The Lineman's and Cableman's Handbook

About the Authors

Thomas M. Shoemaker (deceased) was a Life Senior Member of the Institute of Electrical and Electronics Engineers, a consulting engineer, and former Manager of the Distribution Department at the Iowa-Illinois Gas & Electric Company.

James E. Mack was employed at MidAmerican Energy Company. He is a Senior Member of the IEEE, a Member of the NSPE, a Member of the ISPE, and an Informational Member of ASTM International.

The Lineman's and Cableman's Handbook

Thomas M. Shoemaker, P.E., B.S.E.E.
*Formerly: Consulting Engineer; Manager, Distribution Department,
Iowa-Illinois Gas & Electric Company; Member, Transmission and
Distribution Committee, Edison Electric Institute;
Captain, Signal Corps, U.S. Army; Life Senior Member,
Institute of Electrical and Electronics Engineers*

James E. Mack, P.E., B.S.E.E., M.B.A.
*MidAmerican Energy Company (retired); Senior Member, Institute of Electrical and
Electronics Engineers; Member, National Society of Professional Engineers;
Member, Illinois Society of Professional Engineers;
Informational Member, ASTM International*

Fourteenth Edition

New York Chicago San Francisco
Athens London Madrid
Mexico City Milan New Delhi
Singapore Sydney Toronto

The Lineman's and Cableman's Handbook, Fourteenth Edition

1 2 3 4 5 6 7 8 9   DSS   28 27 26 25 24 23

Library of Congress Control Number: 2022035430

ISBN 978-1-264-26818-4
MHID 1-264-26818-1

Sponsoring Editor
Lara Zoble

Editorial Supervisor
Patty Mon

Project Manager
Nitesh Sharma,
KnowledgeWorks Global Ltd.

Acquisitions Coordinator
Elizabeth M. Houde

Copy Editor
Girish Sharma,
KnowledgeWorks Global Ltd.

Proofreader
KnowledgeWorks Global Ltd.

Indexer
James E. Mack

Production Supervisor
Lynn M. Messina

Composition
KnowledgeWorks Global Ltd.

Illustration
KnowledgeWorks Global Ltd.

Art Director, Cover
Jeff Weeks

Contents

Preface

This Handbook is written for the apprentice, the lineman, the cableman, the foreman, the supervisor, and other employees of electric line construction contractors and transmission and distribution departments of electric utility companies. It is primarily intended to be used as an apprenticeship textbook and a home-study book to supplement daily work experiences. This Handbook has 50 chapters; 11 chapters are devoted to a general understanding of electricity, electrical terms, and electric-power systems; 31 chapters are devoted to actual construction of overhead and underground distribution and transmission lines and to maintenance procedures; and 7 chapters are expressly devoted to safety guidelines. The final chapter can be used by the reader as a self-examination of the information presented.

All chapters in this Fourteenth Edition have been revised where necessary to be consistent with the newest equipment, techniques, and procedures. A special effort was made to present all discussions clearly and in simple language. As in previous editions, a large number of illustrations showing the construction and maintenance processes are provided to assist the reader in a better understanding of the text. The illustrations clarify many details that would require additional words to express. Many of the photographs were taken specifically for use in this edition. They portray the practices in use by some of the foremost electric utility and contracting companies in the United States. Methods of transmission-, distribution-, and rural-line construction have become quite standardized since the First Edition of the Handbook was published in 1928. The construction procedures described and illustrated are in most instances representative of general practice. While each operating company has its own standards of construction to which its linemen and cablemen must adhere, the procedures described explain why things are done in a given way. Such basic knowledge will be helpful to the lineman or cableman who is interested in learning the whys and wherefores of doing things one way or another.

Safety is emphasized throughout this book. Of course, understanding the principles involved in any operation and knowing the reasons for doing things a given way are the best aids to safety. The opinion has become quite firmly established that a person is not a good lineman unless he does his work in accordance with established safety procedures and without injury to himself or others. It is necessary for those engaged in electrical work to know the safety rules and the precautions applicable to their trades, as specified in the *National Electrical Safety Code*, Occupational Safety and Health Act (OSHA) standards, and their employers' safety manuals and standards, and to make the observance of safety rules and procedures an inseparable part of their working habits.

This Handbook places emphasis on the *National Electrical Safety Code*, OSHA standards, American National Standards Institute (ANSI) standards, and ASTM International (formerly known as the American Society for Testing and Materials) standards. Important requirements of all of these are discussed, but the *National Electrical Safety Code*, OSHA standards, ANSI standards, and ASTM standards should be studied for detailed work procedures. Many applicable codes and standards are referenced throughout the text to assist the reader.

The lineman and the cableman must become acquainted with the minimum construction requirements and maintenance and operating procedures in the various codes and standards to ensure the safety of the public and all workers. It is necessary that all linemen and cablemen know the information in the *National Electrical Safety Code* (ANSI C2) and that they adhere to the rules and procedures while performing their work assignments.

The *National Electrical Code* details the rules and regulations for electrical installations, except those under the control of an electric utility. It excludes any indoor facility used and controlled exclusively by a utility for all phases from the generation through distribution of electricity and for communication and metering, as well as outdoor facilities on a utility's own or a leased site or on public or private (by established rights) property.

Reference material for the interested reader includes *Standard Handbook for Electrical Engineers*, edited by Donald G. Fink and H. Wayne Beaty and published by McGraw Hill; *National Electrical Safety Code (NESC) Handbook* edited by David J. Marne, P.E. and published by McGraw Hill; *IEEE Standard Dictionary of Electrical and Electronic Terms*; *Power-lineman Magazine*; and the Electric Power Research Institute's (EPRI) publications.

The editor is well aware that one cannot become a competent lineman or cableman from a study of the pages of this book alone. However, diligent study along with daily practical experience and observation should give the apprentice an understanding of construction and maintenance procedures—and a regard for safety—that should make his progress and promotion rapid.

In conclusion, proper recognition to the lineman profession has been extremely well chronicled in a film documentary series *Storm Soldiers I* and *Storm Soldiers II* and the book *The American Lineman: Honoring the Evolution and Importance of One of the Toughest, Most Admired Professions, Alan Drew*. In 2006, the International Lineman's Museum, Shelby, North Carolina established the Lineman's Hall of Fame.

James E. Mack

Acknowledgments

I wish to express my sincere appreciation to the many companies and their representatives who kindly cooperated in supplying illustrations, data, publisher permission, and valuable suggestions. The editors are also indebted to the Institute of Electrical and Electronics Engineers, Inc., publishers of the National Electrical Safety Code, the Rural Electrification Administration, and the Edison Electric Institute for permission to reprint various items from the literature.

James E. Mack

Lineman working from insulated bucket. The energized primary line conductor is covered with line hose. The lineman has positioned himself for convenient access to the work being completed. (*Courtesy Pepco Technologies.*)

The terms *lineman* and *cableman*, long-established and still current in the industry, are beginning to be replaced by non-gender-specific titles in official documents and government publications. Both men and women are employed in these capacities in the military and in the industry. To avoid awkwardness, this Handbook uses the masculine pronoun, but it in no way implies that the jobs involved are held only by men.

Introduction

Linemen and cablemen construct and maintain the electric transmission and distribution facilities that deliver electrical energy to our homes, factories, and commercial establishments. They provide important skilled services to the electrical industry—important because the health and welfare of the public are dependent on reliable electric service.

When emergencies develop as a result of lightning, wind, or ice storms, linemen and cablemen respond to restore electric service at any time of the day or night.

An understanding of electrical principles and their application in electrical construction and maintenance work is essential to completing the work safely, efficiently, and reliably. New equipment and the public's increased dependence on continuous electric service require that all linemen and cablemen be highly skilled.

The time and effort spent studying these pages will increase the reader's knowledge of electric distribution and transmission facilities and improve his skills. Every

A hot crew with their hot wagon. This photograph is from the 1930s when live line work was in its infancy and was specialized at many companies. Lineman working on these crews developed many effective methods and tools that are still used today.

lineman, cableman, and groundman should develop the highest possible level of skills so that he will be able to meet the challenges of the work and be qualified for promotion when the opportunity becomes available.

As a testament to the significance of the role that the lineman has had in the United States, the following summary was submitted by Alan Drew, Northwest Lineman College and author of *The American Lineman: Honoring the Evolution and Importance of One of the Toughest, Most Admired Professions*.

Evolution of the Lineman

In the 1840s, the use of the telegraph as a means of communication started in the United States. To obtain the benefits of this enhanced method of communication it meant that lines would have to be constructed and maintained. It was found that telegraph lines could be strung on trees if they were available but that wood poles provided the best method of supporting the lines. The expansion of the telegraph system required men who could set poles and then climb them to string the wire. The term "lineman" quickly evolved as a title for those who worked on telegraph lines.

In the late 1870s, the telephone was invented and it began to replace the telegraph as a means of communication. The telephone also needed lines to be constructed, which were similar to the telegraph lines except they utilized more wires for the needed circuits. The term "lineman" was well established and carried on into the telephone era.

In the late 1890s, electric power started to become a useful form of energy and power plants and lines were built. This new form of energy immediately proved to be considerably more hazardous than the telegraph or telephone systems. A new breed of the "lineman" was now needed to work on these lines. These linemen took, and were expected to take, many risks in working on power lines and equipment. A high level of injuries occurred because of the limited training, and lack of proper equipment, construction standards, and safety rules. Although records are sketchy, in some areas it has been said that one of every three linemen was killed on the job, and mostly from electrocution.

As the power system evolved across the country there were many large line building projects which resulted in linemen "booming around" from project to project. These early linemen quickly established a reputation as individuals who worked hard, took many risks, played hard, and took pride in their work.

In the late 1930s, the complexity of the lineman's job was recognized as a good fit for apprenticeship training. This resulted in the establishment of the apprentice lineman, and soon programs started to evolve across the country. This was a significant step in the establishment of more formalized training to develop competent linemen.

The use of electricity had brought with it a higher quality of life at home and the ability to be more productive for businesses. When power outages occurred, the impact on customers started to become significant. Linemen would quickly respond to these outages and make Herculean efforts in all types of weather to restore power. They soon established a reputation as "heroes" in the eyes of many customers.

The early linemen established their reputation while working mainly on wood pole lines. In the late 1950s underground started to evolve, as a popular and reliable way to deliver power and it soon became part of the lineman's work. This added a new dimension and more complexity to the lineman's job.

Today's linemen face similar challenges of the past; however, today customers rely heavily on the continuous delivery of power and are less tolerant of outages.

This results in considerably more work on energized lines than in the past. The lineman of today has a wide array of enhanced vehicles, tools, equipment, and training to meet these challenges. In addition, OSHA rules, standards, and procedures have greatly improved.

Linemen of today take pride in their work and remain heroes in the eyes of many customers when their power is restored after a long outage. As we look back at the vast amount of experience and knowledge that has been gained from the efforts of these pioneering linemen, it is appropriate to commit to maintaining their legacy.

Elementary Electrical Principles

Electron Theory

The basis of our understanding of electricity is the electron theory. This theory states that all matter, that is, everything that occupies space and has weight, is composed of tiny invisible units called atoms. Atoms in turn are subdivided into still smaller particles called protons, neutrons, and electrons. The protons and neutrons make up the central core, or nucleus, of the atom, while the electrons spin around this central core in orbits as illustrated in Fig. 1.1.

The protons and electrons are charged with small amounts of electricity. The proton always has a positive charge of electricity on it, while the electron has a small negative charge of electricity on it. The magnitude of the total positive charge is equal in amount to the sum of all the negative charges on all the electrons. The neutron has no charge on it, either positive or negative, and is therefore neutral.

Atoms differ from one another in the number of electrons encircling the nucleus. Some atoms have as many as 100 electrons spinning around the nucleus in different orbits. The atom of hydrogen gas has only one electron. The atom of lead has 82 electrons.

Positive and negative charges of electricity attract each other; that is, protons attract electrons. But the atom does not collapse because of this attraction. The spinning of the electron around the nucleus causes a centrifugal force that just balances the force of attraction and thus keeps them apart.

Electric Current

The electrons in the outermost orbit of an atom are usually not securely bound to the nucleus and therefore may fly off the atom (Fig. 1.2) and move into an outer orbit of another atom. These relatively free electrons normally move at random in all directions. However, when an electrical pressure (voltage) is applied across a length of wire, the free electrons in the wire give up their random motion and move or flow in one general direction. This flow of free electrons in one general direction, shown in Fig. 1.3, is called an electric current or simply current.[1]

---

[1]By convention, however, the flow of current in a circuit is taken to be opposite in direction to that of the electrons.

FIGURE 1.1   An atom consisting of the nucleus and revolving electrons. The nucleus is composed of protons and neutrons. The protons carry a positive electrical charge, the electrons carry a negative electrical charge, and the neutrons are neutral; that is, they carry neither a positive nor a negative charge.

FIGURE 1.2   Atom showing electron in outer orbit leaving atom. The atom then has more positive charge than negative charge. The nucleus will therefore attract some other free electron that moves into its vicinity.

FIGURE 1.3   Flow of free electrons in a conductor. Only electrons in the outer orbit are free to move from one unbalanced atom to another unbalanced atom. This flow or drift of free electrons is called an electric current.

Conductors and Insulators

Materials having many free electrons, therefore, make good conductors of electricity, while materials having few free electrons make poor conductors. In fact, materials that have hardly any free electrons can be used to insulate electricity and are called insulators. Samples of good conductors are copper and aluminum. Samples of good insulators are glass, porcelain, rubber, paper, polyethylene, and fiberglass.

Electric Circuit Compared with Water Circuit

An electric circuit is the path in which the electric current flows. The flow of electricity in a wire is actually the simultaneous motion of countless free electrons in one direction. It is often compared with the flow of a liquid like water. Electricity can then be said to flow in a wire as water flows in a pipe. A simple water circuit, like the one shown in Fig. 1.4, has a resemblance to a typical electric circuit, shown in Fig. 1.5. The similarity between the water circuit and the electric circuit can give one an understanding of the flow of electric currents.

Figure 1.4 illustrates water that is flowing in the pipe circuit in the direction shown by the arrows. It is evident that this current of water flows because of a pressure exerted on it. This pressure is produced by the rotary pump, often called the centrifugal pump, which is driven by an engine. A water motor is connected on the end of the pipeline, and therefore, all the water that flows around the circuit must pass through the motor. It is plain that it will cause the motor to revolve and, therefore, deliver power to the shaft and the rotating equipment connected by the shaft. Similarly, when an electric

FIGURE 1.4   The water system.

FIGURE 1.5   The electric system.

current flows in a wire, it flows because an electric pressure causes it to flow. Thus, the current in Fig. 1.5 is made to flow because of the electric pressure produced by the dynamo, or electric generator, which is driven by an engine. As the electric current flows along the wire, it will be forced to flow through the electric motor. This motor will begin to revolve as the electricity begins to flow through it and will deliver power to the shaft and the rotating equipment connected to the shaft.

Series Circuit

An electric circuit can be arranged in several ways, as long as the path for the electric current is closed. The simplest arrangement is the series circuit. The series circuit has all the elements of the circuit connected onto each other, end to end, as illustrated in Fig. 1.6. The same current from the battery flows through all the resistors.

FIGURE 1.6   A series circuit. The same current flows through all the resistors.

Parallel Circuit

Another arrangement is the parallel circuit. Each resistor is individually connected across the battery, as shown in Fig. 1.7, instead of all the resistors being connected onto each other, end to end, and then onto the battery. The resistors are now said to be in parallel with each other. The current in the parallel circuit divides, and only a portion of the current flows through each parallel path.

Series-Parallel Circuit

A third arrangement is the combination of the series and the parallel circuits. Part of the circuit is in series and part in parallel, as shown in Fig. 1.8. The same current flows through each element of the series portion of the series-parallel circuit. The current in the parallel part of the series-parallel circuit divides, and only a portion of the current flows through each of the parallel paths.

Electric System

The circuits in Figs. 1.4 and 1.5 can be seen to consist essentially of three main divisions. The section where the electricity is produced is called the generator section. That part of the circuit which furnishes the path for the current from the place where it is generated to where it is ultimately used is the transmission section, and the section where the electricity is distributed and used is the distribution division. In the distribution division, electricity is used to light, heat, cool, and power. An actual electric circuit has three parts: a generation source, a transmission line, and a distribution system. An electric circuit with the different components is called an electric system.

The wires of the system serve to carry the electricity just as highways carry automobiles and railroad tracks carry trains. The reason one does not see the electricity moving along the wire is because it is invisible. The wires and transformers appear lifeless; however, they are very much alive and ready to do almost any work for us.

FIGURE 1.7   A parallel circuit. Each resistor is independent of the other resistors and draws its own current.

FIGURE 1.8   A series-parallel circuit.

One should look on the generation, transmission, and distribution of electrical energy as one does on the manufacture, shipment, and delivery of goods. For example, the customer flips a light switch to the on position or turns on the electric range and an order is flashed back through the distribution system (the retail outlet), the substation (the warehouse), the transmission line (bulk transportation), to the generation source, and the safe and reliable delivery of electricity must be made immediately.

It is important to observe that the current path or the transmission line must have a return wire just as water must have a return pipe. The water passes out along the pipe in one side of the circuit, through the water motor, where it does its work, and then returns to the rotary pump in the other pipe (Fig. 1.4). The electricity passes out along one wire to the motor, does its work, and then returns to the generator in the other wire (Fig. 1.5), a path similar to the water circuit.

Electric Current

It has been pointed out that the flow of electricity in a wire is similar to the flow of water in a pipe. When water flows in a pipe, one speaks of a current of water or a water current. Similarly, when electricity flows in a wire, it is called an electric current.

Ampere (Amp)

Generally, we want to know how much water is flowing in the pipe, and we answer by saying "10 gallons per second (gal/sec)." In the same way, we can express how much electricity is flowing in a wire by saying "25 amperes." The ampere (amp) is the unit of electric current. One can learn how much an ampere is by watching what it can do. An ordinary 60-watt, 120-volt incandescent lamp will require ½ amp. This means that ½ amp is flowing through it all the time that it is glowing.

Ammeter

If we wish to measure the current of water flowing through a pipe, we place a meter right in the pipeline. A meter for measuring the flow of water in a pipe is called a flowmeter. When such a meter is placed in the line, water flows through it, and the meter indicates the number of gallons per second which pass through it. It is clear that the meter must be inserted in the pipe so that the water flows through it. The number of amperes of electric current flowing in a circuit can be measured by connecting a current meter or an ampere meter in the circuit, as shown in Fig. 1.9. Since such a meter is to read amperes, it is called an ampere meter or an ammeter. It should be noted that the ammeter is inserted in the line in order that all the current taken by the motor may pass through the meter.

FIGURE 1.9   Ammeter in an electric circuit.

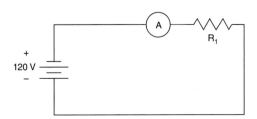

Electric Pressure

We know that a pressure causes a current of water to flow in a pipe. Likewise, in order for current to flow in a circuit, an electric pressure must be present. Voltage is the electric force or pressure that is required. Voltage is determined by the potential difference between any two points in a circuit. Generators and batteries are electrical devices that supply voltage to circuits.

A hand valve in the water circuit will stop the flow of water if it is closed. The water pressure would still be there, but water would not flow through the water motor or the pipe. This can be demonstrated by turning off the water faucet in the kitchen sink. The water flow ceases, but the pressure is still there. A switch placed in the electric circuit will likewise prevent the flow of current if it is in the open position. The electric pressure will still be there with the switch open if the generator is being driven by the engine. Thus, there can be pressure and no current.

Volt

One must be able to talk of its strength to learn something about electrical pressure. This requires a unit with which to measure it; that unit is the volt. One volt will cause 1 amp to flow when impressed across a 1-ohm resistor. We can learn how much a volt is by observing what it can do. We can note, for example, how much pressure or how many volts are required, in general, to force a current through a doorbell, an iron, a washing-machine motor, an electric range, a small factory motor, and a large factory motor. The most common voltage values are as follows:

> An electric doorbell requires 2 to 5 volts.
>
> An electric iron requires 120 volts.
>
> A washing-machine motor requires 120 volts.
>
> An electric range requires 240 volts.
>
> A small factory motor requires 208, 240, or 480 volts.
>
> A large factory motor requires 2,400 to 7,200 volts.

These values simply mean that so many volts are required to push or force the working current through the electrical devices or machines.

One can gain an idea of the strength of electric pressure by observing how much of a shock is received when one puts one's hands across the two wires of a circuit. Under dry conditions, a person cannot detect or feel as little as 5 volts of electrical pressure but can feel 50 volts. At 120 volts, nearly everyone will get a very unpleasant shock, even when a very light and brief contact is made with the wires. It may prove fatal if a firm contact is made with the wires energized at 120 volts. If the hands should be moist or wet and if a firm grasp is made across 120 volts, death is likely to result. All voltages should be considered as dangerous and handled with great care. All higher voltages should be well guarded, and no one except an authorized, fully trained, and qualified person has any business getting near them. System voltages of 600 volts or less are classified as low voltages in *American National Standards Institute* (ANSI) standard C84.1 for Electric Power Systems and Equipment—Voltage Ratings (60 Hz). System voltages of over 600 volts to 69,000 volts are classified as medium voltages. System voltages of over 69,000 volts to 230,000 volts are classified as high voltages. System voltages of over 230,000 volts to 1 million volts are classified as extra-high voltages (EHV). Voltages of 1 million volts or more are classified as ultra-high voltages (UHV).

Voltmeter

If we wish to measure the pressure in a water circuit, all we do is tap a pressure gauge onto the pipeline. Everyone is familiar with such a gauge. The few points to be noted are that the gauge is simply tapped on the pipeline at the point at which the pressure is wanted, so that the pressure at that place can get up into the gauge and make it indicate. It is also evident that flow of water in the pipe is not disturbed by insertion of the gauge.

In the same manner, we can measure electric pressure. Electric pressure or voltage is measured utilizing the voltmeter. The voltmeter measures in volts the potential difference between two points in an electric circuit. We simply connect the two leads from a voltmeter across the line, as shown in Fig. 1.10. The current through the voltmeter will then vary directly with the voltage, and the meter can be made to read volts. It is to be noted that the current which flows to the motor does not flow through the voltmeter. This is because the voltmeter is not a part of the circuit as the ammeter is. Figure 1.11 shows the two meters, the ammeter and voltmeter, properly connected. The ammeter reads the flow of current, and the voltmeter reads the pressure which causes the current to flow.

Water Power

We have compared an electric current to a current of water. When a current of water flows in a pipe in a simple circuit, as shown in Fig. 1.4, power is delivered to the water motor. We know this because the water motor revolves and provides power to downstream connected devices. The power delivered depends on the amount of water

FIGURE **1.10** Voltmeter in an electric circuit.

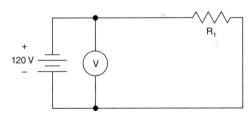

FIGURE **1.11** Ammeter and voltmeter correctly connected.

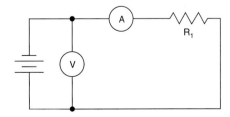

flowing and the pressure under which it flows. This is self-evident, for more power will be developed if 50 gal/sec flows through the water motor than if only 25 gal/sec flows through it. Furthermore, more power will be developed with 100 pounds (lb) of pressure than with only 50 lb of pressure. The power delivered in the pipeline to the motor thus depends on the amount of water flowing and on the pressure.

Electric Power

The amount of power delivered by an electric circuit to an electric motor depends on the number of amperes flowing and the number of volts of pressure in exactly the same way as in water power. The greater the current, the larger the number of amperes, the greater will be the amount of power developed by it; and the greater the pressure, the more effect the current will have. The actual value of power in a direct-current circuit (not true for an alternating-current circuit) is equal to the product of volts times amperes; thus,

$$\text{Power} = \text{volts} \times \text{amperes}$$

Watt

The unit of power in an electric circuit is the watt. An ordinary incandescent electric lamp, when connected to an electric circuit, as in Fig. 1.12, will draw about 150 watts from the circuit. An ordinary iron, when connected to a circuit, will draw about 550 watts of power from it. The motor shown schematically in Fig. 1.13 will draw about 5,600 watts of power. In the case of large machines, the number of watts runs up quickly. Thus, the utilization of decimal system units to quantify amounts is preferred. The decimal system is based on units and powers of 10. The unit for 1,000 is called kilo; therefore, 1,000 watts is called 1 kilowatt (kW). This unit is abbreviated as kW. A kilowatt is equal to about 1.33 horsepower (hp). The motor in Fig. 1.13 thus

FIGURE 1.12   Electric lamp taking 1.25 amps and 150 watts from 120-volt circuit.

FIGURE 1.13   Motor drawing 23.35 amps of current and 5,604 watts of power.

draws 5.6 kW of power (23.35 amps × 240 volts = 5,604 watts or rounded to 5.6 kW). This equates to:

$$5.6 \text{ kW} \times 1.33 \text{ hp/kW} = 7.5 \text{ hp}$$

Wattmeter

We have to measure electric power to know how much power any electrical device or apparatus is drawing from the line. A wattmeter registers watts or kilowatts, and by reading it one can tell how much power any piece of apparatus is consuming. The amount of electric power delivered by a circuit depends on the amperes flowing and the volts of pressure. The meter must be connected so that the entire load current flows through it and the voltage pressure is across it. A wattmeter is essentially a combination of two instruments, an ammeter and a voltmeter. It has an ammeter coil of low resistance, which is connected into the circuit, and a voltmeter coil of high resistance, which is connected across the circuit. A wattmeter will have four terminals or binding posts, two for the current-coil leads and two for the voltage-coil leads.

Electric Energy

Electricity must act for a period of time in order to do useful work. The power expressed in watts tells how much electricity is working, and the hours express the time during which it acts. The product of these two factors gives the amount of work done.

$$\text{Power} \times \text{time} = \text{energy}$$

or

$$\text{Watts} \times \text{hours} = \text{watthours}$$

Likewise, 1,000 watthours equals 1 kilowatthour (kWh). A kilowatthour can thus be thought of as

$$1,000 \text{ watts acting for 1 hour} = 1 \text{ kWh}$$

Any other combination of volts, amperes, and hours whose product is 1,000 watts equates to 1 kWh of energy.

Watthour Meter

The total amount of electrical energy consumed over a period of time, such as a day or month or year, is indicated by a watthour meter. We are all familiar with the common electric house meter. The watthour meter shows how much energy the lights, the television, the stereo, the iron, the toaster, the washing machine, etc. consume in the course of a month. Solid-state watthour meters measure more than just consumption of the kilowatthour. Main printed circuit boards are installed in each solid-state meter and can be programmed to provide integrations and calculations that measure the peak and average kilowatt demands as well as recording voltage swells, voltage surges, and power interruptions.

Conductors

The water pipes merely furnish the path for the flowing water in the water circuit. We would say that the hole in the pipe was the path of the water if we were to be more exact.

We know the water does not flow through the iron shell of the pipe. It actually passes through the opening in the pipe inside the iron shell. The hole or opening in the pipe may then be directly compared to the electric wire. The electricity, however, flows right through the hard wire. It does not need an opening. It passes right through the metal. But we find it can travel with less difficulty through some metals than through others. It travels easily through copper; therefore, we often use copper wire. Copper is therefore called a good conductor. Any metal substance that offers little resistance to the flow of electricity is called a good conductor.

Electricity passes still more easily through silver, but silver is too expensive to be made into wires. Aluminum has been used extensively on long transmission lines because of its rather good conduction and its light weight, making long spans between poles or towers possible. It is in general use in distribution lines as well.

Insulators

It was pointed out earlier that the hole in the pipes serves as the path for water. This does not mean to imply that the shell of the pipe is useless and unnecessary. It is very necessary. The shell serves to hold the water in its path. A pipe without a shell would never transmit any water. There would not be any pipe, and the water would flow everywhere. It is important to remember that the shell keeps the current of water in its path.

In an electric circuit, there must be something to keep the current from leaving the wire. The metal of the wire is its path, but there must be something to keep it from leaving the metal. For overhead installations, bare conductors are separated sufficiently from each other based upon system voltages as defined in the clearance tables of the National Electrical Safety Code (NESC). The air serves as a nonconductor of electricity (a shell, or an insulator, to be more precise). Air has fewer free electrons than the wire conductor and for this reason does not normally conduct electricity. For underground and some overhead applications when the overhead conductors are desired to be placed in closer proximity to one another due to space limitations than the NESC allows, a shell is put around the wire just as in the case of the water pipe. This shell is usually a rubber or polyethylene compound material. The layer of rubber or polyethylene compound material is called insulation, and a wire so covered is called an insulated wire.

Nonmetallic-sheathed cable—which is an assembly of two or more insulated conductors having an outer sheath of moisture-resistant, flame-retardant, nonmetallic material—is usually used for housing wiring where the voltage is 120 or 240 volts. Higher voltages are insulated with rubber or polyethylene.

Wires commonly lay on glass, porcelain, or polymer insulators installed on wooden or porcelain crossarms. The crossarms are attached to poles. Glass, porcelain, and polymers are good insulators of electricity. Bare wires are carefully mounted on glass, "petticoat" (named because of their resemblance to a skirt) porcelain insulators, or polymer insulators to keep the wires from touching each other or the ground, thus preventing the electricity from leaving the wires.

High-voltage transmission lines may be constructed with polymer suspension insulators. The suspension insulators are high strength, light weight, and durable. The greater the length of the insulator, the higher the voltage it can withstand.

Resistance

One can draw up an ice cream soda faster and with less exertion through a short and large straw than through a long and small straw. The water circuit of Fig. 1.4 makes it easy to understand that less water will flow if the pipe is long and has a small opening than if it is short and has a large opening, provided the pressure is the same in both cases. This is to be expected, for in the long pipe the friction is greater because of the greater length, and if the pipe is small in diameter, the friction will be still greater because of the smaller space, through which the water must be forced. It may be said that a long, narrow pipe offers more resistance to the flow of water than a short, wide pipe.

An electric conductor offers a resistance to the flow of current determined by its length and its thickness or diameter. More friction must be overcome if the wire is very long than if it is short; or if the wire is of small cross-sectional area, it will take still more effort to crowd the current through the wire.

Ohm

The unit of resistance for wire is the ohm (Ω). We can picture how much an ohm is by noting how many feet of wire a given size takes to make an ohm of resistance. Wires used for electric purposes are supplied in regular sizes of specified diameter. The no. 10 copper wire has a diameter of about 1/10 inch and has 1 ohm of resistance for each 1,000 feet of length. Thus 5,000 feet of this wire would have 5 ohms resistance. A wire whose cross section is one-half as large would have a resistance of 2 ohms for each 1,000 feet of length.

Ohm's Law

Resistance reduces the amount of current that will flow in a circuit. The number of amperes that will flow in a circuit will not be determined completely by the voltage or pressure which causes the current to flow, but by the amount of friction or resistance in the wires. The greater the resistance of a circuit, the smaller the current will be that flows; and the smaller the resistance, the greater the current will be in a circuit with a constant voltage. This general relation between voltage, current, and resistance is commonly called Ohm's law. It is a law because it has been found to hold true in every case. It is written thus:

$$\text{Current} = \frac{\text{voltage}}{\text{resistance}} \quad \text{or} \quad \text{Amperes} = \frac{\text{volts}}{\text{ohms}}$$

$$\text{Resistance} = \frac{\text{voltage}}{\text{current}} \quad \text{or} \quad \text{Ohms} = \frac{\text{volts}}{\text{amperes}}$$

Voltage (V) = current (I) × resistance (R) or V = IR   or   Volts = amperes × ohms

The law as stated above applies only to direct-current circuits, circuits in which the current continuously flows in one direction in the wire. The law is quite obvious, because it agrees with the common principle with which all of us are familiar, namely, that the result produced varies directly in amount with the magnitude of the effort or force and inversely with the resistance or opposition encountered.

Rheostats

The current in a circuit may have to be reduced or increased. The resistance in the circuit can be increased or decreased with a variable resistance that can be changed as desired. More or less resistance is put into the circuit by turning the handle of the hand-operated resistor, and the current accordingly decreases or increases. Variable resistors are commonly called rheostats and are used to control or regulate the current flowing in a circuit. The most common application utilizing a rheostat is for lighting. The rheostat is used in a dimmer wall switch to control the amount of light emitted by the light source.

Direct Current

A water circuit has been used in which the flow of water was always in the same direction around the circuit. It is evident from Fig. 1.4 that if the engine continues to rotate in the same direction, the rotary pump will rotate in the same direction, and the current of water will continue to flow in the same direction. The electrical equivalent of this current is called a direct current. Direct current flows in the same direction in a circuit. Direct current serves a very minor role in electric distribution. The principal use of direct current at the distribution level is for elevators, electric furnaces, electroplating, etc.

Alternating Current

A rotary engine with a reciprocating pump, like the one shown in Fig. 1.14, will produce a current of water which will first flow in one direction and then, as the piston moves in the opposite direction, flow in the opposite direction. It can be seen in Fig. 1.14 that when the piston is moving upward, the flow of water will be as shown by the arrows in the pipe circuit, and when the piston moves downward, the direction of flow of water will have to reverse. It is noted that in such an arrangement, the engine still keeps on revolving in one direction as before. The pump, however, being of a different type, causes the water current to flow in one direction for a very short time and then reverses it and makes it flow in the other direction for a very short time. This changing of direction continues all the time. The electrical equivalent of the current that flows first in one direction and then in the other direction is said to alternate in direction and is called an alternating current. An alternating current is an electric current which periodically passes through a regular succession of changing values, both positive and negative.

FIGURE 1.14   Alternating-current water pump.

Gasoline engine

Water pipe

Water pump

Water pipe

Cycle

The water pump illustrated in Fig. 1.14 causes the current to change its direction two times for each revolution of the engine. When the piston moves up, the water has a clockwise direction around the circuit, and when the piston moves down, it moves in a counterclockwise direction, opposite to the hands of the clock. When the piston comes up again, the water will then move in a clockwise direction. The water current reverses its direction two times for each complete stroke of the pump. The complete stroke of the pump with its two reversals of water flow is called a cycle. A cycle is, therefore, something which repeats itself.

Frequency

Frequency is the number of complete cycles made per second. If the pump made 60 complete strokes per second, the pump could be said to have a frequency of 60 cycles/sec. Frequency of electricity is expressed in hertz (Hz), which is the unit of frequency equal to 1 cycle/sec. Almost all the electricity generated in the world is of the alternating-current type. Most of the electricity generated and consumed flows first in one direction in the wires and then in the other direction, just as water does in Fig. 1.14. The United States' alternating current frequency is 60-Hz (60-cycle) alternating current. The currents in the house lamp, the toaster, the kitchen range, and the fan motors all flow for a very short time in one direction around the circuit and then quickly change and flow in the other direction. This continues all the time and at a very high rate. These reversals generally take place at the rate of 120 times per second or 7,200 times per minute, making the frequency 60 cycles/sec or 3,600 cycles/min. This is so fast that one cannot notice any flicker in the light of a house lamp. These electric currents are known as 60-Hz (60-cycle) alternating currents.

The prevailing electric frequency is 60 Hz in the United States and just a few other countries, while it is 50 Hz elsewhere in the world. With the higher 60-Hz frequency, output of generator and induction motors will be higher for same size of motor/generator.

Alternating-Current Generator

The discussion of electric pressure or voltage stated that a generator or dynamo produces electric pressure which causes current to flow. A description of the type of generator that produces an alternating pressure will be provided. An electric pressure which causes an alternating current to flow must alternate.

Elementary-Type Alternating-Current Generator

The simplest arrangement for generating an alternating pressure is shown in Fig. 1.15. It consists of a steel horseshoe magnet and a bar of copper to the ends of which is attached a wire, forming the circuit shown. When this bar is moved down between the poles of the magnet so that it passes through the magnetism at the ends of the magnet's poles, an electric pressure or voltage is generated in the wire. A magnet has two poles: north and south. The north pole of a magnet is the magnet end, when freely suspended, that aligns itself toward the earth's north magnetic pole. If the bar is now raised or moved upward so that it passes through the magnetism in the other direction, a voltage will again be generated in the wire, but this time the voltage will be in the opposite direction. Simply stated: electricity can be produced by rotating a wire loop in the presence of a magnetic field. All that is needed to obtain an alternating voltage is a machine in which a wire moves through magnetism first in one direction

FIGURE 1.15 Elements of an alternating-current generator.

and then in the other and arrangements for connecting this moving wire to the out-side circuit. Such a machine is shown in Fig. 1.16 and is called an alternating-current generator. The elementary electric alternating-current generator has a wire bent into the form of a loop or coil so that it can be rotated continuously in one direction instead of having to move up and down. The two sides of the coil will alternately pass through the magnetism at the north pole and through a magnetism at the south pole. The side of the loop marked AB will generate a voltage in a direction from B to A (if this loop is revolved in the direction of the arrow) and the other side of the loop marked CD will generate a voltage in a direction from D to C at the instant shown. A current will flow if the loop is connected to a closed circuit through rings at the ends of the loop by means of sliding-contact brushes. When the loop is in the vertical

FIGURE 1.16 Elementary alternating-current generator.

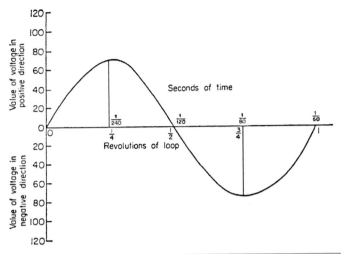

FIGURE 1.17 Alternating-voltage wave. This chart gives the positive and negative values of voltage for each position of the loop. Zero corresponds to vertical position of loop.

position one-fourth of a revolution later, the sides of the loop are not passing through any magnetism, and during this instant there will be no voltage generated and no current flowing. When the loop has advanced one-half of a revolution from the first position, the wires are again passing through magnetism, causing a voltage to be generated and a current to flow if the circuit is closed. This time, however, the current will flow in the opposite direction, because the wires are passing different poles. An alternating current can thus be taken from the rings. The curve shown in Fig. 1.17 results if the values of generated voltage are plotted against the corresponding positions of the coil. This is the familiar sinusoidal waveform of an alternating voltage. Brushes sliding on the collector rings connect the loop to the electric circuit. The machine illustrated in Fig. 1.16 is a generator of alternating currents.

The voltage in the coil will reverse its direction twice for every revolution of the coil, just as the water changed its direction twice for every stroke of the pump. The frequency in cycles per second will then be the same as the revolutions of the coil per second. If the loop revolves 60 times/sec, the frequency of the voltage will be 60 Hz. The time required for each revolution, and for each cycle, will be 1/60 sec.

Single-Phase Alternating-Current Generator

This generator has only two slip rings which connect to a simple circuit of one coiled wire and two connections. A generator with only two slip rings is known as a single-phase generator, single phase meaning that the circuit has only one coiled wire and two connections.

Two-Phase Alternating-Current Generator

A generator with another coil or loop placed on the rotating shaft so that the wires of this loop lie halfway between those of the loop already described and with the ends of the second loop brought out to another pair of slip rings will generate two voltages, one in each loop. One loop would be passing through magnetism when the

FIGURE 1.18   Voltage waves of two-phase generator.

other would be midway between the poles, and the voltage in one would be a maximum while the other was zero. The voltage in one loop would be ahead of the other by the time necessary for the shaft to turn through the space separating the coils or, in this case, one-fourth of a revolution (Fig. 1.18). A machine generating two voltages is called a two-phase generator. The machine would have four connections leading from it to make up the two circuits.

Three-Phase Alternating-Current Generator

Picture three coils or loops equally spaced on the shaft of the machine like the two coils of the two-phase generator. A machine built with three coils and three pairs of collector rings would obtain three alternating voltages in the three circuits, but each of these voltages would be a little ahead of the others (Fig. 1.19). The machine would be called a three-phase alternating-current generator. The cycles per second would again be equal to the revolutions per second if the machine were of the simple type with only two poles. The machine would have six slip rings and six connections leading from it, making up the three circuits.

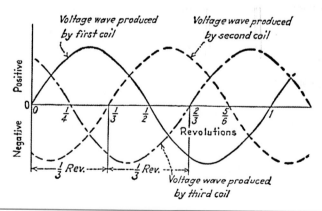

FIGURE 1.19   Voltage waves of three-phase generator. The corresponding zero values are one-third revolution apart.

It has been found that the sum of the currents at any instant in three of the six wires is zero. Some of the currents are flowing away from the generator, others are flowing toward the generator at the same instant, and the net sum of the currents in three wires is zero. This fact makes it possible to do away with three of the six wires, thus making the three-phase circuit a line with only three wires. The generator will have only three collector rings, and only three connections to the machine.

Revolving-Field Alternating-Current Generator

The alternators described have stationary field poles with the conductors arranged to move past the poles. It is better construction to have the armature conductors stationary and the poles revolving, called revolving-field alternators instead of revolving armature alternators. The principal advantages of the revolving field construction are:

1. the armature conductors can be more securely fastened and better insulated and

2. the amount of material in the rotating part is considerably reduced.

Three-Phase Connections for Alternating-Current Generator

The three coils of the three-phase alternator can be connected in two ways, one to form a wye (Y) connection and the other to form a delta (Δ) connection. The wye connection has one set of corresponding ends of each of the three coils of the alternator connected together, as shown in Fig. 1.20. The name "Y connection" is taken from the appearance of the connection when shown as a diagram. The three free ends of the coils are connected to the three-phase line.

The delta connection has the three coils connected in series, as shown in Fig. 1.21. The name is taken from the appearance of the connection when shown as a diagram. The line wires are connected to the junction points or the corners of the delta.

Transmission Lines

Most of the electricity used in the world is three-phase alternating current. Long-distance alternating-current transmission-line towers or poles have three-phase wires and a ground or static wire for lightning protection. In many cases, electric power companies provide lines with two sets of three wires each. Double circuits increase the amount of power transmitted over the line, make more effective use of the right-of-way, and improve the probability of ensuring continuous service.

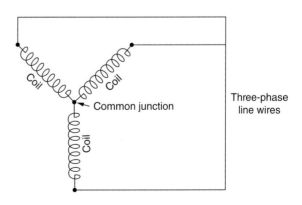

FIGURE 1.20   Wye (Y) connection of three-phase alternating-current generator.

Common junction

Three-phase line wires

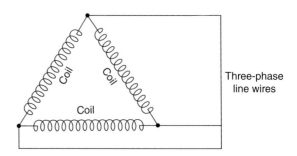

FIGURE 1.21   Delta (Δ) connection of three-phase alternating-current generator.

Three-phase line wires

Power Factor

When an alternating voltage and the current which it causes to flow rise and fall in value together and reverse direction at the same instant, the two are said to be "in phase" and the power factor is unity. This condition is illustrated in Fig. 1.22. The same formula for power as was discussed for direct current holds for alternating current when the power factor is unity and equal to 1:

$$\text{Power in watts} = \text{volts} \times \text{amperes}$$

The current and voltage waves are not in phase in most cases. They do not rise and fall in value together, nor do they change direction at the same instant, but instead, the current usually lags behind the voltage. Figure 1.23 illustrates the usual condition in transmission and distribution circuits. The current and voltage are now said to be "out of phase." The current drawn by idle running induction motors, transformers, or under-excited synchronous motors lags even more than the current shown in the figure.

Occasionally, the current leads the voltage. An unloaded transmission line or an overexcited synchronous motor or a static condenser takes leading current from the line. When the current leads or lags the voltage, the power in the circuit is no longer equal to volts times amperes, but is calculated from the expression:

$$\text{Watts} = \text{volts} \times \text{amperes} \times \text{power factor}$$

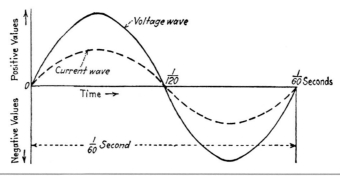

FIGURE 1.22   Voltage and current waves in phase. The power factor under these conditions is unity.

$$\text{Power factor} = \frac{\text{watts}}{\text{volts} \times \text{amperes}}$$

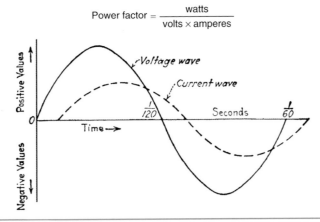

Figure 1.23   Current wave lagging behind the voltage wave. This is the usual condition in transmission and distribution circuits.

From this,

$$\text{Power factor} = \frac{\text{watts}}{\text{volts} \times \text{amperes}}$$

The power factor can thus be defined as the ratio of the actual power to the product of volts times amperes. The latter product is generally called volt-amperes, or apparent power. The value of the power factor depends on the amount the current leads or lags behind its voltage. When the lead or lag is large, the power factor is small, and when the lead or lag is zero, as when the current and voltage are in phase, the power factor is unity and equal to 1. The value 1 is the largest value that the power factor can have. The power factor is usually between 0.70 and 1.00 lagging. An average value often taken in making calculations is 0.80 lagging. Figure 1.24 is an example of an electric circuit with resistive and inductive load and a lagging power factor.

Low Power Factor
The cause of low power factor below 0.80 lagging is an excessive amount of inductive effect in the electric consuming device, be it motor, transformer, lifting magnet, etc. Induction motors, when lightly loaded, exhibit a pronounced inductive effect causing the current to lag the voltage. Idle transformers likewise have a strong tendency to lower the power factor.

Figure 1.24   Electric circuit consisting of a resistive and inductive load with lagging power factor.

Inductance

As the value of current changes in a circuit, properties in the conductor oppose the change in the current and cause a time delay prior to the current reaching a steady state value. This property is known as inductance (L). Inductors store energy in a magnetic field when current flows through it. The transformer and reactor are two electric devices with inductive properties. Inductance is produced when current flows through a wire that is wrapped in a coil around a magnetic core; such as in the transformer. Inductance causes the current to lag the voltage. Inductance is measured in units defined as henrys (H).

Capacitance

Capacitance (C) is the direct opposite of inductance. Capacitance is the property of a capacitor. A capacitor is a combination of metal plates or foil separated from each other by an insulator such as air, paper, or rubber. The capacitance, or the capacity of the capacitor to hold an electric charge in an electric field, is proportional to the size of the plates and increases as the distance between the plates decreases. Capacitance causes the current to lead the voltage. Capacitance is measured in units defined as farads (F).

Power-Factor Correction

One method of raising the low power factor of below 0.80 lagging is to add capacitors to the circuit, since capacitance is the opposite of inductance and since too much inductance is the cause of low power factor. Figure 1.25 is an example of an electric circuit in which the inductive load is offset by the capacitive load and the power factor has been corrected to unity or 1.00. Capacitors are installed in the electric substation or in the electric distribution system on poles. The capacitors can be directly connected to the circuit and switched on and off as needed.

Alternating-Current Generator Review

We have carefully studied how an alternating current of water could be produced by the pump shown in Fig. 1.14. We observed that the current of water circulated around the pipe circuit first in one direction and then in the opposite direction. We observed that the water current reversed its direction twice for each complete stroke of the pump. All this we have carefully compared with the action of a real electric generator and found the processes are quite alike. This much is fundamental in the generation of electricity, whether it is alternating-current or direct-current electricity. To obtain alternating current, we simply fastened slip rings to the ends of the loop and connected those rings to the circuit by means of sliding brushes.

Direct-Current Water Pump

A direct current of water, which is a current that flows continually in one direction, can be obtained by adding a set of valves to our water pump, as shown in Fig. 1.26. The water will not be able to flow backward after it is once forced through the valve, and therefore, the direction of the water flow will always be the same. The piston moves up,

FIGURE 1.25   Electric circuit consisting of a capacitive, a resistive, and an inductive load. The power factor has been corrected to zero.

FIGURE 1.26   Direct-current water pump.

forcing the water through the top valve, and as the piston drops, the top valve closes and the lower valve rises, allowing the water to fill the cylinder. When the piston moves upward again, the cylinder is full of water and more water will be forced into the circuit through the top valve as before.

Elementary-Type Direct-Current Generator

A direct current of electricity can be obtained from the machine shown in Fig. 1.16 by replacing the slip rings with a commutator. The commutator is the electric device which changes an alternating voltage to a direct voltage. It corresponds to the valves in the hydraulic direct-current generator. The process is called commutation. The commutator consists of two bars of copper connected to the two ends of the loop or coil in the simple case. These two copper bars revolve with the shaft, as do the slip rings. These bars have brushes resting on them, which connect the machine to the electric circuit. The general arrangement is shown in Fig. 1.27.

Commutation Direct-Current Generator

From Fig. 1.27, it is clear that the commutator will first connect one side of the loop to one side of the circuit, and then as the shaft revolves, it will connect the other end of the loop to the same side of the circuit. If the brushes are correctly set, this change will take place when both sides of the loop are not generating any voltage, that is, when the coil is midway between the poles. This is the point at which the voltage in the loop changes direction. When A side of the loop is passing the south pole, the voltage is in a given direction and feeds into brush B_1, and D side of the loop feeds into brush B_2. But when the loop has revolved to its midway position, the direction of the generated voltage in the loop is about to change, and just at this instant the ends of the loop are also changed to the opposite sides of the circuit so that when the loop revolves still further, the voltage into the circuit will remain the same. The voltage that alternates in the loop is made to produce a pressure in the same direction in the outside circuit. Such a machine with its commutator is called a direct-current generator.

Commercial Direct-Current Generator

The principal parts of a commercial direct-current generator are the frame, the poles, the armature, the commutator, and the brushes. The purpose of the frame is to support

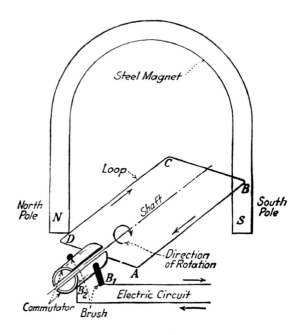

FIGURE 1.27 Elementary direct-current generator.

the north and south poles. The armature supports the conductors which constitute the armature winding. The commutator, a split ring device made up of many copper segments, is also supported by the armature. The commutator keeps the torque on a direct-current motor from reversing every time the coil moves through the plane perpendicular to the magnetic field. A special brush rigging holds the brush holders and brushes.

Direct-Current Circuits

Direct-current (dc) circuits, as we have seen, require only two wires, one for the outgoing current and the other for the return current. Circuits carrying direct current will, normally, consist of only two wires. The use of direct current is generally confined to special applications, such as electroplating, battery charging, and elevator operation, or where fine speed control of industrial motors is required.

High-voltage dc circuits are used to transmit large amounts of power from remote alternating-current (ac) generation sources to populated load areas. High-voltage dc circuits are also used to transmit large quantities of electrical energy by insulated cable circuits. Alternating-current systems that are not synchronized are being interconnected with high-voltage dc circuits. High-voltage dc transmission lines originate and terminate in converter stations that change alternating current to direct current or direct current to alternating current. The converter stations use solid-state electronic devices to convert the energy from alternating current to direct current or direct current to alternating current.

Transformers

The voltages necessary to obtain economical transmission are higher than those which can be directly generated by an alternator. The voltage at which electric power is used in motors and lamps is less than that required for distribution. It is therefore necessary to raise the voltage at the generating station to the value required for transmission and to lower it at the point of consumption to the values required by the motors and lamps on the system. The voltage must be stepped up, in the first case, and it must be stepped down in the latter. The transformer is the apparatus used to make these changes in ac voltage.

Hydraulic Transformer

The water circuit shown in Fig. 1.28 will make it clear how ac voltages are stepped up and down. The water flows back and forth in the circuit since this is an ac circuit. As it does this, it causes the piston in B to move up and down. Piston B is connected to piston C by a rod pivoted at P so that when B moves up and down, C moves down and up. The pressure in B is high and the current of water is small, just like the voltage and current in a high-voltage transmission line. The pressure in cylinder C will be low and the current of water large, because the size of the piston is so much larger. A mechanism such as this could be used to transform, or step down, a water circuit having a high pressure to one having a low pressure. The current on the low-pressure side would be much larger than the current on the high-pressure side.

The frequency of the current has not been changed. The only changes are the decrease in pressure and the increase in current. It should be noted that the total power delivered has not been changed, for under the discussion of power it was pointed out that the power is dependent on the product of voltage and current. The same power can be obtained from a circuit of high voltage and small current as from a circuit that has a low voltage and a large current.

Electric Transformer

An electric transformer operates in the same manner as the hydraulic transformer. The primary, or high-voltage side, of the transformer has a high electric pressure and the current is small. The secondary, or low-voltage side, of the transformer has a low pressure and the current is large. One can usually tell the high-voltage from the low-voltage side of the transformer by observing the size of the insulators or bushings on the top of the transformer case. The high-voltage side must be better insulated than the low-voltage side. The high- and low-voltage leads are brought out on top of the tank.

Figure 1.29 illustrates the elementary parts of an ac electric voltage transformer. Cold-rolled silicon steel strips are so arranged as to form a closed magnetic circuit. These strips have placed on them two coils of insulated wire, one of which has many

FIGURE 1.28 The hydraulic transformer.

FIGURE 1.29   Electric voltage
transformer.

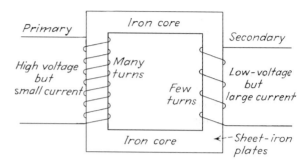

turns of small wire and the other of which has a few turns of heavy, coarse wire. The coil
with the many turns is the high-voltage coil and is called the primary winding in a step-
down transformer. The other coil with the few turns is the low-voltage coil and is called
the secondary winding. Thus, the winding into which current is brought is the primary,
and the winding from which current is taken is the secondary. The coils correspond to
the two pistons in the water transformer, and the magnetic core acts as the coupler
between the two pistons.

The action in the transformer is somewhat as follows: The voltage applied to the
primary causes a current to pass through the primary coil. This current creates a magnetic flux in the core. The flux in the core cuts both the primary and the secondary coils.
This cutting of the primary coil creates a counter voltage in the primary coil which very
nearly equals the primary voltage applied. The current in the primary at no load is only
great enough to magnetize the core at no load. The secondary coil being cut by the flux
will have voltage at no load, but there is no current. Now, let us apply load on the secondary. The load current in the secondary will create a counter flux in the core which
reduces the magnetic flux in the core. The reduction in magnetic flux in the core reduces
the primary counter voltage. The reduction in primary counter voltage increases the difference between the applied voltage and the counter voltage. More current will therefore
flow into the primary, thereby increasing the magnetic flux to its former no-load value.
This again raises the secondary induced voltage to its original value. All these adjustments within the transformer take place automatically and instantaneously.

It should be noted that in a modern transformer the primary and secondary windings are not placed on separate legs, but instead, each winding is generally divided into
two parts, and one-half of each winding is placed on each leg. This gives a more constant voltage with changes in load.

Transformer-Core Construction

The core of a transformer can be built in various shapes. The following types are in
general use:

Core Type

The core is in the shape of a rectangle. Coils are placed on two legs of the core.

Shell Type

The core is rectangular in shape with a central leg in addition to the core first described.
Coils are placed on the central leg.

Cruciform Type

The cruciform core is a modification of the shell type. It is the same as if two shell-type cores were set at right angles, using a common central core for the windings.

Wound-Core Type

The core is made of a ribbon of sheet steel wound in a spiral. After the core is wound and annealed, the coils are wound on the core. This type of core is superseding other core types for small distribution transformers.

Transformer-Core Material

An important consideration for a transformer is the core material because the core of an energized transformer contributes to electric system losses. Transformer losses are classified as either no-load losses or full-load losses. The no-load losses on a lightly loaded transformer are a result of the core material. The energy-efficient cores are constructed of low-loss cold-rolled silicon strips or amorphous steel alloy strips.

Transformer Cooling

When the voltages of the primary and secondary windings are both low and the transformer has a small capacity, the transformer, that is, the steel core with its two windings, is placed in a metal case merely to keep out dirt and moisture. This type of transformer is designated as an air-cooled dry-type transformer. The core and coils of a high-voltage, large-capacity transformer are usually placed in a tank filled with oil. The oil is a good insulator and helps to cool the windings. This type of transformer is designated as oil-cooled. It will perhaps be remembered that whenever current flows in a wire, there is a friction loss in heat. An illustration of this is found in the toaster or the iron. The same is true in a transformer; whenever currents are flowing through the primary and secondary windings, heat is given off which must be carried away so that the transformer may not become too hot and char or burn the insulation on the wire of the coils. If it becomes too hot, the transformer will be ruined, just as a person will die from running a high temperature for a long time.

More power can be safely drawn from a transformer in cold weather than in hot weather because of the lower ambient temperature. More power can be taken from a transformer for short periods of time than can be taken off continuously. A transformer which is overheated because of an overload should be replaced with a larger-capacity rated transformer before it becomes damaged.

Methods of Cooling Transformers

The oil in the tank is the principal cooling agent. It keeps the windings cool by carrying the heat from the coils to the surface of the tank. Many times the transformer case itself is built so that it has much more surface for cooling than if it were smooth. The metal sides of the tank are corrugated, or pipes are connected to the top and bottom of the tank so that whatever oil is in the pipes is exposed on all sides. The object in either case is to increase the surface from which the heat can radiate. This is the same principle as is used in the automobile radiator.

Very large transformers may have coils of pipes placed in the top of the tank through which cold water is circulated by a pump. The cold water in the pipes cools the oil in the tank. The water pipes are placed in the top of the tank because the hottest oil is always at the top.

Transformer Temperature Limits

In general, the maximum safe temperature for such insulating materials as cotton, silk, cambric, tape, fabric, etc., is 105°C, which corresponds to 221°F. The temperature of the oil plus a 10° to 15° correction adder is usually taken as an indication of the temperature of the windings. The temperature of the oil in large power transformers is usually obtained by some form of thermometer.

Distribution transformers may have measuring instrumentation installed which indicates the maximum load on the transformer. The rapid growth of loads on distribution systems makes it necessary, in many cases, to make a check of the loads on transformers in order to determine whether the transformer is still large enough to carry the load. An overload indicator lamp makes use of a bimetallic thermostat to give an indication of the winding temperature. It turns on a red light which is visible from the ground if the transformer is overloaded. A reset switch makes it possible to turn the light off.

Three-Phase Transformer

A three-phase transformer is really three single-phase transformers in one case using a single combined core. The core has much the same shape as the shell-type single-phase transformer. The three-phase transformer, however, has single-phase primary and secondary windings on each leg of the core. The common core acts to supply flux for all three phases.

A three-phase transformer weighs about two-thirds as much as the same capacity in single-phase transformers. Its biggest disadvantage lies in the fact that if one phase fails, the entire transformer must be taken out of service.

Large substation transformers employ forced-air and forced-oil cooling in order to increase the output rating. By forced circulation of the oil through the external radiators and by blowing air against the radiators, the heat is more rapidly removed from the hot oil.

Electronic Circuits

Almost all electronic circuits require dc power for their operation. The ac power delivered from the utility to the electronic equipment is converted to dc by some sort of inverter-rectifier-filter combination contained within the equipment. The inverter is an electrical device that converts dc voltage to ac voltage. The rectifier is an electrical device that converts ac voltage to dc voltage. The inverter is typically built from electronic components called diodes. The rectifier makes use of electrical components like toroidal transformers, diodes, and thyristors. The inverter and rectifier are utilized in uninterruptible power systems (UPS), solar generation, and microgrids. The inverter and rectifier are electric voltage control devices. The inverter and rectifier are built with solid-state components.

Solid-State Circuit Components

Solid-state devices are manufactured from semiconductor materials. Semiconductor materials individually have a high resistance to the flow of electric currents and thus have been identified as semiconductors. Silicon and germanium are semiconductor materials that are commonly used to manufacture solid-state devices. A large, pure crystal of silicon has a high resistance to the flow of electric current. Silicon can be made conductive by adding other materials (referred to as doping) that either add or subtract electrons from the combination material. Materials used to dope the semiconductor materials are boron, phosphorus, gallium, arsenic, and others. A silicon crystal that has been doped with phosphorus, which will donate free electrons to the material, is called an *n*-type semiconductor. An *n*-type semiconductor will have an excess number of electrons. Silicon doped with boron, which takes electrons away, is called a

FIGURE 1.30   Diode semiconductor.

p-type semiconductor. A p-type semiconductor will have a deficient number of electrons. The p-type material with a deficient number of electrons is said to contain holes.

Diode Semiconductors

A semiconductor formed by modifying the silicon material on each end with a junction in the middle separating the p-type and the n-type materials is called a diode (Fig. 1.30).

A diode semiconductor is used in a circuit as a current rectifier. As illustrated in Fig. 1.30, the terminal of the diode on the p-type material is called the anode and the terminal of the diode on the n-type material is called the cathode. When voltage is placed across the diode semiconductor with the positive potential connected to the anode (A) and negative potential connected to the cathode (K), the holes in the p-type material and the electrons in the n-type material will both move toward the junction between the n- and p-type materials. When the impressed voltage is great enough to overcome the resistance of the barrier at the junction, the electrons will combine with the holes and flow externally from the diode, leaving the anode and returning through the cathode (Fig. 1.31a).

The pulsating of the output voltage can be eliminated by adding a capacitor in the circuit connected across the output terminals of the circuit x and y. Full wave rectification can be obtained by adding a transformer on the input to the circuit with a diode connected in series with each of the transformer end terminals and a resistor connected to the cathode of each diode and the neutral terminal on the transformer. Four diodes can be used in a bridge circuit to increase the capacity of the rectified output of the circuit (Fig. 1.31b).[2]

FIGURE 1.31a   Diode semiconductor connected to a 60-Hz ac supply voltage with a resistor in series with it will produce a pulsating dc current through the resistor and a pulsating voltage across the resistor that can be measured from terminals x and y of the circuit.

[2]Additional information on semiconductors and basic solid-state conductor circuits can be obtained from Basic Electronics: Theory and Experimentation by Fredrick W. Hughes (Prentice-Hall, 1984).

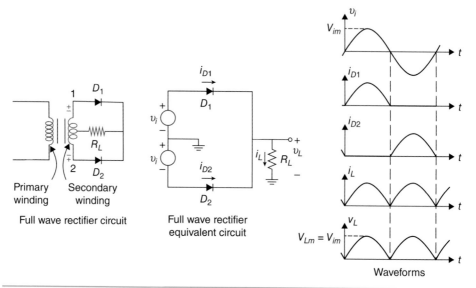

FIGURE 1.31b   Full wave rectifier circuit, full wave rectifier equivalent circuit, and wave forms.

Zener Diode

A diode semiconductor that can be used as a voltage reference in an electric circuit is called a Zener diode. The Zener diode is manufactured by doping the silicon crystal so that when the diode is reversed bias at a voltage greater than the junction breakdown value, a large current will flow through the diode. As the current increases through the diode in the circuit, the voltage across the Zener diode within its rating will remain practically constant (Fig. 1.32).

The operating curve for the Zener diode illustrates that the voltage across the diode, when it is used in a circuit within its rating, varies from +0.7 volt when it is forward biased to –5 volts when it is negatively biased. This characteristic permits its use as a voltage reference when the proper resistance is connected in series with the diode in a circuit with other components. When the voltage applied to the diode increases beyond

FIGURE 1.32   Zener diode symbol and operating-characteristics curve.

the breakdown value of the junction, the high current that flows in the circuit can be used to initiate operation of auxiliary components in the circuit which will correct the voltage variation and provide voltage regulation to the circuit.

Transistor Semiconductor

A semiconductor manufactured with three elements is called a transistor (Fig. 1.33). Transistor semiconductors are constructed with two *n*-type (excess number of electrons) and one *p*-type (deficient number of electrons) or two *p*-type and one *n*-type materials. The current flowing from the emitter (E) to the collector (C) is controlled by the potential applied to the base (B). The base material is very thin.

A transistor can be used in a circuit to amplify a signal. An *npn* transistor connected in a circuit with a collector (C) reversed biased with respect to the base (B) will have a small current flow from the emitter (E) to the base and a large current flow from the emitter to the collector (Fig. 1.34).

When a small ac voltage is connected between the input terminal of the circuit and ground, the bias voltage between the base (B) and the emitter (E) will be increased and

FIGURE 1.33 Transistor semiconductors.

FIGURE 1.34 Schematic diagram of a circuit using a transistor to amplify a voltage.

FIGURE 1.35 Thyristor semiconductor constructed with two p-type and two n-type materials.

Schematic symbol

reduced accordingly. This will vary the current flow from the emitter (E) to the collector (C), which will vary the voltage above and below some positive value between the output terminal and ground. The output voltage will be in proportion to the input voltage and amplified.

Thyristor Semiconductor

A semiconductor consisting of a layer of p-type, a layer of n-type, a layer of p-type, and a layer of n-type materials with an anode (A) connection to one p-type material, and a layer of n-type materials, a gate (G) connection to one p-type material, and a cathode (K) connection to one n-type material is called a thyristor semiconductor (Fig. 1.35).

The thyristor semiconductor is used as a controlled rectifying electrical device. The thyristor center np junction is reverse biased by the operating circuit. A current applied to the gate (G) terminal will reduce the voltage necessary to overcome the reverse bias of the center junction. The operation of the thyristor in a circuit designed to rectify an ac voltage can be controlled by the signal applied to the gate (G) terminal. Controlling the firing or operation of the thyristor is important when changing alternating current (ac) to direct current (dc) and reversing the operation. Thyristors are used in converter stations to convert alternating current to direct current and to convert direct current to alternating current. Thyristors are connected in series in circuits with other circuit components to operate at the high voltages necessary to carry large quantities of electric power over dc transmission lines.

Microprocessor

Solid-state electronic components are also utilized in intelligent electronic devices (IEDs) with microprocessor-based controls. The microprocessor is the brain of an electric system control device. The microprocessor consists of numerous transistors, diodes, and resistors installed on a silicon chip also known as an integrated circuit. Desired actions of the IED occur based upon programed instructions sent by the microprocessor.

Definitions

Definitions of electrical terms used in the text are given as they appear. American National Standard definitions are listed in the *IEEE Standard Dictionary of Electrical and Electronic Terms.*[3]

---

[3]*IEEE Standard Dictionary of Electrical and Electronic Terms.* Wiley-Interscience, New York.

Electric System

Electric System—General

The electric systems in the United States and Canada are interconnected, forming a network of electric facilities across the continent. It is difficult to distinguish the electric system of one company from neighboring utilities. Most electric utility companies will have multiple generating sources supplying a large, complex, and intricate network of transmission and distribution circuits.

The electric system is energized by utility owned and operated nuclear, fossil-fired, and hydroelectric generating stations. Nuclear generating stations use the heat from an atomic chain reaction to generate steam to turn a turbine generator. Fossil-fired steam turbine-driven generating stations use coal, natural gas, or oil as the source of fuel. Hydroelectric generating stations using water pressure to turn a turbine have been constructed at most of the sites where there is sufficient waterfall to economically generate electricity.

Renewable energy generation sources are significantly impacting the industry. Large-scale solar and wind generation projects are emerging, as well as several small-scale distributed solar and wind energy systems and independent microgrid installations. Distributed energy generation sources are normally smaller than 50 megawatts. The microgrids are sourced from the larger electric network and in certain situations isolate themselves from the grid and function independently. In order for the power grid to operate efficiently, all generation sources must be properly integrated (meshed together), managed, and monitored to meet customer demand.

Most electric energy is generated at voltages lower than 24,000 volts. The voltage is raised to transmission levels in a transmission substation located at the generating station. Transmission circuits or lines operating at 138,000 volts to 765,000 volts deliver the energy from the generating source to the load area. Transmission substations adjacent to the load area reduce the voltage to subtransmission levels 34,500 volts to 161,000 volts. The subtransmission circuits or lines extend through metropolitan areas to distribution substations located in the area of the load to be served. Distribution substations reduce the voltage from subtransmission levels to distribution range usually 4,160Y/2,400 volts, 12,470Y/7,200 volts, 13,200Y/7,620 volts, 24,940Y/14,400 volts, or 34,500Y/19,920 volts.

Distribution circuits radiate from the distribution substations to supply the customer's load. Some large industrial customers may be served directly from the transmission or subtransmission circuits. These customers will normally be metered by a high-voltage or primary-meter installation. The large industrial customer will have a substation or substations to reduce the voltage to the desired level for distribution throughout the plant area. Utility-owned distribution circuits will normally exit from a distribution substation underground and then rise and connect to overhead distribution circuits. Underground exit cables are being extended several thousand feet in some cases to minimize the overhead circuits in the vicinity of a substation. Some companies are building new distribution circuits completely underground.

Branch circuits tap to the main distribution feeder circuits through high-voltage fuses commonly called distribution fused cutouts on overhead lines. The branch circuits may be either overhead or underground, as shown schematically. These circuits are often referred to as primaries.

Distribution transformers connect to the branch circuits or primaries. Pole-type transformers are used on overhead circuits, and pad-mounted, or submersible transformers are used on underground circuits. The distribution transformers reduce the voltage to utilization level, usually 120/240 volts single-phase for residential service and 208Y/120 volts or 480Y/277 volts three-phase for commercial or industrial service. Some small commercial installations may be served by 120/240 volts single-phase. The low-voltage service to the customers may be either overhead or underground connecting directly to a distribution transformer or a secondary extending from the distribution transformers.

The main distribution feeder circuits often have normally open switches connecting to other main distribution feeder circuits to provide an alternative source for emergency operation. Other switches may be installed in main feeder circuits to permit sectionalizing the circuit if a component of the circuit fails electrically.

Figure 2.1 is a schematic drawing illustrating a portion of an electric system pictorially. The generators in a power plant will normally be tied to the transmission lines in a transmission substation located at the generating station. Power transformers in the transmission substation raise the voltage from the generation level to the transmission level. Circuit breakers and other equipment in the transmission substation isolate defective circuits or equipment from the electric system. Substations are also used to connect the transmission lines to the distribution system. The purpose of the distribution substation is to provide equipment to protect the complete electric system from trouble on any distribution feeder. The substation is almost always equipped to isolate the feeder in trouble from the other feeders radiating from the substation. Where load and voltage conditions require it, the substation has an additional function of regulating the voltage on the feeders of the distribution system.

Generating Stations

A generating station is a facility in which the energy of coal, oil, natural gas, water, or atom is changed into electrical energy using generators. Traditional sources of energy are not the only sources of fuel considered. Wood, rubber tires, and garbage are also being utilized successfully in some generating stations. Central generating stations came into existence because it is inefficient and uneconomical for every user of power to generate their own electricity. The deregulation of the electric utility industry has significantly impacted the ownership of generating facilities. The ownership of the stations is no longer solely restricted to investor-owned or cooperative electric utilities. Independent power producers (IPPs) are now contributing a greater share of new plant capacity.

Generating stations produce alternating current (ac) power. The frequency of the ac power generated in the United States and Canada is 60 Hz (hertz). In most other countries, the power is generated at a frequency of 50 Hz. Where direct current (dc) is required, it is usually obtained by converting alternating current to direct current.

FIGURE 2.1 Portion of a typical electric system showing generating station, step-up transmission substation, transmission line, step-down transmission substation, subtransmission line, distribution substation, distribution system, and industrial, commercial, and residential customers.

Generating stations may be any of the following types:

1. Hydroelectric
2. Steam turbine
   a. Fossil-fired
   b. Nuclear
3. Combustion turbine
4. Internal-combustion engine

5. Electric generation with renewable energy resources

6. Geothermal

7. Solar

8. Wind

9. Compressed air

Hydroelectric Generating Stations

A hydroelectric generating station is best located at a waterfall to take advantage of the swiftly moving water to propel the blades of the generator. This is sometimes far removed from the territory to be served.

Such plants cost little to run because they do not burn fuel, but the interest and taxes per year are higher than for steam plants because of the larger investment in a dam, land flooded by backwater, etc.

Fossil-Fired Steam-Turbine Generating Stations

Figure 2.2 is a picture of a typical fossil-fired steam turbine-driven generator power plant. The generating station contains multiple generating units. The generating station is located near a river.

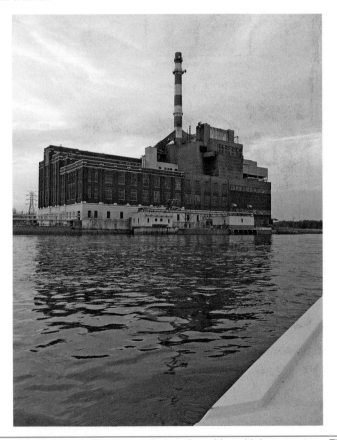

FIGURE 2.2   Fossil-fired steam-turbine generating station with multiple generators. The generating station is located beside a river.

The boilers burn low-sulfur coal. The coal is mined in Wyoming and delivered to the station by 100-car trains hauling 10,000 tons. Thawing sheds, heated electrically, are used to thaw the coal frozen in transit during the period of cold winter weather. The train hopper cars pass over a pit where the bottoms of the cars are opened automatically and unloaded in 2 minutes (min). A conveyor belt transports the coal from the hopper pit to the coal-storage area or the silos in the plant at a rate of 3,000 tons of coal per hour. The coal storage area has a capacity of 800,000 tons.

The boiler for the generator unit develops steam with a pressure of 2,500 pounds-force per square inch (lbf/in^2) and a temperature of 1,005°F. The structure between the boiler enclosure and the 400-foot chimney is the electrostatic precipitator. The precipitator removes 99.2% of the particulates from the boiler gases before they are emitted to the atmosphere through the chimney. The electrostatic precipitator occupies an area about two-thirds the size of a football field.

The turbine generator is located in front of the boiler enclosure. The turbine spindle has rows of blades that stem out from a shaft at the center. Steam from the boiler strikes the blades, like wind on a windmill, to spin the turbine shaft at 3,600 rpm. The turbine shaft turns the generator rotor, a strong electromagnet, to create the electrical energy at 24,000 volts. Water from the river next to the generating station is used to condense the steam after it leaves the turbine. The water condensed from the steam is returned to the boiler feed pumps. The generating station is located on a 122-acre site.

Single steam-turbine generator units are built with a capacity of as much as 1,200 MW. The majority of steam-turbine generators are smaller units with capacity ratings of 25 MW through 1,000 MW.

Nuclear Steam-Turbine Generating Stations

Generating stations using nuclear energy are constructed similar to fossil-fired steam turbine-driven generating stations, except the boiler is replaced with a nuclear reactor. Nuclear plants do not discharge sulfur dioxide and particulates into the atmosphere like plants burning fossil fuel, thus minimizing air pollution.

Radiation, measured in millirems, is present everywhere. Natural background radiation in Iowa and Illinois along the Mississippi River is 100 to 130 millirems every year. If you lived next door to a nuclear power station, you would receive less than 5 millirems of additional radiation annually, about the same amount you would receive on a round-trip coast-to-coast jet flight.

Nuclear fission process is used to produce heat in the nuclear reactor. Fissioning occurs when a subatomic particle, known as a neutron, strikes the nucleus of an atom with sufficient force to split the nucleus. When an atom splits, energy in the form of heat is released. If enough neutrons are released, they will continue to strike and split other atoms, causing a chain reaction. Each separate fission releases only a very small quantity of heat. In a nuclear reactor, many trillions of controlled fissions take place every second. The fissioning process is maintained and controlled in the reactor. Heat from fissioning produces steam, which is piped to large turbines (Fig. 2.3).

Everything in the world is made of atoms—the ground, your body, the food you eat. Certain materials, however, lend themselves to being split, or fissioned, more readily than others. Uranium is such a material, and at the present time it is the most commonly used nuclear fuel.

Large-scale nuclear generating station projects in the United States are stalled. The nuclear power industry is actively developing small-scale nuclear reactor alternatives and is anticipating some growth of nuclear power in the United States energy supply market.

Boiling water reactor (BWR)

FIGURE 2.3   Schematic drawing of large-scale nuclear generating station equipment.

Combustion-Turbine Generating Stations

Generating stations using combustion turbines as prime movers for the generators can be located in an area with other industrial-type installations. Locating the generation near major electrical loads improves reliability and may reduce investment in transmission and distribution equipment.

A combustion-turbine generator facility will include a starting motor, a rotary-type compressor, combustors, ignitors, a multiple-stage turbine, an ac generator, control devices, and necessary accessories. Fuel used in the turbines may be natural gas or oil in crude form or refined state.

The starting motor drives the compressor to produce pressurized air, which is mixed with the fuel in the combustor and ignited. The hot gases flow through the turbine, causing it to accelerate. As the turbine power and speed increase, the starting motor is disengaged and the electric generator is synchronized to the system and connected to the load. The hot gases produced from the burning natural gas or oil expanding through the turbine produce the force to rotate the generator and produce electrical energy.

Combustion-turbine generating stations usually provide peaking power or emergency power. The capacity of the generating units is usually in the range of 15 to 95 MW. The fuel cost for combustion turbines is high, but they have the advantage of low capital cost, automatic operation, fast starting, and minimum electric transmission and distribution investment.

The hot exhaust gases from a combustion turbine can be used to produce steam in a waste-heat boiler. The steam can be used to operate a turbine generator unit, improving the efficiency of the operation and reducing the operating costs. This type of installation is called a combined cycle plant.

Internal-Combustion-Engine Generating Stations

Generating units driven by internal-combustion engines of the diesel type are usually located in remote areas, small towns, and sites requiring standby or emergency power. The engines use oil or natural gas for fuel. The diesel-engine-driven generating units have a low installed cost, can be designed for automatic operation, and can start without an ac source. The fuel cost for the units is high compared to coal- or nuclear-fired stations. Diesel-engine generating stations have a capacity in the range of 500 to 5,000 kW.

Electric Generation with Renewable Energy Resources

Falling water, solar heat, wind, waves and tides, heat from below the earth's surface, agricultural waste combustion, and biomass synthetic fuel are examples of renewable energy sources that can be used to produce electricity. Hydroelectric generation, discussed previously, will continue to produce a significant amount of electricity for consumers.

Geothermal Generating Stations

Geysers located north of San Francisco produce steam that is piped to turbines to generate electricity. New sources of geothermal energy have been discovered in recent years. The geysers (hydrothermal energy) are located principally in California and neighboring states. Geopressured systems use hot water located deep in deposits of sand and sealed off from the surface of the earth by shale. Petrothermal systems require that water or other fluids be injected into molten rock below the earth's surface and pumped out to extract the thermal energy and the dissolved methane. Geopressured and petrothermal energy is generally located throughout the United States. Tests are being conducted with methods to extract the gases from the liquids so that both may be used in a binary-cycle system to capture the geothermal heat and generate electricity.

Solar Generating Stations

The sun can be used to generate electricity by using solar panels. The solar panels contain photovoltaic modules that produce small amounts of electrical energy. An inverter converts the dc generated power to an acceptable utility grade ac power. The solar panels are installed in arrays. The solar panel installations vary from a small-scale residential project of 5 kW to a medium-size commercial project of 75 kW to 10 MW to large-size utility scale solar farms of several hundred megawatts.

Wind Generating Stations

Wind-propelled turbine generator units with a capacity as large as 4.0 MW/unit are operating in regions of the United States where the wind is strong, reliable, and consistent. The wind generators are constructed with propeller rotor blades and a horizontal-axis turbine generator. The equipment is mounted on a steel tower high enough to clear the ground when the propeller blades rotate. The propeller blades vary in diameter, with some as great as 420 ft. The wind-powered generators can produce carbon-free kilowatt-hours, thus replacing consumption of nonrenewable energy sources that are used to generate electricity. Intermittent operation as a result of varying wind velocities limits the dependability of the electric-generating capacity. The ability of wind generation to reduce consumption of oil and natural gas fuels provides a renewable energy alternative to the electric industry. Multiple wind generators are installed at centralized wind farm sites (Fig. 2.4). The wind farms range in size from 100 MW (25 generators) to 500 MW (125 generators).

Compressed-Air Generating Stations

Air can be compressed and stored in natural caverns or underground caverns developed by mechanical methods. The caverns must be sealed to keep the air from escaping. Off-peak electrical energy is used to compress and store the air in caverns.

FIGURE 2.4   Wind generator installed in a central plains region of the United States on a wind farm.

The compressed air can be used to operate a turbine to generate electricity during electrical system peak-demand periods to reduce the base-load generating-equipment requirements.

Alternative Generation Sources

The power system has been positively impacted by the addition of alternative generation sources such as chemical storage of electrical energy and distributed generation.

Chemical Storage of Electrical Energy

Batteries of the lead-acid type and the nickel-cadmium type are commonly used to store small amounts of electrical energy. Zinc-chlorine batteries have been developed that can store large quantities of energy to reduce the peak-load generating requirement of an electric utility. The batteries can be installed in distribution substations close to the point where the electrical energy will be consumed. Installation of the batteries near the consumption location reduces the transmission-capacity requirement as well as the peak electrical-generating requirement.

Distributed Generation

Distributed generation energy sources of 10 MW or less provide power to individual residential or commercial establishments. Solar and wind are the typical sources of distributed generation systems. The small-scale distributed generation system is in direct contrast to the centralized generating station. Distributed generation energy sources primarily supply electricity to the individual establishment and may supply excess power to the electric grid. Distributed generation connects to the grid at the distribution circuit.

Transmission Circuits

A set of conductors energized at voltages over 69,000 volts and transmitting large blocks of electrical energy over relatively long distances are called a transmission circuit or transmission line. Transmission circuits are constructed between transmission substations located at electric generating stations or switching points in the electric system. Transmission circuits may be overhead, with structures supporting conductors attached to insulators, or underground, with conductors surrounded by insulation, shielding, and a sheath or a jacket to form a cable. The distinguishing characteristics of transmission circuits are that they are operated at relatively high voltages, transmit large blocks of electrical power, and extend over considerable distances.

Overhead Transmission Circuits

Transmission lines or circuits are usually constructed for economic and capacity reasons. The lines extend for long distances in rural areas. Alternating-current (ac) transmission circuits generally operate in the voltage range over 69,000 volts to 765,000 volts. High voltages often used for these so-called ac transmission lines are 115,000 volts, 138,000 volts, and 161,000 volts. Transmission circuits energized at 345 kV, 500 kV, or 765 kV are referred to as extra-high-voltage (EHV) lines. Experimental ac ultra-high-voltage (UHV) transmission circuits in the voltage range of 1,100 kV to 1,500 kV are being developed. A 345 kV ac overhead transmission circuit is shown in Fig. 2.5.

Porcelain-disk insulators in strings insulate the two bundled conductors for each phase of the three-phase line from the wood-pole structure. The small wires at the top of the structure are so-called static wires or ground wires. They are not insulated and directly connect to the metal hardware on the structure and the ground. Their purpose is to shield the line conductors from lightning.

Overhead direct-current (dc) transmission circuits are used where it is desired to transmit large amounts of power over long distances or to interconnect two ac systems

FIGURE 2.5   A 345 kV three-phase bundle-conductor transmission line on a wood-pole structure. (*Courtesy Hughes Brothers Co.*)

FIGURE 2.6   Transmission line +250 kV dc, 456 miles long, 1,000 amps, 500 MW capacity located between Minnkota Power Cooperative, Grand Forks, North Dakota, and Minnesota Power, Duluth, Minnesota. (*Courtesy Ulteig.*)

that are not synchronized. The dc circuits are usually several hundred miles in length. A dc circuit in operation in the western United States extends from northern Oregon to southern California, approximately 850 miles, and operates at 400 kV.

Direct-current high-voltage (dc HV) line conductors are either positive or negative polarity. A 400-kV line operates with one conductor 400 kV positive and one conductor 400 kV negative or ±400 kV dc. Some of the voltages used for dc HV circuits are ±250 kV, ±400 kV, and ±500 kV. Figure 2.6 shows a dc HV transmission line tower.

Underground Transmission Circuits

In some areas of large cities, it is not practical nor permissible to build high-voltage overhead lines. All high-voltage electric transmission and distribution circuits must, therefore, be installed underground. Underground transmission circuits may be constructed for ac or dc operation. Alternating-current underground transmission circuits are costly to install and limited in length by charging current. Transmission line charging current is the current that flows in the capacitance of a transmission line when voltage is applied. Circuits over 25 miles long must be compensated by using shunt reactors to absorb the capacitive charging current, or the cable capacity will be used without transmitting any useful power. Direct-current underground transmission circuits are economically attractive for transmitting large amounts of electrical energy where underground construction is required for distances in excess of 25 miles. Underground cables transmit ac power between substations in large metropolitan areas.

Overhead Subtransmission Circuits

Large blocks of power are transmitted from generating stations to one or more transmission substations through high-, extra-high-, or ultra-high-voltage ac lines. Intermediate voltage lines or circuits transmit alternating current from transmission substations to subtransmission or distribution substations. These subtransmission lines carry medium-sized blocks of power and usually form a grid between substations in a metropolitan area. Figure 2.7 illustrates a 69,000-volt three-phase subtransmission circuit constructed along an urban street to provide a source for a distribution substation.

FIGURE 2.7   Modern 69,000-volt
subtransmission line with 13,200Y/
7,620-volt distribution line underbuild.
The picture illustrates armless-type
construction utilized to enhance the
appearance of the line.

FIGURE 2.7   Modern 69,000-volt subtransmission line with 13,200Y/7,620-volt distribution line underbuild. The picture illustrates armless-type construction utilized to enhance the appearance of the line.

Armless-type construction is utilized to gain public acceptance. The lower circuit is a three-phase distribution feeder operating at 13,200Y/7,620 volts.

Substations

A high-voltage electric system facility used to switch generators, equipment, and circuits or lines in and out of the system, change ac voltages from one level to another, and/or change alternating current to direct current or direct current to alternating current is called a substation. Equipment in substations is used to connect electric generators to the system and to terminate and interconnect transmission, subtransmission, and distribution circuits. Power transformers in the substations change the voltage levels to permit interconnecting the generators and circuits forming the electric system.

Substation Types

Transmission substations are constructed at generating stations and at switching points in the transmission system (Fig. 2.1). The transmission substations near generating stations provide a means to connect the electric generators to the system and change the voltage to the appropriate magnitude to transmit the power the required distance. The power is usually generated at 25,000 volts ac or less and stepped up by power transformers to the transmission voltage level of 138,000 volts, 161,000 volts, 230,000 volts, 345,000 volts, 500,000 volts, or 765,000 volts or higher. The power transformers connected to each generator and the transmission lines are switched in and out of service with circuit breakers.

   The transmission substations located at switching points connect various circuits to the network and can serve as a source to subtransmission circuits. Power transformers installed at the substations can change the transmission system voltage from one level to another and the transmission voltage to subtransmission level. The subtransmission circuits originating at a transmission switching substation will serve as a source to distribution substations.

Distribution substations are located near the utilization point in the electric system. The power transformers in the distribution substations change the subtransmission voltage to lower levels and provide a source for the distribution circuits extending short distances to the customer's premises. A circuit switcher installed between the incoming subtransmission line and the power transformer is used to energize and deenergize the power transformer and to isolate the power transformer from the incoming line if a fault develops in the power transformer.

Distribution System

The electric facilities connecting the transmission system to the customer's equipment are called the distribution system. An electric distribution system (Fig. 2.1) consists of several parts: (1) distribution substation, (2) distribution feeder circuits, (3) switches, (4) protective equipment, (5) branch circuits or primaries, (6) distribution transformers, (7) secondaries, and (8) services.

Distribution Feeder Circuits

The conductors for a distribution feeder circuit originate at the terminals of a circuit breaker or a circuit recloser in a distribution substation. The circuits usually leave the substation underground and may be installed entirely underground. The underground cables leaving the substation are commonly called substation exit cables and normally connect to an overhead distribution feeder near the substation. Installing the distribution feeder circuits underground near the substations eliminates multiple circuits on the pole lines adjacent to the substations, thus improving the appearance of the substations and surrounding facilities.

The distribution feeder circuits leaving the substations serve as a source to branch circuits or primary circuits and are often referred to as distribution main-feeder or express feeder circuits. Several distribution feeder circuits will originate at a distribution substation and extend in different directions to serve the customers (Fig. 2.8).

Distribution feeder circuits may have a capacity of 10 MVA or more. The substation exit cables or the underground main-feeder cables for a 13,200Y/7,620-volt, three-phase feeder circuit could have a 750-kCMIL aluminum conductor with solid-dielectric insulation for each phase and a 400-kCMIL bare-copper neutral conductor. The overhead phase conductors might be a 795-kCMIL bare-aluminum or aluminum conductor steel reinforced (ACSR) with a 336.4-kCMIL aluminum or ACSR bare neutral conductor. The substation has a transfer bus, which permits the reclosers to be removed from service for maintenance without deenergizing a feeder circuit. The feeder circuits exit from the substation through underground cables. Figure 2.9 is a typical installation of equipment on a riser pole for substation exit cables connecting to a main-feeder or express-feeder circuit near a substation.

Switches

Distribution feeder circuits, whether overhead or underground, typically have switches installed at strategic locations. These can be used to redirect power flows for load balancing or for sectionalizing to permit repairing of damaged lines or equipment.

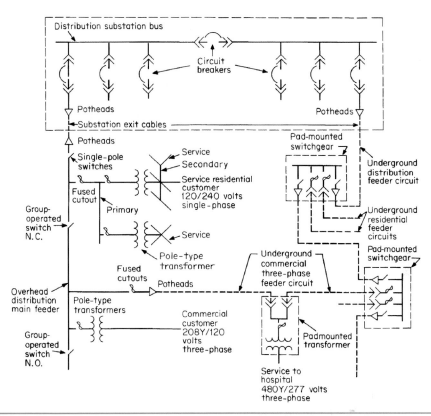

Figure 2.8 Single-line schematic diagram showing a portion of an electric distribution system. Note that distribution feeder circuits originate at the distribution substation bus and extend in different directions from the substation to serve the customers.

Figure 2.9 Riser pole located near a substation with power cables terminated with polymer-body shrink-fit terminators. Lightning arresters protect the cables.

FIGURE 2.10 A distribution fused cutout is being opened using the Loadbuster tool mounted on a universal hot-line stick. In the position shown, Loadbuster has tripped, interrupting the load current. (*Courtesy S & C Electric Co.*)

Single-pole disconnect switches are widely used whenever three-pole switching is not required. Used alone, single-pole disconnect switches without loadbreak interrupters should only be opened when deenergized. However, they can switch load currents to 900 amps when opened with Loadbuster, a portable loadbreak tool (Fig. 2.10).

Three-pole group-operated switches with loadbreak devices are generally installed at appropriate locations in distribution main-feeder circuits. They provide a means for the lineman to sectionalize the feeders or to connect the feeders to other circuits. These switches may be manually operated or power operated and installed on distribution structures or poles or in metal enclosures or vaults. Power-operated switches may be controlled from a remote location.

Protective Equipment

Circuit breakers in the substations serving as a source to distribution circuits are actuated by protective relays that recognize short circuits and grounds on a circuit or line. If a circuit breaker opens, the complete distribution circuit is deenergized, and a large number of customers will have their electric service interrupted.

Lightning arresters and fuses are installed on distribution circuits to minimize the customer outages. Figure 2.11 is a picture of a distribution lightning arrester similar to those mounted on a pole and connected to the top conductor of the circuit in Fig. 2.12. The lightning arresters are designed to bypass electrical surges on the circuits, protecting the insulators and equipment from overvoltages. Fuses are installed in the circuits to isolate branch circuits and equipment from the distribution main-feeder circuit if a fault or short circuit develops on one of the taps to the circuit. It is better to have a fuse blow and interrupt branch circuit service to a small number of customers (100 customers or less) than to have the substation circuit breaker open and deenergize all the customers served from the entire circuit (1,500 customers or more).

Figure 2.13 is a picture of a distribution fused cutout designed to be used on an overhead distribution circuit. The lineman in Fig. 2.13 is wearing a hard hat and safety glasses and is working in the proper position on the pole to open the distribution fused cutout safely. The distribution fused cutouts shown in Fig. 2.13 are located in the circuit in a main feeder where three underground primary conductors connect to the feeder circuit.

FIGURE 2.11   Distribution lightning
arrester equipped with mounting bracket
designed for use on overhead
distribution lines and equipment.
(*Courtesy MacLean Power Systems.*)

FIGURE 2.11   Distribution lightning
arrester equipped with mounting bracket
designed for use on overhead
distribution lines and equipment.
(*Courtesy MacLean Power Systems.*)

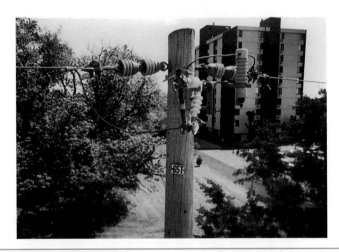

FIGURE 2.12   A fused section of single-phase primary distribution line that is tapped through the lightning arrester prior to the connection to the distribution fused cutout.

Figure 2.14 shows pad-mounted switchgear fuse compartments. The left-hand compartment contains three solid-material feeder circuit power fuses. The right-hand compartment contains three branch circuit fuses. The fuses connect to a bus in the switchgear. The bus is connected through three-pole interrupter switches to the underground distribution main-feeder circuit. The branch circuit conductors are typically smaller than the main feeder circuit conductors.

The feeder circuit fuses can carry up to 600 amps of load current. The solid-material power branch circuit fuses are rated 200 amps or less. Fuses must be selected to carry anticipated loads, interrupt maximum fault currents at that location, and coordinate

FIGURE 2.13   Distribution fused cutout is being opened by the lineman using Loadbuster tool mounted on a universal hot-line stick. The lineman is deenergizing one phase of the three-phase branch circuit that is connected to the main-feeder circuit through distribution fused cutouts. (*Courtesy Northwest Lineman College.*)

FIGURE 2.14   Pad-mounted switchgear showing fuse compartments. Left-hand compartment contains three solid-material feeder circuit power fuses with dual-purpose front barriers removed. Right-hand compartment contains three branch circuit fuses—one dual-purpose front barrier removed, one barrier inserted in the open gap with fuse in the open position, and one barrier hanging in its normal position. (*Courtesy S & C Electric Co.*)

with upstream and downstream protective devices. The fuses connected between the bus and the underground primary feeder cables protect the main feeder circuit from faults on the branch circuits or equipment. Faults or short circuits usually result from cable insulation failures, frequently caused by excavating equipment hitting a cable. After the cause of the fault is isolated, the fuses are replaced and operated by linemen or cablemen using hot-line tools, or "hot sticks." The fuses in the gear will coordinate properly with other fuses in the branch circuit.

Primary Circuits

All high-voltage distribution circuits greater than 600 volts up to 34,500 volts are often referred to as primaries by the lineman. The branch circuits that connect to the main or express feeder distribution circuits are properly called primary circuits. The primary circuits normally connect to the distribution feeder circuit through distribution fused cutouts. The primary circuits may be three-phase (Fig. 2.13) or single-phase circuits (Fig. 2.15).

The size of the fuses protecting the main distribution feeder circuit from a fault on the primary circuit and the size of the primary feeder underground cables are two examples of conditions that will limit the capacity of the circuit.

Figure 2.16 illustrates a three-phase underground primary circuit originating on a riser pole. The underground cables are supplied through distribution fused cutouts connected to the overhead distribution feeder circuit. The branch circuits provide service to small industrial, commercial, and residential customers. The small industrial and commercial customers will normally use three-phase power. The residential areas will normally be provided with single-phase primary service.

Distribution Transformers

The function of the distribution transformer is to reduce the voltage of the primary to the utilization voltage required to supply the customer. The primary voltages vary from 34,500Y/19,920 volts to 4,160Y/2,400 volts. A three-phase circuit with a grounded neutral source operating at 34,500Y/19,920 volts would have three high-voltage conductors and one grounded neutral conductor, a total of four conductors or wires. The voltage

FIGURE 2.15   Single-phase branch circuits are connected to the three-phase distribution main-feeder circuit through distribution fused cutouts. Two of the branch circuits extend away from the reader and one extends toward the reader. A streetlight is mounted on the pole. (*Courtesy S & C Electric Co.*)

FIGURE 2.16   Underground primary cables are terminated in combination distribution fused cutout terminator assemblies called isolating terminators. Lightning arresters are installed on the mounting frame. The arresters provide surge protection for the underground primary cables and the overhead distribution main-feeder circuit. *(Courtesy A. B. Chance Co.)*

between the insulated or phase conductors would be 34,500 volts, and the voltage between one-phase conductor and neutral, or ground, would be 19,920 volts. Similarly, a 13,200Y/7,620-volt distribution feeder would have 13,200 volts between phase conductors and 7,620 volts between a phase conductor and neutral, or ground.

Residential customers are normally provided 120/240-volt single-phase service. The single-phase service would have three wires. Two of the wires require insulation because they will be energized. One of the wires may be uninsulated, or "bare," because it is the neutral wire that is connected to ground. The insulated wires would have 240 volts between the conductors and 120 volts between the insulated wires, or hot wires as they are often referred to by the lineman, and the ground, or neutral wire. Some small older homes may have 120-volt service which would be provided by one insulated, or hot, wire and the ground, or neutral wire. A typical single-phase transformer installed on a pole along the rear lot line of a residential area is shown in Fig. 2.17.

FIGURE 2.17   Typical distribution transformer installation on pole. The primary wire to which the primary winding of the transformer is connected is carried on the insulator on top of the pole. The transformer is a self-protected type complete with a primary lightning arrester, an internal fuse, and a low-voltage, or secondary, circuit breaker. The transformer is rated 7,620-120/240 volts, 15 kVA.

FIGURE 2.18   Single-phase pad-mounted transformer installed on rear lot line of a residential subdivision. Transformer is rated 7,620-120/240 volts, 50 kVA.

The center low-voltage wire connected to the bushing on the side of the transformer tank in Fig. 2.17 is the neutral wire. Pad-mounted transformers are frequently used in underground residential distribution systems. Figure 2.18 shows a single-phase pad-mounted distribution transformer installed along the rear lot line of a residential subdivision. The primary cables connect to the transformer with loadbreak terminators or elbows. In Fig. 2.19, a cableman is preparing the primary cable to install a loadbreak elbow so as to terminate the primary cable on the primary bushing of the single-phase pad-mounted transformer. The single-phase primaries normally feed through the transformer to form a loop. This permits a defective primary cable to be isolated between transformers.

Commercial and light industrial customers normally require three-phase service. The secondary service voltage supplied is usually 480Y/277 volts or 208Y/120 volts. Three-phase pad-mounted transformers (Figs. 2.20 and 2.21) are used with underground primary circuits. The three-phase underground primary cables are installed to form a loop to permit isolating defective underground cables without a long service interruption to customers.

Three single-phase pole-type transformers can be cluster-mounted on a pole to provide three-phase service (Fig. 2.22). The transformers illustrated are single primary bushing transformers connected line to ground.

Network service is provided to many businesses in areas with highly concentrated, large electric loads. Network transformers are designed for installation in a vault usually located under the sidewalk. The network transformers are designed for submersible operation. The primary underground cables terminate in a switching compartment. The primary switch provides a means to deenergize the network transformer without de-energizing the primary circuit that serves as a source to other network transformers. The secondary voltages are usually 208Y/120 volts or 480Y/277 volts. The secondary terminals of the network transformer connect to a network protector. The network protector is a low-voltage circuit breaker and fuse assembly complete with protective relays and control devices designed to isolate the low-voltage network

50 **Chapter Two**

FIGURE 2.19 Cableman is installing terminations on deenergized primary cable in the single-phase pad-mounted transformer. (*Courtesy Northwest Lineman College.*)

FIGURE 2.20 Three-phase pad-mounted transformer. The transformer is rated 13,200Y/7,620-480Y/277 volts, 500 kVA. (*Courtesy ABB Power T&D Company Inc.*)

FIGURE 2.21 Three-phase pad-mounted transformer shown with the door to the primary and secondary compartments in the open position. (*Courtesy ABB Power T&D Company Inc.*)

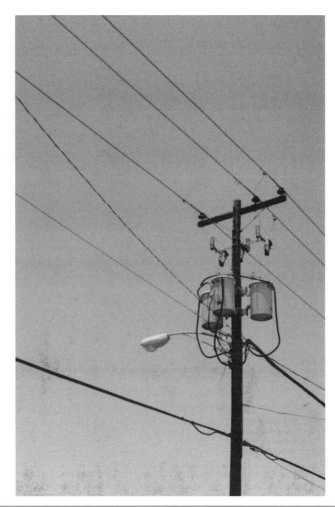

FIGURE 2.22   Three-phase transformer bank rated 13,200Y/7,620-208Y/120 volts, 150 kVA. Primary wires are connected to transformers through distribution fused cutouts. Lightning arresters connect to primary wires to provide surge protection.

from a transformer or primary cable fault. The network is supplied from several underground primary feeder circuits.

Secondaries

The conductors originating at the low-voltage secondary winding of a distribution transformer that extend along the rear lot lines, alleys, or streets past the customers' premises are called secondaries. The secondary conductors for overhead circuits are strung below the primaries and are cabled (Fig. 2.23) or lie in a vertical plane. The secondaries for underground circuits are normally direct buried in the same trench with primary cables. The cables are laid in the trench without separation (Figs. 2.24 and 2.25).

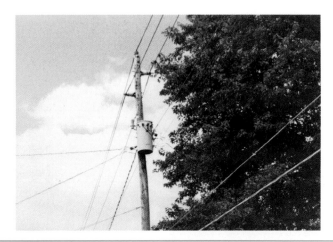

FIGURE 2.23 Completely self-protected (CSP) distribution transformer mounted on pole with primary winding connected to primary, carried on vertical line post three-phase construction. The secondary mains supplied from the transformer are carried on a clevis spool mounted in a vertical plane on the pole.

FIGURE 2.24 Plan-view schematic diagram of an underground residential electric distribution system for installation along a rear lot line. Schematic locates pad-mounted transformers, primary cable, secondary enclosures, secondary cables, and service cables.

Services

The wires extending from the secondary, or distribution transformer, adjacent to the customer's property to the customer's premises are called a service. An overhead triplex cable service extending to a customer's premises is shown in Fig. 2.23. The transformer reduces the voltage from 7,620 volts to 120/240 volts. The triplex service cable has two insulated aluminum conductors and a bare aluminum steel-reinforced neutral wire that provides physical support for the insulated conductors.

Electric System—Emergency Generation

Electric customers offering specialized services may require very reliable electric service. Customers of this type can be provided dual electric service from two distribution feeder circuits. The dual source electric service is normally supplied from two distinct circuits with switching equipment to transfer to the alternative circuit in case of a failure of the principal source of electrical energy. Customers are required to pay the extra

FIGURE 2.25 Sectional view locating primary and secondary cables of an underground distribution system.

cost associated with the extra electrical distribution equipment. Hospitals and similar types of customers will be equipped with emergency electric generators, in addition to dual service from the electric system. The emergency generators are designed to provide sufficient capacity to supply critical loads, such as operating rooms, emergency rooms, and minimum lighting in halls and stairways. The emergency electric generators are designed to start automatically when the normal source of electrical energy is lost. The equipment is installed in a manner to ensure that the generator does not feed back into the electric distribution system. It is essential for the lineman and the cableman to protect themselves from a possible malfunction of the equipment which would permit a feedback of electricity from the generator through the customer's service to the distribution facilities.

Electric System Control

The modern electric system is operated from a control center. Equipment in the control center is used to analyze the electric system operation continuously. The output of the electric generators is varied automatically to match the energy supplied with the customer's demand and the purchases and sales through interconnected electric systems. The control center is located at a strategic point in the electric-system service area. The control-center building is designed to protect the operating personnel and control equipment from severe weather conditions. The operating personnel responsible for the electric system use schematic diagrams displays to give them a picture of the entire electric system.

Computerized systems provide essential information concerning the electric system, such as system net load, generation, net tie-line loads, area control errors, and frequency. Computer workstation monitors display schematic diagrams of the electric system, including generating-station equipment, transmission circuits, substation equipment, and distribution circuits. The monitors connect to digital computers that link up with supervisory control equipment at the generating stations and substations. Electric system service interruptions throughout the electric system are continuously tracked within the electric outage management system. The electrical equipment, such as circuit breakers, is operated by actions contained within the computer software. As the circuits and generators are switched in and out of service, the changes in electrical energy flows are made automatically by the computer to correctly display the information.

The electric-system coordinators use two-way radio equipment to communicate with linemen and cablemen working on the electric transmission and distribution equipment. The electric-system coordinator controls all switching, the placement of hold-off tags, and the installation of protective grounding equipment. Technicians assist the coordinators as needed. Permission to install protective grounding equipment is not given until all switching has been completed to deenergize and isolate the circuit and hold-off tags have been installed. The workers installing the grounds must test the circuit to verify that it is deenergized before the grounding operation is initiated. The correct procedure for installing grounding equipment is described in Chaps. 27 and 46. The workers must report to the system coordinator when the grounding operation is completed. When the work is completed, the lineman must obtain permission from the system coordinator before removing the grounding equipment from the isolated circuit or equipment. The grounds must be removed before the system coordinator authorizes the removal of hold-off tags and the operation of switches to reenergize the circuit. All the switching, tagging, and grounding operations must be completed in accordance with the rules in the *National Electrical Safety Code* (ANSI-C2).

Uninterruptible Power-Supply System

Electric distribution systems experience voltage dips and momentary interruptions as a result of short circuits or electrical surges. The fluctuations on the distribution system are normally caused by lightning discharges, animals bridging an insulator, or switching surges. Lightning may cause an overvoltage which results in an insulator flashover. Squirrels climb poles, get on distribution transformers, and can short out the transformer bushing. Power-system consumers may have equipment such as computers, machine tools, instruments, and other devices that require a continuous and unvarying voltage power supply. An uninterruptible power supply is used to provide a continuous reliable pure sine wave source for critical electrical loads. It must be designed and constructed specifically for the load to be served, and therefore, it is normally owned by the consumer of the electric power. Figure 2.26 is a schematic wiring diagram of an uninterruptible power supply that includes a solid-state rectifier or charger, a battery, an inverter to produce alternating current from the direct current, a solid-state or static switch, a manual transfer switch, and a motor-generator set to provide auxiliary power for the critical electric load. The solid-state rectifier filters out any variations or surges in the ac power source and keeps the battery fully charged. The battery provides power if the distribution circuit is interrupted for a short period of time. A motor-generator set would normally be used in case of a long outage of the electric distribution source circuit. The uninterruptible power supply will operate reliably to supply critical loads.

FIGURE 2.26 Schematic wiring diagram of an uninterruptible power-supply system.

Microgrids

Microgrids are small power systems that are interconnected to the main power system for the emergency supply and delivery of electricity. If service from the main power system is interrupted, the microgrid isolates itself from the main power system and continues to operate independently with no interruption of service. A microgrid utilizes a combination of several types of distributed generation resources (fuel cells, solar panels, gas generators), energy storage devices (batteries, pumped storage hydroelectricity compressed air energy storage), inverters converting dc to ac, transformers and energy control devices.

Electric System—Power Quality

Electric distribution companies have an obligation to provide quality power to their customers. For example, an incandescent light bulb that flickers with the start-up of a large motor is a nuisance to the customer. This problem is resolved by one of several methods. The methods included changing out an undersized transformer, replacing undersized secondary wiring, seeing that the loads were balanced on each phase, or ensuring that some type of reduced voltage soft starting control was installed on the motor. Electric equipment is able to tolerate a few cycles of irregular power. The distribution companies have guidelines regarding transformer sizing, allowable secondary footages, flicker limitations, and limits on allowable starting currents for motors.

With the advent of electronically controlled electrical equipment on the load side of the electric meter, a greater dependence on distortion-free power has emerged. The newer electronic equipment is more sensitive to the voltage provided by the power supply. The electronic components of the devices are dependent upon a pure sine wave voltage supply.

Electric equipment or devices can be of a linear or nonlinear nature. A linear device is one in which the load resembles that of a resistor. The relationship of the current and voltage is predictable in some type of well-defined linear manner. Reactive devices also exhibit linear relationships if the frequency for the device is fixed. Reactance is either inductive or capacitive. Current lags voltage in inductive devices. Current leads voltage in capacitive devices. Linear equipment is not as affected by typical power quality problems.

Electronic equipment contains devices that are of a nonlinear nature. Nonlinear devices contain electronic components that have some inductive or capacitive characteristics that impact the relationship of voltage, current, and frequency. Electronic devices require an alternating current to direct current to alternating current conversion utilizing rectifiers and inverters. The alternating current wave-shape does not mirror that of the voltage wave-shape.

The range of tolerable voltages for computer equipment that has become the industry design guideline for electronic equipment is known as the Information Technology Industry (ITI) Council curve (previously recognized as the CBEMA [Computer and Business Equipment Manufacturer's Association] curve). The horizontal axis of the curve is defined in cycles at 60 Hz or in seconds and the vertical axis is defined in the percentage of rated voltage. Examples of the upper design limit of the curve that should not be exceeded without voltage breakdown concerns are 300% rated voltage at 0.01 cycles, 200% rated voltage at 0.1 cycles, 130% rated voltage at 0.5 cycles, and 106% rated voltage for 1,000 cycles or greater. The minimum lower limit that the equipment will tolerate without noise and voltage breakdown problems is 30% rated voltage at 1.0 cycle, 58% rated voltage at 10 cycles, 70% rated voltage at 100 cycles, and 87% at 1,000 cycles. IEEE Standard 1346-latest revision, "IEEE Recommended Practice for Evaluating Electric Power System Compatibility with Electronic Process Equipment" provides further explanation on the significance of the curve.

According to ANSI-IEEE Standard 84.1, the electric distribution supplier should maintain steady state voltage within ±5% of the nominal voltage. If the nominal voltage is 120 volts, then the maximum delivered voltage is 126 volts and the minimum delivered voltage is 114 volts.

Several types of power quality problems have been defined in IEEE Standard 1159-latest revision, "IEEE Recommended Practice for Monitoring Electrical Power Quality." The problems include: (1) voltage sags due to motor starting, lightning, or fault clearing; (2) voltage surges or voltage swells due to lightning or high-impedance fault currents; (3) flickering due to loose connections at the transformer, in the service lines extended to the meter, or in the electric meter; (4) harmonic waves in the voltage signals due to nonlinear loads such as those of adjustable speed drive motors or high-intensity discharge lights; and (5) momentary interruptions due to faults on the power system causing circuit breaker tripping or relaying.

Harmonics are alternating current and voltage waveforms with frequencies that are multiples of the system frequency. If the system frequency is 60 Hz, the second harmonic is 120 Hz, the third harmonic is 180 Hz, and so on. Harmonic distortion of the voltage and current waveform have a negative effect on equipment. In an ac power system, the odd harmonics are the significant contributors to the total harmonic distortion. Harmonics are addressed in great detail in IEEE Standard 519-latest revision, "Recommended Practices and Requirements for Harmonic Control in Electric Power Systems."

Power quality investigations are required to diagnose power quality complaints and problems. The investigations more often than not involve the lineman. It should be noted that a high percentage of power quality problems, such as poor wiring and improper grounding, are internal to the facility and reside on the load side of the electric meter. Adherence to the NFPA 70 National Electrical Code, latest edition (NEC) Article 250 Grounding and Article 310 Conductors for General Wiring will minimize the likelihood of many of the internal problems. Even so, during the initial stages of the investigation, all possible power quality causes must be reviewed including all of the sources of system impedances that are encountered up to the affected device.

Equipment used to investigate the problem is as follows: true rms (root-mean-square) voltmeters and ammeters, oscilloscopes, and disturbance analyzers. More advanced diagnosis tools include harmonic analyzers, infrared detectors, gauss meters, static meters, and flicker meters.

Equipment used to mitigate the power quality problems includes filters, surge suppressors, isolation transformers, power conditioners, uninterruptible power supplies (UPS), distributed generators, superconducting magnetic energy storage (SMES), mechanical transfer switches, solid state transfer switches, and fuel cells.

The *Power Quality Primer*, Barry Kennedy, McGraw Hill, and *Electrical Power Systems Quality*, Roger C. Dugan, Mark F. McGranaghan, and Wayne H. Beaty, McGraw Hill are excellent reference books on the topic.

Substations

Substation Types

An electrical substation is an installation with electrical equipment capable of interrupting or establishing electrical circuits and changing the voltage, frequency, or other characteristics of the electrical energy flowing in the circuits.

Transmission Substations

Alternating-current (ac) electric transmission 345-161 kV (Fig. 3.1) substation transformers are located at appropriate points in a network of transmission lines. The transmission substation transformers support the transmission system and the smaller subtransmission and distribution substation transformers. The transmission substation contains equipment used to sectionalize the electric transmission system when a fault or short circuit develops on one of the circuits. Circuit breakers in the transmission substation are used to switch generating and transmission circuits in and out of service. The faulted circuit is switched out of service automatically and de-energized, protecting the remaining portion of the transmission network from trouble. The electric transmission system is designed with redundant loops and multiple circuit connections to permit loss of service to any one circuit without interrupting electric service to substations and customers.

Transmission substations located at electric generating stations (Fig. 3.2) have power transformers designed to raise the voltage from generation value, usually in the range of 13,200 volts to 25,000 volts three-phase, to transmission level, 138,000 volts to 765,000 volts or higher.

Subtransmission Substations

Electric substations (Fig. 3.3) with equipment used to convert high voltage, extra-high-voltage (EHV), or ultra-high-voltage (UHV) transmission lines to the intermediate voltage subtransmission lines or to switch subtransmission circuits operating at voltages in the range of 34.5 kV to 161 kV are referred to as subtransmission substations. Subtransmission circuits and substations are located near to high-load concentrations, usually in urban areas. Subtransmission circuits supply distribution substations. One type of specialty subtransmission substation is the gas-insulated substation. Gas-insulated substations (GIS) are installed to provide additional load capacity in heavily populated areas. GIS lines and switchgear are very efficient.

Distribution Substations

Substations located in the middle of a load area are called distribution substations. These substations may be as close together as 2 miles in densely populated areas. The substations contain power transformers that reduce the voltage from subtransmission levels to

FIGURE 3.1   An autotransformer, 560 MVA three-phase, 345-161 kV installed in a transmission substation. (*Courtesy MidAmerican Energy Company.*)

FIGURE 3.2   Power transformers in an ac transmission substation located adjacent to generating station. Power transformers raise the generated voltage from 25 kV to 765 kV. Each single-phase transformer has a capacity of 317 MVA. The three-phase bank capacity is 950 MVA. The fourth transformer serves as a spare.

FIGURE 3.3 Subtransmission ac substation equipped with oil circuit breakers, lightning arresters, disconnect switches, and potential transformers. Substation is located in the fringe area of a large city and serves as a switching center for the subtransmission circuits.

distribution levels, usually in the range of 4.16Y/2.4 kV to 34.5Y/19.92 kV. The transformers are normally equipped to regulate the substation bus voltage. Circuit breakers or circuit reclosers in the distribution substations are installed between the low-voltage bus and the distribution circuits. The distribution circuits vary in capacity from approximately 5 MVA to 20 MVA (1 MVA = 1,000 kVA). Large distribution substations have power transformers with a capacity as large as 100 MVA and serve as a source to as many as 10 distribution circuits. If the power transformers do not have automatic tap-changing underload equipment, it is usually necessary to install voltage regulators (Figs. 3.4 and 3.5).

FIGURE 3.4 A 22 MVA 69-13.8 kV load tap changing substation transformer installed in a distribution substation. (*Courtesy MidAmerican Energy Company.*)

FIGURE **3.5**   Portion of a distribution substation showing power-transformer distribution-voltage bushings, lightning arresters, gang-operated interrupter switch, potential transformers connected to the overhead distribution-voltage bus through solid-material power fuses and three single-phase voltage regulators connected to the bus through bypass switches. (*Courtesy S&C Electric Co.*)

Substation Function

An electric substation is an installation in which the following functions can be accomplished for the power system:

1. Voltage changed from one level to another level
2. Voltage regulated to compensate for system voltage changes
3. Electric transmission and distribution circuits switched into and out of the system
4. Electric power quantities flowing in the transmission and distribution circuits measured
5. Communication signals connected to the circuits
6. Lightning and switching surges eliminated from the electric system
7. Electric generators connected to the transmission and distribution system
8. Interconnections between the electric systems of various companies completed
9. Reactive kilovolt-amperes supplied to the transmission and distribution circuits and the flow of reactive kilovolt-amperes on the transmission and distribution circuits controlled
10. Alternating current converted to direct current or direct current converted to alternating current
11. Frequency changed from one value to another

Substation Equipment

Electric substations contain many types of equipment. The major components of equipment consist of the following:

Power transformers	Shunt reactors
Tap changing equipment	Meters
Buses and steel structures	Substation protective relays
Lightning arresters	Supervisory control
Circuit switchers	Remote terminal units
Circuit breakers	Digital fault recorders
Oil circuit breakers	Capacitors
Air circuit breakers	Voltage regulators
Vacuum circuit breakers	Control house
Disconnect switches	Conduits
Coupling capacitors	Control wires
Potential transformers	Control panels
Current transformers	Power-line carrier equipment
High-voltage fuses	Microwave equipment
Metal-clad switchgear	Batteries

Power Transformers

Transformers are used to change the voltage from one level to another, to regulate the voltage level, and to control the flow of reactive kilovolt-amperes in the power system. Power transformers installed in transmission substations will normally be constructed for and operate at voltages in the range of 138,000 volts to 765,000 volts or higher.

Most substations have three-phase transformers (Fig. 3.6) installed. Some substations will have three single-phase transformers installed in a bank (Fig. 3.7). Often, a fourth single-phase transformer will be located at the substation to serve as a spare. If one of the transformers connected in the bank fails, the spare is energized to expedite restoration of electric service.

The capacity of the transmission substation transformers will normally be in the range of 1,000 MVA to 50 MVA. Power transformers installed in distribution substations will normally be constructed for and operate at the lower voltages and smaller capacities. The voltages will generally be in the range of 161,000 volts to 4,160 volts and the capacities in the range of 50 MVA to 5 MVA. Figure 3.8 is a picture of a power transformer that is part of a distribution substation.

Tap Changing Equipment

An option that may be added to the substation transformer is tap changing equipment. The tap changer feature allows the voltage to be adjusted in increments as system loading conditions require. The controls of the tap changing equipment are installed in a panel mounted on the side of the power transformer. Remote control of the tap changing equipment is usually installed as well. Chapter 19 describes in more detail the tap changing functionality.

FIGURE 3.6   Large power transformer located in a transmission substation. Transformer is connected to the 138 kV and 345 kV buses in the substation. (*Courtesy ABB Power T&D Company Inc.*)

FIGURE 3.7   Transmission substation with three single-phase power transformers connected into a three-phase bank. (*Courtesy Westinghouse Electric Corp.*)

FIGURE 3.8 Distribution power transformer with tap-changing underload equipment in compartment in foreground. The distribution voltage connection to the transformer is completed through the bus duct on the right side of the picture to metal-enclosed switchgear.

Buses and Steel Structures

The substation buses are current-carrying devices that interconnect the individual pieces of substation equipment. The substation bus tubing (or bus bar) is aluminum or copper. Steel structures provide support for insulators, which are used to terminate lines, and support buses. Disconnect switches and other equipment are mounted on the steel structures (Fig. 3.9).

Lightning Arresters

The lightning arresters protect the electric system and the substation equipment from lightning strokes and switching surges. The lightning arresters are installed in a substation near the termination of aerial circuits and close to the more valuable pieces of equipment, such as power transformers. The power transformer in Fig. 3.6 has lightning arresters mounted on the radiators which are connected to both the 345 kV and 138 kV transformer bushings.

FIGURE 3.9 The substation's steel structures provide support for insulators, which are used to terminate lines and support current carrying buses that interconnect the equipment. (*Courtesy MidAmerican Energy Company.*)

Lightning arresters contain semiconductor blocks, which limit the magnitude of high-surge voltages, permit the large-surge currents to pass harmlessly to ground, and interrupt the power-follow current after the surge is eliminated. Metal-oxide surge arresters function like a ceramic capacitor at normal line voltage, limiting the flow of current to ground. When a high voltage begins to build up across the semiconductor blocks, they provide a low-impedance path to ground, which permits the surge current to flow to ground, limiting the voltage buildup and preventing the equipment from being damaged. The semiconductor blocks in the arresters are manufactured using zinc oxide material.

Circuit Switchers

Circuit switchers employ interrupters for switching and protection of transformers, lines, cables, and capacitor banks and have fault-interrupting ratings suitable for use in protecting medium- to large-load transformers (Fig. 3.10). Models are available with and without integral disconnect switches. Operation of circuit switchers is initiated by manually operating a switch, by remote supervisory control equipment, or by substation protective relays that automatically sense abnormal predetermined system or equipment conditions or electrical failures (faults). The circuit switcher shown in Fig. 3.11 is used to energize and de-energize a substation capacitor bank. The substation capacitor bank is used to supply vars to the electric system and to regulate the substation voltage.

Circuit Breakers

Oil, air, gas, and vacuum circuit breakers are used to switch electric circuits and equipment in and out of the system. The contacts of the circuit breakers are opened and closed by mechanical linkages manufactured from insulating materials and utilizing energy from compressed air, hydraulics, or charged springs. Some of the high-voltage circuit breakers utilize compressed air to operate the contacts and interrupt the current flow when the contacts are open.

Figure 3.10 Distribution substation supplied from a 69-kV subtransmission circuit. The substation line terminal structure supports surge arresters. A circuit switcher is used to energize and de-energize the power transformer. The power transformer reduces the subtransmission voltage to 24,940Y/14,420 volts distribution voltage. (*Courtesy S&C Electric Co.*)

FIGURE 3.11 Substation capacitor bank used for power-factor correction and voltage regulation. Capacitors are individually fused. Bank switching is provided by a circuit switcher. Potential transformer connected to bank ungrounded neutral conductor provides sensing voltage for protective control devices. (*Courtesy S&C Electric Co.*)

Operation of the circuit breakers is initiated, utilizing dc circuits, by manually operating a switch, by remote operation of supervisory control equipment, or by relays that automatically recognize predetermined abnormal conditions or electrical failures in the system. Various types of circuit breakers can be seen in Figs. 3.12 to 3.17.

Disconnect Switches

High-voltage equipment is usually installed in substations with disconnect switches in series with it to facilitate isolating the equipment for maintenance. The disconnect switches are normally operated by linkages connected to a lever or gear-type handle or

FIGURE 3.12 Three-phase high-voltage power circuit breaker with vertical break group-operated disconnect switches installed as an isolator. (*Courtesy ABB Power T&D Company Inc.*)

FIGURE **3.13** Cutaway view of one pole of a three-phase high-voltage power circuit breaker showing power interrupter, contact pull rod, bushing through rods, and bushing-mounted current transformers. (*Courtesy ABB Power T&D Company Inc.*)

FIGURE **3.14** A 3,000 amp 345 kV power circuit breaker with insulated composite bushings being installed at an extra-high-voltage (EHV) substation. (*Courtesy ABB Power T&D Company Inc.*)

FIGURE 3.15 West side view of a 3,000 amp 345 kV power circuit breaker installed in an extra-high-voltage (EHV) substation. (*Courtesy MidAmerican Energy Company.*)

a motor operator for remote operation (Figs. 3.18 and 3.19). The switches are designed to interrupt charging current but—unless they are specially constructed—not load current.

The disconnect switches shown in Figs. 3.18 and 3.19 on each side of the circuit breaker can be operated after the circuit breaker contacts are opened to visibly isolate the circuit breaker. The disconnect switches are operated by the motor-driven switch operator with a geared torque-transfer system which moves the interphase rods through a gear arrangement and rotates the insulators, causing the switch blades to move and open or close the switch contacts.

FIGURE 3.16 East side view of a 3,000 amp 345 kV power circuit breaker installed in an extra-high-voltage (EHV) substation. (*Courtesy MidAmerican Energy Company.*)

Figure **3.17** A 2,000 amp 69 kV oil-filled power circuit breaker installed in a 69-13.8 kV distribution substation. (*Courtesy MidAmerican Energy Company.*)

Figure **3.18** Front view of a 3,000 amp 345 kV air break substation disconnect switch installed in a transmission substation. (*Courtesy MidAmerican Energy Company.*)

FIGURE 3.19   Rear view of a 3,000 amp 345 kV substation disconnect switch installed in a transmission substation. (*Courtesy MidAmerican Energy Company.*)

Coupling Capacitors

Communication signals in the form of high-frequency voltages are transmitted to the transmission lines through coupling capacitors. Some of the coupling capacitors are equipped with potential devices which make it possible to measure the voltage on transmission-line circuits. The coupling-capacitor potential devices are accurate enough to be used for supplying voltages to protective relays but—unless they are specifically compensated—not accurate enough to supply voltages for meters designed for billing purposes. Figure 3.20 shows a coupling capacitor.

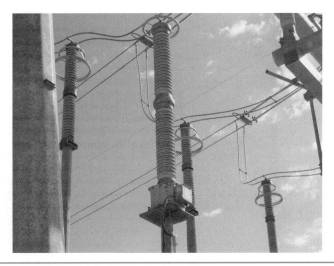

FIGURE 3.20   A 345 kV coupling capacitor voltage transformer (CCVT) device installed in the extra-high-voltage (EHV) substation. (*Courtesy MidAmerican Energy Company.*)

Potential Transformers

High voltages are measured by reducing the voltage proportionately with a potential transformer (Fig. 3.21) which has its high-voltage winding connected to the transmission or distribution circuit and its low-voltage winding connected to a meter or relay or both. Potential transformers are required to provide accurate voltages for meters used for billing industrial customers or connecting utility companies. If single-phase transformers are used, three transformers are generally required to measure power on a three-phase circuit.

Current Transformers

The primary winding of a current transformer is connected in series with the high-voltage conductor. The magnitude of amperes flowing in the high-voltage circuit is reduced proportionately by the ratio of the transformer windings. The secondary winding of the current transformer is insulated from the high voltage to permit it to be connected to low-voltage metering circuits. Current and potential transformers supply the intelligence for measuring power flows and the electrical inputs for the operation of protective relays associated with the transmission and distribution circuits or equipment such as power transformers.

High-Voltage Fuses

High-voltage fuses are used to protect the electrical system from faults in such equipment as potential transformers or power transformers. A group-operated high-voltage disconnect switch and three power fuses are mounted on the terminal structure. The fuses are connected in series with the power transformer that serves as a source for the switchgear.

Metal-Clad Switchgear

Outdoor metal-clad switchgear provides a weatherproof housing for circuit breakers, protective relays, meters, current transformers, potential transformers, bus conductors,

FIGURE 3.21   Attached to the 345 kV substation bus are three single-phase potential transformers (PT). (*Courtesy MidAmerican Energy Company.*)

and other items necessary to provide electric-system requirements. Indoor switchgear must be installed in a building for protection from the elements. For the typical 13.2-7.6 kV distribution substation, the switchgear consists of a 2,000 A bus and six (6) breaker cells. The breaker cells are one 2,000 A main breaker, four 1,200 A feeder breakers, and one 2,000 A future bus-tie breaker that will serve as the mobile sub connection. One of the feeder cells is equipped with a capacitor bank isolation switch.

Circuit Switchers

The importance of the circuit switcher cannot be emphasized enough and thus are discussed once more in this chapter. Circuit switchers serve to provide reliable, economical switching and protection for transformers, single-shunt capacitor banks, line-connected and testing-connected shunt reactors, lines, and cables (Figs. 3.22 and 3.23).

FIGURE 3.22 A 3,000 A 345 kV circuit switcher installed in a transmission substation for a 50 MVA shunt reactor. (*Courtesy MidAmerican Energy Company.*)

FIGURE 3.23 A close up view of the 3,000 A 345 kV vertical break circuit switcher installed in the open position. The downstream device being protected is a 50 MVA shunt reactor. (*Courtesy MidAmerican Energy Company.*)

Shunt Reactors

When the capacitive reactance from extra-high-voltage transmission-line circuits exceeds the ability of the system to absorb the reactive kVA (kilovars), shunt reactors are installed. The shunt reactor shown in Fig. 3.24 is a three-phase single-winding transformer which serves as an inductive reactance on the power system, thus neutralizing the capacitive reactance associated with the long extra-high-voltage transmission line. Shunt reactors are usually located in substations and connected to a transmission-line terminal through a disconnect switch.

Meters

Meters mounted on control panels are used to measure the energy flowing in a circuit or through a piece of equipment. Some of the meters are utilized to bill industrial customers and adjacent utility customers. Interstate flows of electrical energy are measured in the substations that interconnect electric power systems of two different entities.

Substation Protective Relays

Substation protective relays installed on control panels are used to sense electrical failures on transmission and distribution circuits or in pieces of substation equipment, such as power transformers, substation bus, reactors, capacitors, and circuit breakers. Some relays are used to supervise, monitor, and perform predetermined operations as required by system conditions. Relays automatically operate their contacts to properly identify the source of trouble and remove it from the electrical system. Protective relay

FIGURE 3.24   A 50 MVA, 345 kV shunt reactor installed in an extra-high-voltage (EHV) substation. (*Courtesy MidAmerican Energy Company.*)

systems provide overcurrent, differential current, overvoltage, under voltage, over frequency, under frequency, directional, and distance protection.

Supervisory Control
The supervisory control equipment permits the remote control of substations from a system control center or other selected point of control. Distribution substations are normally unattended and operate automatically or are controlled remotely by supervisory control equipment from a central control center. This equipment is used to open and close circuit breakers, operate tap changers on power transformers, supervise the position and condition of equipment, and telemeter the quantity of energy flowing in a circuit or through a piece of equipment.

Remote Terminal Unit
The remote terminal unit (RTU) is the interface between the supervisory control and the substation equipment. The RTU is a monitor and transmitter of substation and electric system data.

Digital Fault Recorder
The digital fault recorder records fault data and provides accurate sequence of events for system disturbances on the electric system.

Capacitors
Reactive kilovolt-amperes can be supplied to the electrical system by connecting banks of capacitors to the distribution circuits in substations or out on the distribution lines to neutralize the effect of customer-inductive loads. Capacitors used in this manner help to control voltages supplied to the customer by eliminating the voltage drop in the system caused by inductive reactive loads. Capacitors in distribution substations are usually mounted in metal cubicles. The capacitors, mounted on the racks in the cubicles, are usually single-phase, single-bushing units rated 100-kVAR through 400-kVAR capacitance, 60 Hz, and a voltage consistent with the distribution system. They are connected between each of the three-phase conductors and ground. Capacitor units for the higher distribution voltages and those connected to conductors energized at transmission voltages are normally mounted on open-type racks. The capacitor units are connected in series and cascaded to provide equipment that can be connected to the higher voltage distribution and transmission systems.

Voltage Regulators
The voltage regulators maintain the system voltage on the distribution circuits. As settings on the voltage regulators are adjusted for various load conditions that occur, the desired voltage is obtained. Chapter 19 describes in more detail the voltage regulators and Chap. 40 describes voltage regulation.

Control House
The substation control house is used to protect the control equipment, including switchboard panels, batteries, battery chargers, supervisory control, power-line carrier, meters, and relays, from the elements. Underground and overhead conduits and control wire are installed to connect the controls of all the equipment in the substation to the control panels normally installed in the control house. Each substation usually contains several thousand feet of conduit and miles of control wire.

FIGURE 3.25   Street view of a decorative wall used to camouflage a distribution substation. (*Courtesy MidAmerican Energy Company.*)

Decorative Walls
Some communities insist that the entire substation, or portion thereof, be hidden from the general public via a decorative wall (Fig. 3.25).

Control Panels
The control panels (Figs. 3.26, 3.27, and 3.28) installed in a control house provide mountings for meters, relays, switches, indicating lights, and other control devices.

FIGURE 3.26   Control panels with meters, relays, and control devices mounted for operation of a substation. (*Courtesy MidAmerican Energy Company.*)

FIGURE 3.27   The 161 kV line protection panel contains two sets of primary and backup relays, one set of primary and backup relays for each 161 kV line. The breaker control switches will also be located on the line panel. (*Courtesy MidAmerican Energy Company.*)

FIGURE 3.28   Substation relay panel controls. (*Courtesy MidAmerican Energy Company.*)

Power-Line Carrier Equipment

The power-line carrier equipment provides high-frequency voltages to be used for transmitting voice communications or telemetering signals on high-voltage transmission line circuits. In the case of voice communications, the sound frequency modulates a high-frequency signal connected to the transmission-line circuit by means of coupling capacitors. This equipment permits using the line conductors for communication, relaying, supervisory control, and metering in addition to the transmission of electric power.

Microwave Equipment

Radio signals used for point-to-point communications between substations or other power-system facilities operating in the megahertz frequencies are called microwave. The frequencies range from 952 MHz to 13,000 MHz or higher. Microwave radio signals are used for communication channels, protective relaying, supervisory control, and remote metering.

Batteries

Control batteries supply energy to operate circuit breakers and other equipment. It is necessary to use dc control systems with a storage battery as a source to make it possible to operate equipment during periods of system disturbances and outage. Battery chargers are used to automatically keep the batteries charged completely to provide sufficient emergency power for all necessary operations.

High-Voltage Direct-Current (HVDC) Substation

HVDC Converter Stations

Converter stations are located at the terminals of dc transmission lines. The converter stations change alternating current to direct current and invert direct current to alternating current using thyristor semiconductor devices. The converter stations can be located at an ac generating station, and the generator can be connected to the converter station power transformer. Converter stations are frequently located at a transmission substation to connect an ac system to a high-voltage dc transmission circuit. Two converter stations can be located adjacent to each other to connect two ac transmission systems together even though they are not synchronized.

Transmission Circuits

Alternating-current (ac) transmission is the movement of ac electrical energy from one location to another, such as from a terminal at a generating station to a substation near a load center.

Direct-current (dc) transmission is the movement of dc electrical energy from one location to another location several hundred miles away. One dc transmission line terminal is at a convertor station of the energy generation source. The second dc transmission line terminal is located in a convertor station located at the load center. Direct current is economically used to transmit large blocks of power long distances.

Transmission Circuit—History

The first electric generators produced direct current. The generators operated at a voltage matched to the load to be served. A Swiss engineer named Thury developed a high-voltage dc transmission system in 1899.

The invention of the ac transformer made it possible to step voltages up and down. Alternating-current generators used with transformers made it practical to locate the generators remote from the utilization area. Polyphase systems were developed because of improved efficiency over single-phase systems and reduced costs for a given quantity of power required to meet the customers' needs. The early systems generated, transmitted, and utilized two-phase ac power. The two-phase system is now obsolete. Three-phase ac systems with a frequency of 60 Hz (hertz) exist throughout the United States and Canada. The development of water power sites made it necessary to transmit electrical energy long distances from the generating stations to the factories located in the cities. High-voltage equipment was developed to permit the transmission of electrical energy from the hydroelectric generating stations to the load.

The development of transmission circuits across the continent has increased the reliability of the electric system, reduced the operating and standby reserve-generating requirements, and made possible both the interchange and sale of electrical energy between utility companies, thus reducing costs and permitting the installation of large, economical generating units. Open access to lines will allow customers to negotiate their own sales independent of the local operating utility, further reducing costs and delaying the need for generation station expansion, as ideally, the power will be supplied by the areas with excess generation capacity to those areas deficient in their ability to meet demands.

The voltages used on transmission circuits have been steadily rising. A 2,000-volt ac transmission line was constructed in Italy in 1886. In 1902, the maximum ac voltage rose to 80,000 volts and remained there for 5 years. In 1913, it reached 150,000 volts and stayed there for 7 years. In 1922, it reached 220,000 volts, which remained the highest

FIGURE 4.1   Transmission-line structure being constructed at 345 kV to provide greater stability to the electric power grid.

voltage until 1936, when the 286,000-volt line from Hoover Dam to Los Angeles was put in operation. In 1952, a 345,000-volt grid was put into operation in Ohio. A typical structure utilized for a 345 kV transmission circuit is shown in Fig. 4.1.

Virginia Electric and Power Company placed in service a 500,000-volt line, 350 miles long designed to transmit 1 million kVA in 1964. American Electric Power Company energized a 765,000-volt circuit (Fig. 4.2) in 1969. American Electric Power Company has constructed approximately 2,000 miles of 765,000-volt ac electric transmission lines across the states of Indiana, Ohio, Kentucky, Virginia, and adjacent areas. Experimental transmission lines have been built and tested to operate at voltages in the range of 1,000 kV to 1,500 kV. China has energized a 1,000 kV ultra-high voltage alternating current (UHV AC) 650 km long transmission line, with an electricity transfer capacity of 5,000 MW from the northern to the eastern and central parts of the country. The Jindognan-Nanyang-Jingmen UHV AC transmission project began commercial operation in January 2009. The UHV AC was built by the country's largest transmission utility, State Grid Corporation of China (SGCC).

Underground transmission circuits have been installed in large metropolitan areas and other places where overhead transmission lines are not practical. The first underground installations were developed for telegraph circuits. A cable insulated with strips of India rubber was used in 1812. Telegraph cables consisting of bare copper conductor

FIGURE 4.2   Transmission-line tower supporting conductors energized at 765 kV. (*Courtesy American Electric Power Service Corp.*)

drawn into glass tubes and joined together with sleeve joints and sealed with wax were used in 1816. Vulcanized-rubber-insulated cables proved to be more successful.

Edison designed a buried system using copper rods insulated with a wrapping of jute. Two or three insulated rods were drawn into iron pipes, and a heavy bituminous compound was forced in around them. The insulated conductors were laid in 20-ft sections and joined together with specially designed tube joints from which taps could be taken if desired. The underground system was used to distribute energy for lighting in New York City. Overhead electric lines are prohibited in sections of Washington and Chicago. New York City enacted a law forcing the removal of all overhead wires from the streets in 1884.

The short life of the early underground systems led to the development of rubber-insulated, lead-covered cables installed in ducts. Cables that were insulated with paper saturated with oil in a lead pipe were developed in England in 1890 and operated at 10,000 volts ac. There was approximately 5,000 miles of underground cables constructed with copper conductor, insulated with paper-saturated oil, and covered with a lead sheath operating at voltages between 6.6 kV and 25 kV in the United States in 1921.

Underground transmission circuits are more expensive than overhead transmission lines. The high cost of underground installations limits their installation. Factors considered in determining the necessity to invest the larger amounts of money for an underground transmission circuit include:

1. City ordinances
2. The availability and costs of rights-of-way for overhead transmission lines compared with underground transmission circuits
3. The need to provide mechanical protection for the circuits
4. The congestion, and the appearance, of the area where the circuit is to be installed

New York City is supplied by an extensive underground transmission system with some cables operating at a voltage of 345 kV ac. Pipe-type cables are used extensively for underground transmission circuits. These cables operate at voltages in the range of 69 kV to 500 kV ac. Research is in progress on cables to operate at 765 kV ac. Charging current as a result of the capacitance between the phase conductors and their shielding conductors limits the ac power-carrying capacity of cable circuits. Uncompensated cables 26 miles long operating at 345 kV ac will have all the cable capacity absorbed by the charging current.

Direct-current transmission was initiated in Germany in 1883. The system consisted of generators operating at 2,400 volts dc transmitting 5 kW over 8 km (kilometers). Transmission of electrical energy by the Thury system was developed in 1902. Generators were connected in series to transmit 150 amps at a maximum voltage of 27,000 volts. The current in the transmission line was kept constant by varying the output voltage of the generators and switching generators in and out of service. The load was varied by bypassing and switching motors, connected in series, in and out of service.

Direct-Current Transmission

Direct-current transmission lines can consist of one conductor with a ground return, called a monopolar system, or two-conductor transmission lines with a ground return, called a bipolar system. The bipolar system operates with one conductor at a positive (+) voltage with reference to ground and one conductor at a negative (−) voltage with reference to ground. The bipolar system can be operated as a monopolar system in an emergency if one of the two line conductors is in good operating condition. The dc transmission lines originate and terminate at stations that convert high-voltage alternating current to direct current and invert direct current to alternating current using rectifier circuits with thyristor semiconductor elements (Fig. 4.3). The converter transformers (3) serve as the power link between ac and dc systems. The power transformers match the ac voltage to the working voltage of the thyristor valves, isolate the dc system from the ac system, and phase shift the ac supply voltage by wye and delta connections to increase the thyristor firing cycles, reducing the harmonic currents generated in the ac system and the voltage ripple in the dc system. The thyristor valves (6) perform the function of converting ac voltage to dc voltage or dc voltage to ac voltage. The valves or converter circuits have thyristors connected in series using the number required to obtain the voltage desired for the system. Resistors and capacitors are connected to the thyristors to equalize the voltage across each of the thyristors in the valve assembly. The thyristor valves are mounted in a structure to form a 12-pulse converter group. Valve thyristors are individually fired at the proper time to provide the proper voltage. A computer develops an electronic pulse at the proper firing time. The electronic pulse is converted to a light pulse, and the light pulse is transmitted to the thyristor valves by light guides. The light pulse is reconverted to an electronic pulse that fires the thyristors. The dc filters (9) attenuate converter harmonic voltages in the dc system to avoid audiofrequency interference on telephone lines adjacent to the high-voltage dc transmission line. Alternating-current filters are installed on the ac systems by the converter equipment. The ac filter equipment is not shown in the schematic diagram. A monopolar system would have only half the equipment shown in Fig. 4.3.

The direction of current flow in the dc lines is determined by the voltage across the line terminals at each station. The current will flow from the terminal with the higher voltage to the terminal with the lower voltage. The magnitude of the current flow is

FIGURE 4.3   Schematic diagram of an ac to dc bipolar convertor station, a dc bipolar transmission line, and a dc to ac bipolar inverter station and a drawing of a typical high-voltage dc bipolar transmission-line structure supporting high-voltage conductors and static wire that serves as a ground conductor. (1) Alternating-current generator; (2) alternating-current air circuit breaker with isolating devices; (3) alternating-current power transformer; (4) gang-operated disconnect switch; (5) lightning and surge arrester; (6) thyristor valve with control connection; (7) reactor; (8) ground or neutral and ground conductor; (9) direct-current filter; (10) direct-current transmission-line conductor; (11) direct-current transmission-line structure; (12) alternating-current system.

Year	Voltage, kV	Country
1950	200	Russia
1954	100	Sweden
1961	±100	England-France (cross-Channel)
1962	±400	Russia
1965	±250	Sweden
1970	±400	United States
1977	±450	Canada
1978	±400	United States
1981	±500	Canada
1986	±600	Brazil

TABLE 4.1   Higher-Voltage dc Three-Phase Transmission Lines

controlled by the magnitude of the voltage difference and the resistance of the line conductors.

Some of the higher-voltage dc transmission lines by operational date and country are as listed in Table 4.1.

Converter stations constructed between 1950 and 1970 used mercury-arc valves. Stations constructed in 1970 and later use thyristor valve assemblies.

Direct-current transmission circuits are economical for installation whenever the overhead lines extend over 400 miles or underground cable circuits extend over 20 miles without a tap to the circuit or to interconnect two ac systems together that are not synchronized.

Additional descriptive information for dc transmission systems can be found in the *Standard Handbook for Electric Engineers*, edited by Donald G. Fink and H. Wayne Beaty, published by McGraw Hill.

Transmission Voltages

Transmission lines are operated at very high voltages to deliver more power with a particular size line conductor. A three-phase ac line operating at 345,000 volts, for example, can carry 6.25 times as much power as a 138,000-volt line. This is so because the power-carrying capacity of a line varies as the square of the line current ($kVA = VI = V^2/R$, where $I = V/R$ and line resistance R remains constant). Raising the line voltage increases the capacity of the line by the square of the ratio between the higher and lower line voltages. Even though the voltage of the 345,000-volt line is only 2.5 times as high as that of the 138,000-volt line, 2.5 squared, that is, 2.5 times 2.5 equals 6.25; therefore, the higher-voltage line can carry 6.25 times as much power as the lower-voltage line. If we assume the 138,000-volt line to be carrying a load of 80,000 kW, the 345,000-volt line can carry 6.25 times 80,000 or 500,000 kW.

In like manner, the three-phase 765,000-volt ac line of the American Electric Power System, having a voltage 5.6 times as great as the 138,000-volt line, can carry 5.6 times 5.6, or 31 times as much power; 31 times 80,000 equals 2,500,000 kW.

Polyphase Transmission

Alternating-current voltages can be changed from one level to another easily by using power transformers. Single-phase ac transmission circuits would be similar to dc circuits except for inductance and capacitance associated with ac transmission. Almost all ac transmission lines, from the smallest to the largest, are three-phase lines.

The two main reasons for building three-phase transmission lines are:

1. A single-phase two-wire transmission line operating at a certain voltage requires a certain amount of conductivity in the wires to carry the load. A two-phase four-wire line would require the same total weight of conductor. The three-phase system requires only three conductors or three-fourths of the total weight of conductor that is required by single- or two-phase systems for the same load. There is a substantial saving in conductors if a three-phase system is selected.

2. Motors operating from a three-phase line are much cheaper and more satisfactory in operation than single-phase or two-phase motors.

This tremendous increase in power-carrying capacity with the higher voltages makes it possible to send more power over the same right-of-way, thereby causing large savings in additional rights-of-way. It also reduces the number of parallel lines needed where heavy loads must be transmitted.

Figure 4.4 visually represents three high-voltage lines operating at 138,000 volts, 345,000 volts, and 765,000 volts, respectively, and their corresponding power-transmitting capacities of 80,000 kW, 500,000 kW, and 2,500,000 kW.

Voltage levels of ac transmission circuits vary from 34,500 volts to 765,000 volts. Voltages originally used for transmission circuits below 34,500 volts have been converted to distribution circuits. Circuits with voltages in the range of 34,500 volts to 161,000 volts have been converted to subtransmission use. In many cases, these circuits will provide high-voltage service to large industrial customers. Extra-high voltages have been used extensively for transmission circuits in the United States.

The determination of the most economical voltage to use requires careful analysis. Usually, several voltages are considered. In each case the total annual operating costs, including interest, depreciation, and taxes, plus power lost in the line, are estimated for

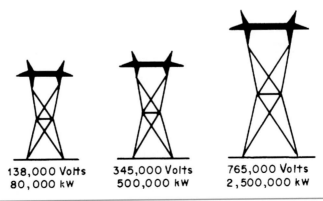

138,000 Volts   345,000 Volts   765,000 Volts
80,000 kW   500,000 kW   2,500,000 kW

FIGURE 4.4   Comparison of power-carrying capability of three-phase ac transmission lines at different voltages. (*Courtesy Edison Electric Institute.*)

each voltage studied. If no special factors are be considered, the voltage which gives the lowest total annual operating cost is the one selected.

The selection of the proper voltage is sometimes also governed by the voltage of existing power lines. If lines are to be connected to the other lines, as they probably will be, it is convenient to have the lines at the same voltage.

Transmission-Line Capacity

Thermal conditions of the conductors are usually the load-limiting factor of transmission circuits. Convection cooling by ambient air allows overhead conductors to carry higher currents than underground conductors. Alternating-current overhead transmission lines are limited in capacity by the thermal expansion and the resulting sag of the conductors. Short overhead transmission lines may be loaded continuously up to their thermal rating. Table 4.2 gives the thermal rating of some typical overhead ac transmission lines installed in the United States.

The current flowing in a transmission-circuit conductor can be calculated if the voltage and the power in kilovolt-amperes (kVA) or megavolt-amperes (MVA) are known. If the power is expressed in kilowatts (kW) or megawatts (MW), the power factor must be known in addition to the voltage. Kilovolt-amperes and kilowatts can be calculated if the current, voltage, and power factor are known.

The current flowing in a single-phase two-wire circuit is given by the formula:

$$\text{Current in amperes} = \frac{\text{apparent power in kilowatts-amperes} \times 1,000}{\text{line voltage in volts}}$$

or

$$\text{Current in amperes} = \frac{\text{power in kilowatts} \times 1,000}{\text{line voltage in volts} \times \text{power factor}}$$

To illustrate the use of these formulas, take the example of a single-phase line carrying a load of 100,000 kVA and operating at 80,000 volts. Substituting in the first formula gives:

$$\text{Current} = \frac{100,000 \times 1,000}{80,000} = 1,250 \text{ amps}$$

Line Voltage, kV	Circuits per Tower	Conductors per Phase	Conductor Size ACSR, kCMIL	Phase Resistance, Ω/ml	Ampacity A/Conductor	Thermal Line Rating, MVA
138	1	1	795	0.140	880	210
138	2	1	795	0.140	880	420
230	1	1	1,431	0.076	1,220	490
230	2	1	1,431	0.076	1,220	980
345	1	2	795	0.070	1,750	1,050
345	2	2	1,590	0.035	2,640	3,160
500	1	2	1,780	0.031	2,850	2,470
765	1	4	954	0.028	3,890	5,150

TABLE 4.2   Capacity of Typical Overhead ac Three-Phase Transmission Lines

The current just calculated is the current in each wire of the single-phase circuit.

If the kilowatts, power factor, and voltage are known instead of kilovolt-amperes and voltage, the second formula is used. For example, if an 80,000-volt circuit delivers 100,000 kW at 80 percent power factor, the current in the line will be given by:

$$\text{Current} = \frac{100{,}000 \times 1{,}000}{80{,}000 \times 0.80} = 1{,}562.5 \text{ amps}$$

If this load of 100,000 kW were at unity power factor, instead of 0.80, the current would be less. The kilovolt-amperes and kilowatts would be the same, and the current would be 1,250 amps.

The current flowing in a three-phase circuit is given by the formula:

$$\text{Current in amperes} = \frac{\text{kilovolt-amperes} \times 1{,}000}{1.73 \times \text{line voltage in volts}}$$

or

$$\text{Current in amperes} = \frac{\text{kilowatts} \times 1{,}000}{1.73 \times \text{line voltage} \times \text{power factor}}$$

To illustrate the use of this formula, take the example of a three-phase line carrying 100,000 kVA of load and operating at 138,000 volts. To find the current flowing in each line conductor, substitute in the formula. Thus:

$$\text{Current} = \frac{100{,}000 \times 1{,}000}{1.73 \times 138{,}000} = 418.9 \text{ amps}$$

If the amperes at unity power factor are known and it is desired to know what the current in amperes would be at any other power factor, the procedure is to divide the current at unity power factor by the power factor at which the current is desired. Thus:

$$\text{Current at any desired power factor} = \frac{\text{current at unity power factor}}{\text{desired power factor}}$$

To illustrate, if the current at unity power factor is 100 amps, the current at 0.8 power factor is:

$$\text{Current at 0.8 power factor} = \frac{100}{0.80} = 125 \text{ amps}$$

The formulas and equations presented are summarized and arranged for ready reference in Table 4.3. It might be pointed out as a matter of interest that power factor does not have to be considered in dc calculations because there is no lag or lead of current in such circuits. In ac circuit calculations, the power factor is important. If its value is not known for a given circuit or line, the value 0.80 is often assumed as a basis for calculation. It will also be noted that the factor 1.73, abbreviated for the square root of 3 (symbolized as $\sqrt{3}$), is used with line voltage for all three-phase calculations.

Underground Transmission-Cable Capacity

Solid-dielectric cables are used extensively for subtransmission and transmission in underground circuits. The insulating compounds used most extensively are crosslinked polyethylene (XLPE) and ethylene propylene rubber (EPR).

To Find	Direct Current	Alternating Current — Single-Phase	Alternating Current — Three-Phase
Amperes when kilovolt-amperes and voltage are known	$\dfrac{\text{Kilovolt-amperes} \times 1{,}000}{\text{volts}}$	$\dfrac{\text{Kilovolt-amperes} \times 1{,}000}{\text{volts}}$	$\dfrac{\text{Kilovolt-amperes} \times 1{,}000}{1.73 \times \text{volts}}$
Kilovolt-amperes when amperes and voltage are known	$\dfrac{\text{Amperes} \times \text{volts}}{1{,}000}$	$\dfrac{\text{Amperes} \times \text{volts}}{1{,}000}$	$\dfrac{1.73 \times \text{volts} \times \text{amperes}}{1{,}000}$
Amperes when kilowatts, volts, and power factors are known	$\dfrac{\text{Kilowatts} \times 1{,}000}{\text{Volts}}$	$\dfrac{\text{Kilowatts} \times 1{,}000}{\text{Volts} \times \text{power factor}}$	$\dfrac{\text{Kilowatts} \times 1{,}000}{1.73 \times \text{volts} \times \text{power factor}}$
Kilowatts when amperes, volts, and power factor are known	$\dfrac{\text{Amperes} \times \text{Volts}}{1{,}000}$	$\dfrac{\text{Amperes} \times \text{volts} \times \text{power factor}}{1{,}000}$	$\dfrac{1.73 \times \text{amperes} \times \text{volts} \times \text{power factor}}{1{,}000}$

TABLE 4.3 Summary of Circuit Calculations

A typical extruded solid-dielectric power-transmission cable consists of a stranded copper or aluminum conductor covered by a thin extruded layer of semiconducting (carbon-black-filled) polyethylene. The semiconducting conductor shield provides a smooth conductor surface for the insulation and excludes air from the interface. The insulation is extruded over the conductor shield. A thin semiconducting polyethylene shield is extruded over the insulation. The semiconducting shields over the conductor and the insulation prevent electric discharges, or corona, that would damage the insulation. A metallic layer on top of the dielectric shield provides a good conductor for the charging current, preventing damage to the semiconducting shield from the heat generated by the charging current. The metallic shield may be copper tapes, concentric wires, tubular corrugated copper, or copper wires embedded in the semiconducting insulation shield. The metallic shield is usually protected by a polyethylene jacket.

Submarine cables constructed with solid-dielectric insulations have been successfully installed and placed in service. The three individual insulated conductors are stranded into a single cable.

For larger ac transmission power, the extruded solid-dielectric cables with a 2,000-kCMIL copper conductor installed direct-buried with the three cables in a flat configuration with one three-phase circuit per trench would have the circuit capacity listed in Table 4.4.

High-pressure oil-filled (HPOF) paper-insulated cables in which the three-phase conductors are enclosed in a steel pipe containing oil are also used for underground cable transmission circuits. The cables are terminated in potheads. Each phase conductor is installed in a nonmetallic pipe to a junction point called a trifurcator where the three-phase cables enter a common pipe. Shipping lengths of cable will normally be approximately 2,500 ft for 138 kV and 230 kV and 1,750 ft for 345 kV and 500 kV. The cables are spliced in manholes. The steel pipe is usually direct buried between the manholes. Oil pressure is maintained in the steel pipe with the use of pumps and reservoirs. The steel pipe is protected from corrosion by cathodic protection equipment. Alarms are installed on the system to detect low oil pressure. The transmission circuit may consist of two cables installed in parallel. The two pipes can be placed in the bottom of the same trench separated as far as possible to minimize the ambient temperature and the installation cost. The dual facilities increase the reliability.

The pressurized oil can be pumped through the pipe for one-cable circuit and returned through the parallel cable-circuit pipe. The capacity of oil-filled pipe-cable systems can be increased significantly by circulating the oil past the cable at several

Voltage, kV	Apparent Power, MVA
138	289
230	470
345	700

TABLE 4.4 Circuit capacity of 2,000-kCMIL copper conductor installed direct-buried with the three cables in a flat configuration with one three-phase circuit per trench for various high voltages

	2,000-kCMIL Conductor				2,500-kCMIL Conductor
Voltage, kV	138	230	345	500	765
Pipesize, nominal, inch	8	10	10	12	14
Insulation thickness, inch	0.505	0.835	1.025	1.25	1.56
Naturally cooled cables Capacity, MVA	224	344	453	571	645
Dielectric losses, watts/cond. ft	0.733	1.39	2.71	3.99	8.70
Resistive losses, watts/cond. ft	7.72	7.02	5.49	4.40	2.30
Pipe temperature, °C	62.0	59.2	58.4	57.1	64.7
Total 3Φ reactive losses per cable mile, MVAR	4.7	8.9	17.3	31.8	69.4
Forced-cooled cables Capacity, MVA	426	688	941	1,352	2,137
Dielectric losses, watts/cond. ft	0.733	1.39	2.71	3.99	8.7
Resistive losses, watts/cond. ft	28.7	28.4	24.2	25.2	24.9
Pipe temperature, °C	50.2	37.8	38.8	32.4	29.1
Total 3Φ reactive losses per cable mile, MVAR	4.7	8.9	17.3	31.8	69.4

TABLE 4.5   Capacities of Three-Phase (3Φ) Oil-Filled Paper-Insulated Pipe-Type Cables

feet per second and cooling it externally by means of an oil-to-air heat exchanger (Table 4.5).

Compressed-gas-insulated or gas-spacer cables are used for high-capacity insulated transmission circuits. The gas-spacer cable transmission system is an ac three-conductor isolated-phase design. The conductor can be copper or aluminum tube with a diameter of approximately 3 inches to 10 inches, depending on the current-carrying capacity required and the operating voltage. The conductor is surrounded by an outer sheath of extruded aluminum tubing, the diameter determined by the space required to provide insulation for the operating voltage. These diameters vary from 8.5 inches at 138 kV up to 20 inches at 500 kV and are approximately 30 inches at 1,000 kV.

The conductor is centered and maintained concentric with the outer enclosure by insulators installed in the enclosure at intervals along the cable. The optimum insulating gas pressure is approximately 50 lbs per square inch (psig). The cable assemblies are manufactured in approximately 40-ft lengths. The conductors are spliced together by plug-in type terminals or welding. The outer enclosures are welded together to form a continuous leak-proof pipe. The cables can be installed on top of the ground, overhead on supporting structures, or direct-buried. Direct-buried circuits would require a trench 4 ft to 7 ft wide, depending on the operating voltage. Thermal sand is used for a back-fill material. The cables are terminated in potheads consisting of an extension of the outer sheath, the inner conductor, and a porcelain insulator. The capacities and charging currents of typical gas-spacer cable installations are itemized in Table 4.6.

Self-contained cables with a central duct using pressurized oil for forced cooling have been used in some locations for underground transmission circuits. The cables can be direct-buried or installed in ducts. The capacity of the circuit is limited severely when these cables are installed in ducts as a result of the inability to dissipate the heat generated in the cable.

Operating Voltage, kV	Capacity, MVA	Charging Current, amps/mi	Uncompensated Maximum Length, mi
138	400	4.0	440
230	700	5.3	350
345	1,200	7.4	295
500	2,200	12.6	250
765	4,200	20.0	
1000	6,500	25.0	

TABLE 4.6   Single-Circuit Three-Phase ac Gas-Spacer Cable Characteristics

Recent breakthroughs with new underground superconducting cables have led to the introduction of high-temperature superconducting (HTS) cable. The superconducting cable has been identified as a significant advancement in the power industry. Southwire has manufactured a unique HTS Triax cable that locates the three electrical phases and neutral conductor concentrically on a single core. The superconducting materials replace the copper or aluminum conductors in conventional cable. This Triax cable delivers up to 500 percent more ampacity.

The first successful underground HTS cable system in a commercial power grid was energized on April 22, 2008 by American Superconductor Corp., Long Island Power Authority, and the U.S. Department of Energy. Other successful HTS projects have been completed.

Direct-current underground transmission-cable circuits have advantages over ac cables since they do not have induction losses or a continuous flow of charging current. Solid-dielectric extruded-polyethylene-insulated cables can be used on voltages up to 300 kV. Pipe-type high-pressure oil-filled cables with paper-tape insulation can be used for circuits with a voltage of ±600 kV. Self-contained oil-filled cables will operate satisfactorily for voltages of ±300 kV.

Direct-current cable insulation must be thick enough to provide a safety factor of 3 to 5 with respect to dc insulation breakdown because of surges. The surges may be caused by malfunction of conversion equipment, operation of switches, accidental polarity reversal, or lightning strokes on overhead lines connecting to the underground cables. Table 4.7 itemizes the capacity of high-pressure oil-filled pipe-type 2,000-kCMIL conductor cable operating at various dc voltages. The high cost of converter stations limits the use of underground dc transmission circuits.

	Voltage dc, kV			
	SUB>±180	±300	±450	±655
Single-conductor loss, kW/mile	42.7	40.7	39.5	33.8
Maximum dc current, amps	1,120	1,095	1,030	1,000
Average electric stress, kV/inch	356	360	440	525
Capacity, MVA	403	657	927	1,307

TABLE 4.7   Data for 2,000-kCMIL Conductor High-Pressure Oil-Filled Pipe-Type Underground Cable Installation Operating at Various dc Voltages

Transmission-Line Environmental Considerations

Underground transmission circuits would eliminate most of the environmental problems associated with electric power transmission. The large investment associated with underground cable installations and technical limitations to the length of ac underground cable circuits prevent using underground installations to solve all the environmental problems.

Rights-of-Way

Transmission-line environmental considerations include visual impact, natural land conditions, electric field effects, land use, corona generation and associated radio and television interference or audible noise, and the limiting of the voltages induced in objects located in the transmission-line right-of-way. Rights-of-way for transmission lines should be selected with the purpose of minimizing the visual impact to the public. Existing rights-of-way should be given priority as the locations for additions to existing transmission facilities.

The first step to be taken prior to the design or construction of any line is to conduct a survey and make a map of the territory where the line is to be constructed. Aerial photography is usually used to survey the land in rural areas for transmission lines. The final location of the line is, of course, not known, and therefore, a wide strip of land must be included on the map. The main points between which the line is to be built will be known. The intervening territory should be laid out on a scale large enough to show clearly all division lines, towns, roads, streams, hills, ridges, railroads, bridges, buildings, and existing communications and power lines.

With the map completed, the following principles should be used as guides in selecting the exact route:

1. *Select the shortest route practicable.* The shortest line naturally is the most economical, other things being equal. This means that the line should be a straight line between the terminals of the line, for a straight line is the shortest distance between any two points.

2. *Parallel streets and highways as much as possible.* This makes the line readily accessible both for construction and for inspection and maintenance. The hauling and delivery of materials will be greatly facilitated if this can be done.

3. *Follow section lines.* Transmission lines located in rural areas should follow section lines. Doing this causes less damage to farmers' property and, therefore, makes it possible to purchase the right-of-way more economically. Paralleling railroads is desirable for the same reason, because the farms have already been cut, and therefore, the additional damage is negligible.

4. *Route in direction of possible future loads.* If there is a possibility of adding power loads, the route selected should be such as to come as near as possible to such locations, provided that the additional cost is not excessive.

5. *Avoid crossing hills, ridges, swamps, and bottom lands.* Hills and ridges subject the line to lightning and storms. Swamps and bottom lands subject the line to floods. Furthermore, delivery of material as well as the construction of the line becomes difficult. Extra guying and cribbing are often necessary in swamps.

6. *Coordinate with communication lines.* High-voltage transmission lines can cause disturbances or interferences in communications lines by induction

and therefore may require transposition of the power conductors. This adds to the cost of the line. In some cases, it is advisable to relocate or place the communications circuits underground. Distribution lines are often constructed on the same poles with communications circuits and/or installed underground in the same trench or conduit system.

7. *Preserve the environment.* Line routes should be selected which are the least visible from scenic locations or other areas where people pass or congregate. A simple test of visibility can be carried out by touring roads and points close to a proposed right-of-way and noting the visual impact of the line prior to final selection.

Width of Right-of-Way

When the route of the line has been selected, permission must be obtained to construct the line or land must be purchased. This is known as the right-of-way. In this connection, it should be pointed out that it is well to select alternate tentative routes so that if difficulties arise in obtaining permits on any one route or the prices demanded become inflated, another route can be chosen. High-voltage lines usually require a continuous strip of land for the right-of-way. Table 4.8 gives a general idea of the widths of rights-of-way necessary for different lines. The width of the right-of-way will vary with the type of construction, the conductor horizontal spacing, the voltage of the line, the terrain, and the area needed for construction and maintenance of the line.

The reason for acquiring a right-of-way much wider than the actual space occupied is to prevent tall trees from being blown into the line and to prevent damage from forest fires. Sometimes it is possible to purchase a narrower right-of-way and obtain permission to clear back the remaining distance beyond the fences.

Electric Field Effects

Electric field effects of transmission lines are not hazardous. Linemen regularly work on energized extra-high-voltage (EHV) and lower-voltage transmission lines using hot-line tools and bare-hand methods with no ill effects. A study of 10 linemen who regularly worked on energized 345-kV transmission lines by personnel of John Hopkins University in Baltimore, MD, over a period of 10 years determined that the linemen were in excellent health.

	Feet
Pole lines	
Single circuit	15–50
Double circuit	30–75
H-frame	60–250
Tower lines	
Single circuit	50–250
Double circuit	75–250

TABLE 4.8   Approximate Range of Widths of Rights-of-Way Required for Various Types of Overhead Lines

Magnetic Effects

A voltage can be induced on a person or conductor under an overhead energized electric transmission-line conductor. The voltage is induced by the magnetic field developed by the current flow in the line conductors. The induced voltage will be insignificant at all times under dc lines and under normal conditions under ac lines. High-fault current on an ac transmission line may increase the induced voltage on a conducting object under the line. The fault current on dc lines is limited by a converter station, which limits induced voltages.

Electric Fields

The magnitude of electric fields at ground level under all overhead high-voltage transmission lines is very low. Static charges on farm fences directly under the lines are usually drained to ground through the metal or wood posts or contacts with grass, weeds, or the ground. Vehicles, such as a tractor or combine, directly under an EHV line may become charged, and a farmer bridging the rubber tires of the vehicle might feel a tingle if the vehicle discharged through his body to ground. The effect could be annoying, but it is not electrically dangerous, as linemen can verify. The annoying shock may startle a person, causing a fall or other act that could result in a personal injury. Vehicles operated under EHV transmission lines or facilities such as fences installed under EHV lines should be grounded to prevent annoying shocks and acts that might lead to injuries. All metal buildings or structures should be grounded, whether they are built near a transmission-line right-of-way or not. Grounding the building will eliminate any static charge.

Linemen use special conductive clothing or electrostatic shields while working on energized ultra-high-voltage (UHV) transmission lines. It is necessary that special precautions be taken when UHV transmission lines are designed and constructed to protect the public from hazardous voltages. The major characteristics affecting a transmission line's electrostatic conditions are voltage, conductor type, phase configuration, phase spacing, and conductor height.

Corona

Ionization of the air around energized high-voltage transmission-line conductors, when the voltage gradient exceeds the air withstand limits, establishes corona conditions. Corona is visible at night as a luminous glow close to the conductors. Corona and resulting radio-interference voltages and audible noise can be controlled by proper transmission-line design and construction precautions. The use of bundled conductors and appropriate hardware on ac EHV and UHV lines keeps corona to acceptable limits if the equipment is installed correctly. Bundling of conductors and selection of conductor hardware will have minimal effect on dc lines because of the space charge surrounding the conductors. The line conductors must be kept clean and free from damage that results in rough or pointed projections. Grease on a conductor can cause the generation of corona. Corona on dc lines is generally caused by the same factors discussed for ac lines. However, corona is generally less on dc lines and does not increase as much during bad weather compared with conditions on ac lines. Radio interference on high-voltage dc transmission lines is caused by the pulses created by the ignition of the thyristor valves in the converter stations, corona on the line, and partial electric discharge over the insulators. The pulses generated in a converter station can be controlled by shielding and proper design and construction of the thyristor valve housing.

Proper design, construction, and maintenance will keep radio interference caused by corona and partial electric discharge over the insulators of both ac and dc high-voltage lines within allowable limits.

Television transmissions are above the 50-MHz frequency. Interference from electric transmission lines above the 50-MHz frequency is very small and does not normally cause problems.

Induced voltages on objects in the right-of-way of ac UHV transmission lines can be controlled by the height of the towers or by shielding. The right-of-way for the transmission line can be purchased and fenced to prevent access to the land. Induced voltages are not a problem on high-voltage dc lines. It is important that transmission lines be designed to minimize the cost and protect the safety of the public.

Individuals opposed to the construction of electric transmission lines have intervened in hearings on the citing of high-voltage lines and have received considerable publicity. The interveners have demonstrated that the weak electric fields under extra-high-voltage ac electric transmission lines will light fluorescent tubes. Electric fields generated by an automobile's ignition system or by a television receiver can make a fluorescent tube glow. A silk cloth rubbed along the surface of a fluorescent tube will cause the lamp to glow. Electric appliances found in the home produce electric fields, allowable by the American National Standards Institute publications, which are greater than those found under high-voltage lines. Research completed by the Electric Power Research Institute, the U.S. Department of Energy, the Minnesota Department of Health, Iowa State University, and other institutions has not identified biologic effects from electric fields that are detrimental to the public.

Long-Range Wireless Transmission Considerations

In areas where the installation of transmission wires is impractical or extremely difficult, the concept of long-range wireless transmission of electricity is being proposed and developed in a New Zealand pilot program. The concept consists of sending the electric power in an electromagnetic waveform between sending and receiving end towers. Multiple relay towers are needed as well. The concept of long-range wireless transmission dates to the 1890s and was initially theorized by Nikola Tesla; however, it was never proven to be feasible until now, due to technology advancements. Lessons learned from the pilot program will be thoroughly studied before widescale deployment of wireless transmission occurs.

CHAPTER 5

Distribution Circuits

Distribution is that system component which delivers the energy from the transmission system to the customers. It includes the substations that reduce the high voltage of the transmission system to a voltage level suitable for distribution and the circuits that radiate from the distribution substation to the customers (Fig. 5.1). The distribution circuits that carry power from the substations to the local load areas are known as express feeders, distribution main feeders, or primary circuits and generally operate at voltages between 2,400 volts and 34,500 volts. The loading of distribution circuits varies with customer load density, type of load supplied, and conditions peculiar to the area served. The nominal capacities of distribution circuits operating at different voltage levels are listed in Table 5.1.

The distribution circuits may be overhead or underground depending on the load density and the physical conditions of the particular area to be served. The overhead circuits are constructed on poles that may also support streetlights, telephone networks, fiber-optic networks, and cable television facilities.

Distribution circuits originate in a distribution substation and include the conductors and equipment necessary to distribute electricity to the customer. The distribution substation may be located at a generating station or at a point remote from the generation and connected to the generation by transmission and subtransmission circuits. Distribution substations are located near the customers utilizing the electricity to ensure good voltage and reliable service.

Distribution transformers are installed in the vicinity of each customer to reduce the distribution primary voltage of between 2,400 volts and 34,500 volts to the desired secondary utilization voltage required by the customer. Secondary circuits and services carry the secondary voltage from the distribution transformers to the customers. Residential customers and some commercial customers utilize a single phase voltage of 120/240 volts. Standard commercial and small industrial customer three-phase voltages are typically 208Y/120 volts or 480Y/277 volts.

Distribution circuits can have a major impact on the customer's voltage and service continuity. The voltage must be maintained within a range that permits the customer's equipment to perform satisfactorily and within limits specified by regulatory agencies. If the customer's base voltage is 120 volts, then the minimum normal service would be 114 volts and the maximum normal service voltage would be 126 volts. It is good practice to maintain the voltage as close to the base voltage as practical and to limit the range of the voltage variations from base.

It is normal to operate the distribution system at higher voltages during peak-load periods to compensate for voltage drop in the distribution equipment and the customer's wiring system. It is important that distribution-circuit equipment be designed and

FIGURE 5.1   Schematic diagram of a distribution system. The distribution circuits may be overhead or underground. The transformers may be pole-type or pad-mounted.

installed in a manner to prevent objectionable light flicker as a result of momentary variations in voltage level.

Alternating-current distribution circuits are commonly operated as three-phase, three-wire (three primary voltage phase wires) or three-phase, four-wire circuits (three primary voltage phase wires and one grounded neutral wire).

Three-Phase, Three-Wire Distribution Circuits

A three-phase circuit can be obtained from either a Y or a delta (Δ) connection of the three coils in the ac generator. Similarly, in distribution systems, the Y or Δ connection

Voltage Class, kV	Capacity, MVA	
	Average	Maximum
4–5	2.5	3.5
12–15	7.5	15.0
20–25	12.0	20.0
30–35	18.0	30.0

TABLE 5.1   Distribution Feeder Capacities of Distribution Circuits

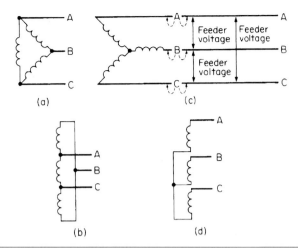

FIGURE 5.2 Three-phase, three-wire distribution circuit. (a) Conventional and (b) practical way of picturing Δ connection of transformer secondaries. (c) Conventional and (d) practical way of picturing Y connection of transformer secondaries.

is obtained by the proper connection of the secondary coils of the transformers in the substation.

Figure 5.2 shows a three-phase, three-wire distribution feeder schematic together with the Y and the Δ connections of the secondary windings of the distribution sub-station step-down transformers. The transformer secondary connections are shown in the conventional and practical way of representation. It should be pointed out here that the Y connection is generally used on distribution feeders when the feeder is operated as a four-wire circuit. The voltage between any two of the three wires for either method of connection is the same; that is, the voltages between conductors A and B, B and C, and C and A are equal. In the Δ connection, this voltage between line wires is the same as the transformer-coil voltage, but in the Y connection, the voltage between line con-ductors is the square root of three ($\sqrt{3}$) times the transformer-coil voltage.

Three-Phase, Four-Wire Distribution Circuits

The three-phase, four-wire feeder circuit is in general use, especially in the larger cities. This circuit can be obtained from a Y connection of the distribution substation trans-former secondary. Figure 5.3 shows this type of circuit. The fourth wire is called the neutral wire because it is connected to the neutral point of the Y connection and is grounded in multiple locations.

It will be noted that the voltage between any two of the line wires is no longer equal to E, the transformer secondary-coil voltage, but instead is $\sqrt{3} \times E$. The voltage E still exists between any line wire and neutral, but the voltage between any two line wires is $\sqrt{3} \times E$ volts. If the voltage E between any wire and neutral is 7,620 volts, then the volt-age between any two-phase wires is equal to $\sqrt{3}\ E$, or $1.732 \times E$, which, in this example, would be $1.732 \times 7,620$, or 13,200 volts. Single-phase voltages equal to 7,620 volts would be available by using one-phase wire and the neutral conductor which is grounded.

The advantages of the four-wire feeder over the three-wire feeder are increased power-carrying capacity and better voltage regulation. Since the line voltage in the

FIGURE 5.3   Three-phase, four-wire distribution circuit. (a) Conventional and (b) practical way of picturing Y connection of distribution substation transformer secondary with grounded neutral brought out.

four-wire system is $\sqrt{3}$ or 1.73 times that of the three-wire system, its power-carrying capacity for the same current in the wires will also be 1.73 times as great, or for the same power transmitted, the current in the four-wire system will only be 58% as large. Moreover, the voltage regulation is also greatly improved. Many large cities have changed over to this system of feeders in order to get the benefit of this increased feeder capacity. Merely stringing the fourth wire and reconnecting the transformer secondaries from Δ to Y make possible an increase in load-carrying capacity of 73%.

In earlier years, 2,400 volts Δ, 4,160Y/2,400 volts grounded wye, and 4,800 volts Δ primary distribution-circuit voltage levels were in wide use, amounting to approximately 60% of the systems in 1955. In 1969, usage of distribution voltages below 5,000 volts was down to about 12%. The next higher levels, 12,470Y/7,200 volts, 13,200Y/7,620 volts, and 13,800Y/7,960 volts, are now the most prevalent primary distribution voltages. Higher primary distribution voltages are in use, including 24,500Y/14,400 volts and 34,500Y/19,920 volts.

From an application standpoint, there are a number of factors to be considered. Factors favorable to the higher primary distribution voltage levels include the steady increase in consumer Kilowatt hours, the permissible use of smaller conductors, better voltage regulation, and the availability of existing industrial feeders or subtransmission circuits at levels that are suitable, or could be made suitable, for the start of the higher-voltage distribution circuits. Some negative factors are the existence of a widespread use of the present levels and the possibility of an alternative solution of using more substations to supply present voltage levels as loads increase. At the increased distribution system voltage levels, the feeders serve a larger number of customers that would be interrupted in case of system faults.

Single-Phase Primary Circuits

A single-phase circuit is generally obtained from a three-phase circuit. It could be obtained from a single-phase ac generator, but such machines are very rare today. Figures 5.4 and 5.5 illustrate how single-phase primaries are obtained from three-phase, three-wire and three-phase, four-wire feeders. These single-phase primaries are tapped to the feeder circuit. By tapping two line wires in the three-wire feeder example of Fig. 5.4, the voltage of the primary will naturally be the same as the voltage across the line wires, 13,200 volts in this example. By tapping from line wire to neutral in the four-wire feeder example of Fig. 5.5, the voltage of the primary line to neutral voltage will be 7,620 volts in this example.

FIGURE 5.4   ABC is a 13,200-volt three-phase primary feeder, and AB, BC, and AC are single-phase 13,200-volt primaries obtained by tapping the 13,200-volt three-phase, three-wire feeder.

FIGURE 5.5   ABC is a 13,200Y/7,620-volt primary feeder obtained by tapping the three-phase primary phase wires, and AN, BN, and CN are single-phase 7,620-volt primaries obtained by tapping one primary phase wire and the neutral wire of the three-phase, four-wire 13,200Y/7,620-volt feeder.

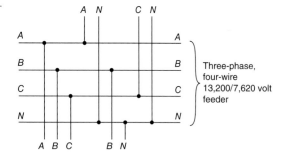

An attempt is always made to keep the loads on the single-phase primaries connected to the three phases of the three-phase line balanced so that the currents in the wires of the three-phase feeder also will be balanced.

Single-Phase Secondary Circuits

Secondary circuits, for residential customers, are generally single-phase, three-wire circuits, although they may be only two-wire circuits. The secondaries originate from the secondary winding of the distribution transformer. Figure 5.6 shows the single-phase, two-wire primary connecting to the primary winding of a single-phase transformer. The secondary winding of a single-phase distribution transformer is nearly always made up in two coils which can be connected either in series or in parallel, as shown. They are most often connected in series. Each coil has induced in it 120 volts so that when they are connected in series, the voltage across the outside wires forms a 240-volt single-phase, two-wire circuit. When the neutral is used, either outside phase wire and the inside neutral wire forms a 120-volt single-phase circuit.

Three-Phase Secondary Circuits

Commercial and small industrial customers usually require three-phase service to operate three-phase motors. Three-phase motors are more efficient and less expensive than single-phase motors. Large single-phase motors cause a sudden drop in the voltage when they start which appears as a voltage flicker that affects incandescent lights and

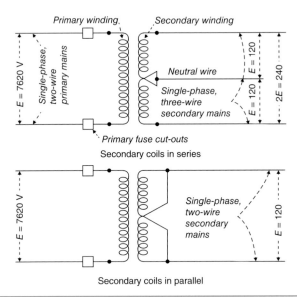

Primary winding.   Secondary winding

Neutral wire

Single-phase,
three-wire
secondary mains

Primary fuse cut-outs

Secondary coils in series

Single-phase,
two-wire
secondary
mains

Secondary coils in parallel

FIGURE 5.6   Standard connections of common step-down single-phase distribution transformer.

distorts television pictures. Three-phase motors draw less current than single-phase motors when they start, thus minimizing voltage flicker. Installation of three-phase service helps to keep the current in each phase wire of the distribution circuit equal. The voltages normally used for three-phase secondary circuits are 208Y/120 volts four-wire, 240 volts three-wire, 480Y/277 volts four-wire, and 480 volts three-wire.

Three-phase, four-wire secondary voltages of 208Y/120 volts or 480Y/277 volts can be obtained by proper selection of pole-type single bushing transformers installed as detailed in Fig. 5.7. The tank of each transformer is connected to the primary neutral wire and ground. It is good practice to ground one of the transformer secondary neutral bushings to limit the secondary voltage rise to ground.

Neutral Conductor

The secondary-neutral conductors in the three-phase, four-wire system and in the single-phase, three-wire system are always grounded, that is, connected to a low-resistance earth terminal. This can be accomplished by connecting the neutral conductor to a ground rod driven deep enough to be in contact with moist earth, as shown in Fig. 5.7 and discussed and illustrated in Chap. 26. The secondary-neutral conductor also must be grounded near the customer's service entrance. This is often accomplished by making a firm connection with water mains embedded in the ground. When this is not sufficient or available, grounds are driven on the customer's premise or metallic grids or plates are embedded in moist earth and an electrical connection is made with them. When only one ground connection is made to the primary-neutral, the neutral conductor is called unigrounded. General practice is to ground the primary neutral conductor at numerous points along its length; it is then called multigrounded. The National Electrical Safety Code requires that multigrounded primary neutral conductors be connected to ground at a minimum of four locations per mile and at each piece of equipment connected to the primary and the neutral conductors along the line. When both primary and secondary circuits are Y-connected

ITEM	No. required
A. Preformed tie wire	7
B. 1″ Pin insulator	4
C. 1″ Pole top pin — 18″	1
D. 5/8 x 10″ Bolt	2
E. 2 1/2″ Curved washer	8
F. 5/8″ Palnut	8
G. Expoxy-rod bi-unit assembly 1/2″ D Rod, 36″ between conductors	1
H. 5/8″ x 12″ bolt	4
I. Stirrup, compression type	3
J. Hot-line clamp	3
K. Transformer secondary lead bracket	2
L. Compression connector	5
M. #6 Split-bolt connector	3
N. 4 Spool rack	1
O. Transformer mounting bracket	1
P. #6 Copper wire	5#
Q. Molding ground wire	30′
R. Staples, fence	10
S. Staples, molding	12
T. 5/8″ x 18′ Ground rod	1
U. Ground-rod clamp	1
V. Transformer lead adapter pin	1
W. 5/8″ x 14″ Double arming bolt	1

Figure 5.7 Three-phase Y-Y pole-type single-primary-bushing transformer installation details showing connections to provide three-phase, four-wire low-voltage service from a three-phase, four-wire primary. Dimensions in illustration are shown for distribution system operating at a primary voltage of 13,200Y/7,620 volts.

103

four-wire circuits and both circuits run along the same route, a single neutral conductor can serve both circuits. Such an arrangement is called a common-neutral system.

In the case of the single-phase secondary, either two- or three-wire, the neutral conductor is grounded at numerous locations, including the consumer's service entrance. The reason for grounding these circuits is primarily for safety.

Distribution Environmental Considerations

The compatibility of distribution systems with the environments in which they are located has become an increasingly important consideration. The location and appearance of both substations and overhead lines have become primary criteria for the planning and design of distribution systems. A strong trend to use underground rather than overhead construction for distribution lines has developed.

Techniques have been developed for improving the appearance of overhead lines by the selection of materials, structural shapes, and colors which are in harmony with the environment. The improved appearance of the overhead lines is sufficient in meeting appearance objectives in many.

The adverse visual impact of the existing overhead lines can be reduced to a significant extent by selective conversion projects of existing overhead distribution lines and associated equipment to underground distribution facilities. The optimal environmental improvement per dollar expended for overhead to underground conversion projects can be obtained by selecting overhead lines that have a particularly adverse impact because of congestion of facilities, location, or exposure to public view.

Efforts to make distribution facilities more compatible with their environments by choice of location, consideration for appearance in the design of visible facilities, and the use of underground construction for new lines where economically feasible is promoted and encouraged by distribution engineering designers.

Harmonic Voltages

Voltage and current nonsinusoidal waveforms with frequencies which are an integer multiple of the base 60 Hz fundamental frequency are referred to as harmonic voltages. For a base fundamental frequency of 60 Hz, the second harmonic frequency is 120 Hz, the third harmonic frequency is 180 Hz, and so on. The measure of harmonic distortion is quantified in units of total harmonic distortion (THD).

These abnormal higher-order harmonic frequencies in distribution circuits can be originated by overvoltages which cause distribution transformers to be saturated. Sources of voltage waveform distortion caused by higher-order harmonic voltage frequencies include:

improperly constructed rotating machinery,

arc furnaces,

arc-welding equipment, and

solid-state devices:

thyristor-controlled equipment,

diode bridges,

rectifiers, and

inverters

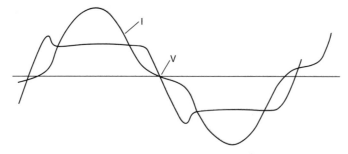

FIGURE 5.8   Plot of voltages and currents in distribution circuit distorted by harmonics generated by an arc furnace.

The installation of customer-owned generating equipment that is not properly manufactured, like wind generators operating in parallel with the distribution circuits, and the proliferation of thyristor-controlled devices have greatly increased the harmonic voltages on the distribution system and the harmonic currents that flow from the voltage sources (Fig. 5.8). Harmonics in the distribution system can damage equipment such as capacitor banks, motors, transformers, and loadbreak devices. Protective relays may malfunction, fuses may melt erroneously, and inductive interference may develop in communication circuits from the harmonic voltages and currents. Harmonics can increase power losses significantly in the distribution circuits, causing a major increase in unaccounted-for energy and operating costs.

Harmonics are corrected by modification of the source equipment with electronic filters consisting of resistors, capacitors, and inductors.

Electrical Interference

High-frequency voltages and currents can be generated by the customer's equipment connected to distribution circuits, by arcing between components of the circuit, and by corona. The electrical interference may interfere with operation of radios or television sets. Radio interference is more common than television interference because of the lower transmitted electrical frequencies.

The best method of eliminating electrical interference is to prevent it by constructing the distribution lines properly. Loose connections, defective insulators, inadequate clearance between high-voltage and ground conductors, and contamination on insulators may cause arcing. Corona occurs on high-voltage transmission circuits more frequently than on distribution circuits. Contaminated distribution-line insulators may permit arcing to be generated during fog or high-humidity conditions.

The customer's equipment, such as thyristor-controlled devices, defective thermostat contacts, and heating devices with a broken wire such as a heating pad or heat tapes, is a common source of electrical interference. Electrical interference from power-line sources will generate a buzzing sound in radio and television audio outputs. Horizontal bands of lines and spots that move up slowly on the television screen can be caused by electrical interference. Arcing-type electrical interference normally causes a raspy intermittent buzz in the audio output and two bands on the television screen that become wider and denser as the intensity of the interference increases.

Observation and information from the public can be helpful in locating the electrical interference. A radio receiver and/or a meter that measures field-strength intensity in the audio and VHF range can be used to good advantage. Special receivers and instruments with a directional antenna are available for locating electrical interference sources.

Neutral to Earth Voltages

Neutral to earth voltage is the voltage measured from the electrical system neutral and/or any structure bonded to the neutral conductor and remote earth which can be established with a ground rod driven in the earth at a location isolated from the neutral system grounds. The surface of the customer's electrical utilization equipment and other metallic surfaces may have voltages impressed on them that can be felt and measured between the metallic surface and a remote ground source. These voltages are usually caused by defective insulation of the customer's appliances or wiring. Unbalanced loads connected to the 120-volt circuits in the customer's premises may create voltages on metallic surfaces, if the electrical equipment grounding wire is not isolated from the neutral conductor beyond the customer's distribution panelboard, as required by the National Electrical Code. When the neutral conductor is connected to the equipment grounding wire, the voltage drop on the neutral conductor will be impressed on the equipment grounding wire and all metallic surfaces connected to it. The primary-neutral conductor on a single-phase electric distribution line normally has a small voltage on it that can be measured with a voltmeter connected to the primary neutral conductor and a remote ground rod. This voltage results from a portion of the return current from the customer's loads flowing through the impedance of the neutral conductor. Adequate size neutral conductors and multiple ground rods connected to the neutral can minimize the level of voltage on the primary-neutral conductor. The voltage on the primary neutral will appear on the customer's neutral conductor, since they are normally connected together at the distribution transformer location.

Stray Voltage

Stray voltage at dairy farms and animal shelters has been defined as low-level voltage present across points (for example, drinking cup to rear hooves) which will cause a current to flow through an animal when the animal simultaneously comes into contact with the points. Stray voltage can be measured by connecting the leads of a voltmeter from the drinking cup or waterline in the barn or milking pallor to a 4-inch square plate on the floor of the barn. The surface of the floor of the barn must be moist and pressure must be applied to the copper plate. A 500-ohm resistor is placed across the voltmeter leads to simulate the resistance of the cow.

Isolation

Isolation of the primary distribution system neutral conductor from the secondary neutral conductor will prevent the primary neutral voltage drop from contributing to stray voltage. However, this will not eliminate stray voltage created by the customer's wiring or equipment. Isolation is permitted by the National Electrical Code as a temporary means of reducing stray voltage. Isolation is undesirable from a safety standpoint because the customer loses the benefits of the low primary ground resistance in parallel with the customer's grounding system. If the isolating device fails, it probably will not be detected for an extended period of time. Failure of the isolating device could permit a high voltage to appear on the customer's wiring if the distribution transformer is

FIGURE 5.9   Schematic diagram for an isolation transformer installation.

damaged by lightning or fails as a result of some other problem such as an overload. Isolation increases the cost of the facilities providing service.

Isolation can be accomplished by installing an isolation transformer with a 1:1 ratio on the secondary of the distribution transformer near the customer's distribution panelboard (Fig. 5.9). The isolation transformer must have adequate capacity for the load on the customer's circuits. An isolation transformer will not eliminate stray voltages if the customer's wiring or equipment is defective.

Isolation can be accomplished with a device containing solid-state electronic components connected between the primary and secondary neutral conductors near the distribution transformer. The primary and secondary neutral conductors must each be connected by insulated wire to a ground rod. The ground rods must be separated by 12 ft or more. The solid-state electronic isolation device becomes conducting if a voltage develops between the primary and secondary neutral conductors that is greater than the isolator's designed isolation level, thus limiting the high voltage that could appear on the low-voltage secondary circuits (Fig. 5.10).

An electronic grounding system has been developed to eliminate neutral-to-earth potentials found to be present in milking parlors of dairy farms. The system can be used for other installations with similar characteristics. The electronic grounding system measures the potential between earth and neutral at the power utilization location. The measured voltage is amplified, and a current proportional to the measured voltage is

FIGURE 5.10   Schematic diagram for installation of solid-state electronic isolation device.

supplied to remote earth, which eliminates the potential between earth and neutral. The electronic grounding system is described fully in a paper entitled "Computation, Measurement and Mitigation of Neutral-to-Earth Potentials on Electric-Distribution Systems," by William K. Dick and David F. Winter, in *IEEE Transactions on Power Delivery*, Vol. PWRD-2, No. 2, April 1987.

The best way to prevent problems as a result of the neutral-to-remote earth voltages in a milking parlor on a dairy farm is to construct the milking-parlor facilities properly with an equipotential plane. The equipotential plane can be established by reinforcing the concrete floor with a 6-inch × 6-inch reinforcing mesh connected together at 3-foot intervals with number 4 AWG or larger copper bonding wire which is connected to the ground rod at the milking-building electric-distribution panel. All metal-conductive equipment that the cows can touch in the milking building, such as metal posts, feeding troughs, curbs, grates, etc., must be connected together with the bonding conductor and connected to the steel-reinforcing wire mesh in the concrete floor.

It is difficult to retrofit the concrete floor for animal confinement areas to establish an equipotential plane. However, if the electrical wiring is installed as specified by the National Electrical Code, it is easy and inexpensive to provide protection for the animals in the confinement area from stray voltage. Painting the concrete floor of the animal confinement areas with a proper paint for application to concrete will provide a moisture seal and insulate the animals from remote earth. A solvent-based sealer formulated with acrylic and silicone resins is recommended. The concrete should receive two coats of the sealer with a skid-resistant additive placed in the second coat.

CHAPTER 6

Construction Specifications

Several transmission and distribution lines are built by powerline infrastructure contracting companies employing linemen and cablemen. A contractual agreement is completed between the electric utility company and the contracting company to specify the responsibilities for completing the project. Construction specifications are included in the contract. The specifications, or parts thereof, are used by linemen and cablemen installing the facilities.

Construction Contracts

The contract describes the scope of work to be completed in the project and would be similar to the following: The contractor agrees to provide all necessary materials, tools, equipment, labor, and supervision and to pay all related costs listed in the contract necessary for the proper clearing of land and for the proper construction and completion of the transmission line described and specified in "Land Clearing Specifications for a New 161 kV Transmission Line" and "Specifications for the construction of a 161-kV transmission line from Substation A to Substation B," which specifications, including all exhibits thereto attached, are made a part of the contract. A hold harmless clause, a schedule for commencing and completing the work, and provisions for payment are usually included in the contract.

Specifications

General specifications consist of the following:

Scope of work

Definitions

Items furnished by contractor

Materials furnished by others

Changed conditions

Changes in work

Inspection and testing of work

Correction of work

Guarantee

Contractor's right to terminate contract

Assignment

Separate contracts

Connection to work of others

Progress of work

Losses to be borne by contractor

Delays

Indemnity against loss of livestock and crop and fence damage

Indemnity against claims by subcontractors or miscellaneous third parties

Performance by owner of contractor's work

Owner's right to terminate contract

Subcontracts

Performance bond

Liens

Patents and royalties

Government labor regulations

Indemnity against loss due to damage or accidents

Labor

Progress payments

Final payment

Payments and payments withheld

Engineer

Specifications

Schedule of work

Overtime

Use of premises

Accidents

Fire prevention

Contractor's liability insurance

Protection of and responsibility for work property

Safety and accident prevention

Right-of-way, easements, and permits

Public regulations, permits, and laws

Cleaning up

Acceptance

Contractor's responsibility

Special Conditions

Such conditions itemized in the specifications would probably consist of these items:

Definitions

Scope

Facilities at site

Standards

Inspection of the site Surveys

Right-of-way

Cooperation with other contractors

Access to the work

Installing gates and maintaining fences

Underground structures

Damages

Loss or damage to livestock

Arrangement with owners of adjacent property

Roads

Use of explosives

Communication equipment

Warning to contractor's employees

Safety precautions

Equal opportunity

Insurance

Withholding of state income tax

Route map

Special provisions

Cost accounting code

Commencement and completion of work

I. Land Clearing Specifications for a New 161-kV Transmission Line Example

Such specifications of land clearing for a new 161-kV transmission line would cover the following:

Scope of Work

The work to be performed under the terms of this specification shall consist of clearing the owner's right-of-way for a new 161kV transmission line. The land to be cleared will be from Substation A to Substation B. The width of the right-of-way is 150 ft.

Conditions for Clearing

The contractor shall clear the right-of-way of all obstructions that will interfere with the operation of the electric transmission line. The clearing to be performed, as a part of these specifications, shall be done in strict accordance with the contract documents. The actual work involved consists of clearing a strip of land 75 ft wide on each side of the centerline of the transmission line, by cutting and/or trimming of all trees and brush in the right-of-way limits. The work includes the removal or trimming of trees which the engineer may classify as "danger trees."

Danger Trees

The contractor must have written permission from the engineer before any danger trees are cut. The trees to be cut shall be marked with a distinctive blaze, by personnel furnished by the contractor, in accordance with the instructions of the engineer.

Clearing in developed areas, such as those planted for orchards or other shade, fruit, or ornamental trees, shall be conducted in accordance with directions given by the engineer. The contractor shall promptly notify the engineer whenever any objections to the trimming or felling of any trees or to the performance of any other work on the land are made by the landowner.

Clearing

All trees, brush, stumps, and other inflammable material, except grass and weeds, shall be removed unless the contractor is otherwise instructed by the engineer. All trees and brush shall be cut 3 inches or less from the ground line (measured on the uphill side). All stumps must be cut so that passage of trucks and tractors will not be hindered.

All trees that are located within highway rights-of-way and are to be removed shall be cut at the ground line or as near to the ground line as possible, in accordance with highway commission practices. The extent of clearing of brush, hedges, and the like within 150 ft of any public road shall be determined by the engineer. Wherever the right-of-way crosses orchards, parks, or indicated special areas, all clearing shall be done specifically as directed by the engineer.

Disposal

The contractor shall dispose of all trees and brush by either chipping and spreading or hauling away unless otherwise specified in the special provisions or as otherwise directed by the engineer. Disposal of any limbs or brush shall be conducted in conformance with state and local laws and applicable regulations and shall be accomplished in such a manner as not to create a hazard or a nuisance. The contractor shall obtain all necessary permits. No damage of any nature shall be inflicted by the contractor upon adjoining property by unwarranted entry or disposal upon said property. All logs and brush shall be the property of the contractor unless otherwise specified in the "special provisions."

Stump Removal

Any stumps that interfere with installation of the structure shall be completely removed and disposed of by the contractor at no extra cost to the owner.

Seeding Right-of-Way

In wooded areas where the contractor may deem it necessary to bulldoze a road for construction purposes or in uncultivated areas where the ground is denuded of vegetation due to construction operations, the denuded areas shall be seeded with a mixture of 2 lb of Alsike clover and 6 lb of White Dutch clover per acre. The seed shall be furnished by the contractor and shall be well raked in to ensure germination. Areas where germination fails shall be reseeded at the contractor's expense.

Maintaining Fences During Clearing

Where the removal of trees requires the removal of fencing along the owner's right-of-way, the contractor shall take steps to provide temporary fencing as directed by the engineer. Such temporary fencing shall remain the property of the contractor and shall be removed when permanent fencing is installed or when directed by the engineer. During disposal and/or removal of trees, the contractor shall keep the fences in good repair where the work has affected their condition and shall keep all gates and temporary openings closed except when in actual use. The contractor shall be responsible for any damage caused by failure to keep fences in good repair or gates closed. On completion of the work, the contractor shall restore to their previous condition all fences that have been damaged during clearing operations. This includes the removal of gates installed for the contractor's use if the landowner wishes those gates removed.

Treatment

Not less than 2 weeks before cutting, all trees and brush shall be given an adequate basal spray with a herbicide that meets with the approval of the engineer. Treatment prior to cutting is required to allow absorption of the herbicide by the tree and roots in order to prevent regrowth. This herbicide shall be furnished by the contractor.

The basal spray shall saturate the bottom 2 ft of the tree or brush. It is important that the entire circumference be treated. Also, it is important that the ground line and all exposed roots be thoroughly saturated. Some species, such as box elder, willow, and Osage orange, are very susceptible to treatment, while others, particularly ash and linden, are tolerant. Because of the difference of susceptibility, some species will require heavier applications than others; among these are ash, linden, maple, hickory, and oak. In the event that the owner is not able to secure the right-of-way to do the work in time to allow the contractor to basal spray all trees and brush prior to cutting, all stumps shall be saturated thoroughly with herbicide, especially at the ground line and on all exposed roots. Spraying after cutting may be done only when approved by the engineer. Care shall be used to prevent any herbicide from contacting any foliage or farm land off the owner's right-of-way. Herbicide spraying shall be prohibited on any state or federal conservation lands.

Staking of Right-of-Way

The owner will stake the right-of-way limits in advance of the contractor's clearing operations. If for any reason the stakes are removed or lost, the contractor shall inform the engineer who will arrange to have the stakes reset.

II. New 161-kV Transmission Line Construction Specifications Example

Such specifications would include the following:

Scope of Work

The work to be performed under the terms of this specification shall consist of installing a new **10-mile** 161-kV transmission line from **Substation** A to **Substation** B. **The designed conductor size and stranding is 556.5 26/7 ACSR (Dove).**

Local Conditions

The contractor must check all local conditions affecting the work and shall make a thorough examination of the route of the line, plans, specifications, and premises in order to be entirely familiar with the details and construction of the installation.

Workmanship

All work shall be executed in a neat and skillful manner as specified or detailed in these specifications as listed herein and in accordance with best construction practice.

Pole Locations

Stakes showing the location of each pole will be set by the utility company. These stakes will be marked with a structure number. The size of poles required, class, height, and other data for each structure number will be shown on supplementary sheets to be furnished to the contractor by the company.

Depth of Pole Setting

All poles are to meet American National Standards Institute specifications: western red cedar, Penta butt-treated with full-length life-span treatment, and shall be set in accordance with the following table except where directed otherwise in writing by the engineer:

Poles shall be set to stand perpendicular and in exact alignment when the line is completed unless otherwise specifically called for. Pole-setting depths shall have a tolerance of not more than ±3 inches except as otherwise directed in writing by the engineer. Poles set partly in earth and partly in rock shall be set to the depth shown for earth. At locations where the poles of any one structure are located at different ground elevations, the bottoms of the poles shall be leveled by deepening the hole on the higher ground, the hole on the lower ground having the depth shown above. In no case shall any pole be set to any depth such that the top of the Penta treatment is less than 9 inches above ground level at the pole. In special cases it may be necessary for the contractor to excavate the shelf in order to meet the above requirements. In special cases a longer pole may be required on the downhill side which shall be set at least as deep as shown in Table 6.1.

Pole Selection

The contractor shall select heavier poles for longer spans and angle points. In addition, on this type of construction, the contractor shall match, as nearly as possible, the ground line and top dimensions of both poles in each structure. A contractor who finds it impossible,

Height of Pole, ft	Depth of Setting	
	In Earth, ft	In Rock, ft
50	7'0"	5'0"
55	7'6"	5'6"
60	8'0"	6'0"
65	8'6"	6'6"
70	9'0"	7'0"
75	9'6"	7'6"

TABLE 6.1   Pole-Setting Depth

because of variation in sizes of poles received, to meet this requirement shall secure the written approval of the engineer before setting poles.

Pole Setting

Holes shall be dug of sufficient size to permit free insertion of tamping bar on all sides of poles after poles are set. In backfilling, tampers shall continually tamp in earth until the hole is completely filled. No earth shall be added until that earth already in place is solid and tight. After hole is filled, excess earth shall be piled up and packed firmly around pole.

Line poles shall be set to stand perpendicular when line is completed. Poles at corners and angles shall be set with a rake of approximately ¼ inch for each foot of height of pole above ground. No change from this specification shall be made except upon the written direction of the engineer. H-type fixtures are all double-armed. Poles shall be set with gains at right angles to line, as no reaming or extra drilling of holes will be permitted to get arm bolts inserted.

Installation of Arms and Braces

Arms on the two-fixtures are to be completely assembled. Poles are to be gained with bolt holes drilled so that the assembled double arm may be spread and slipped over top of pole and bolted in place. This may be done, at contractor's option, either before or after poles are set. Poles are to be drilled for shield-wire ridge iron, which the contractor will bolt in place. Arms must be leveled after installation. This leveling must be done by leveling the bottom of pole holes. It will not be permissible to level arms by drilling additional holes in poles. Poles are to be furnished with top cross-brace bolt holes drilled. The cross braces shall be installed so as to maintain the vertical alignment of the structures without causing undue permanent stresses in either poles or cross braces. In some cases, the company may require the contractor to bore and gain certain poles, in which case contractor will be allowed a unit price shown in the quotation. All field gains and holes shall have brush or spray preservative treatment.

Special Structures

Poles for special structures are to be furnished roofed only, with the contractor to do all boring and gaining. All field gains and holes shall be brush or spray treated with preservative.

Anchors and Guys

Galvanized-steel guy-anchor assemblies are to be used for all heavy guying. Screw anchors are to be used where specifically indicated by the engineer. Anchors shall be installed to the specified depth; anchor rods are to lie in the direction of strain. Thimble-eye guy nuts or guy-anchor assemblies will be installed. Guys shall be installed before wire is pulled and shall be pulled up to sufficient tension to pull pole over slightly. After wire is pulled, where two or more guys are used on one pole, they shall be inspected and equalized if necessary to see that each guy takes its proper portion of the strain. Prefabricated guy grips shall be used for holding wire, and the loose end of the guy wire shall be fastened with a guy-wire clip. Eight-foot guy guards will be installed on all anchor guys. The following numbers of guy clamps are to be used:

3/8″ EHS guy wire with 3 clamps

½″ EHS guy wire with 4 clamps

Insulators

Tangent line structures will be insulated with polymer suspension insulators hanging vertically. Dead-end structures will be insulated with the same polymer suspension insulators doubled up and separated on a yoke plate. All insulators shall be carefully handled to avoid damage, and no damaged insulators shall be assembled on the line. It shall be the contractor's responsibility to inspect all insulators as taken from their crates and to replace all defective units.

Splices

Splices or joints may be made in the conductors and shield wires where necessary, except at crossings over railroads, important highways, and communication lines, and shall be at least 50 feet from the nearest pole or structure. All splices and dead ends shall be made up in strict accordance using steel-core aluminum sleeves and clamps and a hydraulic press. Particular care shall be taken by the contractor in making up all splices or joints in order to see that the proper procedure is followed. The company reserves the right to inspect any or all joints and to reject those which do not meet with the approval of the engineer. All joints shall be made by one foreman and a crew especially trained for this work.

Stringing Conductors and Shield Wires

Conductors and shield wires shall be handled with care and must not be trampled on or driven over. They shall be continuously inspected during installation, and any cuts, kinks, and other injured portions shall be cut out and the wire spliced. Conductors and shield wires may be strung out with reels in a stationary position with conductors supported on the structures in free-running stringing blocks or paying off from reels towed along the line. Conductors or shield wires shall not be dragged out on the ground or dragged over the arms of the structures or over fences or foreign objects which may damage conductors. Each reel shall be equipped with a suitable braking device to keep the conductors or shield wires always under tension and clear of the ground.

Sagging Conductors and Shield Wires

The power-line conductors and the overhead shield wires shall be strung in accordance with sag and tension charts provided by the company. These sag charts are initial stringing sag charts, and no conductors or overhead shield wire shall be prestressed beyond that required by customary practice. The air temperature at time of sagging shall be determined with a certified etched-glass thermometer. All conductor grips, come-alongs, stringing blocks, and pulley wheels shall be of a design which will prevent damaging the conductors or shield wires by kinking, scouring, or unduly bending them during stringing operations.

Sags shall be accurately measured in several spans approximately equal to the ruling span during stringing and checked against the sag charts for the prevailing air temperature. Sagging shall not be done during periods of high wind which, in the opinion of the engineer, might prevent accurate sag measurements. A tolerance of ±½ inch sag per 100 ft of span length, based on the sag and tension charts will be permitted provided that all conductors in the same span assume the same sag and satisfactory clearances are obtained. The engineer will compute and check the sag at all points to be checked, and the contractor shall furnish the necessary crew for signaling and climbing purposes.

The methods for checking sag and the points at which the checks are to be made shall be agreed upon between the engineer and the contractor, the intent of these specifications being that the engineer shall be assured, by means of a sufficient and reasonable number of checks, that proper clearances are obtained at all points, that the proper tensions are being obtained, and that the general appearance of the line shall be satisfactory. The contractor shall make adequate provision to prevent pulling any structure out of line during the stringing and sagging operations.

Dead Ends
All dead ends of the power conductors will be made with compression or bolted dead-end clamps appropriately sized and rated for the conductor being terminated. The shield wires will be dead ended by means of clevis eye, anchor shackle, and 5/8-inch bolt.

Wire Attachment
After sagging and dead ending conductors and shield wire, the contractor shall straighten any poles that may have been pulled out of plumb in the stringing and sagging operation. Conductors and shield wires shall then be securely clamped in their proper suspension clamps.

Pole Grounding
All poles will have a no. 4 bare-copper ground wire installed; the free end to be stapled along the ground section and extending up 12 inches above the ground line. Contractor shall securely install one 7½-inch copper pole-bottom plate on the butts of all of the above poles. The no. 4 bare-copper wire is attached to the bottom plate. A solderless clamp connector shall be installed 6 inches below ground level, connecting the butt section of the ground wire with the down lead from the pole top. The down lead shall be connected to a 5/8-inch × 12-ft copperweld ground rod, and the two ground rods of each structure shall be connected by a no. 4 copper wire buried not less than 12 inches below ground. Eight feet of ground wire molding shall be installed to protect the ground wire and shall extend 6 inches below grade. Ground wire molding shall be stapled to pole with four staples.

CHAPTER 7

Wood-Pole Structures

Structures supporting electric lines must be designed to support conductors, insulators, and shield wires in a manner that provides adequate electrical clearances and ample mechanical strength as specified in the *National Electrical Safety Code* (ANSI C2), published by the Institute of Electrical and Electronics Engineers, Inc. Safe clearances must be maintained when the conductor temperature is elevated as a result of large currents flowing in a circuit and when the conductors are ice-coated and strong winds are blowing. Common types of structures include wood poles, reinforced-concrete poles, steel poles, steel towers, and aluminum towers. The total cost of the installed facilities normally determines the type of structure used. In general, lines built with wood-pole structures have proven to be the most economical where adequate physical and electrical requirements could be satisfied. The American National Standards Institute's publication entitled *Specifications and Dimensions for Wood Poles* (ANSI 05.1) provides minimum specifications for the quality and dimensions of wood poles that are to be used as single-pole utility structures.

Wood-Pole Types

Southern pine, ponderosa pine, western red cedar, and Douglas fir are the most commonly used species of trees used for wood poles. The type of poles used is determined by the physical requirements of the poles needed to construct the line and the cost of the poles as a result of the shipping expense from the location where the species is grown. The suppliers of the poles usually shave and reduce the diameter of the timber to improve the appearance of the poles.

Wood-Pole Preservative Treatment

The poles are given a preservative treatment to prevent deterioration. Southern and Ponderosa pine poles are treated under pressure because of the thick sap wood. Western red cedar poles are normally treated without pressure. Douglas fir poles can be treated with or without pressure. The treatment of poles has been specified by the American Wood Preservers' Association. Experience has shown that the proper preservative treatment of the wood can double the service life of the poles.

Pentachlorophenol (penta) and chromated copper arsenates (CCA) are used commonly to provide a preservative treatment of wood poles.

The Electric Power Research Institute (EPRI) has determined that poles can be treated economically with ammoniacal copper fatty acid. Ammoniacal copper fatty acid treatment is as effective as pentachlorophenol treatment for poles. Poles treated with ammoniacal copper fatty acid have a light color and are easier to climb because the fatty acid softens the wood. The ammoniacal copper fatty acid treatment of poles is acceptable environmentally because it does not contain materials that are toxic to mammals.

The American Wood Preservers' Association has established standards for copper naphthenate treatment of wood poles. Copper naphthenate treatment has environmental advantages.

Pole Classification

Wood poles are commonly classified by length, top circumference, and circumference measured 6 ft from the butt end. The lengths vary in 5-ft steps, and the circumference at the top varies in 2-inch steps. Thus we have lengths, in feet, of 25, 30, 35, 40, etc., and minimum top circumferences, in inches, of 15, 17, 19, 21, etc.

The circumference measured 6 ft from the butt end determines to which class, numbered from 1 to 10 (distribution poles) or designated H1 to H6 (transmission poles), a pole of a given length and top circumference belongs. The classification from 1 to 10 (distribution poles) or H1 to H6 (transmission poles) determines the strength to resist loads applied 2 ft from the top of the pole.

The American National Standards Institute's publication entitled *Specifications and Dimensions for Wood Poles* (ANSI 05.1) provides much technical data for wood utility poles. Pole circumferences are provided in Tables 7.1 and 7.2. The weights of commonly used wood poles are listed in Table 7.3. The horizontal loads used to identify pole classes are referenced in Table 7.4.

Pole Loadings

Wood poles are extremely versatile. Wood poles support wires, crossarms, transformers, and other distribution equipment. Linemen climb the wood poles. In addition, the wood poles are designed to withstand the stresses that will be encountered by wind and ice loads. The National Electrical Safety Code (NESC) C-2 Section 25 Loadings for Grades B and C designates the United States localities classified in the heavy, medium, light, and warm island loading districts (Fig. 7.1).

Stress levels at the point of maximum stress with appropriate safety factors applied determine the pole selected at various locations. For unguyed poles 55 ft or less in length, exception 1 of NESC Rule 261A2a, ground line stress is considered the point of maximum stress not to be exceeded.

Operating Voltages

Wood-pole structures are used for the entire range of distribution-line operating voltages. Wood poles are frequently used for subtransmission lines operating at voltages of 34.5 kV through 161 kV and for transmission lines operating at 138 kV through 345 kV. Some extra-high-voltage (EHV) transmission lines operating at 500 kV are constructed with wood-pole structures.

Distribution Structures

Distribution-circuit structures normally consist of single wood poles using line-post or pin-type insulators with fiberglass support brackets or crossarms. Construction with line-post insulators or fiberglass support brackets with pin-type

		Dimensions of Western Red Cedar[*] and Ponderosa Pine Poles from ANSI 05.1-1992														
Class		H-6	H-5	H-4	H-3	H-2	H-1	1	2	3	4	5	6	7	9	10
Minimum Circumference at Top (inches)		39	37	35	33	31	29	27	25	23	21	19	17	15	15	12
Length of Pole (ft)	**Groundline[†] Distance from Butt (ft)**	Minimum Circumference at 6 ft from Butt (inches)														
20	4.0	–	–	–	–	–	–	33.5	31.5	29.5	27.0	25.0	23.0	21.5	18.5	15.0
25	5.0	–	–	–	–	–	–	37.0	34.5	32.5	30.0	28.0	25.5	24.0	20.5	16.5
30	5.5	–	–	–	–	–	–	40.0	37.5	35.0	32.5	30.0	28.0	26.0	22.0	–
35	6.0	–	–	–	–	48.0	45.5	42.5	40.0	37.5	34.5	32.0	30.0	27.5	–	–
40	6.0	–	–	56.5	53.5	51.0	48.0	45.0	42.5	39.5	36.5	34.0	31.5	–	–	–
45	6.5	64.5	62.0	59.0	56.0	53.5	50.5	47.5	44.5	41.5	38.5	36.0	33.0	–	–	–
50	7.0	67.0	64.5	61.5	58.5	55.5	52.5	49.5	46.5	43.5	40.0	37.5	–	–	–	–
55	7.5	70.0	67.0	64.0	61.0	57.5	54.5	51.5	48.5	45.0	42.0	–	–	–	–	–
60	8.0	72.0	69.0	66.0	63.0	59.5	56.5	53.5	50.0	46.5	43.5	–	–	–	–	–
65	8.5	74.5	71.5	68.0	65.0	61.5	58.5	55.0	51.5	48.0	45.0	–	–	–	–	–
70	9.0	76.5	73.5	70.0	67.0	63.5	60.0	56.5	53.0	49.5	46.0	–	–	–	–	–
75	9.5	78.5	75.5	72.0	68.5	65.0	61.5	58.0	54.5	51.0	–	–	–	–	–	–
80	10.0	80.5	77.0	74.0	70.5	67.0	63.0	59.5	56.0	52.0	–	–	–	–	–	–
85	10.5	82.5	79.0	75.5	72.0	68.5	64.5	61.0	57.0	53.5	–	–	–	–	–	–
90	11.0	84.5	81.0	77.0	73.5	70.0	66.0	62.5	58.5	54.5	–	–	–	–	–	–
95	11.0	86.0	82.5	79.0	75.0	71.5	67.5	63.5	59.5	–	–	–	–	–	–	–
100	11.0	87.5	84.0	80.5	76.5	72.5	69.0	65.0	61.0	–	–	–	–	–	–	–
105	12.0	89.5	85.5	82.0	78.0	74.0	70.0	66.0	62.0	–	–	–	–	–	–	–
110	12.0	91.0	87.0	83.5	79.5	75.5	71.5	67.5	63.0	–	–	–	–	–	–	–
115	12.0	92.5	88.5	84.5	80.5	76.5	72.5	68.5	64.0	–	–	–	–	–	–	–
120	12.0	94.0	90.0	86.0	82.0	78.0	74.0	69.5	65.0	–	–	–	–	–	–	–
125	12.0	95.5	91.5	87.5	83.0	79.0	75.0	70.5	66.0	–	–	–	–	–	–	–

Note: Classes and lengths for which circumferences at 6 ft from the butt are listed in boldface type are the preferred standard sizes. Those shown in light type are for engineering purposes only.

[*] Dimensions of H classes are applicable for western red cedar only.

[†] The figures in this column are intended for use only when a definition of groundline is necessary in order to apply requirements relating to scars, straightness, etc.

TABLE 7.1   Wood-Pole, Minimum Pole Circumferences, Western Red Cedar, and Ponderosa Pine Poles

Dimensions of Douglas Fir (Both Types) and Southern Pine Poles from ANSI 05.1-1992

Class	H-6	H-5	H-4	H-3	H-2	H-1	1	2	3	4	5	6	7	9	10
Minimum Circumference at Top (inches)	39	37	35	33	31	29	27	25	23	21	19	17	15	15	12
Length of Pole (ft) / **Groundline* Distance from Butt (ft)**					Minimum Circumference at 6 ft from Butt (inches)										
20 / 4.0	–	–	–	–	–	–	31.0	29.0	27.0	25.0	23.0	21.0	19.5	17.5	14.0
25 / 5.0	–	–	–	–	–	–	33.5	31.5	29.5	27.5	25.5	23.0	21.5	19.5	15.0
30 / 5.5	–	–	–	–	–	–	36.5	34.0	32.0	29.5	27.5	25.0	23.5	20.5	–
35 / 6.0	–	–	–	48.5	43.5	41.5	39.0	36.5	34.0	31.5	29.0	27.0	25.0	–	–
40 / 6.0	–	–	51.0	51.0	46.0	43.5	41.0	38.5	36.0	33.5	31.0	28.5	–	–	–
45 / 6.5	58.5	56.0	53.5	53.0	48.5	45.5	43.0	40.5	37.5	35.0	32.5	30.0	–	–	–
50 / 7.0	61.0	58.5	55.5	55.0	50.5	47.5	45.0	42.0	39.0	36.5	34.0	–	–	–	–
55 / 7.5	63.5	60.5	58.0	57.0	52.0	49.5	46.5	43.5	40.5	38.0	–	–	–	–	–
60 / 8.0	65.5	62.5	59.5	58.5	54.0	51.0	48.0	45.0	42.0	39.0	–	–	–	–	–
65 / 8.5	67.5	64.5	61.5	60.5	55.5	52.5	49.5	46.5	43.5	40.5	–	–	–	–	–
70 / 9.0	69.0	66.5	63.5	62.0	57.0	54.0	51.0	48.0	45.0	41.5	–	–	–	–	–
75 / 9.5	71.0	68.0	65.0	63.5	59.0	55.5	52.5	49.0	46.0	–	–	–	–	–	–
80 / 10.0	72.5	69.5	66.5	65.0	60.0	57.0	54.0	50.5	47.0	–	–	–	–	–	–
85 / 10.5	74.5	71.5	68.0	66.5	61.5	58.5	55.0	51.5	48.0	–	–	–	–	–	–
90 / 11.0	76.0	73.0	69.5	67.5	63.0	59.5	56.0	53.0	49.0	–	–	–	–	–	–
95 / 11.0	77.5	74.5	71.0	69.0	64.5	61.0	57.0	54.0	–	–	–	–	–	–	–
100 / 11.0	79.0	76.0	72.5	70.5	65.5	62.0	58.5	55.0	–	–	–	–	–	–	–
105 / 12.0	80.5	77.0	74.0	71.5	67.0	63.0	59.5	56.0	–	–	–	–	–	–	–
110 / 12.0	82.0	78.5	75.0	72.5	68.0	64.5	60.5	57.0	–	–	–	–	–	–	–
115 / 12.0	83.5	80.0	76.5	74.0	69.0	65.5	61.5	58.0	–	–	–	–	–	–	–
120 / 12.0	85.0	81.0	77.5	75.0	70.0	66.5	62.5	59.0	–	–	–	–	–	–	–
125 / 12.0	86.0	82.5	78.5	75.0	71.0	67.5	63.5	59.5	–	–	–	–	–	–	–

Note: Classes and lengths for which circumferences at 6 ft from the butt are listed in boldface type are the preferred standard sizes. Those shown in light type are included for engineering purposes only.

*The figures in this column are intended for use only when a definition of groundline is necessary in order to apply requirements relating to scars, straightness, etc.

Table 7.2 Wood-Pole, Minimum Pole Circumferences, Douglas Fir, and Southern Pine Poles

Size/Class	Weight	Material	Size/Class	Weight	Material	Size/Class	Weight	Material
30-6	556	Southern Yellow Pine	55-1	3,258	Southern Yellow Pine	70-5		Douglas Fir
	415	Red Pine		1,937	Red Pine	70-H1	4,278	Douglas Fir
	328	Western Red Cedar		1,674	Western Red Cedar	70-H2	4,784	Douglas Fir
35-4	841	Southern Yellow Pine	55-2	2,826	Southern Yellow Pine	70-H4	5,888	Douglas Fir
	748	Red Pine		1,650	Red Pine	70-H5	6,532	Douglas Fir
	549	Western Red Cedar		1,464	Western Red Cedar	70-H6	7,268	Douglas Fir
35-5	814	Southern Yellow Pine	55-3	2,066	Southern Yellow Pine	75-1	3,981	Douglas Fir
	641	Red Pine		1,443	Red Pine	75-2	3,511	Douglas Fir
	482	Western Red Cedar		1,277	Western Red Cedar	75-3		Douglas Fir
40-2	1,430	Southern Yellow Pine	60-1	2,782	Douglas Fir	75-H1	4,278	Douglas Fir
	1,217	Red Pine		1,912	Western Red Cedar	75-H2	4,784	Douglas Fir
	899	Western Red Cedar	60-2	2,477	Douglas Fir	75-H3	5,244	Douglas Fir
40-4	1,166	Southern Yellow Pine		1,669	Western Red Cedar	75-H4	5,888	Douglas Fir
	791	Red Pine	60-3	2,153	Douglas Fir	75-H5	6,532	Douglas Fir
	675	Western Red Cedar		1,453	Western Red Cedar	75-H6	7,268	Douglas Fir
45-1	1,977	Southern Yellow Pine	60-H1	3,404	Douglas Fir	80-1	4,503	Douglas Fir
	1,661	Red Pine		2,156	Western Red Cedar	80-2	4,042	Douglas Fir
	1,224	Western Red Cedar	60-H2	3,772	Douglas Fir	80-H1	5,244	Douglas Fir
45-2	1,876	Southern Yellow Pine		2,400	Western Red Cedar	80-H2	5,842	Douglas Fir
	1,432	Red Pine	60-H3	4,186	Douglas Fir	80-H3	6,486	Douglas Fir
	1,070	Western Red Cedar		2,643	Western Red Cedar	80-H4	7,176	Douglas Fir

TABLE 7.3 Weights of Commonly Used Wood Poles (Continued)

Size/Class	Weight	Material	Size/Class	Weight	Material	Size/Class	Weight	Material
45-3	1,623	Southern Yellow Pine	60-H6	5,750	Douglas Fir	80-H5	7,968	Douglas Fir
	1,242	Red Pine		3,374	Western Red Cedar	80-H6	8,832	Douglas Fir
	935	Western Red Cedar	65-1	3,150	Douglas Fir	85-1	4,860	Douglas Fir
50-1	2,348	Southern Yellow Pine	65-2	2,806	Douglas Fir	85-2	4,385	Douglas Fir
	1,937	Red Pine	65-H1	3,818	Douglas Fir	85-H1	5,750	Douglas Fir
	1,448	Western Red Cedar	65-H2	4,232	Douglas Fir	85-H2	6,394	Douglas Fir
50-2	2,222	Southern Yellow Pine	65-H3	4,738	Douglas Fir	85-H3	7,084	Douglas Fir
	1,650	Red Pine	65-H4	4,846	Douglas Fir	85-H4	7,866	Douglas Fir
	1,271	Western Red Cedar	65-H5	5,842	Douglas Fir	85-H5		Douglas Fir
50-3	1,925	Southern Yellow Pine	65-H6	6,486	Douglas Fir	85-H6	9,706	Douglas Fir
	1,443	Red Pine	70-1	3,553	Douglas Fir	90-1	5,250	Douglas Fir
	1,098	Western Red Cedar	70-2	3,144	Western Red Cedar	90-H1	6,302	Douglas Fir

TABLE 7.3   Weights of Commonly Used Wood Poles

Pole Class	Horizontal Load, lb
H6	11,400
H5	10,000
H4	8,700
H3	7,500
H2	6,400
H1	5,400
1	4,500
2	3,700
3	3,000
4	2,400
5	1,900
6	1,500
7	1,200
9	740
10	370

Note: 1 lb = 0.4536 kg.

TABLE 7.4 Horizontal loads Used to Identify Pole Classes. (*Courtesy Standards Handbook for Electrical Engineers, H. Wayne Beatty-Donald G. Fink, McGraw Hill.*)

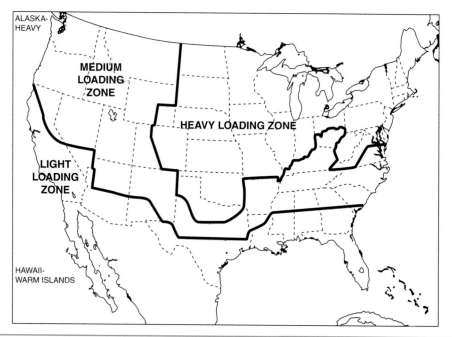

FIGURE 7.1   General loading map of United States with respect to loading of overhead lines. (*Courtesy NESC.*)

FIGURE 7.2   Armless-type construction for 4.16Y/2.4-kV and 13.2Y/7.6-kV distribution-circuit
operation.

insulators is used to create a pleasing appearance for overhead electric distribution
systems (Fig. 7.2). Construction details and materials required for a three-phase,
four-wire 13.2Y/7.6-kV electric distribution line built on wood-pole structures
with fiberglass insulator support brackets to obtain close space construction are
shown in Fig. 7.3.

Many companies are converting 34.5-kV ac transmission circuits to 34.5 Y/19.9-kV
three-phase, four-wire distribution circuits (Figs. 7.4 and 7.5). Three-phase or single-
phase lines are tapped to the converted circuits to supply distribution customers.
The lines tapped to the 34.5Y/19.9-kV circuits in many cases previously operated at
a lower primary voltage. It was necessary to reinsulate, install distribution transformers with the proper voltage rating, and rebuild the lower-voltage lines to operate at the higher voltage. The higher voltage of 34.5Y/19.9 kV has the advantage of
operating at a small amount of current to supply large customer loads. The high
distribution voltage with a small amount of current reduces the required conductor
size and has the advantages of low electrical losses and reduced voltage drop on the
lines.

Subtransmission Structures

Alternating current (ac) subtransmission lines operating at voltages of 138 kV and
below can be built on single wood-pole structures.

Wishbone-type construction on a single-pole structure is popular for use on 69-kV,
115-kV, and 138-kV subtransmission lines. This type of construction provides a symmetrical conductor arrangement which has proved to be economical to build and reliable to operate. Figure 7.6 pictures a single-pole structure with wishbone-type
construction operating at 69 kV. Subtransmission structures have been simplified to
make them as inconspicuous as possible, to improve the aesthetic appearance, and to
preserve the environment. This type of construction minimizes the width of right-of-
way needed and reduces the number of trees that must be removed.

Item	No. required
A. Preformed tie wire size required	7
B. 1″ Pin insulator	6
C. 1″ Pole-top pin	2
D. 5/8″ x 10″ Bolt	3
E. 5/8″ Round washer	2
F. 5/8″ Palnut	9
G. 2 1/2″ Curved washer	1
H. 5/8″ x 12″ Bolt	2
I. Two-phase brackets 48″ spacing	2
J. 5/8″ x 14″ Double arming bolt	2
K. 2 1/4″ Square washer	8
L. #6 Copper wire	1#
M. Split-bolt connector, size required	3
N. Compression connector, size required	1
O. Molding, ground wire	5′
P. Single-clevis insulator	1
Q. Staples, molding	6
R. Staples, fence	6

FIGURE 7.3   Construction drawing and dimensioning of a three-phase 13.2Y/7.6-kV armless-type wood-pole structure (angle pole 6° to 20°).

Conductor Clearances		
Span Length (ft)	Pole Length (ft)	Neutral Location D
200 & less	All	4'0"
201–270	35	6'0"
271–310	40 & greater	7'6"

NOTE: Guy for angles greater than 1°.

Item	Description
A	Insulator, 35 kV, post-type, clamp top.
B	Bracket for post-type insulator.
C	Bolt, 3/4" diameter, length required.
D	Crossarm, wood, 10' long, 4 pin.
E	Washer, round for 5/8" diameter bolt.
F	Bolt, stud, 5/8" diameter, 7" long for line post insulator.
G	Washer, square, flat for 3/4" diameter bolt.
H	Washer, square, curved for 3/4" diameter bolt.
I	Brace, crossarm, wood.
J	Bolt, carriage, 1/2" diameter, 6" long.
K	Washer, round for 1/2" diameter bolt.
L	Insulator, spool, 3" diameter.
M	Clevis for spool insulator.
N	Gain 3" X 4".

FIGURE 7.4 Construction drawing and dimensioning of a 34.5Y/19.9-kV three-phase tangent distribution-circuit structure for installation in a straight portion of the line or where any angle in the line is small.

The structures may use light-colored wood poles and gray-colored post-type insulators (Fig. 7.7). Figure 7.8 is a drawing of a wood-pole tangent structure for a 69-kV three-phase circuit using horizontal post-type insulators with a triangular conductor configuration for installation in a straight portion of the line. Structures with conductor line angles of 45° to 90° angle in the line are double dead ended and appropriately guyed in each direction (Fig. 7.9). Figure 7.10 is a drawing of a 69-kV-circuit single-pole structure designed for installation where it is necessary for the line to change direction. If a line is to be built in an area where appearance is not critical, such as an industrial location, crossarms are frequently used. The use of crossarms reduces the required pole height and permits the addition of a second circuit on the same wood-pole structures if the electrical system requires it at some future time.

NOTES:
1. Omit extension link on single-phase lines.
2. Aluminum conductor, steel-reinforced (ACSR)
3. Stranded aluminum conductor (SAC)

Vertical Construction to	Wire Size	Vertical Spacing D	Max Span
Vertical construction	1/0ACSR 477SAC	5'0"	230'
	795SAC	5'0"	240'
Tangent construction	1/0ACSR 477SAC	4'0"	240'
	795SAC	4'0"	260'

Item	Description
A	Insulator, 4 1/2" diameter, clevis-type.
B	Clamp, dead-end.
C	Link, extension.
D	Bolt, eye, 5/8" diameter, length required.
E	Washer, round for 5/8" diameter bolt.

FIGURE 7.5   Construction drawing and dimensioning of a 34.5Y/19.9-kV three-phase angle distribution-circuit structure.

Figure 7.11 is a drawing of a typical 69-kV-circuit tangent structure with crossarms and provisions for a future second circuit. Figure 7.12 is a drawing of a 7° to 30° angle structure.

Transmission Structures

Two-pole wood structures commonly referred to as H-type structures are frequently used for three-phase ac transmission circuits operating at voltages of 138 kV through

FIGURE 7.6   Wood-pole line using wooden crossarms arranged to provide symmetrical arrangement of the line conductors. The wooden structure is being held in place by the digger-derrick line truck while excavation work around the pole is being completed.

FIGURE 7.7   Post-type insulator in use in horizontal position. Note absence of crossarm and braces and narrow right-of-way required. (*Courtesy Lapp Insulator Co.*)

Item	No. required
A. Ground wire bracket	1
1/2" x 4" Lag screw	1
B. 5/8" x 14" mach. bolt	1
3" Square curved washer/5/8" bolt hole	1
Standard nut	1
Locknut	1
C. 88 kV Horizontal post insulator, gray color	3
D. 3/4" x 12" Machine bolt	2
3/4" x 14" Machine bolt	3
3/4" x 16" Machine bolt	1
3" Square curved washer/3/4" bolt hole	6
Standard nut	6
MF Locknut	6
E. Fiber glass downlead bracket	3
F. 5/8" x 12" Machine bolt	3
Standard nut	3
2 1/4" Square Washer/5/8" bolt hole	6
G. 6A Copperweld groundwire	
H. Two-bolt connector	1
I. Split-bolt Connector	1
J. Ground-wire molding, 1/2" groove	
K. Bronze staple, for 1/2" plastic molding	

Note: Phase spacing (dimension X) varies with span lengths, conductor size, conductor sag, and midspan clearances.

FIGURE 7.8   Construction drawing and dimensioning of a 69-kV single-circuit armless-type tangent structure.

FIGURE 7.9   A 69-kV single-circuit dead-end structure—(90° angle in line).

1'-0"

9'-0"

Ⓐ
Ⓒ

X

Ⓑ

X

30'-6"

65' Pole shown

8'-6"

Plan at conductor

Item	No. Required
A. Pole band	2 Sets (8 sections)
	6 Sets (24 sections)
Stud bolt, 7/8" x 8"	32
Links, 15/16" hole	16 Sets
Guy roller, 15/16" hole	4
7/8" x 3" Machine bolt/MF locknut	4
Shear block	16
7/8" x 19" Threaded rod/2 standard	
nuts, 2 locknuts	2
7/8" x 21" Threaded rod/2 standard	
nuts, 2 locknuts	6
B. Anchor shackle	6
C. Fiberglass strain insulator	6

NOTE: Phase spacing (dimension X) varies with span lengths, conductor size, conductor sag, and midspan clearances.

FIGURE 7.10   Construction drawing and dimensioning of a 69 kV single-circuit dead-end structure (45° to 90° angle in line).

230 kV. Most subtransmission circuits operating in the voltage range of 34.5 kV through 115 kV are constructed on single wood-pole structures; however, a few structures are installed on two-pole H-type wood structures. Extra-high-voltage circuits operating at 345 kV are frequently constructed on two-pole wood structures. A few 500-kV installations have been constructed with wood-pole structures. The wood-pole structures are aesthetically acceptable in wooded areas and can be installed economically in many locations.

Figure 7.13 shows a typical tangent H-type two-pole wood structure commonly used for 161-kV three-phase ac circuits. The structure is usually built on 65-ft poles with some variation due to terrain and has 15½-ft clearance between conductors. A complete structure with crossarms and insulators ready for erection will weigh from 5 to 6 tons. Figure 7.14 is a drawing of a two-pole H-type structure for a 161-kV line. Figure 7.15 is a drawing of a three-pole angle structure. Wood-pole structures in a 161-kV line are shown in Figure 7.16. In Figure 7.17, an east to west 161-kV line crosses over a north to south 69-kV line. As per the NESC, the higher voltage is at the topmost point in the crossing to comply with mandatory clearances.

Guy on slight angles here or below with strain insulator

Tapped lip washer

1/2" Split-bolt assembly

3" Round washer welded to eyebolt

Detail B

6"

5'-0"

3'-2-1/2"

Detail B

4'-6"   4'-6"

3'-0"   3'-0"

Future construction

Standard nut turned 1/4" in pole

Tight hardware at brace & crossarm points

Item	No. Required	
	Single Circuit	Double Circuit
A. Crossarm	2	3
B. Crossarm brace (1 left, 1 right)	2 Pair	3 Pair
C. Shield-wire support	1	1
Washer nut and standard nut	1	1
D. 5/8" x 6 1/2" Eyebolt	3	6
Saddle washer assembly		
Standard nut		
E. 5/8" x 7" Washer head bolt	3	6
Standard nut	3	6
MF locknut	3	6
F. 5/8" x 12" Washer head bolt	1	1
5/8" x 14" Washer head bolt	1	2
Standard nut	2	3
G. 3/4" x 16" Washer head bolt	1	1
3/4" x 18" Washer head bolt	1	2
H. 3/4" Nut and locknut	2	3
Curved guide washer staked	2	3

NOTE: Phase spacing (dimension X) varies with span lengths, conductor size, conductor sag, and midspan clearances. Spacing will be determined in accordance with appropriate design standard.

FIGURE **7.11** Construction drawing and dimensioning of a 69-kV double-circuit crossarm construction tangent structure (0° to 6° angle).

Item	No. Required	
	Single Circuit	Double Circuit
A. Shield-wire support	1	1
Washer nut and standard nut	1	1
B. Pole band	2 Sets (8 sections)	3 Sets (12 sections)
Stud bolt, 3/4" x 6"	8	12
Links	2 Sets	3 Sets
Guy roller, 15/16" hole	2	3
3/4" x 2 1/2" Machine bolt/MF locknut	2	3
C. Double arm crossarm assembly	2	3
D. Crossarm brace (1 left, 1 right)	4 Pair	6 Pair
E. 5/8" x 14" Double arming bolt	2	2
5/8" x 16" Double arming bolt	0	1
4 Standard nuts per bolt		
F. Guy strain insulator, fiber glass	As required	
G. 3/4" x 28" Double arming bolts	2	3
2 Washer nuts and 2 standard nuts per bolt	4	6
3/4" x 28" Double arming bolts	8	12
2 Washer nuts and 6 standard nuts per bolt	16	24
Dead-end Tee	8	12
H. 5/8" x 6" Washer head bolt	8	12
with lip washer		
I. 5/8" x 6" Eyebolt with flange washer	3	6
and standard nut		

NOTE: Phase spacing (dimension X) varies with span lengths, conductor size, conductor sag, and midspan clearances. Spacing will be determined in accordance with appropriate design standard.

FIGURE 7.12   Construction drawing and dimensioning of a 69-kV double-circuit crossarm construction for a 7° to 30° angle structure.

FIGURE 7.13   Two poles combined to make a H-frame structure.

	Item	No. Required
A.	Shield wire support (steel cross angle) with U-bolt, chain link and bonding bolt	1
B.	Double crossarm assembly with 11" spacer fitting	1
C.	Vee brace 9' 1 1/2" C-C role 37 1/2° and 52 1/2° fittings	4
D.	X-brace	2
E.	Center clamp (for X-brace)	1
F.	Pole band	1 Set (4 sections)
	Stud bolt 7/8" x 8"	1 Set (4 sections)
	Links	8
	Guy roller, 15/16" hole	2 Sets
	7/8" x 3" Machine bolt/MF locknut	1
G.	Fiberglass strain insulator	1
H.	7/8" x 12" Machine bolt	2
	Curved washer, 15/16" hole	2
	Standard nut	2
	MF locknut	2
I.	7/8" x 12" Machine bolt	2
	Standard nut	2
	MF locknut	2
J.	7/8" x 23" Threaded rod	2
	Washer nut	4
	Standard nut	4
K.	7/8" x 16" Machine bolt	2
	7/8" x 18" Machine bolt	2
	Curved washer, 15/16" hole	4
	Standard nut	4
	MF locknut	4
L.	Swinging angle bracket	3
M.	Pole shim, 1" or 1 1/2" thick, as required	4

FIGURE 7.14 Construction drawing and dimensioning of a 161-kV H-frame structure for installation where the line has an angle of 1½° to 6°.

136

Item	No. Required
A. Shield wire support	2
B. Double crossarm assembly with filler blocks, 11" separation	2
C. Double crossarm assembly with filler blocks, 11" separation	1
D. Knee brace 8'4" C-C holes 45° fittings	4
E. Diagonal arm assembly	3
F. X-brace	4
G. Center clamp (for X-brace)	2
H. Pole band	3 Sets (12 sections)
Stud bolt 7/8" x 8"	12
Links	5 Pair
Guy roller, 15/16" hole	5
7/8" x 3" Machine bolt/MF locknut	4
Preformed guy grip	
I. 7/8" x 23" Threaded rod	6
Washer nut	12
Standard nut	12
Pole shims as required	
J. 7/8" x 14" Machine bolt	3
Curved washer, 15/16" hole	2
Standard nut	3
MF locknut	3
K. 7/8" x 16" Machine bolt	3
7/8" x 18" Machine bolt	3
Curved washer, 15/16" hole	4
Standard nut	6
MF locknut	6
L. Pole band	3 Sets (12 sections)
Stud bolt 7/8" x 8"	12
Links	3 Pair
Shear block	6
7/8" x 21" Threaded rod/standard nuts and MF locknuts	3
Shackle	3
M. Swinging angle bracket	3
N. Fiberglass strain insulator	3

Rake structure 18° against resultant of angle

Note: Bond all shield wire hardware

3/8" Finished steel

11° separation

Guy front and center poles on angles over 10°

Note: Install extra set of X-braces on poles 90' and over

60' poles shown

36'-0" ground clearance

7'-0"   11'-0"   34'-0"   8'-0"

15'-6"   15'-6"   2'-0"

FIGURE 7.15 Construction drawing and dimensioning of a 161-kV-line three-pole 6° to 17½° angle structure.

137

FIGURE 7.16   Structures in a 161-kV ac transmission line.

Figure 7.18 is a drawing for a double K-type two-pole tangent structure for an EHV ac 345-kV transmission line. A typical line for this use could have 7/16-inch extra-high-strength steel shield wires and two 795-kCMIL 26/7 aluminum conductor steel reinforced (ACSR) bundled conductors per phase, with spans of 1,000 ft between structures. Structure safety factors with ½ inch of radial ice would be 4 for vertical and transverse loads. The structure vertical safety factor for 1½ inches of radial ice would be 1.5, and the structure safety factor for 100 mph winds and bare conductors would be 2.3.

Wood-Substitute Poles

The clear cutting of timber instead of selective harvesting has decreased the supply of wood poles available for use in constructing electric lines. Solid-wood poles are becoming more difficult to obtain and more expensive. Substitute poles made of concrete,

FIGURE 7.17   A 161-kV transmission line traversing from east to west crosses over a north to south 69-kV line.

FIGURE 7.18   Construction drawing and dimensioning of a 345-kV EHV ac transmission line structure.

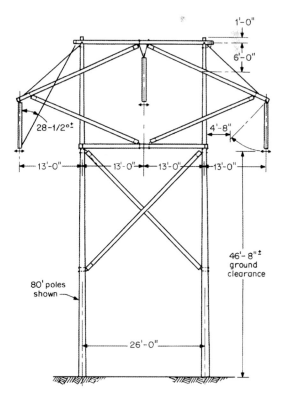

steel, aluminum, and fiberglass are described in Chap. 8. Environmentally friendly laminated wood poles are made by gluing layers of wood together under pressure. This process makes it possible to use smaller trees to manufacture large poles, thus increasing the availability. Laminated poles have been used satisfactorily for use in unguyed, dead end, switch, substation, and tangent applications.

Composite wood poles manufactured utilizing fiber-reinforced polymer (FRP) is an engineered material consisting of reinforcement fibers, polymer resin, and additives. The poles are classified as ANSI 05.1 wood equivalent poles.

The pole dimensions are similar to those for the solid-wood poles they are replacing. The composite-wood poles can be climbed without problems; they are lightweight, smooth, clean, and straight; and they have a satisfactory dielectric strength. The poles should have a long life as demonstrated by accelerated aging tests.

Wood-Pole Fires

Fires on wood poles and crossarms can be initiated by leakage currents. Insulators on wood-pole lines can become contaminated by dust or chemicals in the air during dry weather. Light rain, fog, or wet snow can moisten the insulators, the crossarms, and the poles, causing leakage currents to flow to ground at the base of the pole or a ground wire on the pole. A dry area on the assembly, such as the space between the pole and the crossarm, can impede the flow of leakage current, establishing a voltage across the dry area which may approach the voltage between the line and ground. A small air gap may exist between the pole, the crossarm, and the through bolt securing the crossarm to the pole. An arc can be established across this air gap if the dielectric strength of the insulation of the air in the gap is exceeded. High humidity will reduce the dielectric strength of the air. The arc established across the air gap will ignite the wood in the dry area between the pole and the crossarm. The leakage currents will help to maintain the arc and the fire after it is ignited.

Pole fires can be extinguished with a dry-chemical fire extinguisher designed for use on class C fires (electrical fires) followed by the use of a pressurized-water fire extinguisher on the burnt area after the burning has been minimized. The lineman must be careful to keep the water from the extinguisher clear of the high-voltage circuit while using it to cool the embers to prevent reignition.

Pole fires in locations where experience indicates they may develop can be prevented by cleaning the insulators and/or eliminating air gaps where arcing might develop. Pole-line hardware can be grounded, bridging the air gaps where ignition could occur. Metal shunts can be installed across the dry areas on the wood poles where fires might be started.

Wood Pole Professional Organizations

In addition to the previously mentioned ANSI references, North American Wood Pole Counsel (NAWPC) is the recognized source for information on preservative-treated wood utility poles. In addition, American Wood Protection Association (AWPA), Wood Preservation Canada (WPC) and ASTM International have significantly influenced the utility wood pole industry.

Aluminum, Concrete, Fiberglass, and Steel Structures and Poles

Environmental considerations have increased the use of aluminum, concrete, fiberglass, and steel to support electric transmission and distribution circuits. Wood-pole structures have generally proved to be the most economical for distribution and transmission lines operating at voltages through 345 kV. The design of structures to make them compatible with the environment has resulted in the use of materials and colors to blend together with the natural surroundings. Aluminum, concrete, fiberglass, and steel structures and poles, properly used, may reduce the widths of rights-of ways required for distribution and transmission lines.

Aluminum Structures

Ornamental street lighting is one of the most common applications of aluminum poles (Figs. 8.1 and 8.2). The hollow-tubular street-lighting poles give a pleasing appearance. The poles are lightweight, making them easy to handle and install. The hollow pole facilitates the installation of the low-voltage cables supplying the lamps from underground electric distribution circuits.

Aluminum poles can be used for subtransmission circuits installed along urban streets. Transmission lines operating at voltages through 500 kV have been constructed with H-frame and guyed-aluminum towers. Guyed-aluminum towers have also been fabricated in Y and V shapes. The towers can be completely assembled on the ground at a convenient location and airlifted to the site for erection using a helicopter. The structures can be airlifted, complete with insulators, stringing sheaves, tag lines for the conductor pulling, and guys (Figs. 8.3 and 8.4). Airlifting the structures to the installation site can save time and money in inaccessible locations such as swamps, forest, or mountainous areas. Aluminum towers and poles do not need to be painted for protection, which minimizes the maintenance required.

Concrete Poles

Poles manufactured with concrete are used for street lighting and for distribution (Figs. 8.5, 8.6, and 8.7) and transmission lines (Fig. 8.8). Concrete poles have been used to improve the aesthetics of overhead circuits and are preferred where the life of wooden poles is unduly shortened by decay and woodpeckers. Longer life in certain climates

FIGURE 8.1 Sketch of an ornamental street-lighting pole made of aluminum. (*Courtesy Sentry Electric.*)

FIGURE 8.2 Sketch of an octagonal ornamental aluminum pole. (*Courtesy Sentry Electric.*)

FIGURE 8.3 A 345 kV transmission structure being airlifted by helicopter. (*Courtesy Valmont Industries.*)

and environments and readily available supply are advantages. Compared to wood poles, concrete poles have disadvantages of being more expensive, lower in insulation level, more difficult to climb, heavier to handle, and more difficult to drill when necessary to install equipment.

Hollow-type concrete poles are made by putting the concrete materials and steel reinforcing rods into a cylinder of the desired length and taper. Revolving the cylinder in a lathe-like machine for a period of 10 minutes to 15 minutes then forces the concrete materials to the outside, thereby forming the hollow pole when the concrete cures. The hollow-type pole is lighter than the solid-type pole and provides a means for making connections through the pole to underground cables or services. The hollow-type poles are commonly used for ornamental street-lighting installations served from underground electric circuits.

The hollow-spun prestressed-concrete poles have a high-density concrete shell completely encasing a reinforcing cage containing prestressed high-tensile steel wires. Prestressing produces poles with a high strength-to-weight ratio that can be used for

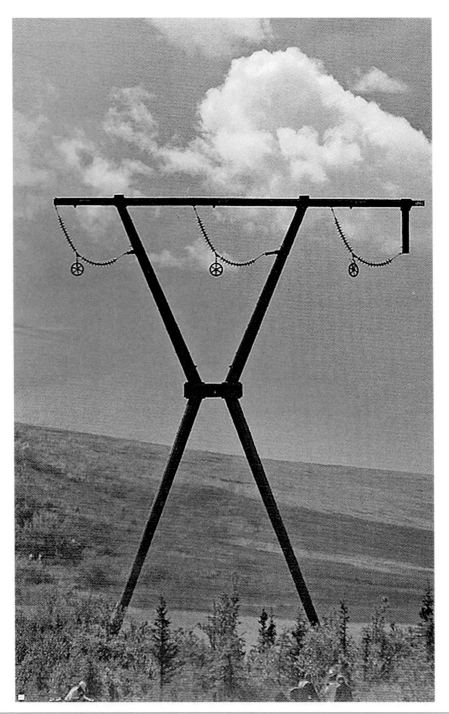

FIGURE 8.4 An airlifted 345-kV structure placed complete with insulators, stringing sheaves, and tag lines for the conductor pulling. (*Courtesy Valmont Industries.*)

FIGURE 8.5   Hollow-type concrete 13.2-7.6 kV three-phase in-line stand-off post insulator pole with three-phase underground riser cables and conduit attached. The lower bracket supports the cable terminators and underground cable lightning arresters. The upper bracket supports the fused cutouts and overhead conductor lightning arresters. (*Courtesy Valmont Industries.*)

FIGURE 8.6   Hollow-type concrete 13.2-7.6 kV three-phase in-line stand-off post insulator pole with three-phase underground riser cables and conduit attached. The bottom bracket supports the cable terminators and underground cable lightning arresters. The middle bracket supports the fused cutouts and the top bracket supports overhead conductor lightning arresters. (*Courtesy Valmont Industries.*)

distribution and transmission lines. Table 8.1 lists typical data for hollow-spun prestressed-concrete poles equivalent to class 1 wood poles.

The tall, long, prestressed-concrete poles are obtained by splicing two sections together. Galvanized-steel davit arms are usually bolted to the poles with a bracket using holes provided in the pole by the manufacturer. Suspension insulators can be fastened to the davit arms, and a crane can be used to set the pole and the assembled equipment attached to it in an augered hole lined with a split-steel casing.

Concrete can be poured around the pole after it has been set and centered in the hole lined with a split-steel casing. The steel casing is removed with the assistance of a crane. Square-tapered prestressed-concrete poles are constructed by placing the stressed reinforcing material in a form and pouring the form full of concrete. The poles may have a center void developed by use of a heavy corrugated cardboard or plastic tube. A copper wire is usually cast in the pole for grounding purposes. A prestressed-concrete 50-ft distribution pole has been found to deflect 2 ft before cracks appeared and 6 ft before failure. Characteristics of square-tapered prestressed-concrete poles are itemized in Table 8.2.

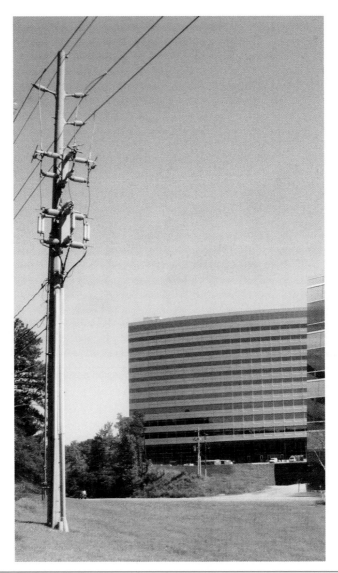

FIGURE 8.7   Hollow-type concrete 13.2-7.6 kV three-phase in-line stand-off post insulator pole with three-phase group operated disconnecting device, three-phase underground riser cables and conduit attached. (*Courtesy Valmont Industries.*)

Large numbers of the prestressed-concrete poles have been used in Florida for street lighting, electric-distribution, subtransmission, and transmission circuits. The installations are clean and neat, satisfying visual environmental requirements.

Florida Power and Light Company has constructed 138-kV transmission lines using line-post insulators on single square-tapered prestressed-concrete poles. Two-pole H-type structures have been used for double-circuit 240-kV transmission lines.

Composite concrete and steel telescoping poles manufactured with a series of nested steel tubes fabricated from high-strength steel and pumped full of concrete can

FIGURE 8.8   A 345-wkV transmission in-line vee-string suspension assembly on concrete pole. (*Courtesy Valmont Industries.*)

be used to support transmission circuits. Pumping concrete into the pole raises it to its full height, ranging up to 150 ft. The assembled telescoping structures are manufactured to be transported to the installation site on a standard trailer. The steel-reinforced concrete foundation is constructed with a tube to accept the telescoping pole. The telescoped pole is installed on the base with crossarms, insulator strings, stringing sheaves, and tag lines in place. The pole is pumped full of concrete, using a commercially available pump and transit-mix concrete. A 100-ft pole can be pumped to its full height in less than 15 minutes.

Fiberglass Composite Poles

Fiberglass is immune to freezing, rotting, the pecking of birds, and damage from nails, all of which are injurious to wood poles. Fiberglass is an excellent insulator, even when wet from rain. Poles made with fiberglass do not require painting. Figure 8.9 shows a fiberglass composite pole installed in a residential area for a streetlight supplied from an underground distribution system.

Linemen use bucket trucks to eliminate the need for pole steps when fiberglass composite poles are located in areas accessible to trucks. Armless construction using fiberglass components is used extensively to enhance the appearance of overhead lines. Fiberglass components are reasonably priced and economically acceptable where the appearance of the facilities is important.

Pole Length, ft			Butt Diameter, inches	Approx. Weight, lb	Moment Capacity, ft-kips
Nominal	Actual	Above Ground			
30	32.8	27.8	14.57	2,300	73
40	39.4	33.4	14.57	2,600	88
45	45.9	39.4	15.75	3,300	105
50	52.5	45.5	16.93	3,900	122
60	59	51	18.11	4,700	138
65	65.6	57.1	19.29	5,500	155
70	72.2	63.2	20.47	6,300	172
80	78.7	68.7	21.65	7,200	188
85	85.3	74.8	22.83	8,200	205
90	91.8	80.8	24.01	9,200	222
100	98.4	86.4	26.38	10,400	237
105	105	92.5	27.56	11,600	255
110	111.5	98.5	28.74	12,800	271
115	118.1	104.6	28.74	14,000	289

Notes:
1. Equivalents are based on using a 2.5 overload factor for prestressed concrete and a 4.0 overload factor for wood.
2. Pole taper is 0.18 in/ft.
3. Poles longer than 90 ft are usually two pieces.
4. Moment capacity is ultimate moment required at ground-line.
5. Ft-kips is 1,000 ft-lb.
Source: PowerSpan, Division of Valmont Industries, Inc.

TABLE 8.1   Characteristics of Hollow-Spun Prestressed-Concrete Poles Equivalent to Class 1 Wood Poles

Length, ft	Nominal Square Pole Top, inches	Minimum Breaking Strength, lb	Minimum Load, with No Cracks, lb	Repetitive Testing Load, lb	Minimum Crack Closure after Repetitive Loading, lb
25	4½	1,200	510	850	255
35	5	1,300	600	900	300
45	6	2,600	1,000	2,400	900
60	9	6,000	2,800	4,200	1,200

TABLE 8.2   Tabulation of Square-Tapered Prestressed-Concrete-Pole Characteristics

FIGURE 8.9   Fiberglass composite street-lighting pole. (*Courtesy The Shakespeare Co.*)

Steel Structures and Poles

There is an increased utilization of steel structures and poles by the industry. Steel poles are utilized for distribution and transmission lines. Steel tower structures have been used extensively to support subtransmission-line and transmission-line conductors. Standard types of steel transmission towers are shown in Fig. 8.10. Towers of this type are required for 765-kV three-phase ac transmission lines. Specially designed steel structures have been created to blend with the natural beauty of the landscape (Fig. 8.11). The size of transmission-line towers should be kept to the minimum feasible to preserve the environment and minimize the conflict with other use of the land. Tubular steel poles continue to be used increasingly to support the subtransmission-line and transmission-line conductors energized at voltages of 345 kV and below (Fig. 8.12). Weathered galvanized-steel structures and poles blend with the sky and reduce the silhouette appearance. Galvanized steel eliminates the need to paint the towers for

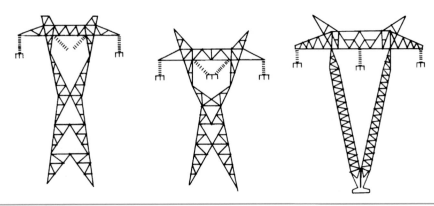

FIGURE 8.10   Standard tower designs.

FIGURE 8.11   Aesthetically designed transmission-line structures.

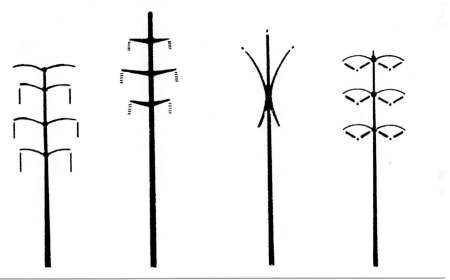

FIGURE 8.12   Tubular steel pole transmission-structure designs.

several years, reducing maintenance costs. Steel is free from damage by woodpeckers and bacteria.

Steel transmission-line towers are of either the rigid or the flexible type. The rigid tower is firm in all directions, whereas the flexible tower is free to deflect in the direction of the line. Rigid towers, in order to be rigid, must have three or more legs, but the flexible construction consists of only two legs. Figure 8.13 shows the four-legged rigid type tower. Rigid towers can resist the same strain in all directions. Flexible towers can resist strain only transversely or across the line and have very little strength in the direction of the line. They will give way or deflect if there is any unbalance in conductor pull; in fact, they depend largely on the conductors to hold them in position.

Flexible H-frame steel structures are usually more expensive than multi-pole wood structures. The steel H-frame structures have many advantages that have increased

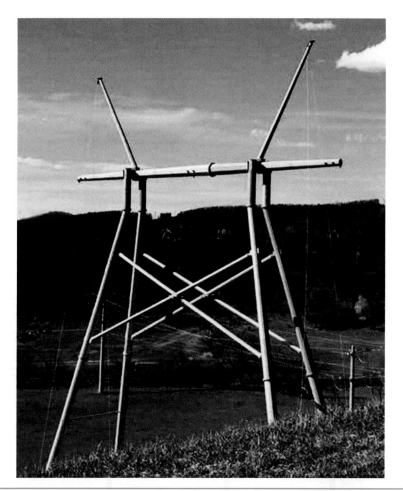

FIGURE 8.13   Four-legged rigid-type 230-kV transmission tower. (*Courtesy Valmont Industries.*)

their use. A 345-kV three-phase ac transmission line constructed across agricultural land in a heavy loading district can have a ruling span of 1,400 ft compared to 850 ft for wood pole structures when the conductor is 954-kCMIL 45/7 ACSR two-conductor bundle. The number of required structures is reduced considerably, minimizing the conflicts with cultivation of the land and reducing the cost of the right-of-way.

The cost of structure assembly of steel H-frame structures is estimated to be approximately 50% of that for standard rigid-latticework steel towers for this type of circuit; it is also less than that of wood-pole structures as a result of the reduced number of structures. The smaller number of structures minimizes the number of insulators and the line hardware required and reduces the labor costs to string, sag, and clip in the conductors. The total cost for the construction of this type of line is approximately the same for rigid and flexible steel towers.

Entire distribution (Figs. 8.14 through 8.19), subtransmission (Figs. 8.20 and 8.21) and transmission (Figs. 8.22, 8.23, and 8.24) lines are being constructed on single steel

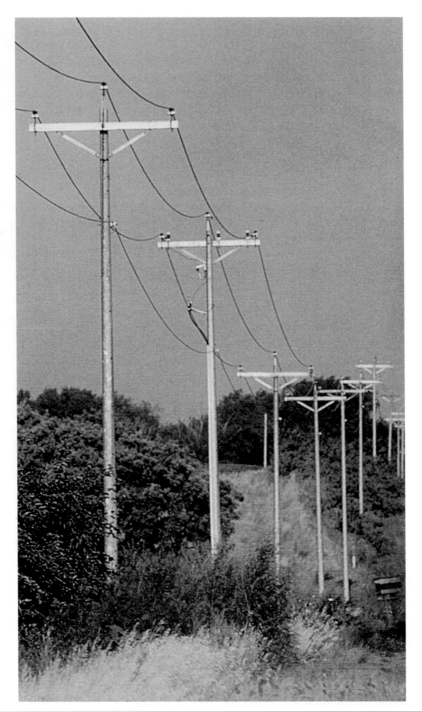

FIGURE 8.14   A 13.2-7.6 kV three-phase 8-ft wooden crossarm in-line steel pole distribution line. (*Courtesy Valmont Industries.*)

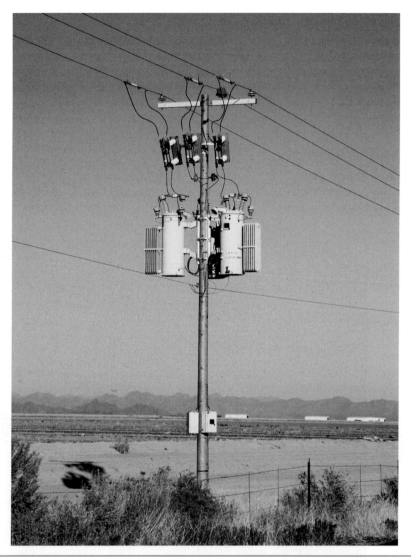

FIGURE **8.15**   A 13.2-7.6 kV three-phase 8-ft wooden crossarm in-line steel pole with a set of three-phase voltage regulators attached. (*Courtesy Valmont Industries.*)

poles. The single steel pole lines are built using narrow profile right-of-way (ROW) footprint single pole construction standards.

The steel pole is manufactured in a single piece or in multiple pieces joined using slip joints or bolted flange joints. The steel utilized to manufacture the poles conforms to ASTM International specifications and standards (Tables 8.3 and 8.4). The tubular distribution steel poles are fabricated with a steel thickness of 0.133 inches to 0.162 inches.

Typically steel poles less than 70 ft in length are shipped by truck. Steel poles 70 ft or greater in length are shipped via escorted truck or rail. The steel pole is handled using

FIGURE 8.16   A 35 kV three phase-steel pole. (*Courtesy Valmont Industries.*)

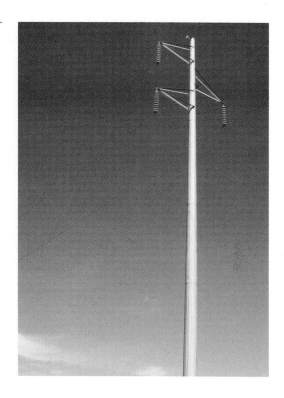

FIGURE 8.17   A double circuit distribution line on COR-TEN finish steel pole with a line angle of 0° to 6°. (*Courtesy Valmont Industries.*)

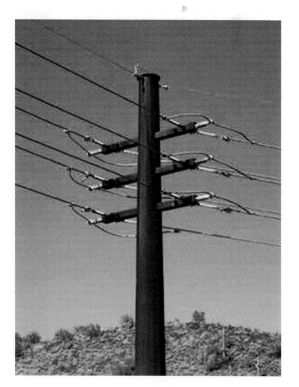

FIGURE 8.18   A three-phase riser steel pole with cutout arrestor bracket. A 69-kV subtransmission line is overbuilt the distribution line. (*Courtesy Valmont Industries.*)

FIGURE 8.19   An in-line double circuit 13.2-7.6 kV distribution line installed on a steel pole. (*Courtesy Valmont Industries.*)

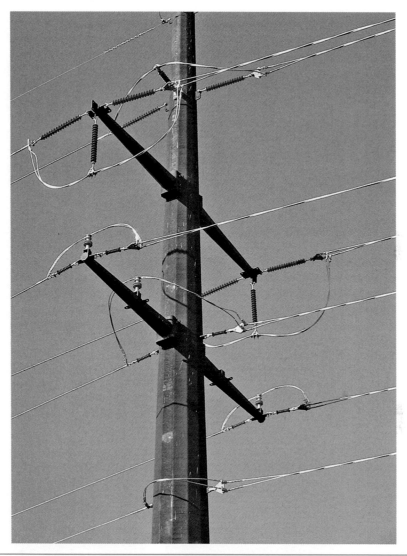

FIGURE 8.20   A 69-kV subtransmission line on a COR-TEN finish steel pole.

nylon straps or rope and not chains or steel cables. Crossarms are mounted on the steel poles using smooth gain plates. The uniformity of the pole minimizes the number and variety of bolt lengths. The steel can be easily drilled. After initial installation, steel pole hardware does not require retightening due to pole shrinkage. When subjected to wind or ice loading, steel poles exhibit tendencies of stiffness and resistance to deflection.

The entire length of the steel pole is coated with either hot-dipped galvanizing, zinc silicate, paints weathered steel, or metalizing. The portion embedded in the ground is layered with an additional urethane coating. The pole is directly embedded in the soil or positioned on a concrete anchor based mounting. The direct embedded poles are backfilled with either the original soil extracted while boring the hole or crushed rock.

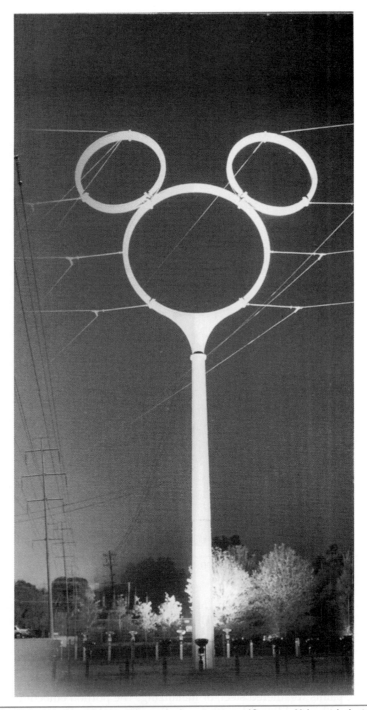

FIGURE 8.21 Unique in-line 69-kV subtransmission structure. (*Courtesy Valmont Industries.*)

FIGURE 8.22   A 230-kV transmission in-line suspension insulator assembly installed on a steel pole. (*Courtesy Valmont Industries.*)

FIGURE 8.23 A 345-kV transmission in-line vee-string suspension insulator assembly installed on a steel pole. (*Courtesy Valmont Industries.*)

Figure 8.24   A 345-kV transmission line using a combination of in-line suspension insulator assembly and vee-string suspension insulator assembly installed on a steel pole.

Pole Length, ft			Butt Diameter, in	Approx. Weight, lb	Moment Capacity, ft-kips
Nominal	Actual	Above Ground			
30	32.8	27.8	14.57	2,300	73
40	39.4	33.4	14.57	2,600	88
45	45.9	39.4	15.75	3,300	105
50	52.5	45.5	16.93	3,900	122
60	59	51	18.11	4,700	138
65	65.6	57.1	19.29	5,500	155
70	72.2	63.2	20.47	6,300	172
80	78.7	68.7	21.65	7,200	188
85	85.3	74.8	22.83	8,200	205
90	91.8	80.8	24.01	9,200	222
100	98.4	86.4	26.38	10,400	237
105	105	92.5	27.56	11,600	255
110	111.5	98.5	28.74	12,800	271
115	118.1	104.6	28.74	14,000	289

Notes:
1. Equivalents are based on using a 2.5 overload factor for prestressed concrete and a 4.0 overload factor for wood.
2. Pole taper is 0.18 in/ft.
3. Poles longer than 90 ft are usually two pieces.
4. Moment capacity is ultimate moment required at ground-line.
5. Ft-kips is 1,000 ft-lb.
Source: PowerSpan, Division of Valmont Industries, Inc.

TABLE 8.3   Characteristics of Steel Poles Equivalent to Class 1 Wood Poles

	Steel Pole LD-1 / Wood Pole Equivalent CL-1			Steel Pole LD-2 / Wood Pole Equivalent H-1			Steel Pole LD-3 / Wood Pole Equivalent H-2			Steel Pole LD-4 / Wood Pole Equivalent H-3		
Total Pole Length (ft)	Bottom Diameter (inches)	Ultimate Moment Capacity @ Groundline* (ft-kips)	Pole Weight (lb)	Bottom Diameter (inches)	Ultimate Moment Capacity @ Groundline* (ft-kips)	Pole Weight (lb)	Bottom Diameter (inches)	Ultimate Moment Capacity @ Groundline* (ft-kips)	Pole Weight (lb)	Bottom Diameter (inches)	Ultimate Moment Capacity @ Groundline* (ft-kips)	Pole Weight (lb)
40	13.20	114	892	13.90	125	920	14.80	141	957	16.20	170	1,046
45	13.72	123	1,024	14.51	136	1,060	15.52	154	1,107	16.98	186	1,209
50	14.25	132	1,162	15.12	148	1,207	16.25	169	1,264	17.75	203	1,381
55	14.77	142	1,304	15.74	159	1,358	16.98	184	1,428	18.52	220	1,561
60	15.30	151	1,450	16.35	171	1,515	17.70	199	1,597	19.26	238	1,739
65	15.83	162	1,603	16.96	184	1,679	18.43	215	1,776	20.07	257	1,944
70	16.35	172	1,763	17.58	197	1,850	19.15	232	1,962	20.85	278	2,145
75	16.88	183	1,927	18.19	211	2,027	19.88	250	2,153	21.62	298	2,356
80	17.40	194	2,096	18.80	225	2,209	20.60	268	2,354	22.40	319	2,574
85	17.93	206	2,270	19.41	240	2,398	21.33	287	2,564	23.17	341	2,801
90	18.45	218	2,450	20.03	255	2,594	22.05	307	2,779	23.95	364	3,032
95	18.98	230	2,634	20.64	270	2,794	22.78	327	3,000	24.72	388	3,274
100	19.50	243	2,832	21.25	287	3,010	23.50	348	3,238	25.50	412	3,524
105	20.03	256	3,026	21.86	303	3,223	24.22	369	3,473	26.27	429	3,781
110	20.55	269	3,225	22.48	320	3,441	24.95	392	3,716	27.05	448	4,046
115	21.07	283	3,543	23.08	338	3,776	25.67	413	4,090	27.82	466	4,454
120	21.60	297	3,753	23.70	356	4,011	26.40	430	4,349	28.60	488	4,734

Table 8.4 Light Duty Single Piece Steel Pole Data

	Steel Pole LD-5			Steel Pole LD-6			Steel Pole LD-7			Steel Pole LD-8		
	Wood Pole Equivalent H-4			Wood Pole Equivalent H-5			Wood Pole Equivalent H-6					
Total Pole Length (ft)	Bottom Diameter (inches)	Ultimate Moment Capacity @ Groundline* (ft-kips)	Pole Weight (lb)	Bottom Diameter (inches)	Ultimate Moment Capacity @ Groundline* (ft-kips)	Pole Weight (lb)	Bottom Diameter (inches)	Ultimate Moment Capacity @ Groundline* (ft-kips)	Pole Weight (lb)	Bottom Diameter (inches)	Ultimate Moment Capacity @ Groundline* (ft-kips)	Pole Weight (lb)
40	17.18	189	1,087	17.18	219	1,261	17.06	246	1,428	18.04	273	1,482
45	18.08	209	1,262	18.08	242	1,464	17.94	271	1,656	19.05	303	1,725
50	18.98	230	1,444	18.98	266	1,676	18.82	298	1,898	20.05	336	1,982
55	19.87	251	1,636	19.87	292	1,900	19.71	326	2,151	21.06	370	2,254
60	20.73	274	1,827	20.73	318	2,123	20.55	356	2,403	22.01	406	2,523
65	21.67	298	2,048	21.67	346	2,380	21.47	389	2,693	23.07	443	2,834
70	22.57	322	2,268	22.57	374	2,635	22.35	421	2,981	24.07	482	3,144
75	23.46	348	2,496	23.46	404	2,901	23.24	453	3,281	25.08	523	3,468
80	24.36	375	2,734	24.36	435	3,177	24.12	486	3,593	26.08	565	3,805
85	25.26	403	2,980	25.26	468	3,464	25.00	522	3,917	27.09	609	4,156
90	26.16	425	3,235	26.16	501	3,761	25.88	559	4,250	28.09	655	4,520
95	27.05	446	3,500	27.05	536	4,069	26.77	597	4,597	29.10	702	4,897
100	27.95	468	3,771	27.95	571	4,385	27.65	637	4,956	30.10	751	5,302
105	28.85	493	4,053	28.85	608	4,713	28.53	678	5,327	31.11	801	5,707
110	29.75	517	4,344	29.75	646	5,052	29.41	720	5,710	32.11	854	6,126
115	30.64	550	5,022	30.64	675	5,598	30.29	763	6,329	33.11	901	6,786
120	31.54	575	5,384	31.54	704	5,959	31.18	808	6,736	34.12	963	7,232

*Groundline assumed to be 10% + 2 ft from base

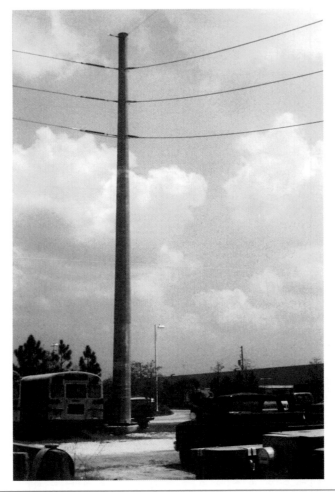

FIGURE 8.25   Galvanized-steel pole used in a 69-kV three-phase subtransmission line to eliminate the need for guys and anchors.

The steel pole is grounded at the base of the pole. The ground rod is placed in undisturbed earth away from the pole base. The ground wire is connected on the pole using stainless steel grounding nuts installed 2 ft above the groundline. A ground wire from the top of the pole running the entire length of the pole is unnecessary.

All steel poles are climbed utilizing steps and the lanyard. Steel poles are not subject to damage from insects, woodpeckers, or fire.

Self-supporting steel poles are used to establish a rigid structure in subtransmission and distribution lines to eliminate the need for guys and anchors (Fig. 8.25). The self-supporting steel poles are frequently installed for angle structures in lines with wood poles used for tangent structures. Elimination of the guys and anchors improves the appearance of the facilities, enhances the environment, and solves right-of-way problems in urban areas. The use of line-post insulators and armless-type construction minimizes the silhouette established.

Distribution Automation and the Smart Grid

To improve reliability, resilience, and power flow, electric distribution systems have become more remotely monitored, controlled, and automated. Controls utilizing intelligent electronic devices (IED) that retrofit existing fault isolation devices, or work in conjunction with the isolating equipment, are installed. Distribution automation (DA) equipment provide the electric system control center personnel (described in Chap. 2) the ability to utilize the supervisory controlled and data acquisition (SCADA) system and distribution management systems (DMS) to not only monitor the system but also provide authorized individuals the ability to communicate with the devices to perform remote site switching or isolating if needed. Personal computers and personal computer software applications have also aided the advancement of distribution automation. A robust communication system must be in place to collect and transmit the signals to and from the equipment.

The Smart Grid

Distribution automation is a key component in a smart grid system. A smart grid network consists of a two-way intelligent device communication system and intelligent electronic hardware that monitors, reports, and controls such that the electric system control center personnel and customers are presented with opportunities to utilize and manage electricity in the most reliable, economical, and efficient manner. Substation automation and automated meter infrastructure (AMI) systems are additional components comprising the smart grid. All smart grid components continuously interact with each other to the benefit of the entire electric system.

The smart grid benefits include a self-correcting network, load management data, emission control, and maintenance planning. A self-correcting network is one that detects and isolates faulted areas of the grid in order to maintain service to as many customers as possible. Load management data provides customers with information to better utilize and consume electricity. Emission control involves the generation of electricity in an environmentally acceptable manner. In the smart grid, device information may be interpreted and utilized to implement proactive maintenance programs.

The benefits of the AMI include automated billing, customer outage detection information, and individual customer load usage profile data.

Monitoring

Past and present distribution automation projects are aimed toward system monitoring of voltage and current. Monitors are placed on the distribution system to sense momentary and continuing power outages (loss of voltage), under- and over-voltages (voltage anomalies), and excessive fault currents (over-currents). The first generation of devices simply logged or displayed status and required field verification to obtain system information. Now utilizing the IED, immediate authorized owner notification is provided. The feedback is communicated by the IED to the owner utilizing a secure communication system.

Circuit fault indicators shown in Fig. 9.1 are installed in either overhead or underground applications. Wireless fault indicators have cellular telephone applications providing authorized individuals the ability to check the device status, receive alarms, and review messages.

An example of an overhead application is the utilization of the IED just beyond a three-phase recloser positioned, as shown in Fig. 9.2. In the event of a single-phase line-to-ground fault, the recloser (reclosers are referenced in Chap. 18) will lock out after progressing through the desired sequence of trips and recloses. The indicator will sense fault current and remain in the faulted position. It is important to note that the fault indicators installed behind automatic reclosing devices must have the inrush current restraint option or the unit will falsely indicate faults. The lineman dispatched to the

FIGURE 9.1 Circuit fault indicators are installed in either overhead or underground systems, and the fault indicators are beneficial in speeding the diagnosis to locate system problems. (*Courtesy Cooper Power Systems, Inc.*)

FIGURE 9.2 Diagram of nine fault indicators positioned on an overhead line beyond a three-phase recloser. An A phase fault has occurred and based on the display of the fault indicators, the lineman will be able to more rapidly track the cause of the fault and restore electric service to the A phase customers without electric service. (*Courtesy Cooper Power Systems, Inc.*)

call to troubleshoot the outage verifies which of the three phases experienced the fault as shown on the face of the IED indicator, and then trace the faulted phase or phases out until he determines the cause of the fault.

Underground fault indicators are beneficial in large underground systems that require troubleshooting. The benefits are briefly described in Chap. 34. Figure 9.3 diagrams a large underground system that has several fault indicators installed and a cable fault on the system has occurred. The lineman troubleshooting a faulted underground section verifies that fault current has flowed through the fault indicator and proceeds to the next location. The fault will be in the cable section between the last fault indicator flagged with a fault and the first fault indicator that has not registered a fault. Fault indicators installed on concentric neutral cable or tape shield neutral cable must be installed in the area of the cable with the neutral either completely removed or doubled back or else the device will not function properly.

Event recorders are available for many distribution devices. Capacitor (capacitors are referenced in Chap. 1) controllers may be monitored with devices, as shown in Fig. 9.4. The monitor records voltage and temperature data. In the event of a system disturbance or fault, the recorder also stores and logs this information. The information may be either retrieved with a laptop computer taken to the worksite, with a telecommunications link to the SCADA system, or with a telecommunications link to a personal computer in the office.

FIGURE 9.3 Schematic of an underground looped distribution system with fault indicators located to aid the lineman to observe the fault indicator target to isolate the problem area. Fault indicators may be installed with visible target mounted flush on the door of the pad-mounted gear or installed so that gear must be opened to observe target. (*Courtesy Cooper Power Systems, Inc.*)

FIGURE 9.4   Lineman interrogating capacitor bank event recorder installed on pole at a height out of reach of typical vandalism. The information is being downloaded into a laptop personal computer that is not shown. (*Courtesy EnergyLine Systems, Inc.*)

Another advantage of monitoring devices is their ability to provide warning indications of malfunctioning equipment prior to the equipment being called on to operate. For example, monitors installed on the capacitors indicate status points or error conditions (Fig. 9.5). If an error condition is detected, the equipment may be maintained as soon as the error message appears and the capacitor will be available as needed.

Conventional voltage sensing controls should be served whenever possible from a dedicated or a lightly loaded transformer. The controls have over- and under-voltage overrides as well as voltage recording capabilities. If it is connected to the secondary of a transformer serving other customers, the controls will be directly affected by those customers' loads instead of the main line voltage, and the voltage data will be corrupted. A "fixed load" transformer is manufactured specifically for this purpose.

Automated Switching, Sectionalizing, and Fault Isolation

Three-phase, group-operated disconnect switches are ideal isolation devices offering the opportunity to automate the distribution system. Motorized operators control the

```
 1 EnergyLine Prog. Cap. Control - Basic Version
 2 Phys. Location:            C-746 13-47-1, Rt 84 & 14 St, Rapid City, IL
 3 Program ID:          C2RP15X1
 4 Upload Time          10/27/94  2:38:11 pm
 5 P.C. Upload Time          2:42:58 pm 10/27/94
 6 Bank Status at upload ...
 7 Voltage:             123.4
 8 Temperature:              62
 9 Cap. Bank State:     Out of Circuit/Switch Open
10 Control Mode:        Automatic Control
11 Voltage Override Status:         None
12 Error Conditions:    All Sensors O.K.
13 Auto. Control Strategy:       Temperature w. Voltage Override
14 Softw. Man. Oper.:       Disabled
15 Softw. Man. Oper. Requested:    Open Switch
16 Softw. Remote/Local Ena.:       Remote
17 Hardw. Man. Override Sw.:       Auto
18 Hardw. Remote/Local Sw.:        Local
19 Low and High Voltage Overrides:       116.0    126.0
20 Bank Voltage Rise Plus Hysteresis:     3.8
21 High and Low Comm. Addr.:       00           00
22 Summer and Winter Control Strategies:     Temperature          Temperature
23 Start., End Months of Summer:     May              September
24 Daily Operating Schedules ...
25 Sunday          Saturday              8    00    23    00
26 (n.a.)          (n.a.)                0    00     0    00
27 Automatic Holidays:          Enabled
28 Holidays by date ...
29 January              1
30 July                 4
31 December            25
32 (unused)            --
33 (unused)            --
34 (unused)            --
35 (unused)            --
36 (unused)            --
37 (unused)            --
38 (unused)            --
39 (unused)            --
40 (unused)            --
41 (unused)            --
42 (unused)            --
43 (unused)            --
44 High and Low Temp. Setpts:          85    82
45 Temp. Change Time Threshold:     10 Min
46 Under/Over Voltage Override Time Threshold:     60 Sec
47 Max. Auto. Control Cycles/Day:          4
48 Cap. Bank Motor Pulse Time:      7 Sec
49 Cap. Bank Minimum Sw. Voltage:     100.3
50 Daylight Sav. Time Auto. Changeover:     Enabled
51 Total Sw.Cycles Since Installation:     217
52 Daily Low, High Volts/Temps, and Cycles Follow ...
53    101.5   123.6     52     71     1
54    117.4   122.6     50     65     1
55    116.4   124.5     49     66     1
56    117.4   124.6     52     67     1
57    116.8   124.9     51     72     1
58    117.4   124.5     52     78     1
59    116.1   124.0     56     80     1
60    117.0   124.3     46     56     1
61    108.4   122.9     39     64     4
62    115.9   124.3     35     65     1
63    117.2   124.3     35     69     1
64    117.2   124.0     37     71     1
```

FIGURE 9.5   Sample sheet of capacitor bank controller event recorder. (*Courtesy MidAmerican Energy Co.*)

FIGURE 9.6 Automated three-phase, group-operated switch. Note the current transformers mounted on the top of the rack at the same level as the disconnect blades. (*Courtesy Kearney-National Company.*)

operation of the switch (Figs. 9.6 and 9.7). Switch monitors provide a system status signal to a remote terminal unit (RTU). A radio link to the SCADA system communicates the RTU information and alerts the control room personnel of a change in system status. The control room personnel may then send a signal back to the RTU. The RTU

FIGURE 9.7 DC motor-driven remote operator unit controlling the operating mechanism of a three-phase, group-operated disconnect switch. (*Courtesy Kearney-National Company.*)

and motorized operator are linked via plug connectors. The operating mechanism changes the status of the switch.

Pad-mounted switchgear equipment with two incoming sources may be controlled such that if power is lost on the incoming primary source, the backup primary feed will be switched to serve the load. Thus, an automatic load transfer has occurred. The switching transfer will not occur if the control device senses a fault on the load side of the line.

In the completely automated disconnect switch distribution system, the system senses a fault has occurred and the isolating devices are activated without manual intervention to reconfigure the system to isolate the least amount of customers affected by the fault from the rest of the system. The customers without electricity will have service restored once the cause of the fault on the isolated section is investigated.

Automated Var Control

Capacitor Switching

A switched capacitor bank is energized or connected to the line on a given schedule. It is sometimes switched in a number of steps, because of its size. The schedule is determined by time, temperature, circuit voltage, or other considerations. It is switched on and off to meet circuit requirements. The switching can be done manually or automatically. A nonswitched capacitor bank is connected to the line at all times and is disconnected only for maintenance purposes.

Capacitor Connections

Figure 9.8 shows a connection diagram for a typical three-phase capacitor bank. The oil switch indicated can be a three-pole device or three single-pole devices that can be controlled automatically to switch the bank in or out to control power factor or to regulate system voltage (Fig. 9.9). Figure 9.10 shows a capacitor application using remote controlled capacitor switches and Figs. 9.11 (a) and (b) show a three-phase pad-mount capacitor bank utilizing remote controlled capacitor switches.

Capacitor banks are turned on and off by oil or vacuum switches. The switches may be activated based upon a clock setting, voltage, or radio controls. The clock settings are established based upon engineering system studies. The clocks may be adjusted in the spring and fall for standard time zone changes. Capacitor banks may be

FIGURE 9.8 Typical connection diagram of three-phase capacitor bank.

FIGURE 9.9   Lineman is adjusting automatic control devices for a capacitor-bank installation. Upper control on panel senses loss of individual capacitor units and initiates bank switching before damaging overvoltages occur. Lower control senses line voltage and initiates switching for voltage regulation. (*Courtesy S&C Electric Co.*)

FIGURE 9.10   Pole-top microprocessor-controlled capacitor bank. (*Courtesy Maysteel Electric Utility Products & Enclosures.*)

brought on line based upon the voltage on the circuit. Sensors monitor the voltage of adjacent customers and supply the desired reactive power flow required to compensate the system. Radio-controlled banks are activated based upon MW, MVAR, and MVA readings in the SCADA system. A radio signal is sent to the switch to alter the status of the capacitor bank.

Static Var Compensators

The static var compensator is a device utilized when a simple electric distribution system capacitor will not meet customer power quality acceptance. Due to harmonic and switching transient issues associated with capacitors, static var compensators are utilized for power factor correction. The static var compensator (Figs. 9.12 and 9.13) stabilizes line and system voltage instantaneously as the reactive component of customer load varies.

(a)

(b)

FIGURE 9.11   Pad-mount microprocessor-controlled capacitor bank: (a) front panel removed to display the remote controlled switches and capacitor units and (b) dead-front and clear see-through barriers installed. (*Courtesy Maysteel Electric Utility & Enclosures.*)

800/900-MHz Radio Systems

Upgraded communications systems are replacing limited low-band radio systems. The high-band systems provide users greater flexibility, clearer reception, and more channel options.

Users of low-band systems had limited range and had to compete for air time. Limited range users often had communications that were walked on or cut out by other users not

FIGURE 9.12   The static var compensator has arrived at the utility service center pending installation at the job site. (*Courtesy Power Quality Systems.*)

FIGURE 9.13   Platform mounted static var compensator installation. (*Courtesy Power Quality Systems.*)

realizing the channel was being utilized. The high-band systems have provided users increased flexibility as the users may be divided into call groups. Each call group is assigned a designation on the radio. Unlike the low-band system in which only limited frequency and corresponding channels were available, the call groups access the radio system and the systems hunt to find an open channel among the several channels available. Thus, the users have freer access to a radio channel to communicate vital information.

The 800/900-MHz system requires multiple antenna sites that are linked utilizing fiber-optic or microwave communication systems. The linking of antennas provides simultaneous radio conversation broadcasting thus linking wide areas of the communications operating territory. Distance limitations of the low-band systems were on the order of 50 to 60 miles. The distance limitation of the high-band communications is bound only by the positioning of antenna sites.

Personal Computer Applications

System data may also be gathered and monitored on personal computers. Computer software vendors are marketing programs to improve electric system reliability, resilience, and performance. Examples of software programs available are listed in Table 9.1. The data is then utilized to study the system to determine required corrective action.

Laptop personal computers are also becoming a part of the lineman/cableman's work tools. Laptops are stationed in the truck. Outage orders are dispatched to the truck via wireless communication from the central office mainframe computer to the truck-mounted computer terminal. The lineman/cableman proceeds to each order noting specifics that provide estimated restoration times for the customer service representatives to relay to all customers calling in reference to that specific order. Upon completion of the order, the lineman/cableman indicates causes of the outage and denotes a completed status and then proceeds to the next order in his assigned responsibility area.

Fiber-Optic Possibilities

Fiber-optic systems offer unlimited potential to distribution companies to provide interactive two-way communication service to their customers. The companies may implement new functions to improve system restoration, such as immediate detection of customer outages, fault detection and immediate system isolation, and improved power quality.

Communication systems have been greatly enhanced by the development and advancement of fiber-optic cable. The fiber-optic cable system allows greater amounts of communication data to be passed via light waves versus an electrical signal sent on a traditional twisted paired copper wire system (one wire to send the signal, the other to receive the signal). Another advantage of fiber-optic systems is that the quality of service is not affected by the outside influence of electromagnetic fields.

Utilities are installing fiber-optic communication systems in a variety of ways that also incorporate the benefits of fiber-optic technology. Some fiber-optic systems are being installed as a part of the overhead transmission shield wire. The shield wire cable utilized is called optical ground wire (OPGW) (Fig. 9.14). Other overhead fiber-optic systems are being installed in the traditional communication zone as defined by the *National Electrical Safety Code* (NESC) on the distribution pole. All dielectric, self-supporting (ADSS) cables

Outage Management Systems (OMS)
Transformer Load Management
Fault Trackers
Voltage Control Systems
Protective Device Coordination
Distribution Automation Systems
Distribution Energy Resource Management System
Distribution Grid Management System

TABLE 9.1 Electric Power System Monitoring Software Examples

Optical Ground Wire—OPT-GW

Aluminum clad (AW), galvanized, aluminum alloy, or combined constructions

Mechanical and electrical characteristics similar to conventional ground wire

Fiber count tailored to individual needs

AW wires

Aluminum pipe

Sheath: Heat-resistive plastics and filler if required

Central strength member

Optical fiber (or fillers)

FIGURE 9.14   Optical ground wire (OPGW) cable utilized as the shield wire of a transmission line. (*Courtesy Alcoa Fujikura Ltd.*)

are used for this application. ADSS cables by definition do not carry electric current, radiate energy, or produce heat. Finally, in the cases of underground systems, fiber cables are being installed in high-density polyethylene (HDPE) conduits.

Each fiber conductor in the cable transmits beams of light from the source end to the receiving end. The fiber conductor is made of glass, silica, or plastic. A thin coating is added to the fiber conductor to keep the light contained. Fiber-optic sensors convert the light signals for processing.

Of course, if the utility installs the fiber-optic system it must design, construct, operate, and maintain the system. The new challenges of the fiber-optic system present the lineman opportunities to learn new skills, such as splicing, fusing, and terminating the fiber. New safety precautions unique to the fiber-optic cable are required. Lasers, glass scraps, and chemical cleaners are all items for the lineman to be aware of.

Power-Line Carrier

Power-line carrier systems are in place to provide channel capabilities for transmission line relaying (refer to power-line carrier equipment in Chap. 3). Utilization of the power-line carrier communication medium for distribution automation purposes is being explored. The amount of data transmitted per second is slow and voice communication is limited, yet this communication medium performs well over short distances. The power-line carrier transmitters and receivers are capacitively coupled to the transmission line. Reliability is not a concern as backup power is provided when the main power source is out to keep the system in operation.

Automated Meter Reading and Automated Meter Infrastructure Systems

Automated meter reading (AMR) systems have existed for many years. AMR allows an electric utility to collect customer meter consumption data at predetermined times without physically having to visit each individual customer site. AMR is significantly less expensive than an automated meter infrastructure (AMI) system.

AMI systems are two-way interactive sources for electric consumption communication and data sharing. Customers are provided many opportunities to consider and select the most efficient way to utilize their electricity. The largest expense involving the AMI is the repeaters and collectors needed to transmit and gather the data. Additional benefits include outage detection, power quality, and load control.

Emergency System Restoration

C ustomers expect reliable electric service due to their ever-increasing dependence upon and utilization of electric equipment and electronic devices. Lifestyle conveniences are associated with reliable electric service. Electric service interruptions negatively affect reliability. Seldom is the lineman more visible to the community than during the repair of service interruptions to the electric transmission and distribution system. The service interruptions occur for a variety of reasons; for instance, wind, ice, animals, and trees are typical interruption causes. However, emergency restoration situations caused by major storms generate the most publicity.

Storms vary in terms of magnitude and severity. Electric system damage may be localized or widespread. Storms may affect several hundred customers or several million customers. Storm damage cleanup and electric system restoration may require only local labor, equipment, and materials for repair or the storm may require both local and mutual assistance emergency manpower response, equipment, and materials. Storms are caused by winds, lightning, ice, or other naturally occurring weather events. There are several recent examples of the devastating damage and destruction created by intense wildfires and storms throughout the United States and Canada (Table 10.1).

The rescue and recovery of human life maintains top priority in electrical emergency restoration situations. Successful electric system restoration depends upon recognizing and assessing the magnitude of the situation, directing resources safely and efficiently, acquiring materials, and keeping everyone informed as to the details of the situation. While the exact date and time of emergency electric system situations cannot be typically predicted, they can be prepared for and thus dealt with in a methodical manner in order to minimize service disruption to customers.

The lineman's performance is critical in the restoration process. Sixteen-hour work days on a continuous basis are required. The lineman relies upon many factors in order to be successful. This section will delve into several of the aspects of the restoration process required in emergency conditions.

Emergency Conditions

Linemen are expected to respond to electrical system emergencies caused by storms, sabotage or terrorism (both to equipment and computer systems), or catastrophic equipment failure (fires or explosions). Other types of emergencies include serious

Date	Event	Location
August 2021	Hurricane Ida	Louisiana
August 2020	Deriecho	Iowa, Illinois, and Indiana
October 2018	Hurricane Michael	Southeastern United States
Multiple events each year since 2017	Wildfires	California
Spring and Summer 2017	Hurricanes Irma, Harvey, Jose, and Maria	Texas to the Caribbean
October 2012	Hurricane Sandy	East Coast

TABLE 10.1   Examples of a Few Recent Significant Newsworthy Electric System Damaging Events

injuries to employees or danger to the public including downed wires (Fig. 10.1), downed poles (Figs. 10.2 and 10.3), and broken pole top hardware (Fig. 10.4), or environmental emergencies (oil spills and fuel leaks). Storm damage is typically more widespread and thus requires more flexibility in terms of the response to the event as opposed to the other emergencies that are more localized in nature.

Business continuity plans, disaster recovery plans, mutual assistance plans, storm plans, and general switching procedures are established to ensure organizations are prepared for the unexpected. The plans and procedures include strategies to be followed to resume business activities.

Pre-event Planning

Frequently, the electrical emergency response plans and procedures are reviewed and practiced in advance of a new event with employees for clarity and understanding. If needed, new rapid response procedures are developed.

FIGURE 10.1   Several spans down of a 161 kV transmission line. The conductor of the 161 kV transmission line was discovered with ice measuring 5 inches in diameter. The devastation resulted from the ice and wind conditions. (*Courtesy L. E. Meyers Co.*)

FIGURE 10.2   A downed pole of a 161 kV transmission line with under built 13.2-7.6 kV distribution line and a 25 kVA overhead transformer attached. (*Courtesy MidAmerican Energy Co.*)

FIGURE 10.3   A view of straight line wind damage to a 345 kV transmission line structure. (*Courtesy MidAmerican Energy Co.*)

Mock storms utilizing past storm data are scheduled. Pre-event risks and risk mitigation solutions are identified so as to minimize the impact of an event. Ideal storm event storeroom inventories of materials and supplies are evaluated. Key supplier agreements are transacted.

Emergency Switching Procedures

Critical to the lineman is the assurance that upstream protective devices are controlled such that a portion of the deenergized system will not become energized unexpectedly. At all times, the lineman expects the system to be properly monitored and controlled, as well as, trouble orders efficiently dispatched. All conductors are treated as energized unless confirmed as isolated and grounded. Be aware of the possibility of back feed from generators (Fig. 10.5).

FIGURE 10.4   A 13.2-7.6 kV distribution facility with a broken crossarm damaged in a storm. (*Courtesy MidAmerican Energy Co.*)

FIGURE 10.5   This 6 kW electric start generator, if wired incorrectly, may be a source of back feed to the electric distribution system. (*Courtesy MidAmerican Energy Co.*)

If mutual assistance emergency response linemen from other companies are involved, the review of the general switching procedures occurs. Topics of discussion include hot line warning procedures, clearance procedures, and hold order procedures. Safety rules are also addressed (Fig. 10.6).

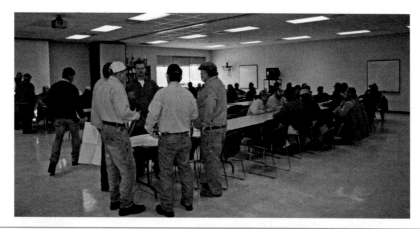

FIGURE 10.6   Emergency response linemen in the show up room prior to the general switching procedure review and assignments being issued. (*Courtesy MidAmerican Energy Co.*)

Storm Plan

Storm plans provide individuals guidance, direction, and expectations regarding defined roles that each person performs in order to maximize the utilization of everyone involved in the recovery effort. Plans emphasize restoring electric service as safely and quickly as possible, emphasizing public health and safety. Environmental concerns are considered as well. Coordination with city, state, and federal authorities is addressed. While the plans are well-defined, flexibility to adapt to each unique case is permitted.

Many individuals with varying backgrounds and past experiences are utilized to support the lineman for storm recovery. People are needed to:

Answer phones

Assess storm damage

Supervise crews

Dispatch orders

Investigate wire down orders

Order materials

Support computer systems

Assist and guide mutual assistance crews

Consult industrial customers

Perform system planning analysis studies

Coordination of the many tasks is the responsibility of storm directors, managers, and supervisors.

Storm plans address logistics concerning mutual assistance emergency response manpower, tree trimmers, fleet, materials, lodging, meals, and communications (phones, computers, radios, fax machines, and printers). The first few hours of response are critical as the situation is properly assessed to ensure appropriate action.

Assessment

Initial assessment data includes:

Number of locked out circuits

Number of trouble orders

Number of customers out

Number of broken poles

Number of wires down calls

Number of police/fire personnel standing by calls

Number of tree damage calls

Number of spans of downed conductor

Number of damaged transformers

The amount of local manpower available to assist

Forecasted weather conditions

Input from other emergency agencies

Based upon the initial assessment of the situation, accurately determining sufficient staffing requirements related to field crews and how to fill the staffing needs are determined. In many cases, manpower is supplemented with mutual assistance emergency response crews (Figs. 10.7, 10.8, and 10.9). Manpower needs include:

Linemen

Servicemen

Groundmen

FIGURE 10.7   Vehicles used by mutual assistance emergency response crews responding to request for mutual assistance. (*Courtesy MidAmerican Energy Co.*)

FIGURE 10.8   An emergency response crew on site at the location of damaged electric distribution facilities. (*Courtesy MidAmerican Energy Co.*)

FIGURE 10.9   An assortment of digger derrick and bucket trucks to be used by the emergency response crews. (*Courtesy MidAmerican Energy Co.*)

Substation electricians

Metermen

Tree trimmers

Mutual assistance emergency response crews

Downed wire watchers

The tree trimmers are required to clear debris from the facilities. In some cases, the tree trimmers are sent out to some areas in advance of the linemen and in other instances, the tree trimmers are sent out with the linemen. The downed wire watchers insure that the affected area is barricaded and made safe for the general public until qualified personnel arrive on-site to repair the downed wire situation.

Logistics involved with mutual assistance emergency response crews involves staging areas for assignments, work and rest hours, lodging, meals, refueling of vehicles, first aid treatment facilities, and laundry.

A various assortment of fleet are considered for utilization:

Single-bucket aerial devices

Double-bucket aerial devices

Material handlers

Digger derricks

Dump trucks

Pickup trucks

All-terrain vehicles

Cranes

Trenchers

Backhoes

Working height requirements of the aerial devices range from 25 ft to over 150 ft.

Material needs are addressed (Fig. 10.10). Adequate reserves of poles, conductor, crossarms, and connectors are typically on-hand. However, as the assessment of the situation continues, material requirements become more defined and arrangements are made so that repair crews are well-stocked. Customers are not interested in excuses if the power is not restored due to a shortage of a given piece of material.

Material needs include standard sizes and sufficient quantities of:

Poles

Overhead transformers

Overhead conductors, both primary and secondary

Crossarms, both in-line and dead-end arms

Braces

Insulators

Steel pins, and pole top pins

Top ties, and spool ties

Dead-end clamps, hot line clamps and bails, and service wedge clamps

Machine bolts, oval eye bolts, and double-arming bolts

Washers and nuts

Cutouts

Arresters

FIGURE 10.10   Boxes of materials in the storeroom to be used for the emergency response restoration effort. (*Courtesy MidAmerican Energy Co.*)

Fuse links

Compression connectors

Sleeves and splices

Guy wires and associated guying materials

Not to be forgotten in the assessment portion of the recovery operation is the need of the environmental service division and of the legal department. Having both of these areas plugged in to the effort pays dividends in the future.

Repair

After all of the information has been processed, the recovery moves from an assessment stage to one of repair (Figs. 10.11, 10.12, and 10.13). People and systems (including computer systems and radio and data communication systems) are in place to support the

FIGURE 10.11 Linemen preparing a pole to be installed replacing a damaged facility. (*Courtesy MidAmerican Energy Co.*)

FIGURE 10.12 The digger derrick truck is being used to assist the lineman restoring a 161 kV structure. (*Courtesy MidAmerican Energy Co.*)

FIGURE **10.13**   A supply truck in route to the jobsite to deliver more materials required for the repair of the damaged 161 kV transmission line. (*Courtesy MidAmerican Energy Co.*)

lineman. Decisions are made regarding the staffing of the office support teams. Round the clock coverage occurs, both for the office staff and for the lineman. The lineman responds to prioritized trouble orders (Figs. 10.14, 10.15, and 10.16). Orders are prioritized based upon public health and safety issues. Critical transmission and substation infrastructure issues receive the next level of attention followed by locked out circuits. To isolate the affected areas in order to restore service, the lineman is directed utilizing switching procedures.

Job briefing conferences are conducted at defined intervals, typically at the start of each designated shift to review current status of customer interruptions, goals for the period, work assignments, and safety reminders. At the job site, the crew personnel job briefings identify possible risks and hazards, and agreed to resolutions.

FIGURE **10.14**   A 13.2-7.6 kV distribution line being repaired. Several new 45-4 poles were required for the repair. New 4/0 ACSR was installed as well. (*Courtesy MidAmerican Energy Co.*)

FIGURE 10.15   The digger derrick vehicle holds a leaning 161 kV transmission line pole while repairs are being completed. The pole also supports 13.2-7.6 kV distribution under build facilities. (*Courtesy MidAmerican Energy Co.*)

FIGURE 10.16   A 161 kV structure being positioned complete with insulators, stringing sheaves, and tag lines for the conductor pulling. (*Courtesy MidAmerican Energy Co.*)

As the recovery effort unfolds, it is important to communicate both internally and externally to address how things are going. Predetermined arrangements are worked out with the public officials and the press so that press briefings share critical announcements with everyone. Press announcements inform the public as to the status of the recovery process. Electrical safety reminders and tips addressing downed or fallen wires, proper utilization of standby generators, and operating specialized equipment are provided as well. In addition, some customers have special individual issues that require immediate resolution.

Post-Storm Assessment

Post-storm assessments provide feedback regarding processes that require revising. The post-storm assessment involves representatives from multiple areas in order to provide varied viewpoints. Success of the recovery operation is measured after the storm in terms of how quickly and safely the system was restored to normal and by the amount of complaints received by the host utility or local media.

Mock Storms

Successful recovery plans are well-developed and practiced prior to actual implementation. The practice sessions require all of the individuals involved in the response plan available to practice and utilize all processes and equipment identified so that everyone is familiar with them.

Outage Management Systems

The outage management system (OMS) is a valuable tool in terms of analyzing all of the outage data and projecting abnormal protective devices based upon customer calls or protective device status information provided from intelligent electronic devices (IEDs). The OMS serves as an outage data collector. The system provides dispatchers with targeted outage, crew, restoration, and network information. Outage data collected includes start date and time, location, and descriptions of the specific events.

OMS displays the information in both text and pictorial form (Figs. 10.17 and 10.18). The data is sorted by state, district, service center, substation, circuit, and customer.

FIGURE 10.17   Incident summary screen of known problems within the service territory. (*Courtesy CGI Information Systems and Management Consultants Inc.*)

FIGURE 10.18 Geographical view of an area that has experienced an interruption of service. *(Courtesy CGI Information Systems and Management Consultants Inc.)*

Statistical snapshots of the information provide information as to the scope of the situation.

The OMS contains maps based upon connectivity models of the entire distribution system. The maps are in several formats; single-line maps, individual circuit maps, district maps, layered informational maps, and combinations of most of the desired geographical features are available. The hierarchy of the distribution system exists in the connectivity model. Protective devices and system isolation points are modeled. Network model changes completed in the field are stored and displayed. Each and every customer meter is associated with a transformer. The transformers are in turn associated to line fuses. The line fuses are associated to higher-level protective devices such as a recloser or a set of mainline fuses. The highest level of upstream protective devices is typically the circuit breaker.

As customer calls are received and entered into OMS, the system groups the calls based upon the connectivity trees of the electric system, diagnoses the information, and targets the most likely protective devices in the electric system where the assessment needs to begin.

The work orders are assigned in a prioritized methodology to restoration crews. The crew status is maintained on the orders.

Crew statuses logged include:

Assigned

Dispatched

En route

Arrived on site

Estimated time to repair

Service restoration

Job completed

Crew management is accomplished using the OMS. Tracking of employee hours is logged in the system.

OMS is capable of creating switch plans and tracking clearances utilized in the isolation of damaged facilities while returning unaffected areas to normal.

Radio Communications Systems

In order for orders to be properly dispatched and switching procedures to be followed, reliable communication networks are established for both company and mutual assistance emergency response crews. In some cases, company personnel (guides) are assigned to each crew. In other cases, mobile phones or handheld devices are utilized. In order to ensure safety for everyone involved, radio procedures and guidelines are reviewed and incorporated.

Summary

Rather than being reactive, a proactive approach to system emergencies will minimize disruption of service to the customer. Customers expect well-coordinated and expertly managed disaster recovery operations. In the majority of cases, due to the dedication of the lineman and of the support teams, exceptional response is what the customers receive.

CHAPTER **11**

Unloading, Hauling, Erecting, and Setting Poles

Unloading Wood Poles

Poles are usually shipped on truck flatbed trailers or railroad flatcars (Fig. 11.1). Delivery by truck allows for more flexible delivery options as poles may be brought to the service center or directly to the job site. In some cases, the delivery unit contains self-unloading pole equipment. Great care must be exercised while unloading poles to avoid all possibility of an accident. Before starting the unloading operation, the delivery vehicle's brakes should be set to prevent the delivery unit from moving. The flatbed trailer wheels should be chocked as an added precaution. Flags and barricades should be set up to safeguard the area. Mechanized equipment reduces the time required and the number of workers needed to unload the poles safely (Fig. 11.2).

The lineman should inspect the stakes holding the poles to be sure the poles are secure before cutting the banding wires. The banding wires should be cut from the end of the poles cautiously by the lineman on the ground. After the banding wires are cut, equipment such as the skid steer loader (Figs. 11.2 and 11.3) can be used to safely remove several poles at a time and place them on the pole storage pile (Figs. 11.4 through 11.6).

Roofing Poles

Most poles are completely impregnated with wood preservatives, eliminating the need to roof them. If poles require roofing, the work is normally completed in the pole yard, where the poles can be readily handled with mechanized equipment.

The roofing operation consists of forming a roof on the top of the pole. The roof can be simple slant cut as in Fig. 11.7(a) or a gable roof as in Fig. 11.7(b). Either type prevents water, snow, or ice from collecting on the top of the pole and causing decay. If poles are completely impregnated with wood preservatives, a flat roof is adequate, as in Fig. 11.7(c). To cut a gable roof, the pole should be supported in a raised position. The pole should then be turned so that the heaviest sag or curve will be nearest to the ground. This is done by having one person turn the pole at the butt end with a cant hook while another sights along the pole and decides when the pole is in the right position. By putting the pole in this position, the direction of the roof will be in the same orientation as

FIGURE **11.1** Flatbed trailer loaded with steel poles. Stakes and binding wires hold the poles in place for shipment. (*Courtesy Steel Market Development Institute.*)

FIGURE **11.2** Skid steer loader being used by to unload steel poles. The hydraulically controlled clamping device holds the poles firmly in place on the forks. (*Courtesy Steel Market Development Institute.*)

FIGURE **11.3** Skid steer loader hauling steel poles to the storage crib. (*Courtesy Steel Market Development Institute.*)

FIGURE 11.4 Poles have been unloaded from the flatbed trailer. The steel poles are still stacked and packaged together as shipped. (*Courtesy Steel Market Development Institute.*)

FIGURE 11.5 Poles are unbanded and are placed in a storage crib. (*Courtesy Steel Market Development Institute.*)

FIGURE 11.6 A storage crib area for the poles. (*Courtesy Steel Market Development Institute.*)

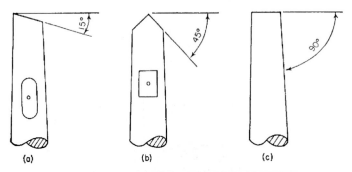

FIGURE 11.7   Three types of pole roofs: (a) slant; (b) gable; (c) flat.

the pole line. When this pole is erected in the line, the pole will appear straight. Figure 11.8 is a sketch of a pole provided with roof and gains and gives the correct relation of these to the crook or bend of the pole. The side of the pole which has the longest concave sweep between the ground line and the point of the load is the face of the pole, as illustrated in Fig. 11.8. The pole supplier normally marks the pole length and class on the face side of the pole with a branding iron. The face of the pole should be determined before the pole is set and before a crossarm or fiberglass insulator supports and insulators are installed. The face of the pole may be different from the side of the pole branded by the supplier as a result of the way the pole was stored in a pole pile.

The gable roof is made by sawing the two sides of the pole top (Fig. 11.9) at an angle of 45° with the pole. One way of obtaining the 45° angle is to measure along the length of the pole a distance equal to one-half the pole-top diameter and to saw accordingly. This roof will shed the rain and prevent any accumulation of water, thereby decreasing possible decay.

FIGURE 11.8   Sketch of completely framed pole showing relation of parts to curvature of pole.

FIGURE 11.9   Pole top showing pole roof cut at 45° and two gains with crossarm through-bolt holes. (*Courtesy Wisconsin Electric Power Co.*)

Pole Gains

A gain is the notch cut into a pole into which the crossarm is placed when mounted. Gains are cut to provide a flat surface to help maintain the arms in alignment.

To cut a gain, one must first saw along the upper and lower edges of the gain with a handsaw and then chisel out a round portion, thereby making a flat recess. It is good practice to hollow the gains out slightly in the center to ensure a snug fit of the crossarm, thereby preventing its rocking from side to side.

The first crossarm gain is usually cut 12 inches from the top of the pole. The succeeding gains are generally spaced 24 inches, 30 inches, or 36 inches apart. The spacing selected depends largely on whether the pole is to carry buck arms or not. When buck arms must be provided, these are placed between the line arms at right angles. In order to permit the use of 28-inch crossarm braces, the spacing between the line arm and buck arm should be at least 18 inches. Gains should be square with the axis of the pole and about ½ inch deep. The height depends on the height of the crossarm. The crossarm should fit snugly.

If several gains are to be cut, a gaining template (Fig. 11.10) is a great convenience. In using the template, the point of the roof of the template is placed over the roof of the pole and the template is shifted until its centerline lies in line with a string stretched from the point of the roof to the center of the butt indicated by a nail at the ground line. Then the outlines of the gains are marked with a sharp tool by marking along the side pieces of the template.

The vertical spacing of the crossarms, and consequently the gains, depends on the nature of the circuits and the voltages of the circuits carried by the arms. The engineering specifications and electric construction standards for the type of construction to be used will show the details and dimension required for gaining the poles.

One of the difficulties encountered in cutting more than one gain on a pole is to keep the flat surfaces of the gains parallel. This can be checked by a resting or hanging T-squares, one in each gain (Fig. 11.11). By sighting over them it is easy to see whether the gains are properly leveled. The square also can be used to advantage in getting buck arms at right angles with the line gains.

FIGURE 11.10   Template for use in marking gains.

FIGURE 11.11   Method of determining whether gains are cut parallel. Test is made by sighting over edges of squares. If gains are parallel, the edges of the T-squares will be in line.

The next operation consists of boring the holes for the crossarms. The size of the hole for the standard crossarm through bolt is 11/16 inch. This is for the 5/8-inch or 10/16-inch through bolt. This operation is generally performed with an electrically or hydraulically driven power drill.

A scheme often used for locating the center of the crossarm bolt hole is to draw two diagonal pencil lines across the gain. The intersection of these lines determines the center of the bolt hole. In case the pole surface is not smooth, it is best not to guess at the pole center but to measure for the vertical center.

Another scheme which takes less time is the use of a template. This consists of a short length of crossarm in which the hole has been bored. This is placed centrally over the pole in the gain, and the center is marked with a punch.

A galvanized crossarm grid gain (Fig. 11.12) with a curved faced eliminates the need of cutting the gain into the pole. A slotted hole on the gain allows for proper alignment on the pole and with the crossarm.

Hauling Poles

Loading of poles and hauling them from the pole storage site to the work site are tasks requiring the skills and techniques of a qualified lineman and groundman necessary to avoid hazards that may be encountered while loading the poles and then while in route to the final destination. Moving poles short distances in the pole yard can be completed efficiently with mechanized equipment (Fig. 11.13). Poles are usually loaded on a trailer, and the trailer is pulled by a line truck to the work site (Fig. 11.14). The trailer is placed near the pile for convenience in loading the poles. The brakes on the trailer should be set and the wheels chocked while the poles are loaded on the trailer with the boom of the line truck. Lightweight steel poles may be lashed to the steel channels of a digger derrick truck (Fig. 11.15).

If one of the poles is to be used as the tongue for the trailer, the pole selected should be of sufficient size and strength to carry the load. The drawbar is chained to the tongue pole (Fig. 11.16). The eye of the drawbar is engaged in the pintle hook on the back of the truck, and safety chains on the truck are fastened to the tongue pole and drawbar (Fig. 11.17). The poles should be blocked into position as they are loaded on the trailer to keep them from shifting. The lineman must be careful not to overload the trailer or balance the load of poles so that the load is tongue heavy.

Figure 11.12   A galvanized crossarm grid gain with a curved face.

FIGURE 11.13   Storeroom worker loading poles for crews. (*Courtesy MidAmerican Energy Co.*)

FIGURE 11.14   Pole on trailer connected to the back of line truck. Flag on the end of the pole is displayed to warn the public. (*Courtesy MidAmerican Energy Co.*)

FIGURE 11.15   Pole is being transported to the worksite by a lineman driving the digger derrick vehicle. (*Courtesy Steel Market Development Institute.*)

FIGURE **11.16**   Drawbar with eye attached to pole.

FIGURE **11.17**   Drawing eye coupled to truck.

When all the poles have been loaded, they must be secured with cables or chains, or both, to keep the load from shifting in transit. Flags should be installed on the end of the poles during daylight hours and warning lights should be used after dark. The lineman must take care to protect the public while hauling poles. The driver should signal turns and stops early to warn other drivers while hauling poles on a highway. The poles may swing into adjacent traffic lanes when short turns are necessary. It is desirable for a lineman to use flags to warn the public if the area is congested.

When the terrain is rugged and access roads are not available, poles can be transported by helicopter (Fig. 11.18). The specialized helicopters are designed to lift and carry the heavy structures. Structures are completely or partially assembled, depending upon weight, and then delivered to the job site. Pole holes are dug or foundations bases installed and correspondingly identified so that each structure is delivered to the correct location and lowered to the ground by the helicopter pilot to complete the installation.

FIGURE **11.18**   Helicopter transporting preassembled four-pole 69-kV wood structure. (*Courtesy MidAmerican Energy Co.*)

Setting Poles

The specifications and drawings will give the location and size of each pole to be set. The pole locations are usually identified by stakes before the construction work is started. If one of the stakes has been removed or destroyed, the proper location can probably be determined by measuring and aligning from adjacent stakes. If several stakes have been lost or destroyed, it may be necessary to ask the foreman or engineer for assistance in resetting the stakes for the poles.

The holes for the poles are normally dug by a mechanical digger, sometimes referred to as an earth-boring machine. The mechanical digger consists of an auger, or helix, operated by a hydraulic system. The mechanical digger used to dig holes for large transmission-line poles is a complete unit normally mounted on a flatbed truck. The casing around the auger-operating shaft can be equipped with a pulley, for a winch line and pole grabber, to make it possible to set poles up to 65 ft long, which are commonly used for distribution lines, with the mechanical digger unit.

The standard way of digging pole holes for the smaller poles used for single phase construction or street and security lights is to use a derrick- or boom-mounted auger. An auger attached to the boom of a corner-mounted derrick line truck is easily aligned to dig the pole hole in the proper location. A line truck, equipped with an auger, increases the versatility of the work that can be performed by the line crew.

Pole-Hole Diameter

The diameter of the hole is determined by the size of the pole to be set. The hole should be large enough to allow plenty of space on each side of the pole for the free use of tamping equipment, which requires at least 3 inches of free space all around the pole. The diameter of the hole at the top should not be greater than at the bottom, but rather it should be larger at the bottom than at the top. A slight increase at the bottom is often necessary to allow for the shifting of the pole in lining-in and to accommodate the larger diameter of the pole at the butt.

Pole-Hole Depth

The depth of setting is determined by the length of the pole and by the holding power of the soil or earth. The required average setting depths have been determined by experience. The recommended depths of setting in soil and rock are given in Table 11.1 and Fig. 11.19 for various pole lengths from 25 ft to 135 ft. It will be noted that a pole set in rock need not be set as deep as a pole set in soil.

A rule often followed for determining the setting depth in soil is to take 10% of the pole length and add 2 ft, with a minimum of 5 ft. The setting depth of a 40-ft pole would thus be:

$$10\% \text{ of } 40 \text{ ft pole} = 4 \text{ ft} + 2 \text{ ft} = 6 \text{ ft}$$

This matches the table. The pole-hole depths in Table 11.1 are for poles to be set in the straight portion of the line. Holes for poles that are to be set at angle or corner points in the line should be at least ½ ft deeper. If soil is soft or marshy, dig the pole holes deeper. Holes for poles set on a slope should be ½ ft deeper than normal, measured from the lowest point on the ground.

Crushed rock is often used for backfill around transmission-line poles and distribution poles requiring extra stability (Fig. 11.20). The rock must be well-tamped to obtain the desired strength.

Length of Pole, ft	Setting Depth in Soil, ft	Setting Depth in Rock, ft
25	5.0	3.5
30	5.5	3.5
35	6.0	4.0
40	6.0	4.0
45	6.5	4.5
50	7.0	4.5
55	7.5	5.0
60	8.0	5.0
65	8.5	6.0
70	9.0	6.0
75	9.5	6.0
80	10.0	6.5
85	10.5	7.0
90	11.0	7.5
95	11.0	7.5
100	11.0	7.5
105	12.0	8.0
110	12.0	8.0
115	12.0	8.0
120	13.0	8.5
125	13.0	8.5
130	13.0	8.5
135	14.0	9.0

Source: Commonwealth Edison Company.

TABLE 11.1   Recommended Pole-Setting Depths in Soil and Rock for Various Lengths of Wood Poles

Poles set in sandy or swampy ground should be set considerably deeper or be supported by guys, braces, or cribbing. One form of cribbing uses a bottomless 55 gallon steel drum into which the pole is set. The steel drum is then filled with concrete or small stones to make the pole secure. When exceptional pole stability is required, an artificial foundation of expanding high-density polyurethane water-blown foam may be placed around the pole base (Fig. 11.21).

Poles set on curves, on corners, or at points of extra strain should be set at least 6 inches deeper than the values given in Table 11.1. The side strain, however, should always be taken up with terminal, side, or line guys, leaving the pole to carry the vertical load only.

Poles subjected to slight side pulls can be kept erect by the use of pole keys. Some companies use treated timber to key poles. Expandable steel-fabricated steel keys have proved to have a longer life and to be more economical to install. Streetlights

FIGURE 11.19   Graph of recommended pole-setting depths in soil and rock.

FIGURE 11.20   Crushed rock used as backfill around pole to provide stability.

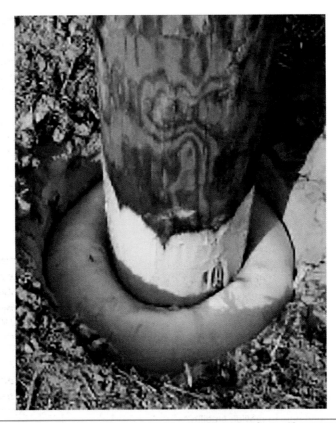

FIGURE 11.21   High-density polyurethane water-blown pole setting foam. (*Courtesy Osmose.*)

with a long mast arm supporting the light fixture, very small angles in the line, and service drops at right angles to the line produce slight side pulls on a pole. When the poles are set in firm soil, one pole key placed 1 ft below ground surface on the strain side of the pole is sufficient. In loose soil, however, a second pole key placed on the opposite side at the butt of the pole is required (Figs. 11.22 and 11.23).

Pole-Hole Digging

Mechanical diggers minimize the physical effort, and time, required to dig holes for poles. The lineman should locate the site where the pole is to be set and determine the best route to get the truck to the location. Special precautions must be taken if the hole is to be dug near an energized circuit. If adequate clearance cannot be maintained, protective equipment must be used to insulate the energized conductors to prevent accidental contact. If adequate protection cannot be obtained from existing conductors, the circuit should be deenergized, isolated, and grounded before the work is started. The truck should be located properly to permit accurate location of the auger.

It may be necessary to start angle-structure holes by hand. A post-hole-type auger, a digging bar, a 5-ft shovel, a 9-ft shovel, and a 9-ft spoon should be available. The power-digging equipment should be inspected before the auger is operated.

FIGURE **11.22** Expandable steel pole key has been placed in the hole adjacent to the butt of the pole by the workman on the side of the pole opposite the overhead strain. (*Courtesy A. B. Chance Co.*)

FIGURE **11.23** Workman expands steel pole key with an expanding and tamping bar. (*Courtesy A. B. Chance Co.*)

The lineman should insure that all personnel maintain adequate clearance of the auger to prevent clothing from being caught by the auger or other rotating devices associated with the mechanical equipment.

When pole holes have been dug along paved streets, near farm buildings, or on any premises used by the public and poles are not set at once, the holes should be barricaded. If a hole is to be left overnight without setting the pole, the hole should be completely covered with a plank and dirt, and the area barricaded.

Pole-Hole Power Borer

Pole holes are dug with a power earth borer. This greatly reduces the physical effort required and speeds up the work. The borer can be tilted up while moving between work sites.

When a hole is to be dug, the truck is driven over the pole location and placed so that the borer is directly over the pole stake. The truck outriggers are lowered to stabilize the vehicle. The table supporting the earth-borer motor and mechanism is moved, or rotated, to place the point of the auger over the stake when it is lowered (Fig. 11.24). The boring equipment must be positioned to maintain proper alignment while the pole hole is dug. Then the operator lowers the auger (Fig. 11.25) while it rotates until it has taken a full "bite." Then the auger is raised until it is above ground and away from the hole and cleaned of the soil (Figs. 11.26 and 11.27). The auger is lowered again, and another bite is taken. Again, the auger is raised and cleaned. This is repeated until the desired depth of hole has been reached.

FIGURE 11.24   Pole-hole power borer in position to start digging. Truck outriggers are down, and point of auger is touching ground at location established by a stake. (Courtesy Northwest Lineman College.)

FIGURE 11.25 Auger lowered into hole for another bite of ground. Note loose dirt around the hole. (*Courtesy Northwest Lineman College.*)

FIGURE 11.26 Auger raised with full bite of ground. (*Courtesy Northwest Lineman College.*)

FIGURE 11.27 Auger has been positioned away from hole and dirt is being knocked from the auger. (*Courtesy Northwest Lineman College.*)

The completed hole will have a ridge of loose dirt completely around it. The loose dirt should be moved away from the edge of the hole to prevent it from falling into the hole. All loose dirt should be removed from the bottom of the hole by use of hand tools.

Holes of different diameters are obtained by the use of augers of different diameters. Augers having different diameters can be mounted on a digging bar. The change from one to the other can be made in a few minutes. The 9-inch auger size, the smallest, is generally used for digging guy-anchor holes. The 16-inch, 24-inch, 36-inch, and 48-inch auger sizes are carried on the truck and are used for different sizes of poles that must be set.

Pole-Hole Derrick Auger

Derrick-mounted power pole-hole augers (Fig. 11.28) are commonly used for digging distribution-pole holes. This type of power hole digger increases the flexibility of the distribution-line crew in utilizing the vehicle and equipment efficiently. The procedures and precautions are similar to those described for the pole-hole power boring equipment. The derrick-mounted auger can be used in locations several feet from the truck by extending the derrick, or boom, on which it is mounted. The auger is driven by the hydraulic system that operates the boom. The point of the auger is lowered to the ground at the pole location, usually identified by a stake. It is important to locate the auger accurately. The lineman must be sure the boom will move easily to keep the pole hole plumb and to prevent damage to the equipment. The auger should be started to rotate slowly to prevent tilting.

The auger rotation should be stopped when the auger worm is full of dirt to avoid damage to the equipment. The rotation of the auger must be stopped before the auger

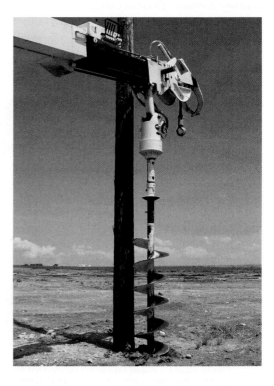

FIGURE **11.28**   Drilling for pole with power auger mounted on boom of truck. (*Courtesy Northwest Lineman College.*)

worm gets stuck. The dirt is removed from the hole by raising the derrick and reversing the rotation. If the derrick will not raise the auger, reverse the rotation of the unit to free it from the earth. A short-handled shovel can be used to remove the dirt from the auger.

Pole-Hole Digging in Rock

The rock-cutting drill is mounted on the boom of a standard digger-derrick line truck. The drill cuts a hole in the rock at the rate of 16 inches to 20 inches (41 cm to 51 cm) an hour. The drill cuts a circular hole in the rock, leaving a core. The rock dust is vacuumed out of the hole by the mechanism as the rock is penetrated. The rock core remaining after the drill reaches the proper depth for the pole to be set is broken at the base with a hydraulic wedge and removed in one piece. The rock drill can eliminate the need to use explosives to form a hole in rock for a pole. The use of explosives is hazardous and requires special training to ensure their safe use.

Pole-Hole Digging by Hand

Digging holes with shovel and spoon is resorted to when it is impossible to use a power digger. The shovel is a straight-bladed tool built as a spade (Fig. 11.29), but the spoon has the blade at an angle (Fig. 11.30). The shovel is used to loosen the soil, and the spoon is used to remove soil from the hole.

FIGURE 11.29   Straight shovel used to loosen soil in digging a pole hole. (*Courtesy Leach Co.*)

FIGURE 11.30   Spoon used to lift soil out of a pole hole while digging. (*Courtesy Leach Co.*)

An experienced hole digger will dig the hole largely without the aid of the spoon under ordinary conditions of soil and moisture. A certain amount of practice is required to do this. One must know just how to strike the soil with the shovel so that the soil will break loose from its position and yet stick to the blade of the shovel. In case the soil is very dry, the shovel can be used to loosen the soil and the spoon used to lift the loose soil from the hole.

Sometimes a bar must be used to break the earth loose if there is frost, rock, or hard clay. One end of the bar should be sharpened to a blunt chisel point and the other end to a round point similar to the sharpened end of a pencil. Such a digging bar is shown in Fig. 11.31.

The usual procedure to be followed is to outline the hole around the stake. The diameter of this circle should be at least 6 inches greater than the diameter of the pole at the butt. For the first 2 ft or 3 ft of digging, it is advisable to use a short-handled shovel (Fig. 11.32), as it is more convenient to hold. The spoon will not be necessary as the earth

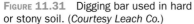

FIGURE 11.31   Digging bar used in hard or stony soil. (*Courtesy Leach Co.*)

FIGURE 11.32   Short D-handled shovel often used in digging the first few feet of a pole hole. (*Courtesy Leach Co.*)

can be lifted out with the shovel. For the remainder of the hole, the long-handled shovel and spoon are used in the usual manner. Where quicksand or swamp soil is encountered and cave-ins occur, a bottomless 55 gallon steel drum is lowered into the hole as it is dug. As the hole is dug deeper, the steel drum is forced down, thus preventing the side of the hole from falling in. The steel drum will also provide a greater bearing surface for the pole when it is set and tamped in the inside of the drum.

Setting Poles with Derrick

The most common method of setting poles is with a truck derrick. The truck should be equipped with a derrick (boom), winch line, and pole grabber (claws). Hand tools required include a cant hook, pike poles, and rope lines. Protective equipment required includes rubber gloves, hard hats, safety glasses, and polyethylene pole covers, or epoxy-rod pole guards.

The line truck should be positioned appropriately to dig the hole, if a boom-mounted auger is to be used, and to erect the pole. The crew should be informed how the work is to be performed. The hazards should be identified so that all personnel will understand the safe way of performing the work (Fig. 11.33). The truck outriggers must be positioned in the proper manner to stabilize the vehicle when the pole is raised. Wood poles are heavy and could tip a line truck over if proper precautions are not taken.

If the pole is to be set near an energized line, the pole should be covered with protective equipment with proper insulation for the voltage of the energized circuit. The winch line is attached to the pole slightly above the midpoint so that the butt end outweighs the top end (Fig. 11.34). The truck derrick is used to move the pole from the pole trailer to the proper position for setting in the hole. The lineman, wearing the appropriate personal protective equipment, may have to balance the pole and guide it to the proper position (Fig. 11.35). The lineman should not stand or pass under the pole while it is suspended by the winch cable and derrick. The pole is gradually raised until it is in

FIGURE 11.33 Line truck with poles to be set on trailer behind truck. Crew leader is holding tailboard conference to discuss the hazards and to give directions for completing the job. (*Courtesy Northwest Lineman College.*)

FIGURE 11.34   Pole is insulated with polyethylene pole covers designed for use when setting a pole near 13.2Y/7.62-kV three-phase circuit. Winch line is attached to pole so that the pole is a little butt-heavy. (*Courtesy Northwest Lineman College.*)

FIGURE 11.35   Linemen guiding and balancing pole while it is supported by derrick and winch line. (*Courtesy Northwest Lineman College.*)

a near vertical position and the butt end is clear of the ground. Lineman wearing rubber gloves guide the pole as it is raised (Fig. 11.36).

The crew leader, or foreman, directs the operation, using standard hand signals for line work. As the pole reaches vertical position, the pole claws are used to grab the pole and stabilize it (Fig. 11.37). When the pole claws have grabbed the pole, the derrick and winch line are manipulated as directed by the crew leader utilizing a plumb bob to set the pole in the hole (Figs. 11.38 to 11.40). If the pole is to be set in an existing energized line, it may be necessary to hold the energized conductors apart to have adequate clearance. This can be done with hot-line tools and rope lines, as shown in Fig. 11.41.

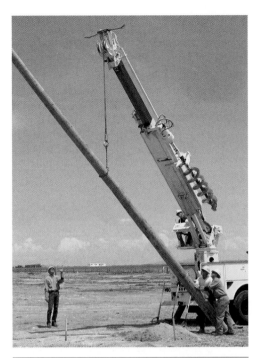

FIGURE 11.36   Linemen, wearing rubber gloves and protectors, guide pole as it is raised. Top of the pole is insulated with polyethylene pole covers. (*Courtesy Northwest Lineman College.*)

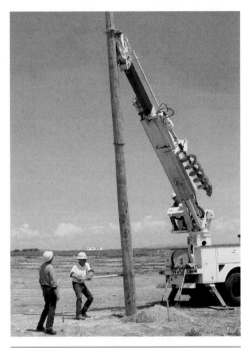

FIGURE 11.37   Pole claws are used to grab and stabilize pole. (*Courtesy Northwest Lineman College.*)

FIGURE 11.38   Crew leader uses standard hand signal to direct lineman operating derrick to set pole in hole. (*Courtesy Northwest Lineman College.*)

FIGURE 11.39   Position of lineman when using plumb bob. Lineman stands at 90 to 180° from derrick regardless of line route and vehicle location. This ensures that straightening the pole in one direction has no effect when aligning the pole in the other direction. (*Courtesy Northwest Lineman College.*)

FIGURE 11.40   Straightening a pole with a plumb bob. String is first aligned with center of pole butt. Equipment operator is then directed to move derrick as needed to align center of pole top with plumb bob string. (*Courtesy Northwest Lineman College.*)

FIGURE 11.41   Energized conductors held in the clear with roller link-stick hotline tools and rope lines. (*Courtesy A. B. Chance Co.*)

FIGURE 11.42   Lineman and groundman backfilling and tamping dirt in hole around pole. (*Courtesy Northwest Lineman College.*)

FIGURE 11.43   Lineman at base of steel pole that is being set. Additional tamping at the base of the pole is required. (*Courtesy Steel Market Development Institute.*)

After the pole is set in the hole and properly aligned in accordance with the specifications, the hole around the pole is backfilled and tamped (Figs. 11.42 and 11.43). After the pole is secured by backfilling and tamping, the lineman removes the temporary protective covering used to set the pole and install the equipment on the pole necessary to complete the project (Fig. 11.44).

Transmission structures are usually assembled on the ground complete with insulators, hardware, conductor stringing blocks, and small rope lines (Fig. 11.45). The structures are set in place with caterpillar-mounted power derricks or mobile cranes.

Setting Poles with Helicopter

Erecting single-pole lines and transmission structures using conventional construction methods in mountainous, swampy, muddy, or snow and ice conditions is difficult, slow, and very costly. Single poles are often transported by helicopter to mountainous locations where land transportation is difficult or practically impossible. The poles are normally

FIGURE 11.44 Lineman in bucket mounted on top of boom attaching equipment and conductor to pole after it is set. Note protective equipment on the lineman and the energized conductor. (*Courtesy Altec Industries.*)

FIGURE 11.45 Two-pole transmission structure assembled on airport concrete apron. (*Courtesy L.E. Myers Co.*)

airlifted with crossarm, insulators, and hardware in place. When the poles arrive at the site, they can usually be lowered into the pole hole directly. Instead of unloading the pole, the helicopter hovers over the assigned pole hole and slowly lowers the pole into it. One or two crewmen are usually needed to guide the butt of the pole into the hole.

Transmission structures can be assembled on the ground at a convenient marshaling yard (Fig. 11.46).

The helicopter hovers over the assembled transmission structure at the marshaling yard while the ground crew attaches the structure to the cables extended from a winch installed on the helicopter. Communications are executed by using two-way radios and standard hand signals for line work. The helicopter airlifts the structure to the site where it is to be installed. The helicopter pilot is directed from the ground by two-way radio. Sometimes considerable saving in time and money can be realized by the use of a helicopter for transport and for erection.

Assembled two-pole transmission structures laid out on airport concrete apron in rotation so that when the helicopter picks them up they are headed in the proper direction. Poles and timbers for additional structures are located adjacent to the assembled structures. (*Courtesy L. E. Myers Co.*)

Setting Poles Along Rear Lot Lines

Powered dollies, equipped with hydraulically operated derrick and auger, are available for installing poles or pad-mounted transformers in inaccessible locations. The poles are hauled to the site on a trailer pulled by a line truck. The boom of the line truck is used to unload the pole from the trailer and place it on the two-wheeled dolly. The two-wheeled dolly with pole is pulled to the work site by the powered dolly. The hydraulically powered auger, supported by the hydraulically operated derrick on the powered dolly, is used to dig the hole for the distribution-line pole. The auger is removed from the boom, and the derrick is attached to the pole, using a nylon sling to set the pole. The pole is raised by the power unit and guided to the hole where it is to be set by the linemen and groundmen. The installation is completed in the same manner as previously described. The power-operated dolly eliminates the need to set a distribution pole manually, thus saving effort and manpower and reducing the cost of installation. The customer's lawn is protected from damage that would result if the line truck was driven to the rear lot-line location.

Setting Poles Manually

The piking method is the oldest method of raising poles. It is a manual method. It gets its name from the tool called a "pike pole" (Fig. 11.47), employed after the pole is lifted beyond the reach of the crew raising the pole. The piking method is used only where a derrick cannot be brought in. The use of a power-operated boom is a faster and more economical method as only a few workers are needed.

The size of the piking crew depends upon length and the weight of the pole to be raised. It varies from 5 individuals for a 25-ft pole to 10 individuals for a 50-ft pole, as shown in Table 11.2. The size of the crew required is not absolutely fixed, but the numbers indicated in Table 11.2 are the size of crews generally employed. As will be noted, one of the crew is a jennyman, one is stationed at the butt of the pole near the hole, and the remaining crew members are called the "pikers."

The first step in the procedure of raising a pole by the piking method is to lay the butt end of the pole over the hole against a bump board or bar, as shown in Fig. 11.48. The use of the board or bar protects the walls of the hole and prevents them from being caved in by the butt of the pole as the pole is raised. The upper end of the pole is raised and placed on the pole support (Fig. 11.49). Pole supports are made in various forms,

FIGURE 11.47   Pike pole used in raising pole beyond the reach of the crew. Note steel spike on end of pole.

Pole Length, ft	Size of Crew	Number of Pikers	Number of Jennymen	Number of Men at Butt
25	5	3	1	1
30	6	4	1	1
35	7	5	1	1
40	8	6	1	1
45	10	8	1	1
50	10	8	1	1

TABLE 11.2   Average Size of Crew Required to Manually Raise Poles of Different Lengths

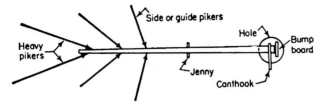

FIGURE 11.48   Plan view of pole-piking method of pole raising.

FIGURE 11.49   Second step in raising a pole. Pole is being lifted from ground into jenny.

two of which are shown in Fig. 11.50. The main duty of the man at the butt is to keep the pole from rolling. This is done by means of a cant hook, illustrated in Fig. 11.51. The men stand side by side on both sides at the top end of the pole (Fig. 11.52). They then push toward each other and up by use of their arms, with the jennyman catching the weight between lifts. They move along the pole until the pole is high enough to permit the use of pikes. The fourth step (Fig. 11.53) is to punch the pikes into the pole and prepare to raise the pole. As the pole is raised, the man carries the jenny forward, always ready to support the pole if need be. The raising continues until "High pike!" is called by one of the men. This means that his pike is too high to be effective if raised any farther. The low man brings his pike down first, because he does not have to lower his pike through the rest of the pikes, but has a clear path. The other men follow in order until all the pikes are lowered. The pole is raised in this manner until it drops into the hole. Figure 11.54 shows the pole almost raised and ready to slide into the hole. Figure 11.54 shows very clearly the duty of each one of the crew. The five pikers raise the pole, the jennyman keeps the jenny always snugly under the pole, and the buttman guides the bottom of the pole and keeps the pole from rolling.

FIGURE 11.50   Two forms of pole supports. The one on the left shown open is called the jenny, and the one on the right is known as the mule.

It will be noted that the pikers are not directly underneath the pole. At least one individual should be well out on each side to guide the pole as well as to help lift it.

Another thing worth noting is the manner of holding the pike pole after the pole is partly raised. It is held in the palm of the hand with the other hand underneath, both arms extended downward. If the pike pole is held in this manner, the worker is in a comfortable position, and if the weight of the pole should be suddenly thrown on the pike pole, he is in a good position for grounding the pike pole or moving into the clear.

FIGURE **11.51**   Cant hook used by man at butt of pole to keep pole from rolling and to turn pole if necessary.

FIGURE **11.52**   Third step in raising a pole. Pikers have raised onto jenny and are now raising pole as high as they can without use of pike poles.

If the pole is very long, the pike pole can be supported on the shoulder as shown in Fig. 11.54. The pike still rests in the palm of the piker's hand. In this way the piker can exert a strong push on the pike.

If the pole to be raised is a very large, heavy pole and the crew is small, it is well to "trench" the hole, that is, to cut a ditch back from the hole. The pole is then placed in this trench. This allows the pole to begin to slide into the hole earlier than if the pole lay

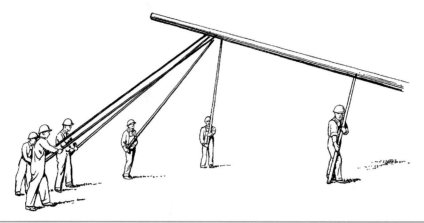

FIGURE 11.53   Fourth step in raising pole. Pikers have placed pike poles against pole and are raising it into vertical position. Note how pikers are spread out.

FIGURE 11.54   Last step in raising a pole. All pikes have been lowered. Pole is about to slide into hole. Note jennyman and buttman with cant hook. Also note manner of pushing on pike poles.

flush on the ground. Furthermore, it allows the weight of the lower end of the pole to balance a portion of the weight of the pole above the point of the trench on which the pole is resting.

Facing the Poles

Facing the poles means turning the poles in the proper direction after they have been placed in the ground. The side of the pole which has the longest concave sweep between the ground line and the top is the face side (Fig. 11.55). Poles are normally branded on the face side by the supplier. The pole's curvature should be checked with respect to the brand after it is stored and before a gain is cut for a crossarm or line hardware is installed. The crossarms and other material must be in the proper position to permit the installation of the conductors.

On straight lines, it is customary to set adjacent poles with the gains facing in opposite directions (Fig. 11.56). This system is sometimes called gain to gain or back to back. It provides for the maximum strength in the line.

Crossarms on poles on each side of the center of a curve should face the center of the curve (Fig. 11.57). The pole at the center of the curve is often equipped with double arms.

Poles next to angles should face the angle, as shown in Fig. 11.58. In case of large angles, the angle pole is generally double-armed or dead-ended.

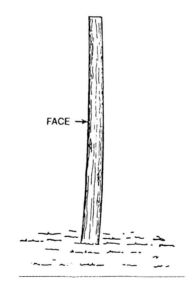

FIGURE 11.55 Concave side of pole is the face. The brand markings should have been applied on the face of the pole by the supplier.

FIGURE 11.56 Facing crossarms on straight lines. Every second pole has crossarm facing in the same direction.

FIGURE 11.57   Facing crossarms on curve. Note that the three poles on each side of the center of the curve face the curve.

Poles next to corners should face the corner (Fig. 11.59). In case of urban distribution, the arms on the corner pole are generally double arms, as illustrated.

Poles on steep grades should face up the grade, as shown in Fig. 11.60.

Poles next to street crossings should face the street (Fig. 11.61). The poles at the intersections are generally double armed as illustrated.

Poles next to poles supporting long spans should face away from the long span, thereby being in a better position to carry part of the load (Fig. 11.62). The poles on each side of a long span are generally of special construction.

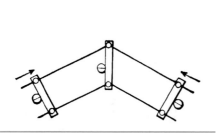

FIGURE 11.58   Facing crossarms at an angle. Adjacent poles face the angle.

FIGURE 11.59   Facing crossarms at a corner. Adjacent poles face the corner.

FIGURE 11.60   Facing poles on steep grade. All poles face upgrade.

FIGURE 11.61   Facing poles at a crossing. Adjacent poles face the crossing.

FIGURE 11.62   Facing poles on long spans. Adjacent poles face away from long span.

FIGURE 11.63 Facing poles at terminals. Adjacent poles face terminal.

Poles immediately preceding end poles shall face toward the end of the line (Fig. 11.63). The end pole, being a terminal pole, is usually of special construction. Sometimes it is good to have the last two poles face the terminal pole.

Poles set at line terminals, curves, corners, and other points of abnormal stress should be given a slight rake against the direction of the pull. This should be sufficient to allow for the change in pole position caused by the continued pull of the load and the normal creepage of the anchor.

Backfilling and Tamping Pole Hole

After the pole is properly faced and lined in, the hole is backfilled and tamped (Fig. 11.64).

Tamping of the dirt around the pole is completed manually with a tamping bar (Fig. 11.65) or with a hydraulic-powered tamper.

Too much stress cannot be laid on proper tamping, as a poorly tamped hole will not hold the pole in alignment. The earth should be backfilled slowly and each layer thoroughly tamped until the tamp makes a solid sound as the earth is struck. In general, the tamping should be done so thoroughly that no dirt need be hauled away.

If small stones or gravel are readily available, these should be used in backfilling, as a better foundation can thus be obtained. Care should be taken that plenty of earth is tamped in to fill the spaces between the stones. The backfill should be piled well up around the base of the pole to allow for settling. An examination should be made 1 or 2 months later and backfilling added if it has settled below the ground level.

FIGURE 11.64 Crew backfilling and tamping.

FIGURE 11.65   Wood and steel tamper bars used in pole setting.

Pulling Poles

The boom of the line truck should not be used to pull poles out of the ground. The strain placed on the boom if it is used to pull poles will exceed the capacity of the boom; and if the boom is rocked to free the pole, the boom can be seriously damaged. A hydraulic pole puller powered by the truck hydraulic system or a portable hydraulic pump driven electrically or with a gasoline engine can be used successfully (Fig. 11.66). The hydraulic pole puller consists of a hydraulic cylinder, a swivel chain hook assembly, a baseplate,

The new TIGER LINE of utility equipment introduces a lightweight pole puller cylinder weighing only 61 lb, yet producing over 46,000 lb of lift at 2,600 psi. Designed by a lineman for linemen; this unit operates directly off the truck hydraulic pto, or a portable gas over hydraulic pump for rear to use.

This compact 4 hp pump delivers over 1.2 gpm at 3,600 rpm. Maximum pressure is 2,600 psi. Weight is approximately 68 lb with a two man carrying handle for easy portability.

FIGURE 11.66 Pole puller in position to raise pole. The pole puller can receive its power from the line truck's hydraulic system or a portable power-driven hydraulic pump. (*Courtesy Thiermann Industries, Inc.*)

6 ft of ½-inch alloy chain, and carrying handles. The puller will lift the pole approximately 16 inches. The line truck boom is used to secure the pole while it is being pulled. If the pole is near energized conductors, it should be covered as shown in Fig. 11.44. The personnel guiding the pole should wear rubber gloves. When the hydraulic cylinder reaches the maximum position, the chain around the pole is released, the cylinder is lowered, the chain is refastened, and the pole is lifted farther. This process is continued until the pole is free from the earth. The boom of the line truck is used to lower the pole to the ground. The groundmen and/or the linemen guide the pole while it is being lowered.

Guying Poles

A *guy* is a brace or cable fastened to the pole to strengthen it and keep it in position. Guys are used wherever the wires tend to pull the pole out of its normal position and to sustain the line during abnormal loads caused by sleet, wind, and cold. Guys counteract the unbalanced forces imposed on the poles. The guy should be considered as counteracting the horizontal component of the forces with the pole or supporting structure as a strut resisting the vertical component of the forces.

Anchor or Down Guys

An *anchor* or *down guy* consists of a wire running from the attachment near the top of the pole to a rod and anchor installed in the ground (Fig. 12.1). This type of guy is preferable if field conditions permit its installation since it transfers the unbalanced force on a pole or structure to the earth without intermediate supports.

An anchor or down guy used at the ends of pole lines in order to counterbalance the pull of the line conductors is called a *terminal guy* (Figs. 12.2 and 12.3). All 90° corners in the line are considered as dead ends. They should be guyed the same as terminal poles except that there will be two guys, one for the pull of the conductors in each direction (Fig. 12.4).

Where the line makes an angle, a side pull is produced on the pole. *A side guy* should be installed to balance the side pull (Figs. 12.5 and 12.6).

Where a branch line takes off from the main line, an unbalanced side pull is produced. A side guy should be placed on the pole directly opposite to the pull of the branch line. Anchor or down guys are installed at regular intervals in transmission lines that extend long distances in one direction to protect the line from excessive damage as a result of broken conductors (Fig. 12.7). Guys installed to protect the facilities and limit the damage if a conductor breaks are called *line guys* or *storm guys*.

An anchor with a horizontal strut at a height above the sidewalk to clear pedestrians on the sidewalk is referred to as a *sidewalk guy*. Figure 12.8 illustrates a sidewalk guy installed at a pole.

Span Guy

A *span* or overhead *guy* consists of a guy wire installed from the top of a pole to the top of an adjacent pole to remove the strain from the line conductors. The overhead or span guy transfers the strain on a pole to another structure. This may be to another line pole or to a stub pole on which there is no energized equipment. A span guy is always installed to extend from the strain pole to the same or lower level on the next line pole.

225

FIGURE 12.1   Anchor or down guy assembly. (1) Galvanized machine bolt with nut; (2) locknut; (3) square curved galvanized washer; (4) galvanized steel guy wire; (5) porcelain guy strain insulator; (6) prefabricated guy dead-end grip; (7) prefabricated guy dead-end grip; (8) plastic guy guard; (9) angle thimble eye; (10) eyenut; (11) steel anchor rod; (12) power-installed screw anchor.

Head Guy

A guy wire running from the top of a pole to a point below the top of the adjacent pole is called a *head guy* (Fig. 12.9). Lines on steep hills are normally constructed with head guys to counteract the downhill strain of the line.

Electric lines crossing railroad tracks must be reinforced by the use of double arms and head guys as shown in Fig. 12.10. These guys run from the top of the pole adjacent to the track to a point about halfway down on the next pole.

Arm Guys

A guy wire running from one side of a crossarm to the next pole is called an *arm guy.* Arm guys are used to counteract the forces on crossarms which have more wires dead-ended on one side than on the other (Fig. 12.11).

FIGURE 12.3 Terminal pole example with the distribution line terminating at the pole labeled as 1 and the guy extending from the pole to the anchor labeled as 2.

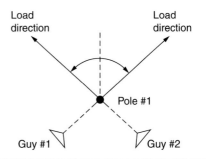

FIGURE 12.4 Terminal guys installed at the corner of the line at pole 1. The down guys labeled guy 1 and guy 2 must counteract the forces that occur in the line direction.

FIGURE 12.2 Guy wires installed on terminal or end pole. Guy wires are connected to system-neutral conductor to provide a low-resistance ground, eliminating the need for a guy insulator.

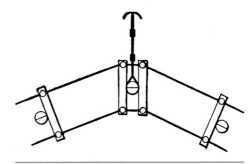

FIGURE 12.5 Guy installed at angle in line. The guy opposes the side pull of the line conductors.

FIGURE 12.6 Down guys installed in line to balance side pulls caused by angles in the line.

FIGURE 12.7   Line guys installed on three-phase transmission line. Note that poles are guyed in both directions, primarily installed to protect line from broken conductors.

FIGURE 12.8   Vertical or "sidewalk" guy completely installed with strain insulator and guy guard. This type of guy is used when the anchor must be set within 5 ft of the pole. The horizontal strut shown consists of a 2-inch diameter galvanized iron pipe and should have a length equal to the distance between the anchor and the pole. The horizontal strut is installed at a height of 8 ft or greater from ground level. This guy is called a sidewalk guy because the pole is set on one side of the sidewalk and the anchor is set on the opposite side. (*Courtesy Wisconsin Electric Power Co.*)

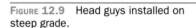

FIGURE 12.9   Head guys installed on steep grade.

FIGURE 12.10   Line reinforcement by use of double arms and head guys at a railroad crossing.

FIGURE 12.11   Arm guy. Guy opposes unbalanced pull on crossarm.

Stub Guys

A guy wire installed between a line pole and a stub pole on which there is no energized equipment is called a *stub guy* (Fig. 12.12). An anchor or down guy is used to secure the stub pole. This type of guy is often installed to obtain adequate clearance for guy wires extending across streets or highways (Fig. 12.13). When transmission or distribution lines parallel to streets or roads must be guyed toward the street or road, the necessary clearance can be obtained by the use of a pole stub on the opposite side.

FIGURE 12.12   Use of pole stub to secure clearance over highway. Pole stub is then guyed in usual manner.

FIGURE 12.13   Stub guy used to obtain clearance of guy wire over highway. (*Courtesy Copperweld Steel Co.*)

Push Guys

A pole used as a brace to a line pole is often referred to as a *push guy* (Fig. 12.14). A push brace or guy is used where it is impossible to use an anchor or down guys. When it is impossible to obtain sufficient right-of-way for a down guy, a push brace can usually be installed (Fig. 12.15).

Direction of pull

Hinged bracket

5"

6"

5/8" machine bolt and square washer

Brace may be cut from sound section of old pole.

Distance to be not less than 1/3 length of brace above ground.

Where additional stability is required install bracing

Approx. 3'-6"

Preformed concrete pad, Use in soft ground or for heavy strains in firm ground. Use rocks if available.

FIGURE 12.14   Push brace or guy.

FIGURE 12.15   Single-pole brace used in place of stub guy.

FIGURE 12.16   Seven-strand steel guy cable. (*Courtesy Joslyn Mfg. and Supply Co.*)

Guy Wire

The wire or cable used (Fig. 12.16) in a down guy is 3-strand, 7-strand, 19-strand, or 37-strand steel wire. The steel wire strands are coated with a layer of zinc or aluminum to prevent corrosion. Guy wire is provided in various sizes with galvanized steel wire strand diameters from 1/8 inch to 1¼ inch, aluminum-coated steel wire strand diameters from 3/16 inch to 1½ inch, and aluminum-clad steel wire strand diameters from 0.220 inch to 1.270 inches. The minimum breaking strength of the guy wire for the various coating types and sizes is listed in Tables 12.1 through 12.4.

Nominal Diameter of Strand, inch	No. of Wires in Strand	Minimum Breaking Strength of Strand, lb				
		Utilities Grade	Common Grade	Siemens-Martin Grade	High-Strength Grade	Extra High-Strength Grade
1/8	7	—	540	910	1,330	1,830
5/32	7	—	870	1,470	2,140	2,940
3/16	7	—	1,150	1,900	2,850	3,990
3/16	7	2,400	—	2,340	—	—
7/32	3	—	1,400	2,560	3,500	4,900
7/32	7	—	1,540		3,850	5,400
¼	3	3,150	1,860	3,040	4,730	6,740
¼	3	4,500	—	—	—	—
¼	7	—	1,900	3,150	4,750	6,650
9/32	3	—	2,080	3,380	5,260	7,500
9/32	7	4,600	2,570	4,250	6,400	8,950
5/16	3	6,500	2,490	4,090	6,350	9,100
5/16	7	—	3,200	5,350	8,000	11,200
5/16	7	6,000	—	—	—	—
3/8	7	8,500	3,330	5,560	8,360	11,800
3/8	7	11,500	4,250	6,950	10,800	15,400
7/16	3	18,000	5,700	9,350	14,500	20,800
½	7	25,000	7,400	12,100	18,800	26,900
½	7	—	7,620	12,700	19,100	26,700
9/16	7	—	9,600	15,700	24,500	35,000
9/16	19	—	9,640	16,100	24,100	33,700
5/8	7	—	11,600	19,100	29,600	42,400
5/8	19	—	11,000	18,100	28,100	40,200
¾	19	—	16,000	26,200	40,800	58,300
7/8	19	—	21,900	35,900	55,800	79,700
1	19	—	28,700	47,000	73,200	104,500
1	37	—	28,300	46,200	71,900	102,700
1-1/8	37	—	36,000	58,900	91,600	130,800
1-1/4	37	—	44,600	73,000	113,600	162,200

(Courtesy Hubbell Power Systems)

TABLE 12.1   Galvanized (Zinc-Coated) Steel Wire (*Sizes and Strengths of Galvanized Strand*)

Nominal Diameter of Strand, inch	No. of Wires in Strand	Minimum Breaking Strength of Strand, lb				
		Utilities Grade	Common Grade	Siemens-Martin Grade	High-Strength Grade	Extra High-Strength Grade
3/16	7	—	1,150	1,900	2,850	—
3/16	7	2,400	—	—	—	—
¼	3	3,150	—	—	—	—
¼	3	4,500	—	—	—	—
¼	7	—	1,900	3,150	4,750	6,650
9/32	7	4,600	—	—	—	—
5/16	3	6,500	—	—	—	—
5/16	7	—	3,200	5,350	8,000	11,200
5/16	7	6,000	—	—	—	—
3/8	3	8,500	—	—	—	—
3/8	7	11,500	4,250	6,950	10,350	15,400
7/16	7	18,000	5,700	9,350	14,500	20,800
½	7	25,000	7,400	12,100	18,300	26,900

(Courtesy Hubbell Power Systems)

TABLE 12.2   Aluminum-Coated Steel Wire (*Sizes and Strengths of Aluminum-Coated Strand*)

Anchors

The common types of manufactured anchors consist of the single-helix anchor (Fig. 12.17), the double-helix anchor (Fig. 12.18), the multiple-helix anchor (Fig. 12.19), the expansion anchor (Fig. 12.20), the screw-type anchor (Fig. 12.21), the cross plate-type anchor (Fig. 12.22), the rock anchor (Fig. 12.23), and the Tough One anchor (Fig. 12.24). Each anchor type has a specific utilization application dependent upon the soil conditions encountered. For example, the helix anchor shown in Fig. 12.18 has a torque rating of 8,000 ft pounds and is acceptable for use in soft and medium hard soils. Figure 12.19 illustrates a multiple-helix anchor. Helixes are stacked on the steel shaft. Each combination of either the 8-inch or 10-inch helix acts essentially as a separate anchor for increased holding power. If rock is located close to the surface, expanding rock-type anchors are used.

Anchor Rods

The anchor rods serve as the connecting link between the anchor and the guy cable. The rod must have an ultimate strength equal to, or greater than, that required by the down guy assembly. The ultimate strength of commonly used anchor rods is provided in Table 12.5. Anchor rods vary in diameter from ½ inch to 1¼ inches and in length from 5 ft to 12 ft. If a greater length is needed, rods can be coupled to reach greater depths.

Anchor rods and extension rods for power-installed helix anchors are fabricated with a connecting device designed to fit over the 1½-in sq drive hub (Figs. 12.25 and 12.26). Galvanized steel eyenuts are installed on power-installed screw anchor rods to provide a connecting device for the guy-strand cable.

Designation No. and Size of Wire, AWG	Nominal Diameter, inch	Minimum Breaking Strength of Strand, lb				
		3 Wire	7 Wire	19 Wire	37 Wire	M-Strand
3#10	0.220	4,532	—	—	—	—
3#9	0.247	5,715	—	—	—	—
3#8	0.277	7,206	—	—	—	—
3#7	0.311	8,621	—	—	—	—
3#6	0.349	10,280	—	—	—	—
3#5	0.392	12,230	—	—	—	—
7#12	0.242 (¼)	—	6,301	—	—	—
7#11	0.272	—	7,945	—	—	—
7#10	0.306 (5/16)	—	10,020	—	—	—
7 #9	0.343 (11/32)	—	12,630	—	—	—
7#8	0.385 (3/8)	—	15,930	—	—	—
7#7	0.433 (7/16)	—	19,060	—	—	—
7#6	0.486 (½)	—	22,730	—	—	—
7#5	0.546	—	27,030	—	—	—
19#10	0.509	—	—	27,190	—	—
19#9	0.572	—	—	34,290	—	—
19#8	0.642	—	—	43,240	—	—
19#7	0.721	—	—	51,730	—	—
19#6	0.810	—	—	61,700	—	—
19#5	0.910	—	—	73,350	—	—
37#10	0.718	—	—	—	52,950	—
37#9	0.801	—	—	—	66,770	—
37#8	0.899	—	—	—	84,180	—
37#7	1.010	—	—	—	100,700	—
37#6	1.130	—	—	—	120,100	—
37#5	1.270	—	—	—	142,900	—

(Courtesy Hubbell Power Systems)

TABLE 12.3 Aluminum-Clad Steel Wire (*Sizes and Strengths of Aluminum-Clad Strand*)

Guy Insulators

The *National Electrical Safety Code* (NESC) requires that ungrounded guys attached to supporting structures that carry open-supply conductors of more than 300 volts, or guys that are exposed to such conductors, be insulated. Otherwise, the guys must be effectively grounded. It may be desirable to install insulators in guys that are effectively grounded to maintain adequate electrical clearances and provide safe working space for the linemen. Linemen must recognize the grounded guy. If work is to be done on a grounded guy assembly, electrical continuity must be maintained. The guy should be

Designation No. and Size of Wire, AWG	Minimum Breaking Strength of Strand, lb					
	Nominal Diameter, inch	3 Wire	7 Wire	19 Wire	37 Wire	M-Strand
4M	0.220	—	—	—	—	4,000
5M	0.247	—	—	—	—	5,000
6M	0.242	—	—	—	—	6,000
7M	0.277	—	—	—	—	7,000
8M	0.272	—	—	—	—	8,000
10M	0.306	—	—	—	—	10,000
12.5M	0.343	—	—	—	—	12,500
14M	0.363	—	—	—	—	14,000
16M	0.386	—	—	—	—	16,000
18M	0.417	—	—	—	—	18,000
20M	0.444	—	—	—	—	20,000
25M	0.519	—	—	—	—	25,000

(Courtesy Hubbell Power Systems)

TABLE 12.4   Aluminum-Clad Steel Wire M-Strand (*Sizes and Strengths of Aluminum-Clad M-Strand*)

FIGURE 12.17   Single-helix anchor. The anchor's sloped lead point improves penetration and helps soil flow from below the hub to flow above the anchor. (*Courtesy A. B. Chance Co.*)

FIGURE 12.18   Double-helix anchor. (*Courtesy A. B. Chance Co.*)

FIGURE 12.19 Multiple-helix anchor designed for heavy guy loading. (*Courtesy A. B. Chance Co.*)

FIGURE 12.20 Expansion anchor. (a) Expanded position and (b) closed position. The anchor is inserted in the hole in the closed position. Then the top of the anchor is struck several blows with a ram to cause the leaves to expand into solid undisturbed earth at the bottom of the hole.

FIGURE 12.21   Screw-type anchor.

FIGURE 12.22   Cross-plate anchor.
(*Courtesy A. B. Chance Co.*)

FIGURE 12.23   Rock anchor. The greater
the strain, the more firmly it is wedged
in the rock. (*Courtesy A. B. Chance Co.*)

Closed   Open

FIGURE 12.24   Tough One Helix Assembly designed for heavy guy loading.

TOUGH ONE® ANCHOR
TORQUE RATINGS: 15,000 ft.-lb.
LARGE HUB (2½" square inside)

Use high-strength 15,000 ft-lb. Tough One in very hard soils short of solid rock.

Use 8,000 ft.-lb. TOUGH ONE® anchor in soft and medium-hard soils.

Nominal Rod Diameter, inch	Ultimate Strength, lb
5/8	16,000
3/4	23,000
1	36,000
1, high strength	50,000

(Courtesy A.B. Chance Co.)

TABLE 12.5   Anchor Rod Strength

FIGURE 12.25   Square Shaft Screw Guy Adapter.

Guy adapter

Extension

Figure 12.26   Square Shaft Screw Anchor Extension.

Figure 12.27   A fiberglass guy strain insulator assembly. The strandvise links the guy strain insulator to the guy wire. (*Courtesy A. B. Chance Co.*)

treated as any other grounded device. Proper protective equipment must be installed on the guy in the area of the work while the lineman is working on or near energized conductors. Ungrounded guys are insulated with fiberglass strain insulators. When higher values of insulation are needed in a guy, longer fiberglass strain insulators are used (Figs. 12.27 and 12.28). Fiberglass strain insulators with three different end style fittings depicted (Figs. 12.29 through 12.31) are impervious to moisture. The fiberglass strain insulators can withstand a direct stroke of lightning to the pole structure without bursting.

Guy Size Determination[1]

Factors to be considered in guying pole lines are the weight of the conductor or cable, the size and weight of crossarms and insulators, wind pressures on poles and conductors, strains due to the contour of the earth, line curvatures, pole heights, and dead-end loads, plus the vertical load due to sleet and ice. To reduce unbalanced stresses to a

---

[1]Material under this heading is from *Encyclopedia of Anchoring*. Copyright A. B. Chance Co. Used by permission.

FIGURE 12.28   Fiberglass guy strain insulator. (*Courtesy A. B. Chance Co.*)

FIGURE 12.29   Insulator, fiberglass, guy strain, clevis end fitting. (*Courtesy A. B. Chance Co.*)

FIGURE 12.30   Insulator, fiberglass, guy strain, thimble eye. (*Courtesy A. B. Chance Co.*)

FIGURE 12.31   Insulator, fiberglass, guy strain, eye end fitting. (*Courtesy A. B. Chance Co.*)

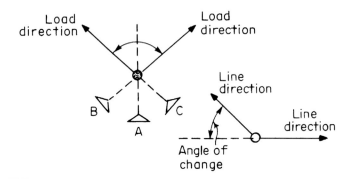

FIGURE 12.32   Proper location of guys and line angle as illustrated.

minimum, correct angling and positioning of guy wires are essential. Where obstructions make it impossible to locate a single guy in line with the load or pull, two or more guys can be installed with their resultant guying effect in line with the load.

Where lines make an abrupt change in direction, the guy anchor is normally placed so that it bisects the angle formed by the two conductors, as shown by anchor A in Fig. 12.32. Under heavy load conditions, it may be necessary to use two anchors, each dead-ending a leg of the line load, as shown by anchors B and C. Long straight spans require occasional side and end guys to compensate for heavy icing and crosswind on conductors and poles.

These, and all other factors that might make it advisable to use guys, should be carefully considered in initial designs for line construction.

To compute the load on the guy, the line load must be determined first. When the line is dead-ended, the line load can be calculated by multiplying the ultimate breaking strength of the conductor used, S, by the number of conductors, N.

For example, if three 1/0 ACSR conductors are dead-ended on a pole, the line load will be 12,840 lb:

$$S \times N = \text{Line load}$$

$$4{,}280 \times 3 = 12{,}840$$

The ultimate breaking strength of aluminum-stranded conductor and aluminum conductor steel-reinforced (ACSR) is found in Table 12.6.

To determine the line load to be guyed when the line is dead-ended, refer to Table 12.7.

The guy wire angle from the attachment at the top of the pole determines the loading. The manner of measuring the height and lead of a guy on sloping ground and of a stub pole guy is illustrated in Fig. 12.33. On sloping ground the lead is the horizontal distance from the guy to the pole, and the height is the distance above this horizontal line, as shown in Fig. 12.33. On a pole stub guy the lead is the horizontal distance from the pole stub to the line pole, and the height is the vertical distance above this horizontal line, as shown in Fig. 12.34. The tension in the guy wire for a given line-conductor load depends on the distance the anchor is installed from the base of the pole or the guy angle (Fig. 12.35).

Aluminum Stranded					ACSR			
Circular Mils or AWG	No. of Strands	Diameter, inch	Ultimate Strength, lb		Circular Mils or AWG	No. of Strands	Diameter, inch	Ultimate Strength, lb
6	7	0.184	528		6	6 × 1	0.198	1,170
4	7	0.232	826		6	6 × 1	0.223	1,490
3	7	0.260	1,022		4	6 × 1	0.250	1,830
2	7	0.292	1,266		4	7 × 1	0.257	2,288
1	7	0.328	1,537		3	6 × 1	0.281	2,250
1/0	7	0.368	1,865		2	6 × 1	0.316	2,790
2/0	7	0.414	2,350		2	7 × 1	0.325	3,525
3/0	7	0.464	2,845		1	6 × 1	0.355	3,480
4/0	7	0.522	3,590		1/0	6 × 1	0.398	4,280
266.8	7	0.586	4,525		2/0	6 × 1	0.447	5,345
266.8	19	0.593	4,800		3/0	6 × 1	0.502	6,675
336.4	19	0.666	5,940		4/0	6 × 1	0.563	8,420
397.5	19	0.724	6,880		266.8	18 × 1	0.609	7,100
477	19	0.793	8,090		266.8	6 × 7	0.633	9,645
477	37	0.795	8,600		266.8	26 × 7	0.642	11,250
556.5	19	0.856	9,440		300	26 × 7	0.680	12,650
556.5	37	0.858	9,830		336.4	18 × 1	0.684	8,950
636	37	0.918	11,240		336.4	26 × 7	0.721	14,050
715.5	37	0.974	12,640		336.4	30 × 7	0.741	17,040
715.5	61	0.975	13,150		397.5	18 × 1	0.743	10,400
795	37	1.026	13,770		397.5	26 × 7	0.783	16,190
795	61	1.028	14,330		397.5	30 × 7	0.806	19,980
874.5	37	1.077	14,830		477	18 × 1	0.814	12,300
874.5	61	1.078	15,760		477	24 × 7	0.846	17,200
954	37	1.124	16,180		477	26 × 7	0.846	19,430
954	61	1.126	16,860		477	30 × 7	0.883	23,300
1033.5	37	1.170	17,530		556.5	26 × 7	0.914	19,850
1033.5	61	1.172	18,260		556.5	26 × 7	0.927	22,400
1113	61	1.216	19,660		556.5	30 × 7	0.953	27,200
1272	61	1.300	22,000		605	24 × 7	0.953	21,500
1431	61	1.379	24,300		605	26 × 7	0.966	24,100
1590	61	1.424	27,000		605	30 × 19	0.994	30,000
1590	91	1.454	28,100		636	24 × 7	0.977	22,600
					636	26 × 7	0.990	25,000
					636	30 × 19	1.019	31,500
					666.6	24 × 7	1.000	23,700
					715.5	54 × 7	1.036	26,300
					715.5	26 × 7	1.051	28,100
					715	30 × 19	1.081	34,600
					795	54 × 7	1.093	28,500
					795	26 × 7	1.108	31,200
					795	30 × 19	1.140	38,400
					874.5	54 × 7	1.146	31,400
					900	54 × 7	1.162	32,300
					954	54 × 7	1.196	34,200
					1033.5	54 × 7	1.246	37,100
					1113	54 × 19	1.292	40,200
					1272	54 × 19	1.382	44,800
					1431	54 × 19	1.465	50,400
					1590	54 × 19	1.545	56,000

(Courtesy Hubbell Power Systems)

TABLE 12.6   Aluminum-Stranded Conductor and Aluminum Conductor Steel-Reinforced Ultimate Breaking Strengths

Line Load, lb	Guy Angle from Pole in Degrees															
	10	15	20	25	30	35	40	45	50	55	60	65	70	75	80	85
1,000	5,759	3,864	2,924	2,366	2,000	1,743	1,556	1,414	1,305	1,221	1,155	1,103	1,064	1,035	1,016	1,004
2,000	11,518	7,727	5,348	4,732	4,000	3,487	3,111	2,828	2,611	2,442	2,309	2,207	2,128	2,071	2,031	2,008
3,000	17,276	11,591	8,771	7,099	6,000	5,230	4,667	4,243	3,916	3,662	3,464	3,310	3,193	3,106	3,046	3,011
4,000	23,035	15,455	11,695	9,465	8,000	6,974	6,223	5,657	5,222	4,883	4,619	4,414	4,257	4,141	4,062	4,015
5,000	28,794	19,319	14,619	11,831	10,000	8,717	7,779	7,071	6,527	6,104	5,774	5,517	5,321	5,276	5,077	5,019
6,000	34,553	23,182	17,543	14,197	12,000	10,461	9,334	8,485	7,832	7,325	6,928	6,620	6,385	6,212	6,093	6,023
7,000	40,311	27,046	20,467	16,563	14,000	12,204	10,890	9,899	9,138	8,545	8,083	7,724	7,449	7,247	7,108	7,027
8,000	46,070	30,910	23,390	18,930	16,000	13,948	12,446	11,314	10,443	9,766	9,236	8,827	8,513	8,282	8,123	8,031
9,000	51,529	34,773	26,314	21,296	18,000	15,691	14,002	12,728	11,749	10,987	10,392	9,930	9,578	9,317	9,139	9,034
10,000	57,588	38,637	29,238	23,662	20,000	17,434	15,557	14,142	13,054	12,208	11,547	11,034	10,642	10,353	10,154	10,038
11,000	63,346	42,501	32,162	26,028	22,000	19,178	17,113	15,556	14,359	13,429	12,702	12,137	11,706	11,388	11,170	11,042
12,000	69,105	46,364	35,086	28,394	24,000	20,921	18,669	16,971	15,665	14,649	13,866	13,241	12,770	12,423	12,185	12,046
13,000	74,864	50,228	38,009	30,761	26,000	22,665	20,224	18,385	16,970	15,870	15,011	14,344	13,834	13,459	13,201	13,050
14,000	80,623	54,092	40,933	33,127	28,000	24,408	21,780	19,799	18,276	17,091	16,166	15,447	14,898	14,494	14,216	14,053
15,000	86,382	57,956	43,367	35,493	30,000	26,152	23,336	21,213	19,581	18,312	17,321	16,551	15,963	15,529	15,231	15,057
16,000	92,140	61,819	46,781	37,859	32,000	27,895	24,892	22,627	20,887	19,532	18,475	17,654	17,027	16,564	16,247	16,061
17,000	97,899	65,683	49,705	40,225	34,000	29,639	26,447	24,042	22,192	20,753	19,630	18,757	18,091	17,600	17,262	17,065
18,000	103,658	69,547	52,628	42,592	36,000	31,382	28,003	25,456	23,497	21,974	20,785	19,861	19,155	18,635	18,278	18,069
19,000	109,417	73,410	55,552	44,958	38,000	33,125	29,559	26,870	24,803	23,195	21,939	20,964	20,210	19,670	19,293	19,073
20,000	115,175	77,274	58,476	47,324	40,000	34,869	31,114	28,284	26,108	24,415	23,394	22,068	21,284	20,706	20,309	20,076
21,000	120,934	81,138	61,400	49,690	42,000	36,612	32,670	29,698	27,414	25,636	24,249	23,171	22,384	21,741	21,324	21,080
22,000	126,693	85,001	64,324	52,056	44,000	38,356	34,226	31,113	28,719	26,857	25,403	24,274	23,412	22,776	22,339	22,084
23,000	132,452	88,865	67,248	54,423	46,000	40,099	35,782	32,527	30,024	28,078	26,558	25,378	24,476	23,811	23,355	23,088
24,000	138,210	92,729	70,171	56,789	48,000	41,843	37,337	33,941	31,330	29,299	27,713	26,481	25,540	24,847	24,370	24,092
25,000	143,969	96,593	73,095	59,155	50,000	43,586	38,893	35,355	32,635	30,519	28,868	27,584	26,604	25,882	25,386	25,095
26,000	149,728	100,456	76,019	61,521	52,000	45,330	40,449	36,770	33,941	31,740	30,022	28,688	27,669	26,917	26,401	26,099

Guy Angle from Pole in Degrees

Line Load, lb	10	15	20	25	30	35	40	45	50	55	60	65	70	75	80	85
26,000	149,728	100,456	76,019	61,521	52,000	45,330	40,449	36,770	33,941	31,740	30,022	28,688	27,669	26,917	26,401	25,099
27,000	155,487	104,320	78,943	63,887	54,000	47,073	42,005	38,184	35,246	32,961	31,177	29,791	28,733	27,952	27,417	27,103
28,000	161,246	108,184	81,867	66,254	56,000	48,817	43,560	39,598	36,551	34,182	32,332	30,895	29,797	28,988	28,432	28,107
29,000	167,004	112,047	84,790	68,620	58,000	50,560	45,116	41,012	37,857	35,402	33,486	31,998	30,861	30,023	29,447	29,111
30,000	172,763	115,911	87,714	70,986	60,000	52,303	46,672	42,426	39,162	36,623	34,641	33,101	31,925	31,058	30,463	30,115
31,000	178,522	119,775	90,638	73,352	62,000	54,047	48,227	43,841	40,468	37,844	35,796	34,205	32,990	32,094	31,478	31,118
32,000	184,281	123,639	93,562	75,718	64,000	55,790	49,783	45,255	41,773	39,065	36,950	35,308	34,054	33,129	32,494	32,122
33,000	190,039	127,502	96,486	78,085	66,000	57,534	51,339	46,669	43,078	40,286	38,105	36,441	35,118	34,164	33,509	33,126
34,000	195,798	131,366	99,409	80,451	68,000	59,277	52,895	48,083	44,384	41,506	39,260	37,515	36,182	35,199	34,525	34,130
35,000	201,557	135,230	102,333	82,817	70,000	61,021	54,450	49,497	45,689	42,727	40,415	38,618	37,246	36,235	35,540	35,134
36,000	207,316	139,093	105,257	85,183	72,000	62,764	56,006	50,912	46,995	43,948	41,569	39,722	38,310	37,270	36,555	36,138
37,000	213,075	142,957	108,181	87,549	74,000	64,508	57,662	52,326	48,300	45,169	42,724	40,825	39,375	38,305	37,571	37,141
38,000	218,883	146,821	111,105	89,916	76,000	66,251	59,116	53,740	49,605	46,389	43,879	41,925	40,439	39,340	38,586	38,145
39,000	224,592	150,684	114,028	92,282	78,000	67,994	60,673	55,154	50,911	47,610	45,033	43,032	41,503	40,376	39,602	39,149
40,000	230,351	154,548	116,952	94,648	80,000	69,738	62,229	56,569	52,216	48,831	46,188	44,135	42,567	41,411	40,617	40,153
41,000	236,110	158,412	119,876	97,014	82,000	71,481	63,785	57,983	53,522	50,052	47,343	45,238	43,631	42,466	41,632	41,157
42,000	241,868	162,276	122,800	99,380	84,000	73,225	65,340	59,397	54,827	51,273	48,497	46,342	44,695	43,482	42,648	42,160
43,000	247,627	166,139	125,724	101,747	86,000	74,968	66,896	60,811	56,133	52,493	49,652	47,445	45,760	44,517	43,663	43,164
44,000	253,386	170,003	128,647	104,118	88,000	76,712	68,452	62,225	57,438	53,714	50,807	48,549	46,824	45,552	44,679	44,168
45,000	259,145	173,807	131,571	106,479	90,000	78,455	70,008	63,640	58,743	54,935	51,962	49,632	47,888	46,387	45,694	45,172
46,000	264,903	177,730	134,495	108,845	92,000	80,199	71,563	65,054	60,049	56,156	53,116	50,755	48,952	47,623	46,710	46,176
47,000	270,662	181,594	137,419	111,211	94,000	81,942	73,119	66,468	61,354	57,376	54,271	51,859	50,016	48,658	47,725	47,180
48,000	276,421	185,458	140,343	113,578	96,000	83,685	74,675	67,882	62,660	58,597	55,426	52,962	51,081	49,693	48,740	48,183
49,000	282,180	189,321	143,266	115,944	98,000	85,429	76,230	69,296	63,965	59,818	56,580	54,066	52,145	50,729	49,756	49,187
50,000	287,939	193,185	146,190	118,310	100,000	87,172	77,786	70,711	65,270	61,039	57,735	55,169	53,209	51,764	50,771	50,191

TABLE 12.7   Guy Load Reference—Line Is Dead-ended (*When the Line Is Dead-ended*)

FIGURE 12.33   Method of measuring height and lead to determine guy angle: Sloping ground.

FIGURE 12.34   Method of measuring height and lead to determine guy angle: Stub guy.

FIGURE 12.35   Sketch of guy installation where H is height of guy attachment to pole, L is lead or distance from pole to anchor rod.

After the line load is known, Table 12.7 is used as a quick reference for determining the load on the guy at different angles of pull. To use the table, determine the line load, and using the line load values in the first column and the guy angle from the pole at the top of the chart, follow the vertical value until it intersects the line indicating the angle of the proposed guy. The figure at this point will be the guy load in pounds. For example, assuming a line load of 10,000 lb and a guy angle at the top of pole of 30°, follow the line at 10,000 lb across the chart until it intersects the line indicating the 30° angle. Reading across the chart to the right, the figure obtained is 20,000 lb, which is the guy load.

To determine the line load to be guyed on a single anchor where the line changes direction, again multiply the ultimate breaking strength of the conductor used (S) times the number of conductors used (N), then refer to Table 12.8. Locate your line load in pounds. Read across the table until it intersects the value of the angle of change of line direction. The result is the line load to be guyed in pounds. For example, assume a line load of 10,000 lb and an angle of change of line direction of 15°. The line load to be guyed is 2,610 lb.

In using the breaking strength of conductor, it should be considered that a conductor properly sagged (according to the NESC) will not exceed 60% of its breaking strength when fully loaded. This automatically allows a safety factor of 1–2/3. However, additional safety factors will be required on important crossings especially over highways, railroads, or rivers, where safety factors of 2 and 3 are generally used. After the guy load has been found, select an anchor with holding power in the soil class, allowing for the desired safety factor. The anchor, rod, and guy strand must be selected with proper size and strength to coordinate with the anchor maintaining the desired safety factor.

Guy Construction

The installation of a guy divides itself naturally into the five steps outlined:

1. Digging in the anchor
2. Inserting the strain insulator in the guy wire or grounding the guy wire or both

Angle Change of Line Direction, Degrees

Line Load, lb	15	20	25	30	35	40	45	50	55	60	65	70	75	80	85	90
1,000	261	347	433	518	601	684	765	845	924	1,000	1,075	1,147	1,218	1,286	1,351	1,414
2,000	522	694	866	1,036	1,202	1,368	1,530	1,690	1,848	2,000	2,150	2,294	2,436	2,572	2,702	2,828
3,000	783	1,041	1,299	1,554	1,803	2,052	2,295	2,535	2,772	3,000	3,225	3,441	3,654	3,858	4,053	4,242
4,000	1,044	1,388	1,732	2,072	2,404	2,736	3,060	3,380	3,696	4,000	4,300	4,588	4,872	5,144	5,404	5,656
5,000	1,305	1,735	2,165	2,590	3,005	3,420	3,825	4,225	4,620	5,000	5,375	5,735	6,090	6,430	6,755	7,070
6,000	1,566	2,082	2,598	3,108	3,606	4,104	4,590	5,070	5,544	6,000	6,450	6,882	7,308	7,716	8,106	8,484
7,000	1,827	2,429	3,031	3,626	4,207	4,788	5,355	5,915	6,468	7,000	7,525	8,029	8,526	9,002	9,457	9,898
8,000	2,088	2,776	3,464	4,144	4,808	5,472	6,120	6,760	7,392	8,000	8,600	9,176	9,744	10,288	10,808	11,312
9,000	2,349	3,123	3,897	4,662	5,409	6,156	6,885	7,605	8,316	9,000	9,675	10,323	10,962	11,574	12,159	12,726
10,000	2,610	3,470	4,330	5,180	6,010	6,840	7,650	8,450	9,240	10,000	10,750	11,470	12,180	12,860	13,510	14,140
11,000	2,871	3,817	4,763	5,698	6,611	7,524	8,415	9,295	10,164	11,000	11,825	12,617	13,398	14,146	14,861	15,554
12,000	3,132	4,164	5,196	6,216	7,212	8,208	9,180	10,140	11,088	12,000	12,900	13,764	14,616	15,432	16,212	16,968
13,000	3,393	4,511	5,629	6,734	7,813	8,892	9,945	10,985	12,012	13,000	13,975	14,911	15,834	16,718	17,563	18,382
14,000	3,654	4,858	6,062	7,252	8,414	9,576	10,710	11,830	12,936	14,000	15,050	16,058	17,052	18,004	18,914	19,796
15,000	3,915	5,205	6,495	7,770	9,015	10,260	11,475	12,675	13,860	15,000	16,125	17,205	18,270	19,290	20,265	21,210
16,000	4,176	5,552	6,928	8,288	9,616	10,944	12,240	13,520	14,784	16,000	17,200	18,352	19,488	20,576	21,616	22,624
17,000	4,437	5,899	7,361	8,806	10,217	11,628	13,005	14,365	15,708	17,000	18,275	19,499	20,706	21,862	22,967	24,038
18,000	4,698	6,246	7,794	9,324	10,818	12,312	13,770	15,210	16,632	18,000	19,350	20,646	21,924	23,148	24,318	25,452
19,000	4,959	6,593	8,227	9,842	11,419	12,996	14,535	16,055	17,556	19,000	20,425	21,793	23,142	24,434	25,669	26,866
20,000	5,220	6,940	8,660	10,360	12,020	13,680	15,300	16,900	18,480	20,000	21,500	22,940	24,360	25,720	27,020	28,280
21,000	5,481	7,287	9,093	10,878	12,621	14,364	16,065	17,745	19,404	21,000	22,575	24,087	25,578	27,006	28,371	29,694
22,000	5,742	7,634	9,526	11,396	13,222	15,048	16,830	18,590	20,328	22,000	23,650	25,234	26,796	28,292	29,722	31,108
23,000	6,003	7,981	9,959	11,914	13,823	15,732	17,595	19,435	21,252	23,000	24,725	26,381	28,014	29,578	31,073	32,622
24,000	6,264	8,328	10,392	12,432	14,424	16,416	18,360	20,280	22,176	24,000	25,800	27,528	29,232	30,864	32,424	33,936
25,000	6,525	8,675	10,825	12,950	15,025	17,100	19,125	21,125	23,100	25,000	26,875	28,675	30,450	32,150	33,775	35,350

Table 12.8 Angle Load Reference Based on Angle Change of Line Direction (When Guy and Anchor Bisects the Angle Formed) (continued)

Line Load, lb	15	20	25	30	35	40	45	60	55	60	65	70	75	80	85	90
									Angle Change of Line Direction, Degrees							
26,000	6,786	9,022	11,258	13,488	15,626	17,784	19,890	21,970	24,024	26,000	27,950	29,822	31,668	33,436	35,126	36,764
27,000	7,047	9,369	11,691	13,986	16,227	18,468	20,655	22,815	24,948	27,000	29,025	30,969	32,886	34,722	36,477	38,178
28,000	7,308	9,716	12,124	14,504	16,828	19,152	21,420	23,660	25,872	28,000	30,100	82,116	84,104	36,008	37,828	39,592
29,000	7,569	10,068	12,557	15,022	17,429	19,836	22,185	24,505	26,796	29,000	31,175	33,263	35,322	37,294	39,179	41,006
30,000	7,830	10,410	12,990	15,540	18,030	20,520	22,950	25,350	27,720	30,000	32,250	34,410	36,540	38,580	40,530	42,420
31,000	8,091	10,757	13,423	16,058	18,631	21,204	23,715	26,105	28,644	31,000	33,325	35,537	37,758	39,866	41,881	43,834
32,000	8,352	11,104	13,856	16,576	19,232	21,888	24,480	27,040	29,568	32,000	34,400	36,704	38,976	41,152	43,232	45,248
33,000	8,613	11,451	14,289	17,094	19,833	22,572	25,245	27,885	30,492	33,000	35,475	37,851	40,94	42,438	44,583	46,662
34,000	8,874	11,798	14,722	17,612	20,434	23,256	26,010	28,730	31,416	34,000	36,550	38,998	41,412	43,724	45,034	48,076
35,000	9,135	12,145	15,155	18,130	21,035	23,940	26,775	29,575	32,340	35,000	37,625	40,145	42,630	45,010	47,285	49,490
36,000	9,396	12,492	15,588	18,648	21,636	24,624	27,540	30,420	33,264	36,000	38,700	41,292	43,848	46,296	48,636	50,904
37,000	9,657	12,839	16,021	19,166	22,237	25,308	28,305	31,265	34,188	37,000	39,775	42,439	45,066	47,582	49,987	52,318
38,000	9,918	13,186	16,454	19,684	22,838	25,992	29,070	32,110	35,112	38,000	40,850	43,586	46,284	48,868	51,338	53,732
39,000	10,179	13,533	16,887	20,202	23,439	26,676	29,835	32,955	36,036	39,000	41,925	44,733	47,502	50,154	52,689	55,146
40,000	10,140	13,880	17,320	20,720	24,040	27,360	30,600	33,800	36,960	40,000	43,000	45,880	48,720	51,440	54,040	56,560
41,000	10,701	14,227	17,753	21,238	24,641	28,044	31,365	34,645	37,884	41,000	44,075	47,027	49,939	52,726	55,391	57,974
42,000	10,962	14,574	18,186	21,756	25,242	28,728	32,130	35,490	38,808	42,000	45,150	48,174	51,156	54,012	56,742	59,388
43,000	11,223	14,921	18,619	22,274	25,843	29,412	32,895	36,335	30,732	43,000	46,225	49,321	52,374	55,298	58,093	60,802
44,000	11,484	15,268	19,050	22,792	26,444	30,006	33,660	37,180	40,656	44,000	47,300	50,468	53,592	56,584	59,444	62,216
45,000	11,745	15,615	19,485	23,310	27,045	30,780	34,425	38,025	41,580	45,000	48,375	51,615	54,810	57,870	60,795	63,630
46,000	12,006	15,962	19,918	23,828	27,646	31,464	35,190	38,870	42,504	46,000	49,450	52,762	56,028	59,156	62,146	65,044
47,000	12,267	16,309	20,351	24,846	28,247	82,148	35,955	39,715	43,428	47,000	50,525	53,909	57,246	00,442	63,497	66,458
48,000	12,528	16,656	20,784	24,864	28,848	32,882	36,720	40,560	44,352	48,000	51,600	55,056	58,464	61,728	64,848	67,872
49,000	12,789	17,003	21,217	25,382	29,449	33,516	37,485	41,405	45,276	49,000	52,675	56,203	59,682	63,014	66,199	69,286
50,000	13,050	17,350	21,650	25,900	30,050	34,200	38,250	42,250	46,200	50,000	53,750	57,350	60,900	64,300	67,550	70,700

TABLE 12.8 Angle Load Reference Based on Angle Change of Line Direction (When Guy and Anchor Bisects the Angle Formed)

3. Fastening the guy wire to the pole

4. Tightening the guy wire and fastening it to the anchor

5. Mounting the guy-wire guard

Digging in the Anchor

The manner of digging in the guy anchor depends on the type of anchor. See Fig. 12.36 for various types of guy anchors in general use. The screw-type anchor is screwed into the earth in the manner shown in Fig. 12.37.

The patented cross-plate anchor is installed in the manner shown in Fig. 12.38. For the driving of the rod, a special maul fitted with oil-soaked hickory inserts is used. The inserts prevent damage to the anchor rod. For hanging of the plate an installing bar fitted with a special hook is employed. The same bar is also used for tamping and for

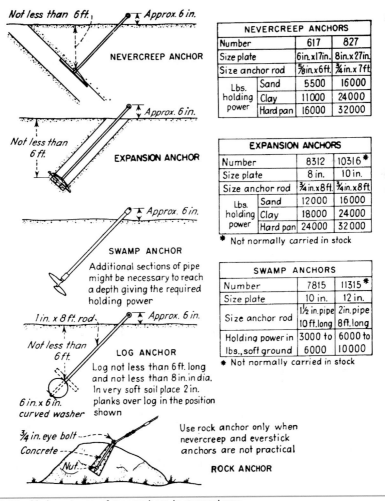

NEVERCREEP ANCHORS		
Number	617	827
Size plate	6in.x17in.	8in.x27in.
Size anchor rod	⅝in.x6ft.	¾in.x7ft.
Lbs. holding power — Sand	5500	16000
Clay	11000	24000
Hard pan	16000	32000

EXPANSION ANCHORS		
Number	8312	10316*
Size plate	8 in.	10 in.
Size anchor rod	¾in.x8ft.	¾in.x8ft.
Lbs. holding power — Sand	12000	16000
Clay	18000	24000
Hard pan	24000	32000

* Not normally carried in stock

SWAMP ANCHORS		
Number	7815	11315*
Size plate	10 in.	12 in.
Size anchor rod	1½ in.pipe 10ft.long	2in.pipe 8ft.long
Holding power in lbs.,soft ground	3000 to 6000	6000 to 10000

* Not normally carried in stock

FIGURE 12.36 Various types of guy anchors in general use.

FIGURE 12.37 Steps in installation of manual screw-type anchor. (a) Digging small hole with bar; (b) screwing in anchor.

(a) Bore the hole

(b) Drive the rod

(c) Hang the plate

(d) Tighten the guy

(e) Fill the hole

FIGURE 12.38 Steps in the installation of the never-creep anchor. Locate the spot desired for the anchor rod, measure from that point back from the pole the length of the rod, and start the hole at this point. A boring-machine auger of almost any size can be used. (a) Bore the hole as nearly at right angles to the line of strain as conditions will allow. Then proceed with steps b, c, d, and e as shown. (*Courtesy A. B. Chance Co.*)

retrieving the anchor in case the guy is no longer needed. The expansion anchor is installed by inserting the anchor into the bottom of the hole and then causing the blades to expand into the solid ground.

The anchor is an important part of a line, for if the anchor fails, the guy fails, and if the guy fails, the pole fails, and if the pole fails, the corner fails, and if the corner fails, the line fails. In other words, the chain is no stronger than its weakest link, and the line is no stronger than its weakest angle or corner. The importance of proper guying has led to the expression, "A line well guyed is half built, but a line poorly guyed is never built."

Digging the Anchor Hole with a Power Digger

Most anchor holes are dug with a power digger. Only isolated holes or holes in locations inaccessible to a digger are dug by hand. The usual diameter of a hole is 9 inches. The length of the hole should be such, as a general rule, that the anchor itself will be not less than 6 ft below the surface of the ground.

Power-Drive Screw Anchor

Screw anchors may be installed by use of a power drive. Tools required for the installation of power-installed screw anchors are depicted in Figs. 12.39 through 12.41. A special tubular wrench (Fig. 12.42) slipped over the anchor rod transmits the torque from the power drive directly to the hub of the anchor wing. Since no torque is transmitted through the anchor rod, the rod has to be only heavy enough to withstand the pull of the guy. Moreover, a large-diameter wing can be readily installed in this manner, thus providing greater holding power.

The steps in the installation procedure are as follows:

1. Remove the eye from the anchor rod.
2. Slide a tubular wrench over the anchor rod to engage the square hub shank of the helix. The wrench is held in position by spring-loaded dogs and a hex nut.
3. Attach the wrench to the driving bar of the power digger.
4. Drive the anchor down with the power digger, which feeds as well as screws the anchor into the earth. The feed should match the natural penetration of the helix so that a minimum of earth is disturbed. Driving at proper revolutions per minute (rpm) and feed eliminates churning the earth, which often results from hand turning, as the anchor then must be drawn down by the ability of the soil to resist crumbling. Churning is similar to stripping a thread.
5. Release the spring-loaded dogs and remove the wrench from the anchor rod.
6. Replace the eye on the anchor rod.

Besides eliminating the hard work of installation when done manually, the use of the power drive saves time. The average time required for actual installation is about 5 min.

Guy Assembly

The tools needed to assemble a guy include bolt cutters, wrenches, pliers, hoist, wire grips, and pulling eye for preformed grips. A typical guy assembly is illustrated in Fig. 12.43. The material required to connect the anchor rod to the pole includes guy wire, hardware, insulators, and preformed guy-wire grips. The lineman assembling a guy

FIGURE 12.39   PISA anchor accessories. *(Courtesy A. B. Chance Co.)*

KELLY BAR
ADAPTER

LOCKING
DOG
ASSEMBLY

EXTENSION
WRENCH

DRIVE-END
WRENCH

FIGURE 12.40   PISA anchor locking dog accessory. *(Courtesy A. B. Chance Co.)*

Locking dog
assembly

FIGURE 12.41 Additional PISA anchor accessories. (*Courtesy A. B. Chance Co.*)

FIGURE 12.42 Special tubular wrench used with power drive to rotate screw anchor. Turning torque is applied on square hub shank next to helix.

Extension

Wrench

10'-0" Max.
8'-0" Min.

10'-0" Max.
8'-0" Min.

Lead dim.

Lead dim.

FIGURE 12.43 Stub guy and down guy assembly. (1) Pole stub; (2) anchor, expansion; (3) guy attachment, thimble eye; (4) guy attachment, eye type; (5) bolt, standard machine, with standard nut; (6) clamp, guy bond, for ¾-inch rod, twin eye; (7) locknut, ¾-inch galvanized; (8) protector, guy; (9) rod, twin eye, ¾-inch × 9 ft, with two standard nuts; (10) washer, 4-inch × 4-inch curved; (11) insulator, strain, porcelain; (12) insulator, strain, fiberglass guy; (13) wire, guy; (14) grip, guy wire; (15) pole topper.

FIGURE **12.44** Prefabricated guy grip partially installed on end of guy strand. (*Courtesy Preformed Line Products Co.*)

must determine the length of the guy wire needed to cut the wire accordingly. All guy wire, whether galvanized, copperweld, or alumoweld has steel core strands; the wire is springy and difficult to handle. Before a guy wire is cut, it is good practice to wrap tape around the wire on each side of the place where the cut is to be made. This will keep the strands from fraying or unlaying after the cut is made. When the guy wire is cut, hold both sides of the cut firmly so that the free ends will not fly into your face or into a live conductor. A good procedure is to place both feet on the guy wire and make the cut between them. After the cut has been made, each end can then be carefully released.

When the guy wire has been cut to the proper length, the insulators should be installed in accordance with the specifications, using prefabricated guy-wire grips (Fig. 12.44). An insulator properly assembled in series with the guy strand is shown in Fig. 12.28.

Guy wires should be placed on the poles and "pulled" before the line conductors are placed on the pole. If the line wires were placed first, the poles would be pulled out of position.

A guy wire is pulled up or tightened from the ground by means of a hoist and approved grips. These grips (Fig. 12.45) are also called *come-alongs*. Come-alongs will grip a straight wire at any point without slipping. The guy is drawn up until the pole is pulled over slightly toward the guy.

Prefabricated guy dead-end grips can be used to terminate the guy strand to the eye on the anchor rod (Fig. 12.46). The guy stand is pulled taut with a hoist by lineman (Fig. 12.47), then the guy strand is wrapped with tape and cut to the proper length. The prefabricated grip is threaded through the eye on the anchor rod, and the legs of the grip are wrapped around the guy strand. The ends of the legs of the grip are snapped into position, and the hoist is removed, completing the installation. Guy-grip dead-ends

FIGURE **12.45** Improved "Chicago" steel wire grip or come-along. (*Courtesy American Safety Utility Corporation.*)

FIGURE 12.46 Prefabricated guy dead-end partially installed. The legs of the grip must be completely wrapped around the guy strand, and the ends must be snapped in place before the hoist is removed. Snapping the ends of the grip into place ensures that the grip will have rated holding strength and prevents the grip from unwrapping. (*Courtesy Preformed Line Products Co.*)

FIGURE 12.47   Link chain hoist used to pull the guy wire taut. (*Courtesy American Safety Utility Corporation.*)

¾-Ton 1-Ton   1½-Ton   2-Ton

may be removed and reapplied three times within 3 months after installation, if necessary, to retension the guys. Guy-grip dead-ends that were removed after their original installation should not be reused.

Figure 12.48 shows the use of a dynamometer in pulling a down guy to the proper tension.

Guys should not be installed where they will interfere with traffic. Figure 12.49 shows a typical guy protector. The ground end of anchor guys exposed to pedestrian traffic must be provided with a substantial and conspicuous marker not less than 8 ft long. Figure 12.50 shows a lineman installing a plastic guard. The guard should be light-colored so that it will be visible at night. The guard should extend from a point near the ground to 8 ft above the ground. A guard installed on a vertical guy is shown in Fig. 12.51.

Pole keys can be used to prevent poles with an unbalanced load from leaning. Street lights installed on poles often cause the poles to lean as a result of the side strain, especially if the pole is set in soft ground. The pole key (Fig. 12.52) is lowered into the pole

FIGURE 12.48 Showing use of come-along, dynamometer, and chain jack in pulling down guy to proper tension. (*Courtesy Coffing Hoist Co.*)

FIGURE 12.49 Typical guy guard.

FIGURE 12.50 Lineman installing plastic guy guard. (*Courtesy MidAmerican Energy Co.*)

FIGURE 12.51 Plastic guy guards.

FIGURE 12.52   An expansion metal pole key. (*Courtesy A. B. Chance Co.*)

FIGURE 12.53   Groundman unscrewing and removing the anchor rod used to lower the key into the pole hole prior to filling the hole with dirt. (*Courtesy A. B. Chance Co.*)

FIGURE 12.54   Pole key has been expanded, Wood pole key is being installed adjacent to the pole just below the surface of the ground by a groundman. (*Courtesy A. B. Chance Co.*)

hole alongside the pole and expanded with a ram. The anchor rod used to lower the pole key into the pole hole is unscrewed and removed (Fig. 12.53). The pole hole is filled with earth and thoroughly tamped. The finished installation will prevent the pole from leaning due to an unbalanced load if the pole key has been properly installed. A treated-wood key can be installed just below the ground surface on the opposite side of the pole from the metal key to provide additional stability (Fig. 12.54).

Insulators, Crossarms, and Conductor Supports

Overhead conductors are supported on the distribution pole using insulators, crossarms, or conductor supports. The National Electrical Safety Code (NESC) establishes the criteria to be followed in the design, construction, and operation of the electric distribution system. The requirements for insulators, crossarms, and conductor supports vary depending upon the grade of construction as defined in the NESC.

Insulators

An insulator prevents the flow of an electric current. The insulator supports electrical conductors. The function of an insulator is to separate the line conductors from the pole or tower. Insulators are designed to withstand electrical stresses, mechanical stresses, and environmental stresses. Insulators are fabricated from porcelain, glass, fiberglass, and polymers.

Porcelain insulators are manufactured from clay. Special clays are selected and mixed mechanically until a plastic-like compound is produced. The clay is then placed in molds to form the insulators. The molds are placed in an oven to dry the clay. When the clay is partially dry, the mold is removed and the insulator drying process is complete. When the insulator is dry, it is dipped in a glazing solution and fired in a kiln. The glaze colors the insulator and provides a glossy surface. This makes the insulator surface self-cleaning.

Large porcelain insulators are made up of several shapes cemented together. Care must be taken when cementing the insulators together to prevent a chemical reaction on the metal parts, causing cement growth. Cement growth can cause stresses on the porcelain great enough to crack the porcelain.

Glass insulators are made from sand, soda, ash, and lime. The materials are properly mixed and melted in an oven until a clear-liquid plastic-like material is produced. The plastic-like compound is placed in a mold and allowed to cool. After the glass insulator is cool, it is placed in an annealing oven.

Fiberglass insulators are manufactured with rods of fiberglass treated with epoxy resins. The rubber-like epoxy compounds are applied to the rods to fabricate suspension, dead-end, and post-type insulators.

Polymer insulators are formed by using various sizes of silica bound together chemically with a resin. The compound's composition is approximately 90% silica. Polymer insulators are replacing porcelain, glass, and epoxy insulators. Polymer insulators have excellent mechanical and dielectric strength. The insulators are lightweight, virtually nonbreakable and much easier to handle than any of the other types of insulators. Polymer insulators are less costly than porcelain, glass, or epoxy insulators because of the ease of manufacture. Field tests and current installation success of polymer insulators have proven that they are durable and that they will provide many years of service in an adverse atmosphere.

The galvanized end fitting of the polymer insulator is crimped to the insulator's fiberglass rod. Polymer material covers the rod. The greatest amount of electrical stress occurs at the interface of the end fitting and the rod.

Conductor Dead-End Insulators

The primary conductor is terminated at the crossarm, at the conductor support, or on the pole using a conductor dead-end shoe and porcelain dead-end bell-shaped insulators or a conductor dead-end shoe and a one-piece polymer dead-end insulator. The porcelain dead-end bells and one-piece polymer dead-end insulator are shown in Fig. 13.1. The length and size of the insulator required is dependent upon the voltage class of the distribution system (4 kV, 15 kV, 25 kV, or 35 kV) and the basic insulation level (BIL) for the system. As defined by IEEE, BIL is the reference insulation level expressed as the impulse crest value of withstand voltage of a specified full impulse voltage wave.

Pin-Type Insulators

The pin insulator gets its name from the fact that it is supported on a pin. The pins are mounted on the crossarm. The pin holds the insulator, and the insulator has the conductor tied to it. Pin insulators are made of glass, porcelain, or polymers (Figs. 13.2 and 13.3). Pin insulators are used on distribution lines and seldom used on transmission lines. The glass pin insulator is principally used on low-voltage circuits.

The polymer pin insulator has unique applications associated with spacer cable. When using the polymer insulator with a bare conductor, a bare tie wire should be used (Fig. 13.4). When using the polymer insulators with covered conductors, a covered tie wire is required to avoid degradation of the conductor covering due to a voltage gradient that exists between the covered conductor, a bare tie wire, and the insulator (Fig. 13.5). Thus, a covered conductor, bare tie wire, and porcelain insulator are not used unless the covered conductor is skinned of the covering at the point of interconnection.

FIGURE 13.1   Porcelain dead-end bells (top) and one-piece polymer dead-end insulators (bottom) for a typical 25-kV class distribution system with a design BIL of 110 kV. (*Courtesy A. B. Chance Co.*)

FIGURE 13.2   Porcelain top-groove pin insulators mounted in position on double crossarm. Note the use of a crossarm double-arming bolt, flat washers, and nuts to maintain crossarm separation.

FIGURE 13.3   A porcelain pin insulator.

FIGURE 13.4   Polymer pin insulator with bare tie wire for use with bare conductor (*Courtesy Hendrix Wire & Cable Inc.*)

HPI insulator with bare conductor and bare tie wire

FIGURE 13.5   Polymer pin insulator with covered tie wire for use with covered conductor. (*Courtesy Hendrix Wire & Cable Inc.*)

HPI insulator with covered conductor and covered tie wire

The smaller insulators are threaded to fit on a 1-inch pin, and the larger units are threaded to fit on a 1-3/8-inch pin. Transmission and distribution pin insulator neck letter designations and diameters are A: 1¾ inch, C: 2¼ inch, F: 2-7/8 inch, J: 3½ inch, K: 4 inch, and N: 6 inch.

Insulator Pins

The function of an insulator pin is to hold the insulator mounted on it in position. The insulator pins are made of iron or steel. Steel pins are in general use. A steel pin provided for use on a wooden crossarm is depicted in Fig. 13.6. The steel pin is usually provided with 1-inch nylon alloy threads for pin-type insulators that prevent any localized pressure on the insulator. The insulator pin has a broad base that rests squarely on the crossarm. An angle pin bracket for use on wooden crossarms is shown in Fig. 13.7.

The spacing of the pins is generally suited to the voltage of the circuit in accordance with the NESC. The spacing should provide sufficient working space for the lineman. The end pins are generally spaced 4 inches from the end of the arm.

Pins that clamp to the crossarm are mounted on the crossarm, as shown in Fig. 13.8. Since these pins do not require pin holes, their use avoids weakening the crossarm.

FIGURE 13.6   Steel pin with 1-inch nylon alloy threads. (*Courtesy MacLean Power Systems.*)

FIGURE 13.7   Angle pin bracket with 1-inch nylon alloy threads for use on wooden crossarms. (*Courtesy MacLean Power Systems.*)

FIGURE 13.8   Clamp style nylon alloy threaded insulator pin. (*Courtesy MacLean Power Systems.*)

FIGURE 13.9   An apitong conductor support installed on a wood pole. The apitong conductor support is 48 inches in length.

Crossarms and Conductor Supports

Overhead conductors rest on insulators supported by crossarms or conductor supports. The crossarms and conductor supports are designed to support the load that they will bear plus appropriate overload capacity factors due to stresses caused by ice or wind. NESC table 261-2 contains the dimensions of a crossarm section of selected Southern Pine or Douglas fir woods. The crossarm dimensions are based upon the grade of construction; either Grade B or Grade C. The conductor support is a more compact style of close-space construction utilizing apitong wood. The apitong supports are 48 inches in length (Fig. 13.9).

Single arms are used on straight lines where no excessive strain needs to be provided for. As already mentioned in Chap. 11, every other crossarm faces in the same direction.

Double arms are frequently used at line terminals, at corners, at angles, or at other points where there is an excessive strain (Fig. 13.10). Where lines cross highways or railroad crossings, double arms are used, as more than ordinary safety as required by the NESC is required at such points.

Buck arms may be used at corners and at points where branch circuits are taken off at a right angle to the main line (Fig. 13.11).

Side arms are used in alleys or other locations where it is necessary to clear buildings, etc. (Fig. 13.12)

Fiberglass crossarms can be used at points where the distribution lines dead-end or at locations where extra safety is necessary, instead of double-wood crossarms (Fig. 13.13).

The fiberglass crossarms are gray and have a small cross-sectional area that minimizes the mass at the top of the pole. The fiberglass crossarms are often installed where

Figure 13.10 Double-crossarm construction details. The steel bracket installed between the crossarms and the pole eliminates the need to cut a gain on the pole for the crossarms. (*Courtesy Hughes Brothers.*)

the appearance of a pole line is important. The bracket between the pole and the fiberglass crossarm eliminates the need for braces.

Galvanized steel crossarms may be used in place of double-wood crossarms or fiberglass crossarms to obtain the safety factor required (Fig. 13.14).

Crossarm braces are used to give strength and rigidity to the crossarm. Metal crossarm braces are made of either galvanized flat bar or light angle iron. The size used varies with the size of the arm and the weight of the conductors. The usual flat-strap brace for ordinary distribution work (Fig. 13.15) is 38 inches long and ¼-inch by 1¼-inches. One end is attached to the crossarm by means of a carriage bolt and the other to the pole by means of a lag screw. One brace extends to each side of the arm. Angle-iron braces are made in one piece and bent into the shape of a V, as shown in Fig. 13.16. These braces are fitted to the bottom of the crossarm instead of the side as is the flat type.

Wooden crossarm braces are used extensively for medium voltages. Their use increases the insulation of the line. A set of wooden braces is illustrated in Fig. 13.17, and their use is illustrated in Figs. 13.9 to 13.11.

Item Description	Number Required
A. Aluminum Tie Wire	As Needed
B. Straight-Line Dead-end Strain Clamp (primary)	3
C. Straight-Line Dead-end Strain Clamp (neutral)	1
D. Aluminum Ampact Connector (primary)	6
E. Aluminum Ampact Connector (neutral)	2
F. Double-Arm Dead-end Assembly	1
G. Pin, 1" x 6" ; 3/4" x 6-1/2" shank	4
H. Insulator, 1" x 5-1/2", pin-type	4
I. Clevis-Type 4-1/2" Suspension Insulator; 5000-lb working load	6
J. Eye Bolt: 5/8"	1
K. Square Washer: flat	7
L. Single Top Tie: formed	3
M. Spool Tie: Formed	1
N. Douglas Fir Wood Arms	1
O. Insulated-Secondary Clevis	1
P. Wood Crossarm Braces	2
Q. Machine Bolts: 1/2 "	2
R. Round Washer	2
S. Machine Bolts: 5/8"	3

FIGURE 13.11   Three-phase, 13,200Y/7,620-volt distribution line constructed with 10-ft crossarms. Double crossarm is used to support conductors for a tap to the main circuit on a tangent pole in the line. A down guy would be installed to counter the pull of the tapped-line conductors.

Where crossarms extend to one side of the pole only, as is sometimes necessary in alley construction, special braces are required. The brace must be longer and more rigid. A single brace made of galvanized angle iron 1¾ inches by 1¾ inches by 3/16 inch and 5 ft to 7 ft long is generally used for distribution lines and is shown in Figs. 13.12 and 13.18. The brace is provided with a step near the middle for use of the lineman.

A variety of bolts and screws are used in line construction, and only the common ones are illustrated. Figure 13.19 shows a typical crossarm through bolt used for fastening the crossarm to the pole. Its common size is 5/8 by 12 inches to 16 inches. A bolt for attaching two crossarms, one on each side of the pole, is shown in Fig. 13.20.

FIGURE 13.12 Alley-arm brace used with side arm to clear obstructions. Note pole step on brace for use by lineman.

Figure 13.21 illustrates a carriage bolt used for attaching the brace to the crossarm. The typical bolt for this purpose has a diameter of 3/8 inch and is 4 inches, 4½ inches, or 5 inches long. Figure 13.22 shows a lag screw used to fasten the crossarm braces to the pole. The recommended size is ½ inch by 4 inches. Lag screws are supposed to be screwed into place either into a small bored hole or after being started by hammering. They should be screwed into place to keep the threads from injuring the fibers of the wood.

The complete assembly of the pieces of pole hardware described above is shown in Fig. 13.23. This figure also lists the standard parts and illustrates their correct use.

Crossarms may be mounted before or after the pole is erected; the practice varies. If the pole is located where it is accessible from a bucket truck and if the pole has not been preframed and drilled, the lineman frames and drills the pole while working from an insulated bucket. The pin-type insulators are installed on the crossarm by a groundman or lineman working on the ground (Fig. 13.24). The groundman hands the crossarm, complete with pins, insulators, and crossarm braces, to the lineman in the bucket of the truck. The lineman installs the crossarm on the pole, using a through bolt while working from the insulated bucket on the truck (Fig. 13.25).

In case the crossarm is mounted after the pole is in place and the pole is located in an area inaccessible to a bucket truck, the lineman climbs the pole, carrying his small tools (pliers, knife, and connectors), hand line, hammer, and lag wrench, and fastens the pulley of the hand line to the pole. The crossarm is then pulled up by the groundman with a handline Fig. 13.26. Various methods of tying the hand line to the crossarm are shown in Fig. 13.27.

FIGURE 13.13 Fiberglass crossarms are used on distribution pole of 13,200Y/7,620-volt three-phase line with a three-phase primary circuit tapped to the conductor. The line is constructed with close spacing to meet environmental requirements. Solid blade disconnect isolation switches are also installed. Plastic line conductor guards and rubber insulating blankets are installed on lower line conductors. The lineman is working from an insulated platform. (*Courtesy A. B. Chance Co.*)

FIGURE 13.14 Galvanized steel insulator supports are used for 13,200Y/7,620-volt three-phase line constructed with close spacing for appearance reasons. A galvanized steel crossarm is used to support dead-ended conductors tapped to main circuit. (*Courtesy Sherman & Reilly, Inc.*)

FIGURE 13.15   Standard flat-strap crossarm brace.

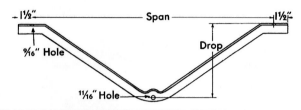

FIGURE 13.16   A V-shaped angle-iron crossarm brace. (*Courtesy Joslyn Mfg. and Supply Co.*)

FIGURE 13.17   Wooden crossarm brace used to increase insulation properties of line. (*Courtesy Joslyn Mfg. and Supply Co.*)

FIGURE 13.18   Alley-arm brace.

FIGURE 13.19   Crossarm bolt.

FIGURE 13.20   Bolt for double crossarms.

FIGURE 13.21   Carriage bolt for fastening brace to crossarm.

FIGURE 13.22   Lag screw for fastening crossarm brace to pole.

FIGURE 13.23   Sketches showing standard sizes and use of bolts, lags, and washers in single- and double-crossarm assembly.

FIGURE 13.24   Lineman and groundman installing pin-type insulator with steel pin on crossarm. (*Courtesy Northwest Lineman College.*)

FIGURE 13.25   Lineman completing installation of crossarm. Pin-type insulator for center-phase conductor is supported by special pole-top steel pin fabricated so it can be bolted to pole. (*Courtesy Northwest Lineman College.*)

FIGURE 13.26   A hand line utilized to transport materials, such as a crossarm, to the lineman at the top of the pole. (*Courtesy American Safety Utility Corporation.*)

The crossarm is attached to the pole by means of a 5/8-inch galvanized machine bolt driven from the back of the pole. A 2¼-inch square washer is placed under the head and the nut of the bolt, and then the bolt is drawn up.

After the through bolt is in place, the crossarm braces are attached. It is necessary first to sight the crossarm, that is, make sure it is horizontal. In case more conductors are to be carried on one side than on the other, that side can be higher, as it will settle later. The braces are usually 1¼-inch by ¼-inch by 28 inches. Sometimes the braces are bolted to the crossarm before the arm is pulled up. The end of the brace attached to the

4 Pin (Single hitch)

6 Pin (Double hitch)

10' Crossarm

This hitch is put on close to end of arm so that the lineman does not have to hold the weight of the arm any longer than necessary. After he removes the top hitch, the groundman pulls the arm up the rest of the way.

FIGURE 13.27   Methods of tying hand line to crossarm for purpose of raising.

crossarm is held in place by means of a 3/8-inch by 4-inch galvanized carriage bolt, and the ends at the pole are fastened by means of a ½-inch by 5-inch lag screw. The braces are attached on the back of the crossarm. When the standard 28-inch braces are used, the hole in the crossarm should be located 19 inches from the middle of the crossarm.

The appearance of the pole line can be improved by the use of armless-type construction. Pin-type insulators can be installed on epoxy rods to provide proper clearances without the use of crossarms. The epoxy rods are available in single, double, and triple assemblies (Fig. 13.28).

FIGURE 13.28   Pin-type insulators mounted on epoxy rod single- and bi-unit assemblies to eliminate use of crossarm. Eyes on bi-unit assembly are for mounting stringing equipment. (*Courtesy A. B. Chance Co.*)

Post-Type Insulators

Post-type insulators are used on distribution, subtransmission, and transmission lines and are installed on wood, aluminum, concrete, fiberglass, and steel poles. The line-post insulators are manufactured for vertical or horizontal mounting (Figs. 13.29 and 13.30).

The line-post insulators are usually manufactured as one-piece solid porcelain units (Fig. 13.31) or fiberglass epoxy-covered rods with metal end fittings and rubber weather sheds (Fig. 13.32). In some instances, the fiberglass rod is combined with a porcelain insulator unit (Fig. 13.33). The insulators are fabricated with a mounting base for curved or flat surfaces, and the top is designed for tying the conductor to the insulator or fitted with a clamp designed to hold the conductor.

Line-post insulators designed for vertical mounting are mounted on crossarms (Fig. 13.34). This type of construction is often used for long-span rural distribution circuits. Figure 13.35 illustrates a distribution circuit constructed with porcelain horizontal line-type post insulators. Armless construction using post-type insulators permits the construction of subtransmission and transmission lines on narrow rights-of-way and along city streets. A 69-kV subtransmission line constructed with porcelain line-post insulators with conductor clamp top is pictured in Fig. 13.36. A 230-kV transmission line constructed with steel poles and line-post-type insulators is illustrated in Fig. 13.37.

Figure 13.29   Vertical pole top post insulator. (*Courtesy Hubbell Power Systems.*)

FIGURE 13.30   Horizontal line post insulator. (*Courtesy Hubbell Power Systems.*)

Characteristics			
Typical application		kV	69
Flash-over voltage	Impulse critical 1.2 x 50 mu-sec wave	Positive	kV	330
		Negative	kV	425
	60 Hz	Dry	kV	200
		Wet	kV	180
Standard glaze color			Sky tone
Radio influence voltage	Test voltage to ground	kV	44
	Maximum RIV at 1000 kHz	Micro-volts	200
Leakage distance		Inches	53
Dry arcing distance		Inches	19-1/4
Cantilever		Pounds	2800

FIGURE 13.31   An 88-kV porcelain horizontal post insulator.

Mechanical characteristics (ultimate ratings)	
Vertical cantilever, lbs	1,500
Longitudinal cantilever, lbs	1,000
Tensile, lbs	4,000
Compression, lbs	1,500
Electrical characteristics	
60 Hz flashover, dry, kV	140
60 Hz flashover, wet, kV	120
Impulse flashover, positive, kV	210
Impulse flashover, negative, kV	320

FIGURE 13.32   A 15-kV tapered-skirt standoff insulator.

FIGURE 13.33 Fiberglass rod and pin insulator used to support top conductor and fiberglass rod and porcelain assemblies to support side conductors. This arrangement permitted increasing the voltage on the circuit from 4.16 kV to 13.2 kV without replacing the pole. (*Courtesy A. B. Chance Co.*)

FIGURE 13.34 Post-type insulator in use in vertical position. Note special supporting hardware used in place of conventional pin. (*Courtesy Lapp Insulator Co.*)

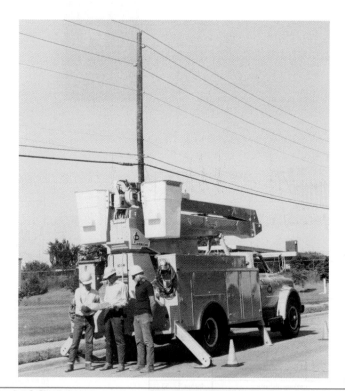

FIGURE 13.35 A 13.2Y/7.6-kV three-phase four-wire distribution circuit constructed with sky-gray line-post insulators to enhance the appearance. Phase conductors are 477ACSR. Neutral conductor, 4/0 ACSR, is supported by porcelain spool-type insulator fastened to pole with a metal bracket. Crew is holding tailgate safety meeting before starting to work on assignment.

FIGURE **13.36**   Post-type insulator in use in horizontal position. Note narrow right-of-way required.

FIGURE **13.37**   A 230-kV transmission line located along a city street. Line uses porcelain post-type insulators for the transmission circuit and the underbuilt distribution circuit on steel poles.

Post-type insulators have been used to construct circuits that operate at voltages through 345 kV. Lines constructed with post-type insulators are clean-looking and aesthetically acceptable to the public (Fig. 13.37). The higher transmission voltages require more than one segment of porcelain to be bolted together (Fig. 13.38).

Suspension Insulators

The suspension insulator, as its name implies, is suspended from the crossarm and has the line conductor fastened to the lower end (Fig. 13.39). The suspension-type insulator was developed when voltages were increased above 44 kV.

The suspension-type insulator is manufactured from porcelain (Fig. 13.40), glass (Fig. 13.41), and epoxy glass rod and rubber composite or silicone materials (Figs. 13.42 to 13.44). Economics, voltage of the circuit, width of right-of-way, size of the conductor, length of the span, and clearances required all enter into the decision regarding the type and material of the insulators to be used.

The diameter of the porcelain disks varies from 3-3/8 inches to 11½ inches or more. The 10-inch disk is perhaps the most common size. The number of units to use in a string depends largely on the voltage of the line. Other factors, such as climate, type of construction, and degree of reliability required, also enter into the consideration. Table 13.1 gives a general idea of the usual number of units employed for the various standard distribution and transmission voltages.

Strain insulators are used where a pull must be carried as well as insulation provided. Such places occur wherever a line is dead-ended—at corners, at sharp curves, or at extralong spans, as at river crossings or in mountainous country. In such places the insulator not only must be a good insulator electrically, but also must have sufficient mechanical strength to counterbalance the forces due to the tension of the line conductors.

230 kV
Steel or concrete structures

138 kV
Steel or concrete structures

5"

138 kV
Wood structures

Line Voltage	Lapp Catalog no.	Unit length (L), inches	Ultimate strength	
			Cantilever, lb	Tension, lb
138 kV	70148	53.0	2800	5,000
	79135	58.0	4000	10,000
	301543	56.75	2800	5,000
	305071	60.50	4000	10,000
230 kV	301372	101.50	3500	10,000
	90974	89.0	5800	10,000

FIGURE 13.38   Porcelain horizontal line-post insulators. (*Courtesy Lapp Insulator Co.*)

Strain insulators are built in the same way as suspension insulators except that they are made stronger mechanically. Furthermore, when a single string is not able to withstand the pull, two or more strings are arranged in multiple.

Glass suspension insulators have been used on the circuits in many cases. Glass is used more commonly in countries other than the United States. Glass insulators can be used interchangeably with polymer insulators wherever their strength and voltage characteristics are adequate.

The development of extra-high-voltage (EHV) bundle-conductor transmission circuits resulted in heavy and cumbersome installations when porcelain suspension insulators were used. The polymer suspension insulator is used to address the issue. The polymer suspension insulator consists of a long epoxy-fiberglass rod of extremely high mechanical and dielectric strength with polymer weather sheds covering the rod. A heavy silicone grease is used as an interface between the rod and the weather sheds. Forged-steel end attachments are swaged to the rod under pressure. The insulators

FIGURE 13.39   Single-circuit 345-kV three-phase high-voltage transmission line. Note the long string of suspension insulators needed to insulate the conductors. Also note two ground wires at extreme top of the tower. These are used to shield the line from lightning.

Characteristics				
ANSI class (Std. C.29.2-1971)				52.8
Flash-over voltage	Impulse critical 1.2 x 50 mu-sec wave	Positive	kV	125
		Negative	kV	130
	60 Hz	Dry	kV	80
		Wet	kV	50
Low-frequency puncture voltage			kV	110
Radio influence voltage	Test voltage to ground		kV	10
	Maximum RIV at 1000 kHz		Micro-volts	50
Leakage distance			Inches	13
Dry arcing distance			Inches	7.75
Section length			Inches	5.75
Porcelain disc diameter			Inches	10
Strength rating	M & E rating		Pounds	40,000
	ANSI M & E category		Pounds	36,000
Recommended maximum sustained load			Pounds	20,000
Impact strength			Inch Pounds	90
Color				Skytone

FIGURE 13.40   Porcelain suspension insulator, ball-and-socket type.

Characteristics				
ANSI class (Std. C29.2-1962)				52-5
Flash-over voltage	Impulse	Positive	kV	125
		Negative	kV	125
	Low frequency	Dry	kV	70
		Wet	kV	55
Low-frequency puncture voltage			kV	130
Radio influence voltage	Test voltage to ground		kV	10
	Maximum RIV at 1000 kHz		Micro-volts	50
Leakage distance			Inches	11-1/2
Impact strength			Inches Pounds	400
Routine proof test			Pounds	15,000
Strength ratings	M & E rating		Pounds	30,000
	Maximum sustained load		Pounds	16,500

FIGURE 13.41   Toughened glass suspension insulator, ball-and-socket type.

FIGURE 13.42   Epoxy glass rod, with rubber weather sheds, suspension insulator. *(Courtesy NGK-Locke Inc.)*

ITEM No.	DESCRIPTION	REQD.
①	ANCHOR SHACKLE	1
②	Y–SOCKET CLEVIS	1
③	YOKE	1
④	Y–90° CLEVIS EYE	2
⑤	SUSPENSION CLAMP	2

FIGURE 13.43   Two conductor, single suspension insulator string material list for a vertical hanging suspension insulator string installation. *(Courtesy NGK-Locke Inc.)*

are manufactured with maximum design tension ratings of 10,000 lb, 20,000 lb, 40,000 lb, and 80,000 lb. A single polymer suspension insulator can replace four parallel strings of porcelain or glass suspension insulators. The synthetic-material insulators weigh only 5% to 10% of the weight of the porcelain suspension insulators they replace. A 500-kV polymer insulator weighs 40 lb, and a 765-kV polymer insulator weighs 65 lb. The synthetic-material insulators are less susceptible to mechanical damage.

Structures can be assembled on the ground and erected without danger of destroying the insulator. The polymer insulators have a high ratio of strength to weight, resistance to vandalism (gunfire), a life expectancy equal to that of porcelain, freedom from tracking and deterioration caused by surface leakage currents in contaminated environments, and a low cost. Corona rings are typically installed on the polymer suspension insulators when the operation voltage is greater than 100 kV (Figs. 13.45 and 13.46).

ITEM No.	DESCRIPTION	REQD.
①	ANCHOR SHACKLE	1
②	Y–SOCKET CLEVIS	1
③	YOKE	1
④	Y–90° CLEVIS EYE	2
⑤	SUSPENSION CLAMP	2

FIGURE 13.44 Two conductor, single suspension insulator string material list for an angled hanging suspension insulator string installation. Note the unique shape of the yoke for the angled installation. (*Courtesy NGK-Locke Inc.*)

Line Voltage, Volts	Number of Suspension Units Required in String	Line Voltage, Volts	Number of Suspension Units Required in String
13,200	2	138,000	8, 9, or 10
23,000	2 or 3	154,000	9, 10, or 11
34,500	2 or 3	230,000	12 to 16
69,000	4 or 5	345,000	18 to 20
88,000	5 or 6	500,000	24 to 28
110,000	6, 7, or 8	765,000	30 to 35

TABLE 13.1 Approximate Number of 10-inch Porcelain or Glass Disk Insulators Required in Suspension String for Various Line Voltages

FIGURE 13.45 Corona ring installed on the suspension insulator string at the end near the energized conductor attachment. (*Courtesy NGK-Locke Inc.*)

4" DISC

FIGURE 13.46 Corona disk, factory installed, on the suspension insulator string at the end near the crossarm attachment. (*Courtesy NGK-Locke Inc.*)

Spacer Cable Spacers

In some instances, bundled, covered, primary conductor is installed. The covered cable is often referred to as "spacer" cable. The messenger supported, bundled configuration has a few advantages, most notably:

- Less tree trimming issues with less tree contacts exposure
- Less animal contact exposure
- Ability to install multiple circuits in congested areas
- Narrower right-of-way requirements
- Less need for overhang easements due to the compactness of the installation

The spacer cable spacers support the conductors and maintain the phase balancing (Fig. 13.47). The spacers are installed in 30 ft intervals. The spacers provide strength and flexibility.

The spacers are a gray, track-resistant, high-density polyethylene with long leakage distance (Figs. 13.48 and 13.49). A self-washing shed design is featured.

Low-Voltage Insulators

Porcelain spool insulators are used for secondary-voltage circuits operating at voltages below 600 volts. Open-wire secondaries of bare conductor or conductor covered with weather-proof material are usually supported by porcelain spool insulators mounted on secondary steel racks (Fig. 13.50).

Porcelain spool insulators supported by steel clevises are often used to support neutral conductors for three-phase, four-wire high-voltage distribution circuits or the bare neutral conductor of insulated twisted-multicable secondary circuits. Porcelain insulators reinforced by a one-piece pressed-steel housing with attached galvanized steel screws, often referred to as house knobs, are available for supporting low-voltage circuits.

Strain insulators used in guy wires are described and illustrated in Chap. 12.

Figure 13.47   Mid-span spacer installed in the three-phase spacer cable installation. (*Courtesy Hendrix Wire & Cable Inc.*)

FIGURE 13.48   Spacer cable spacer installed and hanging from the pole mounted bracket. The spacers are a gray, track-resistant, high-density polyethylene with long leakage distance. (*Courtesy Hendrix Wire & Cable Co.*)

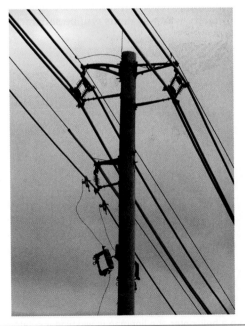

FIGURE 13.49   A fused cutout and cutout bracket installed with spacer cable construction. (*Courtesy Hendrix Wire & Cable Co.*)

FIGURE 13.50 Open-wire bare secondary wires supported by porcelain spool insulators on a galvanized steel rack. Wires are secured to insulators with preformed distribution grips. (*Courtesy Preformed Line Products Co.*)

Insulator Washing

Insulators can be cleaned by hand wiping and by washing. Insulators, hand-wiped or newly installed in areas with an atmosphere that might cause insulator contamination, are usually sprayed with a silicone gel to extend the cleaning cycle and to make it easier to remove contaminants from the insulator's surface. Contaminated insulators may cause excessive leakage current to flow, causing pole fires or insulator flashover and interruptions to electric service.

Insulators can be washed safely, while energized, if proper precautions are taken (Fig. 13.51). The water must have a minimum resistance of 1,000 ohms/cm³. Normally water obtained from city water systems will be satisfactory. The hose nozzle must be

FIGURE 13.51 From a helicopter, the lineman is washing insulators on energized transmission-line structure to remove contamination.

Voltage (Phase-to-Phase), kV	Length of Stream Distance, ft	
	15/64-inch or ¼-inch Nozzle	5/16-inch Nozzle
4–12	7	10
13–23	10	13
24–70	12	15
71–115	15	18
230	15	20
500	20	20

Source: Pacific Gas and Electric Company.

Table **13.2**   Standard Minimum Length of Stream Distances When "Hot Washing" with a Minimum Dynamic Pressure of 550 psi at the Nozzle for Various Voltages

grounded, normally to the tower, static wire, or neutral conductor. The water pressure must be maintained high enough to prevent a solid stream of water flowing from the hose nozzle. The hose nozzle must be kept at a safe distance from the energized conductors (Table 13.2).

The washing of energized insulators should start on the bottom and work up to prevent dirty water from falling on contaminated insulators. Overspray should be avoided to prevent inadequate water volume from contacting insulators and causing flashover.

Figure **13.52**   Cycloaliphatic epoxy-resin insulators are used for the applications illustrated. (*Courtesy S&C Electric Co.*)

Insulators Used in Pad-Mounted Switchgear

Pad-mounted switchgear used with underground circuits requires the use of space-saving insulators to keep the equipment compact.

Cycloaliphatic epoxy-resin insulators, sold under the trade name Cypoxy, are specially formulated to provide the electrical and mechanical strength characteristics detailed in industry standard for porcelain insulators.

Cypoxy insulators and insulating components have given trouble-free service in a wide range of environments including pad-mounted switchgear. The ability to mold Cypoxy into intricate shapes with light weight and high strength has made it possible to develop space-saving switch designs not possible with porcelain insulation. These insulators are nontracking, self-scouring, and nonweathering. Various ways in which this insulating material can be used are illustrated in Fig. 13.52.

CHAPTER 14
Line Conductors

A conductor is a material that readily permits the flow of an electric current. The wires and cables over which electric energy is transmitted are made of copper, aluminum, steel, or a combination of copper strands supported with steel strands or aluminum strands supported with steel strands. Most metals can be used for conductors in an electric circuit. Metallic elements, other than those mentioned, will conduct electricity; however, they are not generally used to make wires and cables because of economic or physical reasons.

Copper Conductors

Copper is a commonly used line conductor. It conducts electric current very readily, ranking next to silver. It is very plentiful in nature, and therefore its cost is comparatively low. It can be easily spliced.

Three kinds of copper wire are in use: hard-drawn copper, medium-hard-drawn copper, and annealed copper, also called *soft-drawn*. Copper wire is hard-drawn as it comes from the drawing die. To obtain soft or annealed copper wire, the hard-drawn wire is heated to a red heat to soften it.

For overhead line purposes, hard-drawn copper wire is preferable on account of its greater strength. Annealing, or softening, reduces the tensile strength of the wire from about 55,000 to 35,000 lb/inch^2. Because of this, it is not good practice to have any soldered splices when using hard-drawn wire, as the soldering anneals the wire near the joint, thereby reducing its strength. Joints in hard-drawn wire should, therefore, be made with splicing sleeves. Annealed or soft-drawn copper wire is used for ground wires and special applications where it is necessary to bend and shape the conductor. Medium-hard-drawn copper is used for distribution especially for wire sizes smaller than American Wire Gauge (AWG) No. 2. Soft copper wire has a yield point elongation less than one-half that of medium-hard copper and hence stretches permanently with a correspondingly lighter loading of ice and wind.

Aluminum Conductors

Aluminum is widely used for distribution and transmission-line conductors. Its conductivity, however, is only about two-thirds that of copper. Compared with a copper wire of the same physical size, aluminum wire has 60% of the conductivity, 45% of the tensile strength, and 33% of the weight. The aluminum wire must be $100/60 = 1.60$ times as large as the copper wire in cross section to have the same conductivity. An aluminum wire of this size will have 75% of the tensile strength and 55% of the weight of the equivalent copper conductor.

FIGURE 14.1   Relative cross sections of copper, aluminum, and ACSR conductors having equal current-carrying capacities.

←.580"→   ← .742" →   ← .826" →

Copper   Aluminum   ACSR

Figure 14.1 shows cross sections of copper, aluminum, and aluminum-conductor steel-reinforced (ACSR) conductors, which have equal conductivity or, what amounts to the same thing, equal current-carrying capacities.

The tensile strength of aluminum is only about one-half that of hard-drawn copper, namely, 27,000 lb/inch². But since the cross section of aluminum wire must be about twice that of copper to have the same conductivity, the actual breaking strength of the aluminum conductor is about the same as that of copper. The weight per foot of aluminum is still only 55% as great, as pointed out above, despite being twice as large in cross section. This fact makes aluminum conductors preferred in many cases, because the lighter weight permits longer spans and, therefore, fewer towers and insulators. This is a decided advantage where long spans are common. The larger size helps to keep corona loss down.

When an aluminum conductor is stranded, the central strand is often made of steel, which serves to reinforce the cable. Such reinforcement gives great strength for the weight of conductor. ACSR cable is especially suited for long spans.

Steel Conductors

Steel wire is used to a limited extent where minimum construction expense is desired. Steel wire, because of its high tensile strength of 160,000 lb/inch², permits relatively long spans, therefore requires few supports. Bare steel wire, however, rusts rapidly and is therefore very short-lived. Steel is also a poor conductor compared with copper, being only about 10% to 15% as good. Galvanized steel conductors are commonly used for shield or static wires in transmission and subtransmission lines. Most guy wires are galvanized steel-stranded cables.

Copperweld Steel Conductors

The disadvantages of short life and low conductivity of steel led to the development of the copperweld steel conductor. In this conductor, a coating of copper is securely welded to the outside of the steel wire. The copper acts as a protective coating to the steel wire, thus giving the conductor the same life as if it were made of solid copper. At the same time, the layer of copper greatly increases the conductivity of the steel conductor, while the steel gives it great strength. This combination produces a very satisfactory yet inexpensive line conductor. Its chief field of application is for overhead primary conductors in rural lines, for guy wires, and for overhead ground wires. The conductivity of copperweld conductors can be raised to any desired percentage, depending on the thickness of the copper layer. The usual values of conductivity of wires as manufactured are 30% and 40%. Figure 14.2 shows a stranded copperweld conductor.

Alumoweld Steel Conductors

Steel wire can be covered with aluminum to prevent the steel from rusting as well as to improve its conductivity. Figure 14.3 shows such an aluminum-covered stranded steel

FIGURE 14.2   Stranded copperweld conductor. (*Courtesy Copperweld Steel Co.*)

cable used for guys. By its use the guys can be expected to last as long as modern treated wooden poles.

Figure 14.4 shows an ACSR conductor in which one conductor is a steel strand covered with a welded layer of aluminum and the other conductors are aluminum strands. The layer of aluminum increased the conductivity of the steel strand in addition to preventing it from rusting. All the strands of the cable, the alumoweld strand, and the aluminum strands thus have the same life.

FIGURE 14.3   Alumoweld guy strand. Each strand is covered with a welded layer of aluminum to protect the steel core from rusting. (*Courtesy Copperweld Steel Co.*)

FIGURE 14.4   ACSR cable with steel strand covered with layer of welded aluminum. The layer of aluminum on the steel strand prevents electrolytic corrosion between strands, prevents rusting of the steel strand, and increases the conductivity of the steel strand by a substantial amount. (*Courtesy Southwire Co.*)

Conductor Classes

Conductors are classified as solid or stranded. A solid conductor, as the name implies, is a single conductor of solid circular section. A stranded conductor is composed of a group of wires made into a single conductor. A stranded conductor is used where the solid conductor is too large and not flexible enough to be handled readily. Large solid conductors are easily damaged by bending. The AWG size No. 0 wire is the approximate dividing line between solid and stranded conductors. The strands in the stranded conductor are usually arranged in concentric layers about a central core (Figs. 14.5 and 14.6). The smallest number of wires in a stranded conductor is 3. The next number of strands is 7, then 19, 37, 61, 91, 126, etc.

Conductor Covering

Conductors on overhead transmission lines are bare conductors; that is, they are not covered. Conductors on overhead distribution circuits, below 5 kV, are usually covered. The common covering is crosslinked polyethylene. Figure 14.7 shows duplex, triplex, and quadruplex conductors used for streetlights and secondary services.

Wire Sizes

Wire sizes are ordinarily expressed by numbers. There are, however, several different numbering methods, so that in specifying a wire size by number it is also necessary to state which wire gauge or numbering method is used. The most used wire gauge in the United States is the *Brown and Sharpe* gauge, also called *American wire gauge* (AWG).

American Wire Gauge

The American standard wire gauge is shown in Fig. 14.8. This cut is full size, so that the widths of the openings on the rim of the gauge correspond to the diameters of the wires whose numbers stand opposite the openings. The actual size of these wire numbers is perhaps better illustrated by Table 14.1. This table gives, for wire sizes from No. 0000 to No. 8, the diameter of the wire in inches and a full-size end and side view of the wire

6/1   7/1   12/7   36/1   24/7   26/7

30/7   30/19   54/7   54/19

FIGURE **14.5**   Sketches showing arrangement of ACSR wire stand conductors. The wires are arranged in concentric layers about a central core. (*Courtesy Southwire Co.*)

FIGURE **14.6**   Sketches showing arrangement of ACAR wire strand conductors. (*Courtesy Southwire Co.*)

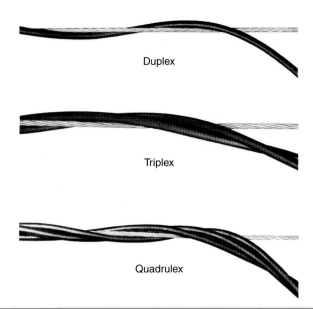

FIGURE 14.7   Secondary service cables. The duplex cable typically used for streetlights, triplex cable for residential and commercial customer single phase service, and quadruplex cable to serve commercial and industrial three phase service. (*Courtesy Kaiser Aluminum.*)

FIGURE 14.8   American standard wire gauge. Cut is full size. Widths of openings on rim correspond to diameters of wires whose numbers are opposite the openings. (*Courtesy L. S. Starrett Co.*)

Gauge Number	Diameter, inch	Full-Size End View	Full-Size Side View
8	0.1285		
7	0.1443		
6	0.1620		
5	0.1819		
4	0.2043		
3	0.2294		
2	0.2576		
1	0.2893		
0	0.3249		
00	0.3648		
000	0.4096		
0000	0.460		

TABLE 14.1   American Wire Sizes (Bare Conductor)

corresponding to each number of the gauge. This table should make it easy to get an idea of the actual physical size of some of the wires commonly used in distribution work. Note that the greater the number of the wire, the smaller the wire is. That came about by numbering the wire by the number of steps required in the wire-drawing process. The smaller the wire, the greater the number of steps required to draw it.

Physical descriptive details for solid copper wire are itemized in Table 14.2. Note in Table 14.2 that with every increase of 3 in the gauge number, the cross section and weight are halved and the resistance is doubled. An increase of 10 in the gauge number increases the resistance 10 times and cuts the weight and cross section to one-tenth. Thus, for example, the No. 5 wire is three sizes smaller than the No. 2 wire. Its cross section is one-half that of the No. 2, for 33,090 circular mils is approximately one-half of 66,360 circular mils, and its resistance is twice that of the No. 2, for 0.3260 Ω is twice 0.1625 Ω ('Ω' for ohms). Furthermore, the resistance of the No. 12 is 10 times that of the No. 2 for 1.65 Ω is nearly 10 times 0.1625 Ω. This general relation between gauge numbers is good to remember. Physical properties of solid round 1350 aluminum wire are listed in Table 14.3.

Size AWG	Area			Weight	Hard		Medium		Soft-Drawn Annealed	
	Diameter inch	Cir Mils	inch²	lb per 1,000 ft	Breaking Strength, Minimum, lb	DC Resistance at 20° (68°F) Maximum,* Ohms per 1,000 ft	Breaking Strength, Minimum, lb	DC Resistance at 20°C (68°F) Maximum,* Ohms per 1,000 ft	Breaking Strength, Maximum,† lb	DC Resistance at 20°C (68°F) Maximum,* Ohms per 1,000 ft
4/0	0.4600	211,600	0.1662	640.5	8,143	0.05045	6,980	0.05019	5,983	0.04901
3/0	0.4096	167,800	0.1318	507.8	6,720	0.06362	5,666	0.06330	4,744	0.06182
2/0	0.3648	133,100	0.1045	402.8	5,519	0.08021	4,599	0.07980	3,763	0.07793
1/0	0.3249	105,600	0.08291	319.5	4,518	0.1022	3,731	0.1016	2,985	0.09825
1	0.2893	83,690	0.06573	253.3	3,688	0.1289	3,024	0.1282	2,432	0.1239
2	0.2576	66,360	0.05212	200.9	3,002	0.1625	2,450	0.1617	1,928	0.1563
3	0.2294	52,620	0.04133	159.3	2,439	0.2050	1,984	0.2039	1,529	0.1971
4	0.2043	41,740	0.03278	126.3	1,970	0.2584	1,584	0.2571	1,213	0.2485
5	0.1819	33,090	0.02599	100.2	1,590	0.3260	1,265	0.3243	961.5	0.3135
6	0.1620	26,240	0.02061	79.44	1,280	0.4110	1,010	0.4088	762.6	0.3952
7	0.1443	20,820	0.01635	63.03	1,030	0.5180	806.7	0.5153	605.1	0.4981
8	0.1285	16,510	0.01297	49.98	826.1	0.6532	644.0	0.6498	479.8	0.6281
9	0.1144	13,090	0.01028	39.61	660.9	0.8241	513.9	0.8199	380.3	0.7925
10	0.1019	10,380	0.008155	31.43	529.3	1.039	410.5	1.033	314.0	0.9988
11	0.0907	8,230	0.00646	24.9	423	1.31	327	1.30	249	1.26
12	0.0808	6,530	0.00513	19.8	337	1.65	262	1.64	197	1.59
13	0.0720	5,180	0.00407	15.7	268	2.08	209	2.07	157	2.00
14	0.0641	4,110	0.00323	12.4	214	2.63	167	2.61	124	2.52
15	0.0571	3,260	0.00256	9.87	170	3.31	133	3.29	98.6	3.18
16	0.0508	2,580	0.00203	7.81	135	4.18	106	4.16	78.0	4.02

*Based on nominal diameter and ASTM resistivities.

†No requirements for tensile strength are specified in ASTM B3. Values given here are based on Anaconda data.

Source: Southwire Overhead Conductor Manual.

TABLE 14.2 Solid Copper Wire—Weight, Breaking Strength, DC Resistance (Based on ASTM Specifications B1, B2, and B3)

Solid Round 1350 Aluminum Wires					
Wire Size, AWG	Wire Diameter, inches	Area		DC Resistance at 20°C (68°F),* Ohms per mile	Weight, lb per 1,000 ft
		Circular Mils	inch²		
4/0	0.4600	211,600	0.1662	0.4228	194.4
3/0	0.4096	167,800	0.1318	0.5333	154.2
2/0	0.3648	133,100	0.1045	0.6723	122.3
1/0	0.3249	105,600	0.0829	0.8476	97.00
1	0.2893	83,690	0.0657	1.069	76.91
2	0.2576	66,360	0.0521	1.348	60.98
3	0.2294	52,620	0.0413	1.700	48.36
4	0.2043	41,740	0.0328	2.144	38.35
5	0.1819	33,090	0.0260	2.704	30.40
6	0.1620	27,220	0.0206	3.409	24.12
7	0.1443	20,820	0.0164	4.297	19.13
8	0.1285	16,510	0.0130	5.419	15.17
9	0.1144	13,090	0.0103	6.837	12.03
10	0.1019	10,380	0.00816	8.617	9.542
11	0.0907	8,230	0.00646	10.88	7.559
12	0.0808	6,530	0.00513	13.70	5.999
13	0.0720	5,180	0.00407	17.26	4.764
14	0.0641	4,110	0.00323	21.78	3.776
15	0.0571	3,260	0.00256	27.44	2.996
16	0.0508	2,580	0.00203	34.67	2.371
17	0.0453	2,050	0.00161	43.60	1.886
18	0.0403	1,620	0.00128	55.09	1.492
19	0.0359	1,290	0.00101	69.42	1.184
20	0.0320	1,020	0.000804	87.38	0.941
21	0.0285	812	0.000638	110.2	0.746
22	0.0253	640	0.000503	139.8	0.588
23	0.0226	511	0.000401	175.2	0.469
24	0.0201	404	0.000317	221.5	0.371
25	0.0179	320	0.000252	279.3	0.294
26	0.0159	253	0.000199	353.9	0.232
27	0.0142	202	0.000158	443.7	0.185
28	0.0126	159	0.000125	563.6	0.146
29	0.0113	128	0.000100	700.7	0.117

*Electric resistance is based on the electric conductivity = 61.2% IACS at 20°C.
Source: Southwire Overhead Conductor Manual.

TABLE 14.3   Solid Round 1350 Aluminum Wire—Dimensions, Weight, DC Resistance

FIGURE 14.9 Area of circle whose diameter is 1/1,000 inch is 1 circular mil.

Circular Mil

The circular mil is the unit of cross-sectional area customarily used in designating the area of wires. A *circular mil* is the area contained in a circle having a diameter of 1/1,000 inch (Fig. 14.9). A mil is a thousandth of an inch. The use of the circular mil as the unit of area came about because of the custom of using a square as the unit of area in measuring the areas of squares or rectangles. Naturally, when the areas of circles had to be measured, a circle having unit diameter suggested itself as the natural unit of measurement.

The circular mils of cross section in a wire of any diameter are obtained by multiplying the diameter in thousandths of an inch by itself. Thus, the number of circular mils of cross section in the No. 10 wire is $102 \times 102 = 10{,}404$. The circular-mil cross section of the AWG No. 2 wire is $257.6 \times 257.6 = 66{,}370$.

If one can remember the relations pointed out above and the data for the AWG No. 10 copper wire, one can build up the remainder of the table any time. The approximate data for the AWG No. 10 copper wire are easily remembered because its resistance per 1,000 ft is 1 ohm (Ω), its diameter is 1/10 inch, and its circular-mil cross section is 10,000. Its weight per 1,000 ft is 31.4 lb. With these figures as a basis, one can calculate the data for the AWG Nos. 7, 4, 1, 13, 16, 19, etc. wire numbers by using the factors 2 and ½, remembering that for every increase in wire number of three sizes, the cross section and weight are halved and the resistance is doubled. In other words, the larger the number of the wire, the smaller the size that wire will be.

To obtain the data for wires one number larger, the factors are 1.25 and 0.80, and for two numbers larger they are 1.60 and 0.625. Thus, the data for the AWG No. 11 wire are obtained by multiplying the resistance of the AWG No. 10 wire by 1.25, that is, 1.0×1.25 equals 1.25, and dividing the cross-sectional area of 10,000 by 1.25, which equals 8,000. The results obtained in this way are, of course, approximate, but they are close enough to give one a fair idea of the wire.

Stranded Conductor Data

Physical descriptive data for bare copper cable is itemized in Table 14.4. Table 14.5 provides physical characteristics of copperweld/copper composite bare stranded conductors.

Table 14.6 lists physical properties of all-aluminum conductors (AAC). ACSR cable physical and electrical descriptive data are itemized in Table 14.7.

Conductor Selection

The size of a line conductor depends on a number of factors. The most important are the line voltage, the amount of power to be transmitted, and mechanical strength.

Conductor Size (AWG or kcmil)	No. of Strands	Stranding Class	Strand Type	Temper	Nom. OD (mils)	Nominal Total Weight (lb/1,000 ft)	Allowable Ampacity	Rated Strength (lb)	DC Resistance at 20°C (ohms/1,000 ft)
14 AWG	1	A	SOL	SD	64	12		124.2	2.525
12 AWG	1	A	SOL	SD	80	20		197.5	1.588
10 AWG	1	A	SOL	SD	101	31		314	0.999
8 AWG	1	A	SOL	SD	128	50	95	479.8	0.628
8 AWG	7	B	CONV	SD	148	51	95	499	0.6408
6 AWG	1	A	SOL	SD	161	79	125	762.9	0.395
6 AWG	7	B	CONV	SD	183	81	130	794	0.403
4 AWG	1	A	SOL	SD	204	126	170	1,213	0.249
4 AWG	7	B	CONV	SD	231	129	170	1,320	0.2534
3 AWG	1	A	SOL	SD	229	159	195	1,530	0.197
3 AWG	7	B	CONV	SD	260	162	200	1,670	0.201
2 AWG	1	A	SOL	SD	257	201	225	1,929	0.156
2 AWG	7	B	CONV	SD	292	205	230	2,110	0.1578
1/0 AWG	7	A	CONV	SD	368	326	310	3,221	0.1002
1/0 AWG	19	B	COMBUL	SD	360	338	355	4,024	0.07949
2/0 AWG	7	A	CONV	SD	413	411	355	4,062	0.07949
2/0 AWG	19	B	COMBUL	SD	404	411	355	4,024	0.07949
3/0 AWG	7	A	CONV	SD	473	518	410	5,118	0.06304
3/0 AWG	19	B	COMBUL	SD	454	520	410		0.06431
4/0 AWG	7	A	CONV	SD	521	653	480	6,459	0.04999
4/0 AWG	19	B	COMBUL	SD	510	653	480	6,453	0.04999
250 KCMIL	19	A	CONV	SD	573	772	530	7,627	0.04231

Size									
250 KCMIL	37	B	CONV	SD	575	772	530	7,940	0.04231
300 KCMIL	37	B	CONV	SD	630	926	590	9,160	0.03526
350 KCMIL	19	A	CONV	SD	678	1,081	650	10,680	0.03022
500 KCMIL	37	B	CONV	SD	813	1,544	810	15,240	0.02116
750 KCMIL	61	A	CONV	SD	998	2,316	1,040	22,890	0.0141
1000 KCMIL	61	B	CONV	SD	1,152	3,087	1,240	30,500	0.1058

Specifications:

- ASTM B1 Hard-Drawn Copper Wire
- ASTM B2 Medium Hard Copper Wire
- ASTM B3 Soft or Annealed Copper Wire
- ASTM B8 Concentric-Lay-Stranded Hard, Medium-Hard, or Soft Copper Conductors
- ASTM B33 Tinned Conductors
- ASTM B787 19 Wire Combination Unilay-Stranded Soft Copper Wire

All dimensions are nominal and subject to normal manufacturing tolerances.

*Ampacity based on 75°C conductor temperature; 2 ft/sec. wind and sun.

Source: Southwire Overhead Conductor Manual.

TABLE 14.4   Bare Copper Cable, Classes AA, A, B—Diameter, Weight, Allowable Ampacity, Breaking Strength, DC Resistance

Size of Conductor	Type of Conductor	Diameter (inch)	Stranding		Copper Eqvlt Ohms/ft @68°F	Minimum Breaking Load, lb	Weight, lb per 1,000 ft	Cross Section Area (in²)
			No. and Diam. of Wires, EHS 30% Copperweld Wires	No. and Diam. of Hard-Drawn Copper Wires				
350,000	E	0.788	7 × 0.1576	12 × 0.1576	0.0302	31,474	1,403	0.3706
350,000	EK	0.735	4 × 0.1470	15 × 0.1470	0.0302	23,141	1,238	0.3225
300,000	E	0.729	7 × 0.1459	12 × 0.1459	0.0353	26,974	1,203	0.3177
300,000	EK	0.680	4 × 0.1361	15 × 0.1361	0.0353	20,338	1,061	0.2764
250,000	E	0.666	7 × 0.1332	12 × 0.1332	0.0423	23,226	1,002	0.2648
250,000	EK	0.621	4 × 0.1242	15 × 0.1242	0.0423	17,302	884.2	0.2302
4/0	E	0.613	7 × 0.1225	12 × 0.1225	0.0500	20,114	848.3	0.2239
4/0	EK	0.571	4 × 0.1143	15 × 0.1143	0.0500	14,881	748.4	0.1950
4/0	F	0.550	1 × 0.1833	6 × 0.1833	0.0500	11,593	710.2	0.1847
2/0	K	0.534	4 × 0.1780	3 × 0.1780	0.0795	17,071	645.9	0.1742
2/0	J	0.494	3 × 0.1648	4 × 0.1648	0.0795	13,420	560.6	0.1493
2/0	F	0.436	1 × 0.1454	6 × 0.1454	0.0795	7,686	446.8	0.1162
1/0	K	0.475	4 × 0.1585	3 × 0.1585	0.100	14,064	512.0	0.1381
1/0	J	0.440	3 × 0.1467	4 × 0.1467	0.100	10,640	444.3	0.1183
1/0	F	0.388	1 × 0.1294	6 × 0.1294	0.100	6,252	354.1	0.0921
No. 1	K	0.423	4 × 0.1412	3 × 0.1412	0.127	11,554	406.2	0.1096
No. 1	J	0.392	3 × 0.1307	4 × 0.1307	0.127	8,735	352.5	0.0939
No. 1	F	0.346	1 × 0.1153	6 × 0.1153	0.127	5,067	280.9	0.0731
No. 2	K	0.377	4 × 0.1257	3 × 0.1257	0.159	9,439	322.1	0.0869
No. 2	J	0.349	3 × 0.1164	4 × 0.1164	0.159	7,107	279.5	0.0745
No. 2	A	0.366	1 × 0.1699	2 × 0.1699	0.159	5,703	256.8	0.0680

No. 2	F	0.308	1 × 0.1026	6 × 0.1026	0.159	4,070	222.8	0.0579
No. 4	D	0.348	2 × 0.1615	1 × 0.1615	0.253	7,120	225.5	0.0615
No. 4	A	0.290	1 × 0.1347	2 × 0.1347	0.253	3,809	161.5	0.0428
No. 6	D	0.276	2 × 0.1281	1 × 0.1281	0.403	4,796	141.8	0.0387
No. 6	A	0.230	1 × 0.1068	2 × 0.1068	0.403	2,499	101.6	0.0269
No. 6	C	0.225	1 × 0.1046*	2 × 0.1046	0.403	2,072	97.34	0.0258
No. 8	D	0.219	2 × 0.1016	1 × 0.1016	0.640	3,160	89.24	0.0243
No. 8	A	0.199	1 × 0.1127	2 × 0.07969	0.640	2,211	74.28	0.0200
No. 8	C	0.179	1 × 0.08081*	2 × 0.08336	0.640	1,214	60.67	0.0160

*High Strength Copperweld, 40% conductivity.
Source: Copperweld Bimetallics, LLC.

TABLE 14.5 Copperweld/Copper Conductors, Physical Characteristics

Code Word	Conductor Size, AWG or kcmil	No. of Wires	Diameter			Area, Square inch	Weight, lbs/1,000 ft	Rated Strength, lbs	Ampacity @ 75°C Amps
			Class	Individual Wire, inch	Complete Cable, inch				
Peachbell	6	7	A	0.0612	0.184	0.0206	24.6	563	103
Rose	4	7	A	0.0772	0.232	0.0328	39.2	881	138
Iris	2	7	AA, A	0.0974	0.292	0.0522	62.3	1,350	185
Pansy	1	7	AA, A	0.1093	0.328	0.0657	78.5	1,640	214
Poppy	1/0	7	AA, A	0.1228	0.368	0.0829	99.1	1,990	247
Aster	2/0	7	AA, A	0.1379	0.414	0.1045	124.9	2,510	286
Phlox	3/0	7	AA, A	0.1548	0.464	0.1317	157.5	3,040	331
Oxlip	4/0	7	AA, A	0.1739	0.522	0.1663	198.7	3,830	383
Sneezewort	250	7	AA	0.1890	0.567	0.1964	234.7	4,520	425
Valerian	250	19	A	0.1147	0.574	0.1963	234.6	4,660	426
Daisy	266.8	7	AA	0.1953	0.586	0.2097	250.6	4,830	443
Laurel	266.8	19	A	0.1185	0.593	0.2095	250.4	4,970	444
Peony	300	19	A	0.1257	0.629	0.2358	281.8	5,480	478
Tulip	336.4	19	A	0.1331	0.666	0.2644	316.0	6,150	513
Daffodil	350	19	A	0.1357	0.679	0.2748	328.4	6,390	526
Canna	397.5	19	AA, A	0.1447	0.724	0.3124	373.4	7,110	570
Goldentuft	450	19	AA	0.1539	0.770	0.3534	422.4	7,890	616
Cosmos	477	19	AA	0.1584	0.793	0.3744	447.5	8,360	639
Syringa	477	37	A	0.1135	0.795	0.3743	447.4	8,690	639
Zinnia	500	19	AA	0.1622	0.811	0.3926	469.2	8,760	658
Hyacinth	500	37	A	0.1162	0.813	0.3924	469.0	9,110	658
Dahlia	556.5	19	AA	0.1711	0.856	0.4369	522.1	9,750	703
Mistletoe	556.5	37	AA, A	0.1226	0.858	0.4368	522.0	9,940	704
Meadowsweet	600	37	AA, A	0.1273	0.891	0.4709	562.8	10,700	738
Orchid	636	37	AA, A	0.1311	0.918	0.4995	596.1	11,400	765
Heuchera	650	37	AA	0.1325	0.928	0.5102	609.8	11,600	775
Verbena	700	37	AA	0.1375	0.963	0.5494	656.6	12,500	812
Flag	700	61	A	0.1071	0.964	0.5495	656.8	12,900	812
Violet	715.5	37	AA	0.1391	0.974	0.5622	672.0	12,800	823

Code Word	Size (kcmil)	Strands	Alloy	Individual Wire Diameter (in.)	Cable Diameter (in.)	Weight (lb/ft)	Rated Strength (lb)	Ampacity	Ampacity 75°C
Nasturtium	715.5	61	A	0.1083	0.975	0.5619	671.6	13,100	823
Petunia	750	37	AA	0.1424	0.997	0.5892	704.3	13,100	847
Cattail	750	61	A	0.1109	0.998	0.5892	704.2	13,500	847
Arbutus	795	37	AA	0.1466	1.026	0.6245	746.4	13,900	878
Lilac	795	61	A	0.1142	1.028	0.6248	746.7	14,300	879
Cockscomb	900	37	AA	0.1560	1.092	0.7072	845.2	15,400	931
Snapdragon	900	61	A	0.1215	1.094	0.7072	845.3	15,900	931
Magnolia	954	37	AA	0.1606	1.124	0.7495	895.8	16,400	948
Goldenrod	954	61	A	0.1251	1.126	0.7498	896.1	16,900	948
Hawkweed	1,000	37	AA	0.1644	1.151	0.7854	938.7	17,200	982
Camellia	1,000	61	A	0.1280	1.152	0.7849	938.2	17,700	983
Bluebell	1,033.5	37	AA	0.1672	1.170	0.8124	970.9	17,700	1,010
Larkspur	1,033.5	61	A	0.1302	1.172	0.8122	970.6	18,300	1,011
Marigold	1,113	61	AA, A	0.1351	1.216	0.8744	1,045	19,700	1,031
Hawthorn	1,192.5	61	AA, A	0.1398	1.258	0.9363	1,119	21,100	1,032
Narcissuus	1,272	61	AA, A	0.1444	1.300	0.999	1,194	22,000	1,079
Columbine	1,351.5	61	AA, A	0.1489	1.340	1.062	1,269	23,400	1,124
Carnation	1,431	61	AA, A	0.1532	1.379	1.124	1,344	24,300	1,169
Gladiolus	1,510.5	61	AA, A	0.1574	1.417	1.187	1,419	25,600	1,212
Coreopsis	1,590	61	AA	0.1615	1.454	1.250	1,489	27,000	1,333
Jessamine	1,750	61	AA	0.1694	1.525	1.375	1,641	29,700	1,408
Cowslip	2,000	91	A	0.1482	1.630	1.570	1,873	34,200	1,518
Sagebrush	2,250	91	A	0.1572	1.729	1.766	2,128	37,700	1,612
Lupine	2,500	91	A	0.1657	1.823	1.962	2,365	41,800	1,706
Bitter root	2,750	91	A	0.1738	1.912	2.159	2,602	46,100	1,793
Trillium	3,000	127	A	0.1537	1.996	2.356	2,840	50,300	1,874
Bluebonnet	3,500	127	A	0.1660	2.158	2.749	3,345	58,700	2,024

Notes:

(1) Data based on a nominal cable manufactured in accordance with ASTM B 399.

(2) Ampacity based on an aluminum alloy (AAAC/6201) conductivity of 52.5% IACS at 20°C.

(3) Ampacity based on a 75°C conductor temperature, 25°C ambient temperature, 2 ft/sec wind, in sun, with an emissivity of 0.5 and a coefficient of solar absorption of 0.5, at sea level.

Source: Southwire Overhead Conductor Manual.

TABLE 14.6   All-Aluminum Conductor (AAC)-Diameter, Weight, Rated Strength, Ampacity

Single-Layer Aluminum Conductor, Steel Reinforced (ACSR)											
			Cross Sectional Area, Sq in		Diameter					Weight/1,000 ft		
					Individual Wires		Steel Core, in	Complete Cable, in				
Code Word	Conductor Size, AWG or kcmil	Stranding (Al/St)	Aluminum	Total	Aluminum, in	Steel, in			Alum., lbs	Steel, lbs	Total, lbs
Turkey	6	6/1	0.0206	0.0240	0.0661	0.0661	0.0661	0.198	24.5	11.6	36.1
Swan	4	6/1	0.0328	0.0383	0.0834	0.0834	0.0834	0.250	38.9	18.4	57.3
Swanate	4	7/1	0.0328	0.0411	0.0772	0.1029	0.1029	0.257	38.9	28.0	66.9
Sparrow	2	6/1	0.0522	0.0609	0.1052	0.1052	0.1052	0.316	61.9	29.3	91.2
Sparate	2	7/1	0.0522	0.0655	0.0974	0.1299	0.1299	0.325	61.9	44.7	106.6
Robin	1	6/1	0.0657	0.0767	0.1181	0.1182	0.1182	0.355	78.1	37.0	115.1
Raven	1/0	6/1	0.0830	0.0968	0.1327	0.1327	0.1327	0.398	98.5	46.6	145.1
Quail	2/0	6/1	0.1045	0.1219	0.1489	0.1489	0.1489	0.447	124.1	58.7	182.8
Pigeon	3/0	6/1	0.1317	0.1537	0.1672	0.1672	0.1672	0.502	156.4	74.0	230.4
Penguin	4/0	6/1	0.1662	0.1939	0.1878	0.1878	0.1878	0.563	197.4	93.4	290.8
High Mechanical Strength											
Grouse	80	8/1	0.0628	0.0847	0.1000	0.1670	0.1670	0.367	75.0	73.9	148.9
Petrel	101.8	12/7	0.0799	0.1265	0.0921	0.0921	0.2763	0.461	95.9	157.9	253.8
Minorca	110.8	12/7	0.0870	0.1378	0.0961	0.0961	0.2883	0.481	104.4	171.9	276.3
Leghorn	134.6	12/7	0.1057	0.1674	0.1059	0.1059	0.3177	0.530	126.8	208.7	335.5
Guinea	159.0	12/7	0.1249	0.1977	0.1151	0.1151	0.3453	0.576	149.7	246.6	396.3
Dotterel	176.9	12/7	0.1389	0.2199	0.1214	0.1214	0.3642	0.607	166.6	274.3	440.9
Dorking	190.8	12/7	0.1499	0.2373	0.1261	0.1261	0.3783	0.631	179.7	296.0	475.7
Brahma	203.2	16/19	0.1596	0.3020	0.1127	0.0977	0.4885	0.714	191.4	483.2	674.6
Cochin	211.3	12/7	0.1660	0.2628	0.1327	0.1327	0.3981	0.664	199.0	327.8	526.8

Notes:

1. Data based on a nominal cable manufactured in accordance with ASTM B 232.

2. Ampacity based on an aluminum conductivity of 61.2% IACS at 20°C and a steel conductivity of 8% IACS at 20°C.

3. Ampacity based on a 75°C conductor temperature, 25°C ambient temperature, 2 ft/sec wind, in sun, with an emissivity of 0.5 and a coefficient of solar absorption of 0.5 at sea level.

4. Rated strengths based on Class A galvanized steel core wire in accordance with ASTM B 498.

Source: Southwire Overhead Conductor Manual.

TABLE 14.7   ACSR Conductor—Physical and Electrical Characteristics

Single-Layer Aluminum Conductor, Steel Reinforced (ACSR)											
Metal Content, % Weight			Approximate Resistance				Reactance @ 1 ft Spacing, 60 Hz				Ampacity @ 75°C, Amps
			dc @ 20°C, Ω/mi	ac-60 Hz			Approximate Inductive, Xa			Capacitive, X'a, MΩ-mile	
Alum., %	Steel, %	Rated Str, lbs		25°C, Ω/mile	50°C, Ω/mile	75°C, Ω/mile	25°C, Ω/mile	50°C, Ω/mile	75°C, Ω/mile		
67.9	32.1	1,190	3.386	3.454	3.860	4.257	0.634	0.734	0.760	0.1423	105
67.9	32.1	1,860	2.129	2.172	2.459	2.719	0.608	0.694	0.723	0.1354	140
58.1	41.9	2,360	2.105	2.147	2.459	2.743	0.598	0.654	0.688	0.1345	140
67.9	32.1	2,850	1.339	1.366	1.583	1.753	0.580	0.652	0.674	0.1285	184
58.1	41.9	3,640	1.324	1.351	1.594	1.787	0.574	0.621	0.637	0.1276	184
67.9	32.1	3,550	1.062	1.083	1.276	1.414	0.564	0.629	0.646	0.1250	211
67.9	32.1	4,380	0.841	0.859	1.034	1.145	0.549	0.601	0.614	0.1216	242
67.9	32.1	5,310	0.668	0.681	0.843	0.932	0.537	0.590	0.599	0.1182	276
67.9	32.1	6,620	0.530	0.541	0.692	0.762	0.524	0.572	0.578	0.1147	315
67.9	32.1	8,350	0.420	0.429	0.573	0.628	0.509	0.553	0.556	0.1113	357
High Mechanical Strength											
50.4	49.6	5,200	1.090	1.112	1.362	1.553	0.553	0.596	0.607	0.1240	204
37.8	62.2	10,400	0.836	0.853	1.102	1.322	0.538	0.613	0.677	0.1173	237
37.8	62.2	11,300	0.768	0.783	1.028	1.241	0.537	0.621	0.670	0.1160	248
37.8	62.2	13,600	0.632	0.645	0.880	1.076	0.527	0.606	0.648	0.1131	273
37.8	62.2	16,000	0.535	0.546	0.773	0.958	0.517	0.590	0.628	0.1107	297
37.8	62.2	17,300	0.481	0.491	0.714	0.892	0.512	0.582	0.617	0.1091	312
37.8	62.2	18,700	0.446	0.455	0.675	0.847	0.505	0.577	0.607	0.1079	324
28.4	71.6	28,400	0.403	0.411	0.642	0.820	0.493	0.553	0.592	0.1043	341
37.8	62.2	20,700	0.403	0.411	0.627	0.793	0.499	0.567	0.596	0.1064	340

The line voltage is a very important factor, for it will be recalled that for a given amount of power, the greater the voltage, the less the current. That is the reason for using as high a line voltage as possible. The current in turn determines the size of the conductor.

Equally important is the magnitude of the load to be transmitted. For a given voltage, the larger the load, the larger the current. That a conductor must also have sufficient mechanical strength to carry its own weight and any load due to sleet or wind is obvious if the line is to give reliable and uninterrupted service. Other factors are length of line, power factor of the load, length of span, etc. Table 14.8 gives the current-carrying capacities of copper and aluminum wires for the wire sizes used frequently in outdoor circuits. These values of current were obtained by experiment and are of such magnitude that the heat produced by them in the conductor will bring it up to a temperature of 75°C (167°F). The air is assumed to have a temperature of 25°C (77°F), and a steady wind of 1.4 miles/hour (2 ft/sec) is blowing over the conductors. The frequency is 60 Hz. If ability to carry current were the only requirement, one could easily pick from this table the right size of conductor to be used for any given current.

Copper		Aluminum	
Cir Mils or AWG	Amps	Cir Mils or AWG	Amps
		3,364,000	2,150
		3,156,000	1,580
		1,780,000	1,460
1,000,000	1,300	1,590,000	1,380
900,000	1,220	1,431,000	1,300
800,000	1,130	1,272,000	1,210
700,000	1,040	1,113,000	1,110
600,000	940	954,000	1,000
500,000	840	795,000	897
400,000	730	636,000	776
300,000	610	477,000	646
No. 4/0	480	336,400	514
No. 3/0	420	266,800	443
No. 2/0	360	No. 4/0	381
No. 1/0	310	No. 3/0	328
No. 1	270	No. 2/0	283
No. 2	230	No. 1/0	244
No. 3	200	No. 1	209
No. 4	180	No. 2	180

TABLE 14.8   Current-Carrying Capacities of Copper and Aluminum Conductors When Used Outdoors

While the values of current shown for the different conductors could be carried in emergency operation, they are far too high for normal loading. Normal values would be approximately one-half of the values shown. Furthermore, current-carrying capacity is not the only consideration in determining conductor size. Power loss and voltage regulation are equally important considerations.

If no other factor must be considered, a good rule to follow for copper conductors is to allow 1,000 circular mils/amp. For example, a current of 300 amps would thus require a copper conductor having 300,000 circular mils of area.

The conductor, besides requiring adequate current-carrying capacity, must be of sufficient size and strength to support itself and any additional load due to ice, sleet, and wind. The weight of the conductor for a given span can be easily computed, but the magnitude of the additional load due to sleet and wind cannot be accurately determined. For this purpose, the entire area of the United States has been divided in to conductor loading zones based upon historical weather conditions, as shown in Fig. 14.10. Table 14.9 provides the values for ice, wind, and temperature.

In the heavy loading zone, the resultant loading on the conductor is assumed to be that due to weight of the conductor plus the added weight of a layer of ice with a radial thickness of 12.5 mm or 0.5 inch (½ inch), combined with a transverse horizontal wind pressure of 4 lb/ft^2 on the projected area of the ice-covered conductor at −20°C or 0°F.

In the medium loading zone, the loading is assumed to be that caused by 6.5 mm or 0.25 inch (¼ inch) of radial thickness of ice combined with a transverse wind pressure of 4 lb/ft^2 at −10°C or 15°F.

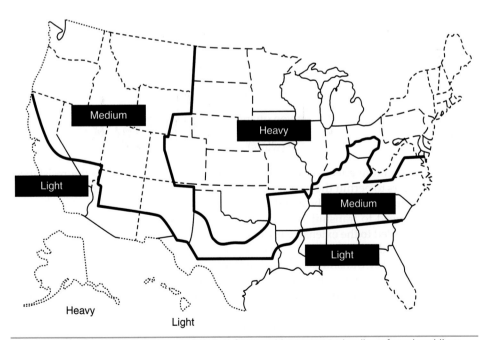

Figure 14.10 General loading map of United States with respect to loading of overhead lines. (*Courtesy NESC.*)

		Loading Zone			
				Warm Islands	
	Heavy	**Medium**	**Light**	**Altitudes Sea Level to 2,743 m (9,000 ft)**	**Altitudes above 2,743 m (9,000 ft)**
Radial thickness of ice (mm)	12.5	6.5	0	0	6.5
Radial thickness of ice (inch)	0.50	0.25	0	0	0.25
Horizontal wind pressure (Pa)	190	190	430	430	190
Horizontal wind pressure (lb/ft²)	4	4	9	9	4
Temperature (°C)	−20	−10	−1	+10	−10
Temperature (°F)	0	+15	+30	+50	+15

Courtesy of National Electrical Safety Code (NESC).

TABLE 14.9   Conductor Loading Conditions

In the light loading zone, the loading assumed is that produced without any ice on the conductor but with a wind pressure of 9 lb/ft² at −1°C or 30°F.

In addition to the ice load, the horizontal pressure due to wind on the conductor and ice must be taken into account, as mentioned above. The forces that cause tension on the conductor, namely, weight of the conductor itself, the weight of the probable ice load, the horizontal force due to probable high winds, the shortening of the conductor, and consequent increase of tension due to extreme cold must be accounted and designed for.

The electrical losses of a transmission line are equal to the current in amperes squared, multiplied by the resistance of the conductors.

$$\text{Losses (watts)} = \text{I}^2(\text{amps}) \times \text{R (ohms)}$$

The current required for a given quantity of power decreases as the voltage increases.

$$\text{P (watts)} = \text{I (amps)} \times \text{V (volts)}$$

It is much more effective to reduce the current than to decrease the resistance of the conductors to permit the transmission of energy economically. If the losses are too great, the conductor will overheat and fail mechanically.

The need to transmit large quantities of electric energy over circuits for long distances has led to the extensive use of extra-high-voltage (EHV) and experimental development of ultra-high-voltage (UHV) transmission lines. The selection of conductors for EHV and UHV transmission lines is influenced by line losses, current-carrying capacity, conductor material costs, and environmental considerations.

Aluminum conductor with steel reinforcement is normally selected for transmission-line conductors because of the good current-carrying capacity and minimum conductor and structure cost. Steel reinforcement of the aluminum conductor does not affect the electrical characteristics of the conductor appreciably. The ACSR conductor will permit higher current densities without loss of the strength since the sag of the conductors is limited by the steel strands as the aluminum expands. Limiting the sag of the aluminum conductor permits the use of longer spans and shorter and fewer supporting structures.

Alternative aluminum conductor types to ACSR for EHV exist and are available to meet industry requirements. All-aluminum conductors (AAC) are used when line designs do not require the strength provided by ACSR. Table 14.10 provides the physical properties of AAC and Table 14.11 contains the electrical properties of AAC.

Aluminum-conductor steel-supported (ACSS) is constructed like ACSR and is the same in appearance except that the aluminum wires are factory-annealed to provide very low yield strength. When tension is applied to ACSS, the aluminum elongates, forcing a steel core to carry the conductor load. ACSS can operate continuously at higher temperatures than ACSR, thus allowing it to carry more current. Physical and electrical properties are shown in Tables 14.12 and 14.13. ACSS will sag significantly less at high temperatures than will other types of conductors when maximum tension under ice and wind loading is the same. ACSS has a high capability for the damping of mechanical oscillations, such as aeolian vibration, which is caused by the winds blowing across the conductors. All-aluminum-alloy conductor (AAAC) has a high strength-to-weight ratio as depicted in Table 14.14. The utilization of the alloy contributes to better sag performance. Aluminum-conductor alloy-reinforced (ACAR) conductor (Tables 14.15 and 14.16) is utilized where current-carrying capacity, high strength, and low conductor weight are desired. Aluminum-conductor composite-reinforced (ACCR) transmission conductor is an ultra-high ampacity, low sag, lightweight conductor.

Aluminum-conductor composite-core (ACCC) cable (Fig. 14.11) has been introduced to the industry as a solution to address transmission constraint issues. The cables utilize composite metallic materials with properties that are of lighter weight but with increased ampacity capabilities. Initial tests have demonstrated that the ACCC conductor operates with cooler temperature performance and lower sag characteristics. As the technology proves itself in trial installations, greater industry acceptance and utilization will occur.

Self-damping (SD) conductor provides a means of controlling aeolian vibration (Fig. 14.12). The conductor is manufactured with trapezoid-shaped aluminum strands surrounding a stranded steel core. The spaces between the conductor layers allow the conductor components to move, generating friction and impact forces which absorb energy and counter aeolian vibration (Fig. 14.13). The use of self-damping aluminum conductor permits higher stringing tensions, longer spans, and/or shorter towers. The aluminum self-damping conductor can be installed in accordance with the procedures for ACSR; however, the control of aeolian vibration permits conductor tensions to be increased to the maximum value specified by the NESC.

Environmental considerations involve visual impact and voltage gradient. Steel reinforcement of the conductor results in shorter structures, or fewer structures, or both, thus improving the appearance of the facilities. The voltage gradient is an electrical characteristic determined by the physical size, arrangement, and surface condition of the conductors. The voltage gradient determines corona loss, radio and TV interference, and audible noise level. The voltage gradient should be minimized. This is accomplished by increasing phase-to-phase spacing and the size and/or number of conductors per phase. The increasing cost of installing generation equipment requires the use of larger conductors to reduce the cost of supplying the losses. This has resulted in the selection of conductors with high ampere capacity, lower voltage gradients, and higher supporting structure costs.

Code Word	Conductor Size: AWG or kCM IL	Cross-Sectional Area, in²	Stranding: Class	Number & Dia. of Strands (Inch)	Nominal Conductor Diam., inch	Rated Strength, lb	Nominal Weight, lb/1,000 ft
PEACHBELL	6	0.0206	A	7 × 0.0612	0.184	563	24.6
ROSE	4	0.0328	A	7 × 0.0772	0.232	881	39.1
IRIS	2	0.0522	AA, A	7 × 0.0974	0.292	1,350	62.2
PANSY	1	0.0657	AA, A	7 × 0.1093	0.328	1,640	78.4
POPPY	1/0	0.0829	AA, A	7 × 0.1228	0.368	1,990	98.9
ASTER	2/0	0.1045	AA, A	7 × 0.1379	0.414	2,510	124.8
PHLOX	3/0	0.1317	AA, A	7 × 0.1548	0.464	3,040	157.2
OXUP	4/0	0.1662	AA, A	7 × 0.1739	0.522	3,830	198.4
SNEEZEWORT	250.0	0.1964	AA	7 × 0.1890	0.567	4,520	234.4
VALERIAN	250.0	0.1963	A	19 × 0.1147	0.574	4,660	234.3
DAISY	266.8	0.2095	AA	7 × 0.1952	0.586	4,830	250.2
LAUREL	266.8	0.2095	A	19 × 0.1185	0.593	4,970	250.1
PEONY	300.0	0.2358	A	19 × 0.1257	0.629	5,480	281.4
TULIP	336.4	0.2644	A	19 × 0.1331	0.666	6,150	315.5
DAFFODIL	350.0	0.2748	A	19 × 0.1357	0.679	6,390	327.9
CANNA	397.5	0.3124	AA, A	19 × 0.1447	0.724	7,110	372.9
GOLDENTUFT	450.0	0.3534	AA	19 × 0.1539	0.769	7,890	421.8
COSMOS	477.0	0.3744	AA	19 × 0.1584	0.792	8,360	448.8
SYRINGA	477.0	0.3744	A	37 × 0.1135	0.795	8,690	446.8
ZINNIA	500.0	0.3926	AA	19 × 0.1622	0.811	8,760	468.5
DAHLIA	556.5	0.4369	AA	19 × 0.1711	0.856	9,750	521.4
MISTLETOE	556.5	0.4368	A	37 × 0.1226	0.858	9,940	521.3
MEADOWSWEET	600.0	0.4709	AA, A	37 × 0.1273	0.891	10,700	562.0
ORCHID	636.0	0.4995	AA, A	37 × 0.1311	0.918	11,400	596.0
HEUCHERA	650.0	0.5102	AA	37 × 0.1325	0.928	11,600	609.8

VERBENA	700.0	0.5494	AA	37 × 0.1375	0.963	12,500	655.7
VIOLET	715.5	0.5623	AA	37 × 0.1391	0.974	12,800	671.0
NASTURTIUM	715.5	0.5619	A	61 × 0.1083	0.975	13,100	671.0
PETUNIA	750.0	0.5893	AA	37 × 0.1424	0.997	13,100	703.2
ARBUTUS	795.0	0.6245	AA	37 × 0.1466	1.026	13,900	745.3
LILAC	795.0	0.6248	A	61 × 0.1142	1.028	14,300	745.7
COCKSCOMB	900.0	0.7072	AA	37 × 0.1560	1.092	15,400	844.0
MAGNOLIA	954.0	0.7495	AA	37 × 0.1606	1.124	16,400	894.5
GOLDENROD	954.0	0.7498	A	61 × 0.1251	1.126	16,900	894.8
HAWKWEED	1,000.0	0.7854	AA	37 × 0.1644	1.151	17,200	937.3
BLUEBELL	1,033.5	0.8114	AA	37 × 0.1671	1.170	17,700	968.4
LARKSPUR	1,033.5	0.8122	A	61 × 0.1302	1.172	18,300	969.2
MARIGOLD	1,113.0	0.8744	AA, A	61 × 0.1351	1.216	19,700	1,044
HAWTHORN	1,192.5	0.9363	AA, A	61 × 0.1398	1.258	21,100	1,117
NARCISSUS	1,272.0	0.9990	AA, A	61 × 0.1444	1.300	22,000	1,192
COLUMBINE	1,351.5	1.061	AA, A	61 × 0.1488	1.340	23,400	1,266
CARNATION	1,431.0	1.124	AA, A	61 × 0.1532	1.379	24,300	1,342
COREOPSIS	1,590.0	1.248	AA	61 × 0.1614	1.454	27,000	1,489
JESSAMINE	1,750.0	1.375	AA	61 × 0.1694	1.525	29,700	1,641
COWSLIP	2,000.0	1.570	A	91 × 0.1482	1.630	34,200	1,873
LUPINE	2,500.0	1.962	A	91 × 0.1657	1.823	41,800	2,365
TRILLIUM	3,000.0	2.356	A	127 × 0.1537	1.996	50,300	2,840
BLUEBONNET	3,500.0	2.749	A	127 × 0.1660	2.158	58,700	3,345

Courtesy of ALCAN CABLE.
ASTM: B230. Specification for Aluminum 1350-H19 Wire for Electrical Purposes.
B231. Specification for Concentric-Lay-Stranded Aluminum 1350 Conductors.

TABLE 14.10 All Aluminum Conductor—Physical Properties

| | Size & Stranding | | Resistance | | | | 60-Hz Reactance 1-ft Equivalent Spacing | | |
| | | | DC 20°C (Ohms/ 1,000 ft) | AC-60-Hz | | | Capacitive (Megohms 1,000 ft) | Inductive (Ohms/ 1,000 ft) | GMR (ft) |
Code Word	AWG or kCMIL	Number or Strands		25°C (Ohms/ 1,000 ft)	50°C(Ohm/ 1,000 ft)	75°C (Ohms/ 1,000 ft)			
PEACHBELL	6	7	0.6593	0.6725	0.7392	0.8059	0.7660	0.1193	0.00555
ROSE	4	7	0.4144	0.4227	0.4645	0.5064	0.7296	0.1140	0.00700
IRIS	2	7	0.2602	0.2655	0.2920	0.3182	0.6929	0.1087	0.00883
PANSY	1	7	0.2066	0.2110	0.2318	0.2527	0.6716	0.1061	0.00991
POPPY	1/0	7	0.1638	0.1671	0.1837	0.2002	0.6550	0.1034	0.0111
ASTER	2/0	7	0.1299	0.1326	0.1456	0.1587	0.6346	0.1008	0.0125
PHLOX	3/0	7	0.1031	0.1053	0.1157	0.1259	0.6188	0.0961	0.0140
OXLIP	4/0	7	0.0817	0.0835	0.0917	0.1000	0.6029	0.0955	0.0158
SNEEZEWORT	250.0	7	0.0691	0.0706	0.0777	0.0847	0.566	0.0934	0.0171
VALERIAN	250.0	19	0.0691	0.0706	0.0777	0.0847	0.566	0.0922	0.0181
DAISY	266.8	7	0.0648	0.0663	0.0727	0.0794	0.581	0.0926	0.0177
LAUREL	266.8	19	0.0648	0.0663	0.0727	0.0794	0.561	0.0915	0.0187
PEONY	300.0	19	0.0575	0.0589	0.0648	0.0705	0.570	0.0902	0.0198
TULIP	336.4	19	0.0513	0.0527	0.0578	0.0629	0.560	0.0888	0.0210
DAFFODIL	350.0	19	0.0494	0.0506	0.0557	0.0606	0.560	0.0883	0.0214
CANNA	397.5	19	0.0435	0.0445	0.0489	0.0534	0.549	0.0869	0.0228
GOLDENTUFT	450.0	19	0.0384	0.0394	0.0434	0.0472	0.539	0.0854	0.0243
COSMOS	477.0	19	0.0363	0.0373	0.0409	0.0445	0.533	0.0848	0.0250
SYRINGA	477.0	37	0.03363	0.0373	0.0400	0.0445	0.533	0.0845	0.0254
ZINNIA	500.0	19	0.0346	0.0356	0.0390	0.0426	0.531	0.0843	0.0256
DAHUA	556.5	19	0.0311	0.0320	0.0352	0.0383	0.522	0.0830	0.0270
MISTLETOE	556.5	37	0.0311	0.0320	0.0352	0.0383	0.522	0.0826	0.0275
MEADOWSWEET	600.0	37	0.0288	0.0297	0.0326	0.0356	0.516	0.0818	0.0285

ORCHID	636.0	37	0.0272	0.0282	0.0300	0.0335	0.511	0.0811	0.0294
HEUCHERA	650.0	37	0.0266	0.0275	0.0301	0.0324	0.510	0.0808	0.0297
VERBENA	700.0	37	0.0247	0.0256	0.0280	0.0305	0.504	0.0799	0.0308
VIOLET	715.5	37	0.0242	0.0252	0.0275	0.0299	0.502	0.0797	0.0312
NASTURTIUM	715.5	61	0.0242	0.0252	0.0275	0.0299	0.502	0.0795	0.0314
PETUNIA	750.0	37	0.0230	0.0251	0.0263	0.0286	0.498	0.0792	0.0319
ARBUTUS	795.0	37	0.0217	0.0227	0.0248	0.0269	0.494	0.0780	0.0328
LILAC	795.0	61	0.0217	0.0227	0.0248	0.0269	0.494	0.0784	0.0331
COCKSCOMB	900.0	37	0.0192	0.0201	0.0220	0.0239	0.484	0.0771	0.0349
MAGNOLIA	954.0	37	0.0181	0.0191	0.0208	0.0227	0.479	0.0763	0.0360
GOLDENROD	954.0	61	0.0181	0.0191	0.0208	0.0227	0.479	0.0763	0.0362
HAWKWEED	1,000.0	37	0.0173	0.0192	0.0199	0.0216	0.476	0.0759	0.0368
BLUEBELL	1,033.5	37	0.0167	0.0177	0.0193	0.0210	0.473	0.0756	0.0374
LARKSPUR	1,033.5	61	0.0167	0.0177	0.0193	0.0210	0.473	0.0754	0.0377
MARIGOLD	1,113.0	61	0.0155	0.0165	0.0180	0.0195	0.467	0.0744	0.0391
HAWTHORN	1,192.5	61	0.0145	0.0155	0.0169	0.0183	0.462	0.0737	0.0405
NARCISSUS	1,272.0	61	0.0136	0.0146	0.0159	0.0173	0.457	0.0729	0.0418
COLUMBINE	1,351.5	61	0.0128	0.0138	0.0151	0.0163	0.452	0.0722	0.0431
CARNATION	1,431.0	61	0.0121	0.0132	0.0143	0.0155	0.447	0.0715	0.0444
COREOPSIS	1,590.0	61	0.0109	0.0120	0.0130	0.0141	0.439	0.0705	0.0468
JESSAMINE	1,750.0	61	0.0099	0.0111	0.0120	0.0129	0.432	0.0693	0.0490
COWSLIP	2,000.0	91	0.0087	0.0099	0.0107	0.0115	0.421	0.0677	0.0525
LUPINE	2,500.0	91	0.0070	0.0084	0.0091	0.0097	0.404	0.0652	0.0588
TRILLIUM	3,000.0	127	0.0058	0.0074	0.0079	0.0084	0.389	0.0629	0.0646
BLUEBONNET	3,500.0	127	0.0050	0.0068	0.0072	0.0076	0.378	0.0612	0.0697

Courtesy of ALCAN CABLE.
Notes: DC resistance is based on 16.946 ohm-cmil/ft @20°C, 61.2% IACS with stranding increments as per ASTM B231.

TABLE 14.11 All Aluminum Conductor—Electrical Properties

| | | Stranding Number & Diameter | | Complete Diameter (inch) | Steel Core (inch) | ACSS/MA Rated Strength (lb) | Weight (lb/1,000 ft) | | |
Code Word	kCMIL	Aluminum	Steel				Total	Aluminum	Steel
PARTRIDGE/ACSS	266.8	26 × 0.1013	7 × 0.0788	0.642	0.236	8,880	366.9	251.3	115.8
LINNET/ACSS	336.4	26 × 0.1137	7 × 0.0884	0.720	0.265	11,200	462.0	316.6	145.4
ORIOLE/ACSS	336.4	30 × 0.1059	7 × 0.1059	0.741	0.317	14,800	526.4	317.7	208.7
IBIS/ACSS	397.5	26 × 0.1236	7 × 0.0961	0.783	0.288	13,000	546.0	374.1	171.9
LARK/ACSS	397.5	30 × 0.1151	7 × 0.1151	0.806	0.345	17,500	621.8	375.3	246.5
FLICKER/ACSS	477	24 × 0.1410	7 × 0.0940	0.846	0.282	13,000	613.9	449.4	164.5
HAWK/ACSS	477	26 × 0.1354	7 × 0.1053	0.858	0.316	15,600	655.3	449.0	206.3
HEN/ACSS	477	30 × 0.1261	7 × 0.1261	0.883	0.378	21,000	746.4	450.4	296.0
PARAKEET/ACSS	556.5	24 × 0.1523	7 × 0.1015	0.914	0.304	15,200	716.1	524.3	191.8
DOVE/ACSS	556.5	26 × 0.1463	7 × 0.1138	0.927	0.341	18,200	765.2	524.2	241.0
EAGLE/ACSS	556.5	30 × 0.1362	7 × 0.1362	0.953	0.409	24,500	870.7	525.5	345.2
PEACOCK/ACSS	605	24 × 0.1588	7 × 0.1059	0.953	0.318	16,500	778.8	570.1	208.7
ROOK/ACSS	636	24 × 0.1628	7 × 0.1085	0.977	0.326	17,300	818.2	599.1	219.1
GROSBEAK/ACSS	636	26 × 0.1564	7 × 0.1216	0.990	0.365	20,700	874.2	599.0	275.2
EGRET/ACSS	636	30 × 0.1456	19 × 0.0874	1.019	0.437	28,000	987.2	600.5	386.7
FLAMINGO/ACSS	666.6	24 × 0.1667	7 × 0.1111	1.000	0.333	18,200	857.9	628.2	229.7
STARLING/ACSS	715.5	26 × 0.1659	7 × 0.1290	1.051	0.387	23,300	983.7	674.0	309.7
REDWING/ACSS	715.5	30 × 0.1544	19 × 0.0926	1.081	0.463	30,800	1,109.3	675.3	434.0
TERN/ACSS	795	45 × 0.1329	7 × 0.0886	1.063	0.266	14,200	895	748.6	146.4
CUCKOO/ACSS	795	24 × 0.1820	7 × 0.1213	1.092	0.364	21,700	1,023	748.8	274.2
CONDOR/ACSS	795	54 × 0.1213	7 × 0.1213	1.092	0.364	21,700	1,022	748.4	273.6
DRAKE/ACSS	795	26 × 0.1749	7 × 0.1360	1.108	0.408	25,900	1,093	749.1	343.9
MALLARD/ACSS	795	30 × 0.1628	19 × 0.0977	1.140	0.489	34,300	1233.9	750.7	483.2
RUDDY/ACSS	900	45 × 0.1414	7 × 0.0943	1.131	0.283	15,800	1,013	847.4	165.6
CANARY/ACSS	900	54 × 0.1291	7 × 0.1291	1.162	0.387	24,600	1,158	847.7	310.3

308

RAIL/ACSS	954	45 × 0.1456	7 × 0.0971	1.165	0.291	16,700	1,074	898.5	175.5
CARDINAL/ACSS	954	54 × 0.1329	7 × 0.1329	1.196	0.399	26,000	1,227.1	898.3	328.8
ORTOLAN/ACSS	1,033.5	45 × 0.1515	7 × 0.1010	1.212	0.303	18,100	1,163	972.8	190.2
CURLEW/ACSS	1,033.5	54 × 0.1383	7 × 0.1383	1.246	0.415	28,200	1,329	972.8	356.2
BLUEJAY/ACSS	1,113	45 × 0.1573	7 × 0.1049	1.259	0.315	19,500	1,254	1,048.7	205.3
FINCH/ACSS	1,113	54 × 0.1436	19 × 0.1862	1.293	0.431	30,400	1,430	1,053.9	376.1
BUNTING/ACSS	1,192.5	45 × 0.1628	7 × 0.1085	1.302	0.326	21,400	1,142	1,123.0	219.0
GRACKLE/ACSS	1,192.5	54 × 0.1486	19 × 0.0892	1.338	0.446	32,600	1,531	1,128.6	402.4
BITTERN/ACSS	1,272	45 × 0.1681	7 × 0.1121	1.345	0.336	22,300	1,432	1,197.7	234.3
PHEASANT/ACSS	1,272	54 × 0.1535	19 × 0.0921	1.382	0.461	34,100	1,634	1,204.3	429.7
DIPPER/ACSS	1,351.5	45 × 0.1733	7 × 0.1155	1.386	0.347	23,700	1,521	1,272.9	248.1
MARTI N/ACSS	1,351.5	54 × 0.1582	19 × 0.0949	1.424	0.475	36,200	1,735	1,279.2	455.8
BOBOUNK/ACSS	1,431	45 × 0.1783	7 × 0.1189	1.427	0.357	25,100	1,611	1,347.5	263.5
PLOVER/ACSS	1,431	54 × 0.1628	19 × 0.0977	1.465	0.489	38,400	1,838	1,354.6	483.4
LAPWING/ACSS	1,590	45 × 0.1880	7 × 0.1253	1.504	0.376	27,900	1,790	1,498.1	291.9
FALCON/ACSS	1,590	54 × 0.1716	19 × 0.1030	1.545	0.515	42,600	2,042	1,505.0	537.0
CHUKAR/ACSS	1,780	84 × 0.1456	19 × 0.0874	1.602	0.437	35,400	2,072	1,685.5	386.5
BLUEBIRD/ACSS	2,156	84 × 0.1602	19 × 0.0961	1.762	0.481	42,100	2,508	2,040.4	467.6
KIWI/ACSS	2,167	72 × 0.1735	7 × 0.1157	1.735	0.347	29,000	2,301	2,051.4	249.6
THRASHER/ACSS	2,312	71 × 0.1744	19 × 0.814	1.802	0.407	35,600	2,523	2,187.9	335.1

Courtesy of ALCAN CABLE.

ASTM: Conductors manufactured to ASTM B856, Specification for Concentric-Lay-Stranded Aluminum Conductors, Coated Steel Supported 3ACSS.

ACSS: Aluminum conductor steel-supported.

ACSS/MA, supported with ZN-5A1-MM-coated steel core wire, coating Class A in accordance with Specification ASTM B802.

TABLE 14.12 Aluminum Conductor Steel-Supported—Physical Properties

Code Word	Size & Stranding		Resistance (Others per 1,000 ft)			Reactance		GMR (ft)	Ampacity
	kCMIL	Stranding	DC @ 20°C	AC 60 Hz @ 25°C	@ 75°C	Capacitive (Megohms/ 1,000 ft)	Inductive (Ohms/ 1,000 ft)		
PARTRIDGE/ACSS	266.8	26/7	0.0619	0.0633	0.0760	0.565	0.088	0.0217	814
LINNET/ACSS	336.4	26/7	0.0492	0.0502	0.0605	0.549	0.085	0.0244	943
ORIOLE/ACSS	336.4	30/7	0.0488	0.0498	0.0600	0.544	0.084	0.0255	955
IBIS/ACSS	397.5	26/7	0.0416	0.0426	0.0513	0.539	0.084	0.0265	1,051
LARK/ACSS	397.5	30/7	0.0413	0.0422	0.0507	0.533	0.082	0.0277	1,069
FLICKER/ACSS	477	24/7	0.0348	0.0357	0.0429	0.524	0.082	0.0283	1,180
HAWK/ACSS	477	26/7	0.0347	0.0356	0.0428	0.522	0.081	0.0290	1,186
HEN/ACSS	477	30/7	0.0344	0.0352	0.0424	0.517	0.080	0.3040	1,200
PARAKEET/ACSS	556.5	24/7	0.0298	0.0305	0.0368	0.512	0.080	0.0306	1,300
DOVE/ACSS	556.5	26/7	0.0296	0.0305	0.0366	0.510	0.080	0.0313	1,316
EAGLE/ACSS	556.5	30/7	0.0291	0.0302	0.0362	0.505	0.079	0.0328	1,336
PEACOCK/ACSS	605	24/7	0.0274	0.0282	0.0339	0.505	0.079	0.0319	1,378
ROOK/ACSS	636	24/7	0.0260	0.0269	0.0322	0.502	0.079	0.0327	1,429
GROSBEAK/ACSS	636	26/7	0.0259	0.0267	0.0320	0.500	0.078	0.0335	1,438
EGRET/ACSS	636	30/19	0.0258	0.0265	0.0319	0.495	0.077	0.0351	1,449
FLAMINGO/ACSS	666.6	24/7	0.0249	0.0256	0.0307	0.498	0.078	0.0335	1,473
STARLING/ACSS	715.5	26/7	0.0231	0.0237	0.0285	0.490	0.077	0.0355	1,550
REDWING/ACSS	715.5	30/19	0.0229	0.0235	0.0283	0.486	0.076	0.0372	1,568
TERN/ACSS	795	45/7	0.0210	0.0219	0.0261	0.488	0.076	0.0952	1,636
CUCKOO/ACSS	795	24/7	0.0209	0.0217	0.0260	0.484	0.076	0.0361	1,647
CONDOR/ACSS	795	54/7	0.0209	0.0216	0.0259	0.484	0.076	0.0368	1,650
DRAKE/ACSS	795	26/7	0.0210	0.0216	0.0259	0.482	0.076	0.0375	1,660
MALLARD/ACSS	795	30/19	0.0207	0.0214	0.0255	0.477	0.074	0.0392	1,693
RUDDY/ACSS	900	45/7	0.0186	0.0194	0.0232	0.479	0.076	0.0374	1,766

CANARY/ACSS	900	54/7	0.0185	0.0191	0.0230	0.474	0.074	0.0392	1,780
RAIL/ACSS	954	45/7	0.0172	0.0180	0.0215	0.474	0.075	0.0385	1,844
CARDINAL/ACSS	954	54/7	0.0174	0.0181	0.0217	0.470	0.076	0.0404	1,856
ORTOLAN/ACSS	1,033.5	45/7	0.0162	0.0170	0.0203	0.468	0.074	0.0401	1,933
CURLEW/ACSS	1,033.5	54/7	0.0160	0.0167	0.0202	0.464	0.073	0.0420	1,936
BLUEJAY/ACSS	1,113	45/7	0.0151	0.0158	0.0189	0.462	0.073	0.0416	2,025
FINCH/ACSS	1,113	54/19	0.0150	0.0156	0.0190	0.458	0.070	0.0436	2,020
BUNTING/ACSS	1,192.5	45/7	0.0140	0.0148	0.0177	0.456	0.072	0.0431	2,116
GRACKLE/ACSS	1,192.5	54/19	0.0140	0.0147	0.0177	0.452	0.071	0.0451	2,122
BITTERN/ACSS	1,272	45/7	0.0131	0.0140	0.0166	0.451	0.072	0.0445	2,223
PHEASANT/ACSS	1,272	54/19	0.0131	0.0138	0.0166	0.447	0.070	0.0466	2,221
DIPPER/ACSS	1,351.5	45/7	0.0123	0.0132	0.0157	0.447	0.071	0.0459	2,303
MARTI N/ACSS	1,431	54/19	0.0123	0.0130	0.0156	0.442	0.070	0.0480	2,313
BOBOLINK/ACSS	1,431	45/7	0.0117	0.0125	0.0149	0.442	0.070	0.0472	2,383
PLOVER/ACSS	1,590	54/19	0.0117	0.0123	0.0148	0.438	0.069	0.0495	2,395
LAPWING/ACSS	1,590	45/7	0.0105	0.0114	0.0135	0.434	0.069	0.0498	2,559
FALCON/ACSS	1,590	54/19	0.0105	0.0113	0.0136	0.430	0.068	0.0521	2,543
CHUKAR/ACSS	1,780	84/19	0.0094	0.0103	0.0122	0.424	0.067	0.0534	2,749
BLUEBIRD/ACSS	2,156	84/19	0.0078	0.0087	0.0103	0.409	0.065	0.0588	3,089
KIWI/ACSS	2,167	72/7	0.0078	0.0089	0.0104	0.411	0.068	0.0570	3,089
THRASHER/ACSS	2,312	76/19	0.0073	0.0084	0.0098	0.405	0.065	0.0600	3,227

Courtesy of ALCAN.
CABLE Notes:
1. Ampacity based on a 200°C conductor temperature, 25°C ambient temperature, 2 ft/sec, wind, in sun, with emissivity of 0.5 and a coefficient of solar absorption of 0.5, at sea level.
2. Resistance and Ampacity based on 63% IACS Al and 8% IACS steel core wire @20°C.

TABLE 14.13   Aluminum Conductor Steel-Supported—Electrical Properties

Code Word	Conductor Size, kCMIL	Conductor Area, inch²	Stranding Number and Diameter, inch	Nominal Diameter, inch	ACSR with Equal Diameter		Rated Strength, lb	Nominal Weight, lb/1,000 ft
					Size	Stranding		
AKRON	30.58	0.024	7 × 0.0661	0.198	6	6/1	1,110	28.5
ALTON	48.69	0.0382	7 × 0.0834	0.250	4	6/1	1,760	45.4
AMES	77.47	0.0608	7 × 0.1052	0.316	2	6/1	2,800	72.2
AZUSA	123.3	0.0968	7 × 0.1327	0.398	1/0	6/1	4,270	114.9
ANAHEIM	155.4	0.1221	7 × 0.1490	0.447	2/0	6/1	5,390	144.9
AMHERST	195.7	0.1537	7 × 0.1672	0.502	3/0	6/1	6,790	182.5
ALLIANCE	246.9	0.1939	7 × 0.1878	0.563	4/0	6/1	8,560	230.2
BUTTE	312.8	0.2456	19 × 0.1283	0.642	266.8	26/7	10,500	291.6
CANTON	394.5	0.3099	19 × 0.1441	0.721	336.4	26/7	13,300	367.9
CAIRO	465.4	0.3655	19 × 0.1565	0.783	397.5	26/7	15,600	433.9
DARIEN	559.5	0.4394	19 × 0.1716	0.858	477	26/7	18,800	521.7
ELGIN	652.4	0.5124	19 × 0.1853	0.927	556.5	26/7	21,900	608.3
FLINT	740.8	0.5818	37 × 0.1415	0.991	636	26/7	24,400	690.8
GREELEY	927.2	0.7282	37 × 0.1583	1.108	795	26/7	30,500	864.6

TABLE 14.14 All-Aluminum-Alloy Conductor—Physical and Electrical Properties

Code Word	Conductor Size, kCMIL	Stranding Number and Diameter, inch	Approx. AAC Size of Equivalent Resistance	Resistance DC-20°C, Ohms/1,000 ft	Resistance AC-60-Hz 25°C, Ohms/1,000 ft	Resistance AC-60-Hz 50°C, Ohms/1,000 ft	Resistance AC-60-Hz 75°C, Ohms/1,000 ft	GMR, ft	Neutral 60-Hz Reactance 1-ft Spacing Inductive, Ohm/1,000 ft	Neutral 60-Hz Reactance 1-ft Spacing Capacitative, Megohms/1,000 ft
AKRON	30.58	7 × 0.0661	6	0.6589	0.670	0.727	0.784	0.00599	0.118	0.751
ALTON	48.69	7 × 0.0834	4	0.4138	0.420	0.456	0.492	0.00756	0.112	0.715
AMES	77.47	7 × 0.1052	2	0.2600	0.265	0.288	0.311	0.00954	0.107	0.678
AZUSA	123.3	7 × 0.1327	1/0	0.1635	0.166	0.180	0.195	0.0120	0.102	0.642
ANAHEIM	155.4	7 × 0.1490	2/0	0.1297	0.132	0.143	0.155	0.0135	0.0989	0.624
AMHERST	195.7	7 × 0.1672	3/0	0.1030	0.105	0.114	0.123	0.0152	0.0963	0.606
ALLIANCE	246.9	7 × 0.1878	4/0	0.0816	0.0831	0.0902	0.0973	0.0170	0.0936	0.588
BUTTE	312.8	19 × 0.1283	266.8	0.0644	0.0657	0.0712	0.0769	0.0202	0.0896	0.567
CANTON	394.5	19 × 0.1441	336.4	0.0511	0.0523	0.0566	0.0610	0.0227	0.0870	0.549
CAIRO	465.4	19 × 0.1565	397.5	0.0433	0.0443	0.0481	0.0517	0.0247	0.0851	0.536
DARIEN	559.5	19 × 0.1716	477.0	0.0360	0.0369	0.0400	0.0431	0.0271	0.0829	0.522
ELGIN	652.4	19 × 0.1853	556.5	0.0309	0.0318	0.0345	0.0371	0.0292	0.0812	0.510
FLINT	740.8	37 × 0.1415	636.0	0.0272	0.0280	0.0305	0.0328	0.0317	0.0793	0.499
GREELEY	927.2	37 × 0.1583	795.0	0.0217	0.0225	0.0244	0.0263	0.0354	0.0768	0.482

Courtesy of ALCAN CABLE.
ASTM: B398, Specification for Aluminum Alloy 6201-T81 Wire for Electrical Purposes.
B399, Specification for Concentric Lay-Stranded Aluminum Alloy 6201-T81 Conductors.
Note: DC resistance is based on 19.755 ohm-cmil/ft, @20°C (68°F) 52.5% IACS. With standard stranding increment of 2%.

Size, kCMIL	Stranding Number & Diameter, inch		Cross-Sectional Area, inch²			Nominal Conductor Diameter, inch	ACSR with Similar Diameter	Nominal Weight, lb/1,000 ft	Rated Strength, lb
	1350	6201	1350	6201	Total				
503.6	15 × 0.1628	4 × 0.1628	0.3122	0.0833	0.3955	0.814	PELICAN	473	10,500
587.2	15 × 0.1758	4 × 0.1758	0.3641	0.0971	0.4612	0.879	OSPREY	551	12,200
649.5	18 × 0.1325	19 × 0.1325	0.2482	0.2620	0.5102	0.927	DOVE	608	15,600
653.1	12 × 0.1854	7 × 0.1854	0.3240	0.1890	0.5130	0.927	DOVE	612	15,400
739.8	18 × 0.1414	19 × 0.1414	0.2827	0.2983	0.5810	0.990	GROSBEAK	693	18,800
853.7	30 × 0.1519	7 × 0.1519	0.5437	0.1268	0.6705	1.063	TERN	801	17,500
853.7	24 × 0.1519	13 × 0.1519	0.4349	0.2356	0.6705	1.063	TERN	800	19,300
927.2	24 × 0.1583	13 × 0.1583	0.4723	0.2559	0.7282	1.108	DRAKE	869	20,900
1,024.5	30 × 0.1664	7 × 0.1664	0.6524	0.1522	0.8046	1.165	RAIL	961	20,900
1,024.5	24 × 0.1664	13 × 0.1664	0.5219	0.2827	0.8046	1.165	RAIL	961	23,100
1,080.6	24 × 0.1709	13 × 0.1709	0.5505	0.2982	0.8487	1.196	CARDINAL	1,013	24,400
1,080.6	18 × 0.1709	19 × 0.1709	0.4129	0.4358	0.8487	1.196	CARDINAL	1,012	27,200
1,109.0	30 × 0.1731	7 × 0.1731	0.7060	0.1647	0.8707	1.212	ORTOLAN	1,041	22,700
1,109.0	24 × 0.1731	13 × 0.1731	0.5648	0.3059	0.8707	1.212	ORTOLAN	1,040	25,000
1,172.0	30 × 0.1780	7 × 0.1780	0.7465	0.1742	0.9207	1.246	CURLEW	1,100	24,000
1,172.0	18 × 0.1780	19 × 0.1780	0.4479	0.4728	0.9207	1.246	CURLEW	1,098	29,500
1,198.0	30 × 0.1799	7 × 0.1799	0.7626	0.1779	0.9405	1.259	BLUEJAY	1,124	24,500
1,198.0	24 × 0.1799	13 × 0.1799	0.6101	0.3304	0.9405	1.259	BLUEJAY	1,123	27,100
1,277.0	54 × 0.1447	7 × 0.1447	0.8880	0.1151	1.0031	1.302	BUNTING	1,199	24,600
1,277.0	42 × 0.1447	19 × 0.1447	0.6907	0.3124	1.0031	1.302	BUNTING	1,198	28,400
1,361.5	54 × 0.1494	7 × 0.1494	0.9466	0.1227	1.0693	1.345	BITTERN	1,278	26,300
1,534.4	42 × 0.1586	19 × 0.1586	0.8297	0.3754	1.2051	1.427	BOBOLINK	1,439	33,800
1,703.0	48 × 0.1671	13 × 0.1671	1.0527	0.2851	1.3378	1.504	LAPWING	1,598	34,600
1,798.0	42 × 0.1717	19 × 0.1717	0.9725	0.4399	1.4124	1.545	FALCON	1,686	39,600
1,933.0	42 × 0.1780	19 × 0.1780	1.0452	0.4728	1.5180	1.602	CHUKAR	1,813	42,500
2,338.0	42 × 0.1958	19 × 0.1958	1.2646	0.5721	1.8367	1.762	BLUEBIRD	2,214	51,500
2,338.0	48 × 0.1958	13 × 0.1958	1.4453	0.3914	1.8367	1.762	BLUEBIRD	2,215	47,500
2,493.0	54 × 0.1655	37 × 0.1655	1.1617	0.7959	1.9576	1.821	KINGFISHER	2,358	57,600
2,493.0	72 × 0.1655	19 × 0.1655	1.5489	0.4087	1.9576	1.821	KINGFISHER	2,362	50,400

TABLE 14.15 Aluminum-Conductor Alloy-Reinforced—Physical and Properties

Size, kCMIL	Stranding 1350	Stranding 6201	DC-20°C, Ohms/1,000 ft	Resistance AC-60-Hz 25°C, Ohms/1,000 ft	50°C, Ohms/1,000 ft	75°C, Ohms/1,000 ft	60-Hz Reactance 1-ft Equivalent Spacing Capacitive, Megohms 1,000 ft	Inductive, Ohms/1,000 ft	GMR, ft
503.6	15 × 0.1628	4 × 0.1628	0.0354	0.0364	0.0398	0.0433	0.531	0.0841	0.0257
587.2	15 × 0.1758	4 × 0.1758	0.0303	0.0312	0.0342	0.0371	0.518	0.0824	0.0277
649.5	18 × 0.1325	19 × 0.1325	0.0287	0.0295	0.0322	0.0349	0.509	0.0812	0.0292
653.1	12 × 0.1854	7 × 0.1854	0.0279	0.0288	0.0315	0.0341	0.509	0.0811	0.0293
739.8	18 × 0.1414	19 × 0.1414	0.0252	0.0259	0.0283	0.0307	0.499	0.0793	0.0317
853.7	30 × 0.1519	7 × 0.1519	0.0208	0.0216	0.0235	0.0257	0.488	0.0777	0.0340
853.7	24 × 0.1519	13 × 0.1519	0.0213	0.0222	0.0242	0.0262	0.488	0.0777	0.0340
927.2	24 × 0.1583	13 × 0.1583	0.0208	0.0216	0.0236	0.0252	0.482	0.0767	0.0355
1,024.5	30 × 0.1664	7 × 0.1664	0.0173	0.0182	0.0199	0.0215	0.474	0.0756	0.0373
1,024.5	24 × 0.1664	13 × 0.1664	0.0178	0.0186	0.0203	0.0219	0.474	0.0756	0.0373
1,080.6	24 × 0.1709	13 × 0.1709	0.0168	0.0176	0.0192	0.0208	0.470	0.0750	0.0383
1,080.6	18 × 0.1709	19 × 0.1709	0.0172	0.0181	0.0196	0.0213	0.470	0.0750	0.0383
1,109.0	30 × 0.1731	7 × 0.1731	0.0160	0.0169	0.0184	0.0199	0.468	0.0747	0.0388
1,109.0	24 × 0.1731	13 × 0.1731	0.0164	0.0172	0.0187	0.0203	0.468	0.0747	0.0388
1,172.0	30 × 0.1780	7 × 0.1780	0.0152	0.0160	0.0174	0.0189	0.463	0.0740	0.0399

Courtesy of ALCAN CABLE.

ASTM: B230, Specification for Aluminum 1350-H19 Wire for Electrical Purposes.

B398, Specification for Aluminum Alloy 6201-T81 Wire for Electrical Purposes.

B524, Specification for Concentric Lay-Stranded Aluminum Conductors, Aluminum Alloy Reinforced (ACAR, 1350/6201).

Notes:

1. DC resistance is based on 16.946 ohm-cmil/ft for 1350 wires at 61.2% IACS and 19.755 ohm-cmil/ft for 6201 wires at 52.5%.
2. IACS @20°C, with standard increments as per ASTM B524.

TABLE 14.16 Aluminum-Conductor Alloy-Reinforced—Electrical Properties (continued)

| Size, kCMIL | Stranding | | Resistance | | | | 60-Hz Reactance 1-ft Equivalent Spacing | | |
| | 1350 | 6201 | DC-20°C, Ohms/1,000 ft | AC-60-Hz | | | Capacitive, Megohms 1,000 ft | Inductive, Ohms/ 1,000 ft | GMR, ft |
				25°C, Ohms/ 1,000 ft	50°C, Ohms/ 1,000 ft	75°C, Ohms/ 1,000 ft			
1,172.0	18 × 0.1780	19 × 0.1780	0.0159	0.0166	0.0181	0.0195	0.463	0.0740	0.0399
1,198.0	30 × 0.1799	7 × 0.1799	0.0149	0.0155	0.0170	0.0184	0.462	0.0738	0.0403
1,198.0	24 × 0.1799	13 × 0.1799	0.0152	0.0159	0.0173	0.0188	0.462	0.0738	0.0403
1,277.0	54 × 0.1447	7 × 0.1447	0.0138	0.0149	0.0161	0.0174	0.456	0.0729	0.0419
1,277.0	42 × 0.1447	19 × 0.1447	0.0142	0.0152	0.0165	0.0178	0.456	0.0729	0.0419
1,361.5	54 × 0.1494	7 × 0.1494	0.0129	0.0138	0.0151	0.0163	0.451	0.0721	0.0433
1,534.4	42 × 0.1586	19 × 0.1586	0.0118	0.0127	0.0139	0.0152	0.442	0.0708	0.0459
1,703.0	48 × 0.1671	13 × 0.1671	0.0105	0.0115	0.0125	0.0135	0.434	0.0696	0.0484
1,798.0	42 × 0.1717	19 × 0.1717	0.0101	0.0110	0.0119	0.0128	0.430	0.0690	0.0497
1,933.0	42 × 0.1780	19 × 0.1780	0.0094	0.0102	0.0113	0.0122	0.424	0.0682	0.0515
2,338.0	42 × 0.1958	19 × 0.1958	0.0078	0.0089	0.0096	0.0103	0.409	0.0660	0.0567
2,338.0	48 × 0.1958	13 × 0.1958	0.0077	0.0088	0.0095	0.0102	0.409	0.0660	0.0567
2,493.0	54 × 0.1655	37 × 0.1655	0.0074	0.0087	0.0093	0.0100	0.404	0.0652	0.0587
2,493.0	72 × 0.1655	19 × 0.1655	0.0072	0.0085	0.0090	0.0098	0.404	0.0652	0.0587

Courtesy of ALCAN CABLE.
ASTM: B230, Specification for Aluminum 1350-H19 Wire for Electrical Purposes.
B398, Specification for Aluminum Alloy 6201-T81 Wire for Electrical Purposes.
B524, Specification for Concentric Lay-Stranded Aluminum Conductors, Aluminum Alloy Reinforced (ACAR, 1350/6201).
Notes:
1. DC resistance is based on 16.946 ohm-cmil/ft for 1350 wires at 61.2% IACS and 19.755 ohm-cmil/ft for 6201 wires at 52.5%.
2. IACS @20°C, with standard increments as per ASTM B524.

TABLE **14.16** Aluminum-Conductor Alloy-Reinforced—Electrical Properties

FIGURE 14.11   Aluminum-conductor composite-core (ACCC) conductor. (*Courtesy Composite Technology Corporation.*)

FIGURE 14.12   Construction details of aluminum self-damping conductor. (*Courtesy Alcoa Conductor Products Company Division of Aluminum Company of America.*)

FIGURE 14.13   Diagram of forces on aluminum self-damping conductor as a result of wind. (*Courtesy Alcoa Conductor Products Company Division of Aluminum Company of America.*)

Aluminum clamp

Lubricated aluminum bolt

Aluminum cap

Sun shield

Cast zinc mass

Elastomeric
damper elements (4)

Drain hole

FIGURE 14.14   Cutaway view of the elastomeric damper for use on static or ground wires. (*Courtesy Alcoa Conductor Products Company Division of Aluminum Company of America.*)

The proper conductor selection is a compromise between economics associated with the cost of electric generation, conductor material, and supporting structures; mechanical strength requirements; current-carrying capacity required; and environmental impact.

Vibration Dampers

Line conductors vibrate as a result of wind-induced motion known as aeolian vibration. The vibration has been found to vary from 3 Hz to 150 Hz. Vibration dampers are installed on the conductors to control the vibration and prevent conductor damage. The Stockbridge-type dampers have been used for many years. The elastomeric dampers developed by Alcoa are smaller, lighter, and more efficient than Stockbridge dampers (Fig. 14.14). The elastomeric damper can protect long span lengths from vibration without causing corona on extra-high-voltage transmission lines.

Conductor Spacers

Transmission lines with bundled conductors, that is, two or more conductors per phase, use spacers to keep the conductors from wrapping together. Spacer dampers are designed with damping mechanisms capable of reducing vibration by converting the mechanical energy of the vibrating conductors into heat in the damping element (Figs. 14.15 and 14.16). Spacer dampers prevent conductor clashing and wind-induced aeolian vibration.

Conductor Reels

All types and sizes of wire are shipped on reels normally. Limited amounts of small wire can be shipped in coils wrapped with paper. Figure 14.17 illustrates reel construction and size for all-aluminum and ACSR wire.

FIGURE 14.15 Vibration damper used on transmission-line phase conductors. (*Courtesy Alcoa Conductor Products Company Division of Aluminum Company of America.*)

FIGURE 14.16 Transmission-line bundle conductor spacers with vibration damping mechanisms. (*Courtesy Alcoa Conductor Products Company Division of Aluminum Company of America.*)

3" Diam.

Approximate dimensions, in							Approximate weight, lb	
Diameter						Lagging thickness		
A	B	C	D	E	F		With lags	Without lags
69	43½	46	30	38	2¾	1¾	900	—
69	32½	35	30	27	2¾	1¾	720	—
65	32½	35	30	27	2¾	1¾	650	—
58	32½	35	30	27	2¾	1¾	540	—
52	31½	33	23½	26	2¾	1¾	420	—
46	31½	33	23½	26	2¾	1¾	380	—
44	22½	24½	22½	18	2¼	1¾	205	140
40	22½	24½	22½	18	2¼	1¾	185	125
36	22½	24½	22½	18	2¼	—	—	100

FIGURE 14.17 Reel for all-aluminum and ACSR cables.

Fiber-Optic Cables

Overhead ground and phase wires can be fabricated with optical fibers as an integral part of the conductor. Phase conductors on high-voltage lines require the use of an insulator constructed with optical fibers in the core.

Ground or static wire with the optical fibers fabricated in the conductor is the method frequently used to establish a fiber-optic communication system (Fig. 14.18).

FIGURE 14.18   Alumoweld ground or static wire with fiber-optic cable placed in the core. (*Courtesy Aluminum Company of America.*)

	Standard Ground Wire Reference	OPT-GW Equivalent	
		Diameter	Strength
Size, reference	½ inch EHS	75 mm AW	99 mm AW
Nominal diameter, inch	0.495	0.5	0.55
Stranding (no. × diam., inch)	7 × 0.165	8 × 0.1358	28 × 2165
Aluminum pipe, OD/ID,	—	0.2283	0.2165
inch		0.1339	0.126
Area, inch²			
Strand	0.1497	0.1159	0.1533
Pipe		0.026	0.0244
Total	0.1497	0.1419	0.1777
Rated strength, lbf	26,900	20,380	27,310
Weight, lb/1,000 ft			
Strand	517	334.42	443.7
Pipe		30.53	28.6
Fiber		(10.1)	(10.1)
Total	517	375	482.4
Strength-to-weight ratio	52	54.5	56

TABLE 14.17   Physical Characteristics of Standard ½-inch Extra-High-Strength Ground Wire and Equivalent Fiber-Optic Ground Wire

The optical ground wire has three main components in the completed assembly: the optical fiber cable assembly, an aluminum pipe, and strand members consisting of aluminum-clad, alloy, or galvanized steel wires.

The optical fiber cable assemblies can be classified as either tight buffered or loose buffered. The tight-buffered cable has the fibers stranded in close contact around a relatively stiff central strength member. All expansions and contractions resulting from external forces and temperature changes are directly supported by the central strength member. The central strength member protects the optical fibers from tensile stress and strain and contraction of the fiber coating material. The loose-buffered cable has the fibers placed in a spacer or tube to make the movement of the whole cable and the optical fibers independent of each other. The aluminum pipe around the optical fibers provides protection while the conductor is being strung and keeps the optical fibers dry while the assembly is in service. Table 14.17 provides data for standard ground or static wire and fiber-optic ground wire (OPT-GW).

Distribution Transformers

T he purpose of a distribution transformer is to reduce the primary voltage of the electric distribution system to the utilization voltage serving the customer. A distribution transformer is constructed with two or more windings used to transfer alternating-current electric power by electromagnetic induction from one circuit to another at the same frequency but with different values of voltage and current.

Distribution transformers available for use for various applications, as shown, include pole-type (Figs. 15.1 and 15.2), pad-mounted (Figs. 15.3 and 15.4), vault or network type (Fig. 15.5), submersible (Fig. 15.6), and step-up/step-down (Fig. 15.7).

The distribution transformer in Fig. 15.1 is self-protected. It is equipped with a lightning arrester, a weak-link or protective-link expulsion-type fuse (installed under oil in the transformer tank), a secondary circuit breaker, and a warning light. The transformer primary bushing conductor is connected to one phase of the three-phase primary circuit through a partial-range current-limiting fuse. The transformer tank is grounded and connected to the primary and secondary common-neutral ground wire. The self-protected transformer contains a core and coils, a primary fuse mounted on the bottom of the primary bushing, a secondary terminal block, and a low-voltage circuit breaker.

Conventional overhead transformers are also available. The key component distinguishing the conventional transformer from the self-protected transformer is the lack of internal fusing in the conventional model. An external fused cutout and lightning arrester are mounted between the distribution primary and the conventional transformer. A conventional unit is depicted in Fig. 15.2.

Pad-mounted transformers are used with underground systems. The single-phase pad-mounted transformers are used for underground residential installations (Fig. 15.3). Three-phase pad-mounted transformers are used for commercial installations, as illustrated in Fig. 15.4. The vault-type transformer (Fig. 15.5) may be installed in a vault under a sidewalk or in a vault inside of the building. They are often used in underground electric network areas.

Subsurface single-phase distribution transformers (Fig. 15.6) are used in some underground systems installed in residential areas. Corrosion problems and the high cost of installation have minimized the use of subsurface transformers. The pole-mounted step-up/step-down transformers (Fig. 15.7) are installed for islanded pockets that are in distribution systems with lower voltages that are being converted to a higher value. For example, step-down transformers are used to convert 13.2/7.6 kV to 4.2/2.4 kV as the area is converted to the higher voltage.

FIGURE 15.1   A 25 kVA, 13.2 GrdY/7.62 kV to 208Y/120 volt, single phase, single bushing, self-protected, overhead transformer. The unit is equipped with a surge arrester and a low-voltage circuit breaker. (*Courtesy Howard Industries, Inc.*)

FIGURE 15.2   A 15 kVA, 13.2 GrdY/7.62 kV to 208Y/120 volt, single phase, two bushing, conventional overhead transformer. The unit is equipped with a surge arrester and a low-voltage circuit breaker. (*Courtesy Howard Industries, Inc.*)

Distribution Transformer Operation

The basic principle of operation of a transformer is explained in Chap. 1. A schematic drawing of a single-phase distribution transformer appears in Fig. 15.8. The single-phase distribution transformer consists of a primary winding and a secondary winding wound on a laminated steel core. If the load is disconnected from the secondary winding of the transformer and a high voltage is applied to the primary winding of the transformer, a magnetizing current will flow in the primary winding. If we assume the resistance of the primary winding is small, which is usually true, this current is limited by the counter voltage of self-induction in the highly inductive primary winding. The windings of the transformer are constructed with sufficient turns in each winding to limit the no-load or exciting current and produce a counter voltage approximately equal to the applied voltage. The exciting current magnetizes, or produces a magnetic flux, in the steel transformer core. The magnetic flux reverses each half cycle as a result

FIGURE 15.3   Single phase, 25 kVA, 13.2 GrdY/7.6 kV to 240/120 V, pad-mount transformer. (*Courtesy Howard Industries, Inc.*)

FIGURE 15.4   Three-phase 500 kVA, 13.2 GrdY/7.62 to 208Y/120 Volt, pad-mount transformer. (*Courtesy Howard Industries, Inc.*)

FIGURE 15.5   Vault-type distribution transformer 1,000 kVA, three-phase 13,200 GrdY/7,620 volts to 208Y/120 volts. (*Courtesy Howard Industries, Inc.*)

of the alternating voltage applied to the primary winding. The magnetic flux produced cuts the turns of the primary and secondary windings. This action induces a counter voltage in the primary winding and produces a voltage in the secondary winding. The voltages induced in each turn of the primary and secondary winding coils will be approximately equal, and the voltage induced in each winding will be equal to the voltage per turn multiplied by the number of turns.

Voltage V = volts per turn × number of turns, N

Voltage primary winding = V_p

Voltage secondary winding = V_s

Volts per turn = V_t

Number of turns of primary winding = N_p

Number of turns of secondary winding = N_s

Therefore, the voltage of the primary winding is equal to the applied voltage and equal to the number of turns in the primary winding multiplied by the volts per turn.

$$V_p = V_t \times N_p$$

Likewise, the voltage of the secondary winding will equal the volts per turn multiplied by the number of turns in the secondary winding.

$$V_s = V_t \times N_s$$

FIGURE 15.6 A 50 kVA, 13.2 GrdY/ 7.62 kV to 208Y/120 volt, single phase, single bushing, subsurface, overhead transformer. (*Courtesy Howard Industries, Inc.*)

FIGURE 15.7 A 500 kVA, 13.2 GrdY/7.62 kV to 4.16 GrdY/2.4 kV, single phase, two bushing, step-up/step-down, overhead transformer. (*Courtesy Howard Industries, Inc.*)

FIGURE 15.8 Schematic drawing of single-phase distribution transformer.

If we use basic algebraic operations and divide both sides of an equation by the same number, we can obtain the following equation:

$$\frac{V_p}{N_p} = \frac{V_t \times N_p}{N_p}$$

$$\frac{V_p}{N_p} = V_t$$

and

$$\frac{V_s}{N_s} = \frac{V_t \times N_s}{N_s}$$

$$\frac{V_s}{N_s} = V_t$$

Quantities equal to the same quantity are equal, so

$$\frac{V_p}{N_p} = \frac{V_s}{N_s}$$

Both of the above quantities are equal to V_t. If we multiply both sides of the equation by the same quantity N_p, we obtain

$$\frac{V_p}{N_p} \times N_p = N_p \times \frac{V_s}{N_s}$$

or

$$V_p = \frac{N_p}{N_s} \times V_s$$

Likewise,

$$V_s = \frac{N_s}{N_p} \times V_p$$

Therefore, if we know the source voltage, or primary voltage, and the number of turns in the primary and secondary windings, we can calculate the secondary voltage. For example, if a distribution transformer has a primary voltage winding with 7,620 turns and a secondary winding of 120 turns and the source voltage equals 7,620 volts, the secondary voltage equals 120 volts.

$$V_s = V_p \times \frac{N_s}{N_p}$$

$$= 7{,}620 \times \frac{120}{7{,}620}$$

$$= 120 \text{ volts}$$

When a load is connected to the secondary winding of the transformer, a current I_s will flow in the secondary winding of the transformer. This current is equal to the secondary voltage divided by the impedance of the load Z.

$$I_s = \frac{V_s}{Z}$$

Lenz' law determined that any current that flows as a result of an induced voltage will flow in a direction to oppose the action that causes the voltage to be induced. Therefore, the secondary current, or load current, in the distribution transformer secondary winding will flow in a direction to reduce the magnetizing action of the exciting current or the primary current. This action reduces the magnetic flux in the laminated steel core momentarily and, therefore, reduces the counter voltage in the primary winding. Reducing the counter voltage in the primary winding causes a larger primary current to flow, restoring the magnetic flux in the core to its original value. The losses in a distribution transformer are small, and therefore, for all practical purposes, the volt-amperes of the load, or secondary winding, equal the volt-amperes of the source, or primary winding.

$$V_p \times I_p = V_s \times I_s$$

If we divide both sides of this equation by the same quantity V_p, we obtain

$$\frac{\cancel{V_p} \times I_p}{\cancel{V_p}} = \frac{V_s \times I_s}{V_p}$$

or

$$I_p = \frac{V_s}{V_p} \times I_s$$

Therefore, if the primary source voltage of a transformer is 7,620 volts, the secondary voltage is 120 volts, and the secondary current, as a result of the load impedance, is 10 amps, then the primary current is 0.16 amp.

$$I_p = \frac{120}{7,620} \times 10 = 0.16 \text{ amp}$$

A three-phase transformer is basically three single-phase transformers in a single tank in most cases. The windings of the three transformers are normally wound on a single multi-legged laminated steel core.

Distribution Transformer Core and Coil Construction

Two basic types of transformer core and coil construction are the *core type* and the *shell type*. In the core type, the metal core is in the form of a rectangular frame with the coils placed on the two vertical sides (Fig. 15.9). The coils are cylindrical and relatively long. They are divided, part of each primary and secondary being on each of the two vertical legs. In the shell type (Fig. 15.10), the core surrounds the coils, instead of the coils surrounding the core. The coils in the shell type are generally flat instead of cylindrical, and the primary and secondary coils are alternated. The core-and-coil assembly of the distribution transformer is completely immersed in insulating oil or other insulating liquids to keep the operating temperatures low and to provide additional insulation for the windings.

Most distribution transformer cores are manufactured utilizing a silicon steel metal. The silicon steel metal is defined by lamination thickness. The thinner the lamination thickness, the higher the core efficiency. Some distribution transformers are being manufactured with amorphous metal cores. The amorphous metal reduces the core loss in a distribution transformer, thus increasing the transformer efficiency. The core losses of distribution transformers built with an amorphous metal core are approximately 60% less than the core losses of distribution transformers with a laminated silicon steel metal core.

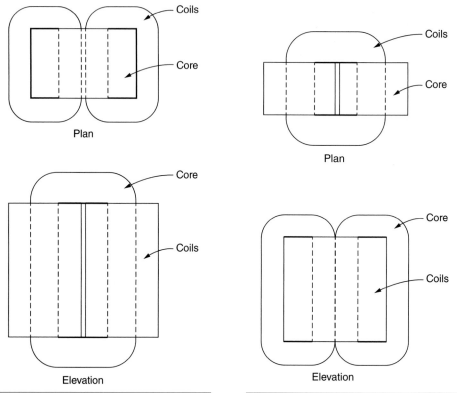

FIGURE 15.9   Views illustrating core-type transformer construction. (*Courtesy ABB Power T&D Company Inc.*)

FIGURE 15.10   Views illustrating shell-type transformer construction. (*Courtesy ABB Power T&D Company Inc.*)

Distribution Transformer Ratings

The capacity of a distribution transformer is determined by the amount of current it can carry continuously at rated voltage without exceeding the design temperature. The transformers are rated in kilovoltamperes (kVA) since the capacity is limited by the load current which is proportional to the kVA regardless of the power factor. The standard kVA ratings are itemized in Table 15.1.

Distribution Transformer Connections

Single-phase distribution transformers are manufactured with one or two primary bushings. The single-primary-bushing transformers can only be used on grounded wye systems. The two-primary-bushing transformers can be used on three-wire delta systems or four-wire wye systems if they are properly connected. Figure 15.11 illustrates schematically the connections of a single-phase transformer to a three-phase 13,200-volt three-wire ungrounded delta primary voltage system to obtain 120-volt single-phase two-wire secondary service. The connections for similar systems operating at other primary distribution voltages such as 13,800, 24,940, and 34,500 would be identical.

Overhead Type		Pad-Mounted Type	
Single–Phase	Three–Phase	Single–Phase	Three–Phase
5	15	25	75
10	30	37½	112½
15	45	50	150
25	75	75	225
37½	112½	100	300
50	150	167	500
75	225		750
100	300		1,000
167	500		1,500
250			2,000
333			2,500
500			

TABLE 15.1   Standard Ratings of Distribution Transformers, kVA

Figure 15.12 illustrates the proper connections for a single-phase transformer to a three-phase three-wire ungrounded delta primary voltage system to obtain 120/240-volt single-phase three-wire service. Normally the wire connected to the center low-voltage bushing will be connected to ground. Grounding the wire connecting to the center bushing limits the voltage above ground to 120 volts even though the wires connecting to the outside secondary bushings have 240 volts between them.

FIGURE 15.11   Single-phase overhead or pole-type distribution transformer connection for 120-volt two-wire secondary service. Transformer secondary coils are connected in parallel.

FIGURE 15.12 Single-phase distribution transformer connected to give 120/240-volt three-wire single-phase service. Transformer secondary coils are connected in series.

Figure 15.13 illustrates schematically the single-phase distribution transformer connections to a three-phase four-wire wye grounded neutral primary system rated 13,200Y/7,620 volts to obtain 120/240-volt single-phase secondary service. The three-phase four-wire wye grounded neutral system has voltages between phases equal to the line to neutral voltage multiplied by 1.73. In Fig. 15.13 the primary system line to neutral voltage is 7,620 volts, and the voltage between phases is 1.73 × 7,620, or 13,200 volts. This system is designated as a 13,200Y/7,620-volt system. Other standard three-phase four-wire wye grounded neutral primary system voltages are 8,320Y/4,800, 12,470Y/7,200, 13,200Y/7,620, 13,800Y/7,970, 22,860Y/13,200, 24,940Y/14,400, and 34,500Y/19,920.

The transformer connections to obtain 120, 120/240, or 240-volt secondary service are normally completed inside the tank of the transformer (Fig. 15.14). The transformer nameplate provides information necessary to complete connections. The voltages should be measured when the transformer is energized, to ensure that the connections are correct before the load is connected to the transformer.

Single-phase pole or overhead-type transformers can be used to obtain three-phase secondary service.

There are four normal connections: delta-delta (Δ-Δ) (Fig. 15.15); wye-wye (Y-Y) (Fig. 15.16); delta-wye (Δ-Y) (Fig. 15.17); wye-delta (Y-Δ) (Fig. 15.18).

Figure 15.15 illustrates the proper connections for three single-phase distribution transformers connected to a three-phase three-wire ungrounded delta primary-voltage system to obtain three-phase three-wire delta secondary service. The illustration is for a 13,200-volt primary system and 240-volt secondary service. Other similar systems operating at other primary voltages would be connected in the same way. Single-phase transformers with secondary windings constructed for voltages of 240/480 could be used to obtain 480-volt three-phase secondary service.

Figure 15.16 illustrates the proper connections for three single-phase distribution transformers connected to a three-phase four-wire grounded neutral wye primary-voltage

FIGURE 15.13 Single-phase distribution transformer connected to provide 120/240-volt three-wire single-phase service. Primary winding is connected line to neutral or ground.

FIGURE 15.14 The left view shows connections for 120/240-volt three-wire service. For 240-volt two-wire service, the neutral bushing conductor is not connected to the jumper between terminals B and C inside the transformer tank. For 120-volt two-wire service, the connections are made as shown in the right view. All connections are completed inside the transformer tank. (*Courtesy ABB Power T&D Company Inc.*)

system to obtain 208Y/120- or 480Y/277-volt secondary service. Figure 15.17 illustrates the proper connections for three single-phase distribution transformers connected to a three-phase three-wire ungrounded neutral delta primary-voltage system to obtain 208Y/120- or 480Y/277-volt secondary service. Figure 15.18 illustrates the proper connections for three single-phase distribution transformers connected to a three-phase four-wire wye primary-voltage system to obtain three-phase three-wire delta secondary service.

Transformers must be constructed with the proper windings for the primary-voltage system and the desired secondary voltage. Properly connected distribution transformers can be wired wye-delta if it is desired to obtain three-wire three-phase secondary voltages from a three-phase four-wire grounded neutral wye primary-voltage system.

The standard secondary system three-phase voltages are 208Y/120, 240, 480Y/277, and 480 volts.

If one of the transformers from a delta-connected bank is removed, the remaining two transformers are said to be open-delta-connected. The remaining two transformers will still transform the voltages in all three phases and supply power to all three phases of the secondary mains. The proper connections, using two transformers, to obtain

Three-phase three-wire delta primary system

13,200 volts

13,200 volts

13,200 volts

240 volts

240 volts

240 volts

Three-phase three-wire delta secondary system

FIGURE 15.15 Three single-phase distribution transformers connected delta-delta (Δ-Δ).

three-phase service, for a delta primary circuit are shown in Fig. 15.19. The capacity of the two transformers is now, however, only 58% instead of 66-2/3% of what it would appear to be with two transformers.

The open-delta connection is often used where an increase in load is anticipated. The third unit is added when the load grows to the point at which it exceeds the capacity of the two transformers. Furthermore, if one transformer of a three-phase bank should become defective, the defective transformer can be removed and the remaining two transformers continue to render service to at least part of the load.

Distribution Transformer Polarity

In connecting transformers in parallel or in a three-phase bank, it is important to know the polarity of the transformer terminals or leads. In building transformers in the factory, the ends of the windings can be connected to the leads extending out through the case such that the flow of current in the secondary terminal with respect to the corresponding primary terminal is in the same direction or in the opposite direction. When the current flows in the same direction in the two adjacent primary and secondary terminals, the polarity of the transformer is said to be *subtractive*, and when current flows in opposite directions, the polarity is said to be *additive*.

Polarity may be further explained as follows: Imagine a single-phase transformer having two high-voltage and three low-voltage external terminals. Connect one high-voltage terminal to the adjacent low-voltage terminal and apply a test voltage across the two high-voltage terminals. Then if the voltage across the unconnected high-voltage and low-voltage terminals is less than the voltage applied across the high-voltage terminals, the polarity is *subtractive*; while if it is greater than the voltage applied across the high-voltage terminals, the polarity is *additive* (Fig. 15.20).

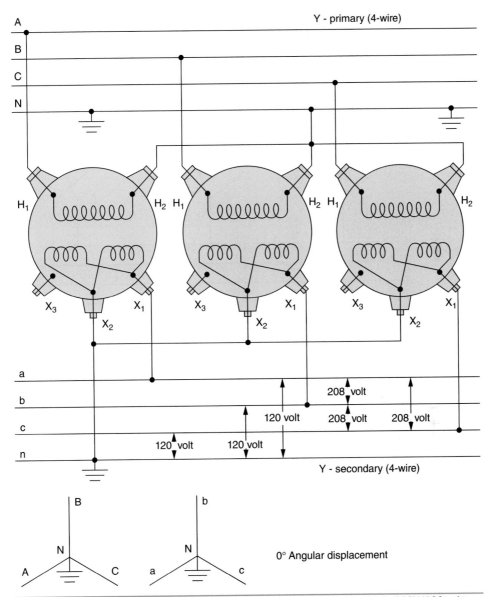

FIGURE 15.16   Three-phase Y-Y pole-type transformer connections used to obtain 208Y/120 volts three-phase from a three-phase, four-wire primary. If single-primary-bushing transformers are used, the H₂ terminal of the transformer primary winding would be connected to the tank of the transformer.

From the foregoing, it is apparent that when the voltage indicated on the voltmeter is greater than the impressed voltage, it must be the sum of the primary and secondary voltages, and the sense of the two windings must be in opposite directions, as in Fig. 15.21. Likewise, when the voltage read on the voltmeter is less than the impressed voltage, the voltage must be the difference, as illustrated in Fig. 15.22. When the terminal markings are arranged in the same numerical order, H_1H_2 and X_1X_2 or H_2H_1 and X_2X_1,

Three-phase three-wire ungrounded delta primary system

13,200 volts   13,200 volts   13,200 volts

Three-phase four-wire grounded-neutral secondary system

FIGURE 15.17   Three single-phase distribution transformers connected delta-wye (Δ-Y). Windings must be rated for the primary system and the desired secondary voltage.

on each side of the transformer, the polarity of each winding is the same (subtractive). If either is in reverse order, H_2H_1 and X_1X_2 or H_1H_2 and X_2X_1, their polarities are opposite (additive). The nameplate of a transformer should always indicate the polarity of the transformer to which it is attached.

Additive polarity is standard for all single-phase distribution transformers 200 kVA and below having high-voltage ratings of 9,000 volts and below.

Subtractive polarity is standard for all single-phase distribution transformers above 200 kVA irrespective of voltage rating.

Subtractive polarity is standard for all single-phase transformers 200 kVA and below having high-voltage ratings above 9,000 volts.

Standard low-voltage terminal designations are illustrated in Fig. 15.23.

Paralleling Single-Phase Distribution Transformers

If greater capacity is desired, two transformers of the same or different kVA ratings may be connected in parallel. Single-phase transformers of either additive or subtractive polarity may be paralleled successfully if connected, as shown in Fig. 15.24, and if the following conditions are met:

1. Voltage ratings are identical.
2. Tap settings are identical.
3. Percentage of impedance of one is between 92½ and 107½% of the other.
4. Frequency ratings are identical.

FIGURE 15.18   Three single-phase distribution transformers connected wye-delta (Y-Δ). The primary neutral bushings are tied together and floated.

Distribution Transformer Tapped Windings

Windings of distribution transformers can be manufactured with various taps to broaden the range of primary voltages suitable for a given installation. The standard taps available are on the primary winding of the transformer and are designed to provide variations in the number of winding turns in steps of 2½% of the rated voltage. The distribution transformers may be designed with four 2½% taps above rated voltage, two 2½% taps above and two 2½% taps below rated voltage, or four 2½% taps below rated voltage. Tap changers on distribution transformers are designed to be operated while the transformer is deenergized. The operating handle (Fig. 15.25) can be located outside of the tank or inside the tank above the oil level. The transformer nameplate provides information concerning the taps available on the distribution transformer.

Figure 15.19   Two single-phase transformers connected into a bank to transform power from three-phase primary voltage to 240 volts three-phase. Primaries and secondaries are open-delta connected.

Figure 15.20   Polarity tests.

Figure 15.21   Sketch showing polarity markings and directions of voltages when polarity is additive.

Figure 15.22   Sketch showing polarity markings and directions of voltages when polarity is subtractive.

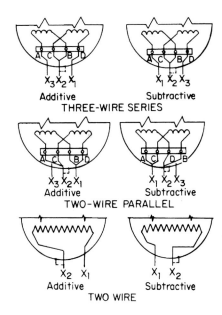

FIGURE 15.23 Standard low-voltage connections for distribution transformer.

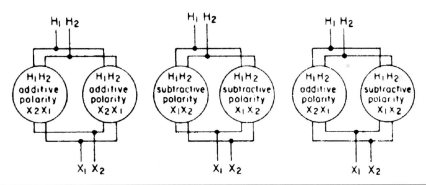

FIGURE 15.24 Schematic diagram showing proper connections for single-phase distribution transformers connected in parallel.

Distribution Transformer Standards

The distribution transformer standards are published by American National Standards Institute (ANSI). The most applicable standards on this topic are:

- ANSI Standard C57.12.20 "Requirements for Overhead-Type Distribution Transformers, 67,000 Volts and Below; 500 kVA and Smaller."

- ANSI Standard C57.12.25 "Requirements for Pad-Mounted Compartmental-Type Single-Phase Distribution Transformers with Separable Insulated High-Voltage Connectors, High-Voltage, 34,500 GrdY/19,920 Volts and Below; Low-Voltage, 240/120; 167 kVA and Smaller."

FIGURE 15.25   No-load tap changer for distribution transformer. (*Courtesy ABB Power T&D Company Inc.*)

- ANSI Standard C57.12.26 "Requirements for Pad-Mounted Compartmental-Type Self-Cooled, Three-Phase Distribution Transformers for Use with Separable Insulated High-Voltage Connectors, High-Voltage, 24,940 GrdY/14,400 Volts and Below; 2,500 kVA and Smaller."
- IEEE C57.12.10 "General Requirements for Liquid-Immersed Distribution, Power, and Regulating Transformers."

A typical specification for conventional single-phase overhead distribution transformers (Fig. 15.26) would read as follows:

This standard covers the electrical characteristics and mechanical features of single-phase, 60-Hz, mineral oil-immersed, self-cooled, conventional, overhead-type distribution transformers rated: 15, 25, 50, 75, 100, and 167 kVA; with high voltage: 7,200/12,470Y, 7,620/13,200Y, or 7,970/13,800Y volts with voltage taps on the high-voltage windings; and low voltage: 120/240.

All requirements shall be in accordance with the latest revision of ANSI C57.12.00 and ANSI C57.12.20, except where modified by this standard.

The transformer shall be bar coded in accordance with the proposed EEI "Bar Coding Distribution Transformers: A Proposed EEI Guideline." The manufacturer and serial number shall be bar coded on the nameplate. The serial number and manufacturer shall be bar coded on a temporary label and should be placed so it is visible on the front side of the shipping pallet.

The transformer shall meet interchangeability dimensions for single position mounting, as shown in Figures 1 and 9 of ANSI C57.12.20.

The two primary bushings shall be wet process porcelain cover-mounted. The primary connectors shall be tin-plated bronze of the eye-bolt type. The H1 bushing is to be supplied with a wildlife protector and an eyebolt connector. The eyebolt connector shall accept a 5/16 inch diameter knurled stud terminal with the

FIGURE **15.26** Diagram of a conventional single-phase, overhead distribution transformer. (*Courtesy MidAmerican Energy Co.*)

wildlife protector in place. The secondary bushings shall be wet process porcelain or epoxy, mounted on the tank wall. Secondary connectors shall be per ANSI C57.12.20, section 6.1.2.

The cover of the transformer shall be coated for wildlife protection. The coating shall have a rating of at least 10 kV AC rms. The insulating finish shall be capable of withstanding 10 kV at a 200 volt/sec rate of rise, tested per ASTM D149-87, using electrode type 3 (¼-inch cylindrical rods).

The transformer tank and cover shall be designed to withstand an internal pressure of 20 psig without tank rupture or bushing or cover displacement, and shall withstand an internal pressure of 7 psig without permanent distortion. The transformer tank shall retain its shape when the cover is removed.

Support lugs for single-position mounting shall be provided and shall be designed to provide a static-load safety factor of 5. Lifting lugs shall be provided, with location and safety factor as specified in ANSI C57.12.20. The tank and cover finish shall be light gray No. 70 (Munsell 5BG7.0/0.4). The interior shall be coated with a durable protective finish from the top to at least below the top liquid level.

The transformer shall be equipped with a tank ground per ANSI C57.12.20, paragraph 6.5.4.1 with eyebolt to accept an AWG No. 6 solid copper ground wire; a tank ground for ground strap connection; and a ground strap, properly sized to carry load and fault current, connected between the secondary neutral terminal and the tank ground. The grounding strap shall be supplied with a connection hole at each end. "J" hook or slotted connection points shall not be allowed.

The transformer shall be equipped with neoprene or Buna-N gaskets on all primary and secondary bushings and tank fittings.

A Qualitrol 201-020-01 automatic pressure relief device, with operation indication, or approved equivalent; or a cover assembly designed for relief of excess pressure in accordance with ANSI C57.12.20, section 6.2.5.2, shall be supplied on the transformer.

Final core and coil assembly shall result in a rigid assembly that maintains full mechanical, electrical, and dimensional stability under fault conditions.

The grade of the silicon core steel shall comply with the U.S. Department of Energy's Office of Energy Efficiency and Renewable Energy Conservation Program for Commercial Equipment: Distribution Transformers Energy Conservation Standards. The rule sets minimum energy-efficiency standard levels for liquid-immersed distribution transformers. The coil insulation system shall consist of electrical kraft paper layer insulation coated on both sides with thermosetting adhesive. Curing of the coils shall cause a permanent bond layer to layer and turn to turn.

The internal primary leads shall have sufficient length to permit cover removal and lead disconnection from the high-voltage bushings. These leads shall be adequately anchored to the coil assembly to minimize the risk of damaging the coil connection during cover removal.

The internal secondary leads shall be identified by appropriate markings, embossed or stamped on a visible portion of the secondary lead near the end connected to the bushing. These leads shall have sufficient length to permit series or parallel connection of the secondary windings. If the internal secondary leads are soft aluminum, a hard aluminum-alloy tab shall be welded to the bushing end of the lead for connection to the bushing.

The transformer should be designed so that the voltage regulation shall not exceed 2.5% at full load and 0.8 power factor for all transformers covered in this specification.

The transformer shall be provided with taps on the high-voltage windings that are: 7,970/7,620/7,430/7,200/7,020 volts. They shall be externally operable as set forth in ANSI C57.12.20, section 6.2.1.1.

The unit shall be of "conventional" transformer design. There shall be no integral fuse, secondary breaker, or overvoltage protection.

All transformers shall be filled with Type I insulating oil meeting the following requirements:

    a. All applicable requirements of ASTM D 3487, paragraph 82a.

    b. Minimum breakdown value of 28 kV per ASTM D 1816, paragraph 84a using a 0.4-in gap.

    c. The PCB content shall be less than 2 parts per million which shall be noted upon the nameplate of each unit. In addition, the manufacturer shall provide and install a "No PCB" identification on the tank wall, near the bottom of the tank centered below the low-voltage bushings. The decal shall be not less than 1 in by 2 in in size and shall indicate "No PCB's" in blue lettering on a white background, and be sufficiently durable to remain legible for the life of the transformer.

A typical specification for self-protected single-phase overhead distribution transformers (Figs. 15.2 and 15.3) would read as follows:

> This specification describes the electrical characteristics and mechanical features of single phase, 60 Hz, mineral oil-immersed, self-cooled, self-protected, with primary weak-link fuse and secondary circuit breaker; overhead type distribution transformers; rated 15 kVA, 25 kVA, 50 kVA, 75 kVA, 100 kVA, and 167 kVA capacity; 13,200 GrdY/7,620-120/240 volts; 95-kV BIL; red overload warning light; no taps; one surge arrester; one primary bushing.

> All requirements shall be in accordance with the latest revision of ANSI Standard C57.12.00 and C57.12.20 except where modified by this specification.

> The primary weak-link fuse, secondary circuit breaker, and red overload warning light shall be coordinated and compatible with the nameplate ratings.

> The arrester shall be a 9/10 kV, direct-connected lead-type distribution class surge arrester (in accordance with the latest revision of ANSI Standard C62.2), with ground lead disconnector, and wildlife protector. The arrester primary lead shall have a minimum length of 24 in with a weatherproof jacket, for connection to the line side of a current-limiting fuse to be installed on the primary bushing in the field.

> The transformer shall be bar coded in accordance with the proposed EEI "Bar Coding Distribution Transformers: A Proposed EEI Guideline." The manufacturer and serial number shall be bar coded on the nameplate. The serial number and manufacturer shall be bar coded on a temporary label and should be placed so it is visible on the front side of the shipping pallet.

> The transformer shall meet interchangeability dimensions for single position mounting, as shown in Figures 1 and 9 of ANSI C57.12.20.

> The primary bushing shall be wet process porcelain cover-mounted. The primary connector shall be tin-plated bronze of the eye-bolt type. The H1 bushing is to be supplied with a wildlife protector and an eyebolt connector. The eyebolt connector shall accept a 5/16 inch diameter knurled stud terminal with the wildlife protector in place. The secondary bushings shall be wet process porcelain or epoxy, mounted on the tank wall. Secondary connectors shall be per ANSI C57.12.20, section 6.1.2.

The cover of the transformer shall be coated for wildlife protection. The coating shall have a rating of at least 10 kV AC rms. The insulating finish shall be capable of withstanding 10 kV at a 200 volt/sec rate of rise, tested per ASTM D149-87, using electrode type 3 (¼-inch cylindrical rods).

The transformer tank and cover shall be designed to withstand an internal pressure of 20 psig without tank rupture or bushing or cover displacement, and shall withstand an internal pressure of 7 psig without permanent distortion. The transformer tank shall retain its shape when the cover is removed.

Support lugs for single-position mounting shall be provided and shall be designed to provide a static-load safety factor of 5. Lifting lugs shall be provided, with location and safety factor as specified in ANSI C57.12.20. The tank and cover finish shall be light gray No. 70 (Munsell 5BG7.0/0.4). The interior shall be coated with a durable protective finish from the top to at least below the top liquid level.

The transformer shall be equipped with a tank ground per ANSI C57.12.20, paragraph 6.5.4.1 with eyebolt to accept an AWG No. 6 solid copper ground wire; a tank ground for ground strap connection; and a ground strap, properly sized to carry load and fault current, connected between the secondary neutral terminal and the tank ground. The grounding strap shall be supplied with a connection hole at each end. "J" hook or slotted connection points shall not be allowed.

The transformer shall be equipped with neoprene or Buna-N gaskets on all primary and secondary bushings and tank fittings.

A Qualitrol 201-020-01 automatic pressure relief device, with operation indication, or approved equivalent; or a cover assembly designed for relief of excess pressure in accordance with ANSI C57.12.20, section 6.2.5.2, shall be supplied on the transformer.

Final core and coil assembly shall result in a rigid assembly that maintains full mechanical, electrical, and dimensional stability under fault conditions.

The grade of the silicon core steel shall comply with the U.S. Department of Energy's Office of Energy Efficiency and Renewable Energy Conservation Program for Commercial Equipment: Distribution Transformers Energy Conservation Standards. The rule sets minimum energy-efficiency standard levels for liquid-immersed distribution transformers. The coil insulation system shall consist of electrical kraft paper layer insulation coated on both sides with thermosetting adhesive. Curing of the coils shall cause a permanent bond layer to layer and turn to turn.

The internal primary leads shall have sufficient length to permit cover removal and lead disconnection from the high-voltage bushings. These leads shall be adequately anchored to the coil assembly to minimize the risk of damaging the coil connection during cover removal.

The internal secondary leads shall be identified by appropriate markings, embossed or stamped on a visible portion of the secondary lead near the end connected to the bushing. These leads shall have sufficient length to permit series or parallel connection of the secondary windings. If the internal secondary leads are soft aluminum, a hard aluminum-alloy tab shall be welded to the bushing end of the lead for connection to the bushing.

The transformer should be designed so that the voltage regulation shall not exceed .5% at full load and 0.8 power factor for all transformers covered in this specification.

All transformers shall be filled with Type I insulating oil meeting the following requirements:

    a. All applicable requirements of ASTM D 3487, paragraph 82a.

    b. Minimum breakdown value of 28 kV per ASTM D 1816, paragraph 84a using a 0.4-in gap.

    c. The PCB content shall be less than 2 parts per million which shall be noted upon the nameplate of each unit. In addition, the manufacturer shall provide and install a "No PCB" identification on the tank wall, near the bottom of the tank centered below the low-voltage bushings. The decal shall be not less than 1 in by 2 in in size and shall indicate "No PCB's" in blue lettering on a white background, and be sufficiently durable to remain legible for the life of the transformer.

A typical specification for a single-phase 50-kVA pad-mounted distribution transformer (Figs. 15.27 and 15.28) would read as follows:

This specification defines the electrical characteristics and mechanical features of a pad-mounted distribution transformer; single phase, 60 Hz, mineral oil-immersed, self-cooled, loop-feed primary, dead front primary; 50-kVA capacity; 13,200 GrdY/7,620-120/240 volts; and the no-load tap-changer switch is to be located on the primary winding.

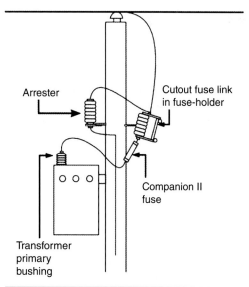

Figure 15.27 A partial-range Companion II current-limiting fuse installed on the source side of the fused cutout. (*Courtesy Cooper Power Systems, Inc.*)

Figure 15.28 A partial-range Companion II current-limiting fuse installed on the load side of the fused cutout. (*Courtesy Cooper Power Systems, Inc.*)

All requirements shall be in accordance with the latest revisions of ANSI Standard C57.12.25 and ANSI Standard C57.12.28, except where modified by this specification.

The transformer housing shall be designed to be tamper-resistant to prevent unauthorized access in accordance with the latest revision of ANSI C57.12.28 "For Switchgear and Transformers—Pad-Mounted Equipment—Enclosure Integrity."

The transformer shall be furnished with a coating system using munsell green color 7GY3.29/1.5 that conforms to the proposed "EEI Finishing Guidelines for Pad-Mounted Equipment."

The transformer shall be furnished with Type 2 interchangeability as per ANSI C57.12.25, Figure 2(a).

The transformer shall be equipped with two primary bolted-on bushing wells rated 8.3 kV and 200 amps in accordance with the latest revision of ANSI/IEEE Standard 386. The bushing wells shall be equipped with 8.3/14.4-kV 200-amps load break bushing inserts to accommodate load break elbows for sectionalizing cable in a loop-feed primary system. A static drain wire shall be attached from each bushing insert to the bushing well clamping flange.

The secondary terminals are to be equipped with bolted ANSI standard copper-threaded studs only.

A ground pad shall be provided on the outer surface of the tank. The low-voltage neutral shall have a fully insulated bushing. A removable ground strap sized for the rating of the transformer shall be provided and connected between the low-voltage neutral bushing and the transformer tank ground pad.

The high-voltage and low-voltage ground pads inside the transformer compartment shall be equipped with eyebolt connectors to accept at least No. 6 solid copper through No. 1/0 bare stranded copper wire.

The transformer shall be equipped with a Cooper Power Systems Bay-O-Net, or approved equivalent, externally replaceable overload sensing primary fuse for coordinated overload and secondary fault protection coordinated and in series with an 8.3-kV partial-range current-limiting fuse. The current-limiting fuse shall be sized so that it will not operate on secondary short circuits or transformer overloads. The current-limiting fuse shall be installed under oil inside the transformer tank. Each load-sensing and current-limiting fuse device shall be capable of withstanding 15-kV system recovery voltage across an open fuse.

The transformer shall be equipped with an oil drip shield directly below the Bay-O-Net fuse holder to prevent oil dripping on the primary elbows or bushings during removal of the fuse.

Final core and coil assembly shall result in a rigid assembly that maintains full mechanical, electrical, and dimensional stability under fault conditions.

The grade of the silicon core steel shall comply with the U.S. Department of Energy's Office of Energy Efficiency and Renewable Energy Conservation Program for Commercial Equipment: Distribution Transformers Energy Conservation Standards. The rule sets minimum energy-efficiency standard levels for liquid-immersed distribution transformers. The coil insulation system shall consist of electrical kraft paper layer insulation coated on both sides with thermosetting adhesive. Curing of the coils shall cause a permanent bond layer to layer and turn to turn.

The transformer shall be designed so the average sound levels do not exceed values specified in NEMA TR1, paragraph 0.05.

The transformer shall be provided with taps on the high-voltage windings that are 7,970/7,620/7,430/7,200/7,020 volts.

The transformer shall be bar coded in accordance with the proposed EEI "Bar Coding Distribution Transformers: A Proposed EEI Guideline." The manufacturer and serial number shall be bar coded on the nameplate. The serial number and manufacturer shall be bar coded on a temporary label and placed so it is visible on the front side of the shipping pallet.

All transformers shall be filled with Type I insulating oil meeting the following requirements:

 a. All applicable requirements of ASTM D 3487, paragraph 82a.

 b. Minimum breakdown value of 28 kV per ASTM D 1816, paragraph 84a using a 0.4-in gap.

 c. The PCB content shall be less than 2 parts per million which shall be noted upon the nameplate of each unit.

A typical specification for a three-phase pad-mounted distribution transformer (Fig. 15.4) would read as follows:

 This specification defines the electrical characteristics and mechanical features of a 300-kVA, three-phase pad-mounted distribution transformer; three-phase; 60-Hz, mineral oil-immersed, self-cooled, loop-feed primary; dead front primary; 300-kVA capacity; high-voltage windings rated 13,800 GrdY/7,970 volts; low-voltage windings rated 208 GrdY/120 volts; the high-voltage winding shall have four 2½% taps below rated voltage.

 All requirements shall be in accordance with the latest revision of ANSI Standard C57.12.26 and ANSI C57.12.28, except where modified by this specification.

 The transformer housing shall be designed to be tamper-resistant to prevent unauthorized access in accordance with the latest revision of ANSI C57.12.28 "For Switchgear and Transformers—Pad-Mounted Equipment—Enclosure Integrity."

 The door sill must be removable to allow the transformer to be slid into place over cables. The door sill must not interfere with the operation of the primary separable connectors.

 The transformer shall be furnished with a coating system using munsell green color 7GY3.29/1.5 that conforms to the proposed "EEI Finishing Guidelines for Pad-Mounted Equipment."

 The transformer shall have specific dimensions for Type B transformers for loop-feed systems as defined in the latest revision of ANSI Standard C57.12.26. The transformer primary bushing wells and parking stands shall be located, as shown in Figure 6(a) of the Standard. The secondary bushing configuration and placement shall conform to Figure 8(a)—"Staggered Low-Voltage Terminal Arrangement."

 The transformer shall be equipped with six primary bolted-on bushing wells rated 8.3 kV and 200 amps in accordance with the latest revision of the ANSI/ IEEE Standard 386, including bushing well interface Fig.#1 of that Standard.

The bushing wells shall be externally clamped to the transformer tank. The bushing wells shall be equipped with 8.3/14.4-kV 200-amps load break bushing inserts to accommodate load break elbows for sectionalizing cables in a loop-feed primary system. A static drain wire shall be attached from each bushing insert to the bushing well clamping flange.

The transformer secondary bushings are to be externally clamped to the tank and conform with ANSI C57.12.26. All spades shall be tinned copper or bronze and the 10-hole spade terminals shall be furnished with additional support designed by the manufacturer and attached to the top side of the spade terminal without interfering with the use of the spade.

The high-voltage neutral (H0) shall be securely grounded to the tank internally and be independent of all other connections. The low-voltage neutral (X0) shall be a fully insulated bushing. A ground pad shall be provided on the outer surface of the tank and a removable ground strap sized for the rating of the transformer shall be provided and connected between the low-voltage neutral bushing and the transformer tank ground pad.

The high- and low-voltage bushings of the transformer shall be separated by a metal barrier having a gap of at least 2 inches between the bottom of the barrier and the transformer pad to allow ground connections between compartments.

The high-voltage compartment and low-voltage compartment ground pads shall be ANSI standard and equipped with a transformer grounding lug to accept at least No. 6 solid copper to No. 2/0 bare-stranded copper wire.

The transformer shall be equipped with Cooper Power Systems Bay-O-Net, or approved equivalent, externally replaceable overload-sensing primary fuses, for coordinated overload and secondary fault protection, coordinated and in series with 8.3-kV partial-range current-limiting fuses. The current-limiting fuses shall not operate on secondary short circuits or transformer overloads. The current-limiting fuses shall be installed under oil inside the transformer tank. Each load-sensing and current-limiting fuse device shall be capable of withstanding 15-kV system recovery voltage across an open fuse.

The Bay-O-Net fuse holder shall be of the Cooper Power Systems flapper design. The fuse holder must provide a positive tank seal and provision must be made for equalizing any tank pressure before withdrawing the fuse.

The transformer shall be equipped with oil drip shields directly below the Bay-O-Net fuse holders, to prevent oil dripping on the primary elbows or bushings during removal of the fuses.

The transformer tank shall have a removable cover or handhole of the tamper-resistant design.

A Qualitrol 201-020-01 automatic pressure relief device with an operation indication, or approved equivalent, shall be supplied on the transformer.

The transformer tank shall be equipped with a 1-in oil drain valve located in the primary compartment, with a built-in oil sampling device.

The primary and secondary windings shall be connected grounded wye-grounded wye (Grd Y-Grd Y) and wound on a four-legged, five-legged, or triplex core to prevent excessive tank heating under single-phase supply or unbalanced load conditions.

The grade of the silicon core steel shall comply with the U.S. Department of Energy's Office of Energy Efficiency and Renewable Energy Conservation Program for Commercial Equipment: Distribution Transformers Energy Conservation Standards. The rule sets minimum energy-efficiency standard levels for liquid-immersed distribution transformers.

The percentage impedance voltage as measured on the rated voltage connection shall be between 1.2 and 6.00.

The transformer shall be designed so the average sound levels do not exceed values specified in NEMA TR1, paragraph 0.05.

The transformer shall be provided with taps on the high-voltage windings that are 13,800/13,200/12,870/12,470/12,160. The no-load tap changer switch is to be located in the high-voltage compartment.

The transformer shall be bar coded in accordance with the proposed EEI "Bar Coding Distribution Transformers: A Proposed EEI Guideline." The manufacturer and serial number shall be bar coded on the nameplate. The serial number and manufacturer shall be bar coded on a temporary label and should be placed so it is visible on the front side of the shipping pallet.

All transformers shall be filled with Type I insulating oil meeting the following requirements:

a. All applicable requirements of ASTM D 3487-82a.

b. Minimum breakdown value of 28 kV per ASTM D 1816-84a using a 0.4-in gap.

c. The PCB content shall be less than 2 parts per million which shall be noted upon the nameplate of each unit.

The above specifications can be changed to meet the kVA capacity and voltages required. Changes in the high-voltage requirements also require revisions in the related items to obtain the proper insulation and surge-withstand capabilities.

Safety Precautions for Distribution Transformers

Distribution transformers must be protected from violent explosions. When a distribution transformer is removed from service as a result of an automatic operation of the transformer protective devices, the transformer should be tested to be sure it does not have an internal fault before it is reenergized by a lineman. Linemen have closed a fused cutout on a distribution transformer, which had an internal fault, and have been injured by the explosion and oil fire that resulted. An internal protective link fuse, commonly installed in self-protected distribution transformers, has a limited interrupting rating and cannot be depended on to operate in a manner to prevent catastrophic failure of the transformer. The interrupting rating of a fused cutout, in series with the primary winding of the transformer with or without an internal weak-link fuse, may exceed the available fault current from the electric system and fail to prevent the explosion of the transformer as a result of an internal arcing fault. The operation of the cutout, with an expulsion fuse, may take one-half

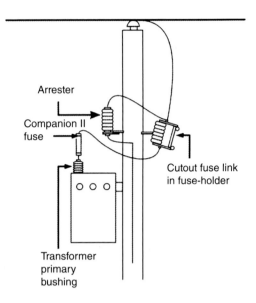

FIGURE 15.29   A partial-range Companion II current-limiting fuse installed on the terminal on the primary bushing of an overhead distribution transformer. (*Courtesy Cooper Power Systems, Inc.*)

cycle, permitting internal arcing under oil in the transformer, which can cause a violent explosion and fire.

Current-Limiting Fuses

A current-limiting fuse can be installed in series with the primary winding of a distribution transformer to prevent violent transformer failures as a result of an internal arcing fault. The current-limiting fuse can be installed external to the transformer or inside the tank under oil. The current-limiting fuse is usually installed externally either on the source side of the fused cutout (Fig. 15.27), on the load side of the fused cutout (Fig. 15.28), on the bushing of a pole-mounted distribution transformer (Fig. 15.29), or internally under oil in a pad-mounted transformer. The current-limiting fuse must be the proper sized to coordinate with the other transformer protective devices. Current-limiting fuses are discussed further in Chap. 17.

Lightning and Surge Protection

Lightning

Lightning flashes are currents of electricity flowing from one airborne cloud to another or between the cloud and earth. They always flow by the path of least resistance, just as water takes the path of least resistance when it flows down the hillside. Such paths down the hillside are not generally straight but, on the contrary, are very crooked. The same is true of the path taken by a bolt of lightning; this fact explains the zigzag path of the flash. Trees, high buildings, towers, transmission and distribution lines, and poles are paths of less resistance than air, and therefore lightning often strikes them and flows through them to ground, most times with damaging results (Figs. 16.1 and 16.2).

Overhead lines have charges of electricity induced on them when charged clouds pass over them. Induced charges trapped on electric circuits can create abnormal voltages on the transmission and distribution lines when lightning strokes travel to ground, discharging the clouds. The abnormal voltages placed on the line will travel along the line until dissipated by surge arrester operation, attenuation, and/or leakage or by failure of insulators, transformers, or other apparatus connected to the line.

Lightning-producing storms are common events. It is estimated that 44,000 thunderstorms rage daily over the earth, producing 100 flashes of lightning every second. Figure 16.2 shows graphically the approximate number of lightning-producing storm days in various parts of the continental United States during a year. In some areas, the number is as low as five electrical storm days, and in other parts it is as high as 90 electrical storm days. Thunderstorms bring rain, strong winds, tornadoes, and lightning. All these can cause major damage to electric facilities.

Surge

Large overvoltages that develop suddenly on electric transmission and distribution circuits are referred to as surges or transients. The surges can be caused by direct lightning strikes or induced charges as a result of lightning strokes to ground, as described, or by circuit-switching operations as well as the operation of devices connected to the lines. The large overvoltages may damage electrical equipment connected to the lines and line insulators, even though they are transient or quick to disappear.

FIGURE 16.1   The damaging result to the underground electric distribution system due to a nearby lightning strike on the overhead distribution system.

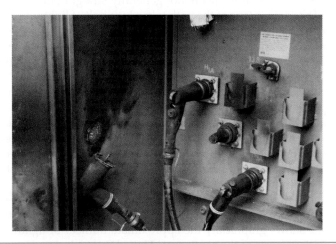

FIGURE 16.2   The load break elbow was violently separated from the single phase pad-mount transformer bushing by the lightning impulse surge.

Shield Wire

A ground wire, or static wire, mounted above the other line conductors and strung along the line, the same as other line conductors, can provide protection or shielding to the circuit from direct lightning strokes (Figs. 16.3 and 16.4). The shield wire on an overhead line may be a galvanized steel, copper-covered steel, or aluminum-covered steel cable.

The shield wire on lines supported by metal towers is grounded by a connection to each tower, with a metal clamp which fastens it to the tower and the ground conductors and ground rods installed at the base of the tower.

The shield wire on lines supported by wood poles is connected to ground at frequent intervals by a ground conductor extending from the metal clamp, which supports the shield wire at the top of the pole, to a ground rod adjacent to the pole just below the ground surface. The grounds may be installed on each pole or may be as far apart as every fifth pole. If the pole line has a common-neutral ground wire on the support structures for a distribution circuit, the ground wire must be connected to the common-neutral conductor on the structure. The National Electrical Safety Code requires that the common-neutral for primary and secondary circuits be grounded at each piece of equipment connected to the line and have at least four connections to ground per mile. A properly installed shield wire is highly effective, but its effectiveness depends on low ground resistance. The shield wire virtually brings the ground or earth potential above the transmission and/or distribution line. The low-resistance shield wire attracts the lightning stroke and diverts the impact of the corresponding overvoltage. The stress on the line and equipment insulation due to lightning direct strokes is greatly reduced by the shield wire.

Figure 16.4   A two-pole 161-kV transmission line is protected with an overhead shield wire. *(Courtesy MidAmerican Energy Co.)*

Lightning or Surge Arresters

American National Standards Institute's Surge Arresters for Alternating-Current Power *Circuits* (ANSI/IEEE C62.1) provides specifications for surge arresters. ANSI C62.2, *Guide for the Application of Valve-Type Surge Arresters for Alternating-Current Systems,* provides proper methods to apply them. A surge arrester is a device that prevents high voltages, which would damage equipment, from building up on a circuit, by providing a low-impedance path to ground for the current from lightning or transient voltages, and then restores normal circuit conditions.

The surge arrester is installed in strategic locations of the transmission and distribution system to protect equipment such as transformers (Fig. 16.5), conductor dead-end insulators, reclosers, sectionalizers, and group-operated disconnect switches. The arrester is installed at transition points from overhead to underground systems (Fig. 16.6). Finally, the arresters are installed at some double-dead-end conductor locations.

Lightning arrester

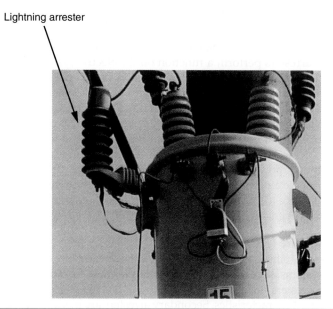

FIGURE 16.5   Lightning arrester installed to protect the overhead distribution transformer.

Lightning arrester

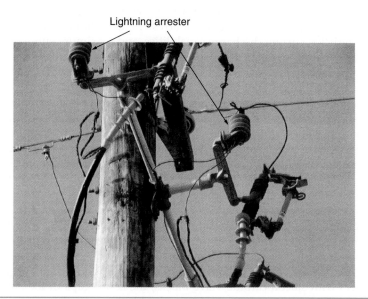

FIGURE 16.6   The transition from the overhead system to the underground system is protected from overvoltages by the installation at the riser pole of the lightning arrester.

The arrester is installed such that when the system is operating at the designed voltage, the arrester is nonconducting. When an overvoltage is induced on the system, the arrester diverts the overvoltage to ground, protecting the system.

Surge arresters perform a function on the electric system similar to that of a safety valve on a steam boiler. A safety valve on a boiler relieves high pressure by blowing off steam until the high pressure is reduced to normal. When the pressure is reduced to normal, the safety valve closes again and is ready for the next abnormal condition. When a high voltage (greater than the normal line voltage) exists on the line, the arrester immediately furnishes a path to ground and thus limits and drains off the excess voltage. Furthermore, when excess voltage is relieved, the action of the arrester must prevent any further flow of power current. The function of a surge arrester is, therefore, twofold—first, to provide a point in the circuit at which the overvoltage impulse can pass to earth without damage to line insulators, transformers, or other line connected equipment and, second, to prevent any follow-up power current from flowing to ground.

Classes of Valve-Type Surge Arresters

Four classes of silicon-carbide valve-type surge arresters are manufactured for alternating-current systems: secondary, distribution, intermediate, and station. These four classes of surge arresters have similar components, but the surge arresters are manufactured with different characteristics. Secondary arresters are manufactured with voltage ratings of 175 volts and 650 volts. Distribution surge arresters for the standard voltage distribution systems are 3 kV, 6 kV, 9 kV, 10 kV, 12 kV, 15 kV, 18 kV, 21 kV, 25 kV, 27 kV, and 30 kV. Intermediate arresters are manufactured with voltage ratings of 3 kV through 120 kV. Station arresters are manufactured with voltage ratings of 3 kV and above. These four different classifications of arresters have different overvoltage conducting characteristics, current discharge capabilities, and maximum surge discharge voltages. The cost of surge arresters increases as the capacities are increased.

Secondary arresters are used on service and other low-voltage alternating-current circuits. Distribution arresters are used on primary distribution systems to protect insulators, distribution transformers, power cables, and other equipment. Intermediate-type surge arresters are often used on substation exit cables and other locations in the distribution system needing a higher level of lightning and surge protection. Station-type surge arresters are used in substations and generating stations to provide a highest level of surge protection for the major pieces of equipment. Surge voltages are generated by operating switches and other equipment connected to the electric transmission system as well as by lightning.

The surge arresters with silicon-carbide valve blocks have gaps in a sealed porcelain casing and grading resistors or capacitors to control the sparkover voltages of the gaps. The silicon-carbide valve blocks function as a nonlinear resistor that passes large surge currents and limits the power follow current.

When a surge voltage of high magnitude develops on a circuit protected with surge arresters with silicon-carbide valve blocks, the series gaps in the arresters will spark over, permitting the nonlinear resistor valve block to pass the surge current to ground. The surge voltage will be limited by the voltage drop across the arrester and the connecting wires ($V = I \times Z$). The resistance of the silicon-carbide valve blocks, while conducting the large surge current, is very low, keeping the voltage across the surge arrester low. This keeps the voltage across the equipment, connected in parallel with the surge arrester, below the value that would damage the insulation of the equipment. After the surge current has been bypassed to ground, power follow current will flow through the surge arrester to ground. The nonlinear resistor silicon-carbide valve block develops a high resistance to the 60-Hz current, preventing a short circuit to ground on the line. When the instantaneous value of the 60-Hz power follow current goes through zero, the arrester gaps interrupt the current, returning the circuit conditions to normal.

The metal oxide varistor (MOV) surge arrester is now the industry standard (Figs. 16.7 through 16.12). First generation MOV surge arresters, constructed with zinc-oxide valve blocks, do not need gaps. The zinc-oxide valve blocks in an arrester effectively insulate the electrical conductors from ground at normal electric system operating voltages, limiting the leakage current flow through the arrester to a small value. When a surge voltage develops on the electric system protected by metal-oxide surge arresters, the arrester valve blocks conduct large values of current to ground, limiting the $I \times R$ voltage drop across the arrester. If the connections to the arrester have a low impedance, the stress on the electric system is kept below the *basic insulation level* (BIL) or critical insulation level of the equipment. The zinc-oxide valve blocks or disks start conducting sharply at a precise voltage level and cease to conduct when the voltage falls below the same level. The arrester zinc-oxide valve blocks have the characteristics of a nonlinear resistance.

The metal-oxide gapless arresters are smaller and lighter than valve-type arresters with gaps. Elimination of the gaps reduces the frequency of failure of arresters as a result of surface contamination, which can cause arrester gap spark over. Metal-oxide valve-type arresters without gaps act quickly to limit the surge voltage, improving the electric system protection.

FIGURE 16.7   Sectional view of metal-oxide valve (MOV) distribution-class surge arrester with zinc-oxide valve blocks encased in a polymer to prevent sharp flying porcelain fragments if the arrester should fail. (*Courtesy Ohio Brass Co.*)

FIGURE 16.8   A 8.4-kV MCOV/10-kV-rated overhead distribution MOV surge arrester. (*Courtesy Ohio Brass Co.*)

Copper
Terminal Stud

Protective Cap

Pulling Eye

Probe

Shield Grounding Tab

Peroxide Cured EPDM Housing

Metal Oxide Varistor

Epoxy-Fiberglass Wrap

Non-Fragmenting (MOV)

Stainless Steel Cap

FIGURE 16.9   Sectional view of a dead-front MOV elbow surge arrester utilized in an underground distribution system. (*Courtesy Ohio Brass Co.*)

FIGURE 16.10   A 8.4-kV MCOV/10-kV-rated underground distribution dead-front MOV elbow surge arrester. *(Courtesy Ohio Brass Co.)*

FIGURE 16.11   Surge arrester utilized on a three-phase primary riser pole. The lightning arrester is installed on the load side of the fused cutout.

FIGURE 16.12   Station-class metal-oxide surge arresters installed in a substation. (*Courtesy Ohio Brass Co.*)

MOV surge arresters produced by some manufacturers have been designed with a combination of metal oxide and gapped arrester technologies. For the gapped MOV arrester, during steady-state conditions, the line voltage is shared by both the gap structure and the MOVs. When surges occur, the gaps spark over, leaving only the MOVs in the circuit. The result is a lower discharge voltage and increased arrester TOV withstand ability than exhibited by the gapless MOV surge arrester.

Surge Arrester Application

Application of surge arresters requires knowledge of the maximum phase-to-ground system voltage, maximum discharge voltage of the arrester, and BIL of the equipment to be protected. The BIL for distribution systems is provided in Table 16.1. The basic impulse insulation level is the crest value of voltage of a standard full-impulse voltage wave that the insulation will withstand without failure. The surge arrester must be selected so that it will not conduct current as a result of maximum system operating voltage and will have a maximum-surge discharge voltage approximately 20% less than the basic impulse insulation level of the equipment to be protected. Typical descriptive data for valve-type surge arresters are given in Table 16.2 for distribution class, in Table 16.3 for intermediate class, and in Table 16.4 for station class.

The metal-oxide valve arresters with zinc-oxide blocks have a very small leakage current at normal system line-to-ground voltage. The zinc-oxide disks go into conduction sharply at a precise voltage level and cease to conduct when the voltage falls below this level. Valve arresters with either zinc-oxide or silicon-carbide valve blocks with the same rating will provide adequate protection for the line and its equipment with approximately the same stress on the protected facilities as a result of the arrester discharge voltage. However, the uniformity of the operation of the zinc-oxide disks

Voltage	Basic Insulation Level (BIL)
5 kV	70
15 kV	95
25 kV	125
35 kV	150

TABLE 16.1   Basic Insulation Levels for Distribution Systems

Arrester Rating Max. Line to Ground, kV rms	Minimum 60-Hz Sparkover, kV rms	Maximum 1½ × 40 Wave kV Crest	Maximum Impulse Sparkover ANSI Front-of-Wave kV Crest, Maximum	Discharge Voltage in kV Crest for Discharge Currents of 8 × 20 Microsecond Waveshape with Following Maximum Crest Amplitudes				
				1,500-amp kV Crest, Maximum	5,000-amp kV Crest, Maximum	10,000-amp kV Crest, Maximum	20,000-amp kV Crest, Maximum	65,000-amp kV Crest, Maximum
3	6	14	17	10	12.4	13.8	15.5	19
6	11	26	33	19	23	26	29	35
9	18	38	47	30	36.5	41	46	56
10	18	38	48	30	36.5	41	46	56
12	23.5	49	60	38	46	52	58	71
15	27	50	75	45	55	62	70	85
18	33	58	90	54	66	74	83	103

TABLE 16.2 Typical Application Data for Silicon Carbide Distribution-Class, Nonlinear Resistor Valve-Type Lightning Arresters with Gaps (Electrical Characteristics)

Arrester Rating kV rms Maximum Valve-Off or Maximum Reseal Rating	Maximum Circuit Voltage Phase to Phase, kV rms		Maximum Front-of-Wave Impulse Sparkover, kV Crest	Maximum 100% Impulse Sparkover 1.2 × 50 Wave, kV Crest	Minimum 60-Hz Sparkover, kV rms	Maximum Discharge Voltage with Discharge Current, 8 × 20 Wave, kV Crest				
	Ungrounded Neutral 100% Arrester	Effectively Grounded Neutral 80% Arrester				2	3 kA	5 kA	10 kA	20 kA
3	3	3.75	11	11	4.5	5.2	6	6.6	7.5	8.7
4.5	4.5	5.63	16	15	6.8	7.8	9	9.9	11.3	13.1
6	6	7.50	21	19	9	10.4	11.9	13.2	15	17.4
7.5	7.5	9.38	26	23.5	11.3	13	14.9	16.5	18.8	21.8
9	9	11.25	31	27.5	13.5	15.6	17.9	19.8	22.5	26.1
10	10	12.50	35	31	15	17.5	20.0	22.0	25.0	29.0
12	12	15.00	40	35.5	18	20.8	23.8	26.4	30	34.8
15	15	18.75	50	43.5	22.5	25.9	29.7	32.9	37.5	43.5
18	18	22.50	59	51.5	27	31.1	35.7	39.5	45	52.2
21	21	26.25	68	59	31.5	36.3	41.6	46.1	52.5	60.9
24	24	30.00	78	67	36	41.5	47.6	52.7	60.0	69.6
27	27	33.75	88	75	40.5	46.7	53.5	59.2	67.5	78.3
30	30	37.50	97	81	45	51.8	59.4	65.8	75	87
36	36	45.00	116	95	54	62.2	71.3	79	90	104.4
39	39	48.75	126	102	58.5	67.4	77.3	85.5	97.5	113.1
45	45	56.25	144	116	67.5	77.7	89.1	98.7	112.5	130.5
48	48	60.0	154	123	72	82.9	95.1	105.3	120	139.2
60	60	75.00	190	153	90	103.6	119	131.2	150	174
72	72	90.00	228	180	108	124.3	142.6	158	180	208.8
78	78	97.50	245	195	117	134.7	154.5	171	185	226.2
84	84	105.00	262	209	126	145	166.4	184.2	210	243.6
90	90	112.50	282	223	135	155.4	178.2	197.3	225	261
96	96	120.00	300	236	144	165.7	190.1	210.5	240	278.4
108	108	135.00	335	263	162	186.5	213.9	237	270	313.1
120	120	150.00	370	290	180	207.2	238	263	300	347.9

TABLE 16.3 Typical Application Data for Silicon Carbide Intermediate-Class, Nonlinear Resistor Valve-Type Lightning Arresters with Gaps (Electrical Characteristics)

Arrester Rating, Maximum Permissible Line-to-Ground kV rms (Maximum Valve-Off or Maximum Reseal Rating)	Minimum 60-Hz Sparkover, kV rms	Maximum Switching Surge Sparkover, kV Crest	Maximum 100% 1½ × 40 Impulse Sparkover, kV Crest	Maximum Impulse Sparkover, ASA Front-of-Wave, kV Crest	Maximum Discharge Voltage for Discharge Currents of 8 × 20 Microsecond Waveshape with the Following Crest Amplitudes					
					1,500 amp, kV Crest	3,000 amp, kV Crest	5,000 amp, kV Crest	10,000 amp, kV Crest	20,000 amp, kV Crest	40,000 amp, kV Crest
3	5	8	8	12	7	8	8.5	9	10	11.5
6	11	17	17	24	15	16	17	19	20	23
9	16	25	24	35	21	23	24	26	28	31.5
12	22	34	32	45	28	30	32	35	38	42
15	27	42	40	55	35	38	40	44	47	52.5
21	36	56	55	72	47	52	55	60	65	73
24	45	70	65	90	56	61	65	71	76	84
30	54	85	80	105	69	75	80	87	94	105
36	66	104	96	125	83	90	96	105	113	126
39	72	113	104	130	90	98	104	114	123	137
48	90	141	130	155	112	122	130	142	153	168
60	108	169	160	190	137	150	160	174	189	210
72	132	206	195	230	167	184	195	212	230	252
90	160	242	228	271	206	226	240	262	283	315
96	175	257	247	294	222	242	258	280	304	336
108	195	294	266	332	244	264	282	316	333	378
120	220	323	304	370	275	301	320	350	378	420

TABLE 16.4 Typical Application Data for Silicon Carbide Station-Class, Nonlinear Resistor Valve-Type Lightning Arresters with Gaps (Electrical Characteristics)

without the need for series gaps permits the selection of a lower-rated voltage arrester, which will provide better protection for the line and its equipment. When a metal-oxide arrester with zinc-oxide blocks is chosen, it is necessary to be sure the arrester's maximum power-frequency continuous operating voltage (MCOV) will not be exceeded. Typical descriptive data for MOV arresters is provided in Tables 16.5 through 16.8. Table 16.9 lists the commonly applied voltage ratings of distribution surge arresters. The MOV arresters are defined as (a) normal duty, (b) heavy duty, (c) riser pole, and (d) station class. Note the difference in the characteristics in the normal and heavy-duty arresters described in Tables 16.5 and 16.6. Specifically, the normal-duty distribution arrester IEEE Standard C62.11 design and test requirements are: (a) low current, long duration; twenty surges with 2,000 μs duration and 75 amps magnitude, (b) high current, short duration; two surges with 4/10 μs duration and 65 kA magnitude, and (c) duty cycle, twenty-two 5 kA discharges. The heavy-duty distribution arrester IEEE Standard C62.11 design and test requirements are: (a) low current, long duration; twenty surges with 2,000 μs duration and 250 amps magnitude, (b) high current, short duration; two surges with 4/10 μs duration and 100 kA magnitude, and (c) duty cycle, twenty 10 kA discharges and two 40 kA discharges.

A few surge arrester application examples are as follows:

Surge arrester selection criteria

Maximum phase-to-ground system voltage

Maximum discharge voltage

Maximum continuous operating voltage

Basic impulse level of the equipment to be protected

Insulation Protective Margin Equation

$$\text{Protective margin} = \left[\left(\frac{\text{insulation withstand}}{V_{max}}\right) - 1\right] \times 100$$

The equation is taken from IEEE Standard C62.2, *Application Guide for Surge Arresters*. The minimum recommended margin is 20%.

Example   Determine the margin of protection for both the normal-duty and heavy-duty MOV arrester on a 15-kV class, 13,200/7,620 volt, four wire, overhead system utilizing the 20 kA 8/20 kV discharge voltage.

Solution   From Table 16.1, the 15-kV equipment class BIL is 95, from Table 16.9, a 10-kV-rated arrester is utilized. The corresponding maximum discharge voltage is 41.5 for the normal-duty arrester (Table 16.5) and 37.5 for the heavy-duty arrester (Table 16.6).

$$\text{Protective margin} = \left[\left(\frac{\text{insulation withstand}}{V_{max}}\right) - 1\right] \times 100$$

Normal-duty arrester protective margin = $[(95/41.5) - 1] \times 100 = 129\%$

Heavy-duty arrester protective margin = $[(95/37.5) - 1] \times 100 = 153\%$

Not included are the effects of faster rates of current rise on the discharge voltage, reduced insulation levels due to various factors, and line and ground leads.

Rated Voltage, kV	MCOV, kV	0.5 μsec 5 kA Maximum IR, kV	500 A Switching Surge Maximum IR, kV	8/20 Maximum Discharge Voltage, kV					
				1.5 kA	3kA	5kA	10 kA	20 kA	40 kA
3	2.55	12.5	8.5	9.8	10.3	11.0	12.3	14.3	18.5
6	5.1	25.0	17.0	19.5	20.5	22.0	24.5	28.5	37.0
9	7.65	33.5	23.0	26.0	28.0	30.0	33.0	39.0	50.5
10	8.4	36.0	24.0	27.0	29.5	31.5	36.0	41.5	53.0
12	10.2	50.0	34.0	39.0	41.0	44.0	49.0	57.0	74.0
15	12.7	58.5	40.0	45.5	48.5	52.0	57.5	67.5	87.5
18	15.3	67.0	46.0	52.0	56.0	60.0	66.0	78.0	101.0
21	17.0	73.0	49.0	55.0	60.0	64.0	73.0	84.0	107.0
24	19.5	92.0	63.0	71.5	76.5	82.0	90.5	106.5	138.0
27	22.0	100.5	69.0	78.0	84.0	90.0	99.0	117.0	151.5
30	24.4	108.0	72.0	81.0	88.5	94.5	108.0	124.5	159.0
36	29.0	134.0	92.0	104.0	112.0	120.0	132.0	156.0	202.0

Source: Ohio Brass Co.

TABLE 16.5   Normal-Duty Polymer Housed Metal-Oxide Valve Distribution Class Surge Arrester Electrical Characteristics

Rated Voltage, kV	MCOV, kV	0.5 µsec 10 kA Maximum IR, kV	500 A Switching Surge Maximum IR, kV	8/20 Maximum Discharge Voltage, kV					
				1.5 kA	3kA	5kA	10 kA	20 kA	40 kA
3	2.55	12.5	8.0	9.5	10.0	10.5	11.0	13.0	15.3
6	5.1	25.0	16.0	19.0	20.0	21.0	22.0	26.0	30.5
9	7.65	34.0	22.5	24.5	26.0	27.5	30.0	35.0	41.0
10	8.4	36.5	23.5	26.0	28.0	29.5	32.0	37.5	43.5
12	10.2	43.5	28.2	31.0	32.9	34.8	38.5	43.8	51.5
15	12.7	54.2	35.0	38.4	41.0	43.4	48.0	54.6	64.2
18	15.3	65.0	42.1	46.0	49.1	52.0	57.5	65.4	76.9
21	17.0	69.5	44.9	49.2	52.5	55.7	61.5	69.9	82.2
24	19.5	87.0	56.4	61.6	65.8	69.6	77.0	87.6	103.0
27	22.0	97.7	63.2	69.2	73.9	78.2	86.5	98.4	115.7
30	24.4	108.4	70.0	76.8	82.0	86.8	96.0	109.2	128.4
36	29.0	130.0	84.2	92.0	98.2	104.0	115.0	130.8	153.8

Source: Ohio Brass Co.

TABLE 16.6 Heavy-Duty Polymer Housed Metal-Oxide Valve Distribution Class Surge Arrester Electrical Characteristics

Duty Cycle Voltage Rating, kV	MCOV, kV	0.5 µsec 10 kA Maximum IR, kV	500 A Switching Surge Maximum IR, kV	8/20 Maximum Discharge Voltage, kV					
				1.5 kA	3 kA	5 kA	10 kA	20 kA	40 kA
3	2.55	8.7	5.8	6.5	7.0	7.4	8.1	9.0	10.6
6	5.1	17.4	11.7	13.0	14.0	14.7	16.2	18.1	21.1
9	7.65	25.7	17.5	19.3	21.0	21.9	24.0	27.0	31.6
10	8.4	28.5	19.2	21.2	23.0	24.0	26.5	29.8	34.8
12	10.2	34.8	23.3	25.9	28.0	29.4	32.3	36.2	42.2
15	12.7	43.1	29.1	32.3	35.0	36.6	40.2	45.1	52.7
18	15.3	51.4	34.9	38.6	41.9	43.8	48.0	54.0	63.2
21	17	57.6	38.7	42.8	46.4	48.6	53.6	60.2	70.5
24	19.5	68.8	46.6	51.6	55.9	58.5	64.2	72.1	84.3
27	22	77.1	52.4	57.9	62.9	65.7	72.0	81.0	94.8
30	24.4	85.5	57.6	63.5	69.0	72.0	79.5	89.4	104.4
36	29	102.8	69.8	77.2	83.8	87.6	96.0	108.0	126.6

Source: Ohio Brass Co.

TABLE 16.7   Riser Pole Polymer Housed Metal-Oxide Valve Distribution Class Surge Arrester Electrical Characteristics

Voltage Rating (kV rms)	MCOV (kV rms)	Equivalent F.O.W.* (kV Crest)	Maximum Discharge Voltage (kV Crest) Using an 8/20 μs Current Wave				
			1.5 kA	3 kA	5 kA	10 kA	20 kA
3	2.55	12.9	10.6	11.2	11.7	13.0	15.7
6	5.1	25.8	21.3	22.3	23.5	25.9	31.5
10	8.4	38.7	31.9	33.5	35.2	38.9	47.2
12	10.2	51.6	42.5	44.8	46.9	52.0	62.9
18	15.3	71.1	58.5	61.4	64.6	71.3	86.2
24	19.5	96.8	79.7	84.0	87.9	97.5	117.9
27	22.0	110.0	90.0	94.9	99.7	110.1	133.7

*The equivalent front-of-wave is the maximum discharge voltage for a 5 kA impulse current wave which produces a voltage wave cresting in 0.5 μs.
Source: Joslyn Manufacturing Co.

TABLE 16.8   Dead-Front Elbow Arrester Electrical Characteristics

System Voltage (kV rms)		Recommended Arrester Rating per IEEE C62.22 (kV rms)		
Nominal	Maximum	Four-Wire Wye Multi-Grounded Neutral	Three-Wire Wye Solidly Grounded Neutral	Delta and Ungrounded Wye
2.4	2.54	—	—	3
4.16Y/2.4	4.4Y/2.54	3	6	6
4.16	4.4	—	—	6
4.8	5.08	—	—	6
6.9	7.26	—	—	9
8.32Y/4.8	8.8Y/5.08	6	9	—
12.0Y/6.93	12.7Y/7.33	9	12	—
12.47Y/7.2	13.2Y/7.62	9	15	—
13.2Y/7.62	13.97Y/8.07	10	15	—
13.8Y/7.97	14.52Y/8.38	10	15	—
13.8	14.52	—	—	18
20.78Y/12.0	22Y/12.7	15	21	—
22.86Y/13.2	24.2Y/13.87	18	24	—
23	24.34	—	—	30
24.94Y/14.4	26.4Y/15.24	18	27	—
27.6Y/15.93	29.3Y/16.89	21	30	—
34.5Y/19.92	36.5Y/21.08	27	36	—
46Y/26.6	48.3Y/28	36	—	—

TABLE 16.9   Commonly Applied Voltage Ratings of Arresters

Example   Determine the margin of protection for a 15-kV underground system utilizing only a riser pole arrester and then utilizing both a riser pole arrester and an elbow arrester at the open point. Assume arrester lead length to be 5 ft and utilize the 20 kA 8/20 kV discharge voltage.

Solution

$$\text{Protective margin} = \left[\left(\frac{\text{insulation withstand}}{V_{max}}\right) - 1\right] \times 100$$

Protective Margin Calculation—Arrester Only at the Riser Pole

In the case of utilizing only an arrester at the riser pole, V_{max} consists of both the arrester discharge voltage and the arrester lead length calculated at 1.6 kV/ft. V_{max} is doubled due to the reflection of the wave at the open point in the underground system.

$V_{max} = 2 \times$ (discharge voltage of the arrester + voltage drop due to arrester lead length)

From Table 16.7, the 10 kV, 20 kA, 8/20 maximum discharge voltage is 29.8. The voltage drop due to a 5-ft lead length is 5×1.6 or 8.0 kV.

$$V_{max} = 2 \times (29.8 + 8.0) = 75.6 \text{ kV}$$

$$\text{Protective margin} = \left[\left(\frac{\text{insulation withstand}}{V_{max}}\right) - 1\right] \times 100$$

$$\text{Protective margin} = [(95/75.6) - 1] \times 100 = 25.7\%$$

Protective Margin Calculation—Arrester at Both the Riser Pole and the Open Point

In the case of utilizing an arrester at the riser pole and at the open point of the underground system, V_{max} consists of the arrester discharge voltage, the arrester lead length calculated at 1.6 kV/ft and some percentage of the front of wave voltage of the open point arrester. (The percentage is typically between 15% and 40%. For this calculation we will use a value of 30%.)

$V_{max} =$ (discharge voltage of the arrester + arrester lead length voltage drop + 0.3 × open point front of wave voltage)

From Table 16.7, the 10 kV, 20 kA, 8/20 maximum discharge voltage is 29.8. The voltage drop due to a 5-ft lead length is 5×1.6 or 8.0 kV. Finally, from Table 16.8, the open point front of wave voltage of a 10-kV arrester is 38.7. For purposes of the calculation, 30% of 38.7 equals 11.61.

$$V_{max} = (29.8 + 8.0 + 11.61) = 49.41 \text{ kV}$$

$$\text{Protective margin} = \left[\left(\frac{\text{insulation withstand}}{V_{max}}\right) - 1\right] \times 100$$

$$\text{Protective margin} = [(95/49.41) - 1] \times 100 = 92.3\%$$

Fuses and
Fuse Coordination

Fuses are overcurrent protection devices connected into circuits to open the circuit and deenergize the apparatus to prevent or limit damage due to an overload or a system fault. Fuses are used to protect the electric system or source from interruption or damage when a short circuit or overload occurs on a circuit beyond the fuse. A fuse is an intentionally weakened spot in an electric circuit. It utilizes an element made of silver, tin, lead, copper, or an alloy such as tin-lead, which will melt at a predetermined current maintained for a predetermined length of time. When the current through such a metal becomes excessive, the resistance offered by the metal to the flow of current develops enough heat to melt the metal, thereby opening the circuit before abnormal current can damage the electric source or any connected apparatus.

Fuses are usually enclosed to prevent the molten metal from flying and doing damage or causing a fire. Enclosing the fuse aids in quenching the arc. The melting of the fuse is often accompanied by a puff of smoke and vaporized metal. This action is sometimes referred to as the *blowing* of the fuse.

Fuses can be broadly classified into low- and high-voltage fuses. The low-voltage fuses are of the plug or cartridge type. High-voltage fuses, commonly used on an electric distribution system, are the expulsion, open-link, current limiting, and solid material types.

Plug-Type Fuses

The low-voltage plug-type fuse can be used for residential services and for ordinary lighting branch circuits on panel boards. The fuse consists of a small cup of solid insulating material within which the fusible wire connects the center contact with the outer metal screw shell. The plug is the same size as the ordinary incandescent-lamp base.

Cartridge Fuse

Cartridge fuses are used on circuits rated 600 volts and below. The fuses are manufactured differently for use on 240- and 480-volt circuits. A fuse of the proper voltage and current rating should be used for the circuit application. In the cartridge fuse the fusible element is contained within a completely enclosed insulating tube. This tube is called the *cartridge*, because of its resemblance to an ordinary shotgun cartridge. Cartridge fuses in which the element is not replaceable after the fuse has blown are called *nonrenewable fuses*. Cartridge fuses in which the fusible element may be readily replaced by the user with suitable renewable elements supplied by the manufacturer are called *renewable fuses*.

The cartridge or tube may thus be used repeatedly with the new renewable elements. In this way, a considerable saving is achieved, especially in the larger fuse sizes.

Distribution Cutouts

A distribution cutout provides a high-voltage mounting for the fuse element used to protect the distribution system or the equipment connected to it. Distribution cutouts are used with installations of transformers, capacitors, underground cable circuits (Fig. 17.1), and sectionalizing points on overhead circuits (Fig. 17.2). Enclosed, open, and open-link cutouts are used for different distribution-circuit applications. Cutouts normally use an expulsion fuse. An expulsion fuse operates to isolate a fault or overload from a circuit because of the arc from the fault current eroding the tube of the fuse holder, producing a gas that blasts the arc out through the fuse tube vent or vents. The mechanical differences between enclosed, open, and open-link cutouts are in their external appearance and method of operation. Enclosed cutouts have terminals, fuse clips, and fuse holders mounted completely within an insulating enclosure. Open cutouts have these parts completely exposed as the name indicates. Open-link cutouts

FIGURE 17.1   Distribution cutouts installed in transition point from an overhead electric distribution system to an underground system.

FIGURE 17.2   A 200-amp cutout fused at 40T for protection of the downstream overhead distribution system.

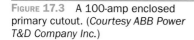

FIGURE 17.3 A 100-amp enclosed primary cutout. (*Courtesy ABB Power T&D Company Inc.*)

have no integral fuse holder; the arc-confining tube for the cutouts is incorporated in the fuse link.

The construction of the cutout fuse holder provides for a dropout operation. Fuses are manufactured with an expendable cap.

Enclosed Distribution Cutout

An enclosed distribution fuse cutout is one in which the fuse clips and fuse holder are mounted completely within an enclosure. A typical enclosed fuse cutout, as shown in Fig. 17.3, has a porcelain housing and a hinged door supporting the fuse holder. The fuse holder is a hollow vulcanized-fiber expulsion tube. The fuse link is placed inside the tube and connects with the upper- and lower-line terminals when the door is closed. When the fuse blows, or melts because of excessive current passing through it, the resultant arc attacks the walls of the fiber tube, producing a gas which blows out the arc. The melting of the fusible element of some cutouts causes the door to drop open, signaling to the lineman that the fuse has blown.

Each time the fuse blows, a small amount of the vulcanized fiber of the expulsion tube is eroded away. The larger the value of the current interrupted, the more material is consumed. In general, a hundred or more operations of average current values can be successfully performed before the cutout fails.

Enclosed cutouts can be arranged to indicate when the fuse link has blown by dropping the fuse holder. The enclosed cutout is designed and manufactured for operation on distribution circuits of 7,200 volts and below. The standard current ratings of the cutouts are 50, 100, or 200 amps.

Open Distribution Cutout

Open-type cutouts are like the enclosed types except that the housing is omitted (Figs. 17.4 and 17.5). The open-type cutout is designed and manufactured for all distribution-system voltages. The open-type cutout is made for 100- or 200-amp operation. Some open-type cutouts can be uprated from 100 to 200 amps by using a fuse tube with dual venting action rated for 200-amp operation. The open-type cutouts are manufactured with a porcelain insulator body (Fig. 17.4) or a polymer insulator body (Fig. 17.5). The polymer body is much lighter and thus easier to handle during installation. The open-type cutout is also designed for load break operation with a hot stick utilizing an arc chute cutout and a blade assembly fuseholder (Figs. 17.6 and 17.7).

FIGURE 17.4   A 200-amp body, 100 amp fuseholder, porcelain body cutout (*Courtesy Hubbell Power Systems, Inc.*)

FIGURE 17.5   A light weight 200-amp, 100-amp fuseholder polymer body cutout. (*Courtesy Hubbell Power Systems, Inc.*)

FIGURE 17.6   A 100-amp open porcelain body insulator distribution cutout with arc chute designed for load break operation. (*Courtesy Hubbell Power Systems, Inc.*)

FIGURE 17.7   The blade assembly fuseholder used for the 100-amp and 200-amp arc chute load break fused cutout. (*Courtesy Hubbell Power Systems, Inc.*)

100A   200A

Fuseholders

Open-Link Distribution Cutout

This type of cutout differs from the open cutout in that it does not employ the fiber expulsion tube. The fuse link is supported by spring terminal contacts. An arc-confining tube surrounds the fusible element of the link. During fault clearing, the spring contacts provide link separation and arc stretching. The arc-confining tube furnishes the necessary expulsion action for circuit interruption. The open-link primary cutout cannot interrupt large values of fault current, and therefore its use is limited.

Primary Fuse Links

A primary fuse link consists of the button, upper terminal, fusible element, lower terminal, leader, and sheath (Figs. 17.8 and 17.9). The button is the upper terminal and the leader is the lower terminal. Fuse links for open-link cutouts are similar except that they have pull rings at each end (Fig. 17.10). In either case, the sheath aids in the interruption of low-value faults, and it provides protection against damage during handling.

Standards specify the size of the fuse holder into which the link must fit freely. Links rated 1 to 50 amps must fit into a 5/16-inch diameter fuse holder, 60- to 100-amp links must fit into a 7/17-inch diameter fuse holder, and 125- to 200-amp links must fit into a ¾-inch diameter fuse holder. Links must withstand a 10-lb pull while carrying no load current and are generally given a 25-lb test.

Fusible elements are made in a wide variety of designs, one group of which is shown in Fig. 17.11. Most silver-element fuse links utilize the helically coiled construction shown in Fig. 17.9. This construction permits the fusible element to absorb, without damage, vibration as well as thermal shock due to current surges and heating and cooling throughout the daily load cycle. The fusible elements shown in Fig. 17.11 are made in four designs:

1. High-surge dual element for links rated 1 to 8 amps
2. Wire element for links rated 5 to 20 amps
3. Die-cast tin element for links rated 25 to 100 amps
4. Formed strip element for links rated over 100 amps

Link designs are further distinguished as strain or spring type. The fusible element of a strain-type link incorporates a high-resistance strain wire for strengthening purposes, while the fusible element of a spring-type link has, in addition to the wire strain, a pretensioned spring to stretch the arc, thereby aiding circuit clearing after the fusible element has melted.

FIGURE 17.8   A 65T primary fuse link that will be installed in an arc-confining fuseholder tube. (*Courtesy Cooper Power Division.*)

**Standard Speed Fuse Links
Rated Less than 6 Amperes;
"N" Speed and "QR" Speed Fuse Links
Rated Less than 10 Amperes**

**Standard Speed, "K" Speed,
and "T" Speed Fuse Links
Rated 6 Amperes through 100 Amperes;
"N" Speed and "QR" Speed Fuse Links
Rated 10 Amperes through 100 Amperes**

Button

Corona shield

Fuse element
(nickel-chrome)

Washer

Swaged upper terminal

Swaged lower terminal

Swaged cable section prevents splaying of strands and wadding of cable . . . ensures smooth withdrawal from sheath during fuse-link operation

High-strength sheath provides superior interrupting performance, especially with respect to low-current transformer secondary-side faults

Tinned copper stranded cable resists corrosion . . . cable diameters selected large enough to resist corona . . . small enough to ensure smooth withdrawal of cable from sheath

Button

Fuse element—silver or silver-copper eutectic—helically coiled to absorb mechanical vibration and thermal shock

Strain wire restrains static force of flipper spring

FIGURE 17.9 Typical primary cutout fuse link showing component parts: button, upper terminal, fuse element, lower terminal, cable (leader), and sheath. Helically coiled fuse element is typical of silver-element fuse links. (*Courtesy S&C Electric Co.*)

FIGURE 17.10 Open primary cutout fuse link provided with pull rings for handling with hot-line tools. Hooks are inserted in rings for placement in cutout. (*Courtesy Cooper Power Division.*)

FIGURE 17.11 Primary cutout fuse links having different fusible elements. (a) High-surge dual element (1 to 8 amps); (b) wire element (5 to 20 amps); (c) die-cast tin element (25 to 50 amps); (d) die-cast tin element (65 to 100 amps); (e) formed strip element (140 to 200 amps).

(a)   (b)   (c)   (d)   (e)

The high-surge-element link of fuses rated 1 to 8 amps (Fig. 17.11a) is a spring-type link having two heater wires in series with a reinforced alloy solder joint. The solder joint is the central component of the element. One wire is fastened to the shank, and the other wire is fastened to the leader.

Links rated 5 to 20 amps (Fig.17.11b) are also of the spring type, consisting of a fusible tin wire and a separate strain wire. Both wires are soldered to the shank and leader.

Links rated 25 to 100 amps (Fig. 17.11c and 17.11d) are the strain type with diecast elements, whereas links rated over 100 amps (Fig. 17.11e) are the strain type with formed strip elements. The strip is lap-soldered to the shank and the leader, and the strain wire is securely fastened to the shank and leader.

Fuse Link Operation

When a fault occurs, the fusible element is melted by the excessive current, and an arc forms across the open gap. The arc is temporarily sustained in a conducting path of gaseous ionized arc products. Gas pressure builds rapidly, and this pressure, acting in conjunction with a spring-loaded flipper at the lower end of the fuse tube, rapidly ejects the fuse link leader, lengthening and cooling the arc. For low values of fault current, the arc acts on the fuse link sheath, generating considerable amounts of deionizing gases. When the current passes through the next zero value, as it changes the direction of flow, the arc is interrupted (Fig. 17.12). As the voltage increases again across the opening in the fuse link, the arc attempts to reestablish itself. A restrike, however, is prevented by

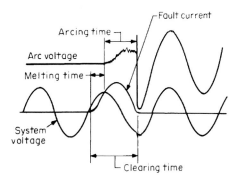

FIGURE 17.12 Diagram of voltages, current, and timing reference recorded with an oscillograph to show fuse operation.

the deionizing gases which will have rebuilt the dielectric strength of the open gap. For large values of fault current, the sheath is rapidly destroyed, and the arc erodes fiber from the inner wall of the cutout tube, generating large amounts of deionizing gas. During the fault interruption process, the cutout expels large amounts of gas under extremely high pressure as well as some fuse link particles. There may also be a very loud noise reported. Accordingly, when one is working near a fused cutout, care should be taken to stay clear of the exhaust path. In addition, when closing a cutout, it is good practice to look down, away from the discharge path, since there is always the possibility of closing into a short circuit (Fig. 17.13).

FIGURE 17.13 Lineman preparing to close cutout. Note that lineman is wearing appropriate protective equipment, is positioned to close the fuse cutout with a final vigorous thrust while looking down and away, and is located well away from the cutout exhaust path, recognizing that the cutout will operate if closed into a fault. (*Courtesy Northwest Lineman College.*)

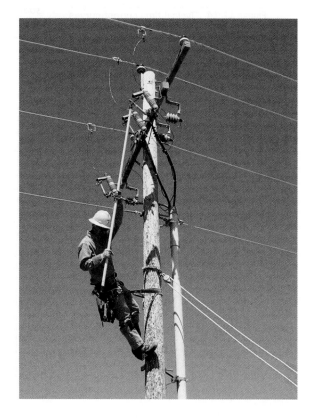

Time-Current Curves

A *time-current curve* is a curve that is plotted between the magnitude of a fault current and the time required for the fuse link to open the circuit. It is obvious that the greater the current, the faster the fuse melts and the shorter the time required for it to blow. Figure 17.14 is a typical time-current curve for a 10 K fuse link (normal current rating of 10 amps). Thus, a current of 10 amps would not cause the fuse to blow; 20 amps, which is twice normal, would require 300 sec or 5 min; 30 amps would require 3 sec; and 100 amps would require about 0.15 sec. These time-current curves are useful in providing required coordination of fuses, reclosers, and sectionalizers.

Fuse links are identified by their amperage ratings and by letter designations such as K, T, N, H, and QR. Links of the same current rating have about the same 300-sec points on the time-current curve, but they have widely varying shapes of time-current curves, depending on the type of metal used for the fusible element and the link's letter designation. For example, Fig. 17.15 shows the time-current curves for a few fuse links that could be used to protect a specific capacitor bank. This plot illustrates graphically why it is important to replace a blown fuse link with the same current rated link or one that is called for in the system fuse coordination plan. Failure to do so may result in miscoordination with other protective devices and needless electric system outages and damage to equipment. Careless storage and handling of fuse links may also result in incorrect operation.

FIGURE 17.14   Typical time-current curves for 10K fuse link. Fault currents are plotted along the abscissa, and time to blow is along the ordinate.

FIGURE 17.15 Total clearing time-current characteristic curves for fuse links recommended for a 1,200-kvar grounded wye-connected capacitor bank rated 13.8-kV three-phase, with two 7.97-kV, 200-kvar capacitor units per phase. (*Courtesy S&C Electric Co.*)

Fuse links designated as K (fast) and T (slow) with the same current rating have the same 300-sec points, but have noticeably different time-current curves, as illustrated in Figs. 17.15 and 17.16. It can be seen in both figures that the K link operates faster than the T link on the higher-current end of the curves by an amount of time equal to the vertical difference between the two curves for a given fault current. Figure 17.16 illustrates that for an overcurrent of 100 amps the 15 K fuse will operate in 0.5 sec while the 15 T fuse will require 1.5 sec, a difference of 1.0 sec.

Recommended Size of Primary Fuse

Table 17.1 gives the recommended size of primary fuse to use with different transformer voltage and kilovoltampere ratings. The table also gives the normal full-load primary current rating of the transformer.

It is general practice not to protect distribution transformers against small overloads. To do so would cause unnecessary blowing of fuses and frequent interruption of

FIGURE 17.16   Curves showing melting times of fast (K) links and slow (T) links having the same 15-amp rating.

the service, both of which are undesirable. It is therefore customary to provide fuses which have a higher current rating than the current rating of the transformer. This is clearly shown by a comparison of the normal current-rating column with the fuse-rating column in Table 17.1.

Solid-Material Power Fuse

The solid-material (SM) power fuses are required when the limits of the fused cutouts and fuse links have been exceeded due to higher fault currents or higher loads. Type SM power fuses are offered with maximum continuous current ratings of 200, 300, 400, and 720 amps and are available with fault-interrupting ratings from 28,000 amps RMS asymmetrical at 34.5 kV to 60,000 amps RMS asymmetrical at 4.16 kV. The solid-material fuses have silver or nickel-chrome fusible elements and utilize a solid-material low-arc energy technique of fault interruption. In some instances, a stainless steel outdoor silencer is utilized (Fig. 17.17). The solid-material power fuse and fuse mounting bracket are shown in Fig. 17.18. The SM power fuse is utilized in either overhead or underground applications. Type SM power fuses are especially suited for protecting transformers, capacitor banks, and cables in outdoor distribution substations on systems rated up through 34.5 kV. Pad-mounted gear protecting distribution systems serving industrial and large commercial load is the equipment where the solid-material fuses are installed (Fig. 17.19).

Transformer Size, kVA	2,400 Δ				4,160Y/2,400		4,800 Δ				8,320Y/4,800	
	Figures 1 and 2		Figure 3		Figures 4, 5, and 6		Figures 1 and 2		Figure 3		Figures 4, 5, and 6	
	Rated amp	Link Rating	Rated amp	Link Rating	Rated amp	Link Rating	Rated amp	Link Rating	Rated amp	Link Rating	Rated amp	Link Rating
3	1.25	2H	2.16	3H	1.25	2H	0.625	1H†	1.08	1H	0.625	1H†
5	2.08	3H	3.61	5H	2.08	3H	1.042	1H	1.805	3H	1.042	1H
10	4.17	6	7.22	10	4.17	6	2.083	3H	3.61	5H	2.083	3H
15	6.25	8	10.8	12	6.25	8	3.125	5H	5.42	6	3.125	5H
25	10.42	12	18.05	25	10.42	12	5.21	6	9.01	12	5.21	6
37.5	15.63	20	27.05	30	15.63	20	7.81	10	13.5	15	7.81	10
50	20.8	25	36.1	50	20.8	25	10.42	12	18.05	25	10.42	12
75	31.25	40	54.2	65	31.25	40	15.63	20	27.05	30	15.63	20
100	41.67	50	72.2	80	41.67	50	20.83	25	36.1	50	20.83	25
167	69.4	80	119.0	140	69.4	80	34.7	40	60.1	80	34.7	40
250	104.2	140	180.5	200	104.2	140	52.1	60	90.1	100	52.1	60
333	138.8	140	238.0		138.8	140	69.4	80	120.1	140	69.4	80
500	208.3	200	361.0		208.3	200	104.2	140	180.5	200	104.2	140

*Reprinted with permission.
†Since this is the smallest link available and does not protect for 300% load, secondary protection is desirable.

TABLE 17.1 Suggested Transformer Fusing Schedule (Protection between 200 and 300% Rated Load)* (*Link sizes are for EEI-NEMA K or T fuse links except for the H links noted.*) (*continued*)

Transformer Size, kVA	7,200 Δ				12,470Y/7,200		13,200Y/7,620		12,000 Δ			
	Figures 1 and 2		Figure 3		Figures 4, 5, and 6		Figures 4, 5, and 6		Figures 1 and 2		Figure 3	
	Rated amp	Link Rating	Rated amp	Link Rating	Rated amp	Link Rating	Rated amp	Link Rating	Rated amp	Link Rating	Rated amp	Link Rating
3	0.416	1H†	0.722	1H†	0.416	1H†	0.394	1H†	0.250	1H†	0.432	1H†
5	0.694	1H†	1.201	1H	0.694	1H†	0.656	1H†	0.417	1H†	0.722	1H†
10	1.389	2H	2.4	5H	1.389	2H	1.312	2H	0.833	1H†	1.44	2H
15	2.083	3H	3.61	5H	2.083	3H	1.97	3H	1.25	1H	2.16	3H
25	3.47	5H	5.94	8	3.47	5H	3.28	5H	2.083	3H	3.61	5H
37.5	5.21	6	9.01	12	5.21	6	4.92	6	3.125	5H	5.42	6
50	6.49	8	12.01	15	6.94	8	6.56	8	4.17	6	7.22	10
75	10.42	12	18.05	25	10.42	12	9.84	12	6.25	8	10.8	12
100	13.89	15	24.0	30	13.89	15	13.12	15	8.33	10	14.44	15
167	23.2	30	40.1	50	23.2	30	21.8	25	13.87	15	23.8	30
250	34.73	40	59.4	80	34.73	40	32.8	40	20.83	25	36.1	50
333	46.3	50	80.2	100	46.3	50	43.7	50	27.75	30	47.5	65
500	69.4	80	120.1	140	69.4	80	65.6	80	41.67	50	72.2	80

*Reprinted with permission.
†Since this is the smallest link available and does not protect for 300% load, secondary protection is desirable.

TABLE 17.1 Suggested Transformer Fusing Schedule (Protection between 200 and 300% Rated Load)* (Link sizes are for EEI-NEMA K or T fuse links except for the H links noted.) (continued)

TABLE 17.1 Suggested Transformer Fusing Schedule (Protection between 200 and 300% Rated Load)* (Link sizes are for EEI-NEMA K or T fuse links except for the H links noted.)

Transformer Size, kVA	13,200 Δ				14,400 Δ				24,900Y/14,400	
	Figures 1 and 2		Figure 3		Figures 1 and 2		Figure 3		Figures 4, 5, and 6	
	Rated amp	Link Rating	Rated amp	Link Rating	Rated amp	Link Rating	Rated amp	Link Rating	Rated amp	Link Rating
3	0.227	1H†	0.394	1H†	0.208	1H†	0.361	1H†	0.208	1H†
5	0.379	1H†	0.656	1H†	0.347	1H†	0.594	1H†	0.374	1H†
10	0.757	1H†	1.312	2H	0.694	1H†	1.20	2H	0.694	1H†
15	1.14	1H	1.97	3H	1.04	1H	1.80	3H	1.04	1H
25	1.89	3H	3.28	5H	1.74	2H	3.01	5H	1.74	2H
37.5	2.84	5H	4.92	6	2.61	3H	4.52	6	2.61	3H
50	3.79	6	6.56	8	3.47	5H	5.94	8	3.47	5H
75	5.68	6	9.84	12	5.21	6	9.01	12	5.21	6
100	7.57	8	13.12	15	6.94	8	12.01	15	6.94	8
167	12.62	15	21.8	25	11.6	12	20.1	25	11.6	12
250	18.94	25	32.8	40	17.4	20	30.1	40	17.4	20
333	25.23	30	43.7	50	23.1	30	40.0	50	23.1	30
500	37.88	50	65.6	80	34.7	40	60.0	80	34.7	40

*Reprinted with permission.
†Since this is the smallest link available and does not protect for 300% load, secondary protection is desirable.

FIGURE 17.17   A stainless steel outdoor silencer utilized with the solid-material power fuse. (*Courtesy Federal Pacific.*)

FIGURE 17.18   Bracket supporting the solid-material power fuse. (*Courtesy S&C Electric Co.*)

PMH Models—Fuse★ and Switch Compartments

S&C's Uni-Rupter™ provides the ultimate in single-pole live switching . . . just a firm opening pull on the fuse interrupts the circuit with a quick-break action. Uni-Rupter self-resets for the next opening operation. Separate current-carrying and fault-closing contacts are provided

Corrosion-resistant non-ferrous door hinges and hinge pins

Insulated roof—"no-drip" compound on underside of roof guards against formation of condensation that could drip onto energized parts

Dual-purpose front barriers of GPO3-grade fiberglass-reinforced polyester guard against inadvertent contact with live parts when in the normal vertical position. Inserted into the open gap of a switch or fuse, barrier provides isolation from bus and upper contacts

Grappler™ takes the work out of fuse handling . . . provides sure grip, perfect balance, and ready control that operators appreciate

Segregated circuits—full-length steel barriers separate side-by-side compartments; GPO3-grade fiberglass-reinforced polyester barriers separate front compartments from rear compartments and isolate the tie bus

S&C's Ultradur® Finishing System provides superb corrosion resistance

Also available:
- **Optional stainless-steel enclosure**
- **Optional key-interlock schemes**
- **Optional inner barrier panel**—meets Rural Utilities Service "dead-front" requirements
- **Optional Canadian Standards Association Listing**

Optional base spacer increases elevation of the gear up to 24 inches, to provide additional cable training space or additional distance above the ground

Cypoxy®—S&C's cycloaliphatic epoxy resin system . . . insulates all live parts from ground

Ground studs for fuse terminals, switch terminals, and the ground pad in each compartment

Optional cable guides assist in cable training and provide protection against damage from excessive cable or foundation movement

★ Fuse compartments illustrate non-standard combination of Fault Fiter® Electronic Power Fuses (on left) and SML-20 Power Fuses (on right) for comparison only.

FIGURE 17.19 Electronic power fuses installed on the left and solid-material power fuses installed on the right-hand side of the pad-mounted switchgear. Protective barriers isolate the fuses from other live parts. (*Courtesy S&C Electric Co.*)

Current-Limiting Fuse

The *current-limiting fuse* (CLF) is a non-expulsion fuse consisting of a silver fuse element wound on a central core enclosed in a tube filled with high-purity silica sand. Under fault conditions the silver element melts and establishes an arc. The fuse element is burned back by the arc. The heat of the arc melts the adjacent sand to produce a glass-like substance known as *fulgurite*. The rapid heat absorption of the sand, together with the rapid burn back of the fuse element, acts to insert a high resistance into the arc path that prevents the current from reaching its natural crest. This produces the fuse's current-limiting action and lowers the energy let-through during faults. In addition, the current-limiting fuse has an interrupting rating in the 50,000-amp symmetric range and a less than ½-cycle clearing time. The current-limiting fuse is produced in two versions: the backup (or partial range) and the general purpose (or full range).

The backup CLF is constructed as described with the central core generally being of ceramic construction. No gas is produced during fault clearing; therefore, the fuse can be hermetically sealed, making it suitable for submersion in oil-filled apparatus. This type of fuse is not capable of interrupting low-magnitude fault currents. It will only interrupt a range of fault currents from stated minimum interrupting rating (usually 3 to 7 times the continuous current rating) to the maximum interrupting rating. Currents below this range but above the continuous current rating of the fuse will melt the fuse element but will not have enough energy to cause interruption. These currents must be interrupted by a series-connected interrupting device (Fig. 17.20). If the fuse is exposed to low-fault currents, the continued arcing will cause the fuse to burn up or explode.

The general-purpose CLF adds modifications to the basic CLF design to make the fuse capable of clearing any current greater than the continuous rating and less than the maximum interrupting rating of the fuse. In this fuse, part of the fuse element is constructed from a solder alloy with a melting temperature one-third of the 960° melting temperature of the silver portion of the fuse element. The central core of the fuse is constructed from a material which evolves gas when heated by the element melting.

FIGURE **17.20** Partial-range current-limiting fuse is installed on top of distribution cutout with expulsion fuse. Current-limiting fuse will operate for high-fault-current insulation failures in the distribution transformer. Operating characteristics of the current-limiting fuse will prevent destruction of the transformer and the lid from blowing off if a fault develops inside the transformer tank. Expulsion fuse will operate for low-primary current or secondary faults. (*Courtesy A. B. Chance Co.*)

FIGURE 17.21 Partial-range current-limiting fuse connected in series with draw-out type of expulsion fuse. Both fuses would be installed in the tank of the pad-mounted transformer or switching equipment. Draw-out feature facilitates replacement of expulsion fuse which is designed to operate for most common electrical faults. (*Courtesy A. B. Chance Co.*)

The lowered melting temperature and the gas-evolving core make the general-purpose CLF capable of clearing the lower range of currents that the backup CLF cannot clear. High-current operation is the same.

Current-limiting fuses are expensive compared to the cost of expulsion fuses. The use of the partial-range current-limiting fuse in series with an expulsion fuse minimizes the operating costs since the expulsion fuse will operate for the most common faults that occur most frequently. Figure 17.21 illustrates an expulsion fuse in series with a partial-range CLF mounted on a metal sheet, as it would be installed in a pad-mounted transformer or pad-mounted switchgear used with underground circuits.

Current-limiting fuses applied as illustrated in Figs. 17.20 and 17.21 and properly sized can prevent catastrophic failure of a transformer. Catastrophic failure of a distribution transformer can result from a high-current arc formed under oil. The failure can result in rupture of the transformer tank and an oil fire that is dangerous to the public and to linemen working near the transformer. When a lineman reenergizes a transformer that has been removed from service as a result of operation of the transformer protective equipment, it is important that special precautions be taken to protect him. The transformer should be tested and/or energized through a current-limiting fuse to prevent a high-fault current from being generated under oil in the transformer.

Electronic Fuses

The *electronic fuse* integrates state-of-the-art electronics with an advanced-design high-current fuse. The electronic componentry provides current sensing, time-current characteristics, and control power for the fuse. High-speed interruption of fault currents to 40,000 amps symmetric is provided by the high-current fuse section (Fig. 17.22). The fuse is rated 600 amps continuous with voltage ratings of 4.16 through 25 kV.

The electronic fuse has a control module and an interrupting module (Fig. 17.23). The control module has an integral toroidal current transformer, which provides line current sensing and control power for the electronic circuitry. The current transformer also provides the energy to operate the interrupting module in the event of a fault. Electrical output from the current transformer is processed by the electronics located inside

FIGURE 17.22   Electronic fuse with interrupting module on top and control module connected to and mounted below the interrupting module. Live switching up to 400 amps is provided by interrupter mounted on assembly above fuse unit. (*Courtesy S&C Electric Co.*)

FIGURE 17.23   Cutaway view of electronic fuse. Toroidal current transformer, far left in control module, provides line current sensing and control power for electronics in the control module. Tripping signal from control module activates a gas-generating power cartridge in the interrupting module, causing the main current path to be separated at two points. Fault current is then diverted through the sand-embedded fusible elements. Fault current is interrupted in a current-limiting mode. (*Courtesy S&C Electric Co.*)

the factory-sealed cast-aluminum control module housing, which serves as both a path for continuous current and a Faraday cage to shield sensing circuits against interference from external electric fields. When a fault occurs, the electronics within the control module initiates a "trip" signal in accordance with electronically derived time-current characteristics. The signal to trip is delivered to the interrupting module through a low-resistance contact. The interrupting module of the fuse assembly has a centrally positioned main current section which carries load current during normal operating conditions. The main current section in the interrupting module is rapidly opened when a trip signal is received from the control module by the action of a gas-generating power cartridge and associated insulating piston. The fault current is shunted into the circuit-interrupting section and the coaxially wound fusible elements, as a result of the opening of the main current path through the interrupting module. The fault current flowing through the fusible element causes the element to open, and the silica sand in the interrupting module limits the magnitude of the energy generated because of the short circuit.

The electronic fuses equipped with inverse-curve type of time-current characteristics are ideally suited for service-entrance protection and coordination because these fuses incorporate unique time-current characteristics designed for coordination with source-side overcurrent relays and in-plant load-side feeder fuses. When used for service-entrance protection and coordination, the electronic fuses provide a buffer to prevent in-plant problems on the customer's system from affecting service on the utility company system. In addition, the electronic fuses provide protection for the switchgear bus against damage from high-current bus faults, as well as backup protection for the load-side feeder fuses (Figs. 17.24 and 17.25).

FIGURE 17.24 Fault-Fiter fuses with electronic control for an underground subloop type of time-current curve (TCC) to provide fault protection for transformers, elbow connectors, and underground cable. Coordination is readily achieved with both the transformer's weak-link fuses and the utility's substation circuit breakers. (*Courtesy S&C Electric Co.*)

Protective Overcurrent Coordination

The coordination of overcurrent protective devices involves their selection and use in such a manner that temporary faults will be quickly removed and permanent faults will be restricted to the smallest section of the system possible.

The locations of the protective devices are known as *coordinating points*. Coordinating points are usually established at the substation, at positions along the feeder, in branch lines off the feeder, and on the primary side of distribution transformers. In the single-line diagram of Fig. 17.26, representing an elementary distribution system, points A, B, C, E, F, H, and I are coordinating points. Each device must be selected so that it can carry the normal line or load current and will respond properly to an excessive fault current. Device A in the substation nearest the power source is normally a circuit breaker; devices B, H, I, E, and F are fused cutouts or power fuses; and device C could be a circuit recloser. Devices A and C could be adjusted to open before the fuses melt, to clear temporary faults and to reclose to restore service to all customers. Devices A and C are adjusted to open after time delay if the first operation fails to clear the fault. A permanent fault, as shown in Fig. 17.26, would be isolated or cleared from the source by fuse E after device C failed to clear the fault when it opened and reclosed. After device E has isolated the branch with the fault, normal load current will continue to flow in the remainder of the system.

The minimum fault current on each circuit must be determined and compared with the minimum operating current of the protective device, to be sure the protective device will operate to clear the fault. The minimum fault current calculations must allow for a high-impedance fault that may occur if an energized wire is lying on the ground. The maximum fault current must be determined to ensure that each protective device has adequate interrupting capacity to deenergize the faulted circuit without a catastrophic failure of the protective equipment.

Relay play an integral role as a part of the overcurrent protection in the electric power system. The relay protects the system as it monitors, alarms, and signals. In the

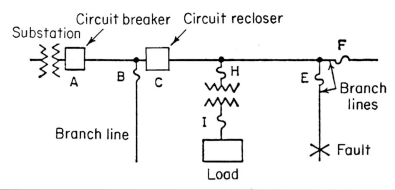

FIGURE 17.26 One-line diagram of an elementary distribution system. Overcurrent protective devices are shown as A, B, C, E, F, H, and I. Fault is shown as X. Device E should disconnect the branch line and permit the remainder of the system to operate.

electric distribution system, the relay signals the circuit breaker to operate. The circuit breaker is the circuit's protective interrupting device. Relays are coordinated with other protective devices such as fuses and reclosers.

Relay to Fuse/Recloser Coordination

Distribution circuit breaker overcurrent relays at the substation coordinate with distribution line fuses and reclosers to achieve optimum fault clearing. The overcurrent substation relay protects for phase or ground faults. The circuit breaker oftentimes initially opens and recloses instantaneously before fully locking open. The purpose of the instantaneous trip is to either allow the fault to clear itself or to allow a downstream fuse to blow. Properly coordinated protective devices isolate the event to the zone being protected.

Electro-mechanical overcurrent relays have been used for many years in substations. This style of relay incorporates an electro-magnet, stationary and moving contacts, and an induction disk as the core components. The electro-mechanical overcurrent substation relays are being replaced by microprocessor-based overcurrent relays. These types of relays have sophisticated hardware and software components. The advantage of the microprocessor-based overcurrent relay is the added functionality offered by the device. In addition to overcurrent protection, the relay records voltage swell, sag, and interruption information. The relay contains advanced communication tools. The relay software has sequence coordination logic used to coordinate relay protection to downstream devices. System reliability is enhanced with the improvements that have occurred in microprocessor-based relays.

Recloser to Load Side Fuse Coordination

Fuse links downstream of the recloser must coordinate such that the load side fault current does not cause the recloser to lock open when the fault is downstream of the fuse link.

Protective Fuse-to-Fuse Overcurrent Coordination

Figure 17.27 shows a simple diagram of a distribution system with two fuses in series that require a fuse-to-fuse overcurrent protection scheme. Figure 17.28 shows the time-current characteristics (TCC) curve of the protecting downstream fuse and the

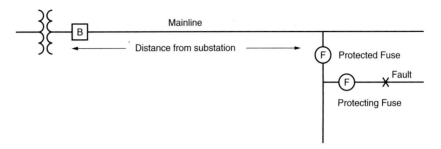

FIGURE 17.27   The two fuses are in series with one another. The fuses must be properly coordinated. For the two fuses depicted, the protecting fuse must melt before the protected fuse in the situation of a fault at location *x*.

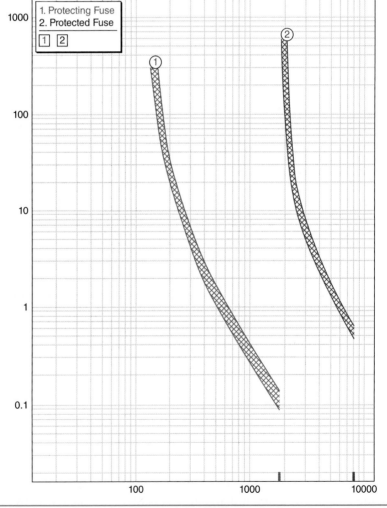

FIGURE 17.28   Sample time-current characteristics (TCC) curve for the protecting fuse and the protected fuse depicted in Fig. 17.27. The maximum clearing time of the protecting fuse (curve 1) occurs prior to the minimum melting time of the protected fuse (curve 2).

protected upstream fuse. The rule of thumb is that 75% of the minimum melting time curve of the protected fuse must not overlap the maximum clearing time curve of the protecting fuse. In some cases, overlap of the curves is unavoidable (Fig. 17.29). If the fault current does not exceed specified ampere values, coordination will occur (Table 17.2).

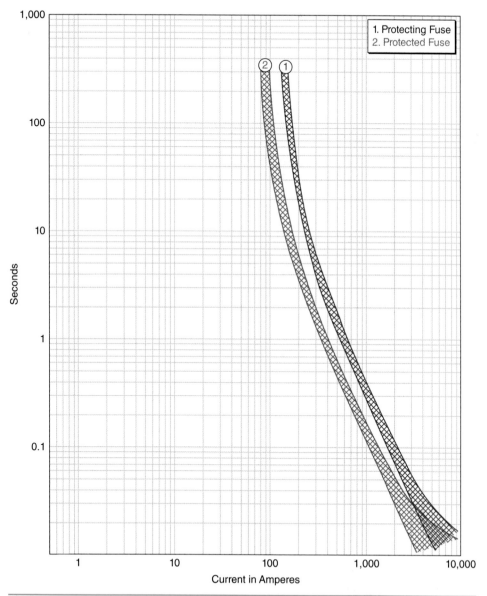

FIGURE 17.29 There is overlap of the two fuse TCC curves. The fuses will coordinate as long as the fault current does not exceed values listed in Table 17.2. The maximum clearing time of the 40T protecting fuse (curve 1) occurs prior to the minimum melting time of the 65T protected fuse (curve 2) with overlap.

Protecting Fuse-Link Rating—Amperes	Protected Link Rating—Amperes													
	8T	10T	12T	15T	20T	25T	30T	40T	50T	65T	80T	100T	140T	200T
	Maximum Fault-Current Protection Provided by Protecting Link—Amperes													
6T		350	680	920	1,200	1,500	2,000	2,540	3,200	4,100	5,000	6,100	9,700	15,200
8T			375	800	1,200	1,500	2,000	2,540	3,200	4,100	5,000	6,100	9,700	15,200
10T				530	1,100	1,500	2,000	2,540	3,200	4,100	5,000	6,100	9,700	15,200
12T					680	1,280	2,000	2,540	3,200	4,100	5,000	6,100	9,700	15,200
15T						730	1,700	2,500	3,200	4,100	5,000	6,100	9,700	15,200
20T							990	2,100	3,200	4,100	5,000	6,100	9,700	15,200
25T								1,400	2,600	4,100	5,000	6,100	9,700	15,200
30T									1,500	3,100	5,000	6,100	9,700	15,200
40T										1,700	3,800	6,100	9,700	15,200
50T											1,750	4,400	9,700	15,200
65T												2,200	9,700	15,200
80T													7,200	15,200
100T													4,000	13,800
140T														7,500

This table shows maximum values of fault currents at which EEI-NEMA Type T fuse links will coordinate with each other. The table is based on maximum clearing-time curves for protecting links and 75% of minimum melting-time curves for protected links. (*Courtesy of Cooper Power Systems, Inc.*)

TABLE 17.2   EEI-NEMA Type T Fuse Link Coordination Chart

Additional detailed information on fuses can be obtained from the ANSI C37.40, Service Conditions and Definitions for High Voltage Fuses, Distribution Enclosed Single-Pole Air Switches, Fuse Disconnecting Switches and Accessories; ANSI C37.42, Specifications for High Voltage Distribution Cutouts and Fuse Links; ANSI C37.46, Specifications for High Voltage Power Fuses and Fuse Disconnecting Switches; ANSI C37.47, Specifications for High Voltage Distribution Fuse Disconnecting Switches, Fuse Supports, and Current-Limiting Fuses and IEEE C37.48 IEEE Guide and Tutorial for the Application of High-Voltage (>1000 V) Fuses and Accessories.

CHAPTER 18

Switches

Aswitch is a current-carrying device installed in an electric circuit. The switch can be opened or closed to connect or disconnect a circuit. A switch is a transitioning device linking circuits together in the closed position or an isolation device separating circuits in the open position.

Non-load break switches are not rated to break load current. Conversely, switches that interrupt current serving load are called load break switches. If the switch is closed and the circuit is energized and current is flowing through the switch when it is operated, load break interrupters are installed on the switch contacts to open the switch safely and stop the current flow.

High-voltage switches are operated with either a switch handle-linked operating mechanism, a hot-line switch stick tool (Fig. 18.1), or a motor-operated controller. The motor-operated mechanism may be manually operated locally by a lineman or remotely operated by an electronic control signal initiated from a control center location (Fig. 18.2). The signals are sent to the motor-controlled operating mechanism at the switch.

Switches may be divided into four general classes based upon the means of interruption:

1. Air switches

2. Oil switches

3. Vacuum switches

4. Sulfur hexafluoride gas (SF_6) switches

Switching devices include circuit breakers, circuit switches, disconnect switches, reclosers, and sectionalizers.

The following documents provide reference information on switches:

ANSI/IEEE C37.30, Definitions and Requirements for High Voltage Air Switches, Insulators, and Bus Supports

ANSI C37.35, Application, Installation, Operation, and Maintenance of High-Voltage Air Disconnecting and Load Interrupter Switches

ANSI C37.45, Specifications for Distribution Enclosed Single-Pole Air Switches

ANSI/IEEE C37.60, Requirements for Overhead, Pad-Mounted, Dry Vault, and Submersible Automatic Circuit Reclosers and Fault Interrupters for AC Systems

ANSI/IEEE C37.61, Guide for the Application, Operation and Maintenance of Automatic Circuit Reclosers

FIGURE 18.1 Installation with six 200-amp, 15-kV, single pole hookstick disconnect switches on the line and load side of the pad-mount switchgear.

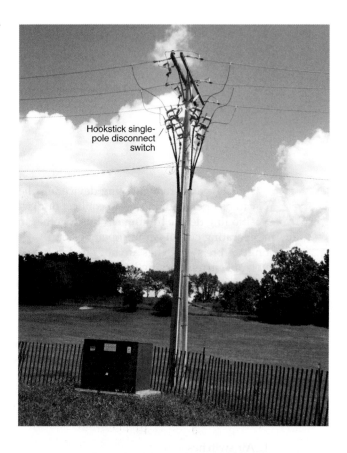

FIGURE 18.2 A remotely controlled 600-amp, three-pole, 69-kV air break switch.

Air Switches

As their name implies, air switches are switches whose contacts are opened in air. Air switches are further classified as air circuit breakers, air break switches, and disconnects.

Circuit Breaker

A substation device used to complete, maintain, and interrupt currents flowing in a circuit under normal or faulted conditions is called a *circuit breaker*. The circuit breaker has a mechanism that mechanically, hydraulically, or pneumatically operates the circuit breaker contacts. Insulating oil, air, compressed air, vacuum, or sulfur hexafluoride gas is used as an arc-interrupting medium and a dielectric to insulate the contacts after the arc is interrupted. If it is desired to open the circuit automatically during overload or short circuit, the circuit breaker is equipped with a tripping mechanism to accomplish this. Circuit breakers at the generating stations and substations are used to control the circuit, as well as, provide protection from overloads, short circuits, and other abnormal conditions.

Air Break Switch

The *air break switch* can have both blade and stationary contacts equipped with arcing horns, as shown in Fig. 18.3. These horns are pieces of metal between which the arc forms when a circuit-carrying current is opened. As the switch opens, these horns spread farther and farther apart, thereby lengthening the arc until it finally breaks. Favorable oblique winds kept the arcs from short-circuiting across phases. Wind also has a cooling effect which aids arc extinction.

Air break switches are usually mounted on substation structures or on poles and are operated from the ground level. In a three-phase circuit all three switches, one in each phase, are opened and closed together as a "group," as the system is called. The switches can be operated by a handle connected to the rod extending from the switch to the base of the pole. Many air break switches are operated by mechanized equipment connected to the switch operating rod. The automation of distribution circuits has resulted in the installation of many sophisticated automated air break switches functioning as a team. The automated switches sense, control, and communicate with one another in order to configure the circuit according to ideal system conditions.

FIGURE 18.3   Air break switch blade with an arc interrupter. (*Courtesy Southern States, Inc.*)

FIGURE 18.4   Dead-front pad-mounted switchgear unit utilized in underground distribution systems. The switchgear unit features include cable termination bushings, switches, and fuses. The unit is contained within the enclosure and is packaged to minimize electrical exposure during switching and fusing operations. (*Courtesy Federal Pacific.*)

Switches for underground distribution circuits are usually installed in pad-mounted switchgear (Fig. 18.4). Protective barriers are installed between each phase of the switch and at the switch opening (Fig. 18.5). The switches are operated with the cabinet doors closed to provide protection for the lineman or cableman.

Customers with essential loads may require dual electric sources; a primary source and a back-up secondary source, with automatic transfer from one source to the other. If the primary electric source is interrupted, the customer load transfers to the back-up secondary electric source. These switches are power-operated and controlled by an automatic control device for source transfer. The underground load feeder circuits are connected to the switchgear bus through power fuses and air break switches with interrupter contacts to permit energized operation while carrying load. The bus-tie switch will permit the load to be served from either source. The power source circuits are monitored by an automatic control device which in turn controls and sequences the power operation of the incoming circuits and bus-tie switches.

FIGURE 18.5   The 600-amp, 15-kV disconnect switches displayed in a pad-mounted switchgear unit. (*Courtesy Federal Pacific.*)

Upon loss of service from either of the two power sources—for a preset time sufficient to confirm permanent loss of service—the interrupter switch associated with the deenergized source automatically opens and the bus-tie switch closes to restore service to the affected feeders. When the normal power source is returned to service the affected feeder bus is retransferred to its normal power source, either automatically or manually depending on the operating mode which has been selected. Overcurrent relays are used to block closing of the bus-tie switch in the event of a bus fault. Paralleling of power sources is precluded by mechanical and electrical interlocks. The equipment can be manufactured and equipped to permit operation of the switches from a remote control point, if that type of operation is desired, instead of the automatic mode.

Solid Blade Disconnect Switch

A *solid blade disconnect switch* is an air break switch not equipped with arcing horns or other load break devices. The solid blade disconnect switch cannot be opened until the circuit in which it is connected is interrupted by some other means, such as a "load buster" attached to a hot-line tool, illustrated and discussed in Chap. 2, or a circuit breaker, illustrated and discussed in Chap. 3. If a disconnect switch is opened while current is flowing in the line, an arc is likely to be drawn between the blade and stationary contact which might easily jump across to the other conductor or to some grounded metal and cause a short circuit. The hot arc could also melt part of the metal, thereby damaging the switch.

Solid blade disconnect switches are used to complete a connection to or isolate the following:

1. Two energized transmission or distribution lines

2. Transmission or distribution lines from substation equipment

3. Substation equipment

4. A distribution feeder circuit and a branch circuit

Solid blade disconnect switches are frequently used to isolate a line or a piece of apparatus such as a circuit breaker to complete maintenance work. In most circumstances, it is necessary to test the equipment for high voltage and, if proved deenergized, to ground it before the maintenance work is performed.

Oil Switch

An oil switch is a high-voltage switch whose contacts are opened and closed in oil. The switch is actually immersed in an oil bath, contained in a steel tank. The reason for placing high-voltage switches in oil is that the oil may help to break the circuit when the switch is opened. With high voltages, a separation of the switch contacts does not always break the current flow, because an electric arc forms between the contacts. If the contacts are opened in oil, however, the oil will help to quench the arc since oil is an insulator. Furthermore, if an arc forms in the oil, the arc will evaporate part of the oil, because of the arc's high temperature, and so will partially fill the interrupters surrounding the switch contacts with vaporized oil. This vapor develops a pressure in the interrupters which assists in quenching or breaking the arc by elongating the arc.

The three lines of a three-phase circuit can be opened and closed by a single oil switch. If the voltage is not extremely high, the three poles of the switch are generally in the same tank, but if the voltage of the line is high, the three poles of the switch are placed in separate oil tanks. The poles are placed in separate tanks to make it impossible for an arc to form between any two phases when all three phases of the switch are opened or closed. An arc between the phases in a high voltage application in a single tank might cause a short circuit across the line and may severely rupture the tank.

When an oil switch is to open the circuit automatically because of overload or short circuit, it is provided with a trip coil. This trip coil consists of a coil of wire and a movable plunger. In low-voltage circuits carrying small currents, this coil is connected in series with the line. When the current exceeds its permissible value, the coil pulls up its plunger. The plunger trips the mechanism, and a spring opens the switch suddenly.

In high-voltage circuits or in circuits carrying large currents, a current transformer is connected into the line and the secondary leads from this transformer supply the current to the trip coil of the oil switch. Since there is a fixed ratio between primary and secondary currents of the current transformer, the coil can be adjusted to trip at any predetermined value of current in the line.

The use of the current transformer on such circuits serves the double purpose of providing a small current for operating the tripping coil and of insulating the coil from the high voltage of the line.

Oil Circuit Recloser

An oil circuit recloser is a type of oil switch designed to interrupt and reclose an alternating-current circuit automatically (Fig. 18.6). It can be made to repeat this cycle several times. Reclosers are designed for use on single-phase circuits (Fig. 18.7) or on three-phase circuits (Fig. 18.8).

A recloser opens the circuit in case of fault as would a fuse or circuit breaker. The recloser, however, recloses the circuit after a predetermined time (for hydraulically controlled reclosers about 2 sec). If the fault persists, the recloser operates a predetermined number of times (1 to 4) and "locks out," after which it must be manually reset before it can be closed again. If, however, the fault was of a temporary nature and cleared before lockout, the recloser would reset itself and be ready for another full sequence of operations.

FIGURE 18.6 Three single-phase oil circuit reclosers installed in 13,200Y/7,620-volt three-phase line. Reclosers have a current rating of 560 amps and interrupting rating of 8,000 amps. Also note installation of lightning arresters.

FIGURE 18.7   Single-phase oil circuit recloser removed from oil tank. Recloser shown has a maximum voltage rating of 14.4 kV and a maximum current-interrupting rating of 1,250 amps.

FIGURE 18.8   Oil circuit recloser with tank removed. Recloser is rated 15-kV, three-phase, 560 amps continuous and 16,000 amps interrupting capacity.

Temporary faults arise from wires swinging together when improperly sagged, from tree branches falling into the line, from lightning surges causing temporary flashover of line insulators, and from animals on the conductors short-circuiting the insulators.

A recloser is unlike a fuse link because it distinguishes a temporary from a permanent fault. A fuse link interrupts temporary and permanent faults alike. Reclosers give temporary faults repeated chances, usually four, to clear or be cleared by a subordinate device like a fuse or sectionalizer. If the fault is not cleared after four operations, the recloser recognizes it as a permanent fault and operates to lock out leaving the line open.

A recloser can be magnetically operated by a solenoid connected in series with the line. Minimum trip current is usually twice the normal load current rating of the recloser coil. The operations are performed by a hydraulic mechanism and a mechanical linkage system.

When the fault current reaches twice the normal line current, the increased magnetic field pulls the plunger down into the coil. As the plunger moves downward, the lower end trips the contact assembly to open the contacts and break the circuit. As soon as the contacts are open, there is no more current in the coil to hold them open, so a spring closes the mechanism and reenergizes the line.

This and succeeding operations are depicted on the oscillogram in Fig. 18.9 as it cycles through to lockout. Note that the line is held open for approximately 60 cycles between operations. This provides time for sectionalizers to operate while no current is flowing in the line. During lockout the contacts are held open until the recloser is reset manually. If a temporary fault clears before the recloser locks out, all mechanical operations cease and the recloser becomes ready to cycle over again when the next fault occurs.

Recloser is set for two fast opens and closes of approximately 2 Hz and two delayed opens and closes of approximately 6 Hz.

Examination of the oscillogram (Fig. 18.9) shows that the first two openings of the recloser are faster than the last two. The first two openings occur in about 2 Hz each, whereas the last two occur in about 5 Hz each. This is arranged so that in case the fault takes place on a branch line or tap protected with a fused cutout, the recloser will try to remove the fault before the fuse has time to melt. If the recloser does not clear the fault after two trials, it will give the fuse time to blow on the longer 5-Hz openings and disconnect the faulted branch. The fact that the fault did not clear with two openings indicates that it was of a permanent nature and could be removed only by disconnecting the defective branch line.

The objective of recloser and fuse coordination is to eliminate outages due to temporary faults and, in the case of permanent faults, to restrict outages to a minimum number of customers for the shortest possible time. This requires satisfying the following conditions whenever possible:

1. Fuses should be protected by a recloser fast trip to clear transient faults.

2. Reclosers and/or fuses in series should be sized such that the source device operates before the load device over the entire range of fault currents.

3. Fuses should be sized as large as fault current and source side protective devices will permit in order to avoid blowing fuses on overload.

In determining time-current characteristic (TCC) curves, tests are conducted at 25°C ambient without any preloading. In accordance with National Electrical Manufacturers Association (NEMA) standards, minimum melting curves are determined by taking

FIGURE 18.9   Oscillogram of single-phase recloser operation interrupting 1,260 amps at 14.4 kV.

average test values minus the manufacturer's tolerance, while total clearing curves represent average melting time plus the manufacturer's tolerance plus the arcing time. Since fuse links are thermal devices, it should be evident that preloading and/or ambient temperatures higher than 25°C will result in links blowing faster than indicated on the published curves. To account for operating variables, it is recommended that 75% of the minimum melting curve be used in coordination work. It is generally agreed that for tin fuse links, this 25% margin consists of 10% for preloading, 5% for extraordinary ambient temperatures, and 10% to prevent damaging the fuse link. The concept of damaging a fuse link provides an explanation for fuses blowing on "nice, sunny days for no apparent reason." Fuse link damage is caused by a fault which lasts long enough for the link to reach its melting temperature, but not long enough for the necessary heat of fusion to be added to completely melt the link. When the link cools, the physical dimensions have changed and consequently the time-current characteristics have been altered.

Recloser curves are plotted to average values with manufacturing tolerances resulting in variations of ±10% in current or time, whichever is greater. The hydraulic fluid used provides consistent timing for ambient temperatures above 0°C. Below this, operations may be slightly slower, but coordination with fuses will be maintained since they are also affected by the low ambient temperature.

To examine fuse-recloser coordination, assume the recloser is the source side device and the fuse is the load side or protecting device. The curve shown in Fig. 18.10 will illustrate coordination between a 35-amp type H recloser and a 15T fuse link. Four operations to lockout may be selected using a maximum of two curves. If fast tripping

FIGURE 18.10   Coordination curves for 35-amp recloser and 15T fuse link.

is desired, select at least one operation on the A curve with the remainder on the B or C curve. The C curve is chosen since it allows more room for coordination. To determine the number of fast trips, the following should be considered. First, the purpose of a fast trip is to prevent transient faults from becoming permanent ones and also to prevent fuse blowing due to transient faults. Since the momentary interruptions resulting from using two fast trips should not be objectionable, it should be used to provide better fuse protection. By comparing the maximum clearing time for a recloser fast trip to 75% of the fuse melting time, it is evident that a 35-amp type H recloser can protect a 15T fuse with one fast trip for faults up to 550 amps. For two fast trips it would obviously be too conservative to compare twice the maximum clearing time with 75% of the fuse minimum melting time since the reclosing interval has a cooling effect on the fuse. For a reclosing interval of 1 sec to 2 sec, the cooling factor can be handled by comparing twice the average clearing time for a fast trip to 75% of the fuse minimum melting time. Thus, for two fast trips, a 35-amp type H recloser can protect a 15T fuse for faults up to 350 amps.

Thus, the fault duty at the fuse location determines the degree of protection afforded by the recloser fast trips.

Sectionalizer   A sectionalizer is another type of oil switch designed to isolate faults on distribution circuits in conjunction with reclosers. Sectionalizers are usually installed on taps or branches off main lines. While in appearance a sectionalizer is similar to a recloser, it should not be confused with it because it does not interrupt a fault current. In fact, a sectionalizer waits until the recloser has opened the line and then sectionalizes the faulty line while the line is still open and no current flowing. It will be remembered from Fig. 18.9 that the recloser holds the line open for about 1-sec intervals (60 Hz) between operations. It is during these periods that the sectionalizer functions.

When a fault occurs beyond the sectionalizer, the recloser will operate. If the fault is of a permanent nature, the sectionalizer will count the number of operations of the recloser and trip and lock itself out after a predetermined number, usually three, of recloser operations. The recloser continues on its fourth operation and restores service up to the sectionalizer. A sectionalizer must therefore always be backed up by a recloser of proper size.

The sectionalizer consists of a set of contacts, a spring-controlled trip, and a solenoid operating coil also connected in series with the line. All are immersed in a bath of insulating oil and housed in a tank.

Vacuum Switches

The vacuum switch is a switch whose contacts are opened and closed in a vacuum bottle. The electric arc that forms between the contacts is quenched within the vacuum container.

Vacuum Reclosers

Vacuum reclosers operate in the same manner as oil circuit reclosers and can be used in the electric distribution system in place of oil circuit reclosers. A vacuum recloser installation is illustrated in Fig. 18.11. A vacuum medium has an outstanding dielectric strength and develops rapid recovery of the dielectric strength in the arc path between the separating contacts following a current zero. The vacuum interrupter limits the arc time to a short period, usually interrupting the current at the first current zero. The contact travel is short, minimizing the energy requirements of the operating mechanism.

FIGURE 18.11   Three-phase vacuum recloser installed to protect an electric distribution circuit from failure of an underground cable tap. The distribution circuit operates at 13,200Y/7,620 volts.

The vacuum reclosers will operate for long periods without maintenance. Contact erosion occurs each time the recloser operates. The amount of contact erosion is directly related to the current magnitude when the interruptions occur. When the contacts have eroded approximately 1/8 inch, the vacuum interrupter should be replaced. The recloser mechanism should be kept clean and checked for proper operation at regular intervals. The insulators should be kept clean and inspected for damage when other routine maintenance is performed.

Sulfur Hexafluoride Gas (SF$_6$) Switches

Circuit breakers, circuit switches, and reclosers are manufactured to use SF$_6$ gas as an insulating and interrupting medium. SF$_6$ gas has proved to be a controversial insulating and interrupting medium. SF$_6$ gas is a greenhouse gas and has proven to have harmful effects on the environment. The Environmental Protection Agency (EPA) is strongly encouraging utilities and manufacturers to eliminate SF$_6$ gas equipment and utilize air, oil, or vacuum interruption devices.

Electronic Microprocessor Controlled Reclosers

In the hydraulically controlled three-phase reclosers it is normally necessary to "untank" the unit to change the characteristics, such as number of fast and slow trips, total number of operations, time-current trip characteristics, or the reclose times. However, in the electronically microprocessor controlled reclosers, changing of all the functions of the control is accomplished by utilization of the sophisticated key pad on the face of the controller, as shown in Fig. 18.12. A much finer and repeatable control function is obtained since the temperature of the oil does not have to be taken into account. The electronic microprocessor type control is used on reclosers using oil, vacuum, and air as an interrupting material. Several models of the electronic microprocessor recloser controllers have bi-directional functionality for utilization in distribution automation loop configurations.

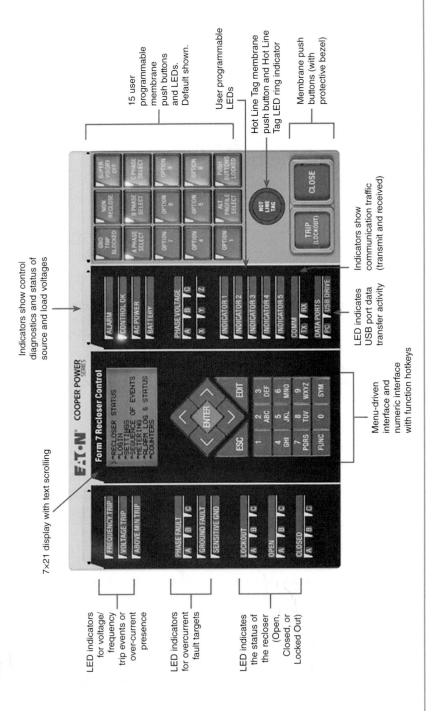

FIGURE 18.12 Electronic controls for recloser. Operating characteristics can be selected by switches and interchangeable components on front of panel. (*Courtesy Cooper Power Systems.*)

CHAPTER **19**

Tap-Changing Transformers and Voltage Regulators

System voltages fluctuate due to the ever-changing load consumption characteristics of customers. Voltage at the customer's premises must be maintained within a range that will permit the customer load to operate properly. The *American National Standard for Electric Power Systems and Equipment—Voltage Ratings (60 Hz)* ANSI C84.1 provides detailed information on standard nominal system voltages and voltage ranges. ANSI C84.1 specifies that 120-volt alternating-current nominal voltage must be maintained between 114 and 126 volts (±5%). There are several options available to utilize to stay within the desired utilization voltage range and compensate for the system load changes. The options include balancing load, increasing feeder conductor size, applying tap-changing transformer equipment, installing capacitors or installing voltage regulators. The concept of voltage regulation is discussed in more detail in Chap. 40. It was stated in Chap. 3 that if the substation power transformer is not equipped with load-tap-changing equipment, it is usually necessary to install voltage regulators. Voltage regulators are used to vary the source voltage to the customer by the proper amount to keep the voltage within the limits desired. Voltage regulators maintain a reasonably constant voltage at the point of utilization. Voltage regulation equipment may be installed in the substation or on a pole out on the distribution circuit.

Tap-Changing Transformers

Distribution substation power transformers or distribution line transformers may be equipped with *tap-changing* equipment. The distribution substation transformers (Fig. 19.1) are normally three-phase with delta-connected primary windings and wye-connected secondary windings to provide a source for three-phase four-wire grounded-neutral alternating-current distribution feeder circuits (Fig. 19.2). The windings of the transformer illustrated are constructed to reduce the subtransmission voltage to the distribution-system primary voltage level. The three distribution voltage windings of the transformer have taps near the neutral, or common, connection. The taps in each winding connect to switch contacts, making it possible to change the number of turns in the transformer secondary windings to vary the substation distribution feeder-bus voltage as desired.

FIGURE **19.1**   Distribution substation transformer with tap-changing underload equipment. Tap-changing switches are located in upper compartment on side of transformer, and control equipment is located in lower compartment.

The tap-changer switches in the low-voltage windings are constructed to permit their operation underload without interrupting the circuits. The high-voltage tap changer is designed for deenergized operation. The tap changers in the high-voltage windings are set to the proper position before the transformer is energized to coordinate with the subtransmission-system voltage level at the substation. The rated voltage for each tap in the low-voltage winding is tabulated in the chart shown in Fig. 19.2. The ratio between high- and low-voltage windings is determined by the rated voltage of the taps at which the transformer operates. For example, if the high-voltage winding is set on no-load tap position 3, rated 69,000 volts, and the low-voltage winding is set on underload tap position lower 4, rated 13,455 volts, the ratio of the windings will be 69,000/13,455 or 5.13/1. If the subtransmission voltage is 68,500 volts, the distribution primary voltage will be 68,500/5.13, or 13,353Y/7,710 volts, for the transformer tap position stated.

The transformer tap changer that operates underload can be controlled manually by using control switches or automatically by voltage-sensitive devices. The automatic regulating equipment can adjust the primary distribution voltage for various distribution feeder line loadings and associated voltage drop with the aid of line-drop compensator circuits. If the distribution primary feeder circuits are short, the substation power transformer tap-changing equipment may be adequate to regulate the voltage. Power transformers with tap-changing underload equipment normally have a capacity to vary the voltage ±10% from rated voltage.

Feeder Voltage Regulators

The function of a feeder voltage regulator is to maintain constant voltage on an alternating-current primary distribution feeder circuit with variations in load. *American National Standard Requirements, Terminology, and Test Code for Step-Voltage and Induction Regulators*

Raise		Reversing switch on 11	
Low voltage	Amp at 33.3 MVA	Indicator position	Contacts on
15180	1268	16	1 + 1
15094	1275	15	2 + 1
15008	1282	14	2 + 2
14921	1290	13	3 + 2
14835	1297	12	3 + 3
14749	1305	11	4 + 3
14663	1313	10	4 + 4
14576	1320	9	5 + 4
14490	1328	8	5 + 5
14404	1336	7	6 + 5
14318	1344	6	6 + 6
14231	1352	5	7 + 6
14145	1361	4	7 + 7
14059	1369	3	8 + 7
13973	1377	2	8 + 8
13886	1386	1	9 + 8
13800	1395	N	9

Lower		Reversing switch on 10	
Low voltage	Amp at 33.3 MVA	Indicator position	Contacts on
13800	1395	N	9
13714		1	9 + 1
13628		2	1 + 1
13541		3	1 + 2
13455		4	2 + 2
13369		5	2 + 3
13283		6	3 + 3
13196		7	3 + 4
13110		8	4 + 4
13024		9	4 + 5
12938		10	5 + 5
12851		11	5 + 6
12765		12	6 + 6
12679		13	6 + 7
12593		14	7 + 7
12506		15	7 + 8
12420	1395	16	8 + 8

FIGURE 19.2 Substation transformer nameplate information containing single line diagram of three-phase distribution substation power transformer windings along with table indicating load tap changer settings. Transformer is rated 20/26.7/33.3 MVA high-voltage winding 69,000 volts, low-voltage winding 13,800Y/7,970 volts.

ANSI/IEEE C57.15 is the industry guideline. The feeder voltage regulator can be three-phase (Figs. 19.3 through 19.6) or single-phase. If the substation power transformer is not manufactured with tap-changing underload equipment, or the primary distribution feeder circuit is long and heavily loaded, a voltage regulator will be installed to maintain proper voltage. In some instances, the voltage regulators are located directly at the substation and other times voltage regulators are located on poles much further out on the distribution circuit away from the substation.

A three-phase voltage regulator basically consists of three single-phase regulators installed in a single tank. Three single-phase regulators are often used to control the voltage on a three-phase distribution circuit. One single-phase voltage regulator will be used on long single-phase taps to the main three-phase distribution circuit.

FIGURE 19.3   Three step-type primary distribution feeder circuit voltage regulators installed on a distribution platform. Regulators are rated 13,800Y/7,970 volts. Each regulator is installed on a different phase. Switches in structure permit regulators to be bypassed and isolated.

FIGURE 19.4   Close up view of three single-phase pole-type distribution feeder voltage regulators and regulator bypass switches. The regulator is rated 167 kVA.

FIGURE 19.5   Three-phase step-type primary distribution feeder circuit voltage regulator platform structure base.

FIGURE 19.6   Single-phase pole-type distribution feeder voltage regulator controllers installed at the base of the distribution pole.

The single-phase feeder voltage regulator, or step voltage regulator, is essentially a transformer having a low-voltage secondary winding connected in series with the line and arranged so that the number of turns in the winding can be varied. The voltage is changed by changing the number of turns in the secondary winding. This is accomplished by means of a rotary tap-changing switch. A schematic diagram of the connections is shown in Fig. 19.7. The primary winding is connected across the line, while the secondary winding with its taps is connected in series with the line. When the "A" switches are closed, the series winding will extend the primary winding and add to the line voltage in an amount depending on which transfer switches are contacted by "x" and "y." When the "B" switches are closed, the voltage in the series winding will be opposite in direction to the primary and so reduce the load voltage.

Regulators generally provide for a variation in load from 10% below to 10% above normal line voltage. This is usually accomplished in 32 steps of 5/8% each. The tap-changing switch is motor driven and is immersed in oil.

FIGURE 19.7   Connection diagram of step voltage regulator. *A* and *B* are reversing switches, and 1 to 9 are transfer switches which contact *x* and *y*. When *A* switches are closed, the line voltage is raised, and when *B* switches are closed, the line voltage is lowered.

The reversing switch provides the 10% regulation on either side of the neutral position. The use of vacuum switches while changing taps on the voltage regulators has reduced maintenance. A tap changer, using a vacuum interrupter, can operate a million times while carrying large current before it is necessary to complete major maintenance. Power thyristors are being installed in some step voltage regulators to electronically switch the number of turns in the winding to completely eliminate arcing in the insulating oil. The use of silicon controlled rectifiers (SCRs), which permit an arcless changing of the taps in the voltage regulator, eliminates the need for periodic maintenance of the equipment.

Single-phase step voltage regulators are manufactured in pad-mounted tanks with provision for cable elbow connectors providing dead-front construction (Fig. 19.8).

The control transformer illustrated in Fig. 19.9 provides a voltage source for the regulator control circuits. The control transformer is an autotransformer that supplies voltages above and below ground to the control circuits (Fig. 19.10).

Two diode rectifiers are connected to the center-tapped supply; one supplies power to the circuit, and the other supplies a sensing voltage through a resistive divider.

Reference voltages are generated from a Zener reference and a resistive divider. These are applied to one input each of two detectors (comparators). The difference in the two reference voltages is the basis of the bandwidth of the control. The sensed voltage is filtered and applied to the other input of both detectors.

One output of each of the detectors goes low if the sensed voltage is out of band on the low side. This initiates the timer and turns on the OUT-BAND light-emitting diode (LED). The other output of the two detectors goes high when the voltage is out of band on the high side. This is ANDED with the output of the timer which goes high after the predetermined delay and the associated Triac fires. The term ANDED is a statement command meaning all specified conditions need to be true.

When the control selector switch is in the automatic position, the tap changer motor (Fig. 19.11) is connected to the control output Triacs. When the selector switch is in either the manual-raise or manual-lower position, the tap-changer motor is disconnected from the automatic circuit and the proper motor circuit is connected to ground to drive the tap changer in the direction selected.

FIGURE 19.8 Pad-mounted primary distribution feeder voltage regulator with compartment door open. High-voltage cables connect to the regulator with elbow connectors. Voltage-sensitive control device is open, displaying voltage control dials.

FIGURE 19.9 Single-phase regulator power circuit schematic diagram.

FIGURE 19.10   Regulator control circuit schematic diagram.

FIGURE 19.11   Regulator motor circuit schematic diagram.

In automatic operation, the voltage from the potential winding is compared with the voltage-level setting on the control. If the voltage falls outside the band, the time-delay sequence is initiated. At the end of the delay time (30 sec is standard), the proper Triac is fired, driving the tap changer in the direction that will correct the voltage. The block diagram of the control (Fig. 19.10), in conjunction with the motor circuit diagram (Fig. 19.11), shows this sequence of operation.

Time delay, which can be altered by changing a resistor, is set by an R/C timing circuit. A capacitive discharge circuit rapidly discharges the capacitor if the input to the timer goes high, resetting the timer. Thus, the timer is quickly reset when voltage goes back in band. Time delay must be continuous without resetting before a tap change is accomplished.

A holding switch—either LHS or RHS (Fig. 19.11)—seals in shortly after the motor starts to run, thus ensuring the completion of the tap change. Resistors in the Triac return from the motor limit Triac current to an acceptable level under all conditions. The Triacs only have to initiate a tap change, at which time the motor torque requirement of the tap changer is low. The holding switch takes over almost immediately after the start of a tap change and carries the motor current for the higher torque required to complete the operation.

FIGURE 19.12 Regulator indicator circuit schematic diagram.

A pair of microswitches (Fig. 19.12) are actuated by the tap-changer reversing switch. When the reversing-switch actuating segment is centered (the tap changer is in the neutral position), the potential voltage is fed through a series connection of the normally closed contacts of the microswitches to the neutral lamp mounted on the tank wall. As the tap changer moves off the neutral position, the reversing-switch segment actuates one of the microswitches (which microswitch depends on whether the segment is moving in a lower or raise direction). On one side, the supply is half-wave-rectified by a diode to form positive pulses and, on the other side, to form negative pulses. These positive and negative pulses are fed through a current-limiting resistor to a pair of parallel reverse-connected LEDs: positive pulses actuate one; negative pulses actuate the other. Thus, one LED—designated *L*—remains on as long as the tap changer is in the lower position. The other LED—designated *R*—remains on as long as the tap changer is in the raise position.

Voltage Regulator Bypass Switches

Bypass switches for voltage regulators are shown in Fig. 19.13. The voltage regulators must be in the neutral position to operate the bypass switches with the circuit energized.

FIGURE 19.13 Voltage regulator bypass switch mounted on distribution structure above voltage regulators. The switch bypasses voltage regulators and interrupts magnetizing current in the correct sequence ... all with one pull of a hookstick. (*Courtesy S&C Electric Co.*)

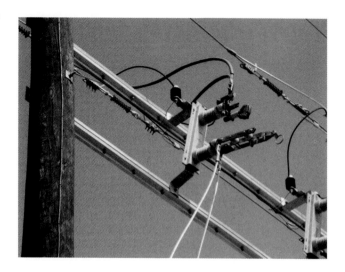

When the voltage regulator is in the neutral position, the series winding output voltage is zero. When the lineman operates the bypass switch to deenergize the voltage regulator, the switch contacts short-circuit the series winding of the voltage regulator to maintain the continuity of the feeder circuit and then in sequence open the contacts in series with the primary winding, interrupting the magnetizing current. If the switches are operated without the voltage regulator in the neutral position, zero voltage output from the series winding, a short circuit will develop, creating a dangerous situation.

Summary

Tap-changing equipment and voltage regulators are widely used within the industry to maintain voltage ranges within acceptable levels. Chapter 40 explains in more detail the concept of voltage regulation.

Transmission Tower Erection

T he general procedure for erecting transmission towers is similar to that for poles, but towers present more problems. The towers require foundations; they are higher and heavier and are therefore more difficult to erect; the conductors are larger and the spans longer, making wire stringing a more difficult job.

The order of the operations in tower erection may be briefly outlined as follows:

1. Clearing right-of-way for the line

2. Installing tower footings

3. Grounding tower bases

4. Erecting transmission towers

5. Insulator installation

These operations will be illustrated in the pages that follow, and comments will be made on the operations performed.

Clearing Right-of-Way

The right-of-way is cleared of all obstructions that will interfere with the operation of the electric transmission line. In scenic and residential areas, clearing of natural vegetation is limited. Trees, shrubs, grass, and topsoil that are not cleared are protected from damage during the construction of the tower line. At road crossings, or other special locations of high visibility, right-of-way strips through forest and timber areas are cleared with varying alignment to blend with the topography of the terrain. Where rights-of-way enter dense timber from a meadow or other clearing, trees are feathered in at the entrance of the timber for a distance of 150 yards to 200 yards. Small trees and plants are used for transition from natural ground cover to larger areas (Fig. 20.1).

A strip of land is cleared on each side of the centerline of the transmission line by cutting and/or trimming the trees and brush. All "danger trees," trees considered to be hazardous to the transmission line, are removed (Fig. 20.2). All trees, brush, stumps, and other inflammable material, except grass and weeds, are removed from the right-of-way. All trees and brush are cut 3 inches or less from the ground line so that the passage of trucks and tractors will not be hindered. The trees and brush cut are disposed of by chipping and spreading, burning, or hauling away. Disposal of the debris by burning, or otherwise, is accomplished in accordance with state and local laws and regulations without creating a hazard or a nuisance.

The right-of-way is treated with a herbicide to retard the growth of brush or trees that if left untreated could endanger the operation of the transmission line.

FIGURE 20.1 Right-of-way cleared for transmission line. The contour of the land and the environment must be taken into consideration by the use of selective clearing techniques. (*Courtesy Asplundh Tree Expert Co.*)

FIGURE 20.2 High-voltage, alternating-current transmission line with right-of-way cleared. (*Courtesy A. B. Chance Co.*)

Installing Tower Footings

Tower sites are graded in accordance with the specifications. Usually the slope of the grade is not greater than 3:1. All topsoil is removed prior to grading the tower location. After the tower construction has been completed and the footings backfilled, the topsoil is replaced. Any excess graded material is removed from the right-of-way.

The excavation is made according to the dimensions and depths shown on the drawings, allowing sufficient space for proper construction of forms and installation of caissons as required (Figs. 20.3 through 20.5).

Excavations may require the use of equipment such as a wellpoint for dewatering the excavation. The excavations are adequately braced and shored to guard against movement or settlement of adjacent structures, utilities, roadways, or railroad facilities. Reinforcing bars and embedded stub angles are installed properly. The reinforcing steel is placed and firmly wired before the concrete pouring is started. Exposed reinforcement steel for bonding future construction is protected from corrosion. Reinforcing steel ties, or spacers, are 1 inch or more from the finished surface of the exterior or exposed portion of the footings. All accessories in contact with the formwork, or the soil, are galvanized steel, plastic, or some other corrosion-resistant material.

Piles are driven to the depth specified in the foundation drawing. Driving is done with fixed leads that hold the pile firmly in position and alignment. Suitable anvils are used to prevent excessive damage to the piles. Driving of the piles is continuous without intermission until the pile has been driven to the specified depth. The tops of the piles are cut off true and level at the proper elevation.

FIGURE 20.3 Various types of tower anchors. Since towers are not massive, footings must be provided to anchor them firmly in the ground. Anchors support the tower and prevent it from blowing over when the lines are subjected to sleet and wind. The anchors shown in (a) and (b) consist of tapering masses of concrete in which anchor bolts or stub angles are embedded. The anchors shown in (c), (d), and (e) are merely extensions of the tower structure embedded in the earth. These extensions are enlarged so that a large bearing surface is presented. (*Courtesy American Bridge Co.*)

Concrete for the footings is thoroughly mixed and is in a uniform workable state when placed. Concrete is placed before the initial set has occurred, which is accomplished by pouring the concrete in the forms within 1¼ hours or before the drum of agitating transport equipment has revolved 300 revolutions. Before placing concrete, the earth foundation is compacted. Granular soil is thoroughly moistened by intermittent sprinkling. The surface, however, is not muddy or frozen when the concrete is poured. Rock surface on which concrete is placed is thoroughly cleaned of dirt, debris, and disintegrated material by a high-velocity jet of air or water. All standing water is removed from the depressions in the area where the concrete is to be placed. The steel-reinforcing rods, embedded steel, and forms are cleaned of all

FIGURE 20.4 Tower anchoring systems. (a) Single-post earth anchor, used with tangent and small-angle towers in line and at sites where earth-augering equipment can economically and feasibly be employed. (b) Triangled earth anchor, used for normal soil conditions with line and angle towers. (c) Quadruped earth anchor, used with dead-end, line, and heavy-angle towers in normal soil conditions. The soil-bearing capacity of this anchor is doubled when it is encased in a concrete footing. (Courtesy Blaw-Knox Co.)

FIGURE 20.5 Tower-anchoring systems. (*a*) Rock anchor, used where welded base assembly is preferred and at sites where elevation of subsurface rock is known before erection. Insertion of the anchor's narrow shaft in drilled rock eliminates rock removal by blasting or other costly means. Swedged anchor bolts in the shaft increase the bond between steel and grout. (*b*) Welded base assembly, an easy-to-erect anchor for use where structural loads require the strength of a welded unit in good soil conditions. (*c*) Adjustable rock and concrete anchor, used for applications where economical concrete anchorage is desirable and where rock may be encountered. (*Courtesy Blaw-Knox Co.*)

425

FIGURE 20.6   Finished concrete foundation with bottom section of steel transmission-line pole set in position.

loose rust scale, paint, mud, dirt, or dried mortar. The forms are oiled before the concrete is poured.

The concrete is conveyed with the proper consistency to the place of final deposit as rapidly as practicable by methods which will prevent segregation, loss of ingredients, rehandling, or premature coating of reinforcing steel or forms. The concrete is consolidated by use of high-frequency internal vibrations and hand spading and rodding while it is poured. The tops of all foundations are troweled smooth, and all irregular projections are removed when the forms have been removed. Any exposed reinforcing steel, broken corners, or edges are cleaned and painted with an epoxy-resin bonding compound. The concrete foundations are properly cured (Fig. 20.6). Curing of the concrete usually requires seven consecutive days, or more, with the temperature of the air in contact with the concrete at 50°F or greater, before the steel for the tower is attached. Backfilling of dirt around the completed tower footing is completed as soon as practicable after the concrete work is finished. Care is taken to exclude large lumps of dirt from the backfill. The backfilling is completed evenly on all sides and properly tamped.

Grounding Tower Base

All steel towers are grounded, usually by means of ground rods. Some tower bases are actually extended to buried steel footers which ground them, while others are connected to buried copper or steel wires. The usual ground rod is a 5/8 inch or ¾ inch. in diameter by 8 ft to 10 ft in length copperweld rod. The rods are driven into undisturbed earth near the tower footing, one rod near each leg. Each tower leg is connected to its own ground rod. If the tower is located near a substation, the four ground rods are often connected together with a heavy copper wire which runs around the tower in a shallow trench. Often the tower ground is also connected to the substation ground by means of the same heavy copper wire. The connection to the tower consists of a heavy clamp bolted to the leg above the ground surface, where it can be removed for testing purposes if need be (Fig. 20.7).

FIGURE **20.7** Ground rod connected to tower leg. (*Courtesy Copperweld Steel Co.*)

Erecting Transmission Towers

The lineman exercises care in unloading, delivering, handling, and erecting towers so as not to damage the finish. Tower members stored pending erection are sorted and neatly piled and supported by suitable blocking. Tower members are not used as unloading or loading skids. Steel erection drawings and structure lists provide the guidelines necessary to assemble the towers. The towers are erected by assembling them in sections on the ground and hoisting the successive sections into place (Fig. 20.8). If the towers are erected by assembling in sections, 50% or more of the bolts are in place on each section before another section is started.

The piece-by-piece method of tower erection is similar to that used in erecting steel buildings or other permanent structures. The method is used when the towers are large and heavy and when the ground is rough.

FIGURE **20.8** Horizontal assembly of the base of a high-voltage transmission-line tower. The first side or panel of the tower has been assembled. Note one anchor or stub in the center foreground. With this type of assembly and modern methods of tower erection, the tower can be assembled on level land, even though it is some distance from the tower footing. Later a crane will pick up the completed tower, carry it to its location, and set it on its foundation. (*Courtesy Wisconsin Electric Power Co.*)

In assembling a tower in place, light members are often simply lifted into place. Sometimes one of the corner legs is used for raising the other members. For heavier towers, a small boom is rigged on one of the tower legs for hoisting purposes or mobile cranes with telescoping or lattice booms on wheeled or crawler-tracked undercarriages are used.

When the tower assembly is complete (Figs. 20.9 through 20.11), all bolts are drawn up tight to specified torque. Palnuts are placed on all bolts and tightened with a special palnut wrench.

The development of the helicopter has made possible its use for transport of poles, tower, and line materials for erection of poles and towers. It is especially suited for transport if the terrain is rugged, if access roads are not available, or if the right-of-way has not been cleared. Figure 20.12 illustrates the helicopter's suitability in the transport and erection of towers in sections or V-type transmission-line towers fully assembled. The delivered towers are anchored in place (Fig. 20.13). Some towers have steps permanently mounted on support pieces for future maintenance activities (Fig. 20.14).

FIGURE 20.9   An in-line, three phase, 345-kV structure height required to maintain proper clearance over a large span length.

FIGURE 20.10 An in-line, three-phase, 345-kV river crossing. The conductors are all placed on one side of the structure in case a second double circuit will be required for the future.

FIGURE 20.11 An interstate highway crossing. In addition to the crossing, the conductors are not in-line but turn an angle.

FIGURE 20.12 Helicopter transporting completely assembled tower including guy wires to line location. (*Courtesy American Electric Power System.*)

FIGURE 20.13 Typical anchor assembly used to anchor the guys on V-shaped, two-legged, aluminum towers. A single center compression foundation, 1 ft by 2 ft, supports the tower. The anchors range in size from 4 ft to 6 ft in diameter. (*Courtesy A. B. Chance Co.*)

FIGURE 20.14 A 765-kV structure with ladders steps permanently affixed for future maintenance activities.

FIGURE 20.15   Dreyfuss-type, Y-shaped steel pole structure used for 500-kV ac transmission line. Note steel wings welded to tower members for lineman to use for climbing tower. (*Courtesy A. B. Chance Co.*)

Steel pole, H-frame structures:, Dreyfuss-type, Y-shaped steel structures (Fig. 20.15); and single steel pole structures are used to improve the aesthetic effects and minimize the right-of-way required. Steel poles eliminate the need for down guys. Single steel poles are used to support transmission-line circuits along city streets. The steel poles are fabricated in a factory and shipped to the site in sections. The foundations are constructed in the same manner as those described for towers. Linemen assemble the steel pole towers (Fig. 20.16) in the field and erect structures with the use of cranes (Fig. 20.17)

FIGURE 20.16   Single steel pole transmission structure assembled on ground prior to installation. Note, brackets on side of pole to facilitate the installation of steps for climbing.

FIGURE 20.17   Crane lifting steel transmission-line pole into position. Insulators are secured to steel arms in a manner to restrain them and prevent breakage. (*Courtesy L. E. Myers Co.*)

and mechanized equipment. Some steel transmission-line poles are fabricated with telescoping construction. The steel poles are shipped to the site, bolted to the concrete foundation, and raised by pumping the interior of the pole full of concrete, using a high-pressure pump. The poles are erected very fast. The concrete reinforces the steel assembly when it has cured.

Insulator Installation

Insulators are handled with care to prevent chipping or cracking the porcelain or glass or damaging epoxy assemblies. Excessive bending strain on the pin shanks or caps is avoided. A cradle, or similar device, is used to lift porcelain or glass insulators. Insulators that are damaged are discarded. The insulators are kept clean and free from grass, twigs, dirt, or other foreign matter. All cotter pins are properly and adequately spread before the insulator strings are attached to the transmission-line structures. The suspension assemblies on all conductors of the line are placed so that all nuts face the center of the structure and cotter pins are parallel to the line with the points of the pins facing inward or toward the space between the conductors.

Stringing Line Conductors

The installation of conductors on poles and towers must be accomplished in a manner to provide an electric circuit that will operate reliably and not endanger the public in accordance with OSHA 29 CFR 1910.269 (q)(2) *Installing and removing overhead lines.* The stringing of transmission lines is accomplished in consideration of IEEE Standard 524 *Guide for the Installation of Overhead Transmission Line Conductors.*

Slack Conductor Stringing

Slack conductor stringing is limited to short lengths of line operating at low voltages utilizing conductors with a weatherproof covering or installations where scratches of the conductor are not important. Scratches on ground or neutral conductors will not cause problems if the strength of the wire is not impaired.

The reels of wire can be mounted on a vehicle in such a manner that the reels are free to rotate. The ends of the conductors are fastened to a pole, tower footing, or other fixed object. The vehicle is then slowly propelled along the route of the line, allowing the conductors to unwind as the reels are moved forward. In this method, the conductors are not dragged over the ground, causing them to become scratched or damaged. They are simply payed out on the ground without being pulled over it. The method, however, cannot be used on one-circuit tower lines or on X-braced, H-frame lines because the center or middle conductor must be placed over the internal framework of the tower or over the X braces of the H frames. This obviously is impossible if the reels are on the lead end of the conductor.

If the reels on which the conductors are wound remain in a fixed location and are raised off the ground or supported in their carriages in such a way that they are free to rotate, the conductors can then be pulled out, thereby rotating the reels and unwinding the conductors.

The conductor should never be payed out from a nonrotating reel or coil, as each turn removed gives the conductor a complete twist which may cause kinks or other damage. The use of a heavy-duty ball bearing swivel assures the conductor remains in the neutral position during stringing (Fig. 21.1).

Unreeling conductors from stationary reels provides two possibilities, one to simply draw the conductors forward, sliding or dragging them over the ground, and the other to keep the conductors suspended in the air in tension so that they will not touch the ground. When the conductors are pulled forward, sliding over the ground, they are apt to become scratched and nicked, especially if the ground is rough and covered with surface stone. Aluminum conductors are more easily injured in this manner than copper. Conductors with surface scratches or defects energized at high voltages will have corona losses and generate radio-interference voltages that will be transmitted through the atmosphere.

Ball bearing swivel

Figure **21.1**   Heavy-duty ball bearing swivel installed in the conductor reel.

Tension Conductor Stringing

The process of installing overhead transmission- or distribution-line conductors in a manner which keeps the conductors off the ground, clear of vehicular traffic and other structures that might damage the conductors, and clear of energized circuits is called *tension stringing*. Tension stringing, to reconductor existing energized lines, provides a means to keep the new conductors under control and prevent them from contacting the energized conductors. The new conductors being installed are effectively grounded for the protection of the workmen. Transmission-line conductor installation requires the linemen to install conductor-stringing blocks on the transmission-line insulator strings to permit pulling the conductors in under tension (Fig. 21.2).

The conductor-stringing blocks are built for the installation of one to four conductors. The sheaves on the stringing blocks are usually lined with a conductive-type neoprene, or urethane, to protect the phase conductors which are usually aluminum or aluminum and steel (ACSR). The blocks used for static, or ground, wires are not lined since the static wires are usually galvanized steel, copperweld, or alumoweld, which are not easily damaged.

The conductive neoprene lining of the stringing blocks permits the conductors to be effectively grounded by a jumper from the stringing block to a ground wire on the metal tower structure. Effectively grounded stringing blocks will eliminate induced static, or impulse voltages that might be present during construction. The stringing blocks, often called *travelers*, can be equipped with outriggers to permit the installation of the pilot line with a helicopter.

When linemen install the stringing blocks on the insulators, they normally place a lightweight rope called a *finger line* over the traveler. The finger line must be long enough to reach the ground on both sides of the traveler. Finger lines are used by the linemen to pull the pilot line through the traveler from the ground. If the pilot line is installed with a helicopter, finger lines are not used. Pilot lines are lightweight rope used to pull pulling ropes through the travelers. A pilot-line winder is used to provide the power necessary to pull the conductor-pulling rope from the reel on the take-up reel stand through the travelers. Tension is kept on the pulling rope, often referred to as the *bull line*, to prevent interference with objects on the ground.

The conductor reels, tensioners, and pulling machines must be in line before the conductor pulling is started (Figs. 21.3 through 21.5). The tensioner, often called a *bull-wheel* tensioner, and the bull-wheel puller should be set up as near to midspan as possible. The slope of the conductors between the equipment and the stringing blocks at the first tower must not be steeper than 5 horizontal to 1 vertical.

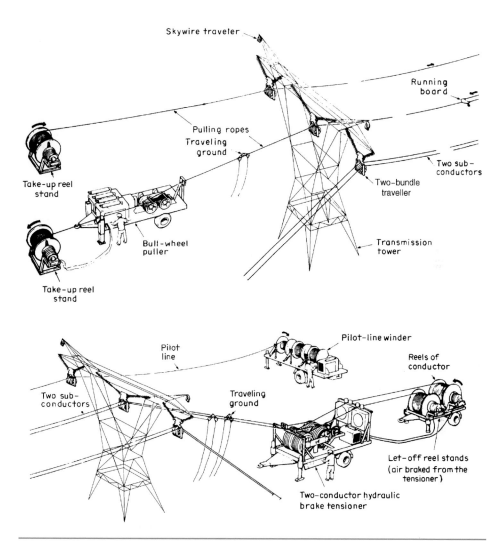

FIGURE **21.2** Typical equipment setup for tension stringing a two-conductor bundle transmission line. (*Courtesy Timberland Equipment Ltd.*)

The transmission-line conductor retarding bull wheels should have a minimum of five turns of the conductor over the bull-wheel and have multiple grooves lined with neoprene or other approved nonmetallic resilient material so that the conductor will cushion into the lining to prevent flattening or otherwise damaging the conductors as they are payed out.

The diameter of the bull-wheel grooves must not be less than the conductor diameter, plus 25%. The bull-wheels should have a minimum bottom grove ("root") diameter of 25 times the conductor diameter.

The conductors should be pulled directly from the cable reel onto the tensioners and into the stringing blocks without touching the ground. As per ANSI 1919.269 (q)(2) (vi), load ratings of stringing lines, pulling lines, conductor grips, load-bearing hardware

FIGURE 21.3 Take-up reel stand in combination with a bull-wheel tensioner/puller.

FIGURE 21.4 Front view of the hydraulic bull-wheel tensioner/puller.

FIGURE 21.5 Equipment in alignment prior to conductor-pulling operation.

and accessories, rigging, and hoists shall not be exceeded. The conductor-pulling tension should not exceed 70% of the conductor-sagging tension. Care must be taken at all times to ensure that the conductors do not become kinked, twisted, abraded, or damaged and that foreign matter does not become deposited on them. Dependable communication must be maintained between the linemen operating the pulling equipment and the tensioning equipment and observers at intermediate points at all times during the wire-stringing operations.

Bundled conductors are pulled simultaneously by one pulling line with the use of a unidirectional articulated running board. Special bundle-conductor-type stringing blocks are required. For stringing two conductors (Fig. 21.6) the block consists of two sheaves and one drum, all of which turn independently of one another. The two outer sheaves support the conductors, and the center portion of the drum carries the pulling line. The drum is designed to accept the running board. For stringing a bundle of three conductors, a block with three sheaves and a two-part drum are used. The pulling line rides in the center groove as does the center conductor. The sheave spacing corresponds to the final spacing of the bundled conductor.

The tensioner for a two-conductor bundle consists of two pairs of bull wheels, one pair for each conductor. The bull wheels are geared together in such a way that a small auxiliary engine can be used to even up the conductors if one should sag more than the other. The engine is also used to air-cool the brakes and to reverse the bull wheels slowly, if this should become necessary.

When a bundle of four conductors is strung, it is good to string the bottom two conductors first and leave them in soft sag. The top conductors are then strung over them.

The conductors are grounded while they are being installed. Traveling grounds can be used at the tensioner and puller sites. The use of travelers with a conductive-type neoprene lining permits grounding the conductors at each structure. The grounding cable must be large enough to adequately conduct fault current to ground without fusing.

When the conductors being installed reach the pulling equipment, they must be secured to prevent them from developing slack, which might let them touch the ground.

FIGURE 21.6   Pulling board about to be drawn over two-conductor stringing block. Central drum carries pulling line, and two outer sheaves carry the two-conductor bundle. The wedge-shaped running board ensures block centering and placing conductors in grooves. (*Courtesy Sherman & Reilly, Inc.*)

New distribution-circuit conductors are installed by using tension stringing methods, as described, except that the equipment for pulling and tensioning the conductors is smaller and lighter (Figs. 21.7 through 21.10). Distribution circuits are shorter in distance than most transmission lines, and the conductors are usually smaller with a limit of one conductor per phase. Finger lines can be used to pull the pilot line through the stringing blocks. However, care must be taken to tie the finger lines off high enough on the pole to keep them out of reach of the public.

FIGURE 21.7   Diagrammatic sketch showing arrangement of tension-stringing equipment to install new line conductor. (a) Conductor reel and carriage; (b) line conductor; (c) tensioner; (d) stringing sheave; (e) pulling grip; (f) pulling rope; (g) puller; (h) take-up reel and carriage; (i) traveling ground; (j) ground rod. The conductor being installed is grounded on the reel, by traveling grounds and by grounds connected to the metal-stringing blocks. (*Courtesy Peterson Engineering Co., Inc.*)

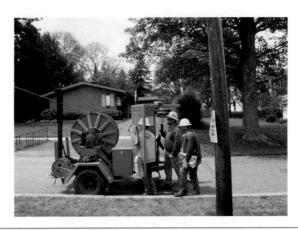

FIGURE 21.8   Lineman is using a two-way radio communication system in order to coordinate pulling and tensioning.

FIGURE 21.9 Bull-wheel tensioner/puller equipment set up for a field demonstration.

FIGURE 21.9 Bull-wheel tensioner/puller equipment set up for a field demonstration.

FIGURE 21.10 The hydraulic bull-wheel tensioner/puller equipment set up with the traveling ground connection.

Tension-stringing procedures are used successfully to reconductor distribution lines. The energized conductors must be kept isolated from the new conductors being installed. The linemen often use hot-line extension arms (Fig. 21.11) to provide clearance for the new conductors. The new conductors being installed adjacent to the energized wires are effectively grounded and controlled to prevent them from being energized by accidental contact with energized wires. In the instances that the new conductors are not effectively grounded, a determination must be made that there is no presence of hazardous induced voltages. Compliance with 1910.269 (q)(2)(iv) as found in Chap. 46 is required.

Pilot lines can be installed for distribution-line tension stringing by using the *spider system*. Pilot-line controllers are installed on the lead pole by a lineman approximately 15 to 20 ft above the ground. A pilot-line storage reel is hung on each controller (Fig. 21.12). The pilot line comes off the storage reel over an arm that releases the brake on the pilot-line controller when the rope is pulled and engages the brake when the rope is slack. This eliminates "backwinding" and holds tension during the installation of the pilot line.

All pilot lines can be carried up each pole and placed in the stringing blocks when the blocks are placed on the pole. As this is accomplished, the pilot lines can be tied up and left in the clear of any obstructions below, such as tree limbs, transformers, and street lights.

FIGURE 21.11   All three energized phase conductors on pole in foreground have been set out in the clear on hot-line extension arms. Stringing blocks for phase conductors are installed on crossarm. Stringing block for neutral-ground wire is clamped to pole below crossarm. The new conductors being installed will be grounded on the reels, by traveling grounds and by grounded stringing blocks. The multigrounded neutral conductor is the best source for the ground connections.

FIGURE 21.12   Spider system of installing pilot line for tension stringing of distribution-circuit conductors. Reels are mounted high enough on pole to prevent pedestrian contact and overnight vandalism. (*Courtesy Sherman & Reilly, Inc.*)

The pulling line is positioned at the opposite end of the spider controller and storage reels and is tied to the end of the pilot line. The partially empty storage reel is then mounted on a pole-mounted winch or a line truck winch. The pole-mounted winch is powered hydraulically or electrically. The line truck's hydraulic system can be used to supply power for the pole-mounted winch.

The winch turns the reel and rewinds the pilot line, thus hauling in the pulling line under tension and eliminating the need for lineman to walk through while carrying the pulling line over each structure.

Fast, efficient, safe, and field-tested, the spider system is used by many utilities. The pilot line is used to pull the pulling line, or bull rope, through the stringing blocks. The pulling rope is connected to the conductor with a conductor grip (Fig. 21.13).

The conductors are effectively grounded while they are being installed (Fig. 21.14). The grounding cable connecting to the running ground must be large enough to safely conduct fault current to the point of ground attachment. The multigrounded neutral conductor provides the best source for a ground, if it is available.

Equipment similar to the spider system described is available that can be used to install small distribution conductors under tension without pulling in a larger pulling rope.

The ropes can be placed in the stringing blocks while the lineman is mounting the block, or finger lines can be installed in the blocks to pull the stringing rope through the blocks at a later time. The conductor tension pulling equipment is powered by the line truck's hydraulic system. The equipment can be pole-mounted or can be installed on a trailer.

FIGURE 21.13 Attaching the Kellum grip on a 795,000-cir-mil ACSR cable. This grip is used for pulling the cable over the stringing sheaves on the structures. When the grip is in place, it is taped down to the conductor to keep it from pulling back and releasing. (*Courtesy Northwest Lineman College.*)

FIGURE 21.14 Running ground device designed to maintain positive contact with the conductor being strung regardless of tension. (*Courtesy Sherman & Reilly, Inc.*)

FIGURE 21.15   In case conductors have to be strung over a live line, guard poles which hold the wires above the live line are used. Shown here are two guard poles carrying three line conductors and two ground wires of a 132 kV line over a 4.8 kV single-phase distribution feeder. (*Courtesy Wisconsin Electric Power Co.*)

FIGURE 21.16   The end result is a properly installed line with proper clearance between the transmission line and the distribution line crossing below.

If distribution- or transmission-line conductors are to be strung over highways or energized circuits, guard poles should be used to protect the public and the workmen (Figs. 21.15 and 21.16).

Self-Contained Continuous Line Puller

The self-contained continuous line puller (CLP) machine is stringing and reconductoring equipment that is being utilized for full tension reconductor line refurbishment. With this line reconductoring equipment, the scrap conductor take-up reels at the collection point are no longer used. Instead, the old conductor is continuously pulled, collected, and cut into sections as the CLP conveyer system feeds the conductor sections into disposal containers to be recycled. This process saves time, increases efficiency, and improves safety.

Sagging Line Conductors

S agging operations begin as soon as the conductor stringing is completed, to establish the proper conductor tension for the conditions at the time the work is performed. The correct conductor sag measurements for control spans are provided in the construction specifications or job details. Conductor is sagged to meet National Electrical Safety Code clearances rules. The conductor tension is established in the control spans. The weather conditions, including temperature and wind velocity, must be taken into consideration to complete the sagging process. Sighting methods using a surveyor's transit or targets can be used to measure the sag in the conductor. The tension in the conductor can be measured with a dynamometer or determined by timing mechanical waves induced into the conductor.

All conductor grips, come-alongs, stringing blocks, and pulley wheels used in the sagging operations must be of a design which will not damage the conductors by kinking, scouring, or unduly bending them. Sag sections must be selected before the operations are started. The lengths of sag sections are limited by dead-ends, angle structures, and the terrain. The sag sections are normally not longer than 4½ miles.

General

Sag sections must have control spans. The control spans will be used to measure the sag for each sag section. If the control spans are properly sagged, the conductors in the sag section should all be at the proper tension. The control spans should be approximately the length of the ruling span. Ruling spans are a calculated dead-end span length assumed to have the same conductor tension as all other spans in the sag section, even though they vary in length. The ruling span is found using the following expression:

Ruling span (ft) = [average span (ft)] + 2/3 [longest span (ft) − average span (ft)].

Control spans should be located away from the dead-end sections and large line angles in a level portion of the line, if possible. If only one span is used for checking the sag, it should be approximately in the middle of the section being pulled to tension. Where two or more spans are used for checking the sag, they should be located approximately equidistant from each other and from each end of the sag section. The sag of spans on both sides of points in the line where the grade varies by more than 10 degrees should be checked.

The use of more than one sag control span in each sag section will eliminate problems due to cumulative sheave effect and human error. One control span is permissible in short sag sections of five spans or less. A minimum of two control spans should be used in sag sections of five to eight spans in length. A minimum of three control spans should be used in sag sections of nine or more spans. Sag sections are normally limited to approximately 20 spans. The control spans selected should be checked to be sure there are no major discrepancies between the actual span lengths and those instructions included with the job specifications.

All wires on new lines being sagged under deenergized conditions should be grounded (Figs. 22.1, 22.2, and 22.3). When a bundle of three or four conductors is sagged, the top conductors are sagged first and bottom conductors are sagged last. The tension in the bottom conductors should be about 3% less than in the top conductors. All wires must be pulled up to the proper sag without exceeding the maximum ultimate conductor tension. The wires must not be pulled up to a tension greater than the sagging tension. The conductors must be sagged to the initial sag specifications.

The temperature of the conductor must be determined with a thermometer that is normally inserted into a piece of conductor approximately 30 inches-long and suspended at the approximate elevation of the wire to be sagged. Two thermometers should be used at locations near sag spans in the sag section. The thermometers should be in position for a minimum of 15 min before the temperature is recorded to determine the proper wire sag. After the conductors have been properly sagged, intermediate spans should be inspected to ensure that the sags are uniform and correct. The wires are sagged from the tensioner end of the sag section to the puller end.

The foreman initiates and supervises the sag operations, maintaining communication with the linemen at each sag span and other points under observation in the sag section between the tensioner end and the pulling end. The sagging operations cannot be

FIGURE 22.1 Ground device installed on moving conductor while sagging and stringing operations are completed. Grounding cable is attached with hot-line tool. Rope secured to pole or tower prevents ground device from moving with conductor.

FIGURE 22.2   Ground device installed on conductor for a field demonstration.

FIGURE 22.3   The ground device is connected to the ground rod.

completed on days with strong winds because of the conductor uplift from wind pressure. The aerodynamic effect of the wind on the conductors can cause sagging errors.

Special precautions must be taken to sag distribution conductors that are near energized circuits or are energized (Figs. 22.4 through 22.7).

FIGURE 22.4 Schematic diagram of distribution line being pulled up to proper sag. Note use of free-running sheave block at every point of support.

FIGURE 22.5 Distribution-phase conductor in stringing block mounted on insulated epoxy stand-off bracket in position for sagging operation. (*Courtesy Sherman & Reilly, Inc.*)

FIGURE 22.6 Distribution-circuit conductors in stringing blocks mounted on line-post insulators. Circuit can be energized in this position while sagging operations are completed if proper precautions are taken. (*Courtesy Sherman & Reilly, Inc.*)

FIGURE 22.7   Come-along grip depends on the wedge action to grip conductor. The greater the pull on the wedge, the harder the conductor is forced against the steel grip insert in the body. Come-alongs can be installed in live-line work with a tie stick, using the lifting ring provided on the top. (*Courtesy James R. Kearney Corp.*)

Sagging Line with Computer Software Sag Tables

Vendor-supplied computer software programs are available that calculate initial and final sag tables for various span lengths of conductor. Tables 22.1 and 22.2 provide example calculations of the initial and final sag in inches for the ruling span of 1/0 ACSR conductor installed at 40% of rated conductor ultimate strength for selected conductor temperatures.

The conductor sag in any span may be determined by the expression:

Span sag (inches) = [span (ft)/ruling span (ft)]² × [ruling span sag (inches)]

Sagging Line with Transit

The method of determining conductor sag used commonly by linemen is to climb one structure and place targets at the specified sag level for the conditions that prevail. The lineman then climbs the structure on the opposite end of the control span and mounts a surveyor's transit, or an especially designed sag scope (Figs. 22.8 and 22.9), at a point equal to the sag distance below the conductor support. The sag scope with a variable mount can be easily leveled under a variety of field conditions. The clamping device for steel, aluminum, or wood-pole structures permits adjustments from 0 inches to 3 inches. The lineman then sights through the transit and observes conductor sag. When the conductor is at the proper level, he tells the foreman supervising the operations to prevent overstressing of the conductor. If the structures on each end of the control span are not the same height or not located on level ground, special adjustments must be made in the target and transit locations to compensate for the differences in elevation.

In the case of an H-frame line, the line conductors are on one or the other side of the poles. A lineman stationed on one pole and looking to the corresponding pole a span length away cannot include the lowest point of the conductor in his line of sight. To get around this difficulty, a transit is securely fastened to the pole at a distance equal to the

Distribution Sag Table Initial Sag								
Span Length (ft)—Sag (inches)				Conductor Raven 1/0 (6/1) ACSR Maximum Tension 40% ULT. = 1,712 lb				
Temp.	100.00	110.00	120.00	130.00	140.00	150.00	160.00	170.00	180.00
−20.00	1.32	1.56	1.92	2.28	2.64	3.12	3.72	4.32	4.92
0.00	1.32	1.68	2.04	2.52	2.88	3.48	4.08	4.68	5.52
10.00	1.44	1.80	2.16	2.64	3.12	3.60	4.32	5.04	5.76
20.00	1.56	1.80	2.28	2.76	3.24	3.84	4.56	5.28	6.24
30.00	1.56	1.92	2.40	2.88	3.48	4.08	4.80	5.64	6.60
40.00	1.68	2.04	2.52	3.00	3.60	4.32	5.16	6.12	7.20
50.00	1.80	2.16	2.64	3.24	3.96	4.68	5.52	6.60	7.80
60.00	1.92	2.40	2.88	3.48	4.20	5.04	6.00	7.20	8.52
70.00	2.04	2.52	3.12	3.84	4.56	5.52	6.60	7.92	9.36
80.00	2.16	2.76	3.36	4.20	5.04	6.12	7.32	8.76	10.44
90.00	2.40	3.00	3.72	4.56	5.52	6.72	8.16	9.84	11.76
100.00	2.64	3.36	4.20	5.04	6.24	7.56	9.24	11.4	13.32
120.00	3.36	4.20	5.28	6.60	8.16	9.96	12.12	14.52	17.40
32.00	7.68	9.48	11.28	13.32	15.60	18.12	20.76	23.64	26.76
+½ ice 2-lb wind									
0.00	10.08	12.12	14.40	16.92	19.68	22.56	25.68	29.04	32.52
+½ ice 4-lb wind									
170.00	8.16	10.08	12.12	14.40	16.92	19.56	22.56	25.68	29.04

Courtesy of MidAmerican Energy Company.

TABLE 22.1   Initial Sag Table for Various Span Lengths of 1/0 ACSR Conductor to Be Installed at 40% of Rated Ultimate Strength

desired sag below the conductor support. Then the transit is leveled. To observe the sag, the transit is sighted at the conductor at midspan and then is swung around until in line with the pole a span length away. The sag is then observed, and the line conductor is drawn up to the specified distance.

A simple and accurate method of measuring the sag is by use of targets placed on the poles below the crossarm, as shown in Fig. 22.10. The targets may be a light strip of wood, like a lath, nailed to the pole at a distance below the conductor when resting on the insulator equal to the desired sag. The lineman sights from one lath to the next. The tension on the conductor is then increased until the lowest part of the conductor in the span coincides with the lineman's line of sight. Careful attention must be paid to the temperature at the time of sagging in of the conductor as the sag varies considerably with temperature. This is illustrated in Fig. 22.11, where typical values of sag are shown for the three values of temperature. The reason, of course, that the sag is greater at the higher temperature is because of the expansion of the metal in the conductor, causing it to be elongated.

	Span Length (ft)—Sag (inches)				Conductor Raven 1/0 (6/1) ACSR Maximum Tension 40% ULT. = 1,712 lb				
Distribution Sag Table Final Sag									
Temp.	100.00	110.00	120.00	130.00	140.00	150.00	160.00	170.00	180.00
0.00	1.44	1.80	2.16	2.64	3.24	3.84	4.68	5.52	6.60
10.00	1.56	1.92	2.40	2.88	3.48	4.20	5.04	6.12	7.32
20.00	1.68	2.16	2.64	3.24	3.84	4.68	5.64	6.84	8.16
30.00	1.92	2.28	2.88	3.60	4.32	5.28	6.36	7.68	9.36
40.00	2.04	2.64	3.24	3.96	4.92	6.00	7.20	8.88	10.68
50.00	2.28	2.88	3.60	4.56	5.52	6.84	8.40	10.20	12.48
60.00	2.64	3.36	4.20	5.16	6.48	7.92	9.72	11.88	14.52
70.00	3.00	3.84	4.92	6.12	7.56	9.36	11.52	14.04	16.92
80.00	3.60	4.56	5.76	7.32	9.00	11.16	13.56	16.44	19.44
90.00	4.32	5.52	7.08	8.76	10.92	13.20	15.96	18.60	21.48
100.00	5.28	6.84	8.52	10.20	12.24	14.40	16.80	19.56	22.44
120.00	6.36	7.80	9.48	11.52	13.56	15.96	18.60	21.48	24.60
32.00	8.64	10.56	12.60	14.88	17.40	20.04	23.04	26.16	29.52
+½ ice 2-lb wind									
0.00	10.08	12.12	14.40	16.92	19.68	22.56	25.68	29.04	32.52
+½ ice 4-lb wind									
170.00	8.88	10.80	12.96	15.36	17.88	20.64	23.76	26.88	30.36

Courtesy of MidAmerican Energy Company.

TABLE 22.2   Final Sag Table for Various Span Lengths of 1/0 ACSR Conductor to Be Installed at 40% of Rated Ultimate Strength

The method of sighting for sag gives very satisfactory results when the sag is not less than 6 inches and when the visibility is good. If the visibility is not good, it is often helpful to provide a suitable target at the distant pole for visual contrast with the conductor. This may be a piece of wood or metal approximately 2 square feet, painted white with a horizontal black line about 1-inch wide across the center. For the sighting end, a device may be used, having a sighting slot in place of the black line (Fig. 22.12). The distant target and the slot are set below the support according to the desired sag, and the conductor sag is varied until the low point of the conductor is in line with the slot and the black line.

Sag Measurement by Timing

Accurate sagging by sighting is difficult if the spans are long or if the sags are quite small. Furthermore, if the poles used in sighting are not at the same level, it is even more difficult. A mechanical wave, instigated near one support, will travel to the next support, be reflected in reverse phase, and pass back and forth repeatedly between the supports. Although the wave is attenuated continuously, the time required for the wave to return is independent of the conductor span length, conductor size, or conductor type.

FIGURE 22.8   Sag scope with zoom capabilities for wire sagging. (*Courtesy Hi-line Industries, Inc.*)

FIGURE 22.9   Sag scope with three times- to nine times magnification-power zoom telescope cross hairs for precision sighting and a two-way bubble level. (*Courtesy Sagline Corp.*)

FIGURE 22.10   Sketch showing lineman sighting for sag by means of two lath strips nailed to poles at proper distance below the conductor when resting on the insulator.

SAG AT 30°F.
SAG AT 60°F.
SAG AT 90°F.

24" 20" 15"

15" 20" 24"

SIGHTING
WIRE SAG

NAIL STICK TO POLE
THE SAG DISTANCE
BELOW WIRE

130 FT. SPAN

FIGURE 22.11 Sketch showing how sag increases with rise in temperature also illustrates the correct manner of sighting for sag.

FIGURE 22.12 Sighting target about 2 square feet and painted white. Black line is about 1-inch wide. Sighting slot is also about 1-inch wide.

Target

Black line

Sighting slot

The sag of the conductor can be determined by timing the wave returns with a standard stopwatch and converting the time measured to sag from the equation

$$\text{Sag } D = 48.3 \, (t/2n)^2$$

where D is in inches, t is time in seconds, and n is the number of return waves counted.

The number of returns depends primarily upon the conductor weight or size and the length of the span being sagged. For long spans and/or larger conductor sizes, the wave energy imparted to the conductor will dissipate more rapidly, and the number of return waves which can be counted accurately is less than for lightweight conductors and/or short spans.

Attach a rope to the conductor about 3 ft from the support. Then give the rope a sharp jerk and at the same time start a stopwatch or note the reading of the second hand on your (pocket) watch. Striking the conductor causes a wave or ripple to travel along the conductor until it reaches the next pole. When it reaches the next pole, it is reflected and returns to the pole where you are stationed. Here it is also reflected, and thus starts its second round trip. This will continue until the energy of the blow has been expended. The time in seconds required for the wave to return to the near support corresponds to a definite sag which can be calculated or read from prepared tables or graphs. The time is independent of the span length or the size or type of the conductor.

Observe the travel of this wave, and count the number of returns until the wave has returned 3, 5, 10, or 15 times. Upon the arrival of the third, fifth, tenth, or fifteenth return wave read the stopwatch or your (pocket) watch and note the time elapsed in

FIGURE 22.13 Graphs giving relation between time in seconds and sag in inches for third to tenth return of wave. (*Courtesy Copper Wire Engineering Association.*)

seconds. A sufficient number of tests should be made until at least three identical readings are obtained. Then refer to Fig. 22.13 and read the corresponding sag opposite the time in seconds. If this is not the desired sag, change the pull on the conductor and repeat the timing operation.

The choice of the number of return waves depends principally upon the span length and the size of the conductor. For long spans and large conductors, the number of return waves which can be accurately counted is less than for short spans and small conductors. The distance of the "jerk" line from the support of a conductor will largely determine the force of the impulse given to the conductor as well as the ease with which the lineman with the stopwatch can feel the returning waves. A distance of 3 ft or more is commonly used. The largest number of return waves which can be observed should be used as this minimizes errors in recording the time.

The return wave may be felt by a man on the pole by placing a finger lightly on the conductor, if the line is deenergized and effectively grounded. Or readings may be made from the ground by throwing a light, dry, and clean polypropylene rope over the conductor about 3 ft from the support. This light rope may also be used to give the impulse initiating the wave. When the line is "hot," that is, energized, the lineman handling the rope must wear rubber gloves while handling the clean, dry, polypropylene rope, which must be constructed and maintained properly for use on energized lines.

Care must be taken not to count 1 when the impulse is given to the line, but to count 1 on the first return of the wave. In other words, count, "hit," 1, 2, 3, etc.

This sagging method is most satisfactory if the line is not in motion. Any vibration of the line, such as might be caused by working on it or by a strong wind, makes it difficult to determine the exact time of the return wave. The conductor should rest on the crossarm or in the stringing block to reflect the wave, although it need not be tied in.

Unisagwatch

The watch (Fig. 22.14) developed by Philip C. Evans takes the place of the stopwatch and sag-time tables and curves. The Unisagwatch carries logarithmic scales calibrated in feet and decimals for both third and fifth returns of a mechanical wave, providing direct

FIGURE 22.14 Unisagwatch calibrated to determine wire sag regardless of span length, temperature, composition, size, or weight of conductor. (*Courtesy Sagline Corp.*)

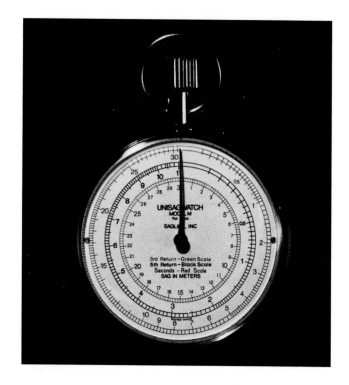

readings of sags up to 100 ft. The watch is calibrated for an accuracy of 1% and has dials for reading sags in feet or meters. Sagging with this instrument eliminates the need for conversion tables and setting and sighting targets. The Unisagwatch is suitable for use in poor visibility conditions due to weather or in a selectively cleared right-of-way. The sag of an energized conductor may be readily checked by an observer on the ground.

Sagging Line with Dynamometer

When conductors are to be strung with unusually small sags, the measuring of the sag may not be as accurate or convenient as measuring the conductor tension. This is done by means of a dynamometer which is inserted in the pulling equipment, as shown in Fig. 22.15. One of the pointers indicates the pull at all times, and the other will remain at the point of maximum load after the tension is released. The pull is increased until the tension is obtained for the desired conductor type and size, span length, loading district, and prevailing temperature.

FIGURE 22.15 Traction dynamometer used with block and tackle in pulling line conductor to proper tension.

Any convenient span in the section of line may be used for measuring the tension, but care should be taken that the tension is uniform in all spans. After stringing, a common practice is to allow the conductor to remain untied for several hours before final measurement and tying. On a long section of line, it is advisable to check the sag or tension at several spans to ensure uniformity of the tension in the line. Figure 22.16 illustrates a tension dynamometer equipped with a swivel hook designed for a maximum conductor tension of 20,000 lb.

A shunt dynamometer can be used to measure conductor tension without being in series with the conductor. The operating principle of the shunt dynamometer is based

FIGURE 22.16 Dynamometer equipped with a swivel hook to prevent conductor twisting. Dynamometer has one hand for continuous load indication and one hand for maximum tension indication. (*Courtesy John Chatillon & Sons.*)

on the relation of the tension in the conductor to the force necessary to displace it in a direction perpendicular to the axes of tension. The use of the shunt dynamometer has been limited because of the cost of the instrument and the complications associated with its successful application.

Clipping in Conductors

Conductors and shield wires should be clipped in as soon as practical after the sagging is completed. In all cases, the conductors and static wires, or ground wires, should be clipped in within 2 days after the sagging is completed. Conductors should be lifted from the stringing sheaves by using a standard suspension clamp or a large plate hook. Rope slings should not be used. The inside diameter of the plate-hook seat should be at least 1/16 inch greater than the diameter of the conductor to be lifted.

The center longitudinal portion of the plate-hook seat must be bent down at substantially the same curvature as found on the suspension clamp. The inside surfaces of the plate-hook must be free of all burrs and rough edges. The conductors must be attached to the insulator assemblies at all suspension and dead-end points in accordance with the specifications or the standards for the line under construction. Armor rods and vibration dampers are usually specified for all high-voltage transmission lines. The dampers should be placed on the conductors as soon as the conductors are clipped in, to avoid the risk of conductor damage as a result of critical wind conditions. Any conductor supported above ground in long spans under relatively high tension is subject to vibration. Such vibration is caused by the passage of air over the conductor at right angles to the line. As the air blows over the conductor, eddy currents are set up on the leeward side. These eddy currents oscillate, first in one direction and then in the other. When the frequency of these eddies corresponds to some natural frequency of the line conductor, a tendency to vibrate exists. Changes in velocity or direction of wind dampen the original vibrations and set up new ones. It is for this reason that a steady gentle breeze may set up far more destructive vibrations than high winds of unsteady force and direction.

Under certain conditions, these vibrations may build up an amplitude which produces alternating stresses large enough to cause fatigue failure. The failure usually takes place at the point of support, very seldom at a dead-end or at a joint. Failure may not be experienced over a period of years, or it may develop in a few months. When a vibration passes through a wire, the conductor tends to conform to the waveform of the vibration. When this vibration encounters some obstacle such as a supporting clamp, it may be reflected back on itself with consequent increase in destructive stresses. More destructive still is the effect of two vibrations of similarly directed bending tendencies meeting at opposite ends of a common supporting clamp.

Vibration fatigue of conductors may be prevented either by reinforcing the conductor at points of stress to resist the effects of repeated bending or by reducing the vibration to negligible amplitudes with damping devices. One form of damper known as the *Stockbridge damper* is illustrated in Fig. 22.17. It consists of a resiliently supported weight with a suitable clamp for attaching it to the conductor. The action of the damper is to absorb continuously the energy of the vibrations started by winds of all kinds and to prevent these vibrations from building up to damaging proportions. The damper is attached a suitable distance ahead and back on the line from various types of conductor supports, dead-ends, etc.

FIGURE 22.17   The Stockbridge damper clamped to transmission-line conductor to reduce conductor vibration by changing natural frequency of line conductor. (*Courtesy Aluminum Company of America.*)

Distribution-circuit conductors are fastened to insulators by the use of either wire ties or clamps. Wire ties are used in distribution systems where the pin insulator is employed. Insulator clamps are used on some high-voltage, pin-insulator lines and entirely on suspension insulators.

Conductors should occupy such a position on the insulator as will produce minimum strain on the tie wire. The function of the tie wire is only to hold the conductor in place on the insulator, leaving the insulator and pin to take the strain of the conductor.

In straight-line work, the best practice is to use a top-groove or *saddle-back* insulator. These insulators all carry grooves on the side as well. When the conductor is placed in the top groove (Fig. 22.18a), the tie wire serves only to keep the conductor from slipping out. The groove practically relieves the tie wire of any strain and allows the entire strain to be placed on the insulator and pin. When side-groove insulators are used in straight-line work, the wires should be placed as shown in Fig. 22.18b. The conductors on the pins nearest to the pole are placed on the side away from the pole. This is to increase the climbing space for the lineman. All other wires are placed on the pole side of the insulator. This prevents the conductors on the end pins from falling in case their tie wires break.

On corners and angles where the wires are not dead-ended, the conductors should be placed on the outside of the insulators (Fig. 22.19) on the far side of the pole and on the inside of the insulators on the near side of the pole, irrespective of the type of insulator employed. This puts the conductor pull against the insulator instead of away from the insulator.

Distribution-circuit conductors are usually secured to the insulators with prefabricated ties (Figs. 22.20 through 22.27). The ties are packaged and identified by the manufacturer to facilitate their correct application. The prefabricated ties can be installed on energized circuits with proper protective equipment in place to insulate the tie from grounded equipment or other energized phase conductors. Hot-line tools or rubber gloves and rubber sleeves are used, depending on the voltage applied to the conductor. Ties used for tying conductors to pin-type insulators should be relatively simple and easy to apply. They should bind the line conductor securely to the insulator and should reinforce the conductor on both sides of the insulator. Any looseness between the conductor, tie wire, and insulator will result in chafing and injury to the conductor. The hand-wrapped ties illustrated in Figs. 22.28 through 22.31 have been widely used and when properly applied give excellent service. These ties should be applied by hand without the use of pliers, and care should be taken to use the proper length and size of fully annealed tie wire specified for each conductor.

FIGURE 22.18   Position of wires on insulator in straight-line work for (a) top-groove insulators, and (b) side-groove insulators. Top-groove insulators are generally used with conductors size 0 and larger and side-groove insulators with size 2 and smaller.

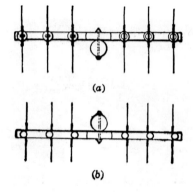

(a)

(b)

FIGURE 22.19   Position of wires on both top- and side-groove insulators at angles. All conductors are placed on side of insulator so that pull is against insulator.

FIGURE 22.20   Prefabricated conductor tie kit for top-tie insulator. (*Courtesy Preformed Line Products Co.*)

FIGURE 22.21   Installation of prefabricated conductor tie on deenergized circuit. The neoprene pad is centered in the insulator groove between the conductor and the insulator. The center of the tie is positioned over the insulator so that both legs are parallel to the conductor. (*Courtesy Preformed Line Products Co.*)

FIGURE 22.22   The prefabricated tie is rotated around the insulator so that one leg is around the neck of the insulator wrapped onto the conductor and the end snapped in place. (*Courtesy Preformed Line Products Co.*)

FIGURE 22.23   Completed application of prefabricated tie with both legs wrapped around the conductor and the ends of the tie snapped in place. (*Courtesy Preformed Line Products Co.*)

FIGURE 22.24   Prefabricated conductor tie for application as a side tie. Neoprene pad is placed around the conductor to prevent conductor contact with insulator. The legs of the tie are squeezed together to enlarge the loop, enabling it to be pushed over the head of the insulator. (*Courtesy Preformed Line Products Co.*)

FIGURE 22.25   The side tie is positioned around the neck of the insulator so that the conductor is between the legs of the tie. (*Courtesy Preformed Line Products Co.*)

In general, the tie wire should be the same kind of wire as the line wire. If the line wire is a bare conductor, the tie wire should be bare also; if the line conductor is covered, the tie wire should also be covered. Copper tie wires should be used with copper line conductors and aluminum tie wires with aluminum line conductors. The tie wires, however, should always be made of soft-annealed wire as the hard-drawn tie wire would be too brittle and cannot be wrapped snugly. A hard tie wire might also injure the line conductor.

A tie wire should never be used the second time.

FIGURE **22.26** The legs of the side tie are pulled firmly and wrapped around the conductor. (*Courtesy Preformed Line Products Co.*)

FIGURE **22.27** Prefabricated conductor side-tie installation completed. When both legs are completely wrapped around the conductor, the ends of the legs must be snapped into position against the conductor. (*Courtesy Preformed Line Products Co.*)

FIGURE **22.28** Single-pin-type insulator tie for copper conductors. (*Courtesy Copper Wire Engineering Association.*)

FIGURE 22.29   Double-pin-type insulator tie for copper conductors. (*Courtesy Copper Wire Engineering Association.*)

FIGURE 22.30   Single-pin-type insulator tie for ACSR and aluminum cables equipped with armor rods. (*Courtesy Aluminum Company of America.*)

FIGURE 22.31 Double-pin-type insulator tie for ACSR aluminum cables equipped with armor rods. (*Courtesy Aluminum Company of America.*)

Good practice is to use No. 6 tie wire for line conductors of sizes No. 4 and smaller, No. 4 tie wire for line conductors No. 1 to No. 4, and No. 2 tie wire for line conductors No. 0 and larger, as shown in Table 22.3.

The length of the tie wire varies. The approximate lengths for various ranges of conductor sizes for some of the common ties are given in Table 22.4.

Size of Line Conductor, AWG	Size of Tie Wire, AWG
No. 4 and smaller	No. 6
No. 1 to No. 4	No. 4
No. 0 and larger	No. 2

TABLE 22.3 Size of Tie Wire to Use with Various Sizes of Line Conductors

Type of Tie	Length of Tie Wire, inches
Western Union	No. 0000 bare cable, 54 2,000,000-cir-mil cable, 87
Bridle	No. 0000 bare cable, 54 2,000,000-cir-mil cable, 87
Horseshoe	All sizes, 39
Armored Western Union	All sizes, 102
Armored top	All sizes, 114
Stirrup	All sizes, two-pieces-each 54

TABLE 22.4 Approximate Length of Tie Wires Required for Different Types of Ties

Rules on good tying practice are as follows:

1. Use only fully annealed tie wire.

2. Use a size of tie wire which can be readily handled, yet one which will provide adequate strength.

3. Use a length of tie wire sufficient for making the complete tie, including an end allowance for gripping with the hands. The extra length should be cut from each end after the tie is completed.

4. A good tie should

   a. provide a secure binding between line wire, insulator, and tie wire.

   b. have positive contacts between the line wire and the tie wire so as to avoid any chafing contacts.

   c. reinforce the line wire in the vicinity of the insulator.

5. Apply without the use of pliers.

6. Avoid nicking the line wire.

7. Do not use a tie wire which has been previously used.

8. Do not use hard-drawn wires for tying or fire-burned wire which is usually either only partially annealed or injured by overheating.

Steps in Making Ties

The simplest tie is the Western Union tie. It is universally used in distribution systems and is made as follows:

Place the middle of the tie wire on one side of the insulator and the cable on the other side, both bearing in the side groove. Each end of the tie wire is then brought around the neck of the insulator, but underneath the cable, then up and around the cable, and served closely for about seven wraps.

The looped Western Union is quite similar except that the start is made on the cable side and the two ends are wound around the insulator once before being wrapped.

Armor Rods

Transmission and distribution-line conductors can be reinforced with armor rods, as illustrated in Figs. 22.32 and 22.33. This armor consists of a spiral of rods surrounding the conductor for a short distance. The attachment of the conductor to its support is made in the middle of this armored length (Fig. 22.34). Since this provides a cable of much larger diameter than the conductor itself, the resistance to bending is greatly increased. This not only reduces the stress by distributing the bending but also strengthens the cable in the region of maximum bending stress.

Clamp-top line-post insulators can be used on distribution circuits to eliminate tying the conductor to the insulator. The conductor can be protected with armor rods if necessary before it is placed in conductor clamp. Dead-end clamps of the bolted, or compression, type are used to dead-end distribution conductors.

FIGURE 22.32   An ACSR line conductor cable reinforced with armor rods. These armor rods reduce vibration and protect against chafing and flashover at point of support. Increased conductor size at support provides better grip for tie wire and also reduces danger of cable slippage. (*Courtesy Aluminum Company of America.*)

FIGURE 22.33   Typical single-phase rural line illustrating the use of armor rods to minimize the effects of conductor vibration. (*Courtesy Ohio Brass Co.*)

FIGURE 22.34   Steps in the application of preformed armor rods. (1) Center preformed rods (maximum two or three) at support of suspension point and wrap on one or two turns on both sides of center. Distortion will be avoided by not applying too many rods at one time. (2) Follow the same procedure with additional rods until all rods are applied at center for several pitches. (3) After all rods are applied several turns at center, the whole group can be rotated until the set is completely installed. (4) Always rotate the rods from center out toward end. They easily fit into position. (5) Never grip preformed rods too far from the point of application, as distortion will usually result. (*Courtesy Preformed Line Products Co.*)

Spacer Installation

Bundled-conductor transmission lines use spacers to maintain uniform distance between the conductors for each phase. The number of spacers to be installed at equal intervals in each span will be detailed in the specifications for the line. Spacers should be installed immediately after clipping-in operations are completed in each pull to minimize conductor damage due to contacts between the subconductors. The bolted spacers should be installed with the conductor clamp free to rotate on one conductor without binding, and then the second clamp should be bolted loosely on the second conductor. After the spacer is in place properly, the lineman should hold the clamp firmly in place and tighten all bolts to the proper tension.

Corona Shield Installation

Extra- and ultra-high-voltage transmission lines require corona shields at each point of conductor support. The corona shields are normally installed after the conductors are clipped in and spacers installed.

CHAPTER 23

Joining Line Conductors

Transmission- and distribution-line conductors must be joined together with full-tension splices if the conductors are under tension. Bolted connectors can be used to join electric conductors at locations where the conductors are slack, such as between conductor dead-ends. Compression tools are used for joining and dead-ending electric distribution conductors (Fig. 23.1).

Compression-, internally fired-, implosive-, and automatic-type splices are used to obtain the strength and electrical conductivity needed to join together conductors in tension. The number of locations where the conductors are joined should be kept to a minimum. All splices in transmission-line conductors should be made at least 50 ft or more from the suspension or dead-end points. Splices should not be made in spans crossing over railroads, rivers, canals, or interstate highways. It is advisable to avoid splices in spans crossing over communications circuits or electric transmission and distribution lines, if possible.

The splices must be applied in accordance with the transmission-line specifications or distribution standards and in accordance with the splice manufacturer's instructions. The conductors of a transmission line requiring a full-tension splice should be laid out straight for 40 ft each way from the joining point and straightened at the ends to prepare the conductors for splicing. The conductors should be supported and if necessary laid on polyethylene sheets to keep them clean if laid along the ground in preparation for splicing.

Compression Splice for ACSR Conductor

A compression splice for an aluminum conductor with steel reinforcing (ACSR) usually consists of two separate splices, one for the steel core and the other for the overall cable (Fig. 23.2). The complete splice, therefore, consists of a small compression slice for the central steel core to establish the tensile strength of the steel core of the spliced conductors and a much larger aluminum compression splice over the outside of the two joined cables to establish good conductivity for the spliced aluminum strands of wire. The plugs are for sealing the holes in the aluminum sleeve after the paste filler has been injected.

The steps for making a compression splice on an ACSR conductor are as follows (Fig. 23.3):

Caution: Before proceeding, make sure that the bores in the two sleeves are perfectly clean.

1. A compression connector shall be used only for the size of conductor for which it is specified. Neither the connector nor the conductor shall be altered to attempt installation of a fitting on a conductor for which it is not designed.

FIGURE 23.1   A battery actuated compression tool used to join and terminate electric conductors. The tool utilizes standard compression dies or the tool may be ordered to use with a die-less tool head. (*Courtesy Burndy Corp.*)

FIGURE 23.2   Parts of compression splice for ACSR conductor. (*Top*) Sleeve for steel core of conductor, flanked by plugs for sealing holes in the large aluminum-alloy sleeve, after paste filler has been injected. (*Bottom*) Aluminum-alloy splicing sleeve for complete cable. (*Courtesy James R. Kearney Corp.*)

(a)

(b)

(c)

FIGURE 23.3   Several steps for completing a compression splice on an ACSR conductor as described in the text are shown. (a) *Step 4*: Aluminum strands cut back a distance equal to one-half the length of the steel sleeve; (b) *Step 6*: Completed compressed sleeve on the steel core; (c) *Step 9*: Completed compression splice on ACSR conductor. Cutaway view shows inner-steel-core compressed sleeve. (*Courtesy Burndy Corp.*)

2. Select the correct press and dies for the conductor and compression connector (Figs. 23.4 and 23.5).

3. Slip the aluminum compression sleeve over one cable end and back it out of the way along the cable.

4. Cut off the aluminum strands from each cable end, exposing the steel core for a distance of a little more than half the length of the steel compression sleeve. Use care not to nick the steel core. Before cutting, secure the aluminum strands of the cable just behind the cut.

5. Insert the steel-core ends into the steel-core compression sleeve, making sure that the ends are jammed against the stop in the middle of the sleeve.

6. Compress the sleeve over its entire length, making the first compression at the center and working out toward the ends, allowing the dies to overlap their previous position.

7. Measure from the center of the steel-core splice along the aluminum wire strands the distance of one-half the length of the aluminum splicing sleeve and

FIGURE 23.4   Compression tool operated by foot, electric, or hydraulic power pump capable of developing 60 tons of compression force. It can be used to install compression sleeves on conductors as large as 1,780,000 circular mils and weighs 38 lb. (*Courtesy Burndy Corp.*)

FIGURE 23.5   Example of a die used in a compression tool to compress splicing sleeve or other connectors of conductors. (*Courtesy Huskie Tools.*)

mark the cable surface. This will facilitate centering the sleeve when it is slipped over the steel-core splice. Brush the aluminum conductor surface with a wire brush to a bright finish. Coat the brushed conductor with an inhibitor-type compound.

8. a. Slip the aluminum splicing sleeve up over the steel-core splice to the mark placed on the outside surface of the conductor, centering the sleeve over the steel-core splice.

   b. Using a caulking gun equipped with a tapered nozzle, inject filler paste through the holes provided in the aluminum splicing sleeve until the paste is visible at the ends of the sleeve. The preferred filler paste contains zinc chromate.

   c. Insert the plugs in the filler holes and hammer them firmly in place.

9. Compress the aluminum sleeve with a hydraulically powered compression tool. Make the first two compressions with the inner edges of the dies matching the positions stenciled on the aluminum sleeve. Make additional compressions advancing to the ends, allowing the dies always to overlap the previous position (Fig. 23.6).

Compression sleeves are manufactured for transmission-line ACSR conductors that contain a gripping unit internal to the aluminum-alloy splicing sleeve. The conductor is prepared for the splice in the same manner as described above. The conductor is inserted into the splicing sleeve so that the steel cores of the conductors that are being spliced are in the gripping unit. The gripping unit establishes the continuity of the tension strength of the steel core of the conductors being spliced. The splice is completed in the same manner as described above for compression of the outer aluminum conductor sleeve. Only one die is required in the compression tool. The two-step operation in which a sleeve is compressed on the steel core before compressing the outer-conductor splicing sleeve is avoided.

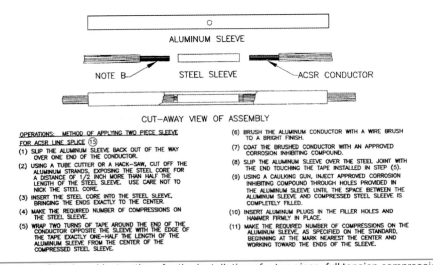

FIGURE 23.6 Sketch and instructions for the installation of a two-piece, full-tension compression splice for use on ACSR transmission line cable. (*Courtesy WE Energy Co.*)

Compression Splices for Distribution Conductors

Single-sleeve full-tension splices are normally installed for aluminum, aluminum-alloy, ACSR, and copper distribution-line conductors (Fig. 23.7). Compression splices for smaller distribution-line conductors can be made with lightweight hand-powered compression tools like those shown in Fig. 23.8. Power-driven and hand-operated, pump-type compression tools can be used for the larger distribution-line splices. Smaller hydraulically powered compression tools using line-truck hydraulic-tool circuits can be used to splice both transmission- and distribution-line conductors. Distribution-line conductors can be spliced and tapped while energized if the truck is properly equipped with an insulating boom and the lineman wears the proper personal protective equipment and follows procedures for working on energized conductors.

Single-sleeve compression splices should be completed in the following manner:

1. Select the proper compression splicing sleeve for the type and size of distribution-line conductor to be spliced.

2. Use the proper compression tools and dies for the sleeve and the wire.

3. Clean the conductor thoroughly to remove corrosion and film. Aluminum conductors should be brushed with a wire brush until the wire has a bright finish.

4. Cover the conductor with an inhibitor compound.

5. Cut off the ends of the conductors to be joined with a tube cutter or a hack saw to obtain flush ends. The splice should be made at least 12 inches from the edge of the nearest tie wire, armor rod, or dead-end clamp on the distribution conductor.

FIGURE 23.7   Compression-type splicing sleeve used for distribution-line conductors. Upper view shows sleeve before compression, and lower view shows sleeve fully compressed. Markings on sleeve are produced by die used in compression. (*Courtesy Copperweld Steel Co.*)

FIGURE 23.8   Hand-operated, die-less compression tool. (*Courtesy Homac Manufacturing Co.*)

6. Support the conductors to be joined in a straight line on each side of the joint to keep the splice straight. The wires should be straight for a distance of 10 ft or more if possible.

7. Wrap two turns of tape around one conductor exactly one-half the length of the sleeve from the end of the conductor if the sleeve has no center stop.

8. If the sleeve is not completely filled with inhibitor compound, apply additional compound to the conductor so that the sleeve will be completely filled when installed.

9. Slip the ends of the conductors into the sleeve with one end of the sleeve touching the tape and the conductor ends touching inside the sleeve or against the stop if one is provided in the sleeve.

10. Make the required number of compressions on the sleeve beginning at the center and working toward the ends of the sleeve.

Compression tools can be used with the proper accessories in a manner similar to the procedures described for splicing conductors to dead-end conductors, complete taps, and install terminals or lugs (Figs. 23.9 through 23.18).

FIGURE 23.9   Sketch of a wedge connection for secondary service wire to house service connection. (*Courtesy WE Energy Co.*)

PROCEDURE:

A. Prepare the conductor and dead end as stated in procedure for splicing ACSR conductors.

B. Slide aluminum body over conductor until sufficient working length protrudes from pad end.

C. Cut back aluminum strands a distance equal to the depth of wire bore in steel barrel plus 1/2 inch. Do not nick steel strand. File off all burs or sharp edges from ends of steel and aluminum strands.

D. Insert steel strands in steel barrel to full depth.

E. Select die size for compressing steel barrel. The die size number on die and die size number for steel barrel marked on steel assembly must be the same.

F. Make the required number of compressions on steel barrel beginning at knurl adjacent to corrugations and working toward mouth of barrel.

G. Coat steel assembly with inhibitor compound up to felt or neoprene washer.

H. Slide aluminum body over conductor and steel assembly until pad end butts solidly against felt or neoprene washer. Washer should butt against shoulder of clevis or eye.

I. Align eye or clevis of the steel assembly so that the pad of the aluminum body will be in the proper position and the eye or clevis will line up properly with the attaching hardware.

J. Select die size for compressing aluminum body. The die size number or die and die size number for aluminum marked on aluminum body must be the same.

K. Hold aluminum body in position against felt or neoprene washer and compress onto the steel assembly by making the required number of compressions beginning at the knurl mark nearest pad and working to the second knurl mark from pad end. No compressions shall be made between the second and third knurl mark from the pad end.

L. Inject Inhibitor compound into filler hole until compound is visible at the end of aluminum body. Insert and drive filler plug into hole and peen edge of hole over top surface of plug.

M. Compress aluminum body onto the conductor by making the required number of compressions beginning at third knurl from pad end and working toward outer end.

N. Finish the installation to provide the compressed portion of the connector with a smooth uniform appearance. Remove any flashing with file or emery cloth and crocus cloth. Wash the connector and adjacent wire with solvent to remove all inhibitor compound or other possible causes of corona.

FIGURE 23.10   Compression dead end for ACSR conductor.

Secondary and service conductors can be spliced with bare or insulated compression connectors, using a lightweight hand-operated compression tool. The tools are designed so that the compression cycle must be completed before the tool can be removed, ensuring proper compression of the connector. The hand tool is used to install compression streetlight tap connectors and pin terminals. Large-size service conductors are spliced with compression connectors using hydraulically operated tools powered by a truck hydraulic system or hand-operated or ac-dc electric-driven hydraulic-fluid pump, previously illustrated. The head of the hydraulically operated tool weighs

PROCEDURE:

A. Prepare the conductor and dead end as stated in procedures for splicing distribution conductors.

B. Coat steel assembly with inhibitor compound up to felt or neoprene washer.

C. Slide aluminum body over steel assembly until pad end butts solidly against felt or neoprene washer. Washer should butt against shoulder of clevis or eye.

D. Align eye or clevis of the steel assembly so that the pad of the aluminum body will be in the proper position and the eye or clevis will line up properly with the attaching hardware.

E. Select die size for compressing aluminum body. The die size number on die and die number for aluminum marked on aluminum body must be the same.

F. Hold aluminum body in position against felt washer and compress onto the steel assembly by making the required number of compressions beginning at the knurl mark nearest pad and working to the second knurl mark from pad end. No compressions shall be made between the second and third knurl mark from the pad end.

G. Insert conductor to the full depth of the bore and mark conductor at end of body. Check by measuring.

H. Inject sufficient inhibitor compound into the aluminum body to ensure that excess compound will be visible when body is completely compressed.

I. Insert clean, brushed end of the conductor into the body to the mark on the conductor.

J. Compress aluminum body onto the conductor. Using same die as used in step E, by making the required number of compressions beginning at third knurl mark from pad and working toward outer end.

K. Finish the installation to provide the compressed portion of the connector with a smooth uniform appearance. Remove any flashing with file or emery cloth and crocus cloth. Wash the connector and adjacent wire with solvent to remove all inhibitor compound or other possible causes of corona.

Dead end installed on conductor

FIGURE 23.11   Compression dead end for aluminum and aluminum-alloy conductors.

approximately 15 lb and can be equipped with a hot-stick adapter for use on energized lines. The hoses connecting the hydraulic-fluid pump to the tool head must be insulated for energized work. Special hydraulic fluid with a high dielectric strength must be used to operate the tool. The tools operate with a hydraulic-fluid pressure of 10,000 lb/in^2. The lineman must protect the equipment from damage to ensure safe operation. Specific dies are needed for different size conductors.

Internally Fired Connector System

Internally fired connectors can be used for both transmission and distribution lines. The connectors are made for joining stranded aluminum, aluminum-alloy, and aluminum-steel composite conductors. The housing of the connectors is made of high-tempered

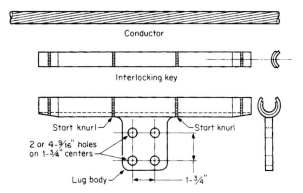

Conductor

Interlocking key

Start knurl

2 or 4-9/16" holes
on 1-3/4" centers

Lug body

Start knurl

1-3/4"

PROCEDURE:

A. Prepare the conductor and tee lug as stated in procedure for splicing distribution conductors.

B. Coat the clean, brushed conductor and inner surface of key and body with sufficient compound so that the groove will be completely filled.

C. Wire sizes smaller than 477 MCM do not have interlocking key.

D. Insert clean, brushed conductor and interlocking key into groove.

E. Select correct die size for tee lugs. The die size number on die and die size number marked on tee lug must be the same.

F. Make the required number of compressions for tee lugs. The center portion which is over the pad is not compressed.

Make initial compression beyond "start" knurl on one end of the barrel of tee lug. Make second compression beyond "start" knurl on opposite end and complete compressions to the end of the barrel of tee lug. Then complete compressions on starting end.

G. Clean contact areas of pad and connecting fitting. Coat with inhibitor compound. Wire brush through compound and assemble joint without removing compound.

H. Finish the installation to provide the compressed portion of the connector with a smooth uniform appearance. Remove any flashing with file or emery cloth and crocus cloth. Wash the connector and adjacent wire with solvent to remove all inhibitor compound or other possible causes of corona.

Tee lug installed on conductor

FIGURE 23.12   Compression tee-lug connector for installation on aluminum conductor.

FIGURE 23.13   Compression straight-lug terminal installed on large aluminum conductor with bucket-truck hydraulic tool circuit or hand-operated hydraulic pump providing force to operate compression tool on energized conductors. Hydraulic hose and fluid have a high dielectric. Electric pump may be used to provide force to operate compression tool on deenergized conductors. (*Courtesy Somerset Division of Homac Manufacturing Co.*)

Lug

2 or 4-9/16" holes
on 1-3/4" center

Conductor

PROCEDURE:

A. Prepare the conductor and lug as stated in procedures for splicing distribution conductors.

B. If lug is prefilled with inhibitor compound, mark conductor end with pencil or crayon at a distance equal to length from barrel end of connector to the knurl at pad end plus 3/8 inch.

C. If lug is not prefilled with inhibitor compound, insert conductor to the full depth of the bore and mark conductor with pencil or crayon at end of barrel. Remove conductor after marking. Inject sufficient inhibitor compound into the barrel to ensure that excess compound will be visible when barrel is completely compressed.

D. Insert clean, brushed end of the conductor into the barrel to the mark on the conductor.

E. Select correct die size for straight lugs. The die size number on die and die size number marked on lug must be the same.

F. Make the required number of compressions for straight lugs. Beginning at pad end and working to outer end.

G. Clean contact areas of pad and connecting fittings. Coat with inhibitor compound. Wire brush through compound and assemble joint without removing compound.

H. Finish the installation to provide the compressed portion of the connector with a smooth uniform appearance. Remove any flashing with file or emery cloth and crocus cloth. Wash the connector and adjacent wire with solvent to remove all inhibitor compound or other possible causes of corona.

Lug installed on conductor

FIGURE 23.14   Compression straight-lug terminal for aluminum conductor.

Notes:

1. Clean all conductors thoroughly with soft wire brush or with No. 70 emery cloth.

2. After the cleaning operation, coat the conductors with corrosion-inhibiting grease before installing connector. Apply grease not later than one hour after the cleaning operation. If connectors are factory-packed with grease, be sure any supplementary grease used on run or tap conductors is similar or compatible.

3. Be sure connector contact surfaces are clean. (Consider connectors factory-packed with grease to have clean contact surface when received.)

4. Always place copper conductor below aluminum conductor.

5. Install connectors directly on line conductor where possible instead of over armor rods.

6. Use compression tool with proper dies to install connector.

FIGURE 23.15   Compression stirrup connector. Stirrup permits use of hot-line taps on distribution line without damaging line conductor.

Figure 23.16 Compression stirrup installed on aluminum distribution line. Hot-line tap installed on stirrup permits readily removable connection for distribution transformer. (*Courtesy Sherman & Reilly, Inc.*)

Figure 23.17 Compression-type tap connector. Connector can be used to make a nontension tap to primary or secondary conductors with all combinations of aluminum-to-aluminum or aluminum-to-copper conductors. (*Courtesy James R. Kearney Co.*)

Figure 23.18 Compression-type tap connector installed to make a copper-conductor connection to an aluminum primary conductor. Conductors and connector must be cleaned with a wire brush and coated with inhibitor compound before connector is compressed. Aluminum conductor must be kept above the copper conductor to prevent copper salts from washing down onto aluminum conductor. (*Courtesy James R. Kearney Co.*)

• Spring-action "C" shaped connector housing generates constant pressure between the conductors and tapered wedge compensating for mechanical fatigue and thermal cycling, which maximizes connector service life.

• Large metal-to-metal contact areas are generated by scrubbing action during connector installation.

• Factory applied AMP Inhibitor Compound seals the contact area during termination, preventing performance-degrading oxidation and the corrosion processes. (Courtesy AMP., Inc.)

Figure 23.19   Spring-action "C"-shaped connector. (*Courtesy Amp., Inc.*)

aluminum alloy, tapered at the ends where the conductors enter (Fig. 23.19). An impact tool contains a specially designed cartridge that is loaded with a fast-burning propellant charge. Igniting the charge creates instantaneous high pressure in the cartridge. This pressure drives cylindrical sets of wedge-shaped serrated aluminum jaws (into which the conductor ends have been inserted) at high velocity into the tapered ends of the housing. The jaws clamp and lock the conductor ends in position, providing the required holding strength and establishing a low-resistance current path across the housing.

The C-member is installed by the lineman hitting the impact tool end with a hammer (Fig. 23.20).

Figure 23.20   Taps utilize wedge pressure technology. Using the impact tool, a wedge is driven into a specially shaped C-member. During installation, several key aspects of utility connector performance are addressed: the connector-to-conductor surfaces are abraded; the conductive surfaces are maximized; an oxidation inhibitor is applied to the surfaces; and spring tension is applied within the connection. (*Courtesy Amp., Inc.*)

The impact tool is convenient for lineman to use to install propelled tap connectors (Fig. 23.21). The tool is loaded with a specially designed color-coded shell that contains a small propellant charge. The tap connector consists of a tapered C-shaped member and a wedge (Fig. 23.22). The lineman hooks the C-member over the main and tap conductors and places the wedge in position. The joint can be completed on energized conductors using hot sticks or with rubber glove and sleeve procedures (Fig. 23.23).

FIGURE 23.21   Detail and sectional view of the impact tool. (*Courtesy Amp., Inc.*)

FIGURE 23.22   The impact tap component parts, assembly, and completed tap joint. (*Courtesy Amp., Inc.*)

FIGURE 23.23   Lineman installing impact tap with hot-line tools. The impact tool is in position to complete the connection. A piggyback hot-line clamp is used to hold the tap conductor in place while regular tap is completed. (*Courtesy Amp., Inc.*)

FIGURE 23.24   Dead-end pole with jumpers connected between conductors with impact taps. (*Courtesy Amp., Inc.*)

The lineman clamps the tool over the tap and detonates the propellant in the tool by striking the end of the tool with a hammer. The tool drives the wedge at high velocity between the main conductor and the tap wire within the support of the C-member. This action spreads the C-member. As it attempts to revert to its original shape, the C-member exercises a permanent retentive force on the electrical contact areas. The locking tap ensures that the wedge remains in position even under severe service conditions (Fig. 23.24).

The impact tool can be used to remove the taps. The propellant reduces the effort needed to remove the tap. The main conductor is free of connector bodies after the tap has been removed.

Implosive Compression Connectors

Implosive connectors are manufactured for splicing and terminating all transmission-line conductors. The implosive compression splice for ACSR conductor consists of a steel sleeve, an inner aluminum sleeve, and an outer aluminum sleeve on which the implosive charge is mounted. The inner aluminum sleeve is used instead of filler compound to fill the central void in the splice. The conductor is prepared for splicing in the same manner as described for a hydraulically compressed splice. The implosive charge is detonated by a number eight strength blasting cap, initiated by a safety fuse or electrical means. The implosion-compressed splice has a strength equal to 95% of the conductor's breaking strength, meeting the requirements for a full-tension splice. The use of a hydraulic press and its accessories is avoided. The finished splice is corrosion-resistant and corona-free. The splices are completed without birdcaging the conductor and without the "banana" look sometimes obtained with a hydraulically compressed splice. The finished splice can be pulled through the stringing blocks without bending and without damaging the conductor. The manufacturer specifies the maximum line angle for various stringing tensions when the splice is to be pulled through the stringing blocks.

Automatic Tension Splice

This type of joint is made with a single-bore sleeve that has internal gripping jaws at each end (Fig. 23.25).

Automatic line splices may be used in any span where the tension during installation exceeds 15% of the rated breaking strength of the conductor. It is necessary to force the teeth of the jaws into the conductor to obtain reliable gripping. This should be accomplished by imposing several severe jerks on the conductor or by exerting enough pulling tension on the conductor to obtain 15% of the rated breaking strength. Automatic line splices cannot be used in taps or jumpers.

Splices should not be made in spans crossing railroads and preferably not in adjacent spans that are depended on for withstanding the longitudinal load of the crossing conductors. Splices should not be installed within 12 inches of a tie wire or armor rod. Automatic line splices must remain free of dirt, which would impair the proper operation of the internal jaws. Splices that have lost their original protective equipment, such as plastic end inserts, must be carefully inspected for the presence of dirt and other contaminants that would impair the operation of the jaws.

During installation, automatic line splices must not be dragged through dirt, mud, ice, or water and must remain free of fouling until the conductor reaches its final installed tension. Any splice with a deformed or dented barrel cannot be used. The preceding conditions interfere with the proper seating of the internal jaws.

The making of a good splice with an automatic line splice requires following these important procedures:

1. Choose the proper size and type of splice for the conductor size and material.

2. Clean the conductor with a wire brush and remove all burrs but *do not grease* (automatic splices are pregreased with a material that is grit-free; the presence of grit prevents proper jaw operation).

3. Be sure the conductor strands are in normal lay and not deformed.

4. Insert the conductor with one sharp, continuous thrust, forcing it in as far as possible.

   a. Aluminum splices are provided with a pilot cup for retaining and guiding the strands into jaws of the chuck. This pilot cup must remain in place and will be driven into the interior of the splice until it strikes the center barrier.

   b. Copper splices are provided with slotted openings which allow visible examination to determine if the conductors are properly inserted into the jaws.

Figure 23.25 Automatic tension splice for use on aluminum, aluminum-alloy, and copper conductors. Ends of conductors are surrounded by free-floating, self-aligning chucks. The spring serves to position the chucks against the conductors. (*Courtesy Hubbell Power Systems Inc.*)

FIGURE 23.26   Bolted jumper connection for a T-2 transmission cable termination at a substation terminal.

5. It is extremely important to set the jaws into the conductor by applying one or more momentary jerks to the conductor, forcing the jaws to bite into the conductor.

When an attempt is made to withdraw the conductors, the jaws clamp down on the conductors because of a taper in the bore of the sleeve and on the jaws. Tension causes a wedging action which increases with the pull applied on the conductor. A loss of tension could cause the jaws to release their grip and allow the conductor to drop out. This type of splice is therefore used where the wires are kept in tension. It is also used where service must be restored quickly. It is especially suited for live-line splicing.

Bolted Joints

Conductors can be joined together with bolted connectors if the wires are not under tension (Fig. 23.26). Bolted connectors are used successfully on both copper and aluminum conductors. The conductors must be cut to the proper length, cleaned, and coated with an inhibitor to obtain a good connection. Bolted, parallel-groove clamps, U-bolt clamps, tag lugs, and dead-end clamps are manufactured for use on a range of conductor sizes. The lineman must use care to obtain the proper tension when tightening the bolts on a bolted connector.

Live-Line Maintenance with Hot-Line Tools

The trend in electric-system management is to do more and more of the testing, repair, and maintenance work while the lines are live (energized). The reason for this is to reduce the number of times electric service is interrupted. Unscheduled interruptions in electric service are referred to as outages. Live-line maintenance has been made possible by the development of special tools and procedures for such work.

Under live-line maintenance, a great variety of work is included. The most common of these live-line operations are as follows:

1. Replacing insulators
   a. Pin
   b. Suspension
   c. Strain

2. Replacing crossarms

3. Replacing poles

4. Tapping a hot line

5. Cutting conductor slack in or out

6. Splicing conductors

7. Installing vibration dampers

8. Installing armor rods

9. Phasing conductors

It will not be the objective here to give detailed instructions covering the procedure of every hot-line-tool live-line maintenance job. This would be impossible because of the great variety of jobs and types of line construction and special conditions. This section is therefore intended to give the reader only a general conception of the methods and procedures employed in the most fundamental operations of live-line maintenance work with hot sticks.

The Institute of Electrical and Electronics Engineers, Inc. (IEEE) issued IEEE Standard 516 entitled *IEEE Guide for Maintenance Methods on Energized Power Lines,* and this standard contains the following information:

1. Minimum tool-insulation distance ac energized-line work
2. Minimum tool-insulation distance dc energized-line work
3. Altitude correction factor
4. Minimum approach distance ac energized-line work air insulation
5. Minimum approach distance dc energized-line work
6. Physical aspects of energized-line work
    a. Electric fields
    b. Field intensities in the work area
        (1) Shielding and fields
        (2) Types of shielding
    c. Effects of fields

The guide should be reviewed for the information listed and other safety precautions required to complete live-line maintenance with hot-line tools.

The maintenance of high-voltage lines while energized, or "hot," may appear on first thought to be hazardous, especially when compared with working on dead, or deenergized lines. However, the work is completed safely when the lineman is continually conscious of the fact that the lines are "hot" and of the need to be careful to follow the correct procedures. Furthermore, there is no possibility of the line being "hot" when it was thought to be dead, as there is when working on deenergized lines. Also, there is no possibility of confusion with live duplicate circuits, for in live-line work every conductor is worked "hot" (energized) and every operation is planned and performed accordingly.

Tools

The insulated-pole portions of the hot-line tools are made from fiberglass or wood. Wood was used originally before fiberglass became available. Fiberglass has proven to be the best material for insulated-pole tools.

If the hot stick pole is made of wood, it must be kept dry and clean. Scratches, scars, or other abrasions on the stock will damage the wood surface and permit moisture and dirt to enter the fibers of the wood. Moisture and dirt will form conducting paths along the hot stick that will render it hazardous for use.

If the pole is made of fiberglass, it will not absorb moisture. Fiberglass is not affected by creosote, gasoline, oil, grease, or most solvents. The smooth pole does not have any grooves to trap dirt or moisture. Fiberglass poles are stronger and a better insulator than wood. They can also be molded in bright colors for improved visibility. However, fiberglass insulated-pole tools should be treated carefully and kept clean and dry. Any material that becomes contaminated and/or wet loses its insulating qualities.

Hot-line tools are of great variety and those most generally used can be classified as follows:

1. Wire tongs
2. Wire-tong supports or saddles
3. Insulated tension links

4. Auxiliary arms

5. Strain carriers

6. Tie sticks

7. Platforms and ladders

8. Insulated hoods (Figs. 24.1 and 24.2)

FIGURE 24.1 A hot-line tool to cover the conductors is being applied by the lineman on the pole. The circuit is operating at 13.8 kV. The second lineman on the pole is the observer.

FIGURE 24.2 The hot-line tool has provided the protection the lineman requires as he completes his work.

9. Insulated tools

10. Temporary jumpers

The *IEEE Guide on Terminology for Tools and Equipment to be Used in Live-Line Working* (ANSI/IEEE Standard 935-1989) provides detailed information on live-line tool nomenclature.

Wire Tongs

A wire tong is a slender insulated pole provided with a hook clamp on one end and a swivel on the other end (Fig. 24.3). Tongs are used primarily in holding live conductors. They are sometimes referred to as lifting sticks or holding sticks. The hook clamp is opened or closed by turning the pole. A live conductor is clamped by placing the hook over the conductor and turning the pole until the jaws of the hook are totally closed. The swivel clevis on the other end is for the application of the lifting tackle.

The wire tong is used to hold the conductor away from the point where work is to be done such as when replacing insulators, crossarms, or transformers. The simplest case is illustrated in the sketch in Fig. 24.4, where the wire tongs are positioned to hold the live-line conductor to one side. A more complicated operation is depicted in the

FIGURE 24.3 Wire tong consisting of pole, conductor clamp, and swivel. *(Courtesy A. B. Chance Co.)*

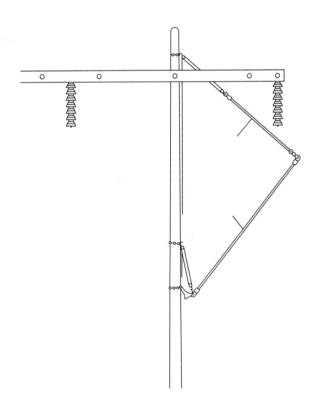

FIGURE 24.4 Sketch of two wire tongs positioned to hold live conductor on one side so as to create needed working space. *(Courtesy A. B. Chance Co.)*

FIGURE 24.5 Sketch of wire tong cluster used to hold each line conductor away from the pole. (*Courtesy A. B. Chance Co.*)

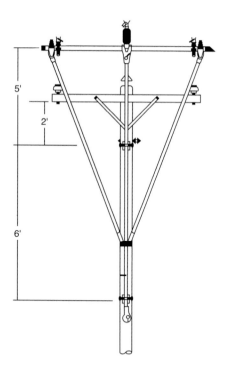

sketch in Fig. 24.5, where a number of wire tongs are used to hold the three conductors of a three-phase line away from the pole.

Wire-Tong Supports

A wire-tong support serves the special purpose of guiding and holding a wire tong in a fixed position after the conductor has been moved. A typical tong support is shown in Fig. 24.6. The support consists essentially of a base or saddle fastened by a chain to the pole or tower. A swivel clamp is mounted on the saddle. The wire tong slides between the jaws of this clamp, while the swivel action permits movement of the tong in all directions. Wing nuts are used to tighten the jaw of the clamp onto the tong.

A multiple form of tong support is illustrated in Fig. 24.7. When a number of saddles are needed on the same side of a pole or tower, a slide type of wing-tong support may be used. This support consists of a round rod secured to the pole at each end by means of a chain. On the rod are placed the tong holders. These holders slide up and

FIGURE 24.6 Typical wire-tong base for anchoring tong to pole. (*Courtesy James R. Kearney Corp.*)

FIGURE **24.7**   Multiple wire-tong support. Each support can be clamped in any position along the rod. (*Courtesy A. B. Chance Co.*)

down the rod freely but can be clamped in any position. One to six tongs can be supported from one fastening. The rod is available in lengths up to 6 ft.

Insulated Tension Link

An insulated tension link is a strain insulator with a conductor clamp on one end and a swivel on the other, as shown in Fig. 24.8. It is commonly used between the line conductor and the rope tackle when pulling conductors in the clear of structures or when changing insulators, crossarms, or poles.

Figure 24.9 shows two tension links in use when changing post insulators. The wire tongs hold the conductor at the desired distance, and the links keep the tongs from swaying toward the pole.

FIGURE **24.8**   Insulated tension link used to spread line conductors for pole installations and removals, pulling conductors in the clear of structures for maintenance work, and handling extremely heavy conductors on H-frame and tower lines. (*Courtesy James R. Kearney Corp.*)

Figure 24.9 Two conductors supported by wire tongs, with two insulated tension links being used in series with ropes to maintain the wires at a proper distance from the pole to complete the maintenance work. (*Courtesy James R. Kearney Corp.*)

Auxiliary Arm

An auxiliary arm is an arm that is used to support the line conductors temporarily while the insulators or crossarms are being replaced. It may be used to raise and support the conductors vertically, or it may be used to support one or more conductors to one side horizontally. An auxiliary arm is especially useful where it is desired to install a pole of greater height or where low-voltage lines must be cleared preparatory to working on higher-voltage lines above them. Clamps are provided on the top of the arm to hold the conductors securely.

Strain Carrier

The strain carrier is used to take the load or strain from suspension insulators while the insulators are changed. Two types of strain carriers are in use, one for light loads up to about 2,000 pounds and one for heavy loads up to about 10,000 pounds.

The light-duty strain carrier consists of two rocker arms hinged on a bar (Fig. 24.10). By bringing together the two rocker arms, the insulator string is compressed, and the insulator units may be removed and replaced.

The heavy-duty strain carrier consists of two parallel insulating tension rods which are set astride the insulators. The strain is relieved by taking up screws on the ends of the rods, thereby compressing the insulator string. The insulator string is supported in a cradle consisting of two insulated rods below the insulator string. The strain carrier may be used on either pole or tower construction.

Figure 24.10 Light-duty strain carrier of the rocker-arm type. Illustration shows strain carrier mounted on a corner pole and used to relieve the insulator string of conductor pull. (*Courtesy James R. Kearney Co.*)

FIGURE 24.11 Tie sticks for use on live lines. (A) Blade type: (1) fixed, (2) rotary. (B) Prong type: (1) two-prong, (2) three-prong, (3) rotary-prong.

Tie Sticks

Tie sticks are used to fasten or unfasten the tie wire which holds the conductor to the pin-type insulator (Fig. 24.11). Also, tie sticks are usually equipped with a small double hook called the *tie assistant*. It is fastened to the tie stick a short distance from the end. It has several uses, such as prying loose ends of old tie wires that may be wrapped close to the conductor, holding the conductor down in the insulator groove while being tied in, or as a means of hanging the tie stick on the conductor when not in use.

Insulated Platforms and Ladders

The insulated platforms and ladders provide the lineman safe and comfortable access for completing rubber-glove work assignments on energized lines. Support assemblies secure the equipment to a pole or tower. Figure 24.12 shows a small utility platform, and Fig. 24.13 shows a more elaborate 6-ft insulated platform designed for distribution hot-line work. Use of insulated platforms (Fig. 24.14) and ladders enables the lineman to reach farther out from a pole or tower and thus perform the work with greater reach and ease. Helicopters transport and deliver the insulated ladder equipment and the lineman to towers located in hard to reach locations. Insulated fiberglass heavy duty hook ladders are available in 8 ft to 30 ft lengths.

FIGURE 24.12 Small utility platform. (*Courtesy A. B. Chance Co.*)

FIGURE 24.13   The lineman is working from the insulated platform designed for live-line maintenance work. (*Courtesy A. B. Chance Co.*)

FIGURE 24.14   Insulated platform mounted to a steel pole. (*Courtesy Steel Market Development Institute.*)

Insulated Hoods

Insulated hoods are available to place over conductors and insulators to add to the safety of lineman doing hot-stick work.

Insulated Tools

Many other tools can be adapted to hot-line work by mounting them on the ends of insulated poles. Among these are pliers, wire cutters, tree trimmers, cotter-key pullers, screwdrivers, fuse pullers, saws, wrenches, brushes, and clamps. Some of these are illustrated in Fig. 24.15.

Temporary Jumpers

The lineman uses flexible insulated temporary jumper cables for hot-line distribution work less than 35 kV. The lineman must be aware of the jumpers rated ampacity prior to installation. These cables are not intended for use on permanent installations.

Required Clearance

Table 24.1 gives the AC live-line work minimum approach distances as specified in the Occupational Safety and Health Act (OSHA) Standards. The *IEEE Guide for Maintenance Methods on Energized Power-Lines* provides additional information on safe working distances that takes into account switching-surge voltages and the altitude at the location where the work is being performed.

Changing Pin Insulators

The principal parts of a pole line are the conductor, insulator, crossarm, and pole. Since the conductor hardly ever fails, live-line maintenance resolves itself into

Figure **24.15**   Heads of hot-line tools used in holding off and repairing lines. Hot-line tools must be inspected and tested periodically for safe use. (*Courtesy American Electric Power System.*)

For Phase to Phase System Voltages, Volts	Distance to Employee					
	Phase to Ground Exposure			Phase to Phase Exposure		
	Feet	Inches	Meters	Feet	Inches	Meters
0 to 50	Not Specified			Not Specified		
51 to 300	Avoid Contact			Avoid Contact		
301 to 750	1	1	0.33	1	1	0.33
751 to 5,000	2	1	0.63	2	1	0.63
5,001 to 15,000	2	2	0.65	2	3	0.68
15,001 to 36,000	2	6	0.77	2	11	0.89
36,001 to 46,000	2	9	0.84	3	3	0.98
46,001 to 72,500	3	5	1.00	3	11	1.20
72,501 to 121,000	3	9	1.13	4	8	1.42
121,001 to 145,000	4	3	1.30	5	5	1.64
145,001 to 169,000	4	9	1.46	6	4	1.94
169,001 to 242,000	6	7	2.01	10	1	3.08
242,001 to 362,000	11	3	3.41	18	1	5.52
362,001 to 420,000	13	11	4.25	23	4	6.81
420,001 to 550,000	16	8	5.07	27	0	8.24
550,001 to 800,000	22	7	6.88	37	4	11.38

TABLE 24.1   AC Live-Line Work Minimum Approach Distance

replacing insulators, crossarms, and poles. To replace one of these materials requires that the conductors be untied from the insulators, removed, and held in the clear. When so held, the insulators or crossarms, or both, can be removed and replaced. If the pole is to be replaced also, the best practice is to set the new pole first and then to mount all wire tongs and saddles and other tools on the new pole. In this way the old pole, which usually is weak, will not have to be specially guyed for this operation.

The steps in the process are performed in the following order:

1. Fasten wire tongs to conductor.

2. Untie conductor from insulator.

3. Move conductor into clear.

4. Remove old insulator, or crossarm, or both.

5. Mount new insulator, or crossarm, or both.

6. Return conductor to insulator.

7. Tie in conductor.

8. Remove wire tongs or auxiliary arm.

Moving Line Conductors

Three methods for moving line conductors into the clear are available:

1. Wire-tong method

2. Auxiliary-arm method

3. Combination of wire-tong and auxiliary-arm methods

Wire-Tong Method

In the wire-tong method, the conductor is moved and supported by the use of two wire tongs, one being used for lifting, called the *lifting tong* or *stick*, and the other for holding, called the *holding tong* or stick. The lifting stick is clamped onto the conductor to be moved, and its saddle is fastened to the pole in such a position that several feet of swivel end of the lifting stick extend through the saddle. The saddle clamp is then tightened onto the stick by turning the wing nut. Next, a set of blocks is attached between the saddle and the swivel on the end of the lifting stick. This block and tackle will later be used to lift and move the conductor out of reach.

Another saddle for the holding stick is fastened to the pole just below the crossarm braces. The holding stick is now clamped onto the conductor and placed in the top saddle.

The conductor may now be untied from the insulator by means of the insulated tie sticks. After the conductor is untied, it is ready to be pushed out of reach. To move the conductor, pull up on the set of blocks, thereby raising the conductor off the insulator. If the outer conductor is being moved, the distance raised should be about 10 inches. If the center conductor is being moved, it should be raised enough to clear the end of the crossarm. When the conductor has been moved outward to a safe working distance, the saddle clamps on both sticks are tightened firmly. The other outside conductor is moved next, and the inside or center conductor is moved last, since it must be moved above one of the outer conductors. The saddles for the sticks supporting the outer conductors may be placed on opposite sides of the poles, while a saddle extension may be used for the saddles supporting the middle conductor.

After the insulators, or crossarms, or both have been replaced, the above procedure is reversed and the conductors are brought back in. As the conductors are returned, they are tied in, using the special tie sticks.

For heavier construction such as the wishbone, H-frame, or at angles in the line, the same procedure may be followed except that an insulated link stick with blocks may be used to help move the conductor away from the insulator. The top holding stick should still be used because it helps to stabilize the conductor on the line stick.

Auxiliary-Crossarm Method

Where construction permits, the work of moving conductors may be simplified by the use of the auxiliary crossarm. The auxiliary crossarm may be used to lift all conductors from a crossarm vertically at one time, or it may be used to hold two or more conductors out on one side of the pole.

Mast Arm

When the auxiliary arm is used as a mast arm, that is, in lifting the conductors up from a crossarm, two saddles are mounted on the pole. The first saddle is mounted just beneath the crossarm braces. The assembled mast arm is then placed in position in this saddle and pushed up against the conductors and clamped into position. The lower saddle is then mounted on the pole below the point where the bracing is fastened to the mast arm, and the main lifting pole is clamped in position. Saddle extensions should be used if necessary.

The conductors are then untied from the insulators with the tie sticks and placed in the holders on the mast arm provided for that purpose. When all the conductors have been transferred, the whole mast crossarm assembly is lifted up until the conductors are in the clear.

If the line makes an angle greater than 5°, it is advisable to side guy the mast arm using an insulated link stick and a rope.

Side Arm

When the auxiliary arm is used as a side arm, it should be placed on the two-conductor side of the pole. One end of the side arm is mounted on a saddle at a point just below the crossarm braces. The conductors are then transferred to the holders on the side arm provided for that purpose with the use of lifting and holding tongs. The outside conductor is moved first. A block and tackle fastened to the upper end of the lifting tongs is also used to assist in moving the conductor.

The third conductor on the opposite side of the pole is moved and supported with wire tongs.

Removing the Tie Wire

The tie wire is untied by starting to unwind the end of the tie wire with the tie-wire blade on the tie stick. If the tie wire has no end which can be easily reached with the blade of the tie-wire stick, an end is made by cutting the tie wire at some convenient point with a hack saw. The tie wire is then unwound by pulling the end of the tie wire if it points away from the pole. If the end of the tie wire points toward the pole, it is started by pushing up with the tie-wire stick. If it has a long wrapping, it should be cut with the hack saw as soon as the untied portion becomes difficult to handle. If it gets too long, there is danger of its touching the other conductor and shorting the line.

During the process of removing old tie wires and tying in, the lineman should not be allowed to sit on the crossarm or put his legs or arms through the braces but should be made to stand with his full weight in the climbing hooks with his feet as close together as possible, and he should be instructed to lean back in his safety belt and keep his hands off the pole, crossarms, and crossarm braces. In following this recommendation, he will benefit from the insulation properties of the pole and the crossarm.

Tying in Conductor

Particular mention should be made of the procedure employed in tying the conductor onto the new insulator. Special provision is made for this by providing the new insulator with special tie wires before the insulator is screwed in place.

Each piece of the tie wire is placed in the neck of the insulator and bent until the ends meet (both ends being the same length). They are then given a couple of twists close to the neck of the insulator and allowed to extend horizontally in the same direction as the top groove of the insulator, one pair of tie-wire ends extending out in the opposite direction for the other side of the insulator. These ends are now bent over the top of the insulator so that in screwing the insulator on the pin, the tie-wire ends will not come in contact with any part of the live line. The insulator is then sent up on the pole and screwed onto the pin until it is tight and with the top groove of the insulator parallel to the line conductor. The line conductor is then placed in the top groove of the insulator, care being taken that both ends of the tie wire come on the same side of the line conductor and that the two pairs of tie-wire ends come on opposite sides of the line conductor. When in this position, the insulator is ready to be tied in. The tie stick is used for this purpose. The man making the tie should be well below the crossarm. The tie stick is passed up on the far side of the line conductor from the pair of tie-wire ends on that side of the insulator. One horn of the tie stick is put through the loop in the end of the tie wire nearest the insulator. This end is then pulled downward. The same is then done with the other end of the tie wire. When both ends are then pointing downward, the end of the tie wire which was first pulled down is pushed upward with a stiff punch on the other side of the line conductor; the other tie wire is likewise pushed up. By repeating this operation, both ends of the tie wire are thus served up to the loops in their ends. These loops can then be removed if desired, although there is no reason for removing them, by cutting them with a hack saw and breaking them off or by twisting them with a tie stick until they break off close to the line conductor. Obviously, the same method can be used in tying the line conductor in the neck of the insulator as will be necessary on angles. In the case of double arms, half the above tie is made on one insulator and half on the other insulator.

This is by no means the only tie which can be applied; any desired form of tie can be made. This type of tie is chosen here only because it lends itself to a clear and accurate description and is perhaps the easiest tie to learn.

Note that both wires on one side of an insulator are turned in the same direction and that both wires on the other side of the insulator are turned in the opposite direction. They are started from the top and bottom of the conductor or from opposite sides of the conductor. This provides a self-snubbing action which prevents the conductor from rolling and assists in holding the conductor tightly against the cap of the insulator.

Note that eyes are on all ties. A lightweight stick with prong-type head and this type of tie make a very satisfactory working combination. Some linemen prefer having the tie without eyes. In this case, the blade-type head is more convenient. The use of eyes is especially helpful in untying as it makes it unnecessary to pry up the ends of the tie wire.

A handy tie stick is one with the blade-type head on one end and the prong-type head on the other.

Prefabricated ties manufactured to fit specific insulators and conductors provide a fast, firm, and economical system for securing wires to pin- and post-type insulators. Hot-stick application of a prefabricated tie for a conductor installed on top of a pin-type insulator is illustrated in Figs. 24.16 through 24.23. Application of a prefabricated tie for a conductor installed on side groove of a horizontal line post insulator with hot sticks is shown in Figs. 24.24 through 24.28.

Removing Conductors from Clamp-Top Insulators

In case the pin insulator is of the clamp-top type, removal of the conductor from the insulator is greatly simplified. Likewise, fastening of the conductor is also made easy.

The clamp-top insulator is similar to other pin insulators except that it has a bolted clamp cemented to the top. Two carriage bolts hold the keeper piece over the conductor. To remove the conductor, therefore, it is only necessary to unscrew the two nuts.

FIGURE 24.16   Lineman is installing top tie to secure energized conductor to top of pin-type insulator. The pole is covered with insulated pole guard to provide safe working space. Lineman places unmarked leg of tie in a clamp stick one pitch in from the end and places tie on insulator and conductor. (*Courtesy A. B. Chance Co.*)

FIGURE 24.17   Lineman uses hot stick with tie loop to wrap the marked tie leg to completion while holding the tie loop snug against the indicator with clamp stick. (*Courtesy A. B. Chance Co.*)

FIGURE 24.18   Lineman snaps marked leg of tie securely into place with tie loop on hot stick. (*Courtesy A. B. Chance Co.*)

FIGURE 24.19   Lineman slips the second tie of the set between the wrapped and unwrapped legs of the first tie using clamp stick on the marked tie leg. The marked leg must remain on top of the conductor parallel to the previously installed leg. The tie wires must not cross. A clamp stick is used to hold conductor in position on top of insulator. (*Courtesy A. B. Chance Co.*)

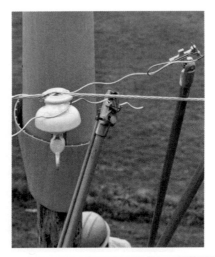

FIGURE **24.20** Lineman removes clamp stick from marked leg and places it on the unmarked leg, holding it while marked leg is wrapped to completion with tie loop. (*Courtesy A. B. Chance Co.*)

FIGURE **24.21** Lineman applies unmarked leg of second prefabricated tie wire with tie loop. (*Courtesy A. B. Chance Co.*)

FIGURE **24.22** Lineman wraps remaining unapplied leg of first tie around energized conductor. (*Courtesy A. B. Chance Co.*)

FIGURE **24.23** Lineman has completed installation of prefabricated tie securing conductor to insulator. Clamp stick holding conductor can now be removed. (*Courtesy A. B. Chance Co.*)

FIGURE 24.24   Lineman places tie over insulator side groove before energized conductor is lowered onto insulator side groove. Note plastic cover-up material on insulator support and pole. (*Courtesy A. B. Chance Co.*)

FIGURE 24.25   Lineman rotates tie on insulator with tie loop on hot stick so legs of tie point skyward. The conductor supported by a clamp stick is lowered onto insulator side groove with conductor between tie legs. (*Courtesy A. B. Chance Co.*)

FIGURE 24.26   Lineman applies left leg one pitch with tie loop on hot stick and holds tie while applying the right-hand leg. (*Courtesy A. B. Chance Co.*)

FIGURE 24.27   Lineman completes application of left leg of tie after making sure right leg is snapped into place. (*Courtesy A. B. Chance Co.*)

FIGURE 24.28   Lineman has completed application of prefabricated tie. Clamp stick supporting conductor and temporary protective equipment can now be removed. (*Courtesy A. B. Chance Co.*)

This is done by means of a special ratchet wrench and socket secured to the end of an insulated handle. Before this is done, the lifting and holding tools are attached to the conductor and secured to the pole.

After the nuts are loosened and backed off about 5/8 of an inch, the conductor is raised slightly, about ½ inch, to take its weight off the clamp seat. The clamp is then lifted from the supporting pintles with a special wire cradle mounted on a hot stick and slid along the conductor away from the insulator. With the clamp slid away from the clamp-top seat, the hot conductor is free and can be moved to any desired position to clear the insulator.

For attaching the conductor to the insulator, the reverse procedure is followed.

Changing Suspension and Strain Insulators

The principal parts of a steel-tower line are the steel tower, steel crossarm, insulator string, and conductor. Suspension insulators are installed on tangent (in-line) towers. At crossings, angles, and dead-ends, strain insulators are used instead of suspension insulators. The most frequently performed live-line maintenance work on tower lines will be the replacement of suspension or strain insulators. The steel towers, steel crossarms, and conductors seldom fail.

In order to replace any defective units in the insulator string, the weight of the conductor must be taken off the string. This is accomplished by the use of the strain carrier. By drawing up on the strain carrier, the insulator string is relieved of the conductor load and can then be disconnected from the conductor clamp. The lineman is then free to remove and replace one or more of the insulator disks. When this is accomplished, the conductor load is transferred back to the insulator string and the strain carrier is removed. These steps can be listed as follows:

1. Attach strain carrier.
2. Draw up on strain carrier.
3. Disconnect insulator string.
4. Replace defective insulator units.
5. Reconnect insulator string.
6. Release tension from strain carrier.
7. Remove strain carrier.

Clearing Working Space

Before proceeding with the work, the lineman should make sure that sufficient working space is available and that the required clearance from live conductors can be maintained. If protective gaps are in the way, they should be pushed out of the way or removed temporarily. Jumpers around dead-end strings also should be pushed out of the way and held away by use of a holding stick in the regular manner.

Replacing Suspension Insulators

Wire-Tong Method

In many cases, especially on light-pole lines, the suspension insulators may be replaced with the use of wire tongs and other hot-line tools.

The procedure is similar to that used with pin-type insulators. The conductor and clamp are secured with a lifting tong. The assembly is then pushed away from the pole or tower a distance of about 6 inches. The lifting tong is then secured in its saddle clamp. The holding tong is fastened to the conductor, and the conductor is then pulled back to the pole or tower. This will cause the conductor to be raised slightly and thus transfer the conductor load onto the lifting tong. The insulator string will now hang limply. The insulator string may now be disconnected from its clamp and the line conductor moved farther out of the way. A link stick may be used in pulling the conductor away or in pulling it back, but the holding tong should always be used as it will help to stabilize the conductor in its position away from the pole or tower. After the defective insulator units are replaced, the conductor is pulled back in and the string is reconnected to the conductor clamp. All tongs and other tools are then removed in reverse order of their installation.

Strain-Carrier Method

The strain carrier is placed in position vertically adjacent to the insulators. As the suspension string hangs vertically, the strain carrier also hangs vertically. A cradle for supporting the insulator string will usually not be required. The fixed end of the carrier is hooked onto the conductor on opposite sides of the clamp. The adjustable or take-up end is fastened to the crossarm, allowing enough slack in the tightening arrangement for taking the strain off the insulator string.

After the strain is taken off the insulator string, the insulators hang limply and may be disconnected from the strain clamp. This requires pulling out a cotter key or pin. This is performed with a cotter-key puller, which consists of a special cotter-key pulling hook fastened to the end of another hot stick. If the cotter key is hard to pull out, the conductor load on the string should be further reduced by additional tightening of the strain carrier.

With the insulator string now free from the clamp, the string can be removed and the defective insulator units replaced. It has been found preferable in most cases to remove the entire string of insulators rather than to replace the insulator disks singly. This is especially so when the insulator string is short, consisting of only two or three disks. The insulator string may be removed by using a shepherd hook, fork stick, lifting yoke, sling, or other suitable device.

After the insulators are replaced, the string is reconnected to the clamp. The cotter key is installed with a special cotter-key replacer. The cotter key is spread by means of a screwdriver fastened to the other end of the cotter-key-replacer hot stick. The strain is now transferred to the insulator string by letting off on the strain carrier. The carrier and other tools should then be removed in the reverse order of their installation.

Replacing Strain Insulators

The strain carrier is first placed in position as with suspension insulators. Since the strain insulator strings are somewhat horizontal, the carrier can be slid over the insulators out to the line conductor. The fixed end of the carrier is attached behind the strain clamp on the span side of the clamp. The adjustable or take-up end of the carrier is fastened to the pole, crossarm, or tower, allowing enough slack in the tightening arrangement for taking the strain off the insulator string. The cradle for supporting the slack string of insulators is then placed in position by hanging the adjustable hook over the conductor and fastening the other end to the crossarm. When the insulator string is short, that is, consists of only two or three disks, the supporting cradle need not be used, as the insulators can then be handled by means of a yoke arrangement on the end of a hot stick.

The next step is to take the conductor load off the insulators. This is accomplished by drawing up on the strain carrier. In the buck-saw type of strain carrier, this is done by pulling together the ends of the two hinged members using a set of blocks. In the cradle type of strain carrier, the conductor load is transferred to the carrier by taking up on the jackscrew at the rear of the two tension members.

After the strain is taken off the strain insulator string, the string may be disconnected from the clamp. The defective insulator disks may now be replaced and the complete string reconnected to the clamp. The conductor load is then returned to the strain insulator string by letting off on the strain carrier. All hot-line tools should now be removed in reverse order of their placement.

Pulling Slack

String Insulator

The same general procedure is followed when pulling or removing slack from a conductor when the line is energized as is used when the line is deenergized. When the lines are energized, properly insulated tools and insulated links must, of course, be used. The order of work is as follows: The pulling grip or come-along is attached to the conductor with a tie stick, holding stick, or other hot stick. An insulated link stick is next attached to the pulling eye of the come-along, pole, crossarm, or tower. A set of blocks or a hoist is placed between the end of the link stick and the pole, crossarm, tower, or conductor (Fig. 24.29). The conductor is then secured by taking up on the set of blocks or hoist. After the conductor is thus secured, the clamp holding the conductor to the

FIGURE 24.29   A hoist with a nylon strap and an insulated handle is connected to an insulated link stick so that slack can be pulled out of a 15-kV distribution-line conductor. (*Courtesy A. B. Chance Co.*)

insulator string is loosened by unscrewing the nuts on the holding bolts by means of a socket wrench using hot-line procedures. The slack is then taken up by pulling further on the blocks or the hoist. When sufficient slack has been removed, the conductor clamp is tightened. If all the unwanted slack cannot be pulled up in one operation, the foregoing procedure is repeated until the desired conductor sag is obtained. When sufficient slack has been pulled up, the blocks or hoist, insulated stick, and pulling grip are removed. The pulling grip is loosened by tripping the releasing lock with a tie stick or other hot stick.

Pin Insulator

The same procedure is followed when pulling slack in a line supported on pin insulators. Instead of unscrewing the nuts on the conductor clamp, the tie wires holding the conductor to the insulator are loosened but not removed. Then the slack is pulled up and held while the insulator ties are remade.

Cutting Out Conductor

If a short length of conductor needs to be cut out of the line, the slack can be pulled in both directions from the same pole or tower. The extra conductor may then be cut out and the conductor spliced and then refastened to the insulators.

In order not to interrupt the flow of current in the line when the conductor is cut, a jumper is placed around the section where the conductor is to be cut out. However, before the ends of the jumper are tapped onto the conductor, the jumper must be firmly held away from the lineman by means of a holding tong or stick securely fastened to the pole or tower. An insulated link stick with rope also may be used for this purpose. The ends of the jumper are then clamped onto the live conductor. The desired length of conductor is cut out of the line, and the ends are spliced. After splicing is completed, the jumper is removed.

If the line conductor is a stranded cable, extreme care must be taken to prevent the cable from unraveling and allowing the wires to get out of control.

Splicing Live Conductors

The procedure for splicing a hot conductor with live-line tools is as follows: The conductors to be spliced are pulled tight with blocks, using link sticks on the come-alongs for insulation. The conductors are connected together with a temporary jumper to bypass the current while the splice is completed. The ends of the conductors are held in position with holding sticks. The splicing connector is next fastened to the end of one conductor. Then the end of the other conductor is brought over and placed in the connector. The connector is now firmly fastened, compressed, or fired, completing the splice. Insulated tools, of course, are used for all the operations.

Tapping Live Line

Taps are easily made to live conductors by the use of appropriate equipment. A stirrup is installed on the conductor to be tapped, as discussed and illustrated in Chap. 23. The conductor to be tapped to the live line is fastened to the body of a special hot-line tapping clamp and is then held with a hot stick and hooked over the live-line conductor stirrup. The clamp is then tightened by turning the screw head of the clamp using the hot stick.

Hot-line clamps are not designed to pick up load current or fault current. If there is load on the line being energized or if there is any doubt as to the condition of the insulation of the line being energized, the line should be energized by a switch designed to energize circuits, carry load current, and withstand the forces associated with fault current.

Load-pickup tools are available to energize circuits operating at voltages of 15 kV or below, which will carry 200 amps of load current (Fig. 24.30). Once the load has been energized successfully through the load-pickup tool, the conductor is connected to the main circuit with the hot-line clamp using a hot stick, and the load-pickup tool is removed.

FIGURE 24.30   Load-pickup tool designed for 15-kV, 200-amp operation. The tool has a one-way mechanism design which precludes its use to interrupt load current. (*Courtesy A. B. Chance Co.*)

Applying Vibration Dampers

Vibration dampers of different types may be applied to lines while energized using live-line tools for their installation.

A special tool simplifies the installation of the Stockbridge type of damper on live conductors. The damper is placed in the head of the tool, as shown in Fig. 24.31. A spring holds the damper clamp in place whether it is open or closed. With the damper clamp in the open position, the damper is placed on the live conductor. The locking nut on the bottom of the clamp, which has previously been placed in a socket, is tightened by turning the hot stick. The tool is removed by pulling down on the hot stick, compressing the spring, and turning the stick a quarter turn to engage the socket in a locked position on a projecting lug on the head of the tool. The tool may then be easily lifted away. A ring is provided on the side of this tool for attaching the link stick and hand line or blocks for aid in handling heavy dampers of this type.

Applying Armor Rods

In order to install armor rods, it is first necessary to remove the conductor from the insulator in the usual manner, and move it with wire tongs to a suitable working position. The conductors are supported on either a clamp or a holding stick. The armor rods are then applied with a special tool (Fig. 24.32).

The tool consists essentially of two split wheels into which different size dies are set. The armor rods are placed into these dies, and the assembly is placed around the conductor by means of holding or clamping sticks. The wheels or holders are rotated in the proper direction depending on the lay of the wire, one holder moving clockwise and the other counterclockwise. Rotation of the holders is accomplished by means of hooks fitted onto hot sticks engaging lugs on the holders. At the same time the holders are rotated, they are moved apart, starting near the center of the armor rods and moving toward the ends. The rotary movement of the holders is continued until the armor rods are completely and tightly twisted into place.

FIGURE 24.31 Vibration damper placed in a special tool preparatory to fastening to a live-line conductor. (*Courtesy A. B. Chance Co.*)

FIGURE 24.32 Special armor-rod wrenches and wrench holder used to apply armor rods while line is live. (*Courtesy A. B. Chance Co.*)

The holders are then pushed off the rods and removed by means of clamp sticks or other hot sticks.

Armor clips are then inserted on each end of the armor rods and clamped into place by tightening the bolt nuts. Special spring sockets mounted on the ends of hot sticks are used to tighten the nuts, the spring being used to thrust the saddle into place as soon as the two halves of the clip are in alignment.

Phasing Out

Phasing out is absolutely necessary when a new line is to be paralleled with another line, new or old, and after repairs or changes have been made on either of two lines which have previously operated in parallel. It is necessary in the latter case because of the possibility of interchanging conductors when making repairs or changes.

The process of phasing out consists of determining whether the phases of a given line or apparatus correspond with the phases of another line with which it is to operate in parallel. This problem arises most frequently at sectionalizing switches. The voltage across corresponding lines or phases should be zero. Therefore, to determine if zero voltage exists across corresponding lines, it is necessary to read the voltage. This is done with a phasing-out voltmeter.

Phasing-Out Voltmeter

A phase tester consists of two high-resistance units connected in series with a milliammeter that is calibrated to read voltage in kilovolts mounted on Epoxiglas housings. The resistors and instrument are connected in series with 20 ft of 15-kV insulated flexible cable coiled on a reel. Universal insulated hot sticks are fastened to the resistor housing to provide the lineman with isolation from the instrument. Only the necessary length of cable should be unreeled to avoid contact with the insulated cable or the cable with a ground that could distort the reading. Resistor units can be attached to the basic phase tester to extend the voltage range of the instrument (Fig. 24.33).

FIGURE 24.33   Resistors have been attached to a basic 15-kV phase tester to increase the voltage range of the instrument to 75 kV. *(Courtesy A. B. Chance Co.)*

Phasing-Out Procedure

Circuits to be connected together or connected in parallel must first be checked for proper phasing to prevent short circuit. The phase tester should be connected between all phases of each line and/or between each phase and ground of both lines to ensure that both lines are energized and that the tester is indicating properly. The tester is then connected between one phase of one line and a phase of the other line to ensure the circuits are properly coupled. If the circuits are properly coupled, the voltmeter will have reading equal to line-to-line voltage. If both lines are energized and coupled together, the tester is connected from a phase of one line to each phase of the other line in turn to determine the corresponding phases of each line. The conductors that have zero voltage between them are in phase and can be connected together for parallel operation. The tester should be removed from the lines as soon as the readings are completed.

Since the operation of high-voltage paralleling is usually done through a circuit breaker, the customary procedure in phasing out is to open the disconnects on one side of the circuit breaker and then close the circuit breaker so that the corresponding blades and contact jaws of the open disconnects are alive from the two sources that are to be paralleled. The instrument is checked for proper operation by connecting it between phases on each side of the open disconnect switch. This procedure verifies that both circuits to be phased out are energized if the instrument works correctly. The phasing-out voltmeter is then used to detect the presence of any difference in potential between a blade and the corresponding contact jaw of each disconnect.

If the voltmeter gives no indication as the two leads are successively applied to each pair of the three disconnect terminals, then no potential exists between each jaw and blade. This then shows that the correct phases are connected to each of the two parts of the respective disconnects. A voltage read between the jaw and hinge of one of the disconnect switches indicates that the circuits are out of phase. The proper phasing can be determined by checking between the jaw of one switch and the hinge of the other switches. The conductors with zero voltage between them have the same phase relationship. The phase conductors must be connected to provide the proper phase relationship before the circuits are paralleled. If no voltage is read across the open disconnects, the high-voltage voltmeter should be connected from the jaw of one switch to the blade of an adjacent switch, verifying that the circuits are similar. Ungrounded neutral or three-wire, three-phase circuits without adequate capacitive coupling will require the use of two high-voltage voltmeters simultaneously on adjacent phases to obtain a proper reading.

Similar tests can be completed for proper phasing with a low-voltage meter if potential transformers or potential devices are connected to the lines on both sides of the separation between circuits. The voltmeter leads are connected to the low-voltage terminals of the potential devices in the same sequence as described for the high-voltage tester on energized primary lines. Transmission-line circuits are usually phased out in substations using the low-voltage terminals of potential devices installed in the substation for metering and relaying circuits.

Insulator Testing

All live-line testing requires knowledge of the voltage distribution over the units of a suspension string or across the shells of a multipart pin insulator. Unfortunately, this distribution is not uniform, which would make the voltage across all units the same. Instead, the distribution is quite uneven; that is, the voltage across the unit nearest the line conductor is several times as great as the voltage across the unit nearest the crossarm or support. In some cases, the voltage across the unit nearest the conductor will be five times as much as across the unit next to the support.

Now, if for any reason one or more units become defective, the distribution curve becomes distorted.

Therefore, if the voltage across each unit of a string can be conveniently measured, defective units can be located easily. Any unit whose voltage reading is lower than the reading for a like unit on a string in good condition would show that the former unit is not withstanding its share of the voltage drop and is therefore defective. Actually, if a unit is very defective, the voltage drop across it will approach zero, as the broken-down insulation tends to act as a conductor. Ordinarily, if a reading is less than 60% of the average reading, the unit is probably defective.

Insulator Tester

One make of tester employing a meter for reading the voltage distribution is illustrated in Fig. 24.34. The meter on which the voltage is read is mounted on the end of the insulated hot stick. Although this meter actually is an ammeter, being actuated by a condenser current, its readings are proportional to the voltage; therefore, it may be looked on as a voltmeter.

In the case of disk-type insulators in a string or insulators in a stack, the two prongs are touched to the metal portions on both sides of the insulator, and readings are taken on the meter. Since the two prongs actually are the leads from the voltmeter, the readings give the voltage drop across the insulator. These readings are then compared with those for the corresponding units on a good insulator string. If any of the readings are much lower, the units are defective.

FIGURE 24.34 Parts of meter-type live-line insulator tester.

In the case of multilayer pin-type insulators, the prong is inserted between the porcelain layers until it makes contact with the cement. This prong is one lead from the voltmeter. The other lead from the meter connects to a condenser made of several metal tubes located in the hollow of the stick. The metal tube is one plate of the condenser, and the earth is the other plate. The condenser is made variable by providing two metal tubes of different length. The longer one is used at all times unless the reading goes off the scale. Then the shorter one is used. On the other hand, if the reading is too small and needs to be increased, both tubes may be joined and used.

On a two-piece insulator, a reading of the upper part will give the condition without testing the bottom part. If the upper part should be bad, the meter reading will drop. If the lower part is bad, the meter reading on the upper part will be above normal.

This type of insulator tester is made with different type heads and a variety of contact tips to fit the many types of insulators in use. The holding stick is furnished in three sections for convenience in reaching the insulator.

The phasing-out high-voltage voltmeter can be used to complete the tests, following the same procedure as described for the insulator tester, if the voltages are within the range of the voltmeter capacity.

A tool has been developed to easily test insulator strings as shown in Fig. 24.35. The tool is lightweight and easy for the lineman to handle with a hot-stick as it is slid along the insulator string.

Hot-Line-Tool Care

The outside surface of the insulated stick portion of hot-line tools must be kept clean and dry. The trailers used to store and transport hot-line tools are usually equipped with electric heaters to reduce the humidity and prevent condensation. The surface of the sticks should be inspected before they are used. Tools that are battered, scratched, or dirty should be maintained before they are used. Hot-line tools are factory-tested by the manufacturer at 100,000 V/ft. Many companies have laboratories for testing protective equipment and hot-line tools. The insulated stick portion of the hot-line tools is normally tested at 50,000 V/ft after delivery to the customer. The tools can be tested in the field using a portable hot-stick tester.

FIGURE 24.35   Close-up of the insulator
tester. (*Courtesy A. B. Chance Co.*)

FIGURE 24.36   Lineman using portable tester to check insulation of hot-line tool. Tester can be used to check all kinds of insulated sticks regardless of the material. (*Courtesy A. B. Chance Co.*)

The tester is energized from a 120-volt ac source. The meter is set to zero without a stick in it while energized to eliminate the effect of leakage current. A test bar is used to ensure proper operation. The tester puts 1,800 volts across the two center electrodes with the two outer electrodes grounded. The outer electrodes shield the inner electrodes to prevent the lineman using the tester from affecting the meter reading while the tester is moved along the surface of the insulated portion of the hot-line tool (Fig. 24.36).

Robotic Hot-Line-Tool Live-Line Maintenance

The three-phase material handler lifter assists the lineman to accomplish hot line work while quickly maneuvering energized phase conductors out of one's way (Fig. 24.37). Tomcat, a remote-controlled manipulator arm, is being used to complete live-line maintenance. Tomcat (teleoperator for operations, maintenance, and construction

FIGURE 24.37   Three-phase material handler lifter in place allowing lineman to make insulator repair on the pole. The line remains energized.

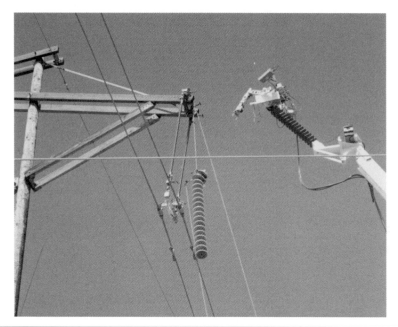

FIGURE 24.38   Tomcat has completed installation of strain carrier and disconnected the insulator string, and the insulators are being lowered to the ground by a groundman. New insulators will be raised to the structure by the groundman, and Tomcat will be used to complete the replacement.

using advanced technology) was developed by Philadelphia Electric Company and the Electric Power Research Institute. Tomcat is operated by a lineman from a position in a van. The lineman uses closed-circuit television and a master-slave control to perform hot-line maintenance work. Fiber-optic cable between the manipulator and the equipment in the van provides the communication link and electrical insulation from the energized line.

Figure 24.38 pictures the insulator string being lowered to the ground by a lineman or groundman after Tomcat has completed the installation of the strain carrier and disconnected the insulator string from the line and its support on the structure. Tomcat can be programmed to complete certain operations automatically, which permits it to function robotically.

Live-Line Maintenance from Insulated Aerial Platforms

T he *IEEE Guide for Maintenance Methods on Energized Power-Lines* (IEEE Standard 516) provides detailed information concerning work by linemen from insulated aerial platforms. The procedures discussed in the standard should be followed while performing work on energized power lines.

Insulated Aerial Lifts

The American National Standard entitled Vehicle-Mounted Elevating and Rotating Devices (ANSI Standard A92-2) provides information on insulated aerial devices commonly used by linemen and cablemen. Mobile extension ladders have long been useful for tree trimming and for the replacement of burned-out lamps in street lighting fixtures. The insulated aerial platform has replaced the insulated extension ladder for most operations. The mobile insulated aerial platform has an insulated section in the upper portion of the boom assembly and an insulated insert in the lower portion of the lifting boom (Fig. 25.1). The fiberglass-insulated section of the upper portion of the boom provides protection from ground for the lineman and the insert near the mounting mechanism for the boom provides isolation for a groundman if the metal in the boom assembly should accidentally contact an energized conductor. The mobile insulated aerial platform is an improved device for streetlight maintenance and a safer platform for tree trimming near energized lines, but its greatest field of usefulness is for live-line maintenance (Fig. 25.2). It can serve as a platform from which to perform live-line operations in three ways:

1. Using rubber gloves on distribution lines

2. Using conventional hot-line tools on distribution and transmission lines

3. Using bare hands on medium- and high-voltage transmission lines

For tree trimming or replacement of street lamps, the platform usually consists of a single bucket or basket with room for only one lineman, while for live-line maintenance, single and double buckets are employed.

FIGURE 25.1   Aerial lift trucks setup for transmission line maintenance. The vehicles have the outriggers extended to stabilize the vehicle. The double bucket aerial device is extended away from the truck. The insulated sleeves in the upper and lower portions of the lift device boom are visible. (*Courtesy Altec HiLine LLC.*)

FIGURE 25.2   Two linemen in a two-man single-basket bucket truck installing line hose. (*Courtesy Altec HiLine LLC.*)

Use of the aerial platform is limited only by accessibility to the work area near the overhead lines. As access is sometimes impossible, the aerial platform can only supplement regular line crews using conventional pole climbing procedures.

Description of Insulated Aerial Lift

The lifting-boom pedestal is placed behind the truck cab. Outriggers from the pedestal base are used to stabilize the boom when it is extended (Fig. 25.3). The boom and baskets are positioned over the cab while traveling. The upper member of the boom and the baskets are made of fiberglass to provide the required insulation from ground. A fiberglass section may be installed in the lower boom to provide protection for groundman. All movements of the lift are controlled by the lineman in the basket. Mobile aerial platforms, baskets, or bucket trucks used for live-line bare-handed work should be tested and specially certified for the voltage of the line to be maintained.

Maintenance with Rubber Gloves

Live-line maintenance from an insulated platform with rubber gloves is limited to the distribution voltages (Figs. 25.4 through 25.6). Some electric companies work on energized lines with voltages through 20 kV to ground or 34.5 kV phase-to-phase; their employees use rubber gloves and sleeves from an insulated aerial bucket or an insulated platform clamped to the structure. It is good practice to test the mobile-insulated aerial platform or bucket truck in the field each day before the work is started and each time the assembly is to be used on a higher distribution voltage or if there is some question about the insulating conditions. The insulating arm of the bucket truck can be tested by raising the bucket with all personnel at ground level until the bucket contacts a conductor energized at the voltage to be worked. Contact should be maintained for 3 minutes or more. If the insulation of the assembly does not give indication of excessive leakage current or electrical failure, the work can proceed. An insulated boom extension on the aerial-lift device is being used to support the conductor while the insulator is being replaced. Aerial-lift platform accessories improve the safety and the efficiency of the linemen performing energized-line maintenance work. Other operations that are usually completed with rubber gloves and rubber sleeves are installing transformers,

FIGURE 25.3 Lineman in the process of lowering the vehicle outrigger. (*Courtesy Steel Market Development Institute.*)

Figure 25.4   Lineman installing line hose from an insulated bucket. (*Courtesy Steel Market Development Institute.*)

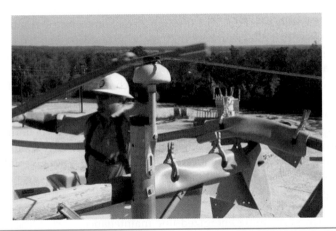

Figure 25.5   Rubber-glove work demonstrated on a 15-kV line. The lineman is installing protective line hose. (*Courtesy Steel Market Development Institute.*)

Figure 25.6   Lineman inspecting the installation of the cover-up prior to starting to work. (*Courtesy Steel Market Development Institute.*)

switches, and cutouts, replacing insulators, repairing conductors, and inspecting equipment. Work on distribution circuits operating at 12,500Y/7,200 volts or higher requires the use of insulated aerial devices or insulated platforms to supplement the rubber gloves and sleeves. Rubber gloves and sleeves must be manufactured and rated for a voltage greater than the distribution voltage to be worked. Distribution voltages lower than 12,500Y/7,200 volts may be worked using rubber gloves while the lineman's gaffs are engaged in wood poles.

Maintenance with Hot-Line Tools

Live-line maintenance from an insulated aerial platform using hot-line tools is more conveniently and more rapidly done than when working from the pole or tower. The ease of reaching elevated positions and the reduction in the required protective covering, as well as the reduced number of linemen required, favor this method. Sometimes a combination of linemen working both from the platform and from the pole or tower is preferred.

Maintenance Using Bare-Hand Method

The insulation of mobile aerial devices and fiberglass ladders to be used for bare-hand live-line work should be tested each time they are used in the field, before the electrical work is started. The devices can be tested by contacting an energized conductor to be worked on with the device and reading leakage current to ground at the ground location. Contact should be maintained for 3 minutes. The leakage-current readings should remain essentially constant. If the leakage current increases or decreases by more than 20% from the initial reading, a longer test period should be used and the cause of the variation should be identified and corrected before the bare-hand electrical work is started. ANSI Standard A92.2 provides detailed testing information. When live-line work is to be done bare-handed from an insulated aerial platform, the lineman in the basket must be at the same potential as the live conductor on which the repair is to be made. This is accomplished by bonding the metal-mesh lining of the basket to the live-line conductor. The mesh lining, also referred to as *basket shielding*, is brought to line potential by means of a bonding lead, one end of which is permanently attached to the shielding while the other end is clamped onto the energized conductor as the basket approaches it. A short hot stick is used to fasten the metal clamp. One lead is employed for each lineman. Since the two leads are electrically interconnected, the second lead serves as a safety measure in case one should accidentally be knocked off. If both were removed accidentally, the lineman would receive a charging-current shock each time he touched or removed his hand from the energized conductor.

The linemen, wearing conductive suits if working on extra-high-voltage (EHV) or ultra-high-voltage (UHV) lines, are also "bonded" to the shielding by means of a spring-type ankle clamp or by the wearing of conductive-soled shoes.

While the linemen and their shielding are thus bonded to the energized conductor, the linemen are maintained at conductor potential. They can move around in the baskets and touch the conductor and any metal accessories which are at the same potential without having any physical awareness of the fact that they are charged at line potential. As long as the lineman is at the same potential as the conductor he is working on, and as long as he is insulated from ground, no current can flow from conductor to man. The lineman can work bare-handed on the conductor or on any metal fittings at the same potential as the conductor without experiencing any discomfort. The rule is,

"Touch only that which you are clipped to." Work is completed safely on transmission lines using proper equipment and procedures.

In many locations it is impossible to position an insulated bucket truck at the work site. Bare-hand live-line work can be performed by linemen in special conducting suits working from special insulated ladders (Figs. 25.7 and 25.8). The conducting suit is connected to the energized conductors.

Touching live conductors with bare hands when insulated from ground is similar to the contact birds make when resting on live conductors, except that the birds would call it the barefoot method. The birds apparently experience no uncomfortable effects or

FIGURE 25.7   Lineman setting up to work from insulated ladders on a 345-kV transmission line.

FIGURE 25.8   Lineman working from insulated ladders on a 345-kV transmission line.

they would leave the conductor. Of course, the birds are thoroughly insulated from ground, since they require no auxiliary support.

Live-line maintenance is completed on EHV circuits constructed with steel towers. The procedures for performing work on 345-kV and 765-kV energized lines are similar. Linemen performing bare-hand maintenance on energized EHV circuits wear a specially designed conductive suit. The conductive suit has a steel mesh that equalizes the charge around a lineman's body as he approaches the energized conductor from an insulated platform. The metal-mesh suit acts like a Faraday cage, shielding the lineman from the electric field. The lineman contacts the energized line conductor with a wire or wand that connects to the conductive suit. The wand draws an arc as it approaches the energized conductor. The connection between the wand and the conducting suit bypasses the charging current, eliminating any sensation the lineman would feel if he contacted the energized conductor directly with his bare hand before being clipped to the conductor.

Live-line bare-hand maintenance is frequently performed with the aid of a helicopter (Fig. 25.9). The lineman is carried to the energized line conductor while on a platform suspended below the helicopter (Fig. 25.10). When the lineman approaches the energized line, he connects the metal platform below the helicopter and his conductive suit to the energized line conductor on which the work is to be performed (Figs. 25.11 and 25.12). The lineman uses a portable two-way radio to communicate with the helicopter pilot. The helicopter makes it possible to perform work in areas where the transmission line is inaccessible for an insulated bucket truck (Figs. 25.13 and 25.14).

FIGURE 25.9 Lineman replacing conductor spacer on 765-kV line using bare-hand techniques while being supported by a helicopter. The lineman wears a conductive suit that is clipped to the energized conductors. (*Courtesy American Electric Power Company, Inc. and Haverfield Helicopters.*)

FIGURE 25.10   Lineman being transported to the worksite via helicopter. (*Courtesy MidAmerican Energy Company.*)

FIGURE 25.11   The lineman has reached the work site destination and is in the process of clipping in. (*Courtesy MidAmerican Energy Company.*)

FIGURE 25.12   Lineman is adjusting a stringing block while working from the platform of the helicopter. (*Courtesy MidAmerican Energy Company.*)

FIGURE 25.13 Lineman being transported to the worksite via helicopter while two other linemen are in position to assist. (*Courtesy MidAmerican Energy Company.*)

FIGURE 25.14 The linemen are working on the transmission structure. (*Courtesy MidAmerican Energy Company.*)

Protective Gaps

Linemen performing live-line maintenance work on EHV and UHV transmission lines may be working in the vicinity of insulator strings with a small number of insulators shorted out. The insulators may be shorted out because they are defective or as a result of the procedures necessary to complete the work. The dielectric strength of the air surrounding the insulators being worked on or between the lineman and the energized conductor or between the lineman in contact with the energized conductor and the transmission tower depends on the air humidity, temperature, density, pressure, and the shape of the electrodes separated by the air gap.

Protective gaps can be installed in the vicinity of the work being performed, usually on an adjacent tower, to protect the linemen performing bare-hand or hot-line-tool work. The gap distances are specified by the organization responsible for the facilities and the work being performed. The linemen performing the work then always maintain an approach distance greater than the gap distance to ensure their safety. The protective gaps ensure that a surge voltage on the energized transmission line will always cause a flashover at the temporary protective-gap installation instead of at the location where the linemen are performing the work.

CHAPTER 26

Grounding

Proper grounding of electrical systems is very important for electric equipment performance and the safety of the linemen, cablemen, groundmen, and the public. Correct operation of protective devices is dependent on the grounding installation. The *National Electrical Safety Code* (ANSI-C2 Standard) specifies the grounding methods for electrical supply and communications facilities. *The National Electrical Code* (NFPA-70 Standard) specifies the grounding methods for all facilities that are within or on public or private buildings or other structures.

The grounded static wires for transmission lines and the grounded neutral wire for primary distribution circuits originate at substations. The static wires for transmission lines will normally originate and terminate at a substation. The substation may be located at a generating station.

The substation grounding system governs the proper functioning of the whole grounding installation and provides the means by which grounding currents are conducted to remote earth. It is extremely important that the substation ground grid has a low-ground resistance, adequate current-carrying capacity, and safety features for personnel. The substation grounding system normally consists of buried conductors and driven ground rods interconnected to form a continuous grid network. The surface of the substation is usually covered with crushed rock or concrete to control the potential gradient when large currents are discharged to ground and to increase the contact resistance to the feet of personnel in the substation. Substation equipment should be connected to the ground grid with large conductors to minimize the grounding resistance and limit the potential between the equipment and the ground surface to a safe value under all conditions. All substation fences should be constructed inside the ground grid and connected to the grid frequently to protect the public and workmen.

Ground Resistance

The electrical resistance of the earth is largely determined by the chemical ingredients of the soil and the amount of soil moisture present. Measurements of ground resistances completed by the Bureau of Standards are summarized in Table 26.1.

Measurement instruments used to gather the readings are voltmeters, ammeters, clamp-on ground resistance testers, or earth ground tester kits.

The resistance of the soils tested varied from 2 ohms to 2,700 ohms. The type of soil, the chemical ingredients, and the moisture level surrounding an electrode determine the resistance, as illustrated in Fig. 26.1. If the soil has uniform resistivity, the greatest resistance is in the area immediately surrounding the electrode which has the smallest cross section of soil at right angles to the flow of current through the soil. The area of the current path, 8 ft to 10 ft from the electrode, is so large that the resistance is negligible

Soil	Resistance, ohms		
	Average	**Minimum**	**Maximum**
Fills and ground containing more or less refuse such as ashes, cinders, and brine waste	14	3.5	41
Clay, shale, adobe, gumbo, loam, and slightly sandy loam with no stones or gravel	24	2.0	98
Clay, adobe, gumbo, and loam mixed with varying proportions of sand, gravel, and stones	93	6.0	800
Sand, stones, or gravel with little or no clay or loam	554	35	2,700
Bureau of Standards Technologic Paper 108.			

TABLE 26.1   Resistance of Different Types of Soil

FIGURE 26.1   Resistance of earth surrounding an electrode. This may be pictured as successive shells of earth of equal thickness. With increased distance from the electrode, the earth shells have greater area and therefore lower resistance. (*Courtesy Copperweld Steel Co.*)

compared to the area immediately surrounding the ground rod. Measurements show that 90% of the total resistance surrounding an electrode is generally within a radius of 6 ft to 10 ft. A variation of a few percent in moisture will make a great difference in the effectiveness of a ground connection made with electrodes of a given size for moisture contents less than approximately 20%. Experimental tests made with red-clay soil indicated that with only 10% moisture content, the resistivity was over 30 times that of the same soil having a moisture content of about 20% (Fig. 26.2).

The soil resistivity also varies greatly with the surrounding soil temperature (Fig. 26.3). The water in the soil freezes when the temperature is below 32°F, causing a large increase in the resistance of a ground connection. Grounding electrodes which do not extend below the frost line where the soil freezes will have a large variation in resistance as the seasons change. The depth of the ground electrode is important to the electrical performance. Driven ground rods should be long enough to reach the permanent moisture level of the soil (Fig. 26.4). Soil is seldom of uniform resistivity at different depths. The first few feet of soil near the surface normally have a relatively high resistance because it is subject to changes in moisture content as a result of rainfall.

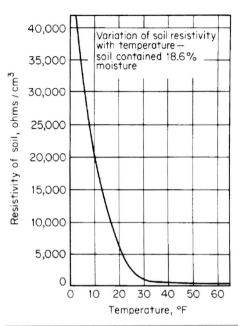

FIGURE 26.2 Variation of soil resistivity with moisture content. (*Courtesy Copperweld Steel Co.*)

FIGURE 26.3 Variation of soil resistivity with temperature. (*Courtesy Copperweld Steel Co.*)

FIGURE 26.4 Chart showing the relation between depth and resistance for a soil having a uniform moisture content at all depths. In the usual field condition, deeper soils have a higher moisture content and the advantage of depth is more pronounced. (*Courtesy Copperweld Steel Co.*)

The ground rod (Fig. 26.5) should be driven deep enough to be in perpetually moist earth. In most localities, a depth of 8 ft to 10 ft is usually necessary to meet this requirement. A ground rod that penetrates perpetually moist earth will have a fairly low resistance. Low values of resistance are desirable and essential for the proper functioning of lightning arresters. The resistance of grounds on lines in service can be measured easily with a clamp-on ground-resistance tester (Fig. 26.6). The ground-wire molding must be removed to gain access to the ground wire near its connection to the ground rod.

If when the rod is driven to its full length, it is found that the resistance is not well below 25 ohms, resort can be had to one of the two following procedures:

1. Extend the length of the rod.
2. Drive additional rods.

A sectional ground rod is one that makes possible adding section upon section in order that the rod can be driven deeper into more favorable moist soil, thereby lowering its ground resistance. Figure 26.7 shows the parts of a typical sectional ground rod and gives details of its construction. The sections are joined by a heavy-tapped bronze coupling. A special impact-resisting steel driving stud is screwed into the top section to take the hammer blows while driving. It protects the threading for subsequent attachment of additional sections of rod if needed. The sections are threaded at each end, one of which is pointed. The usual section length is 8 ft, 10 ft, or 12 ft.

FIGURE 26.5   Copper-bonded steel ground rod. (*Courtesy of Blackburn.*)

FIGURE 26.6   Measuring ground resistance of ground on a distribution circuit without disconnecting the ground connection. (*Courtesy Greenlee.*)

FIGURE 26.7   Typical sectional ground rod showing driving stud, couplings, rod sections, and driving point. (a) Removable driving stud of special steel takes the blows, protects the threads. (b) Cutaway view showing coupling and driving stud installed. (c) The steel core gives strength and rigidity for driving. The molten-welded copper covering provides high conductivity, permanent protection against corrosion. (d) A strong union and tight contact—easily made as successive sections are joined by the heavy bronze coupling. (e) Bottom end of sectional rods are pointed for easier installation. Rods are threaded on both ends. (*Courtesy Copperweld Steel Co.*)

The installation procedure for sectional ground rods is as follows. It is the same whether hand driven or power driven.

1. The coupling and driving stud are screwed on the top end of the sectional rod.
2. This assembly is then driven until the coupling on the rod reaches ground level.
3. The driving stud is then removed.
4. The lower end of the second sectional rod is then screwed into the coupling.
5. Another coupling is screwed onto the top end of the second sectional rod.
6. The driving stud is screwed into the second coupling.
7. Driving is now continued until the second section reaches ground level.
8. This sequence is repeated until resistance measurements indicate that the desired value of resistance has been obtained.

The other means of securing lowered ground resistance is by use of additional rods. When two or more driven rods are well spaced from each other and connected together, they provide parallel paths to earth. They become, in effect, resistances in parallel or multiple and tend to follow the law of metallic parallel resistances. Figure 26.8 shows one, two, three, and four ground rods installed at each pole, respectively. The rods are connected together with wire no smaller than that used to connect to the top of the pole.

If the rods were widely separated, two rods would have 50% of the resistance of one rod, but because they are usually driven only 6 ft or more from each other, the resistance of two rods is about 60% of that of one rod. Three rods, instead of having 33% of one rod, have on an average about 40%. Four rods, instead of having 25% of one rod, have about 33%. These values are shown graphically in Fig. 26.9.

FIGURE 26.8   Sketches showing one ground rod (a) and two, three, and four ground rods at poles (b), (c), and (d), respectively. Rods at each pole are connected together electrically.

FIGURE 26.9   Installations of multiple ground rods. Upper graphs show approximate extent to which ground resistance is reduced by the use of two, three, or four rods connected in multiple.

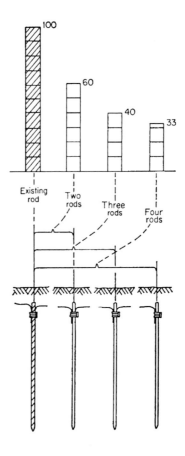

Multiple rods are commonly used for arrester grounds and for station and substation grounds. In addition to lowering the resistance, they provide higher current-carrying capacity and can thus handle larger fault currents.

Substation Grounds

Substations utilize an elaborate interconnected **substation ground grid** for the safety of the workman. The ground grid equalizes and limits step and touch potential differences. Metallic matts, gratings, bare connectors, and ground rods are arranged throughout the substation and are bonded together. Equipment and substation fence grounds are coupled to the substation ground grid.

Transmission-Line Grounds

High-voltage electric transmission lines are designed and constructed to withstand the effects of lightning with a minimum amount of damage and interruption of operation. When lightning strikes an overhead ground or static wire on a transmission line, the lightning current is conducted to ground through the metal tower or the ground wire installed along the pole. The top of the structure is raised in potential to a value determined by the magnitude of the lightning current and the surge impedance of the ground connection. If the impulse resistance of the ground connection is high, this potential may be many thousands of volts. If the potential exceeds the insulation level of the equipment, flashover will result, causing a power arc which will initiate the operation of protective relays and removal of the line from service. If the transmission structure is well grounded and proper coordination exists between the ground resistance and the conductor insulation, flashover can usually be avoided. The transmission-line grounds are installed in accordance with the specifications provided. Ground rods are usually used to obtain a low-ground resistance (Figs. 26.10 through 26.13).

A pole butt grounding plate (Fig. 26.14) or a grounding coil can be used on wood-pole structures. A grounding coil is a spiral coil of bare copper wire placed at the bottom of a pole, as shown in Fig. 26.15. The wire of the coil continues up the pole as the ground-wire lead. Sometimes it is wound around the bottom of the pole a few times as a helix (see Fig. 26.16) to increase the amount of wire surface in contact with the earth. Grounding coils should have enough turns to make good contact with the earth. A coil

FIGURE 26.10   Ground wire connects ground rod to metal tower.

FIGURE 26.11 Shield-wire connection to ground wire on wood pole for transmission line.

having seven turns and 13 ft or more of wire would provide satisfactory ground contact. Number 6 or larger AWG soft-drawn or annealed copper wire should be used. To prevent the coil from acting as a choke coil, the ground lead from the innermost turn is stapled to all the other turns.

If the soil has a high resistance, a grounding system called a counterpoise may be necessary. The counterpoise for an overhead transmission line consists of a special grounding terminal which reduces the surge impedance of the ground connection and increases the coupling between the ground wire and the conductors. Counterpoises are normally installed for transmission-line structures located in areas with sandy soil or rock close to the surface. The types of counterpoises used are the continuous or parallel type and the radial or crowfoot type (Fig. 26.17). The continuous, or parallel, counterpoise consists of one or more conductors buried under the transmission line for its entire length or under sections with high-resistance soils. The counterpoise wires are

Figure 26.12 Ground-wire connection to ground rod for wood-pole transmission-line structure (a) Ground-wire molding; (b) staples, molding; (c) 6A Copperweld wire; (d) staples, fence; (e) ground-rod clamp; (f) 5/8-inch × 10-ft ground rod.

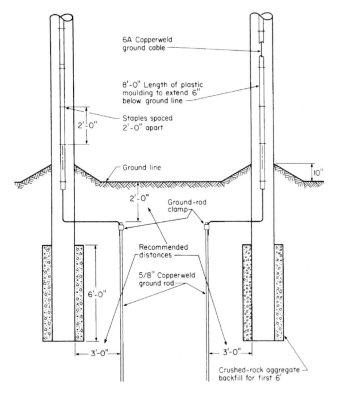

Figure 26.13 Ground-wire connections to ground rods for wood-pole H-frame transmission-line structure.

FIGURE 26.14 Pole butt grounding plate attached to pole of a transmission or distribution line before the pole is set. The grounding plate is made from copper plate with moisture-retaining cups to maintain a good ground. (*Courtesy Homac Manufacturing Co.*)

FIGURE 26.15 Grounding coil in position on the bottom of the pole. Entire weight of pole helps to maintain contact with earth below pole. The spiral coil is stapled to the pole bottom. Wire from inner spiral is stapled to all turns as it crosses them to shunt out the turns.

FIGURE 26.16 Views of grounding coil. Elevation shows six twin helixes with 3 inches between turns. Bottom view shows spiral pancake coil of seven turns.

connected to the overhead ground, or static wire, at all supporting structures. The radial-type counterpoise consists of a number of wires extending radially from the tower legs. The number and length of the wires will depend on the tower location and the soil conditions. The continuous counterpoise wires are usually installed with a cable plow at a depth of 18 inches or more. The wires should be deep enough that they will not be disturbed by cultivation of the land. The utilization of mesh networks and ground plates are alternate solutions when the soil has a high resistance (Fig. 26.18).

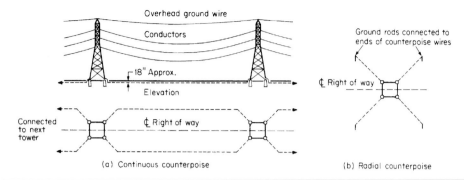

Overhead ground wire

Conductors

18" Approx.

Elevation

Connected to next tower

₵ Right of way

(a) Continuous counterpoise

Ground rods connected to ends of counterpoise wires

₵ Right of way

(b) Radial counterpoise

FIGURE 26.17   Two general arrangements of counterpoise installations. (*Courtesy Copperweld Steel Co.*)

FIGURE 26.18   Mesh networks and ground plates are utilized when the soil has a high resistance. (*Courtesy Fluke.*)

Mesh network.

Ground plate.

Distribution-Circuit Grounds

A multigrounded neutral conductor for a primary distribution line is always connected to the substation grounding system where the circuit originates and to all grounds along the length of the circuit. The primary neutral conductor must be continuous over the entire length of the circuit and should be grounded at intervals not to exceed ¼ mile. The primary neutral conductor can serve as a secondary neutral conductor. If separate primary and secondary neutral conductors are installed, the conductors should be connected together if the primary neutral conductor is effectively grounded. All equipment cases and tanks should be grounded (Figs. 26.19 and 26.20). Lightning-arrester ground terminals should be connected to the common neutral if available and to a separate ground rod installed as close as is practical. Switch handles, streetlight fixtures, down guys, and transformer secondaries must be properly grounded (Fig. 26.21). The common conductor on the transformer secondary for a single-phase 120/240-volt service must be grounded. A three-phase delta- or open-delta-connected transformer secondary for 240-volt or 480-volt service would have the center tap of one of the transformer-secondary terminals grounded. If the center tap of one of the transformer

ITEM

A. Preformed tie wire
B. 1″ × 5 1/2 Pin-type insulator
C. 1″ Pole-top pin
D. 5/8″ Bolt
E. 2 1/2″ Curved washer
F. 5/8″ Palnut
G. Epoxy-rod biunit assembly
   1 1/2″ Diameter rod, 36″ between
   conductors
H. Stirrup, compression type
I. Hot-line clamp
J. Three-phrase cutout-arrester bracket
K. 10 kV arrester
L. 15 kV cutout
M. #6 Copper wire
N. Molding, ground wire
O. Staples, fence

P. Staples, molding
Q. Split-bolt connector
R. 5/8″ × 10′ Ground rod
S. Ground-rod clamp
T. Small epoxy-glass standoff
U. #4 Stranded, copper wire
V. Capacitor bank
W. Partial range current limiting fuse

Detail A

FIGURE 26.19 Capacitor bank installation for 13.8-kV circuit with grounding details illustrated.

ITEM

A. Dead-end clamp
B. 4 1/4" Suspension insulator
C. 5/8" Eyebolt
D. 5/8" Eyenut
E. 2 1/2" Curved washer
F. Compression-type stirrup
G. Hot-line clamp
H. 10 kV arrester
I. Cutout-arrester bracket
J. Compression connector
K. #6 Split-bolt connector
L. Single-clevis insulator
M. 5/8" x 10" bolt
N. Automatic recloser
O. #6 Copper wire
P. Molding-ground wire
Q. Staples, molding
R. 5/8" x 10' Ground rod
S. Ground-rod clamp
T. Staples, fence
U. Preformed tie wire

FIGURE 26.20 Single-phase 15-kV rural distribution automatic recloser installation with grounding details illustrated.

secondaries is not available, one-phase conductor is normally grounded. The neutral wire for three-phase, four-wire Y-connected secondaries for 208Y/120-volt or 480Y/277-volt service is grounded (Fig. 26.22). If the primary circuit has an effectively grounded neutral conductor, the primary and secondary neutrals are connected together as well as to ground. Many times the primary distribution-line grounded neutral conductor is used for the secondary grounded neutral conductor and is called a common grounded neutral conductor. The neutral conductor of the service to the customer must be connected to ground on the customer's premises at the service-entrance equipment. A metal water pipe provides a good ground. Care must be taken to avoid plastic, cement-lined, or earthen-ware piping. Driven grounds may be installed near the customer's service entrance to provide the required ground connection.

ITEM
A. 5/8" x 10" Eyebolt
B. 2 1/2" Curved washer
C. 5/8" Palnut
D. 4 1/4" Suspension insulator
E. Dead-end clamp
F. Stirrup, compression type
G. Hot-line clamp
H. Staples, fence
I. 5/8" Bolt
J. Split-bolt connector
K. Compression-connector cover
L. Aerial-cable clamp
M. Compression connector
N. Partial-range current-limiting fuse
O. #6 Copper wire
P. Molding, ground wire
Q. Staples, molding
R. 5/8" x 10' Ground rod
S. Ground-rod clamp
T. Single-phase self-protected distribution transformer

FIGURE 26.21   Distribution transformer installed on dead-end pole for 15-kV line with grounding details illustrated.

Installing-Driven Grounds

Since the most common ground in distribution systems is the driven ground, its installation will be briefly described. A galvanized-steel rod, stainless-steel rod, or copperweld rod is driven into the ground beside the base of the pole at a horizontal distance of 3 ft to 5 ft in undisturbed earth, as shown in Fig. 26.23. The rod should be driven until the top end is 2 ft below the ground level. After the rod is driven in place, connection is made with the ground wire from the pole. The actual connection to the ground rod is usually made with a heavy bronze clamp, as shown in Fig. 26.24. The clamp is placed over the ground-rod end and the ground-wire end. The size of the ground wire should not be less than No. 6 AWG copper wire. Number 4 wire is generally used. When the setscrew in the clamp is tightened, the ground wire is squeezed against the ground rod. Several precautions are taken in making the actual connection to minimize the possibility

ITEM

A. Preformed tie wire
B. 1″ x 5 1/2″ Pin insulator
C. 1″ Pole-top pin, 18″
D. 5/8″ Bolt
E. 2 1/2″ Curved washer
F. 5/8″ Palnut
G. Epoxy-rod biunit assembly
   1 1/2″ D rod, 36″ between conductors
H. Non-tension sleeve
I. Stirrup, compression type
J. Hot-line clamp
K. Transformer secondary lead
L. Compression connector
M. #6 Split-bolt connector
N. 4 Spool rack
O. Transformer mounting bracket
P. #6 Copper wire
Q. Molding, ground wire
R. Staples, fence
S. Staples, molding
T. 5/8″ x 18′ Ground rod

Detail A

U. Ground-rod clamp
V. Small epoxy-glass standoff
W. Partial-range current-limiting fuse
X. No. 33 Scotch electrical tape
Y. No. 4 Stranded copper wire
Z. Single-phase self-protected
   distribution transformer

FIGURE 26.22   Three-phase Y-Y 15-kV distribution transformer bank with quadruplex secondaries.

FIGURE 26.23   Lineman helper driving the ground connection. The rod is copperweld, has a diameter of 5/8 inch, and is 8 ft long. Lineman helper is using a hydraulic ground-rod driver which slips over the end of the rod. Care must be taken not to lift the driver above the head of the rod and have it slide off, as this might cause injury. (*Courtesy Steel Market Development Institute.*)

FIGURE 26.24   The ground rod clamp is used to connect the ground wire to the ground rod. The clamps shown work on ½ inch, 5/8 inch, and ¾ inch ground rods for wire ranges of No. 8 through 1/0. (*Courtesy Burndy Corporation.*)

of the ground wire being pulled out by frost action or packing of the ground above it. The ground wire may be brought up through the clamp from the bottom. It is then bent over the clamp so that it cannot pull out. The wire is trained loosely so that there is plenty of slack between the ground-wire molding and the ground rod. This connection must be made before the lightning arresters are completely connected up on the pole. If this were the last connection—the one that connects the arrester to ground—the lineman could get a high-voltage electric shock if one of the lightning arresters is defective. The last connection is always made at the lightning arrester by the lineman, either to its top or bottom terminal.

Ground-Wire Molding

The molding protects the wire as well as the lineman. The molding shall extend 2 ft below and at least 8 ft above the surface of the ground. The ground wire is molded to reduce the likelihood of electric shock if the wire becomes energized.

CHAPTER 27

Protective Grounds

Protection of the lineman is most important when a transmission or distribution line or a portion of a line is removed from service to be worked on using deenergized procedures. Precautions must be taken to be sure the line is deenergized before the work is started and remains deenergized until the work is completed. The same precautions apply to new lines when construction has progressed to the point where they can be energized from any source.

The installation of temporary protective grounds and short-circuiting leads at the worksite protect against the hazards of static charges on the line, induced voltages, backfeeds from a customer generators, and accidental energizing of the line. A safe work zone is created with a common work source ground to eliminate hazardous differences of potential. Conductive objects are bonded to the common work source ground.

References include:

- *OSHA Standard 1910.269(n)–Grounding for the protection of employees,*
- *OSHA Standard 1926.962–Grounding for the protection of employees,*
- *Personal Protective Grounding for Electric Power Facilities and Power Lines, Volume 5-1, US Bureau of Reclamation,*
- *ASTM F855–19a Standard Specifications for Temporary Protective Grounds to Be Used on De-energized Electric Power Lines and Equipment*
- *IEEE1048 Guide for Protective Grounding of Power Lines.*

When a deenergized line and an energized line parallel each other, the deenergized line may pick up a static charge from the energized line because of the proximity of the lines. The amount of this static voltage "picked up" on the deenergized line depends on the length of the parallel, weather conditions, and many other variable factors. However, it could be hazardous, and precautions must be taken to protect against it by grounding the line at the location where the work is to be completed. This will drain any static voltage to ground and protect the workman from this potential hazard.

When a deenergized line parallels an energized line carrying load, the deenergized line may have a voltage induced on it in the same manner as the secondary of a transformer. If the deenergized line is grounded at a location remote from where the work is being done, this induced voltage will be present at the work location. Grounding the line at the work location will eliminate danger from induced voltage.

Grounding and short-circuiting protect against the hazard of the line becoming energized from either accidental closing in of the line or accidental contact with an energized line which crosses or is adjacent to the deenergized line.

The procedures established to control the operation of equipment in an electrical system practically prevent the accidental energizing of a transmission or distribution line. Hold-off tagging procedures have proven to be very effective. If a circuit should be inadvertently energized, the grounds and short circuits on the line will cause the protective relays to initiate tripping of the circuit breaker at the source end of the energized line in a fraction of a second and deenergize the hot line. During this short interval of time, the low-impedance grounds and short circuits on the line being worked on will protect the lineman (Fig. 27.1). If it isn't grounded, it isn't dead!

Inspecting Equipment before Grounding

The lineman thoroughly inspects the grounding equipment before every use for defects or damages. The use of a low-impedance grounding jumper is imperative. The grounding jumper components are:

- Insulated grounding cable
- Copper ferrules
- Grounding clamps (C-type, Duckbill, Tower & Flat-Face)

The protective ground sets are tested periodically using a protective grounding set tester.

FIGURE 27.1   Grounding cluster installed on four-wire three-phase line. Line conductor mounted on side of pole is the grounded neutral conductor. The grounding conductors create a low-impedance short circuit around the lineman's work area. (*Courtesy A. B. Chance Co.*)

Testing Line Conductors before Grounding

The lineman must test each line conductor to be grounded to be sure it is deenergized (dead) before the protective grounds are installed. The circuit can be tested with a high-voltage voltmeter (Figs. 27.2 and 27.3). Special testing equipment that produces an audible sound is available for detecting high voltage before applying grounding equipment.

FIGURE 27.2 Analog voltage detector utilized for testing potential on an energized circuit. (*Courtesy A. B. Chance Co.*)

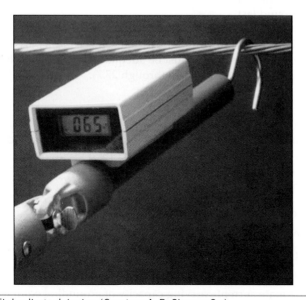

FIGURE 27.3 Digital voltage detector. (*Courtesy A. B. Chance Co.*)

A high-voltage voltmeter can be used to check for static or induced voltage if the lineman is unable to positively determine that the line is safe for grounding. The high-voltage voltmeter connected between ground and conductor energized with a static charge will give an indication of voltage and then may gradually indicate that voltage has reduced to zero, as the static charge is drained off to ground through the high-voltage voltmeter.

Protective Ground Installation

After the testing is completed, the protective grounds should be installed in the following sequence:

1. Connect one end of the grounding conductor to an established ground. The grounded neutral conductor on a distribution line and the grounded static wire on a transmission line are the best grounded sources. A driven ground rod at the worksite connected to the neutral conductor or static wire provides additional protection. If the work is being performed on a wood-pole structure, the installation of a cluster block on the pole below the work area connected to the grounding conductor, as shown in Fig. 27.1, will ensure the establishment of an equipotential zone on the pole for the lineman performing the work.

2. Connect the other end of the grounding conductor, using a hot-line tool, to the bottom conductor on vertical constructions or the closest conductor on horizontal constructions (Fig. 27.4).

3. Install grounds or jumpers from a grounded conductor to the ungrounded conductors in sequence until all conductors are grounded and short-circuited together.

When the work is completed and the protective grounds are to be removed, always remove the grounds from the line conductors in reverse sequence, removing the connection to ground last.

When a connection is made between a phase conductor and ground, the grounding lead, connection to the earth, and the earth itself become a part of the electric circuit. All parts of the grounding circuit must be adequate to provide protection to the workmen under severe conditions (Figs. 27.5 through 27.7).

Protective Ground Requirements

An inadequate or poorly installed grounding jumper can be a safety hazard. It can give a lineman a false sense of security without actually protecting him under severe conditions. The protection that is provided by a grounding jumper is as good (or as bad) as the care that is taken to make sure it is installed properly. The requirements of a good grounding jumper are as follows:

1. A low-resistance path to earth

2. Clean connections

3. Tight connections

4. Connections made to proper points

5. Adequate current-carrying capacity of grounding equipment

FIGURE 27.4 Lineman completing the installation of protective grounds using a hot-line tool. Grounding operation was completed by connecting the cluster block mounted on the pole below the work area to a driven ground rod. Then a grounding jumper was connected from the cluster block to one phase wire. Jumpering was completed in sequence from bottom to top, providing the lineman with a zone of protection as the protective grounds are installed. The grounded static wire at the top of the pole provides an effective ground source. (*Courtesy A. B. Chance Co.*)

A good available ground (earth connection) is the system neutral; however, it usually is not available on a transmission line. However, if there is a common neutral at the location where the line is to be grounded, it must be connected to the grounding system on the structure so that all grounds in the immediate vicinity are connected and bonded together.

If the structure is metal, the neutral should be bonded to the steel, either with a permanent connection or a temporary jumper. If the structure is a wood pole that has a ground-wire connection to the static wire, the neutral should be permanently connected to the ground wire. If there is no permanent connection, a temporary jumper should be installed. If there is no ground wire on the pole, the common neutral should be

FIGURE 27.5 Suggested maximum allowable fault currents for copper grounding cables. Total current-on time, for minimum off time between consecutive current-on periods-reclosures. These values exceed IPCEA recommendations for cable installations by 1.91. (Based on tests of 10-cycle to 60-cycle duration, 30°C ambient.) (*Courtesy A. B. Chance Co.*)

temporarily bonded to whatever temporary grounding connection is used to ground the line. Static wires are connected to earth at many points along the line and provide a low-resistance path for grounding circuits. These are normally available for use as the ground-end connection of the grounding leads. If the line has a static wire, it is usually grounded at each pole. A ground wire runs down the wood pole(s) and connects to the ground rod. All metal poles and towers are adequately grounded when they are installed and serve as a good ground-connection point, provided the connection between the metal structure and the ground rod has not been cut or disconnected. If there is an exposed lead from the structure to the ground, it should be checked for continuity.

If a neutral conductor or static wire grounded at the structure is not available, use any good metallic object, such as an anchor rod or a ground rod, which obviously extends several feet into the ground as a ground for the protective ground source.

If no such ground is available, the lineman must provide a ground by driving a ground rod or screwing down a temporary anchor until it is firm and in contact with moist soil.

FIGURE 27.6   Fusing current time for copper conductors, 40°C ambient temperature. (*Courtesy A. B. Chance Co.*)

Protective Ground Locations

A line is grounded and short-circuited whenever it is to be worked on using deenergized procedures. It is grounded before the work is started and must remain so until all work is completed.

Special devices are used to permit the grounding of conductors during stringing operations (Fig. 27.8). Deenergized lines are grounded for the protection of all the men working on the line. The protective grounds are installed from ground in a manner to short-circuit the conductors so that the linemen and everything in the working area will be at equal potential. This is the "man-shorted-out" concept, since the grounding, short-circuiting, and bonding leads will carry any current which may appear because of potential differences in the work area.

When a line is to be worked on deenergized, the line must be grounded and short-circuited at the work location, even though the work location is within sight of the disconnecting means used to deenergize the line.

When work is to be done at more than one location in a line section, the line section being worked on must be grounded and short-circuited at one location, and only the conductor(s) being worked on must be grounded at the work location.

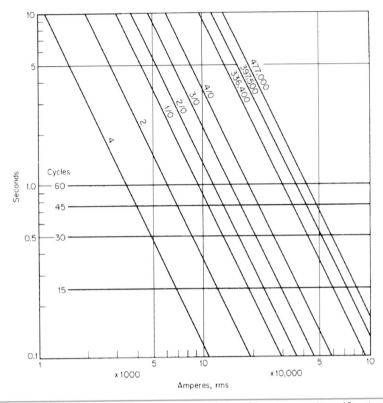

FIGURE 27.7 Fusing currents, aluminum conductor, 40°C ambient temperature. (*Courtesy A. B. Chance Co.*)

FIGURE 27.8 Running ground connection designed to keep the conductor grounded as stringing operations proceed. A rope is used to secure the connector in place. (*Courtesy Sherman & Reilly, Inc.*)

When a conductor is to be cut, or opened, the conductor must be grounded on both sides of the open or grounded on one side of the open and bridged with a jumper cable across the point to be opened (Fig. 27.9).

A cluster block can be installed on a wood-pole structure to facilitate the ground connections and to help maintain the equipotential area on the pole (Figs. 27.10 through 27.12).

Protective grounds installed on a steel transmission-line structure to connect conductors to ground and together are illustrated in Fig. 27.13.

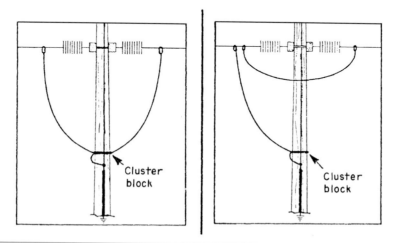

FIGURE 27.9   Method for grounding and bridging conductors across an open point is illustrated.

FIGURE 27.10   Metal cluster block equipped with 4-ft chain binder. The connection bar will accommodate four ground-lead clamps.

FIGURE **27.11**   Grounding and short-circuiting jumper equipped with C-type clamps designed for operation with hot-line tool. Jumper cable is 1/0 copper conductor or larger with a large number of strands to provide flexibility.

FIGURE **27.12**   Protective grounds installed on conductors of a single-circuit wood-pole subtransmission line.

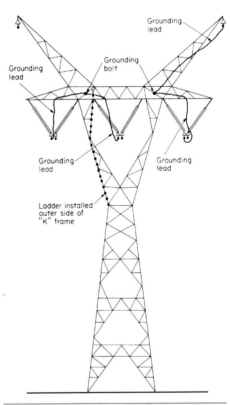

FIGURE **27.13**   Protective grounds installed on conductors of a single-circuit transmission line constructed on steel towers.

Underground-System Protective Grounds

Underground transmission and distribution cable conductors are not readily accessible for grounding. Transmission cables originating and terminating in substations can normally be grounded with portable protective grounding equipment at the cable termination points in the same manner as described for overhead transmission circuits. If the cable termination points are not accessible, special equipment manufactured and provided for grounding the circuit must be used. Underground distribution cables that originate in substation switchgear to supply complete underground circuits must be tested and grounded with special equipment provided at the substation before work is performed on the cable conductors. Underground equipment installed in vaults may be equipped with oil switches which can be used to isolate and ground the cable conductors. Underground cable circuits originating at a riser pole in an overhead line can be grounded at the riser pole with portable protective grounding equipment. Underground cable circuits originating in pad-mounted switchgear illustrated in Fig. 27.14 can be grounded at the switchgear near the cable termination. The switches in the switchgear must be open to isolate the cable circuit. Opening the switch provides a visible break in the circuit. A high-voltage voltmeter can be used by the lineman to be sure the circuit does not have a backfeed from the remote end, proving that the cables are deenergized. A special conductor is available above the cable terminations for the application of portable grounding devices in the switchgear.

Underground cable conductors terminating in pad-mounted transformers can be isolated, tested, and grounded at the transformer location. The lineman opens the pad-mounted transformer-compartment door and installs a feed-through device with a

FIGURE 27.14 Lineman has installed grounding jumpers with special grounding clamps on three-pole interrupter switch in pad-mounted switchgear. Note that insulating front barrier is inserted in open gap of three-pole interrupter switch to isolate incoming cables from bus and upper switch contacts. (*Courtesy S&C Electric Co.*)

FIGURE 27.15   Using hot-line tool to mount feed-through device on bracket in high-voltage compartment of pad-mounted transformer. (*Courtesy MidAmerican Energy Co.*)

hot-line Grip-All clamp stick to prepare for the insulating and grounding operation (Fig. 27.15). The cable to be worked on must be isolated to separate it from a source of energy at the remote end. The lineman or cableman identifies the cable to be worked on at the pad-mount transformer with the markings on the cable and maps of the circuit. When the proper cable has been identified, the elbow cable terminator can be separated from the bushing on the transformer with the Grip-All clamp stick. The lineman then installs the elbow on the bushing on the feed-through device with the hot stick (Fig. 27.16). A high-voltage tester equipped with a special connection can be used to test

FIGURE 27.16   Removing cable to be grounded from bushing with hot-line tool hooked to elbow-connector terminating cable and installing elbow connector on feed-through device. (*Courtesy MidAmerican Energy Co.*)

FIGURE 27.17   High-voltage voltmeter with special probe for testing cable circuit through feed-through device. (*Courtesy MidAmerican Energy Co.*)

the isolated cable through the second bushing on the feed-through device (Fig. 27.17). If the cable is found to be deenergized, the lineman can connect a special grounding conductor to the ground wires in the pad-mounted transformer compartment and place it on the bushing of the feed-through device with a Grip-All clamp stick (Figs. 27.18 through 27.20). The cable is now properly grounded so that the necessary work can be performed safely.

The *IEEE Guide for Protective Grounding of Power Lines* (IEEE Standard 1048) provides additional information about the installation of protective grounds.

FIGURE 27.18   Lineman testing cable circuit with high-voltage voltmeter through feed-through device. Note one lead of the high-voltage voltmeter is connected to grounding conductor in pad-mounted transformer compartment. (*Courtesy MidAmerican Energy Co.*)

FIGURE 27.19   Connecting special grounding jumper with elbow connector on one end and C clamp on the other end to grounded neutral cable in pad-mounted transformer compartment. C clamp is connected to ground lead. (*Courtesy MidAmerican Energy Co.*)

FIGURE 27.20   Lineman installs elbow connector of grounding jumper on feed-through-device bushing, grounding the cable. (*Courtesy MidAmerican Energy Co.*)

CHAPTER 28

Street Lighting

The modern streetlight system reduces traffic hazards, helps to prevent crime, and enhances the beauty of our surroundings. The incandescent streetlight lamp that was commonly used in the early part of the twentieth century was replaced by electric discharge lamps of the fluorescent, mercury-vapor, metal-halide, and high-pressure-sodium types. The mercury-vapor lamp had wide use because of its good efficiency, long life, and good light maintenance. Metal-halide and high-pressure-sodium lamps (Fig. 28.1) had a greater light output per watt of electric power input than the mercury-vapor lamps. Until recently, high-pressure-sodium lamps were the most prominent light source for new street-lighting installations. The street lighting industry is now in the energy efficient light-emitting diode (LED) era. A mass conversion of mercury vapor and high-pressure-sodium street lights to LED street lights is occurring.

The American National Standards Institute's publication ANSI/IES RP-8-21: *Lighting Roadway and Parking Facilities* addresses street-lighting design and installation of roadway lighting systems in the United States.

Street-Lighting Terms

Light. A form of radiant energy. Measurements are based on a unit of luminous intensity equal to the light emitted by a "standard candle" in a horizontal direction.

Lamp. A source of light.

Luminaire. The device which directs, controls, or modifies the light produced by a lamp. It consists of a light source and all necessary mechanical, electrical, and decorative parts.

Candlepower. The amount of light that will illuminate a surface 1 ft distant from the light source to an intensity of 1 footcandle (fc).

Lumen. The unit of luminous flux. It is equal to the flux on a unit surface, all points of which are at unit distance from a uniform point source of one candle. A uniform 1-candlepower source of light emits a total of 12.57 lumens.

Footcandle. The unit of illumination produced on a surface all points of which are 1 ft distant from a uniform point source of 1 candle, or the illumination of a surface 1 ft² in area on which a flux of 1 lumen is uniformly distributed.

Ballast. An auxiliary device used with vapor lamps to provide proper operating characteristics. It limits the current through the lamp and also may transform voltage.

Mast arm. An attachment for a pole used to support a luminaire.

551

FIGURE **28.1** High-pressure sodium streetlight installed on a wood pole in an area served by an overhead electrical distribution system.

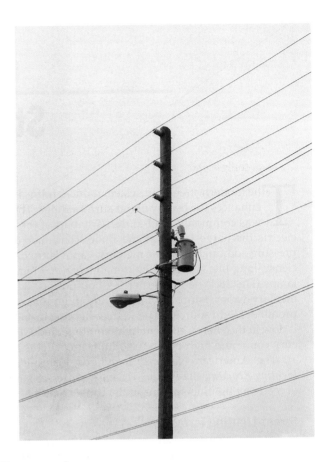

Street-Lighting System Components

A street-lighting system consists of the following:

1. Light-fixture support
   a. Mast arm
   b. Post

2. Streetlight grounding

3. Streetlight Luminaire
   a. Incandescent
   b. Electric discharge lamps
      1. Mercury vapor
      2. Metal halide
      3. Low-pressure sodium vapor
      4. High-pressure sodium vapor
      5. Fluorescent
   c. Light-emitting diodes (LEDs)

4. Light control
   a. Reflector
   b. Refractor
   c. Diffuser

5. Time control
   a. Manual
   b. Clock-operated time switch
   c. Photoelectric cell control
   d. Cadmium-sulfide-cell control
   e. Silicon Light Sensor Electronic Control
   f. Intelligent LED street lighting

Light-Fixture Support

The streetlight fixture can be supported above the street in two different ways:

1. Supported from a bracket attached to a pole, post, or mast arm (Fig. 28.2)

2. Supported on top of a post, streetlights installed in areas served with an overhead electrical distribution system usually are mounted on the poles supporting the overhead lines (Figs. 28.3 through 28.5). The length of the luminaire support or mast arm will determine the location of the luminaire with respect to the edge of the street (Table 28.1).

The mounting height of luminaires, by size, that should be achieved wherever practical:

Luminaire size	Type	Minimum mounting height, ft	Recommended pole height,* ft
100 watts	MV	25	30
175 watts	MV	25	30
250 watts	MV	25	30
400 watts	MV	30	40
1000 watts	MV	35	45
100 watts	HPS	25	30
150 watts	HPS	25	30
250 watts	HPS	30	40
400 watts	HPS	35	45

* For poles supporting street lights only.

Item	Description	Item	Description
A	Wood pole	H	Ground-rod clamp
B	Luminaire	I	Ground-wire moulding
C	Luminaire support (mast arm)	J	Split-bolt connectors
D	Insulated-secondary clevis	K	Machine bolt: 5/8"
E	Insulated cable	L	Ground-wire-moulding staples
F	Polyethylene-covered #6 AWG copper wire	M	Staples
		N	Square washer: flat
G	5/8" x 10' ground rod	O	Lag screws

FIGURE 28.2 Construction standard details for streetlight installed on a wood pole for installation in an area served by an overhead electrical distribution system.

FIGURE 28.3   High-pressure sodium streetlight with a 6-ft mast arm installed on a wood pole supporting a 13.2Y/7.62-kV distribution circuit.

FIGURE 28.4   High-pressure sodium streetlight with a 12-ft mast arm installed on a wood pole supporting a 13.2Y/7.62-kV distribution circuit.

The mounting height of luminaires, by size, that should be achieved wherever practical:

Luminaire size	Type	Minimum mounting height, ft
100 watts	MV	25
175 watts	MV	25
250 watts	MV	25
400 watts	MV	30
1000 watts	MV	35
100 watts	HPS	25
150 watts	HPS	25
250 watts	HPS	30
400 watts	HPS	35

Item	Description	Item	Description
A	Luminaire	F	Split-bolt connectors
B	Luminaire support (mast arm)	G	Machine bolt: 5/8"
C	Insulated-secondary clevis	H	Square washer: flat
D	Insulated cable	I	Lag screws
E	Polyethelyne-covered #6 AWG copper wire	J	Photoelectric control

FIGURE 28.5 High-pressure sodium streetlight construction standard detail for installation on wood pole supporting 13.2Y/7.62-kV distribution circuit.

Mast Arm Length	Street Width	Luminaire Location
3 ft	< 15 ft	0 ft to 6 ft out over the street surface
6 ft	< 40 feet	0 ft to 6 ft out over the street surface
12 ft	> 40 ft	6 ft to 12 ft out over the street surface

TABLE 28.1 Recommended Luminaire Support or Mast Arm Length

If the lighting circuit is placed underground along the street, the most convenient support for the lighting fixture is an ornamental pole or post. The supply circuit for the lamp enters the pole or post underground and passes up inside the hollow post or pole. The light fixture can be mounted directly on the top of the post, as illustrated in Fig. 28.6, or supported by a bracket or mast arm out over the street, as shown in Figs. 28.7 and 28.8. Construction details for various types of ornamental streetlight installations served by underground supply circuits are shown in Figs. 28.9 through 28.11.

FIGURE 28.6   Streetlight fixture mounted on top of an ornamental pole. Power is supplied to the lamp from a buried cable that extends up through the hollow center of the pole to the luminaire.

FIGURE 28.7   Streetlight fixture supported from pole. Power is supplied from underground circuit.

FIGURE 28.8   Streetlight luminaire attached to ornamental pole. Power is supplied from underground circuit.

Item   Description

A Pole, direct embedded, 20'
   luminaire mounting height.
   (1) Square laminated-wood pole,
       24'6" overall length.
   (2) Fiberglass pole, 24' overall
       length.
B Luminaire-mounting-adapter
   bracket for laminated-wood pole.
C Luminaire, colonial style, with
   100-watt, high-pressure sodium
   lamp.
D No. 6 AWG, insulated cable,
   2 conductor.
E No. 6 AWG, polyethylene-
   covered copper wire.
F Copperweld, 5/8" x 10' ground
   rod.
G Ground-rod clamp.
H Crushed rock or concrete.
I Plastic conduit elbow, 9¹/₂"
   radius.

FIGURE 28.9 Ornamental, direct-embedded post streetlight construction standard detail for installation supplied from underground wiring. Luminaire has a high-pressure-sodium lamp.

FIGURE **28.10** A 35-ft steel streetlight pole specification drawing. (*Courtesy American Electric Power System.*)

Area lighting has become common to enhance security and to permit nighttime sports activities. Figure 28.12 is an illustration of an ornamental pole and luminaire installation used to provide light for a parking lot. Figures 28.13, 28.14, and 28.15 show the installation of ornamental luminaires with high-pressure-sodium lamps mounted on ornamental poles.

Streetlight Grounding

All luminaire supports (mast arms) and luminaires installed on wood poles should be grounded to the neutral conductor, if available, or to a driven ground rod located near the base of the pole and connected to the mast arm by a polyethylene-covered No. 6 AWG copper wire. All metal poles, concrete poles, and metal U-guards and conduits must be grounded to a minimum height of 8 ft above ground, as required by the *National Electrical Safety Code* (ANSI C-2).

Supplied with pole   25'

Item	Description
A	Pole, with mast arm, galvanized steel, direct embedded, for 25' luminaire mounting height.
B	Luminaire, with high-pressure sodium lamp.
C	Plastic conduit elbow, 91/2" radius.
D	No. 6 AWG, insulated cable, 2 conductor.
E	No. 6 AWG, polyethlyene-covered copper wire.
F	Copperweld, 5/8" x 10' ground rod.
G	Ground-rod clamp.
H	Pole grounding terminal.
I	Crushed rock or concrete.

Hand hole

2'0"

2' min

Roadway Surface

5' min

FIGURE 28.11 Construction standard detail for installation of a streetlight luminaire mounted on galvanized steel pole with power supplied from underground circuit.

FIGURE 28.12 Ornamental pole and luminaire installed in parking lot of a hospital. High-pressure-sodium lamps are installed in the luminaires.

559

FIGURE 28.13 High-pressure-sodium lamps are installed in ornamental luminaires mounted on ornamental poles to light walkway.

FIGURE 28.14 Ornamental luminaires installed on ornamental poles for lighting the downtown shopping district of a community.

FIGURE 28.15   Ornamental streetlights on a main roadway leading to the downtown area. (*Courtesy MidAmerican Energy Company.*)

Streetlight Luminaires

Luminaires used for street lighting are:

1. Incandescent
2. Electric discharge
   a. Mercury vapor
   b. Metal halide
   c. Low-pressure sodium vapor
   d. High-pressure sodium vapor
   e. Fluorescent
3. LEDs

Incandescent Lamp

In an incandescent lamp, a wire filament is sealed in a glass bulb from which the air is extracted. Electric current flowing through the filament heats it to incandescence and produces light. It is necessary to heat the filament to about 5,000°F to produce full brilliance. The metal tungsten has a melting point of 6,120°F, well above the operating temperature, and is therefore well suited for use in incandescent lamps. The absence of oxygen prevents the filament from burning up. A mixture of inert gases is often put into the bulb. These gases carry any tungsten that evaporates from the filament to the top of the bulb.

Lamps are rated as to light output in lumens. A lumen is defined as one unit of light output. Incandescent lamps are operated at 120 volts and are rated in watts input as well as lumens output. Table 28.2 lists the incandescent lamps most commonly used.

Bulb Shape	Volts	Watts	Initial Lumens	Nominal Lumens	Rated Life, h
PS 25	120	189	2,920	2,500	3,000
PS 35	120	295	4,950	4,000	3,000
PS 40	120	405	6,900	6,000	3,000
PS 40	120	620	10,800	10,000	3,000

TABLE 28.2   Ratings of Incandescent Lamps in Common Use

Electric-Discharge Lamps

In electric-discharge lamps, electrons flow between electrodes through an ionized gas or vapor. All gaseous conduction lamps have a negative-resistance characteristic. This means that the lamp resistance decreases as the lamp heats up. As the resistance decreases, the current increases. In fact, the current will increase indefinitely unless a current-limiting device is provided. All gaseous conduction lamps, therefore, have current limiters called ballasts.

The main advantage of a gaseous-discharge lamp is that it gives out more light than an incandescent lamp of comparable rating. Mercury and sodium are the most economical chemicals used in electric-discharge lamps.

Vapor Lamps

The mercury-vapor lamp contains a small quantity of argon gas, which permits a starting discharge between the special starting electrode and one of the main electrodes. During the starting interval, the liquid mercury gradually vaporizes and eventually the full lamp current flows between the main electrodes, causing the mercury vapor to give off a considerable amount of light—extremely rich in blue, green, and yellow colors.

Where more normal color appearance of people or objects is important, improvement is obtained by the use of mercury lamps having fluorescent materials coated on the inside of the outer bulbs. The phosphors convert some of the invisible ultraviolet energy into visible light, in colors that supplement those from the mercury-vapor discharge.

A typical 400-watt mercury lamp is rated at 135 volts, and 3.2 amps and in its clear design gives out an average of 20,500 lumens initially when operated in a vertical position. When it is operated horizontally, its light output is slightly reduced, but this burning position is commonly used for street lighting because it permits a more effective distribution of light from the fixture. The mercury-vapor lamp has an output of approximately 50 lumens per watt and a life expectancy of more than 24,000 burning hours.

The metal-halide lamp is similar in construction and operation to the mercury-vapor lamp. The electrodes in the lamp are modified to permit use of the metallic halide compounds. Inert gases are placed within the arc tube and halogen compounds are introduced into the mercury, increasing the efficiency of the lamp. Ballasts must be constructed to produce a starting voltage higher than that normally utilized for mercury-vapor lamps. The lamp has an output of approximately 75 lumens per watt and a life of approximately 15,000 burning hours.

The low-pressure sodium-vapor lamp operates in a similar manner electrically. However, its light is limited to an intense yellow-orange color. This lamp type has little present-day use in the United States.

The high-pressure sodium-vapor lamp operates at high pressure and temperature. The lamp is elliptical and smaller than a mercury-vapor lamp of equal lumen output. The lamp is constructed with a ceramic arc tube utilizing sodium and other metallic traces and is contained by a borosilicate glass envelope. High voltage for starting the lamp is generated by high-impulse electronic equipment. The ballast is designed specifically for the high-pressure-sodium lamp and is not interchanged with ballasts for other electric-discharge lamps. The high-pressure sodium-vapor lamp has an output of approximately 100 lumens per watt and a life expectancy of approximately 24,000 burning hours.

Fluorescent Lamp

The fluorescent lamp is built in the shape of a long, slender glass tube with an electrode at each end.

In the fluorescent lamp, the flow of current takes place between electrodes through a mercury vapor as in a mercury-vapor lamp. The inside of the glass tube, however, is coated with a thin visible layer of fluorescent material called phosphor. This material will glow and give off visible light when struck by the electrons flowing through the mercury vapor between electrodes. The light produced is thus a secondary effect of the current flow. A "cool white" color is usually preferred, although other colors can be obtained by the use of other phosphors.

A small measured amount (about a drop) of mercury placed in the arc tube permits the lamp to operate, and a ballast in the form of a choke coil in the circuit limits the current. Lamps require starters to provide a temporary heating circuit. This circuit opens after a few seconds' operation.

The fluorescent lamp has an output of approximately 70 lumens per watt and a life expectancy of 18,000 or more burning hours.

Light-Emitting Diodes

Diode semiconductors were discussed in Chap. 1. The light-emitting diode (LED) is the process where when the electrons pass through the semiconductor junction from the p-type material to the n-type material and combine, photons are produced. The photons are the light source. The LEDs are integrated into the streetlight luminaire. The overwhelming reason for the conversion to LED lighting is economics. The LEDs are extremely energy efficient. The LED has been found to be up to 60% more efficient than the vapor lamps. The projected life of the LEDs is 10 years to 15 years. The failure mode is a noticeable decrease in luminous output. The LEDs immediately start up upon energization or re-energization.

Light Control

Bare lamps emit light in many directions. Light in the skyward direction is largely wasted, and crosswise light on streets is also partly wasted.

Streetlight fixtures (luminaires) at street intersections can be designed to direct their light equally in four directions, and luminaires along the sides of streets should emit light in two principal directions. Such light-distribution patterns are illustrated in Figs. 28.16 and 28.17. Figure 28.16a shows uniform distribution in all directions, and Fig. 28.16b shows a light pattern in four main directions. Figure 28.17 shows light distribution in two principal directions along the street.

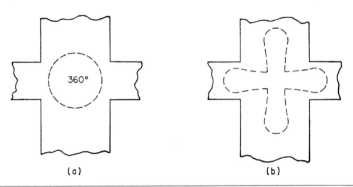

FIGURE 28.16   Light-distribution patterns at street intersection. (a) Uniform, (b) four-way.

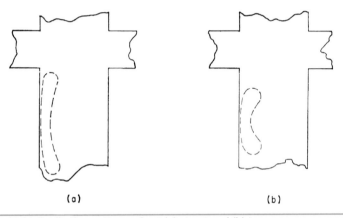

FIGURE 28.17   Light-distribution patterns along (a) street and (b) two-way.

The required control of the light emitted from the lamp is obtained by the use of reflectors, refractors, and diffusing globes or combinations thereof.

A reflector is a surface which stops the light ray and redirects it in the desired direction. Figure 28.18 shows a typical light fixture equipped with a reflector. Such a reflector will stop the light rays shining upward and redirect them downward to the ground. Reflectors may be made of porcelain, metal, or silvered glass. They are made in various shapes to give different types of light distribution.

FIGURE 28.18   Streetlight luminaire with lamp and reflector back of lamp shown. Refractor has been removed from luminaire. (*Courtesy General Electric Co.*)

A refractor is made of a transparent material, usually glass or acrylic plastic, which passes light rays but bends or deflects them in the desired direction. The enclosing globe is actually a series of optical prisms which bend the light rays. It is used with pendant luminaires, while dome refractors are used on upright post luminaires.

Refractors must be positioned correctly to get the desired light distribution in the horizontal plane. They are adjusted vertically with respect to the lamp filament.

Sometimes the enclosing globe is made to diffuse the light, that is, scatter the light in many directions. Diffusing the light reduces the glare by spreading the apparent light over a larger area at lower brightness. The emitted light from the lamp so lights up the globe that the globe itself appears to be the light source.

Special types of light fixtures often combine the several methods of light control in the same fixture. Such a combination fixture is shown in Figs. 28.19 and 28.20.

Figure 28.19 A LED luminaire using reflector and refractor for light control. Note photoelectric control mounted on top of luminaire.

Side view

23.25"
591 mm

4.38"
111 mm

Bottom view

11"
279 mm

Weight: 12.2 lbs
EPA: 0.53 sq. ft.

Figure 28.20 A detailed drawing of the LED luminaire utilizing reflector and diffusing bowl for light control.

Time Control

As the purpose of street lighting is to illuminate streets and intersections when natural lighting from the sun is not sufficient, the lights should be switched on as darkness approaches and off when daylight returns. This can be accomplished either manually or automatically by the following methods:

1. Manual
2. Clock-operated time switch
3. Photoelectric-cell control
4. Cadmium-sulfide-cell control
5. Silicon light sensor electronic control
6. Intelligent LED street lighting

Manual

This method requires that some member of the operating personnel be given the specific task of closing the lighting circuits at dusk and opening them at dawn, as well as at other daylight hours when unusual weather conditions indicate the need for artificial light. On series circuits, the switch on the primary side of the constant-current transformer is operated.

Clock-Operated Time Switch

Clock-operated time switches can be:

1. Motor driven
2. Spring driven, hand rewound
3. Spring driven, motor rewound

In case of a power interruption, the motor-driven time switch stops and has to be reset to the correct time manually. The spring-driven hand-rewound clock has to be rewound regularly. The spring-driven motor-rewound clock rewinds itself automatically. The last arrangement is the most satisfactory when clock-driven time switches are employed.

Time switches are usually equipped with astronomical dials which automatically change the turn-on and turn-off times to conform to the seasonal changes of sunrise and sunset. This gradual change is reflected in the shape of a cam which is driven by a small electric motor. The cam allows the contacts to be made or broken in accordance with the shape of the cam. The dial is adjusted at the factory for the locality in which the time switch is to be used.

Photoelectric-Cell Control

A photoelectric-cell-controlled relay may be employed to control street-lighting circuits by turning on the lights whenever the natural illumination falls below a given level. The photoelectric cell activates a relay which switches the lights on or off.

Cadmium-Sulfide-Cell Control

The cadmium-sulfide cell is commonly used for automatic street-lighting circuit control. This device can be mounted on the luminaire itself, thereby eliminating the need of control circuits wherever secondary voltage is available.

A diagram of the basic circuit for this type of control is shown in Fig. 28.21. Its operation is based on the variable resistance of the cadmium-sulfide cell rather than on the small current-generating characteristic of the photoelectric tube. The cadmium-sulfide cell has a very high resistance (about 15,000,000 ohms) when dark, but when exposed to sunlight, it has a relatively low resistance, less than 1,000 ohms. The cell is placed in series with the holding coil of the relay. Thus in the day-time the resistance of the cell is low, and sufficient current flows through the relay coil to hold the contacts open and keep the lights off. As evening approaches and the resistance of the cell begins to increase owing to the decreasing light level, the current of the holding coil decreases. When the natural light level reaches the control setting, the current of the holding coil lets the relay contacts close and the streetlight is energized.

The next morning, as the light level begins to increase, the resistance of the cell again is reduced. When the light level reaches the control setting, the current in the relay holding coil has become large enough to cause the relay to open the contacts and deenergize the light. The relay holds these contacts open again throughout the day until the following evening.

The controls are manufactured to fail on, as described, or fail off: the streetlight remains off if the control fails. The fail-on control has the advantage of the streetlight being on in daytime, indicating that the control has failed and that there is no interruption of streetlight output at night. The fail-off control has the advantage of eliminating daytime burning of the streetlight, thus conserving energy.

A typical cadmium-sulfide-cell control unit is shown in Fig. 28.22. The light sensor cell is located directly behind the window in the cover.

Silicon Light Sensor Electronic Control

The concept of silicon light sensor electronic control operation is similar to the cadmium-sulfide light sensor cell. The silicon light sensor electronic controls have peak sensitivity to infrared and red light. These electronic controls operate at a cool temperature that extends the life of the device. The control turns on instantaneously. The cell has MOV surge protection.

FIGURE 28.22   Assembled view of cadmium-sulfide cell streetlight control unit. Cadmium-sulfide cell is in lower control location directly behind window in cover. (*Courtesy American Electric Lighting.*)

Intelligent LED Street Lighting

The intelligent streetlight incorporates technology, such as cameras, light-sensing photocells and other sensors, to introduce real-time distribution automation monitoring functionalities. The intelligent streetlights are managed by a central management systems that will connect to the new LED streetlight installations.

Design of Street-Lighting Installations

For information on the design of street-lighting systems, the reader is encouraged to refer to the *Practice for Roadway Lighting* (ANSI/IES RP8), published by the American National Standards Institute (ANSI) under the sponsorship of the Illuminating Engineering Society. This provides information on the design of street-lighting systems and the measurement of their performance. Luminaire spacing factors such as city block lengths, type of roadway, lot lines, luminaire mounting heights are addressed.

Copies of the guide may be obtained from the Illuminating Engineering Society, 345 East 47th Street, New York, N.Y. 10017. An additional design guide is the Edison Electric Institute Street Lighting Manual.

CHAPTER 29

Underground System

Underground transmission and distribution systems are installed when

1. Space is not available for overhead lines, as in the congested downtown areas of large cities.

2. The appearance of numerous overhead lines would be unsightly, as in dense downtown areas or in new or redeveloped districts of cities.

3. Community ordinances, administrative codes, or franchise agreements require the installation of underground electric facilities in defined areas.

Underground lines cost more than the equivalent overhead lines. Underground lines are therefore used only when necessity demands them, as stated above.

Parts of Underground System

An underground system may consist of six parts:

1. Conduits or ducts
2. Manholes
3. Cables
4. Transformer vaults
5. Risers
6. Transformers

The first three parts listed above are illustrated in Fig. 29.1. The conduits or ducts are the hollow tubes connecting the manholes, one with another. The cable is the electrical power circuit placed in the conduit through which power flows, and the manhole is the chamber where cable ends are spliced together to make a continuous circuit. Transformer vaults (Figs. 29.2 and 29.3) are underground rooms in which the power transformers, network protectors, voltage regulators, circuit breakers, meters, etc. are housed. Cables terminate in transformer vaults or customers' substations or connect with overhead lines, at the cable riser.

FIGURE 29.1 Principal parts of an underground conduit system for underground electrical power transmission and distribution.

Conduits

Types

The hollow tubes or ducts running from manhole to manhole can be made of plastic polyvinyl chloride (PVC), fiberglass, concrete, or steel. Direct-buried plastic PVC is used in most cases. Plastic PVC conduit is available for direct-burial installation or encasement in concrete. The conduit may be purchased in 10 ft lengths or purchased as in coilable reels.

Sizes

Ducts are made in varying sizes with inside diameters normally from 2 inches to 6 inches. The size of duct selected depends on the size of the cable to be drawn into it or to be installed in the future. In general, when a single cable is installed in a duct, the diameter of the duct should be at least ½ inch to ¾ inches greater than the diameter of the cable. When more than one cable is drawn into one duct, the diameter of the duct should be at least ½ inch to ¾ inches greater than the diameter of the circle enclosing all cables. The duct opening should always be large enough to make the pulling in of the cable as easy as possible.

Duct runs should preferably be short, straight runs between manholes. This facilitates pulling in the cable. When this is impossible, as in going around obstacles, the bends should be laid out with the greatest possible radius of curvature, thereby avoiding sharp curves.

Duct runs should be slightly graded; that is, they should slope gently toward one or both manholes, thus allowing any water that may seep in to drain to one of the manholes. Figure 29.4 illustrates a duct run on level ground. It has a double slope; that is, half the run drains to each manhole. Good practice requires that the slope be no less than 3 inches for every 100 ft of length. Figures 29.5 and 29.6 illustrate the manner of providing proper grade on moderate and steep slopes. In these cases, the drainage is all in one direction.

In general, the top of a duct run should not be less than 2 ft below the street surface, as indicated in the diagrams.

Manholes

A manhole is the opening in the underground duct system which houses cable splices and which cablemen enter to pull in cable and to make cable splices and to complete cable tests. The manhole is therefore often called a splicing chamber or cable vault.

Floor plan

Section "E-E"

Top Plan

FIGURE 29.2 Typical transformer vault. (*Courtesy MidAmerican Energy Company.*)

Notes:
1. Manhole design shall meet all requirements for application in state and federal highways along with requirements of AASTO H-20 loadings.

2. Concrete anchors are to be placed in walls at these locations. Anchors are to accept 3/8 × 16 threads. Anchors are to be 2" deep.

FIGURE 29.3 Cross-sections of the typical vault in Fig. 29.2. (*Courtesy MidAmerican Energy Company.*)

FIGURE 29.4   Double-slope duct run on level ground. Water drains toward manholes from point midway between manholes.

FIGURE 29.5   Single-slope duct run between manholes on sloping street.

FIGURE 29.6   Single-slope duct run on street with steep grade. Note that duct run leaves at bottom of one manhole and enters near top of next manhole.

Manholes are generally built of reinforced concrete or brick, and the covers are made of steel. The opening leading from the street to the manhole chamber is called the chimney or throat. An opening having a minimum diameter of 32 inches is recommended. This size opening is large enough for a man to enter on a ladder and also to pass the equipment needed for splicing and testing. No manhole is to be entered until testing for combustible gases is complete and the atmosphere considered safe. The specific requirements are detailed in Chap. 46 in OSHA 1910.269 (e), Enclosed spaces.

Manholes are usually located at street intersections so that the ducts can be run in all four directions. The distance between manholes seldom exceeds 500 ft, as this is about as long a cable as can be pulled into the duct. In long blocks, additional manholes are placed in the middle of the block.

Two-Way Manholes

The shape and size of manholes depend on the number and size of cables to be accommodated and the number of directions in which ducts leave the manhole. The simplest shape of manhole is illustrated in Fig. 29.7, where ducts and cables enter and leave in only two directions. Such a manhole is called a two-way manhole. Note in Fig. 29.7b that the cables do not pass through the manhole in a straight line but instead are offset; that is, they enter and leave at opposite corners. This is to provide opportunity to rack the cables around the sides of the manhole. Cables expand and contract in length with

FIGURE 29.7   Plan views of two-way manhole. Note how cable is racked around sides of manhole with double-reverse curve and recommended dimensions shown.

changes in temperature caused by changes in electrical power flow through the cable. The most practical way to take care of this expansion and contraction is to provide a large reverse curve in the cable as it passes through the manhole. This reverse curve, consisting of two large-radius 90° bends, enables the cable to take up the expansion movements, thereby reducing to minimum the danger of damaging the cable.

A dimensioned side view of such a two-way manhole is given in Fig. 29.8. Note that the inside vertical height should not normally be less than 7 ft or more than 8 ft. These

FIGURE 29.8   Side view of two-way manhole. Ducts enter and leave at same level. (*Courtesy MidAmerican Energy Company.*)

FIGURE 29.9   Plan view (a) and side view (b) of manhole when ducts and cables enter and leave manhole at different levels.

dimensions allow for the racking of the cables with their splices around the side of the manhole and sufficient space and headroom for the splicer to work.

In case ducts and cables do not enter and leave the manhole at the same level, the manhole design would appear as in Fig. 29.9. Note the change in level of the cables as they pass through the manhole in the side view and the reverse curve of the cables in the plan view.

A two-way 90° turning manhole with cables at the same level would appear as in Fig. 29.10. Only one curve is needed here to allow for cable expansion. Such a manhole is usually called a corner manhole.

Three- and Four-Way Manholes

Manholes designed for three-way duct and cable lines are illustrated in Fig. 29.11, and manholes designed for four-way lines are illustrated in Fig. 29.12. The headroom in each case need not exceed 8 ft but should not be less than 7 ft.

Manhole Drainage

Drainage should be provided in manholes to keep them free of water. Oftentimes a "dry well" is resorted to. This requires that a part of the manhole floor be left open, the dirt excavated to a depth of a foot or two, and the hole filled with stones. This permits the water to seep away gradually. In case a connection to the city storm sewer line can be made conveniently, this is often done. Figure 29.13 shows a cross section or side view of a typical sewer connection. In areas along waterfronts where the soil is so moist or the water table so high that water will not drain out of a manhole, the only alternative remaining is to build the manhole walls and floor of waterproofed concrete. In addition, the concrete is usually painted with a waterproof paint. In extreme cases, an automatic pump is installed which removes the water from the sump below floor level.

FIGURE 29.10   Plan view (a) and side view (b) of two-way manhole when ducts and cables make a 90° turn. Such a manhole is usually called a corner manhole.

FIGURE 29.11   Possible variations of three-way manhole with recommended dimensions shown.

Distribution Manholes

When only a few trunk ducts are required for primary feeders to serve only the local distribution system, smaller manholes are satisfactory. The cables are usually smaller in diameter and therefore more easily racked. The dimensions for a two-way distribution manhole are given in Fig. 29.14. This manhole could accommodate a maximum of six trunk ducts. Note that the headroom is reduced to 7 ft. Most distribution manholes are made with precast concrete. The precast-concrete manholes are made at a remote location and hauled to the site and set in place. This method of construction for small manholes has proven to be economical.

PLAN

FIGURE 29.12   Possible variations of four-way manhole with recommended dimensions shown.

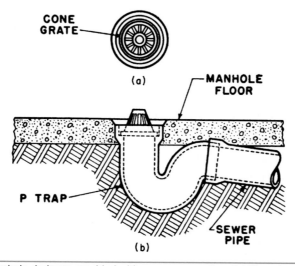

CONE
GRATE

(a)

MANHOLE
FLOOR

P TRAP

SEWER
PIPE

(b)

FIGURE 29.13   Manhole drainage provided with sewer connection. (a) Plan view; (b) elevation.

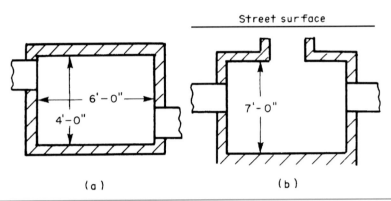

FIGURE 29.14 Two-way distribution manhole with recommended dimensions shown. (a) Plan view; (b) elevation.

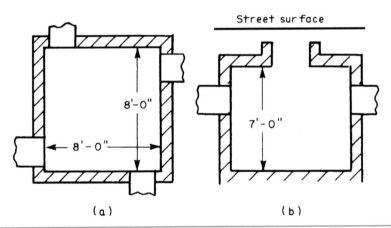

FIGURE 29.15 Four-way distribution manhole with recommended dimensions shown. (a) Plan view; (b) elevation.

A similar distribution manhole providing for primary and secondary circuits radiating in four directions at a street intersection is shown in Fig. 29.15.

The same type of cable manhole may be used for secondary distribution at locations where it is necessary to accommodate several circuits. This occurs frequently in congested areas adjacent to buildings requiring large and multiple service from either or both the primary and secondary circuits.

Handholes

When small splicing chambers are necessary on lateral two-way duct lines, so-called handholes are used. They should be restricted to one-conductor cables only because the cable makes a complete loop once around in the hole, as shown in Fig. 29.16. The handhole cover should be large enough to permit working from above the handhole.

FIGURE 29.16 Distribution handhole with recommended dimensions shown. (a) Plan view; (b) elevation.

Cable

The Insulated Cable Engineers Association (ICEA) is a professional organization dedicated to developing cable standards for electric power, control, and telecommunications industries. On systems operating above 2 kV to ground, the design of the conductors or cables installed in nonmetallic conduit should consider the need for an effectively grounded shield and/or sheath. Bending during handling, installation, and operation should be controlled to avoid damage to any of the components of the supply cable. Pulling tensions and sidewall pressures of cable should be limited to avoid damage to the supply cable. Manufacturer's recommendations may be used as a guide. Ducts should be cleaned of foreign material which could damage the supply cable during pulling operations. Cable lubricants should not be detrimental to cable or conduit systems. On slopes or vertical runs, consideration should be given to restraining cables to prevent downhill movement. Supply, control, and communication cables should not be installed in the same duct unless the cables are maintained or operated by the same utility.

Cable supports should be designed to withstand both live and static loading and should be compatible with the environment. Supports should be provided to maintain specified separation between cables. Horizontal runs of supply cables should be supported at least 3 inches above the floor or be suitably protected. This rule does not apply to grounding or bonding conductors. The installation should allow cable movement without destructive concentration of stresses. The cable should remain on supports during operation. Adequate working space should be provided. Where cable and/or equipment is to be installed in a joint-use manhole or vault, it should be done only with the concurrence of all parties concerned. Supply and communication cables should be racked from separate walls. Crossings should be avoided. Where supply and communication cables must be racked from the same wall, the supply cables should be racked below the communication cables. Supply and communication facilities should be installed to permit access to either without moving the other. Clearances should be maintained as specified in Table 29.1.

Phase-to-Phase Supply Voltage	Inches, Surface to Surface
0 to 15,000	6
15,001 to 50,000	9
50,001 to 120,000	12
120,001 and above	24

TABLE 29.1 Minimum Separation between Supply and Communication Facilities in Joint-Use Manholes and Vaults

These separations do not apply to grounding conductors, and they may be reduced by mutual agreement between the parties concerned when suitable barriers or guards are installed. Cables should be permanently identified by tags or otherwise at each manhole or other access opening of the conduit system. This requirement does not apply where the position of a cable, in conjunction with diagrams or maps supplied to workmen, gives sufficient identification. All identification should be of a corrosion-resistant material suitable for the environment. All identification should be of such quality and located so as to be readable with auxiliary lighting. Where cables in a manhole are maintained or operated by different utilities or are of supply and communication usage, they should be permanently marked as to company and/or type of use.

Insulation shielding of cable and joints shall be effectively grounded. Cable sheaths or shields which are connected to ground at a manhole shall be bonded or connected to a common ground. Bonding and grounding leads shall be of a corrosion-resistant material suitable for the environment or suitably protected.

Although fireproofing is not a requirement, it may be provided in accordance with each utility's normal service reliability practice to provide protection from external fire. Special circuits operating at voltages in excess of 400 volts to ground and used for supplying power solely to communication equipment may be included in communication cables under the following conditions:

1. Such cables shall have a conductive sheath or shields which shall be effectively grounded, and each such circuit shall be carried on conductors which are individually enclosed with an effectively grounded shield.

2. All circuits in such cables shall be owned or operated by one party and shall be maintained only by qualified personnel.

3. Supply circuits included in such cables shall be terminated at points accessible only to qualified employees.

4. Communication circuits brought out of such cables, if they do not terminate in a repeater station or terminal office, shall be protected or arranged so that in event of a failure within the cable the voltage on the communication circuit will not exceed 400 volts to ground.

5. Terminal apparatus for the power supply shall be so arranged that live parts are inaccessible when such supply circuits are energized.

6. Such cables shall be identified, and the identification shall meet the pertinent requirements.

The requirements do not apply to supply circuits of 550 volts or less which carry power not in excess of 3,200 watts.

Cable Construction

Power cables are constructed in many different ways. Cables are generally classified as nonshielded or shielded cables. Nonshielded cables normally consist of a conductor, conductor-strand shielding, insulation, and a jacket. The use of nonshielded power cables is limited to the lower primary voltages, which normally do not exceed 7,200 volts. Shielded power cables are generally manufactured with a conductor, conductor-strand shielding, insulation, semiconducting insulation shielding, metallic insulation shielding, and a sheath. If the sheath is metallic, it may serve as the metallic insulation shielding and be covered with a nonmetallic jacket to protect the sheath. Shielded cables are commonly used on circuits operating at 4,160Y/2,400 volts or higher.

Multiconductor cables used for three-phase power circuits consist of three single conductor cables with a common sheath and/or jacket.

Cable Conductors

Copper and aluminum are commonly used conductor materials. The conductors of large power cables have a stranded conductor to make a cable as flexible as possible, thus facilitating bending. Power cables with a conductor size of No. 2 AWG and smaller may use a solid conductor. Copper conductor has a high conductivity and is easily worked and handled. Aluminum conductor is lighter in weight and is commonly used to reduce the initial installation cost. Aluminum conductors must be larger than copper conductors to supply a given load because of the lower conductivity. Aluminum will "cold flow" if improper connectors are used. The surface of aluminum conductors oxidizes quickly and can be damaged by galvanic action when touching other metals with moisture present. The manufacturer's instructions must be followed carefully to prevent a defective splice or termination of aluminum conductors.

The stranded conductors have several shapes that are used for various power-cable constructions (Fig. 29.17).

Concentric-round-stranded conductors have wires of the same diameter. When six wires of the same size are helically wrapped around a single wire, they lie tightly to the inner wire. This progression of six additional wires in each layer will produce a conductor of any diameter.

The compact-round-stranded conductor has a smaller diameter than the concentric-round conductor but is less flexible. The smaller diameter is obtained by laying all the strands in the same direction and then rolling each layer to eliminate the spaces.

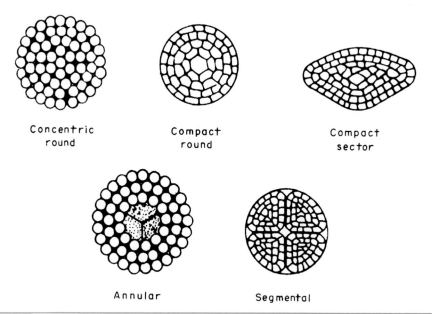

Concentric round   Compact round   Compact sector

Annular   Segmental

FIGURE **29.17**   Diagrams of common shapes of stranded conductors.

A compact-sector-stranded conductor is used to obtain a smaller-diameter cable. It is normally furnished with impregnated-paper or varnished-cambric insulation. Sector conductors are stiffer and more difficult to splice. However, they are cheaper and lighter in weight.

Annular conductors are usually composed of wires stranded concentrically around a ropelike cord. This conductor is generally used for sizes above 1,000 kCMIL to help combat the "skin effect" problem. It has been shown that alternating current flows more densely near the outer surface of a conductor; therefore, increasing the surface area increases the current-carrying capacity.

A segmented conductor is composed of three or four segments which are electrically separated from one another. This reduces the skin effect problem and results in small-diameter cable with high current-carrying ability.

Cable-Conductor Sealing

The space between the conductor strands in cables can be filled with a semiconducting compound during the manufacture of the cable conductor. The filling compound is designed to prevent water from penetrating the stranded-cable conductor during manufacture, storage, installation, splicing, and terminating of the cable.

Water in the interstices of the cable conductor is one of the principal causes of electrochemical trees that cause premature failure of polymer-insulated cables. Compound filling of the spaces between conductor stranding during manufacture of the conductor eliminates the possibility of water being trapped in the conductor.

Cable-Strand Shielding

Strand shielding is applied between the conductor and the cable insulation. It is made from materials such as conductive fibrous tapes, conducting paints, conducting rubber, and graphite compounds. The strand shielding should be securely anchored to the dielectric so that no ionizable air can exist between them if the cable is sharply bent. A semiconducting tape has proven that it can perform this function in addition to providing an easily stripped and clean conductor for splicing or terminating.

Cable Insulation

The two most common materials used as insulation for distribution cables are tree retardant cross-linked polyethylene (TRXLPE) and ethylene-propylene rubber base (EPR).

Tree retardant cross-linked polyethylene cable insulation has compounds consisting of the polyethylene resin, fillers, antioxidants, and vulcanizing agents. It has most of the electrical advantages of polyethylene plus much better mechanical properties such as superior resistance to heat deformation and environmental stress cracking. It should be noted that the insulation thickness used on these cables is dependent upon the distribution voltage and the fault clearing requirements (Table 29.2). Most cross-linked polyethylenes are rated at 90°C at 15 kV with a 100-hour per year overload temperature of 130°C for 5 years of its life. For the same voltage rating, the insulation thickness of the rubber type is generally thickest, cross-linked polyethylene the thinnest, and straight polyethylene somewhere between these two. The specific thickness for rubber and polyethylene are listed in the NEMA-IPCEA manual.

The Association of Edison Illuminating Companies[1] has written Specifications for Thermoplastic and Crosslinked Polyethylene Insulated Shielded Power Cables Rated 5

---

[1]Association of Edison Illuminating Companies, 51 East 42nd Street, New York, NY, 10017.

Distribution Voltage, kV	100% Insulation Level, mil	133% Insulation Level, mil
5	90	90
15	175	220
25	260	345
35	345	420

100% level	Ground fault will be cleared within 1 minute.
133% level	Fault-clearing requirements will not meet 100% criteria but fault will clear within 1 hour.

TABLE 29.2   Insulation Thickness

through 46 kV and Specifications for Crosslinked Polyethylene Insulated Shielded Power Cable Rated 69 through 138 kV. Experimental cross-linked polyethylene insulated cables have been developed for operation at 230 and 345 kV.

Cable Insulation Shielding

Insulation shielding confines the dielectric field within the cable to obtain symmetrical radial distribution of voltage stress within the dielectric. The shielding protects cables connected to overhead lines or otherwise subject to induced potentials and limits radio interference. If the shield is grounded, it reduces the hazard of shock and provides a ground return in case of a cable failure.

Conducting nonmetallic tape is used directly over the insulation of shielded power conductors to provide more intimate contact between the insulation and metallic shield. It is at least 0.00025 inch thick and consists of carbon-black-filled cloth, rubber, or neoprene. Extruded semiconducting polyethylene is used over the insulation of concentric neutral underground cables. The extruded jackets are superior to tapes in both durability and moisture-penetration resistance.

Insulation shieldings are either metal, tape, braid, or wire. In general, rubber-covered cable with a synthetic jacket should be shielded when applied to systems over 3,000 volts. Requirements for shielding are given in the following industrial specification: *Wire and Cable with Rubber, Rubber-like and Thermoplastic Insulations*, issued by IPCEA and NEMA. The metallic tape is generally 5 mils thick and is applied over the conducting nonmetallic tape.

Cable Bedding Tapes

Bedding tapes are generally used in a cable to add mechanical strength. They are located in various places by different manufacturers—under the sheath, under the shielding, and between the conductor and insulation. If one is not sure whether a tape is bedding or conductive, it should always be considered to be semiconductive and handled accordingly.

Cable Filler Tapes

Filler tapes of suitable materials generally are used in the interstices of cables, where necessary, to give the completed assembly a substantially circular cross section. Fillers are normally either treated hemp rope, synthetic rubber compound, or thermoplastic material.

Cable Jackets

The jacket over the cable insulation is a protective covering which provides physical protection.

Strand Filled, Tree Retardant, Cross-Linked Polyethylene Cable Specification

This specification covers a typical 1/0AWG Aluminum, 15-kV single conductor, strand filled, tree retardant cross-linked polyethylene, shielded, jacketed power cable. Application of this cable is for 60-Hz, Y-connected system with grounded neutral operating at 13,200Y/7,620 volts.

> The cable shall be suitable for installing in ducts or direct burial in the earth in wet or dry locations.
>
> Except as modified in this specification, the cable must meet or exceed the latest revision of the AEIC and the ICEA publications.
>
> Maximum conductor temperature shall be 90°C normal, 130°C emergency, and 250°C short-circuit.

Conductor

The phase conductor shall be 1/0AWG, 19 strand, Alloy 1350 aluminum. Stranding shall be class B stranded per ASTM B231 and semicompressed.

The phase conductor shall be strand filled. The strand fill compound shall be fully compatible with the shielding and insulation compounds, and not reduce the ratings of applicable connectors.

Conductor Shield

The conductor shield shall be Dow HFDA-0580, Borealis LE0594, or equivalent. The shield shall be firmly bonded to the insulation but free stripping from the conductor.

Insulation

The insulation shall be 175 mils of Dow HFDB-4202 EC, Borealis LE 4212 or an approved equivalent, extra clean, extruded, tree retardant, cross-linked polyethylene.

Insulation Shield

The insulation shield shall be Dow HFDA-063, Borealis LE310MS or an approved equivalent, semiconducting, extruded, strippable, cross-linked material.

Concentric Neutral

A full size concentric neutral consisting of a sufficient number of bare copper strands shall be spirally wound over the insulation shield, with uniform spacing between wires.

Moisture Barrier

Moisture blocking agents shall be applied in the interstices of the concentric neutral wires. A moisture blocking tape shall be applied over the concentric neutral wires, including two underlying rip cords, located approximately 180° apart, to assist jacket removal. The rip chords shall be located over the concentric neutrals while under the moisture blocking tape.

Jacket

The jacket shall be black, nonconducting linear low-density polyethylene over the concentric neutral and be free stripping. The compound shall be Dow DFDG-6059 or an approved equivalent. The thickness of the jacket over the concentric neutral shall be 70 mil minimum point.

Identification

Cable and center strand identification shall be in accordance with AEIC CS8. An electric cable symbol shall be marked on the outer jacket.

Conductor Tests

The conductor shall be tested in accordance with AEIC and ICEA standards.

Test Reports

Certified test reports of all cable will be required, one copy of test reports to be sent to the ordering party.

Shipping Lengths

The shipping lengths will be specified on the inquiry and purchase order. Shipping lengths shall have a tolerance of ±5%. Unless otherwise specified on the purchase order, the cable shall be furnished in one continuous reel.

Reels

Reels shall be nonreturnable, in compliance with NEMA WC-26, Tables 1-5 and 3-1, with maximum dimensions of 66-inch flange diameter and 36-inch traverse width. The cable shall be protected by a wrapping of corrugated plastic.

Cable Sealing

Durable, weather-tight, heat-shrink seals with internal mastic shall be applied to all cable ends.

Reel Marking

Each reel shall be marked with the purchaser's name, address, the purchase order number, description of the cable starting and ending sequential footage numbers, the net length of the cable, manufacturer's name, gross weight of the reel, and the net weight of the cable all printed on durable weather-resistant tags.

Cable Ampacity Tables

The normal and emergency ampacity ratings for direct-buried primary cable are available from the manufacturer or other published sources. The data in Table 29.3 provides aluminum and copper cable ampacity values that have been calculated based upon a set of given assumptions listed in the table.

Adjustments may be factored in due to temperature differences of the assumed rated cable ampacity values. Table 29.4 provides adjustment factors as only the cable temperature is varied from the assumed value upon which Table 29.3 is based.

Tables 29.5 and 29.6 provide multiplication factors due to various combinations of the ambient earth temperatures or cable temperature deviations from the given values.

Material, Application, & Size	Normal Conditions				Emergency Conditions				
	Max Amps	Maximum MVA at			Max Amps	Maximum MVA at			Assumptions
		12.47	13.2	13.8		12.47	13.2	13.8	
Aluminum									Earth Interface Temperature 60°C
Single-phase application									Normal Cable Temperature 90°C
2	165	3.6	3.8	3.9	190	4.1	4.3	4.5	Emergency Cable Temperature 130°C
1	190	4.1	4.3	4.5	220	4.8	5.0	5.3	Ambient Earth Temperature 20°C
1/0	215	4.6	4.9	5.1	245	5.3	5.6	5.9	Earth Resistivity 90°C-cm/W
2/0	250	5.4	5.7	6.0	290	6.3	6.6	6.9	Depth of Burial 36"
3/0	280	6.0	6.4	6.7	320	6.9	7.3	7.6	Daily Load Factor (Avg/Peak) 80%
4/0	320	6.9	7.3	7.6	370	8.0	8.5	8.8	Three-Phase Load Balanced
Three-phase application (triplexed arrangement)									Single-Phase Application Full Neutral (Concentric)
2	155	3.3	3.5	3.7	180	3.9	4.1	4.3	Three-Phase Application Reduced Neutral (Concentric)
1	175	3.8	4.0	4.2	200	4.3	4.6	4.8	Single-Conductor Cables with XLPE, TR-XLPE, or EPR Insulation
1/0	200	4.3	4.6	4.8	230	5.0	5.3	5.5	
2/0	230	5.0	5.3	5.5	265	5.7	6.1	6.3	
3/0	260	5.6	5.9	6.2	300	6.5	6.9	7.2	
4/0	290	6.3	6.6	6.9	335	7.2	7.7	8.0	
250	320	6.9	7.3	7.6	370	8.0	8.5	8.8	
350	385	8.3	8.8	9.2	445	9.6	10.2	10.6	
400	415	9.0	9.5	9.9	475	10.3	10.9	11.4	
500	460	9.9	10.5	11.0	530	11.4	12.1	12.7	
600	495	10.7	11.3	11.8	565	12.2	12.9	13.5	
750	550	11.9	12.6	13.1	635	13.7	14.5	15.2	
1,000	620	13.4	14.2	14.8	715	15.4	16.3	17.1	

Material, Application, & Size	Normal Conditions				Emergency Conditions				Assumptions
	Max Amps	Maximum MVA at			Max Amps	Maximum MVA at			
		12.47	13.2	13.8		12.47	13.2	13.8	
Copper									
Single-phase application									
2	215	4.6	4.9	5.1	250	5.4	5.7	6.0	
1	245	5.3	5.6	5.9	280	6.0	6.4	6.7	
1/0	275	5.9	6.3	6.6	315	6.8	7.2	7.5	
2/0	315	6.8	7.2	7.5	360	7.8	8.2	8.6	
3/0	355	7.7	8.1	8.5	410	8.9	9.4	9.8	
4/0	410	8.9	9.4	9.8	470	10.2	10.7	11.2	
Three-phase application									
2	200	4.3	4.6	4.8	230	5.0	5.3	5.5	
1	225	4.9	5.1	5.4	260	5.6	5.9	6.2	
1/0	260	5.6	5.9	6.2	300	6.5	6.9	7.2	
2/0	290	6.3	6.6	6.9	335	7.2	7.7	8.0	
3/0	330	7.1	7.5	7.9	380	8.2	8.7	9.1	
4/0	370	8.0	8.5	8.8	425	9.2	9.7	10.2	
250	405	8.7	9.3	9.7	465	10.0	10.6	11.1	
350	480	10.4	11.0	11.5	550	11.9	12.6	13.1	
400	510	11.0	11.7	12.2	600	13.0	13.7	14.3	
500	550	11.9	12.6	13.1	635	13.7	14.5	15.2	
600	580	12.5	13.3	13.9	665	14.4	15.2	15.9	
750	630	13.6	14.4	15.1	725	15.7	16.6	17.3	
1,000	700	15.1	16.0	16.7	805	17.4	18.4	19.2	

Courtesy MidAmerican Energy Company.

A multiplication factor of 1.15 is used to go from normal to emergency conditions. This approximately accounts for the effect of temperature on the conductor material and other cable components including dielectric losses. In accordance with AEIC standards, the 130°C operating temperature should be limited to 1,500 hours for the life of the cable and 100 hours in a 12-month period.

Table 29.3 Cable Ampacity Tables

		Cable								
		1/0 AL Set: Direct		4/0 AL Set: Direct		750 AL Set: Direct		750 AL Set: Duct		
Cable Temp. °C	Adjustment Factor	Amps	% Change	Amps	% Change	Amps	% Change	Amps	% Change	
75	0.908	182		263		499		422		
80	0.941	188	3.6%	273	3.6%	517	3.6%	437	3.6%	
85	0.971	194	3.2%	282	3.2%	534	3.2%	452	3.2%	
90	1.000	200	3.0%	290	3.0%	550	3.0%	465	3.0%	
95	1.027	205	2.7%	298	2.7%	565	2.7%	478	2.7%	
100	1.053	211	2.5%	305	2.5%	579	2.5%	489	2.5%	
105	1.077	215	2.3%	312	2.3%	592	2.3%	501	2.3%	
110	1.100	220	2.1%	319	2.1%	605	2.1%	511	2.1%	
115	1.122	224	2.0%	325	2.0%	617	2.0%	522	2.0%	
120	1.143	229	1.9%	331	1.9%	628	1.9%	531	1.9%	
125	1.162	232	1.7%	337	1.7%	639	1.7%	541	1.7%	
130	1.181	236	1.6%	343	1.6%	650	1.6%	549	1.6%	
135	1.200	240	1.5%	348	1.5%	660	1.5%	558	1.5%	
140	1.217	243	1.5%	353	1.5%	669	1.5%	566	1.5%	
145	1.234	247	1.4%	358	1.4%	679	1.4%	574	1.4%	
150	1.250	250	1.3%	362	1.3%	687	1.3%	581	1.3%	

Courtesy MidAmerican Energy Company.
Ambient Earth temperature   = 20°C
Earth Thermal Resistivity, RHO = 90°C-cm/W
Load Factor   = 100%
Depth of Burial   = 36" for direct-buried, 30" to top of duct

TABLE 29.4   Ampacity Values as the Cable Temperature Is Varied

Aluminum		Ambient Earth Temperature, T_A, °C						
		0	5	10	15	20	25	30
	75	1.060	1.024	0.987	0.948	0.908	0.866	0.821
	80	1.086	1.052	1.016	0.979	0.941	0.901	0.859
	85	1.111	1.078	1.043	1.008	0.971	0.933	0.893
	90	1.134	1.102	1.069	1.035	1.000	0.964	0.926
	95	1.156	1.125	1.093	1.061	1.027	0.992	0.956
	100	1.177	1.147	1.116	1.085	1.053	1.019	0.985
	105	1.197	1.168	1.138	1.108	1.077	1.045	1.012
Cable temperature T_c, °C	110	1.216	1.188	1.159	1.130	1.100	1.069	1.037
	115	1.234	1.207	1.179	1.151	1.122	1.092	1.061
	120	1.252	1.225	1.198	1.171	1.143	1.114	1.084
	125	1.268	1.243	1.217	1.190	1.162	1.134	1.106
	130	1.284	1.259	1.234	1.208	1.181	1.154	1.126
	135	1.300	1.276	1.251	1.225	1.200	1.173	1.146
	140	1.315	1.291	1.267	1.242	1.217	1.192	1.165
	145	1.329	1.306	1.282	1.258	1.234	1.209	1.184
	150	1.343	1.320	1.297	1.274	1.250	1.226	1.201

Courtesy MidAmerican Energy Company.

TABLE 29.5 Multiplication Factors as the Aluminum Cable Temperature and Ambient Earth Temperatures Are Varied from a Standard Cable Temperature of 90°C and an Ambient Earth Standard of 20°C

Copper		Ambient Earth Temperature, T_A, °C						
		0	5	10	15	20	25	30
	75	1.060	1.024	0.987	0.948	0.908	0.865	0.821
	80	1.086	1.051	1.016	0.979	0.940	0.900	0.858
	85	1.111	1.077	1.043	1.008	0.971	0.933	0.893
	90	1.134	1.102	1.069	1.035	1.000	0.964	0.926
	95	1.156	1.125	1.094	1.061	1.027	0.992	0.956
	100	1.177	1.147	1.117	1.085	1.053	1.020	0.985
	105	1.197	1.169	1.139	1.109	1.077	1.045	1.012
Cable temperature T_c, °C	110	1.217	1.189	1.160	1.131	1.100	1.069	1.038
	115	1.235	1.208	1.180	1.152	1.123	1.093	1.062
	120	1.253	1.226	1.199	1.172	1.144	1.115	1.085
	125	1.270	1.244	1.218	1.191	1.164	1.136	1.107
	130	1.286	1.261	1.235	1.209	1.183	1.156	1.128
	135	1.301	1.277	1.252	1.227	1.201	1.175	1.148
	140	1.316	1.293	1.269	1.244	1.219	1.193	1.167
	145	1.331	1.308	1.284	1.260	1.236	1.211	1.185
	150	1.345	1.322	1.299	1.276	1.252	1.228	1.203

Courtesy MidAmerican Energy Company.

TABLE 29.6 Multiplication Factors as the Copper Cable Temperature and Ambient Earth Temperatures Are Varied from a Standard Cable Temperature of 90°C and an Ambient Earth Standard of 20°C

The temperature adjustment formulas are:

Copper conductors

$$I' = I\sqrt{\frac{T'_C - T'_A}{T_C - T_A} \cdot \frac{234 + T_C}{234 + T'_C}} \quad \text{amps} \tag{29.1}$$

Aluminum conductors

$$I' = I\sqrt{\frac{T'_C - T'_A}{T_C - T_A} \cdot \frac{228 + T_C}{228 + T'_C}} \quad \text{amps} \tag{29.2}$$

where the primed (') values are the revised parameters.

Sample Calculation
A 90°C rated copper cable has a published ampacity of 500 amps under a given set of conditions which includes an ambient temperature (T_A) of 40°C. Find the ampacity at a conductor temperature of 80°C and ambient temperature of 50°C.
  Using Eq. (29.1):

$$I' = I\sqrt{\frac{T'_C - T'_A}{T_C - T_A} \cdot \frac{234 + T_C}{234 + T'_C}} \quad \text{amps}$$

$$I' = 500\sqrt{\frac{80 - 50}{90 - 40} \cdot \frac{234 + 90}{234 + 80}} \quad \text{amps}$$

$$I' = (500)\ (0.787) = 393 \text{ amps}$$

Temperature Correction Formula (*Information Courtesy Southwire Company.*)

Cable Riser
When it is desired to connect an underground cable to an overhead line, the cable is brought out of a manhole to a pole through a run of conduit which is usually extended 10 ft or more up the pole in metallic pipe or plastic duct for mechanical protection. If the cable is low voltage, it terminates in a so-called weather head, and if it is a high-voltage cable, it terminates in a so-called pothead. The whole assembly is called a riser. When the cables are single-conductor cables, they are fanned out just below the crossarm and terminate in single-conductor potheads. When the cable is a multiconductor cable, it terminates in a multiconductor pothead like that shown in Fig. 29.18. From the pothead the conductors spread out into the overhead line. The pothead is supported by means of bolts fastened to the pole or crossarm.

Pothead
A pothead (Fig. 29.18) provides for the connection between an underground cable and an overhead wire. It serves to separate the conductors from their close position to one another in the cable to the much wider separation in the overhead line without permitting arcing between them. It also seals off the end of the cable so that moisture cannot enter the insulation of the cable.

Figure 29.18   Three-conductor, cap-nut
porcelain pothead terminator. (*Courtesy
G&W Electric Specialty Co.*)

The pothead consists essentially of a pot, from which it gets its name, and a cover. The pot has a hole in the bottom through which the cable enters. The pot is attached to the sheath of the cable by a wiped joint or clamp. The conductors of the cable are made to flare out to give the necessary separation. The conductors of the cable are connected to terminals encased in porcelain attached to the cover. These projections of porcelain have "petticoats" on them the same as pin insulators and naturally serve the same purpose. The overhead-line conductors are connected to these terminals. The potheads are being replaced by cold shrink terminators.

Cold Shrink and Heat Shrink Terminators

With the introduction of molded rubber and polymer compounds, primary cables at the riser pole or in switchgear are terminated using cold shrink or heat shrink terminators. The cold shrink cable terminations contain one-piece, skirted, silicone rubber terminations. The cold shrink termination assemblies consist of a skirted insulator, high-dielectric constant stress control tube, conformable stress controlling compounds. The insulator is made of a dark gray silicone rubber with excellent tracking resistance and hydrophobic properties. The complete cold shrink assembly is prestretched and loaded onto a removable core. The heat shrink terminator is a heat-shrinkable heavy-wall tubing for insulating and sealing power cables and accessories. The terminator provides the electrical and physical properties of a cable combined with ruggedness and easy installation. On heating, the tubing recovers to a smaller diameter, fitting tightly over a wide range of cable sizes because of its high shrink ratio. The tubing's inner adhesive wall gives a dependable moisture seal.

Gas-Insulated Lines

Gas-insulated lines (GIL) are an integral component of the gas-insulated substation. Environment friendly GILs with a g3 gas mixture are replacing the sulfur hexafluoride (SF$_6$) gas lines. Single- and three-conductor rigid compressed-gas cable lines have been developed for operation at 345 kV ac. The three-phase cable costs less to fabricate and install than the isolated-phase compressed-gas cable. Experimental flexible compressed-gas cable lines for operation at 345 kV ac have been fabricated. The flexible cable line is easier to handle and ship, requires less excavation, and can be routed around obstacles when it is installed. It is estimated that the cost of flexible compressed-gas cable lines will be 25% less than the current costs for fabricating and installing rigid compressed-gas cable lines.

CHAPTER 30

Laying Conduit

A conduit is a structure containing one or more ducts. A duct is a single enclosed runway (hole) for conductors or cables. Conduit systems should be located so as to be subject to the fewest disturbances practical. Conduit systems extending parallel to other subsurface wires, cables, pipes, structures, and casings should not be located directly over or under other subsurface facilities. If this is not practical, the rules on clearances should be followed. Conduit alignment should be such that there are no protrusions which would be harmful to the cable. When bends are required, the minimum radius shall be sufficiently large to prevent damage to cable being installed in the conduit. The maximum change of direction in any plane between lengths of straight, rigid conduit without the use of bends should be limited to 5°.

Routes through unstable soils or highly corrosive soils should be avoided. If construction is required in these soils, the conduit should be constructed in such a manner as to minimize movement and/or corrosion. When conduit must be installed longitudinally under the roadway, it should be installed in the shoulder or, to the extent practical, within the limits of one lane of traffic. The conduit system shall be located so as to minimize the possibility of damage by traffic. It should be located to provide safe access for inspection or maintenance of both the structure and the conduit system. The top of a conduit system should be located not less than 36 inches below the top of the rails of a street railway or 60 inches below the top of the rails of a railroad.

Where unusual conditions exist or where proposed construction would interfere with existing installations, a greater depth than specified above may be required. Where this is impractical, this clearance may be reduced by agreement between the parties concerned. In no case, however, shall the top of the conduit or any portion of the conduit duct bank system extend higher than the bottom of the road bed section of coarse gravel. Submarine crossings should be routed and/or installed so they will be protected from erosion by tidal action or currents. They should not be located where ships normally anchor.

The clearance between a conduit system and other underground structures paralleling it should be as large as necessary to permit maintenance of the system without damage to the paralleling structures. A conduit which crosses over another subsurface structure shall have a minimum clearance sufficient to prevent damage to either structure (Fig. 30.1). These clearances should be determined by the parties involved. When conduit crosses a manhole, vault, or subway tunnel roof, it may be supported directly on the roof with the concurrence of all parties involved. Conduit systems to be occupied by communication conductors shall be separated from conduit systems to be used for supply systems by 3 inches of concrete, 4 inches of masonry, or 12 inches of well-tamped earth. Lesser separations may be used where the parties concur. If conditions require a

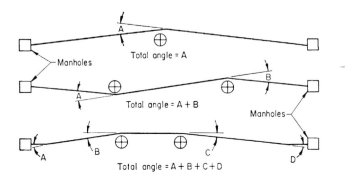

FIGURE 30.1 Typical outlines of duct runs showing manner of avoiding obstructions. Angles *A*, *B*, *C*, and *D* are degree deflections from straight line.

conduit to be installed parallel to and directly over a sanitary or storm sewer, it may be done provided both parties are in agreement as to the method.

Where a conduit run crosses a sewer, it shall be designed to have suitable support on each side of the sewer to avoid transferring any direct load onto the sewer. Conduit should be installed as far as practical from a water main in order to protect it from being undermined if the main breaks. Conduit which crosses over a water main shall be designed to have suitable support on each side as required to avoid transferring any direct loads onto the main. Conduit should have sufficient clearance from fuel lines to permit the use of pipe-maintenance equipment. Conduit and fuel lines shall not enter the same manhole. Conduit should be so installed as to prevent detrimental heat transfer between the steam and conduit systems.

The bottom of the trench should be undisturbed, tamped, or relatively smooth earth. Where the excavation is in rock, the conduit should be laid on a protective layer of clean, tamped backfill. All backfill should be free of materials that may damage the conduit system. Backfill within 6 inches of the conduit should be free of solid material greater in maximum dimension than 4 inches or with sharp edges likely to damage it. The balance of backfill should be free of solid material greater in maximum dimension than 8 inches. Backfill material should be adequately compacted.

Duct material shall be corrosion resistant and suitable for the intended environment. Duct material and/or the construction of the conduit shall be designed so that a cable fault in one duct would not damage the conduit to such an extent that it would cause damage to cables in adjacent ducts. The conduit system shall be designed to withstand external forces to which it may be subjected by the surface loading, except that impact loading may be reduced one-third for each foot of cover, so no impact loading need be considered when cover is 3 ft or more. The internal finish of the duct shall be free of sharp edges or burrs which could damage supply cable.

Conduit, including terminations and bends, should be suitably restrained by backfill, concrete envelope, anchors, or other means to maintain its design position under stress of installation procedures, cable-pulling operations, and other conditions such as settling and hydraulic or frost uplift (Figs. 30.2 and 30.3). Ducts shall be joined in such a way as to prevent solid matter from entering the conduit line. Joints shall form a sufficiently continuous smooth interior surface between joining duct sections that supply cable will not be damaged when pulled past the joint.

Steel coupling
Pipe clamp
Steel nipple
Plastic-to-steel adapter shall be at or below subgrade
Plastic 90° bend
Plastic coupling
Conduit to be encased in sand

FIGURE 30.2 Typical conduit-riser construction details for a conduit system buried directly in the earth.

Rerod welded to coupling
Steel coupling
Plastic-to-steel male adapter
Plastic nipple
Plastic coupling
Plastic 90° bend
1/2" rerod in each corner
Plastic coupling

FIGURE 30.3 Typical conduit-riser construction details for a conduit system encased in concrete. The terminals of some conduit systems buried directly in the earth are secured by encasing the ducts in concrete.

When conditions are such that externally coated pipe is required, the coating shall be corrosion resistant and should be inspected and/or tested to see that it is continuous and intact before backfilling is begun. Precautions shall be taken to prevent damage to the coating when backfilling.

Conduit installed through a building wall shall have internal and external seals intended to prevent the entrance of gas into the building insofar as practical. The use of seals may be supplemented by gas-venting devices in order to minimize building up of positive gas pressures in the conduit. Conduit installed in bridges shall be such as to allow for expansion and contraction of the bridge. Conduits passing through a bridge abutment should be installed so as to avoid or resist any shear due to soil settlement. Conduit of conductive material installed on bridges shall be effectively grounded. Conduit should be installed on compacted soil or otherwise supported when entering a manhole to prevent shear stress on the conduit at the point of manhole entrance.

Laying Out Line

In general, a conduit run should be as straight as possible. When this is impossible due to obstructions, the run is made up of combinations of straight sections and curved sections. The combinations of straight sections and curves can follow the typical outlines shown in Fig. 30.1 depending on the number and position of the obstructions to be avoided. After the outline for a particular set of conditions has been made, the total angular displacement is determined by adding the degrees at angles A, B, C, and D as illustrated in Fig. 30.1; thus

$$\text{Total angle} = A^\circ + B^\circ + C^\circ + D^\circ$$

If there is a displacement in a vertical plane also, this is included in the summation of the total angle. This total angle limits the length of the run between manholes, as shown in Fig. 30.4. The greater the total angle, the shorter the permissible duct length; and the smaller the angle, the longer the permissible length. In general, the maximum length of duct, even when entirely straight, is not much over 500 ft.

The radius of the curve connecting the straight sections should be as large as possible to facilitate the pulling in of the cable (Fig. 30.5). Plastic or polyvinyl chloride (PVC) duct is available with preformed bends and in short lengths with a small amount of curvature. If the curved section is to be made up of short, straight sections, these straight sections should be short enough to approximate a smooth curve. Table 30.1 shows the preferred lengths of sections for various lengths of radii of curvature.

FIGURE 30.4 Chart showing relation between total angular deflection of duct and permissible maximum length of run.

FIGURE 30.5   View showing a conduit run making a turn with a large radius of curvature. Note use of spacers to keep ducts in correct alignment.

Lengths of Radius of Curvature, ft	Lengths of Short Straight Sections of Conduit, inches
10–20	6
20–30	9
30–150	18
150 and over	36

TABLE 30.1   Preferred Lengths of Sections for Various Lengths of Radius of Curvature

Grading the Line

Conduit runs are graded so that any water seeping into the conduit will drain into the manhole. Good practice requires that the minimum grade be no less than 3 inches for every 100 ft of run.

Grades are established by an engineer with a level and usually marked with wood grade stakes driven 5 ft apart on the center line of the bottom of the conduit trench. The top of the grade stakes indicates the top surface of the bottom duct sheathing.

The methods of grading are double slope and single slope. Figure 29.4 illustrates the double-slope grade. This is the preferred method in the level areas as it reduces the distance any seep water needs to drain. The duct run is highest midway between manholes, coming to within 2 ft of the street surface, and grades off in each direction. Figure 29.5 illustrates single-slope grading. The entire conduit run is graded in only one direction. This is the logical method when the area slopes from one manhole to the other.

Manner of grading conduit run in order to pass under obstruction.

Sometimes obstructions also prevent the use of double-slope grading, as illustrated in Fig. 30.6. Here the single-slope grade is increased in order to have the conduit pass under the obstruction.

Arrangements of Ducts

The National Electrical Manufacturers Association (NEMA)[1] publication entitled *User's Manual for the Installation of Underground Plastic Duct* (Bulletin No. TCB2) discusses the procedures in detail. Ducts can be arranged in the trench in various ways, as shown in Fig. 30.7, depending on the number of ducts actually needed and the number of spares to be provided for future load growth. A rule often followed is to double the number initially filled with cable. This would therefore provide for a 100% increase in load.

Since cables carrying electric power become heated from the electrical losses in the conductors, they should be so installed that cooling can take place by radiation. When a large number of cables are close together, the ability to radiate heat is greatly reduced. Hence the number of cables in duct run should be kept low. A recommended maximum of ducts in a single run is nine. When a greater number is used, the heat radiation per cable becomes greatly reduced, and the total power that can be carried in the conduit run is only slightly increased.

Excavating the Trench

The trench for the conduit run is first marked off. If a digging machine is to be used, only one side of the trench need be marked. Cuts in a paved street must conform to the shape of the required excavation. The width of the cut is made equal to the overall width of the conduit run plus necessary clearance for the digging equipment. The sides of the excavation are shored where necessary. The trench is dug true to the marked line with perpendicular sides. The sides are trimmed smooth to provide for a uniform sheath of concrete around the ducts if they are to be encased. The entire trench is opened

Various arrangements of ducts in a conduit run. Interior ducts are sometimes called dead ducts because they do not radiate heat as well as outside ducts. They are therefore generally used for control cables.

[1]National Electrical Manufacturers Association, 155 East 44th Street, New York, N.Y. 10017.

between manholes or termination before any duct is laid in order to determine if any obstructions exist that must be bypassed.

When the required depth of the trench is known for all points along its length, the grade may be established. The bottom of the trench should be relatively smooth, undisturbed earth, well-tamped earth, or sand. Trenches which have been dug too deep at any point are partially refilled and tamped solid.

Directional Boring

In some instances, an open cut trench is not an option. Perhaps the conduit system must cross under a paved roadway, a body of water, or an environmentally sensitive area. In these instances, directional boring techniques are utilized to complete the conduit installation. Boring is accomplished utilizing directional boring construction equipment (Fig. 30.8). The equipment controls a guided boring "missile-shaped" tool that is attached to the lead end of a drill rod. The 10-ft sections of drill rod are continuously connected together as the tool advances beneath the ground's surface, from the starting point to the end point, without disturbing the ground-line between the two points. In some cases, small excavation pits are utilized at either the starting or end points. This technique minimizes the clean up required over the course of the conduit route. High-pressure fluid jets on the head of the directional boring tool provide the equipment operator the ability to control the device as it advances underground. Once the device has completed its desired route, backreaming tools are utilized to solidify the tunnel or place the conduit system. The soil conditions dictate the best directional boring technique to use. In loose soils, it may be necessary to utilize drilling fluids such as bentonite to hold the bored hole to reduce the occurrence of collapsed soil until the conduit system is installed. The bore lengths range in distance from 75 ft to distances as long as 1,200 ft. Prior to beginning directional boring, it is important to determine and locate any known underground facilities or obstructions that are installed and must be avoided.

FIGURE 30.8   A directional boring machine being used to bore a hole from the start point to the end point without disturbing the ground-line in between these two points. Note the ten-foot sections of drill rod on the side of the equipment that are utilized to extend the boring tool. (*Courtesy Vermeer Manufacturing Co.*)

Cutting the Duct

Plastic and fiberglass ducts can be cut with a hand saw, power saw, or rotary cutter. The duct must be cut square. Remove the burrs left by sawing and the ridges left by a rotary cutter with a file or a knife. Sharp edges on the outside or inside of the duct must be removed by beveling the edge of the duct. If the burrs and ridges are not removed, a cable may be damaged when it is pulled into the duct. If burrs and ridges are not removed, a defective joint may result.

Joining the Duct

Surfaces of the duct to be joined must be clean and free from dirt, foreign materials, and moisture. Clean the outside surface of the dust spigot end for the depth of the socket and the inside surface of the socket with a clean, dry cloth.

Fiberglass ducts are manufactured with a bell or socket on one end which is equipped with a gasket. The spigot end of one duct is pushed into the bell end of the next duct. The gasket in the bell end of the fiberglass duct provides the seal.

Plastic (PVC) duct joints are sealed with cement. Solvent cements are specified for each thermoplastic duct material and should not be interchanged between duct materials. Combination cements are available for joining ducts and fittings of different materials. The solvents used in the cements are flammable. Fumes from the solvents can be harmful if adequate ventilation is not provided. Forced ventilating equipment should be used while cementing ducts in manholes, vaults, or enclosed areas. Open flames, smoking, and exposed heating elements must be avoided while applying the cement to the ducts. The cement should be applied with a brush or applicator to easily and rapidly coat the surfaces of the ducts to be joined. While the spigot surface of the duct is still wet with cement, insert the duct spigot with a slight twisting motion into the socket until it bottoms home at the socket shoulder. The bonding of the duct joint begins immediately and should be completed within 15 sec. After assembly, wipe any excess cement from the duct at the end of the socket to prevent excess cement from dripping on and possibly weakening a lower duct.

Laying Ducts

The laying of ducts is begun midway between manholes where duct spacing is standard and then working toward the manholes where the ducts may flare out. If the duct run is located between a termination such as a riser (Fig. 30.2) and a manhole, the laying of ducts is started at the termination and laid toward the manhole. Manufactured plastic spacers are normally used to support the ducts, to give them proper spacing for cooling, and to secure the ducts so that they will not float or move in any direction (Fig. 30.9).

(a)   (b)

FIGURE 30.9   Plastic base spacer (a) and intermediate spacer (b).

FIGURE 30.10   Assembled bank of ducts being installed. Note use of spacers to hold ducts in proper position.

If the ducts are to be encased in concrete, the manufactured spacers provide the proper clearance from the bottom of the trench and the proper spacing between ducts for encasement (Fig. 30.10). The use of bricks or wood is not recommended because these materials may deform the duct wall. The spacers are normally placed approximately 5 ft apart to provide adequate support during the placement of concrete or backfilling of the trench with soil. The bottom layer of ducts is placed between the terminals of the conduit run, which may be any combination of manholes, handholes, vaults, or risers, before the second layer is started. It is recommended that the end ducts that do not terminate in a riser be full length. The ducts should terminate in a manhole, handhole, or vault with end bells that are grouted in place or manufactured duct terminators which are designed for installation into the walls to create a watertight duct-entry system (Fig. 30.11). The spacers normally provide for anchoring the ducts into position after all have been installed. Some manufactured spacers provide vertical openings on either side of the spacers of the duct assembly through which reinforcing rods can be passed and driven into the trench floor. The reinforcing rods will secure the duct assembly while concrete is being poured for encasement or while earth is placed around the ducts. If the duct assembly cannot be secured with reinforcing rods, a trench jack may be placed directly over the spacer location and wedges used between the trench jacks and the spacers to hold the ducts in place while the concrete is poured or the earth placed in the trench. The open ends of the ducts should be closed with plastic plugs to prevent dirt, animals, or other objects from entering and becoming lodged. This should be done whenever the work is halted for any length of time, even for overnight. Figure 30.12 is a cross-sectional view of ducts laid in a trench in preparation for backfilling with earth or encasement with concrete.

FIGURE 30.11 Conduit run terminated in a handhole. The end bells provide an outlet to the handhole with rounded edges to prevent cable damage.

FIGURE 30.12 Ducts resting on spacers installed in a trench.

Burial of Ducts in Earth

Plastic or PVC duct is manufactured in different wall thicknesses and strengths. The extra-strength plastic ducts can be directly buried in the ground in many locations, eliminating the need for and cost of concrete encasements. The ANSI/NEMA publication entitled *Extra-Strength PVC Plastic Utilities Duct for Underground Installation* (ANSI/NEMA TC8) provides detailed information on ducts. The extra-strength plastic ducts are available with inside diameter measurements of 1 inch through 6 inches. The ducts are available in type EB for encased burial in concrete and type DB for direct burial without encasement. A suitable backfill properly placed is critical for a satisfactory direct-buried duct installation. A stable backfill material should be selected. Sand and small gravel are the best materials. Sand and gravel mixed with silts or clays in which the sand and gravel amount to at least 50% of the mixture are satisfactory. Rock-free sandy loam and small crushed quarry stone are suitable. Soils of highly organic content, identified by odor or spongy feel, or highly plastic clays are not suitable for backfilling. Each soil layer should be compacted to 85% to 90% of its maximum density. Care should be taken when compacting to avoid excessive tamping in order to prevent shifting or

Warning tape

6" above conduit

Intermediate spacers

Sand

Conduit spacing will vary with conduit size (this drawing typical for 4" conduit)

Base spacers

FIGURE 30.13 Cross-sectional view of a completed conduit installation with ducts directly buried in the earth.

distorting of the ducts. Vibratory methods are the preferred method of compacting sand and gravel. The fill should be in a nearly saturated condition while the compaction is in progress. After the initial cover of select backfill over the duct bank has been completed, the remaining trench should be backfilled to grade with sufficient depth to obtain the required density when tamped (Fig. 30.13).

Encasement of Ducts in Concrete

The concrete envelope surrounding an encased duct structure is usually 3 inches thick on the top, bottom, and sides. The concrete mixture used is generally one part cement, three parts clean sand, and five parts broken stone or gravel. The maximum size aggregate should be 3/8 inch and the slump should be 7 inches to 9 inches to ensure proper distribution of the concrete around the ducts. The concrete should be well mixed and not too wet. Very wet concrete has a tendency to float the ducts and change the alignment. Care should be taken to limit the fall of the concrete to a minimum height from the chute to the top tier of ducts to minimize flotation effects. It is recommended that the pouring of the concrete start at the point the ducts terminated in a manhole or at another end of the conduit run where the ducts are securely restrained. The concrete is poured around the ducts as soon as possible after they have been placed to protect them from mechanical damage. Continuous spading is necessary to ensure a flow of concrete between and under the individual ducts. A long, flat tool or spatula worked carefully up and down between each vertical line of ducts will help to eliminate voids. Power-driven tampers should not be used for this purpose. The pouring is continued to a height of 3 inches above the top layer of ducts. If a trench jack and wedges were used to secure the ducts, the wedges should be removed before the concrete sets and the voids left by them filled with concrete and spaded.

After the concrete has taken its initial set, the trench is backfilled. The dirt should be placed in the trench in 6-inch to 12-inch layers, and each layer carefully tamped. In case a pavement was cut, this is then replaced. Figures 30.14 and 30.15 are examples of a

FIGURE 30.14   Direct buried
distribution feeders exiting new
switchgear box pads. Five
distribution feeders exiting at a
transition from conduit to direct
buried. Direct buried distribution
feeders are randomly placed in
the trench.

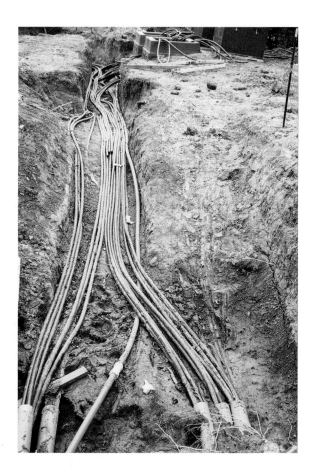

project that was originally installed with direct buried feeder cable for a portion of the
job. Figure 30.16 is the re-engineered project under construction using two distinct
2-duct × 2-duct conduit systems.

Steel Encasement of Ducts

Plastic (PVC) ducts can be installed in driven or bored steel casings. Steel casings
are frequently used when crossing under highways and railroads. A steel casing may be
used to support a conduit run in soft soil provided proper bearing support is available
at the pipe terminals. The steel casing used should be strong enough not to bend
between supports. The extra-strength ducts normally used for direct burial should be
used inside the steel casing.

The space between the ducts and the steel casing must be filled to provide a means
of dissipating the heat in the power cables installed in the ducts. A cement grout used
to fill the voids in the casing contains an additive called Elastizell or Mearlcrete which
reduces the grout density to approximately 75 lb/ft³. The cement grout should have
about an 11-inch slump, which is fluid enough to allow low hydraulic pumping pressure.
The steel casing should be filled from both ends, and the hoses should be withdrawn as

FIGURE 30.15   The trench for the direct buried distribution cable is at a lower depth in and around the switchgear.

FIGURE 30.16   The placement of two distinct 2-duct × 2-duct, 6-inch conduit systems to provide separation for distribution feeder cables.

back-pressure develops while the grout is being pumped. Ducts in a cradle are properly separated to permit the installation of the cement grout. Special care must be taken not to deform the ducts while installing them in the steel casing and while pumping in the grout to fill the casing.

Flexible-Plastic Ducts

Plastic (PVC) utility duct is available in smooth-walled and corrugated form, packaged in coils or on reels. ANSI/NEMA Publication No. TC7 provides detailed information on smooth-walled, coilable polyethylene electrical plastic duct. The smooth-walled, coilable plastic ducts are available with inside-diameter dimensions of ½ inch through 3 inches. ANSI/NEMA Publication No. TC5 provides detailed information on corrugated-polyolefin coilable plastic utilities duct. The corrugated-plastic ducts are available with inside-diameter dimensions of 1¼ inches through 5 inches. The manufacturer's instructions should be followed for proper installation of the flexible-plastic ducts. Duct may be unreeled directly into the trench for installation. The flexibility of the ducts permits their installation around corners without the use of prefabricated bends or the splicing of short sections of duct. The flexible ducts can be installed with plowing equipment if the soil is appropriate.

CHAPTER 31

Manhole Construction

Manholes, handholes, and vaults shall be designed to sustain all expected loads which may be imposed on the structure. The horizontal and/or vertical design loads shall consist of dead load, live load, equipment load, impact load due to water table or frost, and any other load expected to be imposed on and/or occur adjacent to the structure. The structure shall sustain the combination of vertical and lateral loading that produces the maximum shear and bending moments in the structure.

In roadway areas, the live load shall consist of the weight of the moving tractor semi-trailer truck illustrated in Fig. 31.1. The vehicle wheel load shall be considered applied to an area as indicated in Fig. 31.2. In case of multilane pavements, the structure shall sustain the combination of loadings which result in vertical and lateral structure loadings which produce the maximum shear and bending moments in the structure. Loads imposed by equipment used in road construction may exceed loads to which the completed road may be subjected.

In designing structures not subject to vehicular loading, the minimum live load shall be 300 lb/ft². Live loads shall be increased by 30% for impact. When hydraulic, frost, or other uplift will be encountered, the structure shall be either of sufficient weight or so restrained as to withstand this force. The weight of equipment installed in the structure is not to be considered as part of the structure weight. Where pulling-iron facilities are furnished, they should be installed with a factor of safety of 2 based on the expected load to be applied to the pulling iron.

A clear working space sufficient for performing the necessary work shall be maintained. The horizontal dimensions of the clear working space shall not be less than 3 ft. The vertical dimensions shall not be less than 6 ft, except in manholes, where the opening is within 1 ft horizontally of the adjacent interior side wall of the manhole. Where one boundary of the working space is an unoccupied wall and the opposite boundary consists of cables only, the horizontal working space between these boundaries may be reduced to 30 inches. In manholes containing only communications cables and/or equipment, one horizontal dimension of the working space may be reduced to not less than 2 ft provided the other horizontal dimension is increased so that the sum of the two dimensions is at least 6 ft.

Round access openings in a manhole containing supply cables shall be not less than 26 inches in diameter. Round access openings in any manhole containing communication cables only, or manholes containing supply cables and having a fixed ladder which does not obstruct the opening, shall be not less than 24 inches in diameter. Rectangular access openings should have dimensions not less than 26 inches by 22 inches. Openings shall be free of protrusions which will injure personnel or prevent quick egress.

Variable spacing–14 ft. to 30 ft. inclusive

Spacing to be used is that which results in vertical and lateral structure loading which produces the maximum shears and bending moments in the structure

FIGURE 31.1   Vehicle loading requirements for manhole construction.

FIGURE 31.2   Vehicle wheel loading requirements for manhole construction.

When not being worked in, manholes and handholes shall be securely closed by covers of sufficient weight or proper design that they cannot be easily removed without tools. Covers should be suitably designed or restrained so that they cannot fall into manholes or protrude into manholes sufficiently far to contact cable or equipment. Strength of covers and their supporting structures shall be at least sufficient to sustain the applicable loads.

Vault or manhole openings should be located so that safe access can be provided. When in the highway, they shall be located outside the paved roadway when practical. They should be located outside the area of street intersections and crosswalks whenever practical to reduce the traffic hazards to personnel working at these locations. Personnel access openings in vaults or manholes should be located so that they are not directly over the cable or equipment. Where these openings interfere with curbs, etc., they can be located over the cable if one of the following is provided: a conspicuous warning sign, a protective barrier over the cable, or a fixed ladder. In vaults, other types of openings may be located over equipment to facilitate work on this equipment.

Where accessible to the public, access doors to utility tunnels and vaults shall be locked unless qualified persons are in attendance to prevent entry by unqualified persons. Such doors shall be designed so that a person on the inside may exit when the door is locked from the outside. This rule does not apply where the only means of locking is by padlock and the latching system is so arranged that the padlock can be closed on the latching system to prevent locking from the outside. Fixed ladders shall be corrosion resistant.

Where drainage is into sewers, suitable traps or other means should be provided to prevent entrance of sewer gas into manholes, vaults, or tunnels.

Adequate ventilation to open air shall be provided for manholes, vaults, and tunnels having an opening into enclosed areas used by the public. Where such enclosures house transformers, switches, regulators, etc., the ventilating system shall be cleaned at necessary intervals. This does not apply to enclosed areas under water or in other locations where it is impractical to comply.

Supply cables and equipment should be installed or guarded in such a manner as to avoid damage by objects falling or being pushed through the grating.

Manhole and handhole covers should have an identifying mark which will indicate ownership or type of utility.

Description

Manholes are usually made of brick, concrete blocks, reinforced concrete, precast concrete, precast plastic, fiberglass, or a combination of these materials. The manhole cover is made of steel. As a rule, manholes are constructed at the manufacturer. It is more economical and convenient to precast the manhole and transport it to the location in the street. This procedure eliminates the field parking of materials, pouring of concrete, etc., in crowded streets and avenues.

Precast manholes can be fabricated in various configurations such as a straight manhole, an L manhole, and T manhole (Figs. 31.3 through 31.5). Large manholes are usually precast in sections or in pieces (Fig. 31.6). Manholes for direct-buried plastic conduit systems are often fabricated with precast concrete, plastic, or fiberglass pipe with a minimum internal diameter of 60 inches and a minimum depth of 7 ft. Even larger manholes are utilized in heavily congested areas (Figs. 31.7 and 31.8). The plastic conduits entering the manhole are terminated with an end bell and grouted if precast

FIGURE 31.3 Straight-type precast cable manhole. Approximate weights: top section 14,800 lb and bottom section 14,650 lb. Manhole is designed for use with 12-duct conduit run. (*Courtesy Commonwealth Edison Co.*)

FIGURE 31.4 An L-type precast cable manhole. Approximate weight 30,000 lb. Manhole is designed for use with 18-duct conduit run. (*Courtesy Commonwealth Edison Co.*)

FIGURE 31.5 A T-type precast cable manhole. Approximate weights: top section 15,065 lb and bottom section 13,235 lb. Manhole is designed for use with 18-duct conduit run. (*Courtesy Commonwealth Edison Co.*)

28" Dia. Hole

7'-2"

1'-2"

5"

2.5"

5"

6'-0"

4'-0"

7"

5"

2.5"

5"

2'-1"

12" Dia. sump hole

FIGURE 31.7 A 15-ft × 9-ft precast manhole specification with two manhole openings. *(Courtesy Utility Concrete Products.)*

FIGURE 31.8 Cross-section view of the 15-ft × 9-ft precast manhole with two manhole openings. (*Courtesy Utility Concrete Products.*)

613

concrete pipe is used for the manhole. The conduits joined to a plastic or fiberglass manhole must have a watertight connection normally completed with mechanical devices. Precast concrete pipe can be used in an area subject to vehicular traffic by setting it on a 6-inch reinforced poured concrete base with a 6-inch tile, 1 ft in length to serve as a sump. Circular manholes in an area not subject to vehicular traffic can be set on an 18-inch layer of crushed rock. Manholes installed in areas not subject to vehicular traffic must support a minimum load of 300 lb/ft^2.

The area at the manhole installation site must be cleared of obstructions such as gas pipes and water mains. Mechanized equipment such as a backhoe can be used to excavate the area for the manhole. The bottom of the excavation should be firm, undisturbed, or compacted earth, leveled without any large rocks or obstructions that would prevent the manhole from setting properly. The excavation should be shored to protect the worker preparing the excavated area for the manhole installation. The precast manhole can be moved to the site on a lowboy trailer. A mobile crane is normally used to lift the manhole from a trailer and place it in the excavated area.

The hole around the precast manhole must be backfilled and thoroughly compacted. Sand or the excavated material, if it is fine and dry, can be used for backfilling. Manhole frame and cover are installed after the precast manhole is in place (Fig. 31.9). Precast manhole necks can be used to extend the manhole opening to the proper level for the surrounding area (Fig. 31.10).

As manholes differ in size and complexity owing to variations in the number of ducts entering and leaving; whether one-, two-, three-, or four-way; size of cables; length of splices; horizontal offsets of conduits; etc., only general principles of construction can be outlined.

FIGURE 31.9 A precast concrete manhole set on concrete slab. (*Courtesy MidAmerican Energy Company.*)

FIGURE 31.10 Manhole frame and cover in place on precast manhole. A precast concrete manhole neck is used to extend opening proper distance so that cover will be flush with surface of terrain.

Floors

Manhole floors are built of poured concrete approximately 6 inches thick. The mixture for the concrete usually is:

> One part Portland cement Three parts clean sand
>
> Five parts broken stone or gravel

The floor should slope toward the sewer drain or sump with a grade of about 1 inch in 6 ft.

Walls

The conduit runs should terminate at the interior side of the manhole wall in conduit bells. A conduit bell (Fig. 31.11) is a flared opening that makes it easier to pull the cable and eliminates the sharp corner that might injure the cable sheath as the cable is pulled in. It also protects the cable sheath as the cable moves in and out owing to load cycling. The walls will have the duct openings placed to meet project specifications.

Pulling eyes (Fig. 31.12) are usually embedded in the walls opposite the conduit runs to facilitate the pulling in of the cables.

FIGURE 31.11 Typical plastic duct bell used to terminate duct in manhole wall.

FIGURE 31.12 Typical galvanized wrought-iron pulling eye built into manhole wall opposite duct runs.

Vaults

Large manholes are often referred to as vaults. Vaults are used for transformer and network protector equipment protection. The construction of a vault is similar to that of the manhole and must meet the same construction specifications. The vault being lifted in place is shown in Fig. 31.13. Electric equipment being installed in a vault by linemen is shown in Figs. 31.14 and 31.15. The energized equipment as viewed from the manhole vault opening is shown in Fig. 31.16.

Figure 31.13 A precast manhole vault being hoisted from a flatbed trailer to be placed in the ground. (*Courtesy Utility Concrete Products.*)

Figure 31.14 The lineman placing electric equipment in the manhole vault opening. (*Courtesy G & W Electric Co.*)

FIGURE 31.15   Linemen placing network switchgear in vault below sidewalk in a business area. (*Courtesy G & W Electric Co.*)

FIGURE 31.16   The vault cover has been removed and the network switchgear primary cable connections in the vault are visible. (*Courtesy G & W Electric Co.*)

CHAPTER 32

Pulling Cable

Installing Cable Racks

Before any cables are pulled into ducts, provision for supporting the cable ends on racks in the manholes is made. Allowing the cable ends to lie on the manhole floor enables them to be stepped on or damaged by falling objects. The rack support channel is fastened to the manhole wall with expansion bolts. Holes for the expansion bolts are drilled 3 inches deep. Safety goggles are worn when concrete or brick is being drilled.

The spacing between racks adjacent to the proposed cable splice is usually 3 ft. In general, a cable should be supported at points 6 inches from each end of the splice. Additional racks are spaced about 3 ft to 4½ ft apart depending on the nature of the cable bends.

Selecting the Duct

Before the start of cable installation, the duct to be occupied should be selected throughout the entire length of the conduit run. As far as possible, the same relative position in the duct bank should be maintained. In general, the longer cables should be installed in lower ducts, and the shorter cables in upper ducts. In this way the vacant spaces will be left in the upper ducts, where they can be utilized later on without disturbing the cables already installed.

Automated Cable-Pulling Equipment

Portable cable-pulling equipment or its accessories and a winch truck can be used to clean a duct and pull the steel cable contained on the cable puller equipment or the winch line on a truck through the duct simultaneously (Fig. 32.1). The hose from an air compressor is attached to the air inlet on the portable cable-pulling equipment or to the air control valve assembly that can be mounted on a winch truck. The steel cable on the truck or the pulling equipment is fastened to the double-disk birdie and inserted in the duct to be used (Figs. 32.2 and 32.3). A duct-sealing head attached to the outlet air hose is fastened around the steel cable, inserted into the duct on top of the birdie, and locked in place to provide an air seal (Fig. 32.4). The air control valve is operated so that the compressed-air pressure forces the double-disk birdie through the duct (Fig. 32.5). The birdie cleans any dirt out of the duct and pulls the winch line through the duct. When the birdie reaches the end of the duct, the air seal is broken, and the travel of the birdie and the steel winch cable stops. The steel cable is now available to be used to pull in the electric power cable. If the duct has an obstruction in it, the workmen may have to use manual methods to rod the ducts and to clean them.

FIGURE 32.1   Air hose and accessories used with winch truck or portable cable-pulling equipment for automated duct cleaning and cable pulling. (*Courtesy Sherman & Reilly, Inc.*)

FIGURE 32.2   Double-disk birdie with winch line of cable-pulling equipment is being inserted into a duct. The double-disk birdie will clean the duct and pull the winch line to the far end of the duct so that the winch line can be used to pull the power cable being installed through the duct. (*Courtesy Sherman & Reilly, Inc.*)

Rodding the Duct

Rodding is the term used to describe the insertion of a flexible fiberglass or steel rod into a duct to clear or dislodge obstructions. Rodding equipment is used if the automated cable equipment is not successful or if the automated equipment is unavailable. Figure 32.6 shows a steel rod on portable reels. The size of the rod is such as to make it stiff enough for pushing. Coupling of rods is accomplished by use of a leader and pickup fittings on the ends.

FIGURE **32.3** Double-disk birdie fastened to winch line of the cable-pulling equipment is being inserted into a duct for a field demonstration.

FIGURE **32.4** The duct-sealing head with air hose attachment is partially inserted into the duct. The winch line attached to the double birdie is threaded through the head. The head will be locked into place to provide an air seal on the end of the duct. The double birdie will pull the winch line through the duct when compressed air is supplied to the head. (*Courtesy Sherman & Reilly, Inc.*)

FIGURE 32.5   The duct-sealing head with air hose attachment is fully inserted into the duct and in the process of clearing obstructions and cleaning the duct as it travels through the duct system.

FIGURE 32.6   Flexible steel rod on reels. (*Courtesy Superior Switchboard and Devices Division.*)

Cleaning the Duct

Before the cables are installed, each duct must be checked to see that it is clean and free from all obstructions. If the automated procedures are not successful, rodding equipment can be used to ensure a cable is successfully pulled through a clean unobstructed duct as previously described. The duct is cleaned with a mandrel, a circular wire brush, and a swab attached to the cable. The mandrel is about ¼ inch to 3/8 inch smaller in diameter than the duct. A common mandrel is a steel tube rounded and fitted with a pulling eye at each end. It will remove serious obstructions such as concrete which has dripped through a joint by working it back and forth until the obstruction is cleared. The circular wire brush has the same diameter as the duct and is used to remove small pebbles or bits of concrete, etc. The swab is made by stuffing waste into a holding grip.

Selecting the Pulling-In Manhole

The selection of the manhole from which the cable is to be pulled is important. Cables can be pulled on the first pull only as far as the pulling rig will permit (Fig. 32.7). When the pulling eye in the cable end or on the basket grip reaches the sheave, the pulling must stop. The length of cable then in the manhole may not be sufficient to reach the cable rack on which the splice will be supported. Furthermore, additional cable length is needed to train the cable around the walls to the center of the splice on the rack. The shape of the manhole and the position of the splice on the wall will determine the amount of cable length needed. In a manhole where the ducts come in at one corner and leave at the diagonally opposite corner, the splice is always closer to one duct than to the other (Fig. 32.8). When there is considerable difference in duct elevations, the length

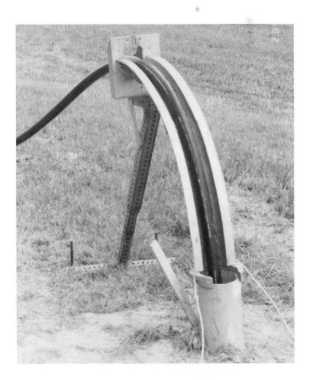

FIGURE 32.7   The electric power cable being pulled through the duct toward the manhole.

Figure 32.8 Types of cable training in manholes of different shapes. (*Courtesy Edison Electric Institute.*)

of cable from the splice to one entrance may be less at one end than at the other. The rule, therefore, is that the manhole requiring the shorter cable length in the manhole to the center of the splice should be the pulling end, as any required length of cable can be left at the feeding end.

Manhole Rigging

The rigging used to pull in cable varies with the type of manhole, length of cable, and conditions prevailing on the street.

The simplest rigging arrangement possible is a pulling eye that has been embedded in the manhole wall (Fig. 32.9). Here the pulling eye is directly opposite the duct, and the snatch block holds the pulling rope in line with the duct. In case the embedded pulling eye is located much below the duct level, a chain or rope of proper length is inserted between the snatch block and the pulling eye, as shown in Fig. 32.10. This is the usual

Figure 32.9 Pulling cable by use of eye embedded in manhole wall opposite duct. Snatch block holds pulling rope in line with duct.

Figure 32.10 Pulling eye located below duct level requires use of short rope or chain between snatch block and pulling eye.

FIGURE 32.11   Double-sheave stand made of two parallel channel irons. Holes in channel members permit placing sheaves at proper levels.

situation in a multi-way manhole, where ducts enter the manhole on all sides and the eye must be placed below the duct level. A sheave in the manhole opening is often used in deep manholes to guide the pulling rope to the pulling winch on the street as shown.

When no embedded pulling eyes are available, the sheaves are supported on a sheave stand. It consists of two timbers, channel irons, or I beams bolted together with a space between them for the sheaves. Holes through the two parallel members (Fig. 32.11) permit placing the sheaves at the desired locations.

Attaching Pulling Rope

The pulling rope is attached to the cable by means of a woven cable grip or by means of a clevis or eye.

The principle of a woven cable grip, or *basket* grip, as it is sometimes called, shown in Fig. 32.12, is that the wires, which are in the form of an open weave, will contract laterally and thus hold fast to the cable jacket when a longitudinal pull is applied.

FIGURE 32.12   Woven cable grip, sometimes called basket grip. Note swivel inserted between pulling rope and cable grip.

FIGURE 32.13   Swivel inserted between pulling rope and cable to prevent twisting of cable. (*Courtesy Edison Electric Institute.*)

When the pull is heavy, the pulling rope must be attached to a clevis or eye. The eye in turn is fastened to the conductors in the cable. Such eyes are usually attached at the cable factory.

Pulling eyes can be attached in the field. The conductor strands are wrapped around the lug of the eye. Since the pulling rope has a tendency to twist during the pull, a swivel (Fig. 32.13) should be inserted between the pulling rope and the cable.

Feeding Tube

To prevent injury to the cable by scraping on the manhole frame or at the duct opening or in passing over other cables, a feeding tube is sometimes used. This guiding tube is usually made of flexible steel tubing about 4 inches in diameter with a flare at the reel end and a removable nozzle fitting at the duct end so that it can be adapted to various size ducts. The tubes vary in length from 10 ft to 20 ft.

Handling Cable

A common cause of underground system outages is damaged cable resulting from improper handling and installation. It is important that extreme care be exercised while handling and installing the cable. Minor physical damage to the cable jacket or installation can lead to premature failure.

Cable packaged on reels should arrive on the site suitably protected against damage. Reels should be upright and with means to prevent cable unrolling or unwrapping. The lineman or cableman should inspect the cable for obvious and hidden damage. If the reels or reel wrap is damaged, the inspector should use extra caution. The cable must arrive at the site with the ends sealed to prevent moisture from entering the cable. The cable must be protected from cold weather prior to installation. If the temperature is less than 14°F, the cable reels should be warmed for 24 hours prior to the starting of the cable-pulling operation.

The lineman or cableman unloading the cable at the installation site should use a cradle to support the reel flanges or a shaft through the arbor hole to lift the reels. If a forklift is used by the workmen, the forks must lift the cable reel at 90° to the flanges and be long enough to make complete lifting contact with both flanges. The cable surface and/or the protective wrap must be protected from damage by the lift forks or other equipment. The cable is often placed on a trailer designed for handling cable at the service center and transported to the installation site. The cable can be installed while the reel is mounted on the trailer, minimizing the handling of the cable and the danger of damage while it is being installed.

Cable Lubrication

To protect the cable from excessive tension during pulling in, the cable is lubricated. This reduces the friction between cable and duct walls, especially on curves in the duct run, by as much as 70%.

The lubricants for cables are greases, oils, soapstone, etc. The lubricant is applied with a paddle or brush just before the cable enters the feeding tube. A coating of about 1/16-inch thick is ample and will amount to 6 lb/100 ft to 8 lb/100 ft on 3-inch diameter cable.

No lubricant should be applied to the first and last 5 ft of cable for convenience and cleanliness in splicing.

Pulling the Cable

The cable is drawn into the duct by means of a winch, or capstan. The winch is usually mounted on a truck or a portable cable puller located near the manhole or the riser conduit at the pulling end.

The reel of cable must be properly placed at the feeding end to cause minimum flexing of the cable. It should always be located on the side of the manhole toward which the cable is to be pulled. This avoids putting a reverse bend in the cable and makes it unwind easily.

The end of the cable is then threaded into the feeding tube extending from above the edge of the manhole down to the end of the duct. The pulling grip is now slipped over the end of the cable, the swivel is attached, and finally the pulling rope is attached. In case a pulling eye is provided on the cable, the swivel and pulling rope are attached to it.

The cable is inspected and greased as it enters the funnel of the feeding tube. On cables having a rubber jacket on the outside, a mixture of soapstone and water or a pulling compound specified by the cable manufacturer is applied to the jacket as the cable is drawn in.

Workers stationed at the cable reel regulate the amount of slack in the cable so that it passes freely into the tube without being loose on the reel and without scraping on the manhole frame.

A man should be stationed in the manhole at the pulling end to signal a stop to prevent pulling the cable over the sheaves. Another should be stationed at the top of the manhole to relay signals to the winch operator.

Excessive pulling tensions that exceed the elastic limit of the conductor or stress the extruded portions of a cable will cause permanent damage to the cable that will eventually cause the cable to fail. The maximum pulling tension should not exceed 0.008 lb/circular mil of copper conductor or 0.005 lb/circular mil of aluminum conductor, when the pulling eye is attached to the cable conductor, with the maximum tension to be limited to 5,000 lb for single-conductor cable and 6,000 lb for multiconductor cables. If a basket is placed over the surface of the cable for pulling, the maximum tension should not exceed 1,000 lb. Sidewall pressure on the insulation at a bend in the duct should not exceed 300 lb/ft of radius of the duct bend.

The speed at which the cable is drawn into the duct varies with conditions and cable size. It is good practice, however, to pull the cable slowly. In a straight duct, a single cable can be pulled in at about 60 ft/min. If a higher speed is used, there is difficulty in greasing properly and in inspecting the cable. An average speed of about 40 ft/min, or 2/3 ft/sec, is preferable when a single cable is being pulled. When two or more cables are pulled into one duct, the speed should be further reduced to about 20 ft/min. A slow speed will prevent the possibility of crossing the conductors as they enter the duct.

The savings in time obtained with higher speeds is negligible, as the time required for drawing in a 400 ft length at 40 ft/min is only 10 minutes. This is a small part of the total time needed to prepare for the pulling-in operation.

Slack Pulling

When the cable end has been drawn up to the manhole rigging, the first pulling operation must stop. If not enough cable length is then available in the manhole to reach to the center of the splice as the cable is trained around the wall, additional pulling is necessary. To obtain this additional slack in the cable, a new cable grip must be applied to the cable the required number of feet from the cable end. A special *basket* grip is used. This grip is open at both ends and can be slipped over the cable to the desired position. Care must be used to avoid injuring the cable.

When the cable has been pulled in as far as necessary, the forward end is placed on the cable supports. No attempt, however, should be made to train the cable into its final position. This will be done by the splicing crew. A check should be made to see that the seal at the end of the cable is still watertight. At the feeding end the cable is cut off with enough slack left to train it properly around the manhole walls. The end of the cable is then also placed on its supports, and the end of the cable is sealed. The end of the cable remaining on the reel is also sealed.

CHAPTER **33**

Splicing and Terminating Cable

Multiple lengths of cable are connected together to form a continuous length. These connections are called joints or splices. Proper cable splices and terminations are necessary to minimize electrical stress areas developed at cable ends. Two cable ends may be spliced together or a cable end may be terminated at the primary cable riser, in a cable pothead, or at a primary cable elbow. Cable ends are to remain clean during the splicing process.

Electric Stress Areas

Utilizing a proper cable splice, the cable's electric field is properly distributed to eliminate any highly concentrated electric stress areas (Figs. 33.1 and 33.2). The highest concentration of electric stress in an underground cable is at the edge of the cable shield. The typical underground distribution cable consists of the following layers:

1. Stranded conductors
2. Semiconducting cross-linked polyethylene strand shield
3. Tree-retardant cross-linked polyethylene primary insulation
4. Semiconducting cross-linked polyethylene insulation shield
5. Bare concentric neutral wires
6. Insulating or semiconducting polyethylene jacket

The cable layers are depicted in Fig. 33.3. The cable must be properly spliced or terminated to ensure the integrity of the cable so that cable performance remains intact.

Prefabricated Cable Inline Splice Installation

Prefabricated inline cable splice kits are normally used on underground distribution 15 kV and above primary cable installations with ethylene propylene rubber or cross-linked polyethylene insulation. The splices may be direct-buried or placed in enclosures to permit access. The prefabricated splice kit consists of a metallic connector, a connector shield, insulation, and an insulation-shielding jacket. The prefabricated splices are cold shrink or heat shrink jacket design. When an unjacketed repair sleeve is uncovered (Fig. 33.4), a wraparound cross-linked polyolefin sleeve with a raised rail profile stainless steel locking channel system is utilized to seal the splice (Fig. 33.5).

629

FIGURE 33.1   Electric stress distribution for insulated shielded high-voltage cable with shield and insulation partially removed from cable conductor. Note high concentration of electrical stress at the edge of the cable shield.

FIGURE 33.2   Schematic of electric stress distribution in a completed prefabricated splice installed on an insulated shielded high-voltage cable. Note electric stress in transition area from cable to splice is evenly distributed.

FIGURE 33.3   The layers of a typical underground distribution cable.

FIGURE 33.4   Prefabricated splice installed on concentric neutral cross-linked polyethylene insulated underground cable. The cable and splice will be encased in a wraparound sleeve and buried in the ground.

FIGURE 33.5 A wraparound cross-linked polyolefin sleeve with a raised rail profile stainless steel locking channel system. (*Courtesy Raychem Corporation.*)

FIGURE 33.6 Adjustable cutting blade cable insulation stripper being set to the correct blade depth by the lineman. (*Courtesy Speed Systems.*)

    The cables are prepared for splicing by removing the cable jacket, laying the concentric neutral wires back, removing the cable insulation semiconducting material, removing the cable insulation, removing the cable strand semiconducting layer, and exposing the conductor strands. Special cable stripping tools are used to prepare the cable ends for the prefabricated splice. Tools used are:

1. End strippers (Figs. 33.6 and 33.7).
2. Encapsulated jacket stripper
3. Adjustable blade semi-con scoring tool (Fig. 33.8)
4. Insulation chamfer tool
5. Penciling tool

FIGURE 33.7 Cable insulation partially cut for removal. (*Courtesy Speed Systems.*)

FIGURE 33.8 Special tool manufactured to score cable semiconducting insulation jacket. (*Courtesy Speed Systems.*)

FIGURE 33.9   Prefabricated splice on unjacketed, insulated, concentric neutral cable. Conductor connector has been crimped. (*Courtesy 3M Co.*)

FIGURE 33.10   Completed installation of prefabricated unjacketed insulated concentric neutral cable splice. (*Courtesy 3M Co.*)

The chamfer tool creates a 45° bevel needed for some pre-molded splices and terminations. The penciling tool provides a smooth insulation taper.

Splice kit instructions include these steps: Apply one half-lapped layer of vinyl tape over the semiconductive jacket of the cables. Apply silicone grease over cable insulation and vinyl tape. Slide splice end caps onto each cable until against concentric wires. Slide the splice body onto one of the cables. Place cable conductors into connector and crimp the connector with the proper tool in the manner specified by the manufacturer (Fig. 33.9). The splice body is slid into final position over cable connector, leaving a small area of insulation exposed at both ends of the splice body (Fig. 33.10). The cable splicer or lineman removes the vinyl tape applied to the semiconductive jacket prior to completing the splice assembly. The end caps are slid onto the splice body, completing the semiconductive shielding jacket over the cable and splice insulation. The concentric neutral wires are extended over the splice and splice insulation. The concentric neutral wires are extended over the splice and connected together with a compression-type connector. The concentric wires are connected to the grounding eyes on the splice body to ensure the splice jacket is grounded. The insulation shield on a prefabricated splice normally consisting of a conductive jacket retains the electric field completely within the spliced cable insulation.

Prefabricated Cable Inline Cold Shrink
Shear Bolt Connector Splice Installation

The aluminum shear bolt cable conductor connector splice (Fig. 33.11) is encased in a cold shrink ethylene propylene diene monomer (EPDM) rubber elastomer housing. The connector shear bolts will shear at the proper torque. The connector eliminates the need for dies and compression tools. A cordless drill and the properly sized drill bit are the tools required. The shearing occurs below surface of the connector such that there are no protruding sharp edges that must be filed.

FIGURE 33.11 Aluminum shear
bolt cable conductor connector
splice. (*Courtesy TE Connectivity
Raychem.*)

FIGURE 33.11 Aluminum shear
bolt cable conductor connector
splice. (*Courtesy TE Connectivity
Raychem.*)

General Splicing and Termination Recommendations

When a cable is cut, in preparation for splicing or for any other reason, it must be protected from moisture and dirt. Cable ends exposed to the atmosphere will collect moisture and contamination. The cable ends should be properly sealed at all times except during the period when the splice is being completed or the cable termination is being installed. The cable ends should be thoroughly inspected before they are trained into final position for splicing.

When underground ducts are used, the splices are normally completed in a manhole. If the splices are to be completed in a manhole, the cables and the splices should be supported on racks mounted on the manhole wall as illustrated in Chap. 29. The joint or splice is supported between two brackets located on the central portion of the manhole wall. Because of the expansion and contraction of the cables, caused by changes in the cable temperature, it is necessary to provide reverse curves or offset bends in the portion of the cable in the manhole. These reverse curves enable the cable to take up the expansion and contraction movements without cracking or buckling of the insulation. Therefore, the training of cables on the manhole wall is not only for making a neat arrangement in the manhole but more importantly for providing the expansion space needed to absorb the movements of the cable.

In general, the bending radius R (Fig. 33.12) of the reverse curves in covered cables must be no less than eight times the overall diameter of the cable. Thus, a 3-inch cable should have a minimum bending radius of 24 inches or more, and a 4-inch cable should have a minimum bending radius of 32 inches or more.

FIGURE 33.12 Sketch showing
minimum bending radius on reverse
curves of cable as it is trained
around wall in manhole; D is cable
diameter, X is number of times D,
and R is bending radius.

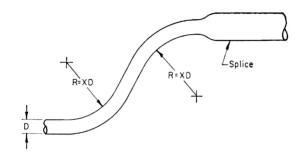

The range of values of R in terms of cable diameter varies from 5 to 12 and depends on the size and number of conductors and whether it is a polyethylene-, thermoplastic-, or ethylene propylene rubber cable, as well as on the rated voltage of the cable.

At least 6 inches of straight cable should extend beyond each end of the splice to provide space for resting on the saddles of the supporting racks. The clips holding the saddles on the bracket should be spread enough to permit the saddle to slide freely along the bracket.

All cables and joints should be so racked in the manhole that they are not directly under the manhole cover.

The exact makeup of a splice depends on the specific cable construction, that is, whether the cable is a single- or multiple-conductor cable; whether the insulation is ethylene propylene rubber, polyethylene, or thermoplastic; or whether the cable insulation has a conducting shield; etc. The cable manufacturer's specific instructions must be followed in all instances. The worker must maintain tools in good condition and keep them clean and dry. The working area and the area of the cable being worked on should be protected from moisture with a rubber blanket or other waterproof material. The cable splicer or lineman performing the work must maintain clean and dry hands.

Splice Jacketing

Once the cable splice of a jacketed cable has been completed, the prefabricated splice body is jacketed as well. Cold or heat shrink jackets are used. The ground wire must also be spliced. The system shown in Figs. 33.13 through 33.16 is cold shrink jacketed with pre-stretched tubing. The tubing shrinks onto the cable as the inner core is unwound. The final product is a tight-fitting sealed cable splice.

FIGURE **33.13** Drawing of 15-kV cable splice and splice jacket kit. (*Courtesy 3M Co.*)

FIGURE **33.14** Drawing of 15-kV cable splice that has been completed and about to be jacketed. (*Courtesy 3M Co.*)

FIGURE 33.15   Drawing of 15-kV cable splice jacket kit inner core being unwound. (*Courtesy 3M Co.*)

FIGURE 33.16   Drawing of completed jacketed 15-kV cable splice. (*Courtesy 3M Co.*)

Cable Riser Pole Terminations

Cable riser pole terminations are completed after preparing the end of the cables in a manner similar to the procedure described for preparing the cable ends for splicing to other cables. Solid-dielectric insulated cables are usually terminated with a molded-rubber terminator (Figs. 33.17 through 33.19). The molded-rubber cable termination is usually delivered in a kit which contains a one-piece molded-rubber stress cone,

FIGURE 33.17   Schematic diagram of a portion of a molded-rubber termination for a cable. (*Courtesy 3M Co.*)

FIGURE 33.18   Cable terminated with a molded-rubber terminator using a cable aerial lug. (*Courtesy 3M Co.*)

FIGURE 33.19   An insulated cable molded-rubber termination with a stem-cable connector on a cable riser pole. The terminator pin is directly into the bottom of the fused cutout.

a stainless steel grounding strap, a molded silicone one-piece skirted rubber insulator, a stem-cable connector or cable aerial lug, a cable support, silicone lubricant, and complete instructions.

Cable Terminations in Pad-Mounted Equipment

Cables are terminated in pad-mounted equipment in an elbow (Fig. 33.20) or with a connector encased in a tee body (Fig. 33.21). The elbow termination and tee body connector are fully shielded and insulated. The steps to build the elbow termination are:

1. Apply mastic tape on the conductor jacket below the portion of the end where the jacket has been removed.
2. Fold concentric neutral strands back.

FIGURE 33.20   Pad-mounted equipment elbow cable termination. (*Courtesy 3M Co.*)

SEMI-CONDUCTING SHIELD
Precision-molded peroxide-cured
semi-conducting shield provides
ground shield continuity and meets the
requirements of IEEE Std 592™ standard.

HEX HEAD
One-inch hex head
is used to tighten the
connection.

INSULATING PLUG CAP
Semi-conducting rubber cap fits over
insulating plug test point for waterproof
seal and deadfront shielding.

INSULATING PLUG
Molded epoxy insulating plug provides
excellent electrical, thermal, and mechanical
reliability.

SEMI-CONDUCTING INSERT
Molded semi-conducting peroxide-cured
insert provides corona-free electrostatic
shielding of the compression connector.

CAPACITIVE TEST POINT
Optional capacitive test point on
molded T-body with snap-on cap
provides a shielded, hotstick-operable
means to determine circuit condition when
used with high-impedance voltage-sensing
devices designed for capacitive test points,
and a place to mount Cooper TPR Series
Fault Indicators.

DRAIN WIRE TAB
Drain wire tab provides
a convenient point to
connect a drain wire to
ensure grounding of the
connector shield.

COMPRESSION CONNECTOR
Aluminum compression
connector is sized to ensure a
cool running connector with
maximum current transfer.

T-BODY
Molded T-body adapts
to all cable sizes and
provides a deadfront shielded
connection.

EPDM INSULATION
High-quality peroxide-cured
EPDM insulation is mixed
and formulated in-house for
complete control of raw
rubber characteristics.

CABLE ADAPTER
Molded cable adapter, sized to fit
cable insulation diameters from
0.640 in to 1.965 in (16.3 to 49.9 mm),
provides stress relief for the terminated
cable.

9.6"
(244 mm)

12.5"
(318 mm)

FIGURE 33.21 Tee body connector. (*Courtesy Cooper Power Systems.*)

Dead-front load break elbow termination detail

Concentric neutral

Mastic (First layer)

Step 1

Overall jacket

Insulation

Conductor

Concentric neutral

Semi-conductive material

Step 2

Mastic
(Second layer-over concentric)

Vinyl tape
(Two half layers over mastic)

Compression lug
supplied with
elbow kit

Step 3

Cold shrink

Step 4

2-ground wires

Step 5

Cold shrink seal

FIGURE 33.22   The steps involved in the make-up of a cold-shrink seal of a dead-front termination. For proper dimensioning of the cable layers, consult the instructions enclosed in the manufacturer's elbow installation kit. (*Courtesy MidAmerican Energy Co.*)

3. Terminate the cable with the compression lug and apply a second layer of mastic tape over the concentric neutral wires.

4. Install the elbow, insert the probe, and slide the cold shrink in position.

5. Attach two ground wire strands to elbow drain wire tab (Fig. 33.22).

The steps to build the tee body connector are essentially the same as building the elbow termination with the exception that the tee body slides over the compression connector.

Underground Distribution

Underground distribution has proven to be more reliable than overhead distribution. Underground distribution lines are relatively immune to some of the major causes of failures in overhead circuits, such as damage from lightning, wind, ice, and snowstorms; or contacts to the wires by trees or other foreign objects. Underground cables have their own unique cable insulation failure problems caused by moisture, corrosion, cable dig-ins, or improper installation of terminations and splices.

Power Source

In most cases, underground distribution systems will be supplied from an overhead main feeder circuit installed adjacent to residential subdivisions or commercial business centers. The customer loads in the area may be served by several underground cable laterals connected to the open-wire overhead system at several cable riser poles. A typical riser pole with cable, cable guard, molded polymer cable terminator, lightning arrester, and fused disconnect is shown in Fig. 34.1. Figures 34.2 and 34.3 illustrate the essential components required to connect the underground single-phase cable to an overhead distribution circuit. A three-phase riser pole is shown in Fig. 34.4.

Total underground distribution circuits have been constructed in some locations. The three-phase main feeder circuit extends from substation to substation underground. The main feeder circuit has a capacity of 10 MVA (megavolt-amperes). The underground cables consist of 750-kCMIL aluminum conductor with cross-linked polyethylene insulation rated 15 kV, shielding consisting of semiconducting tapes and copper tapes and a polyethylene jacket. Taps to the main circuit to serve residential areas or commercial customers are made in switching modules installed in pad-mounted switchgear or a submersible container.

Most of the underground distribution circuits tapping the main feeder are single-phase loop feeds with fuses to protect the main feeder circuit. Extending the three-phase main feeder circuit between two substations makes it possible to isolate a faulted cable section and restore service to customers prior to completing the cable repairs. The loop-type feed serving the underground distribution area makes it possible to isolate faulted cable circuits in order to restore service to the customers in the subdivision.

FIGURE 34.1   Single-phase underground cable riser installed on pole. The single-phase cable is connected to the overhead circuit through a fused cutout and protected by a lightning arrester.

Types of Systems

Direct-buried underground distribution systems can be installed in several different ways. A common method of installation places the primary and secondary cables along front lot lines and utilizes pad-mounted transformers. The pad-mounted transformers would normally serve from 4 customers to 12 customers depending upon the terrain, the customer loads, and the size of transformers installed. This type of installation is illustrated in Fig. 34.5.

It is desirable to install the cables along streets in front of the homes. This type of installation is commonly referred to as front-lot line. The front-lot line type of installation can use pad-mounted transformers, submersible transformers, direct-buried transformers, or unit residential transformers (see Fig. 34.5.). The installation of pad-mounted transformers along the streets may be objectionable from an environmental point of view.

Submersible transformers are usually installed in fiber-material vaults. The submersible transformer must be protected from corrosion. Maintenance of submersible transformers and the cost of installation are usually greater than the costs associated with pad-mounted transformers. Direct-buried transformers have not been used extensively to date. It is necessary to provide corrosion protection for this type of installation.

The unit residential transformer is usually installed adjacent to a house and back from the street, where it is inconspicuous. The unit residential transformer usually consists of a dry-type core and coils installed in a metal housing for mounting on a pad. Each house is served by an individual transformer. Different types of high-voltage switching arrangements have been installed to vary the flexibility and the cost associated with each different type of underground distribution installation.

QUANT.	MATERIAL DESCRIPTION
2	WASHER. SPRING. TWIN COIL. $\frac{5}{8}$ INCH. CARBON
2	WASHER. FLAT SQUARE. GALV. STL. $\frac{13}{16}$ INCH HOLE
1	TERMINATOR. CABLE. OUTDOOR. 15KV. 1/0
50	WIRE. BARE. #6 COPPER. SOLID. DEAD SOFT
15	WIRE. BARE. #2 COPPER 7W. SOFT DRAWN
3	GUARD. CABLE. "U". HMWPE. 2 IN X 10 FT
1	GRIP. CABLE. CLOSED MESH. DBL-EYE. 1.00-1.24
1	ARRESTER. SURGE. RISER CLASS. 8.4KV MCOV
1	CUTOUT. FUSED. NON-LBR. 100A. 7.8/13.8 KV
2	NUT. HEX. ASTM 651 SI BRZ. $\frac{1}{2}$ IN-13 UNC
1	BRACKET. INSULATOR. VERTICAL. GALV. 18 IN
1	CLAMP. HOT-LINE. BRZ. #8-2/0 LN. #8-1/0 TAP
1	ROD. GROUND. COPPERBONDED. SECTIONAL. $\frac{1}{2}$ IN.
5	GUARD. LEAD-WIRE. PLASTIC. SPLIT TUBE
1	BAIL. STIRRUP. #2 SOL COPPER. CRANK-SHAPE
2	BOLT. HEX. CAP. SCREW. SS. $\frac{1}{2}$ INCH X 6 INCH
1	CLAMP. GROUND ROD. BRZ. $\frac{1}{2}$ IN.
1	CONNECTOR. COMPRESSION. C-TAP. CU. .162-.232
1	CONNECTOR. COMPRESSION. C-TAP. CU. .128-.292
42	SCREW. LAG. WASHERHEAD. GALV. $\frac{1}{4}$ IN X 2 IN
1	HOOK. DRIVE. GALV. $\frac{7}{16}$ IN X $4\frac{3}{4}$ IN ←
1	MOLDING. GROUND WIRE. PVC. $\frac{1}{2}$ IN X 8 FT
2	WASHER. BELLEVILLE. SS. $\frac{1}{2}$ INCH
2	WASHER. FLAT. ROUND. SS. $\frac{1}{2}$ INCH HOLE
1	HOLDER. TAG. IDENTIFICATION. 4-TAG. POLY. VERTICAL
1	DECAL. CAUTION. UG ELECT CABLE. ONE CALL
2	BOLT. MACHINE. GALV. $\frac{5}{8}$ INCH X 12 INCH
1	FUSE. LINK. UNIVERSAL
4	TAG. ID. POLY VERTICAL
1	KIT. PREPARATION. CABLE
2	COMPRESSION CONN—#2 CU
1	COMPRESSION CONN—#2 CU
1	COMPRESSION CONN—#6 CU

FIGURE 34.2 Construction standard detail for installation of a 7.62-kV single-phase primary underground distribution riser pole.

Cables

Cable should be capable of meeting high potential dielectric withstand tests applied in accordance with an applicable standard issued by a recognized organization such as the Association of Edison Illuminating Companies, the Insulated Power Cables Engineers Association, the National Electrical Manufacturers Association, or the American Society for Testing and Materials. The design and construction of conductors, insulation, jacket, and shielding shall include consideration of mechanical, thermal, environmental, and electric stresses which are expected during installation and operation. Cable shall be designed and manufactured to retain specified dimensions and structural integrity during manufacture, reeling, storage, handling, and installation.

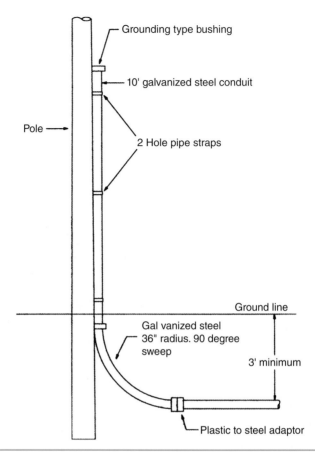

Cable shall be designed and constructed in such a manner that each component is protected from harmful effects of other components. The conductor, insulation, and shielding shall be designed to withstand the effects of the expected magnitude and duration of fault current except in the immediate vicinity of the fault. Sheaths and/ or jackets shall be provided when necessary to protect the insulation or shielding from moisture or other adverse environmental conditions. Conductor shielding should be provided in accordance with an applicable standard issued by a recognized organization such as the Association of Edison Illuminating Companies, the Insulated Power Cable Engineers Association, and the National Electrical Manufacturers Association.

Insulation shielding shall be provided for cable operating at more than 5 kV to ground and is recommended for cables operating above 2 kV to ground. Shielding is not required for short jumpers which do not contact a grounded surface within enclosures or vaults provided they are guarded or isolated. Insulation shielding may be sectionalized provided that each section is effectively grounded. The shielding system

FIGURE 34.4   Three-phase underground cable riser installed on pole. The cables are connected to the overhead circuit through fused cutouts and protected by lightning arresters.

FIGURE 34.5   Single line schematic of an underground distribution front-lot line system with submersible and unit residential transformers.

3-No. 2/0 AWG AL 600 volts cables

3-No. 4/0 AWG AL 600 volts cables

1-No. 2 AWG AL 15 kV concentric neutral cable

URT

▼=Transformer
S=Submersible
URT=Unit residential

may consist of semiconducting materials, nonmagnetic metal, or both. The shielding adjacent to the insulation shall be designed to remain in intimate contact with the insulation under all operating conditions. Shielding material shall be designed to resist excessive corrosion under the expected operating conditions and shall be protected.

Voltages

The primary voltages used for underground circuits vary from 2,400 to 34,500 volts. One of the most common primary voltages is 13,200Y/7,620 volts. Other voltages used extensively are 4,160Y/2,400 volts, 12,500Y/7,200 volts, 13,800Y/7,960 volts, 24,900Y/14,400 volts, and 34,500Y/19,920 volts.

Utilization voltages for residential installations are normally 120/240-volt single-phase three-wire grounded neutral system. Utilization voltages for commercial installations may be 120/240-volt single-phase three-wire grounded neutral system; 208Y/120-volt three-phase four-wire grounded neutral system; 240-volt three-phase three-wire ungrounded delta system; 480Y/277-volt three-phase four-wire grounded neutral system, or 480-volt three-phase three-wire ungrounded delta system. Large industrial customers may receive electric service at voltages commonly used for subtransmission or transmission circuits such as 13,200 volts, 69,000 volts, or 345,000 volts three-phase.

Primary Cable

The high-voltage cables for the underground system main feeder circuit have a 600-amp capacity. The main feeder cables are usually installed direct-buried at a depth greater than required by the National Electrical Safety Code to provide protection from dig-ins. The main feeder circuit cables may be installed in polyethylene plastic pipe.

The primary cables supplying residential areas or commercial customers tapped from the main feeder circuits are usually 200-amp capacity, direct-buried single-conductor concentric neutral type.

The cable layers shown in Fig. 34.6 are:

1. Aluminum or copper conductor
2. Extruded semiconducting polyethylene conductor shield
3. Tree-retardant cross-linked polyethylene or ethylene propylene rubber insulation
4. Extruded semiconducting polyethylene insulation shield

Figure 34.6 Typical 200-amp concentric cable used for underground primary distribution. Aluminum-stranded conductor has a semiconducting extruded shielding, cross-linked polyethylene insulation, semiconducting extruded insulation shielding, surrounded by spiral-wound exposed tinned copper neutral conductor. The copper neutral conductor is often jacketed with polyethylene. (*Courtesy Southwire Corporation.*)

5. Spiral-wound tinned copper neutral strands

6. Polyethylene cable jacket

Low-Voltage Cables

The secondary and service cables for underground electric distribution are generally larger than those applied on overhead secondary systems because of the reduced capacity for a given conductor size as a result of the temperature limitations of the cable insulation, greater load density, and the need to provide for load growth. Aluminum conductors are generally used for underground distribution because of the lower cost compared to copper. The secondary cables often have a conductor of No. 4/0 AWG or 350-kCMIL aluminum. Service cables often have a conductor of No. 2/0 AWG aluminum to provide 200-amp capacity where National Electric Code ratings are not applicable. The underground residential distribution services will vary in size from No. 1/0 AWG to No. 4/0 AWG aluminum conductor. A nominal three-wire 120/240-volt single-phase grounded neutral system is normally installed to serve homes in areas supplied by underground distribution circuits.

A twin-concentric neutral cable is constructed with two 600-volt cross-linked insulated aluminum conductors in parallel configuration with an overall spirally applied concentric neutral of No. 14 AWG, No. 12 AWG, or No. 10 AWG-tinned copper wires in sufficient number to provide the conductivity required. This type of cable construction is used where telephone and power conductors are installed randomly in a single trench. Triplexed cables may be used for secondaries and services. This cable construction consists of three single-conductor aluminum cables with 600-volt cross-linked polyethylene insulation twisted together. This insulated aluminum neutral cable may be color-coded, providing ease of identification and distinguishability from other cables (Fig. 34.7). Secondary and service cables constructed in a flat parallel configuration are used in some installations. This cable assembly consists of two aluminum phase conductors and a neutral conductor, each individually insulated with cross-linked polyethylene and covered with a tight, contour-fitted polyethylene jacket. The jacket assembles the three cables in a flat parallel configuration, providing a web between the conductors for ease of separation when splicing or terminating the cable, without danger of damaging the insulation.

FIGURE 34.7   Reel of low-voltage insulated triplexed cables mounted on back of truck to facilitate installation of underground secondaries and services. Cable is manufactured for direct-buried installation. Light-colored insulation identifies neutral conductor.

Transformers

Many component parts of transformers used for underground electric distribution are similar to those installed in transformers designed for pole mounting. The core and coil assemblies are interchangeable in many instances. The transformers can be equipped with no-load tap changers, high-voltage fuses, low-voltage circuit breakers, and pressure relief devices. The core and coil assembly is normally immersed in insulating oil to provide the major insulation and a cooling medium. A typical core and coils assembly is shown in Fig. 34.8.

Pad-mounted transformers are used more extensively than other types of underground distribution transformers. The transformers are built in different sizes and configurations depending upon the kVA capacity of the transformer, the type of connections, and the accessory equipment installed. The pad-mounted transformers vary in height from 18 inches to 48 inches or more. A low-profile pad-mounted transformer construction standard diagram is shown in Fig. 34.9.

In Fig. 34.10, the lineman is installing a grounded parking stand that is used to ground a high-voltage cable for maintenance work. It is very important that the cables be tested with a high-voltage meter to be sure they are deenergized prior to placing the elbow connector on the grounded device installed on the parking stand bracket. Other devices that may be accessible with the transformer cover open include draw-out load-break fuses, operating handle for low-voltage circuit breaker, and pressure relief device.

Figure 34.10 illustrates a pad-mounted transformer with the hinged cover open. The transformer is constructed with two high-voltage load-break bushings. In Fig. 34.10, one of the high-voltage load-break bushings is marked L_1. The high-voltage bushing is shown in Fig. 34.10, after the connections have been made. The low-voltage bushings are marked X_1, X_2, and X_0. A parking stand bracket for the high-voltage cable connections is located between the two high-voltage load-break bushings.

FIGURE 34.8   Core and coil assembly for installation in transformer. (*Courtesy ABB Power T&D Company Inc.*)

GROUND ROD MUST BE DRIVEN A MINIMUM OF 8 FEET IN THE GROUND

----NOTES----

1. Confirm that a drain wire is installed from each bushing insert to the bushing well clamping flange.
2. Install fault indicator sensor on the load side elbow. Fault indicator sensor installation should conform to prior installations.
3. Attach end of ground bus to the neutral secondary terminator.
4. One copper strand from cable scrap is to be connected between the drain wire lobe and cable neutral. Do not use required concentric neutral for drain wire.

FIGURE 34.9 Single-phase pad-mounted transformer construction standard diagram. (*Courtesy MidAmerican Energy Company.*)

FIGURE 34.10 Close-up view of pad-mounted transformer with the hinged cover open, showing high-voltage and low-voltage terminal equipment and accessories. Lineman is installing a grounding device on high-voltage terminator parking stand bracket with hot-line tool. (*Courtesy A. B. Chance Co.*)

Submersible distribution transformers are similar in appearance to pole-mounted transformers and contain similar equipment (Fig. 34.11). High-voltage cable connections are completed through bushings by using elbow connectors installed on the high-voltage cables. The low-voltage connections are completely insulated. The transformers are designed to operate under water. Tanks of the transformers are made with stainless steel or mild steel and protected by special enamel finishes to prevent corrosion. Special care must be taken to protect the transformer finish from being scratched or damaged while the unit is being stored, transported, and installed. It is desirable to install cathodic protection equipment connected to the submersible transformer bank to prevent electrolysis.

Direct-buried single-phase distribution transformers are available in sizes of 10 kVA through 37.5 kVA. The transformers are usually installed as a part of a front-lot line system with one transformer per customer. The direct-buried transformer reduces the cost of installation by eliminating the vault and minimizes the maintenance compared to the requirements associated with submersible vault-installed transformers. The transformers are constructed with a metallic tank or a polyester tank with a metallic liner. Figure 34.12 illustrates a direct-buried transformer with a metallic tank.

Submersible transformers installed in vaults (see Fig. 34.11) and direct-buried transformers (see Fig. 34.12) must be protected for corrosion. Figure 34.13 is a suggested installation diagram for a cathodic protection system designed to prevent transformer tank corrosion. If the tank of the transformer is connected to the system ground, the copper neutral is part of the galvanic cell, and the protective anode will attempt to protect it as well as the tank; thus, the anode will be depleted rapidly. If the tank of the

Figure 34.11 Cutaway view showing submersible single-phase distribution transformer installed in fiber vault. A steel grate on top of the vault provides protection and ventilation openings. (*Courtesy ABB Power T&D Company Inc.*)

Figure 34.12 Direct-buried transformer with a metallic tank. (*Courtesy General Electric Co.*)

Note: Waterproof all electrical connections   Insulate all bare connectors within 8 ft of transformer

FIGURE 34.13   Sketch of suggested direct-buried transformer cathodic protection system. (*Courtesy General Electric Co.*)

transformer is isolated from the system ground, the anode will protect the transformer tank only, giving satisfactory anode life.

Three-phase pad-mounted transformers are often used to serve shopping centers, schools, hospitals, churches, apartment complexes, and office buildings located in the vicinity of electric underground distribution circuits serving several residential areas. The three-phase pad-mounted transformers are available in sizes of 75 kVA through 2,500 kVA, with primary voltages of 4,160Y/2,400 through 34,500Y/19,920 and secondary voltages of 208Y/120, 240, 480Y/277, and 480. Standard accessories such as high-voltage bushings for use with load-break elbow connectors, fuses, no-load tap changer, secondary breaker, pressure relief device, liquid-level gauge, and lightning arresters are available. A three-phase pad-mounted transformer is illustrated in Fig. 34.14.

Isolating Devices

Equipment used to sectionalize three-phase underground main feeder circuits and provide a means for isolating single-phase and three-phase taps for underground residential and underground commercial circuits may be constructed for pad-mounted or submersible installation. The pad-mounted or direct buried gear may be equipped with air, vacuum, or oil switches. The submersible isolating devices are constructed with vacuum or oil switches. The use of air or vacuum switches minimizes the maintenance of the equipment. The single- and three-phase lateral taps

FIGURE 34.14 Three-phase pad-mounted transformer with underground 7.62-kV primary connecting cables. (*Courtesy MidAmerican Energy Co.*)

originating at the switching cubicles are usually fused to protect the main feeder circuit from an outage as a result of failure of some of the equipment installed as part of one of the lateral circuits.

Installation

An electric underground distribution system serving a residential subdivision normally consists of single-phase primary cable originating on riser poles connected to an overhead distribution circuit through fused cutouts and protected by lightning arresters. Mechanical protection for supply conductors or cables installed on a riser pole shall be provided. This protection should extend least 1 ft below ground level. Supply conductors or cable should rise vertically from the cable trench with only such deviation as necessary to permit a reasonable cable bending radius. Exposed conductive pipes or guards containing supply conductors or cables shall be grounded.

The installation should be designed so that water does not stand in riser pipes above the frost line. Conductors or cables shall be supported in a manner designed to prevent damage to conductors, cables, or terminals. Where conductors or cables enter the riser pipe or connecting bend, they shall be installed in a manner that shall minimize the possibility of damage due to relative movement of the cable and pipe. Risers should be located on the pole in the safest available position with respect to climbing space and possible exposure to traffic damage. The number, size, and location of riser ducts or guards shall be limited to allow adequate access for climbing.

The single-phase primary cables may be supplied by an underground distribution circuit and originating at a sectionalizing switching device similar to those illustrated in Figs. 34.15 through 34.18.

Most electric underground residential distribution systems employ direct-buried cables and pad-mounted transformers installed on transformer pads (Figs. 34.19 and 34.20). Pad-mounted transformers are installed on fiberglass or concrete pads as illustrated in Figs. 34.21 and 34.22.

FIGURE 34.15 Open-door view of one main underground circuit compartment of pad-mounted switchgear. The main circuit cables and terminators have a capacity of 600 amps. Viewing window in the compartment permits the lineman to verify the switch position and check that the switch, when open, has visible open gaps. (*Courtesy S&C Electric Company.*)

FIGURE 34.16 Pad-mounted switchgear showing interrupter switches. The interrupter switches are connected to the main underground distribution circuit to permit sectionalizing the circuit. A bus connected between the two switches in the switchgear permits the feeder circuits and their protective equipment to be tapped to the main circuit between the two switches. (*Courtesy S&C Electric Company.*)

FIGURE 34.17 Closed door pad-mounted switchgear similar to units shown in Figs. 34.15 and 34.16 with power-operated equipment installed on the exterior of the gear to operate the two three-pole interrupter switches. The power-operated equipment permits the interrupter switches to be operated remotely by supervisory control equipment to sectionalize the main underground feeder circuit. (*Courtesy S&C Electric Company.*)

200-amp or 600-amp load break vacuum switch (3Ø, open position)

Separable connectors

Separable connectors

600-amp bushings or 200-amp flower pot universal bushing well for load break or non-load break bushings and elbows

Optional laterals and 1Ø fuse holders

FIGURE 34.18 Schematic diagram of submersible three-phase gang-operated vacuum sectionalizing switch for the underground main feeder circuit.

FIGURE 34.19 Fiberglass pad kept in position with screw anchors being installed by lineman's helper. (*Courtesy A. B. Chance Co.*)

FIGURE 34.20   Fiberglass pad in position with pad-mounted transformer installed. Ground preparation is being completed by lineman and helper. (*Courtesy A. B. Chance Co.*)

FIGURE 34.21   A single-phase pad-mounted transformer concrete pad construction standard diagram. (*Courtesy MidAmerican Energy Company.*)

In rural areas and urban locations free of obstacles underground, cables are frequently installed with plowing equipment. If the cables are to be installed in conduits, cable preassembled in plastic pipe is usually installed by trenching or plowing. Both concentric neutral primary cables and secondary cables are available with this type of construction. The pipe or conduit protecting the cable may be either smooth wall or corrugated wall for greater flexibility.

The distribution facilities should be installed on utility easements or along the streets in public property where the utility has a franchise authorization to make the installation.

Cables should be located to be subject to the least disturbance practical. Cables to be installed parallel to other subsurface structures should not be located directly over or under other subsurface structures; but if this is not practical, the NESC rules on

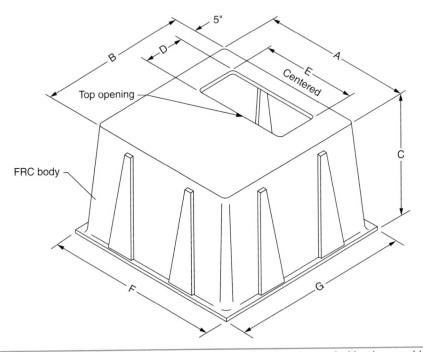

Figure 34.22   A single-phase pad-mounted transformer fiberglass box pad with primary cable basement. (*Courtesy MidAmerican Energy Company.*)

clearances should be followed. Cables should be installed in as straight and direct a line as practical. Where bends are required, the minimum radius shall be sufficiently large to prevent damage to the cable being installed. Cable systems should be routed to allow safe access for construction, inspection, and maintenance. The location of structures in the path of the projected cable route shall, as far as practical, be determined prior to the trenching, plowing, or boring operation.

Routes through unstable soil such as mud, shifting soils, corrosive soils, or other natural hazards should be avoided. If burying is required through areas with natural hazards, the cable should be constructed and installed in such a manner as to protect it from damage. Such protective measures should be compatible with other installations in the area.

Supply cables should not be installed within 60 inches of a swimming pool or its auxiliary equipment. Cables should not be installed directly under building or storage tank foundations. Where a cable must be installed under such a structure, the structure shall be suitably supported to prevent the transfer of a harmful load onto the cable. The installation of cable longitudinally under the ballast section for railroad tracks should be avoided. Where cable must be installed longitudinally under the ballast section of a railroad, it should be located at a depth not less than 60 inches below the top of the rail. Where a cable crosses under railroad tracks, the same clearances shall apply.

The installation of cable longitudinally under traveled surfaces of highways and streets should be avoided. When cable must be installed longitudinally under the road-way, it should be installed in the shoulder or, if this is not practical, within the limits of one lane of traffic to the extent practical. Submarine crossings should be routed and installed so that they will be protected from erosion by tidal action or currents. They should not be located where ships normally anchor.

The horizontal clearance between direct-buried cable and other underground structures shall be controlled at a minimum of 12 inches or larger as necessary to permit access to and maintenance of either facility without damage to the other.

Where a cable crosses under another underground structure, the structure shall be suitably supported to prevent transfer of a harmful load onto the cable system. Where a cable crosses over another underground structure, the cable shall be suitably supported to prevent transfer of a harmful load onto the structure. Adequate support may be provided by installing the facilities with sufficient vertical separation. Adequate vertical clearance shall be maintained to permit access to and maintenance of either facility without damage to the other. A vertical clearance of 12 inches is, in general, considered adequate, but the parties involved may agree to a smaller separation.

If conditions require a cable system to be installed with less than 12 inches of horizontal separation or directly over and parallel to another underground structure, or if another underground structure is to be installed directly over and parallel to a cable, it may be done providing all parties agree as to the method. Adequate vertical clearance shall be maintained to permit access to maintenance of either facility without damage to the other.

Cable should be installed with sufficient clearance from other underground structures, such as steam or cryogenic lines, to avoid thermal damage to the cable. Where it is not practical to provide adequate clearance, a suitable thermal barrier shall be placed between the two facilities.

The bottom of a trench receiving direct-buried cable should be relatively smooth, undisturbed earth, well-tamped earth, or sand. When excavation is in rock or rocky soils, the cable should be laid on a protective layer of well-tamped backfill. Backfill within 4 inches of the cable should be free of materials that may damage the cable. Backfill should be adequately compacted. Machine compaction should not be used within 6 inches of the cable.

Plowing in of cable in soil containing rock or other solid material should be done in such a manner that the solid material will not damage the cable, either during the plowing operation or afterward. The design of cable plowing equipment and the plowing-in operation should be such that the cable will not be damaged by bending, sidewall pressure, or excessive cable tension.

Where a cable system is to be installed by boring and the soil and surface loading conditions are such that solid material in the region may damage the cable, the cable shall be adequately protected.

The distance between the top of the cable and the surface under which it is installed shall be sufficient to protect the cable from injury or damage imposed by expected surface usage. Burial depths as indicated in Table 34.1 are generally considered adequate.

Volts	Depth of Burial, inches
600 and below	24
601 to 22,000	30
22,001 to 40,000	36
40,001 and above	42

TABLE 34.1 Burial Depths for Supply Conductors or Cables

In areas where frost conditions could damage cables, greater burial depths than indicated in Table 34.1 may be desirable. Lesser depths than indicated may be used where supplemental protection is provided. Where the surface under which a cable is to be installed is not to final grade, the cable should be placed so as to meet or exceed the requirements indicated in Table 34.1 both at the time of installation and subsequent thereto.

The conductors or cables of a supply circuit and those of another supply circuit may be buried together at the same depth with no deliberate separation between facilities provided all parties involved agree (Fig. 34.23).

The conductors or cables of a communication circuit and those of another communication circuit may be buried together and at the same depth with no deliberate separation between facilities provided all parties involved agree.

FIGURE 34.23 Underground distribution transformer, pedestal, and cable layout for joint construction of telephone, gas, and 13,200Y/7,620-volt electric distribution circuit.

Supply cables or conductors and communication cables or conductors may be buried together at the same depth with no deliberate separation between facilities provided all parties involved agree and the following requirements are met:

1. Grounded supply systems shall not be operated more than 22,000 volts to ground.
2. Ungrounded supply systems shall not be operated more than 5,300 volts, phase to phase.

A supply facility operating above 300 volts to ground must include a bare, grounded conductor in continuous contact with earth. This conductor, adequate for the expected magnitude and duration of the fault current which may be imposed, shall be one of the following:

1. A sheath and/or shield.
2. Multiple concentric conductors closely spaced circumferentially.
3. A separate bare conductor in contact with the earth and near the cable where such cable or cables also have a grounded sheath or shield not necessarily in contact with the earth. The sheath and/or shield as well as the bare conductor shall be adequate for the expected magnitude and duration of the fault currents which may be imposed.

Where a buried cable passes through a short section of conduit, as under a roadway, the contact with earth of the grounded conductor can be omitted provided the grounded conductor is continuous through the conduit. The bare conductors in contact with the earth shall be of suitable corrosion-resistant material. Cables of an ungrounded system operating above 300 volts shall be effectively grounded concentric shield construction in continuous contact with the earth. Such cables shall be maintained near each other.

More than one cable system buried in random separation may be treated as one system when considering clearance from other underground structures or facilities.

The transformer pads and riser poles should be installed prior to placing the cable in the ground. The trenches for the direct-buried cables are usually dug with a small trencher. Care exercised in handling the cable during installation, as illustrated in Figs. 34.24 and 34.25, will help to avoid trouble later, for damage sustained by the cable during installation has proved to be a major cause of subsequent cable failure.

FIGURE 34.24 Linemen installing electric, telephone, and cable-television cables in a common trench with no deliberate separation between cables. (*Courtesy MidAmerican Energy Co.*)

FIGURE 34.25   Primary and secondary mains buried directly in ground. Primary cable is two-conductor concentric cable having neutral ground spiraled around outside of cable. (*Courtesy Commonwealth Edison Co.*)

The trench should be backfilled carefully. The cable can be best protected by using screened dirt or sand to provide several inches of cover over the cables. The backfill should be tamped to prevent settling. Tamping tools should not be allowed to hit the cables.

The cables installed on riser poles are terminated at the top of the pole with a cable terminator, as shown in Fig. 34.26. The manufacturer's instructions must be followed carefully to prepare the cable for insertion in the terminator body. It is necessary to remove each cable layer the correct distances and insert the cable in the terminator properly.

The cables in switching cubicles and transformer enclosures are normally terminated with load-break elbow terminators. Figure 34.27 is a cross-sectional sketch of a typical load-break elbow terminator illustrating how cable is prepared and installed.

Correct Installation of Termination

FIGURE 34.26   Typical riser pole termination makeup for use with a two-conductor No. 2 AWG concentric neutral single-phase 7,620-volt cable used for underground distribution. (*Courtesy 3M Co.*)

Pulling eye

Male contact

Arc follower

Conductive shielding

Crimp connector

Silicon valve

Cable

Semicon tape, insulating tape, and PVC jacket tape

Cable conductive sheath

FIGURE 34.27 Cross-sectional view of load-break terminator with component parts identified. (*Courtesy Cooper Power Systems.*)

Figures 34.28 and 34.29 show a 7,620-volt load-break elbow terminator being operated by a lineman with a hot-line tool to energize a pad-mounted transformer.

Aluminum and copper cable is available in desired lengths, so that splices are seldom required. Prefabricated high-voltage splices are available. It is necessary to follow the manufacturer's recommendations to properly prepare the cable for installing a prefabricated splice. The semiconducting insulation shielding must be removed from a section of the insulation to permit the insulation buildup at the splice location. Stress cones in the prefabricated splice must contact the insulation shielding to maintain the shielding across the splice. Aluminum connectors and terminating devices should be used with aluminum conductors so as to avoid differential expansion and contraction that could result from the use of dissimilar metals.

FIGURE 34.28 Load-break elbow termination in position to be connected to bushing on a pad-mounted transformer. Lineman holds energized terminator in position with hot-line tool. (*Courtesy I. T. T. Blackburn Co.*)

FIGURE 34.29 Load-break elbow terminator installed or closed in on a bushing of pad-mounted transformer by lineman using hot-line tool. (*Courtesy I. T. T. Blackburn Co.*)

The low-voltage leads may go directly from the transformer to the customer, or secondaries may be installed to service pedestals and the services to the customer extended from the service pedestals. If secondaries are installed, they will be in the same trench with the primary cable in most instances, with no definite separation. Figures 34.30 through 34.32 illustrate the installation of a below-grade service enclosure and the secondary and service wire connections completed in the enclosure.

The service wires originating at a below-grade secondary enclosure or a transformer such as the one illustrated in Fig. 34.33 extend directly to the customer's house or meter pedestal. At the customer's house, the service wires usually enter a conduit and terminate in a meter socket. The diagram (Fig. 34.34) shows a typical method of terminating the service wires at a house. The installation of an underground service conduit and electric meter mounted on the side of the residential structure is shown in Fig. 34.35.

FIGURE 34.30   Typical convergence of secondary and service cables in a trench, below grade, below where the service pedestal is installed. The service pedestal is installed to completely cover all the cables. (*Courtesy I. T. T. Blackburn Co.*)

FIGURE 34.31   The service enclosure (not visible) has been placed in the ground. The lineman has slightly penciled the cable insulation and is slipping sealing sleeves on the cable. (*Courtesy I. T. T. Blackburn Co.*)

FIGURE **34.32** The lineman is applying silicon lubricant to the insulated neck of the bus bar and to the inside of the sealing sleeve to facilitate removal of the sealing sleeve at some future time. (*Courtesy I. T. T. Blackburn Co.*)

FIGURE **34.33** Lineman connecting service wires to the low-voltage terminals of a pad-mounted transformer. (*Courtesy General Electric Co.*)

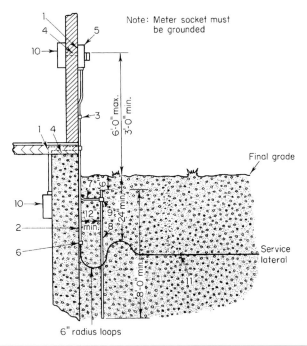

FIGURE **34.34** Typical specification diagram for installing underground service wires at customer's house for 120/240-volt three-wire single-phase supply. (1) Service entrance conductors. Minimum recommended capacity 200 amps. (2) Galvanized steel rigid conduit 2½ inch. (3) Galvanized steel conduit strap 2½ inch. (4) Galvanized steel rigid conduit 2½ inch. (5) Outdoor meter socket with 2½-inch conduit hubs. Minimum recommended capacity 200 amps. (6) End bushing for 2½-inch conduit. (7) Ground wire. (8) Ground rod copperweld ½ inch minimum or galvanized water pipe ¾ inch minimum. (9) Connector. (10) Fused switch, circuit breaker, or fuse box. Minimum recommended size 200 amps. Two possible locations shown. (11) Underground service conductors. Minimum recommended capacity 200 amps.

FɪɢᴜʀE **34.35**   Meter mounted at the customer's house.

Pad-Mount Primary Meter Cabinets

Large industrial or commercial customers served from an underground electric distribution system may have the need for the installation of a pad-mount primary meter cabinet. The meter cabinet contains a primary and secondary compartment. The primary compartment contains the high-voltage primary cable that is landed on the high-voltage bushings. The secondary compartment contains the current transformers and potential transformers with connections and leads to an electric meter.

Pad-Mounted Capacitor

The electric underground distribution circuit may be VAR deficient and require power factor correction. The pad-mounted capacitor is suited to this particular requirement.

Streetlights

Electric underground distribution areas usually have streetlights installed on ornamental poles with underground supply conductors. The standard is usually fabricated from aluminum or precast concrete. The luminaire is generally constructed for use with mercury-vapor or high-pressure sodium lamps. The underground supply cables are connected to the terminals of a transformer or the connectors in a service enclosure.

Fault Indicators

When a fault occurs on an underground circuit, it is important to locate the source of the trouble quickly and restore service with minimum outage time. Fault indicators installed at strategic points directly on the underground distribution cable in the system will aid in identifying the trouble location. The fault indicators are usually equipped with a neon light that glows, indicating fault current passed through the cable. The fault locators are waterproof and reset automatically after normal operating current is detected.

Protection

Supply circuits operating above 300 volts to ground or 600 volts between conductors shall be so constructed, operated, and maintained that, when faulted, they shall be promptly deenergized initially or following subsequent protective-device operation. Ungrounded supply circuits operating above 300 volts shall be equipped with a ground-fault interruption system. Communication protective devices shall be adequate for the voltage and currents expected to be impressed on them in the event of contact with the supply conductors. Adequate bonding shall be provided between the effectively grounded supply conductors and the communication cable shield or sheath, preferably at intervals not to exceed 1,000 ft. In the vicinity of supply stations where large current flow, the effect of these currents on communication circuits should be evaluated before communication cables are placed in random separation with supply cables.

Troubleshooting Underground Distribution Circuits

Underground electric distribution systems have advantages compared to overhead systems and some disadvantages which, if not recognized, could present hazards to linemen, cablemen, and the public. Underground distribution systems are innocent in appearance without exposed bare conductors and insulators to remind the lineman of the presence of primary voltages. Switchgear, transformers, and pedestals are easily accessible at ground level, rather than being located at a height which minimizes the likelihood that unqualified personnel will be in close proximity to them. All ground-level equipment must be securely labeled, locked, and/or protected to prevent the public, especially children, from gaining access to the energized parts. The relatively small dimensions of equipment used in underground distribution systems result in confined working space and close clearances between energized parts. Linemen working on underground distribution systems are generally in contact with ground, providing one necessary ingredient for an accidental electric shock.

The first indication that trouble exists on an underground electric distribution system is generally a call from a customer stating that he has power outage and/or the operation of a substation circuit breaker or circuit recloser. The reclosing of a substation circuit breaker or a recloser, supplying power to an underground circuit, should be kept to a minimum to limit the damage to the equipment. Test reclosing of the source equipment seldom restores service to the customers and may result in high-fault currents that can damage cables or other facilities.

A fuse in series with the failed cable will normally blow to isolate the trouble from the main circuit, thus permitting the restoration of service to most customers.

Information received from the customers' calls can normally provide the approximate location of the trouble. An inspection by the lineman of the route of the underground distribution circuit will often reveal the location of the trouble. An inspection inside the switchgear and pad-mounted devices may reveal to the lineman or cableman damaged components. Cable equipped with fault indicators will help the lineman isolate the faulted cable section. The fault indicators indicate the flow of fault current even after the circuit has been deenergized. The fault indicators reset automatically when the circuit has been repaired and reenergized.

Most underground distribution circuits use looped cable routes to permit the lineman to isolate the failed cable between two transformers and restore electric service to the customers. A secondary cable failure, a transformer failure, or trouble on a radial system may make it necessary to install temporary facilities such as cables across the surface of the ground to restore electric service.

Cable Fault Detection

A faulted cable is isolated by the lineman using a hot stick and properly insulated rubber gloves. The faulted cable is tested to ensure the cable is deenergized. The location of the cable faulted is precisely located using with either:

1. High-voltage impulse fault locator (Thumper),
2. Cable fault finder,
3. Distributed sensor-based approach.

In the case of the cable fault locator method, a signal transmitter is used to impress a power tone signal on the isolated deenergized faulted cable. The cable fault locator is used to follow the signal on the cable to the fault location by observing the cable fault finding instrument or with the use of an earphone headset.

Once the exact location of the cable fault is identified, it will need to be dug up and repaired. The cable failure cause due to corrosion, water trees, mechanical stress, dig-in overheating, animal damage, etc. is recorded and reported.

One Call System

Prior to digging, the location of underground cables and other buried facilities should be determined utilizing one call locating services. In most jurisdictions in the United States, an 811 One Call system is used (Fig. 34.36). The location address, City, County, State, Township, Range and Section Number information will be requested if the global positioning satellite point is unknown. All underground facilities, including the electric, will be precisely located and marked. American Public Works Association (APWA) uniform color code for locating underground facilities is adhered to (see Fig. 34.37). Always remember to "Call before you dig" (Fig. 34.38).

FIGURE 34.36 The 811 reminder telephone number logo for underground locates. (*Courtesy American Public Works Association.*)

APWA UNIFORM COLOR CODE FOR LOCATING UTILITY LINES:

Red: Electric Power Lines, Cables, Conduit and Lighting Cables

Yellow: Gas, Oil, Steam, Petroleum or Gaseous Materials

Orange: Communication, Alarm or Signal Lines, Cables or Conduit

Purple: Reclaimed Water, Irrigation and Slurry Lines

Blue: Potable Water

Green: Sewer and Drain Lines

Pink: Temporary Survey Markings

White: Proposed Excavation

LOCATING AND MARKING BEST PRACTICES

Locating and marking is necessary before excavation can be carried out safely. Without accurate and timely location and effective marking of facilities, excavator damage to underground facilities is more likely and may have severe consequences.

4–1: Locators use available facility records at all times.

4–2: If a facility locator becomes aware of an error or omission, then the facility locator provides information for updating records that are in error or for adding new facilities.

4–3: A uniform color code and set of marking symbols is adopted nationwide.

4–4: A single locator is used for multiple facilities.

4–5: Locators are properly trained. Locator training is documented.

4–6: Locates are performed safely.

4–7: A visual inspection is completed during the facility locating process.

4–8: Facilities are adequately marked for conditions.

4–9: Positive response is provided to facility locate requests.

4–10: Multiple facilities in the same trench are marked individually and with corridor markers.

4–11: Information on abandoned facilities is provided when possible.

4–12: When locating electromagnetically, active/conductive locating is preferable to passive/inductive locating.

4–13: The facility owner/operator is identified.

4–14: Communication is established between all parties.

4–15: Documentation of work performed on a locate is maintained.

4–16: A damaged facility is investigated as soon as possible after occurrence of damage.

4–17: Forecasting/planning for predictable workload fluctuation. A plan is developed for managing unpredictable fluctuations.

4–18: Underground facility owners/operators have a quality assurance program in place for monitoring the locating and marking of facilities.

4–19: All stakeholders adhere to all best practices and the general guidelines prior to, during, and after any trenchless excavation (as applicable).

4–20: Permanent markers are placed as close as practical at the entrance and exit points of facilities located underneath bodies of water where facilities are at risk of being damaged. Temporary markers are placed within the areas of proposed excavations as close as practical over facilities that are submerged in bodies of water where facilities are at risk of being damaged without impeding or creating additional hazards.

4–21: A service line is marked in response to a locate request to the operator who uses the service line to pursue a business that derives revenue by providing a product or service to an end-use customer via the service line or a governmental entity that provides a product or service to an end-use customer via the service line.

4–22: Facility operators ensure that new facilities in areas with continuing excavation activity are marked upon installation to indicate their presence.

Go to **CommonGroundAlliance.com** for the latest version of CGA Best Practices. Download or view online the Practice Description for each Practice Statement above.

RESOURCES

- call811.com
- CommonGroundAlliance.com
- CGAconference.com
- dp-pro.com
- ExcavationSafetyUniverity.com
- ExcavationSafetyGuide.com
- Locaterodeo.com
- LocatorSafety.com
- NULCA.com
- OSHA.gov
- phmsa.dot.gov/pipeline

INTERNATIONAL LOCATE PROFESSIONALS SYMPOSIUM

Held EVERY MARCH in conjunction with the CGA 811 Excavation Safety Conference & Expo, the International Locate Professionals Symposium is an annual event hosted by NULCA that provides a forum for locators around the world to meet and exchange ideas. The symposium includes the New Locate Technology Forum, in-depth workshops, educational sessions specific to the locating industry, and opportunities to network with other locate professionals.

– Learn about current technologies
– Develop best practices
– Improve quality
– Gain insight on damage investigation and claims resolution

LocateSummit.com

EXCAVATION SAFETY UNIVERSITY

Choose from a selection of training DVDs and Field Guides developed to help locators learn in the office AND in the field:

- Basic Locating Theory
- Basic Locating Skills
- Locating Best Practices
- Locating Safely: Avoiding Hazards
- Damage Investigation and Claims Resolution

Locating Buried Utilities: A Professional Approach
A full-length, full-color, illustrated textbook providing comprehensive details on how to locate buried utilities using standard electromagnetic equipment. With chapters specific to every industry.

BE SAFE OUT THERE TODAY

Top 10 Locator Safety Hazards:

10. Confined Space
9. Threats to the Eye
8. Climate and Weather
7. Dog Bites
6. Punctures and Foot Trauma
5. Poison Ivy and Other Skin Threats
4. Insect Bites and Stings
3. Walking, Lifting, Bending and Squatting
2. Slips, Trips and Falls
1. Time on the Road

Learn more about protecting yourself against these threats at **ExcavationSafetyUniversity.com**

MARKS

Membership in your local One Call Center

Accurately locate and mark underground facilities

Respond to locate requests as prescribed by law and in a timely manner

Keep accurate records of underground facility locations

Safety is a shared responsibility

FIGURE 34.37 Best practices used for locating purposes. (*Courtesy American Public Works Association.*)

Figure 34.38   The 911 one-call locate employee prepared to complete an assignment.

When the cable is exposed at the fault location, it should not be handled until it is grounded, ensuring that it is deenergized. Dead end elbows may contain test points that may be utilized. The cable can be grounded by using a "cable spear." The cable spear is operated with a hot-line tool (shotgun). The cable spear has a C-clamp arrangement with a pointed threaded stud. The clamp is placed around the faulted cable thought to be deenergized, and the stud is twisted with the shotgun until the C-clamp is in contact with the concentric neutral conductor and the pointed stud has penetrated the insulation and made contact with the insulated conductor.

The faulted cable may be repaired with a prefabricated splice. If it is necessary to replace the cable, precautions should be taken to minimize the damage to the terrain.

Expedited Interruption Solution

When faulted cable interruptions occur for sensitive customer loads served from underground distribution systems, one expedited solution that is in place at a major Midwest utility is the use of an above ground emergency cable system. The cable system is contained in emergency response trailers for transport to the site (Fig. 34.39).

Typically, restoration times for faulted cable interruptions are 10 to 12 hours or more. The use of the emergency cable system reduces this outage duration to at least half that time duration. The emergency response cables are of great benefit to those customer loads that must be restored immediately.

The underground response trailer is equipped with heavily insulated 15-kV mining cable on rolling designed reels terminated at both the start and end of the cable reel (Fig. 34.40). Other equipment on the trailer include:

- Warning signs
- Primary enclosures
- Secondary enclosures

FIGURE 34.39   Emergency response trailer at a training and demonstration site. (*Courtesy MidAmerican Energy Company.*)

FIGURE 34.40   The lineman is taking an inventory of the heavily insulated 15-kV mining cable reels stored on the trailer. (*Courtesy MidAmerican Energy Company.*)

- Traffic barricade cones
- Cable troughs
- Cable fault locating machine
- Generator
- Feed-through load-break bushings

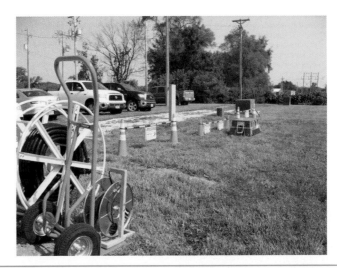

FIGURE 34.41   A view of the trial demonstration site from source underground pad-mounted transformer to the radially feed pad-mounted transformer. The 150 ft reel of the heavily insulated 15-kV mining cable and neutral wire are positioned for use. (*Courtesy MidAmerican Energy Company.*)

- Neutral cable
- Ground wire
- Load-break cutouts
- Fuse links
- Underground cable locator

Once it has been determined, where in the underground distribution cable system that the failed cable section is using the underground cable fault locator or pad mount transformer fault indicators, the failed conductor is isolated from the rest of the system. After isolation of the failed cable, the next step to restore service with equipment from the emergency response trailer without delay is to select the correct reel length for the job and to begin placing materials in position. The simplest scenario may be that only one cable reel will be required between two transformers with load-break elbows on each end. Other situations may require a load-break elbow on one end and a terminator at the other. In yet another possible example, a primary enclosure and two cable reels may be required.

Proper signage of the cable and equipment is a must. The cable guards are placed to allow vehicle traffic without damaging the cable.

Now that materials are in place, the final step is connecting the emergency cable into the system (Fig. 34.41). The grounds and bleed wires are connected first. If required, a primary enclosure is placed along the route. The primary enclosure is equipped with load-break feed-through bushings. The emergency cables are removed after the permanent faulted cables are repaired and put back in service.

Vegetation Management— Tree Trimming

V egetation management is a specialized field that goes far beyond the trimming of trees. Vegetation management involves right-of-way ground clearance as well as line clearance issues. Tree trimming must be completed in a manner that is environmentally acceptable. The reader is encouraged to review the following two industry standards for tree care operations: ANSI A300, *Tree, Shrub, and Other Woody Plant Maintenance,* and ANSI Z133.1, *Pruning, Trimming, Repairing, Maintaining, and Removing Trees, and Cutting Brush—Safety Requirements.*

Line Clearance Objectives

Line clearance is preventive line maintenance to ensure that the utility's service to its customers is not interrupted as the result of interference with conductors or circuit equipment by growing trees. Elements include removal of dangerous trees and overhangs, trimming to clear the conductors, and clearing transmission right-of-way. Natural tree growth and storm-tossed branches can ground or break distribution and transmission lines, interrupting electric service and endangering the public. The trimming process is intended to anticipate such a possibility by removing this hazard. Lines are checked and cleared on a planned time cycle. The amount of clearance sought should provide hazard-free operation for at least 2 years. It should be accomplished while maintaining the health and beauty of the trees involved, the goodwill of property owners and public authorities, and the safety of the trimming crew and the public.

Clearing the Right-of-Way

Practically all overhead lines will traverse through some brush or timberlands. Even lines built in the Middle West, an area generally thought of as level prairie, may pass through forests or hilly country covered with shrubs and underbrush. Lines built in the mountainous country of the West or the hills of the East naturally pass through areas in which the right-of-way must be cleared before construction can be started.

In clearing the right-of-way, all stumps should be cut low. All logs and brush should be removed for a distance of at least 25 ft under each conductor so that there is ample room to assemble and erect poles or towers and later to string the line conductors.

Any underbrush or piles of dead wood should be removed. Otherwise, they may catch fire and burn the line or anneal the conductors and cause them to sag abnormally.

If it is possible, permission should be obtained to cut any extremely tall trees immediately adjacent to the right-of-way. All dead limbs and branches in the adjoining strip also should be cut down since a high wind may blow them into the line. In case of a high-voltage line, it is more important to remove so-called danger trees along the side of the right-of-way.

Trees which would reach within 5 ft of a point underneath the outside conductor in falling are known as danger trees.

Herbicide Brush Control

Herbicides are a means of which the growth of brush may be controlled. The herbicide is sprayed onto the basal area of the shrubs or small trees from the ground line up to a height of 12 inches to 15 inches above ground. The rest of the shrub need not be sprayed. However, complete encirclement of the stem or trunk with the spray solution is necessary. When applied to stumps, the spray prevents resprouting. Herbicides registered by the Environmental Protection Agency have been tested by power companies and found to be effective. The herbicides can be used for basal bark injection or cut-stump treatments.

Factors in Trimming Techniques

There are fundamentals essential to safe and competent trimming operations. Primary fundamentals among these are (1) knowing how to climb and use a rope, (2) knowing how to tie essential knots, (3) knowledge of tree species—their growth characteristics and wood strengths, (4) knowledge of electrical conductors, (5) knowledge of alternative trimming methods, and (6) knowing how to cut limbs and lower them under full control. Ideally, climbers should be trained to recognize structural problems such as weak crotches and chronic-disease symptoms, so that trimming and tree removal decisions result in optimum accomplishment.

Time for Pruning

Trees may be pruned at any time of the year, and utility line clearance operations are maintained throughout the year. In the past, work often was concentrated into a particular season. However, today, it is recognized that there are compensating values and factors, so that trimming operations run year-round except where there are unique, limiting local considerations.

A Basic Set of Pruning Tools

1. One light power chain saw (16-inch cutter bar length)
2. One handsaw and scabbard
3. One 12-ft pole saw
4. One 12-ft pole clip or pruner
5. One ½-inch polyester climbing rope (120 ft)
6. Several hand lines (½-inch polyester)
7. One ¾-inch bull rope (150 ft)
8. One wooden 24-ft extension ladder
9. One dielectric hard hat for each crew member
10. Tree-wound paint in a pot or spray can

11. Belt snaps for handsaw and paint

12. A set of road signs and flags warning of men working in the trees

Some of the pruning tools appear in Figs. 35.1 and 35.2.

Other items should be considered, especially when working in residential areas where cleanup is vital: a pair of loppers, a sledge and wedges, a brush rake, a street broom, and scoop shovel.

Climbing Equipment

For safety and proficiency, climbers should always tie into a tree crotch before beginning to trim. The elements involved are a body sling or saddle and a climbing rope. The climbing rope is a ½-inch polyester rope, 120 ft long, attached to the snap of the sling which is in front at the center and then secured by a taut-line hitch. The method of tying

FIGURE 35.1 Trimming hand tools. Starting at the top, tools are a center-cut bull clip, a side-cut pole clip, a center-cut pole clip, a pole saw, and a pair of loppers. (*Courtesy Asplundh Tree Expert Co.*)

FIGURE 35.2 Two types of handsaws. Upper saw cuts on the push, lower saw cuts on the pull. Use is primarily a matter of personal preference. (*Courtesy Asplundh Tree Expert Co.*)

STANDING ROPE

X

Y

Rope fastened to safety sling
with clove hitch or
a round turn and two
half hitches.

DETAIL - A
METHOD OF TYING

DETAIL - B
COMPLETED HITCH

FIGURE 35.3   Tying a taut-line hitch to a standing rope. (*Courtesy Edison Electric Institute.*)

the rope with a taut-line hitch is covered in detail in Fig. 35.3. This combination permits the climber to move around a tree freely and with safety. Proper use distributes the climber's weight, so he will tire less easily and be able to move onto branches which would not otherwise support his full weight.

Permission

No trimming should be started until permission is obtained from the property owner or the state, county, or municipal agency with jurisdiction. Verbal permission is usually satisfactory for routine trimming, but written permission may be advisable for heavy pruning or tree removals. If the property owner or the public authorities refuse to give permission to trim or remove trees endangering the power lines, a permanent record should be maintained for future reference. The trimming should then be done under the direction of the person who secured the permission, since he can best interpret any special directions or limitations imposed. To ensure goodwill necessary for future trimming cycles, the work must be professional and the cleanup thorough.

Good and Bad Methods

Before providing the details of specific tree-trimming techniques, it will be useful to review good and bad general methods. Pollarding and shearing or rounding over are undesirable because the visual effect is ugly and contrary to normal tree form, and because these methods represent uneconomical line clearance. Many small cuts take unnecessary time, create an unhealthy tree condition, and stimulate rapid regrowth back into the conductors. Natural trimming is desirable. In this method, branches are flush cut at a suitable parent limb back toward the center of the tree. This method is also called *drop-crotching* or *lateral trimming*. It involves fewer but heavier cuts, usually made with a saw, not a pole pruner. This natural pruning generally permits more clearance, so that trimming cycles can be lengthened. In addition, better tree form and value are maintained, and regrowth direction can be influenced. Note the examples in Figs. 35.4 through 35.8.

Removing Large Branches

The procedure in the removal of a large side limb is shown in Fig. 35.9. The first cut is an undercut 10 inches to 12 inches out from the place where the final flush cut is to be made. This cut should be one-quarter to one-half the distance through. The second cut

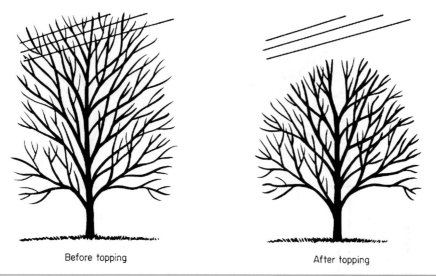

Before topping   After topping

FIGURE 35.4 *Topping.* Topping is cutting back large portions of the upper crown of the tree. Topping is often required when a tree is located directly beneath a line. The main leader or leaders are cut back to suitable lateral. (The lateral should be at least one-third the diameter of the limb being removed.) Most cuts should be made with a saw; the pole pruner is used only to get some of the high lateral branches. For the sake of appearance and the amount of regrowth, it is best not to remove more than one-fourth of the crown when topping. In certain species—sugar maple, for example—removal of too much of the crown will result in the death of the tree.

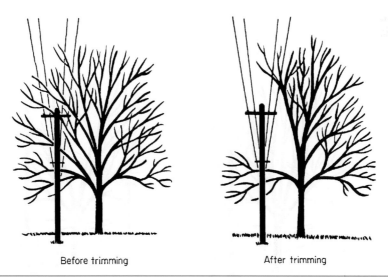

Before trimming   After trimming

FIGURE 35.5 *Side trimming.* Side trimming consists of cutting back or removing the side branches that are threatening the conductors. Side trimming is required where trees are growing adjacent to utility lines. Limbs should be removed at a lateral branch. Unsightly notches in the tree should be avoided if possible. Shortening branches above and below the indented area or balancing the opposite side of the crown will usually improve the appearance of the tree. When trimming, remove all dead branches above the wires, since this dead wood could easily break off and cause an interruption.

Before trimming   After trimming

FIGURE 35.6 *Under trimming*. Under trimming involves removing limbs beneath the tree crown to allow wires to pass below the tree. To preserve the symmetry of the tree, lower limbs on the opposite side of the tree should also be removed. All cuts should be flush to avoid leaving unsightly stubs. The natural shape of the tree is retained in this type of trimming, and the tree can continue its normal growth. Overhangs are a hazard, however, when a line passes beneath a tree. Overhangs should be removed in accordance with the species of tree, location, and the general policy of the utility that you work for. When trimming, remove all dead branches above the wires, since this dead wood could easily break off and cause an interruption. Many utilities have a removal program set up for trees that overhang important lines.

Before trimming   After trimming

FIGURE 35.7 *Through trimming*. Through trimming is the removal of branches within the crown to allow lines to pass through the tree. It is best suited for secondaries, streetlight circuits, and covered cables, although it is often used on primary circuits where there is no other way of trimming the tree. Cuts should be made at crotches to encourage growth away from the lines.

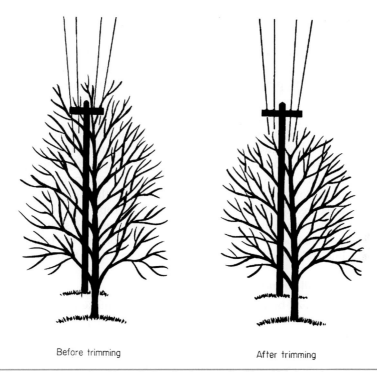

Before trimming

After trimming

FIGURE 35.8 *Natural trimming*. Natural trimming is a method by which branches are cut flush at a suitable parent limb back toward the center of the tree. This method of trimming is sometimes called drop-crotching or lateral trimming. Large branches should be removed to laterals at least one-third the diameter of the branch being removed. Natural trimming is especially adapted to the topping of large trees, where a great deal of wood must be removed. In natural trimming, almost all cuts are made with a saw, and very little pole pruner work is required. This when finished results in a natural-looking tree, even if a large amount of wood has been removed. Natural trimming is also directional trimming, since it tends to guide the growth of the tree away from the wires. Stubbing or pole-clip clearance, on the other hand, tends to promote rapid sucker growth right back into the conductors. The big factor to remember is that natural clearance does work and that two or three trimming cycles done in this manner will bring about an ideal situation for both the utility and the tree owner. Most shade trees lend themselves easily to this type of trimming. Elm, Norway maple, red oak, red maple, sugar maple, silver maple, and European linden are the most common street trees, and these species react especially well to natural clearance methods.

Cut No 2

Cut No 4 (flush with trunk)

Cut No.1

6"

Bark will strip down the trunk here - - - - - - - -

12"

Cut No. 3 (flush with trunk)

For small branches omit cuts No 1 & 2

Right Way

Wrong Way

FIGURE 35.9 Steps in right and wrong removal of a large side limb. (*Courtesy REA.*)

is made about 1 inch to 6 inches farther out on the upper side of the branch. The reason for shifting out a few inches is to prevent the bark from being stripped back beyond the point of the final cut when the limb falls. When the bulk of the weight has been removed, the finish cut is made flush with the parent limb or trunk in two steps. First an undercut to prevent bark stripping is made, and finally the limb is sawed through from above.

Lowering Large Limbs

Large branches that might cause damage to the line conductors or other property in falling should be carefully lowered with tackle. Figure 35.10 illustrates the manner of lowering a large limb which would otherwise fall directly into the line conductors. Before the branch is sawed off, it is supported with two ropes, one at the butt and the other well out toward the end. The ropes are run through crotches in the branches above and secured to the trunk of the tree. A third rope—known as the guide rope—is fastened to the branch at such a point that it can be used to swing the branch out from over the line conductors. After the sawing is completed, the branch is swung out to clear the line and then gradually lowered by means of the two supporting ropes.

Removing Small Branches

In the case of small branches, it is not necessary to make three separate cuts. One cut with the saw close up against the limb while it is held in place, as shown in Fig. 35.11, is

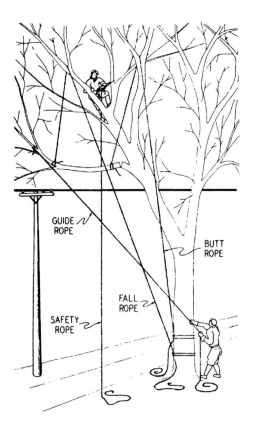

FIGURE **35.10** Lowering large or hazardous limb with rope tackle. The two supporting ropes are called butt and fall ropes; the third rope is the guide rope. (*Courtesy Edison Electric Institute.*)

FIGURE 35.11 How a small branch can be removed with a single cut.

FIGURE 35.11 How a small branch can be removed with a single cut.

all that is necessary. Small branches can also be cut off with a pole pruner. When trees are being trimmed, all dead branches should be pruned out.

Brush Disposal

All brush and wood trimmed in line clearance operations must be disposed of in such a way as to maintain the goodwill of property owners. In residential trimming, it is usually most economical to saw the large limbs and main trunk of the tree into 3 ft to 6 ft lengths for easier handling and disposal. The small limbs, branches, and brush are chipped into small 4-inch wood chips and saw dust. The debris is used as mulch. Note Figs. 35.12 and 35.13 illustrate the woodchipper. Vegetation management personnel exhibit extreme care when feeding limbs, branches, and brush into the woodchipper hopper. The chipper is a powerful tool that only trained personnel utilize. The chipper's feeding mechanism consisting of a hopper with a collar is designed with safety guards.

Hands and feet should always remain clear of the feed chute. A push stick is often utilized during feeding.

Public relations generally require that the work area be raked or swept up. Material suitable for fireplaces should be cut to size and neatly stacked. It should first be offered to the property owner, and if he rejects it, it may be left available for the public. However, some situations may require that the wood be hauled away and dumped (Fig. 35.14).

FIGURE 35.12 A typical woodchipper unit utilized by the tree crew. (*Courtesy Carlton.*)

FIGURE 35.13   A manual climbing crew in operation. Note the groundman chipping brush behind the climbers. The chute of the woodchipper directs the chips into the back of the dump van. (*Courtesy Asplundh Tree Expert Co.*)

FIGURE 35.14   A helicopter is being used to transport large limbs removed from an electric transmission line right-of-way to a central location for disposal. (*Courtesy Asplundh Tree Expert Co.*)

On right-of-way clearance or when trimming in remote areas, it is sometimes possible to pile brush and leave it as game cover. Generally, burning is no longer permitted because of concern about smoke and fly ash.

Power Equipment

Technology responds to operational needs, and various pieces of specialized equipment are designed to solve special problems or to increase productivity (Fig. 35.15).

Trim Lift

Wherever a truck can be driven, an aerial device can be effectively used to position a man for trimming, thus saving the trouble of climbing. The mechanism involves two connected booms, with a work platform or bucket at one end, mounted on a truck chassis. The upper boom is supported by the lower boom through the use of cables. A hydraulic power system runs off the truck engine and lifts the booms, operates the tools, and lifts the dump bodies to discharge chips and other materials gathered in trimming (Figs. 35.16 and 35.17). Mechanical outriggers are used to stabilize the unit during operation. Generally, the booms and the work platform are fully insulated from ground to protect the operator from shock and the truck body from becoming energized if the boom or platform should come in contact with a conductor.

All-Terrain Tree Crane

This type of device is used primarily for taking down trees too tall to reach with a trim lift. The all-terrain equipment utilizes a boom that telescopes and is attached to the tree by a tree crew member so that the crane operator may safely lower the portion

FIGURE 35.15   A large tree limb chipper equipped with a mechanical in-feed system for clearing overgrown ROW around electric transmission and distribution lines. (*Courtesy Asplundh Tree Expert Co.*)

FIGURE 35.16   A trim lift and chipper in use by crew to trim trees and dispose of brush. (*Courtesy Asplundh Tree Expert Co.*)

of the tree that has been cut to the ground (Fig. 35.18). Entire trees, or portions of cut trees, weighing between 3,000 lbs to 6,000 lbs are removed using the crane.

Power Tools

Chain and circular-blade saws are powered by compact gasoline engines and designed in a variety of sizes and weights to adapt to the many conditions encountered, from take-downs to tree trimming to bush cutting (Figs. 35.19 and 35.20). Noise levels generally dictate hearing protection for the operator.

A variety of cutting tools can be hydraulically powered or air-operated by a system connected with the bucket of a trim lift. These include pole chain saws, circular saws, and limb loppers illustrated in Fig. 35.21. Some situations, such as hospital areas, may require the use of electrically powered tools operated by a portable gasoline generator. Power equipment requires systematic inspection and preventive maintenance to ensure crew safety and efficiency.

Poisonous Plants

When trimming trees and removing brush there is always the possibility that a lineman may encounter plants which may cause skin poisoning. These may include the following:

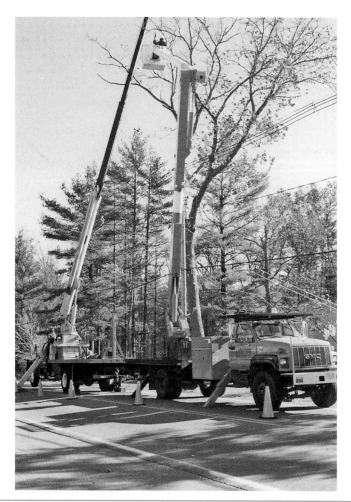

FIGURE 35.17 A specialized right-of-way trim lift with the boom mechanism mounted on a diesel-powered, four-wheel-drive vehicle. (*Courtesy Asplundh Tree Expert Co.*)

Poison Ivy

This plant sometimes climbs trees and poles. It is a crawling plant. Poison ivy has broad and glossy leaves that always grow in clusters of three leaflets, two branching from the stem, 1 inch or less below the center one.

Poison Oak

A low-growing erect plant, poison oak never climbs. The leaves are broad but not always glossy and occur in clusters of three. The dark-green leaves are permanently and very heavily hairy on the undersurface.

Poison Sumac

This shrub or small tree may grow to a height of 20 ft. Its smooth glossy leaves are in the form of 7 oblong leaflets to 13 oblong leaflets (always an odd number). Poison sumac grows in swamps and low areas.

FIGURE 35.18 Attachment device of the all-terrain tree crane being placed in position by a tree crew member working from an aerial lift. (*Courtesy Asplundh Tree Expert Co.*)

FIGURE 35.19 Gasoline-powered chain saw utilized by tree crew personnel. (*Courtesy Stihl.*)

FIGURE 35.20 Large portable power-driven saw. (*Courtesy Asplundh Tree Expert Co.*)

FIGURE **35.21** Hydraulic tools of a type used from a trim lift bucket. (*Courtesy Asplundh Tree Expert Co.*)

The poisoning from these plants is caused by an oily substance which gets on a person's skin. The poison can be conveyed by smoke insects, clothing, and direct contact, and it is more likely to occur when the skin is covered with perspiration. No one should ever consider himself immune to these poisons, even after having had an attack. If a worker suspects that he has been exposed to these plants or their oil, he should wash the exposed skin with warm water and ordinary laundry soap. Another way to prevent poisoning is to wash the exposed skin in rubbing alcohol, rinse in clear water, and dry the skin. No brush should be used, because it would irritate the skin.

Tree Characteristics

Characteristics of trees are given in Table 35.1. The table lists the common tree species, growth form, average annual normal growth, annual sucker growth, mature height, modulus of rupture, and remarks. Tree growth forms are upright, spreading, and horizontal (Fig. 35.22).

The annual sucker growth and total growth of the same species of tree will vary considerably as a result of soil conditions and available moisture.

Modulus of rupture is the breaking point of green wood of the various species. The modulus of rupture test is conducted by placing a piece of green wood over a beam and subjecting each end of the wood to increasing pressure until it finally breaks. These breaking points give a comparison of wood strength. However, there is a difference between a test of strength and actual strength for climbing. Deformities or weak points may cause a premature break on normally strong wood. Table 35.1 should be used as a comparison between species only to arrive at a sound working knowledge. Table 35.2 gives the suggested tree-trimming clearances for some electric distribution system voltages.

Species (Common Name)	Growth Form	Avg. Ann. Normal Growth, inches	Avg. Ann. Sucker Growth, inches	Mature Height. ft	Modulus of Rupture (gm.), lb/inch²	Remarks
Ailanthus (tree of heaven)	S	48	60	50	11,060	Fast-growing, weak-wooded. Remove if possible.
Alder, Red	U	36	84	120	9,800	Weak-wooded.
Ash, Black	U	18	36	80	12,600	Strong wood, but tends to split. Will not stand heavy topping.
Banyan	S	24	60	70		Rather strong branching. Good crotches for lateral trimming.
Basswood (linden)	S	18	27	75	5,000	Wood not very strong. Tends to split. Watch dead wood.
Beech, American	S	12	30	60	8,600	Strong wood. Mature trees do not stand topping well.
Birch, White	S	21	52	50	6,400	Very sensitive. Intolerant of heavy trimming. Form is spoiled by topping.
Box-elder	H	26	72	50	5,000	Weak-wooded. Suckers grow rapidly. Remove if possible.
Catalpa	S	12		60	5,000	Soft, weak wood. Slow-growing.
Cedar, Eastern Red (juniper)	U	15	15	50	7,000	Host to the Cedar-Apple rust.
Cherry, Wild Black	S	14	24	60	8,000	Found mostly in rural areas. Does not stand heavy topping. Dies back easily. Watch dead wood.
Elm, American	U	26	60	85	7,200	Strong, flexible wood. Avoid tight crotches.
Elm, Chinese	U	40	72	65	5,200	Wood rather brittle. Strips easily.
Eucalyptus (blue gum)	U	24	50	125	11,200	
Reus	H	36	60	50	5,800	
Fir, Douglas	U	18	21	125	7,600	
Gum, Sweet	U	12	20	90	6,800	A strong slow-growing tree.
Hackberry	S	18	30	60	6,500	Strong, stringy wood. Similar to elm.
Hickory	U	14	21	65	11,000	Wood very strong and tough. Will not stand heavy topping.

Tree						Remarks
Locust, Black	S	18	80	80	13,800	Normal growth is moderate. Vigorous sucker growth when topped. Remove small trees if possible. Watch thorns.
Locust, Honey	U	22	33	80	10,200	Watch thorns.
Madrona	S	8	24	50	7,600	
Magnolia	U	20	42	90	7,400	Wood splits easily.
Mango	S	18	36	50		
Maple, Big Leaf	S	30	72	60	7,400	Wood tends to split.
Maple, Norway	H	15	35	50	9,400	
Maple, Red	S	18	42	75	7,700	Wide-spreading as a mature tree. Sucker growth rapid. Can stand heavy trimming.
Maple, Silver	U	25	65	65	5,800	Suckers grow rapidly. A rather weak and brittle tree.
Maple, Sugar	S	18	40	75	9,400	Mature trees cannot stand heavy topping.
Melaleuca (punk tree)	U	36	96	50		Softwood. Weak branches. Suckers grow very fast. Remove if possible or top under wires.
Oak, Australian Silk	U	36	72	50		Not too strong. Rather irregular growth.
Oak, Black and Red	S	18	30	85	6,900	Very sturdy trees.
Oak, Live	S	30	45	70	11,900	Very hard strong wood.
Oak, Pin	U	24	36	100	8,300	Strong wood. Topping ruins shape of tree.
Oak, Water	S	30	45	75	8,900	Moderately strong branches.
Oak, White	S	9	18	75	8,300	A sturdy and slow-growing tree. Dies back when topped heavily. Sucker growth thin.
Oak, Willow	H	24	40	50	7,400	Strong. Tough. Can be topped safely.
Palm, Coconut	U	60		90	12,970	
Palm, Queen	U	36		65		

Source: Asplundh Tree Expert Co.

Table 35.1 Characteristics of Various Trees (*continued*)

Species (Common Name)	Growth Form	Avg. Ann. Normal Growth, inches	Avg. Ann. Sucker Growth, inches	Mature Height. ft	Modulus of Rupture (gm.), lb/inch²	Remarks
Palm, Royal	U	48		80		
Palm, Washington	U	36		70		
Pecan	U			140	9,800	
Pine, Australian	U	60	144	100		Has hard wood. Grows rapidly. Watch weak crotches. Suckers profusely.
Pine, Shortleaf	U	36	48	80	7,300	Wood rather strong, but weak branching.
Pithecellobium (ape's earring)	S	24	72	50		Hard brittle wood. Dangerous thorns.
Plane (sycamore)	u	34	72	100	6,500	Rank growth of suckers when topped.
Poinciana, Royal	H	18	36	30		
Poplar, Carolina (cottonwood)	U	52	80	85	5,300	Wood very brittle. Breaks abruptly. Suckers grow at tremendous rate. Remove if possible.
Poplar, Lombardy	U	45	72	60	5,000	Very brittle. Remove if possible. A short-lived tree.
Sassafras		24	36	50	6,000	Does not stand heavy topping. Branches break easily.
Tulip Tree	U	30	52	100	5,400	Wood splits easily.
Walnut, Black	S	20	40	80	9,500	Strong wood. Dies back under heavy topping.
Willow, Black	H	40	70	50	3,800	Very weak and brittle. Breaks easily in storms.
Willow, Weeping	S	48	72	50	3,800	A weak-wooded tree. Remove if possible.

Source: Asplundh Tree Expert Co.

TABLE 35.1 Characteristics of Various Trees

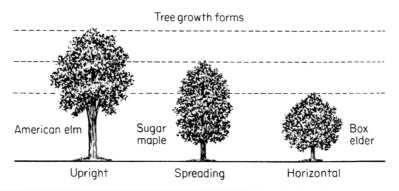

FIGURE 35.22 Tree-form types that are basic in planning and maintaining clearance for overhead lines. (*Courtesy Asplundh Tree Expert Co.*)

Clearances	Secondary, 100–600 Volts ft	Primary, 2,400–4,800 Volts ft	Primary, 7,200–13,800 Volts ft
Topping			
Fast growers	6	8	9
Slow growers	4	6	7
Side			
Fast growers	4	6	8
Slow growers	2	4	6
Overhang			
Fast growers	4	8	12*
Slow growers	2	6	12*

*Remove if possible.
Source: Asplundh Tree Expert Co.

TABLE 35.2 Tree-Trimming Clearances, ft

Tree Growth Retardants

The use of high-volume growth-retardant foliage spray in rural areas has been successful in reducing tree growth without significantly changing the tree characteristics.

Distribution Transformer Installation, Selection, and Utilization

Transformers constructed to reduce the primary voltage down to the customer's utilization voltage are commonly referred to as *distribution transformers*. The distribution transformers may be single-phase, normally used for residential customers, or three-phase, normally used for commercial or industrial customers. The transformers are constructed for mounting on poles or racks, in vaults, or on pads constructed with concrete, plastic, or fiberglass and for direct burial in the earth. The construction, theory of operation, and the method of making connections to distribution transformers are described in Chap. 15, Distribution Transformers.

Distribution Transformer Size

The transformer load capacity in units of kVA required to supply any given load may be estimated as follows:

Residential Loads

Residential loads may be estimated in accordance with Table 36.1, which gives the demand per house. As the table shows, a small house is taken as having an estimated demand of 8,000 watts, a medium-sized house is taken as having an estimated demand of 12,000 watts, and a large house is taken as having an estimated demand of 15,000 watts. A large home utilizing electricity for heating may have a demand of 20,000 watts to 30,000 watts. When several houses are served from the same transformer, there will be diversity between the peak use of energy, as shown in Table 36.2.

The size of distribution transformer required to serve several houses is determined by totaling the estimated watts for each house obtained from Table 36.1, dividing the total by 1,000 to obtain kilowatts, and multiplying by the diversity factor obtained from Table 36.2.

If three small houses and one medium house are to be served from one transformer, the size of the transformer can be determined by calculations itemized in Table 36.3.

The nearest size transformer adequate to serve the four houses with the loads shown in Table 36.3 would be 37.5 kVA. A distribution transformer can carry considerable overload without damage. A distribution transformer with a rated capacity of

Size of House	Demand, Watts
Small	8,000
Average	12,000
Large	15,000

TABLE 36.1 Wattage Demand for Residences to Be Used in Determining Size of Transformer Required

Number of Houses	Diversity Factor
1	1
2	0.9
4	0.8
6	0.75
8	0.7
12	0.67

TABLE 36.2 Diversity Factor for Selecting Transformer Capacity Required to Serve Several Houses

Houses	Watts	Watts
3 ×	8,000 =	24,000
1 ×	12,000 =	12,000
Total		36,000
36,000 W/1,000 =		36.0 kW
36 kW × 0.8 diversity factor =		28.0 kW

TABLE 36.3 Sample Calculation for Determining Proper Size of Transformer

37.5 kVA should be adequate to serve the four houses for several years even with normal load growth.

Commercial Lighting

In determining the size of transformer necessary for commercial lighting customers, such as stores of all kinds, it is first necessary to add the total connected load. This requires making a list of all lights, motors, air conditioners, etc., together with the wattage of each. The sum is the total connected load on the transformer. The average demand on the transformer will be 40% to 60% of the connected load, depending on individual cases. The demand represents the necessary transformer capacity.

Motor Loads

The transformer capacity required for motor loads is best found from the motor ratings. Add the ratings of all the motors to be supplied from one transformer or bank of transformers. Then from Table 36.4 obtain the proper percentage of the number of motors. Multiply the actual connected load by this percentage. The result is the required transformer capacity. Select the next larger standard rating.

If the transformer bank is to be open-delta-connected, multiply the rating as found above by 1.35 before selecting the transformer rating required.

The foregoing load estimates should be used only if there is no local information available for arriving at an approximate transformer rating. Load demands vary considerably

Total Connected Load, hp	Percentages for Calculating Kilovoltampere Demand from the Normal Rated Capacity of Motors					
	Numbers of Motors					
	1	2	3–5	6–10	11–20	20 or more
Less than 10	83	81	75	73		
10 to 50	78	76	70	68	67	65
50 to 100	74	72	65	62	61	60
100 to 300	72	70	61	59	58	57
Over 300	70	69	58	56	55	54

(Percentages for various numbers of motors and magnitude of connected horsepower load.)

TABLE 36.4 Percentages by which to Multiply Total Connected Motor Load to Obtain Required Transformer Kilovoltampere Capacity

for the various classes of customers throughout the country, and therefore local data should be collected to form a basis for estimating.

Load Checks

When a new subdivision in an urban area is first supplied with electric service, the smallest transformer used by the utility is generally installed. This may be a 15-kVA or 25-kVA size. This is completed when the first house is connected to the secondary. When the load has built up to nearly full transformer capacity, load measurements are made on the transformer to see what the actual load is.

The customer's monthly kilowatthour consumption as determined by meter readings made for billing the customer can be analyzed automatically by a computer to determine the distribution transformer loading. If a transformer is over- or underloaded by some predetermined value, the computer provides an exception report. The report can be used to initiate action to change the distribution transformers to the proper size. The distribution transformers must be properly loaded to efficiently utilize the investment and to provide the customer reliable service with good voltage. If a distribution transformer is overloaded, it may be desirable to install an additional transformer and divide the secondary and services to distribute the load properly on each transformer.

Distribution Transformer Location

The distribution transformer should be installed as near as possible to the center of the load area. The transformer secondary leads and services should be kept as short as possible to minimize the voltage drop and line losses. Transformers installed on a rear lot line at the point where the corners of four lots are common permit four houses to be served by services direct from the transformer whether the transformer is mounted on a pole, on a pad, or in a vault. If additional houses are to be served from one transformer, the secondaries can be extended along the rear lot line.

Commercial loads are usually served by an individual transformer or a three-phase transformer bank. The transformers should be installed as close as possible to the load to minimize the secondary and service voltage drop.

Distribution Transformer Grounding

The term *ground* is used in electrical work to refer to the earth as the zero potential. A ground actually consists of an artificial electrical connection to the earth, having a very low resistance to the flow of electric current. To ground a circuit or apparatus, therefore, means to connect it to earth. All distribution transformer tanks and one conductor of all transformer secondary circuits should be grounded to protect life and property. The points to be grounded in the various kinds of circuits are illustrated in Chap. 15, Distribution Transformers.

If the distribution transformer has a two-wire secondary, either of the two wires can be grounded. In a three-phase transformer bank with a delta-connected secondary, the center of one of the transformer windings should be grounded. If the secondaries are Y-connected, the neutral of the Y is grounded.

The ground connection starts at the ground rod, usually driven 8 ft or more into the earth, and the connecting wire extends to the proper secondary terminal of the distribution transformer. The rod must be driven deep enough to penetrate moist soil.

All service neutrals are grounded at the customer's service entrance. The usual ground is the customer's steel water main and/or a driven ground rod. If the water system is not a good ground source, the customer must have a connection to a driven rod at the service entrance.

The low-voltage secondary circuit is grounded to guard against an excessive voltage being impressed on that circuit from an external source. A conductor of the primary feeder may break and fall upon the secondary conductors below it. The greatest danger, however, exists in the breaking down of the insulation in the distribution transformer, thereby bringing the secondary winding in contact with a high-voltage primary winding. If a person should then touch a bare part of the secondary circuit and should also happen to be in contact with a radiator, bathtub, gas range, or water faucet, he would be likely to receive a very severe shock. Furthermore, any excessive voltage on the low-voltage fittings in lamp sockets, etc., is very apt to cause fire. If the secondary is grounded, however, a breakdown in the transformer insulation is very apt to blow the primary transformer fuses, thereby clearing the defective circuit from the source.

Distribution Transformer Mounting

Small distribution transformers normally installed on overhead circuits are often referred to as pole-type transformers. These transformers may be fastened directly to the pole, hung from crossarms, mounted on racks or platforms suspended from poles, or placed on pads. Figures 36.1 and 36.2 illustrate a distribution transformer bolted directly to the pole, Fig. 36.3 illustrates a bank of three transformers supported by means of cluster mount brackets, and Fig. 36.4 illustrates the installation of three transformers mounted on a platform. Platforms are built of any length or size to suit local conditions (Fig. 36.5). If the structure is to be permanent, the supporting members should be galvanized steel. Steel has a good appearance and requires no maintenance. The supporting members should be covered with wood flooring. The development of three-phase pad-mounted transformers has practically eliminated new installations of transformers on platforms supported by two poles. The costs of the two-pole platform transformer installations are excessive, and the platforms are visually unacceptable to the public. Pad-mounted transformers are safer for the linemen to install and maintain, and they reduce the hazard to the public. Very large transformers with

FIGURE 36.1 Distribution transformer mounted directly on pole. Transformer is rated 25 kVA, 7,620-120/240 volts. Note service wires extending from the pole near the transformer to the house in the background of the picture. The service wires are dead ended on the side of the house. The customer's service-entrance wires protruding from the weatherhead and conduit on the side of the house connect to the 120/240-volt single-phase service wires.

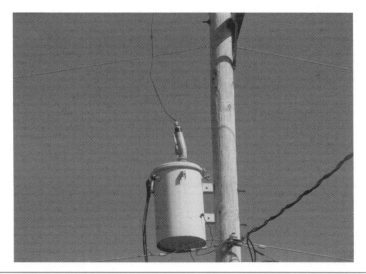

FIGURE 36.2 Close-up view of a distribution transformer mounted directly on the pole. Note the insulated protective wildlife covering installed on the current-limiting fuse and transformer primary bushing.

exposed live parts are set on pads on the ground. A high fence must then be provided to keep persons from coming in contact with the live parts. Small distribution transformers can be bolted directly to the pole by means of brackets. The transformer shown in Fig. 36.6 is direct mounted on the pole. The transformer is equipped with a partial-range current-limiting fuse and a direct-connected surge arrester. The transformer is suspended by a line truck boom-tip winch line which was used to lift the transformer from the material trailer fastened to the back of the line truck.

FIGURE 36.3 Three-phase transformer bank supported on pole by means of cluster mount brackets. The pole must be set 1 ft to 3 ft deeper than normal. Note three-phase four-wire secondary along the line from a secondary bracket on the pole below the transformer bank.

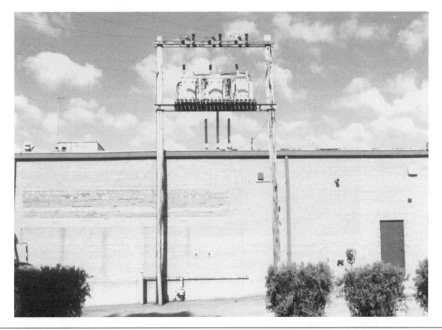

FIGURE 36.4 Three single-phase transformers connected into a three-phase bank and mounted on a platform supported by two poles. The three-phase bank of transformers provides service to a commercial customer.

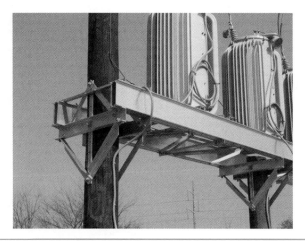

FIGURE 36.5   The platform supports are used for transformers or voltage regulators.

The pole-type transformers are normally raised to the mounting position with a boom and winch on the line truck (Fig. 36.7). Installation of distribution transformers on poles located on rear lot lines in residential areas or other locations inaccessible for line trucks requires the use of a block and tackle or a portable winch to raise the transformers (Figs. 36.8 and 36.9).

The distribution transformer should be handled carefully. Bushings and other equipment on the transformer can be damaged easily. The windings of the transformer may be damaged if the transformer is dropped from the truck or severely jolted.

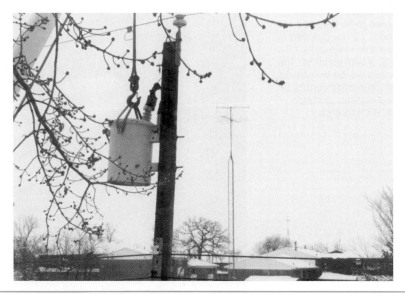

FIGURE 36.6   Distribution transformer designed to be supported by direct-to-pole mounting. Transformer is fastened to pole by bolts passing through brackets and pole.

FIGURE 36.7 Lineman installing distribution transformer on pole. Transformer is lifted in place by boom and winch line of line truck. Transformer is mounted on bracket clamped to the pole. (*Courtesy A. B. Chance Co.*)

FIGURE 36.8 Capstan portable hoist designed to be fastened to a pole. The hoist can be powered by a motor operated electrically, hydraulically, or with gasoline. The electric motors can be constructed for 120-volt alternating current or 12-volt direct current operation. (*Courtesy A. B. Chance Co.*)

FIGURE 36.9 Lineman is using a capstan hoist to raise a distribution transformer to the proper mounting height on a pole with the aid of a block mounted temporarily near the top of the pole. The hoist is used to lift equipment at a location not accessible by a vehicle. (*Courtesy A. B. Chance Co.*)

The lifting equipment, including slings used on the transformer, should be carefully inspected before the operation is started. Linemen and groundmen should stay in the clear while transformers are raised into position. Protective rubber line hose and equipment cover-up should be installed on the pole to make the primary area safe. Lineman must wear rubber gloves, protectors, and other appropriate personal protective equipment (PPE) while working near the primary wires. Figure 36.10 illustrates a complete conventional pole-type single-phase distribution transformer installation. Conventional distribution transformers should always be protected on the primary side with distribution fuses (cutouts) to disconnect the transformer in case of trouble. A current-limiting fuse can be mounted on each cutout to prevent the transformer lid from blowing off if a high-current fault starts to develop in the distribution transformer. A lineman could be seriously injured when working on the transformer pole if a lid blew off the distribution transformer and he was sprayed with burning oil. The cutouts are mounted near the transformer so they will be readily accessible to the lineman.

When the transformer has been energized from the primary feeder, the secondary connections can be installed. The lineman should check the secondary with a voltmeter to be sure it is deenergized before completing the connections. If the transformer is to be connected in parallel with an existing transformer, the secondary circuits must be properly phased out by using the voltmeter to be sure they can be connected without causing a short circuit. The voltmeter should always indicate zero voltage between the transformer low-voltage terminal and the secondary wire before the connections are installed. Linemen must be especially careful to be sure the customer's neutral, or ground, wire is not connected to an energized transformer terminal or secondary wire, to prevent a short circuit and damage to the customer's equipment.

A voltmeter or test lamp can be used to check connections if a single-phase transformer is to be connected in parallel with an existing transformer installation.

Connect the voltmeter or lamp across a live secondary lead to test for proper operation. After the ground or neutral connection has been completed and the transformer

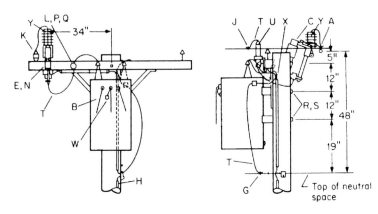

Phase to neutral 2400-7620 volts

Phase to phase 13,200 volts

ITEM

A. Live-line clamp connections	Q. Pin, steel INS, lag screw, for XA
B. Single-phase transformer secondary taps	R. Bolt, machine, 5/8 x 12 in.
C. Cutouts and fuses, 15 kV and below	S. Washer, SQ flat, 2 1/4 in.
E. Arresters	T. Wire, #4 WR SD Sol CU
G. Conn. wedge (#4 Sol CU-neut)	U. Lightning arrester
J. Conn. wedge (4 Sol CU to pri.)	W. Ground transformer case
K. Assembly, insulator, post type, 12.5 kV	X. Conn. wedge, (4 Sol CU — Sol CU)
L. Conductor ties (conventional)	Y. Partial-range
N. Bracket, cutout and arrester, Arm MTG	current-limiting fuse
P. INS, pin, 12 kV top groove	

FIGURE 36.10   Single-phase 25-kVA conventional distribution transformer construction standard installation for 13,200-volt or lower voltage distribution primary feeder circuit.

is energized, test for voltage between one of the remaining secondary mains, as shown in Fig. 36.11. If this is the proper connection, the voltmeter will read zero or very near zero. To avoid confusion, it is best to connect this lead immediately to the secondary main. Now proceed to the last lead and last secondary main. This, of course, should also read zero. After completing this connection, the new transformer is in parallel with the other transformers connected to the same secondary mains and will share the load connected thereto.

FIGURE 36.11 Paralleling two single-phase transformers each for 120/240-volt three-wire service by using voltmeter or test lamp.

FIGURE 36.12 Paralleling a single-phase transformer two-wire, 120 volts with another connected three-wire, 120/240 volts by using voltmeter or test lamp. One of two possible connections.

Figure 36.12 shows the procedure to be followed in connecting a two-wire 120-volt transformer to a three-wire 120/240-volt secondary. All connections should be completed carefully in accordance with the instructions and checked for proper tightness.

Three-phase transformer connections must be completed properly to obtain the voltages desired and to prevent an accidental short circuit. A voltmeter must be used on the wye-wye transformer bank to measure the phase-to-phase voltages for proper magnitude and balance. The primary connections for a wye-delta or delta-delta transformer bank can be installed and the transformer energized without danger of a short circuit providing one corner of the delta secondary is left open. The voltmeter or test lamp must have a voltage rating equal to twice the secondary line voltage to safely check the delta-delta voltage. The voltmeter or test lamp is connected across the open corner of the secondary delta windings (Fig. 36.13). If the connection of the delta is correct, the lamp will not burn, and the voltmeter will read zero or almost zero. But if one of the transformer's connections is reversed, the lamp will burn brightly. If the voltmeter is used instead of the lamp, it will indicate a voltage equal to twice the voltage of one of the secondary windings. Under these conditions the delta must not be closed, for a severe short circuit would result. To correct the connection, reverse the leads from one of the transformer secondaries and repeat the test. If this does not correct the connection, reverse the secondary leads of another transformer and repeat the test. When the lamp fails to burn or the voltmeter reads zero, the delta connection is correct.

On higher-voltage secondaries, a small step-down potential transformer with a lamp or voltmeter connected to its secondary can be used while following the procedure described above.

The CSP Distribution Transformer

The *completely self-protected* (CSP) distribution transformer requires no auxiliary protective equipment except for current-limiting fuses. Primary fuses and lightning arresters are included with the transformer and therefore do not have to be supplied or mounted separately. Figure 36.14 shows an exterior view of a CSP transformer, and Fig. 36.15 is a

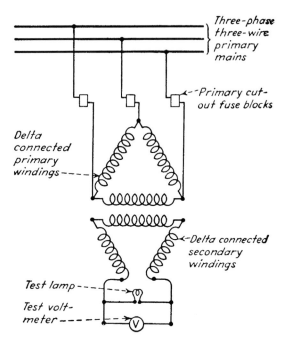

FIGURE 36.13 Use of lamp or voltmeter in testing delta connection on three-phase transformer bank. When the lamp tests dark or voltmeter reads zero, the delta connection is correct.

FIGURE 36.14 Exterior view of a CSP distribution transformer showing primary bushing, lightning arrester, secondary terminals, tank ground connection, and secondary circuit breaker operating handle. (*Courtesy ABB Power T&D Company Inc.*)

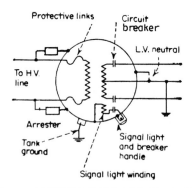

FIGURE 36.15 Schematic diagram of connections of a typical CSP distribution transformer. Note primary protective links, secondary circuit breaker, lightning arresters, signal lamp winding, and circuit breaker handle.

schematic diagram of the transformer and its component parts. Such transformers are equipped with support brackets for direct mounting to a pole; but when they are provided with hanger irons, they can also be hung from crossarms.

The following features can be incorporated in a CSP transformer:

1. Primary-voltage lightning arresters. These are supported externally on the tank.

2. Primary protective links or fuses in series with line leads serve to disconnect the transformer in case of internal transformer fault.

3. Overcurrent protection in the secondary leads provided by a low-voltage circuit breaker. This breaker is mounted above the core and coils but operates under oil.

4. Overload signal lamp which becomes lighted when the maximum safe operating load on the transformer is exceeded or when the secondary circuit breaker has tripped automatically.

5. An external operating handle for opening or closing the secondary circuit breaker or for resetting the signal lamp and breaker-tripping mechanism.

6. No-load tap changer. This tap changer provides four 2½% voltage taps which give a choice of four values of transformation ratio and compensate for voltage drop on long distribution lines. The operating handle is above the oil level, but the taps and contacts are under oil. The taps should be changed only when the transformer is deenergized.

To install a transformer of this type, it is only necessary to ground the neutral conductor, connect the high-voltage terminals to the primary line, and connect the secondary terminals to the secondary mains. Figures 36.16 through 36.18 illustrate typical installation details to obtain single-phase and three-phase secondary service using CSP single-phase distribution transformers manufactured for line-to-ground primary connections.

Phase Sequence

Phase sequence is the sequence or order in which the three voltages of a three-phase system appear. The phase sequence of a three-phase system is often desired in order to:

1. Determine the direction of rotation of polyphase motors

2. Determine the proper connections for paralleling three-phase transformer banks, generators, and power buses

3. Determine that the phase sequence is not changed when a three-phase transformer installation is replaced

4. Determine the proper connections for watthour meters, instruments, and relays

One type of phase-sequence indicator makes use of a small capacitor connected in wye with two small neon lamps. When used, the terminals of the wye are connected to the conductors of the three-phase circuit of which the phase sequence is to be determined. A schematic diagram of the connections is shown in Fig. 36.19.

When the phase-sequence indicator is connected across a three-phase line, as shown in Fig. 36.19, an unequal distribution of voltage occurs in the three arms of the wye network. This makes the voltage across one of the neon lamps considerably greater than

Installation with
current limiting fuse
where required.

ITEM

A. Preformed tie wire	H. Hot-line clamp	O. Compression connector
B. 1″ Pin-type insulator	I. #6 Split-bolt connector	P. Wedge clamp
C. 1″ Pole-top pin, 18″	J. #6 Copper wire	Q. 5/8″ x 10′ Ground rod
D. 5/8″ x 10″ Bolt	K. Single clevis insulator	R. Ground-rod clamp
E. 2-1/2″ Curved washer	L. Terminal pin	S. Squirrel & bird guard
F. 5/8″ Palnut	M. Molding, ground wire	T. Staples, fence
G. Compression-type stirrup	N. Staples molding	U. Partial-range current-limiting fuse

FIGURE 36.16   Construction standard for installation of a single-phase CSP distribution transformer connected line to ground on a single-phase tap to a 13.2Y/7.62-kV three-phase four-wire grounded-neutral primary circuit.

that across the other for a given phase rotation. In fact, if the voltage across the darker lamp falls below the minimum breakdown or ignition value, the lamp will cease to glow completely. Consequently, only one of the neon lamps glows brightly at a time. The one that glows brightest indicates the phase rotation of the line voltages.

Potential transformers or potential devices must be connected in series with the phase-sequence instrument to check phase rotation of high-voltage lines.

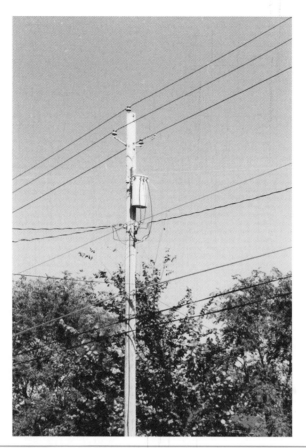

FIGURE 36.17 An installation of a single-phase CSP distribution transformer connected line to ground on a 13.2Y/7.62-kV three-phase four-wire grounded-neutral primary circuit.

Phase sequence or rotation is designated as 1-2-3 or 3-2-1. The first order is engraved on the case under the left-hand (*A*) lamp, and the second order is engraved under the right-hand (*B*) lamp. The leads which are used to measure the phase sequence of the three-phase line are identified with engraved markings of 1, 2, and 3.

To use the phase-sequence indicator, the voltage selector switch is first set to the desired voltage. The three leads of the instrument are then connected to the lines to be measured. One lamp should then glow much more brightly than the other. If the left-hand (*A*) lamp glows brightest, the phase sequence is 1-2-3, and if the right-hand (*B*) lamp glows brightest, the phase sequence is 3-2-1.

In order to check the sequence indicated, interchange any two of the leads. This changes the order, and therefore the first lamp should become dark and the second lamp should become bright. This also serves as a check on the condition of the lamps. If one of the lamps should fail to glow during this test procedure, it should be replaced with a new neon lamp and the checking procedure repeated.

When it is necessary to change out a three-phase transformer installation, the lineman should always check the phase sequence. When the new transformer installation

Detail A

ITEM

A. Preformed tie wire	J. Hot-line clamp	Q. Molding, ground wire
B. 1" x 5-1/2" Pin insulator	K. Transformer secondary lead bracket	R. Staples, fence
C. 1" Pole-top pin, 18"	L. Compression connector	S. Staples, moulding
D. 5/8" Bolt	M. #6 Split-bolt connector	T. 5/8" x 10' Ground rod
E. 2-1/2" Curved washer	N. 4 Spool rack	U. Ground-rod clamp
F. 5/8" Palnut	O. Transformer mounting bracket	V. Small Epoxy-glass standoff
G. Epoxy-rod biunit assembly 1-1/2" D rod, 36" between conductors	P. #6 Copper wire	W. Partial-range current-limiting fuse
H. Non-tension sleeve		X. #33 Scotch electrical tape
I. Stirrup, compression type		Y. #4 Stranded copper wire

FIGURE 36.18   Construction standard for installation of single-phase CSP distribution transformers manufactured for line-to-ground primary connections to obtain three-phase four-wire grounded-neutral secondary voltages by using a wye-wye transformer connection from a 13.2Y/7.62-kV three-phase four-wire grounded-neutral primary circuit.

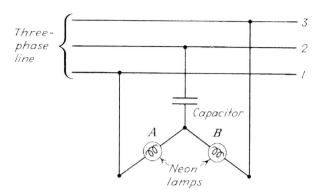

FIGURE 36.19 Circuit diagram depicting phase-sequence indicator connections to a three-phase line. Indicator consists of a capacitor and two neon lamps, *A* and *B*, connected in wye.

is completed, the phase sequence should be rechecked before the load is connected. The phase sequence must not be changed. Reversing the phase sequence would reverse the rotation of the customer's three-phase motors and would probably damage the customer's equipment.

Distribution Secondaries

The low-voltage wires from the distribution transformer low-voltage terminals to the terminal point of the customer's service-wire connections at a pole or an underground system pedestal are called *secondaries* or *secondary mains*. In well-designed distribution systems the voltage drop from the transformer secondary terminals to the end of the secondary in any direction should not exceed reasonable values. For secondary lighting mains, this drop is usually limited to 3% and for secondary power mains to 5%. These percentages apply to the voltage of the circuit. On a 120-volt two-wire circuit a 3% drop would equal 3.60 volts, and on a 240-volt circuit a 3% drop would equal 7.20 volts. A 5% drop would be equal to 12.0 volts on a 240-volt circuit.

The estimating chart for size of three-wire single phase secondary mains shown in Fig. 36.20 can be used to find the size of secondary mains required for residential areas. The chart is based upon a conductor span length of 100 ft. A span is defined as the measure of distance between two points. For this chart, one point is the transformer pole and the second point is pole number 1. Make a secondary layout sketch similar to that shown in Fig. 36.21, and show the kilowatt demand at each pole for the residences served from that pole. For example, pole 1 in either direction serves a load of 12 kW and pole 2 in either direction serves a load of 8 kW.

Then the kilowatt spans are calculated for each pole by multiplying the demand at each pole by the number of pole spans between the pole and the transformer. The number 2 pole in the sketch, for example, has a demand of 8 kW and is 2 spans from the transformer pole. The kilowatt spans, therefore, are 8 × 2 = 16 kW spans. The kilowatt spans should be as nearly equal as possible. This is the same as saying that the transformer should be located as nearly as possible in the center of the load.

Now, to determine the size of wire required for the three-wire secondary mains, refer to the chart of Fig. 36.20 and find the value of kilowatt spans along the horizontal scale. Then move vertically upward on the chart until you reach the line marked "maximum allowable."

FIGURE 36.20   Chart for selecting wire size for 120/240-volt single-phase 60-Hz three-wire secondary mains. Chart is based on the use of conductors spaced 12 inches apart. Assumed power factor is 80%. Span length is 100 ft. Unbalance factor is 1.2.

FIGURE 36.21   Secondary layout sketch showing transformer pole, services at each pole, and demand at each pole. Demand per customer at pole number 2 is assumed to be 8.0 kW.

If the actual average span length varies very much from the assumed span length distance of 100 ft, the total kilowatt spans as calculated should be multiplied by the ratio of the actual average span to 100 ft before using the chart.

Accurate determination of the size of wire required for the secondary mains of lighting circuits can be made by use of the curves in Figs. 36.22 and 36.23. To use these curves, the kilowatt demand for each house and the distance in feet between the house and the transformer must be determined. Then the demand in kilowatts and the distance in feet are multiplied to get the kilowatt distance for that house. The kilowatt distance is likewise obtained for each house and the total for all the houses on one side of the transformer. This total may be called KD. Next add the total kilowatt demand on the same side of the transformer and call it K. Now divide KD by K and get D, which is the equivalent distance in feet to which a concentrated load equal to K may be transmitted. With K and D known, refer to Fig. 36.22 for 120-volt two-wire mains or to Fig. 36.23 for 120/240-volt three-wire mains, and find the intersection of the lines from K and D. If the point of intersection does not fall directly on one of the curves, use the wire size of the curve next above the point. These curves are based on a 3% voltage drop.

In calculating the wire size for three-phase power secondaries, the procedure is the same as for lighting secondaries outlined above. In the case of power loads, however, the load can be more easily determined, as the motors to be supplied and their ratings

FIGURE 36.22 Load curves for 120-volt single-phase lighting secondary. Curves show distance that any load in kilowatts K can be transmitted with 3% drop on lines. Voltage at load, 120 volts; spacing of wires, 8 inches; power factor of load, 95%; 60 Hz current.

FIGURE 36.23 Load curves for 120/240-volt single-phase three-wire lighting secondary. Curves show the distance in feet D that any load in kilowatts K can be transmitted with 3% voltage drop on lines, allowing for unbalance of three-wire circuit. Voltage at load, 120/240 volts; spacing of wires, 4 inches; power factor of load, 95%; 60 Hz current.

FIGURE 36.24 Load curves for 240-volt three-phase power secondary. Curves show distance in feet D that any load in kilowatts K can be transmitted with 5% voltage drop on lines. Voltage at load, 240 volts; equivalent spacing, 5 inches; power factor of load, 80%; 60 Hz current.

are known. The procedure consists in getting the value of each motor load K and the distance D from the transformer. Then obtain the total of KD products, divide by K, and obtain the equivalent distance D. Now refer to Fig. 36.24 for 240-volt three-phase secondaries, and obtain the intersection of K and D. The curve on which this point of intersection falls is the proper wire size to use for a 5% voltage drop.

FIGURE 36.25   Load curves for 240-volt single-phase power secondary. Curves show distance in feet D that any load in kilowatts K can be transmitted with 3% drop on lines. Voltage at load, 240 volts; equivalent spacing of wires, 4 inches; power factor of load, 94%; 60 Hz current.

The procedure for single-phase power secondaries is the same as that already outlined for lighting and three-phase secondaries, except that the curves in Fig. 36.25 must be used. These curves are for 240 volts, single-phase, and allow for a 3% voltage drop.

Installing Overhead Secondary Mains

Overhead secondary mains are usually installed on vertical racks bolted to the poles (Fig. 36.26) or with triplex or quadruplex cable secured to the poles (Fig. 36.27).

In stringing wires on bracket construction, the conductors are unreeled and passed through the rack, as shown in Fig. 36.28a. When the desired number of pole spans has been laid in place, the conductors are drawn up and dead-ended. As many as 10 spans can be drawn up at one time in this manner. Figure 36.29 shows a secondary rack extension bracket used to avoid a surface mounted bracket. Figure 36.30 shows the conductors properly dead-ended on the rack.

The conductors are then lifted up to the insulator groove and tied in. The regular Western Union tie is generally used (see Fig. 36.28b and c). If the conductors are to be tied to the outside of the insulator, the Western Union tie is also used (see Fig. 36.28d).

Telephone cables are attached well below the secondary main and CATV cable is located one foot below telephone cable.

In turning corners and at angles in the line, the position of the line wires on the insulators will be determined by the direction of the strain. They should always be placed so that the conductor is pulled against the insulator and not away from it. Figure 36.31 illustrates the correct positions for corner and angle construction.

Distribution Services

The low-voltage wires that connect to the secondary main conductors and extend to the customer's service entrance on the building are called *services*. The services may be overhead or underground, usually depending on the type of distribution system. However, underground services are frequently installed from an overhead secondary main to the customer's building to eliminate the overhead wires across the customer's property. The size of the wire used for the customer's service is determined by the size of the electrical load. Single-phase 120/240-volt three-wire services are installed to all new homes (Fig. 36.32).

Single-conductor wire may be used with supporting racks. The use of single-conductor wires on racks permits the use of weatherproof wire instead of insulated wire for overhead services, reducing wire costs. However, using triplexed cables with the close conductor spacing reduces the reactance and improves the voltage for the customer.

FIGURE 36.26 Secondary mains are secured to pole with vertical rack. The neutral or grounded conductor of the 120/240-volt three-wire single-phase secondary main is located on the white center insulator.

FIGURE 36.27 A triplex secondary main operating at 120/240 volts single-phase extends from the transformer pole to adjacent poles along the street. The secondary conductors are attached to the pole below the transformer tank. An overhead service and an underground service connect to the secondary conductors on the transformer pole. The services supply power to the residences next to the transformer pole.

(a)   (b)

(c)   (d)

FIGURE 36.28 Stringing secondary mains. (a) Pulling line wires through rack; (b) line wires tied to inside of insulator; (c) top view of line wire tied to inside of insulator; (d) top view of line wire tied to outside of insulator.

FIGURE 36.29 Secondary Rack Extensions are needed where obstructions are to be overcome and where proper clearances are required. The bracket is fastened to the pole for mounting secondary racks.

FIGURE 36.30 Completed dead-end on secondary-rack construction. Two or more through-bolts should be used on each rack.

FIGURE 36.31 Suggested rack construction on corners and angles. Line conductor is always placed on insulator so that the pull is against the insulator as shown.

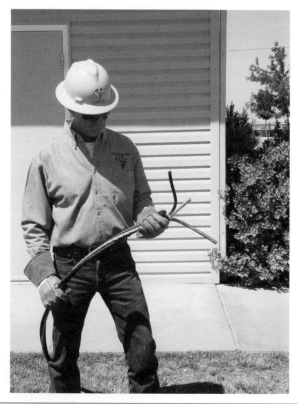

FIGURE 36.32 Lineman holding ends of three-conductor, No. 4 AWG aluminum, polyethylene covered triplex cable. Light-colored conductor will be used for neutral or grounded wire. Similar cable is used for overhead and underground low-voltage services.

Underground services to new homes are usually installed with 2/0 aluminum conductor or larger to provide a minimum capacity of 200 amps. Underground services are installed large enough to prevent replacing them when the customer's load increases. Overhead services can be installed appropriately for the customer's load and replaced with larger conductors when the load increases, minimizing the investment.

Installing Overhead Services

Service brackets are installed on the customer's building, that is, store, or factory, as the case may be, and on the pole to support the service (Figs. 36.33 and 36.34). Service supports should be mounted about 15 ft above the ground. The practice varies slightly with

Four $\frac{3}{8}$" double expansion bolts or three $\frac{5}{8}$" toggle bolts for services up to 3-No. 2 W.P. Copper

3-No. 1/0 W. P. Copper require three $\frac{5}{8}$" through bolts and plates on inside of wall

12' for residences

16' for business buildings

Ground line

3-Wire service for No. 6 to 1/0 copper wire or No. 4 to 3/0 aluminum wire

FIGURE 36.33   Details of service connection for three-wire services for residences and business buildings.

FIGURE 36.34   Typical service-drop cable terminated on the pole near residence with prefabricated grip. (*Courtesy Preformed Line Products Co.*)

different companies. Figure 36.33 shows the practice of one company which requires 12 ft for residences and 16 ft for business buildings, and Fig. 36.35 illustrates the requirement of another company.

The minimum required clearance to ground of the service-drop wires varies with the use made of the ground below the wires. The minimum clearance is 12 ft for voltages from 0 volts to 750 volts in areas restricted to pedestrian traffic. If the service wires cross an alley or street, the minimum ground clearance is 16 ft. The span length is limited to 125 ft.

FIGURE 36.35   Installation of customer's outside meter and service entrance, three-wire 120/240-volt single-phase, from overhead distribution system. (1) Weatherhead must extend above service wires. (2) Building service wires–size of service wires is determined by application of the National Electrical Code. Service wires must extend at least 36 inches from weatherhead. The neutral wire must be identified. (3) Service rack. (4) Service wires. (5) Galvanized rigid conduit (not water pipe). (6) Outdoor meter socket. (7) Galvanized conduit fitting with threaded hubs. (8) Galvanized conduit or water pipe of size to carry neutral wire. (9) A ½-inch or larger copperweld ground rod or ¾-inch galvanized water pipe. The ground rod must be driven a minimum of 8 ft in ground. (10) Solderless connector (copper or bronze) of proper size to connect ground wire to rod or pipe. (11) Galvanized conduit nipple and end bushing. (12) Service-entrance fused switch, circuit breaker, or fuse box. (13) Galvanized conduit straps spaced approximately 4 ft apart. (14) Compression sleeve connector. (*Courtesy MidAmerican Energy Co.*)

Lowest point of attachment on building

Lowest service wire 0 to 750 volts

Lowest service wire 0 to 750 volts

12' min   Building to be served   12' min   12' min

16' min

Meter

High place in contour

Ground line

Sidewalk

Space accessible to pedestrians only

Street, alley or highway

125' max

FIGURE 36.36   Sketch showing minimum ground clearances of service-drop wires, having voltages of 0 to 750 volts, above open spaces, sidewalks, and streets or alleys as required by the National Electrical Safety Code. Note that the span length is limited to 125 ft.

The minimum clearances must be maintained under all loading, wind, and ice conditions. Figure 36.36 illustrates the minimum required ground clearances.

The service connection should be made at the nearest pole in the distribution line (Fig. 36.37). If this distance should exceed 125 ft, an intermediate support pole should be provided.

After the service wires are unreeled, one of the linemen dead-ends the wires onto the service bracket, near the customer's service entrance wiring. As soon as this is done, the other lineman climbs the service pole, pulls up the service wires, and dead-ends them on the distribution pole. These wires need to be pulled up by hand only, as little strain should be placed on the service supports. If the service-drop wires are pulled otherwise than by hand, too great a strain may be put on the service brackets which may cause them to break or be pulled out of the building structure.

If the service drop consists of a self-supporting service-drop cable (Fig. 36.38), only the messenger wire need be dead-ended. Usually a special feed-through dead-end fixture is employed, as shown in Figs. 36.39 and 36.40.

The type of support for the service wires on the distribution pole depends on the type of secondary main construction. If the secondary mains are supported by secondary racks, the same racks may also be used for the support of the service drops, as shown in Fig. 36.41.

The installation of the watthour meter (Fig. 36.42) is the next step in the running of the service. The meter is, as a rule, the property of the company and is therefore installed by the company at the time of making the service connection (Fig. 36.43). The details of the installation of the meter socket will not be discussed here as this work is generally done by an electrician working for the customer.

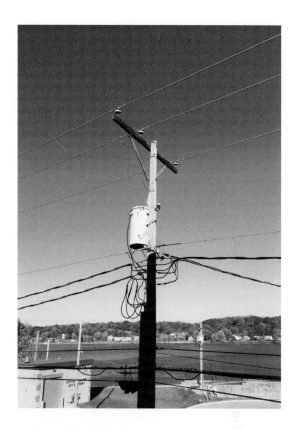

FIGURE 36.37   The distribution transformer serves four triplex 120/240 volt services from this location on the distribution pole.

FIGURE 36.38   Self-supporting service-drop cable insulated with polyethylene or neoprene. Messenger serves as neutral conductor.

FIGURE 36.39   Dead-ending service-drop cable on consumer's building by means of (a) feed-through dead-end fixture attached to eye bolt, (b) feed-through dead-end fixture attached to wire holder.

Feed – through
dead end

FIGURE 36.40 Dead-ending service-drop cable on distribution service pole by means of special feed-through dead-end attached to secondary-rack insulator spool.

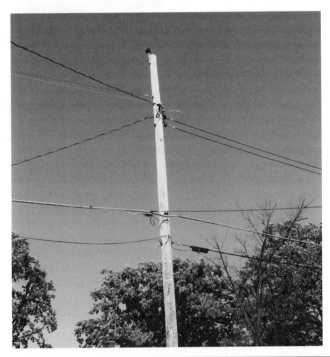

FIGURE 36.41 Service-drop wires supported on secondary main bracket. In this instance neutral conductor of three-wire service is middle conductor. Note cable television (CATV) wires below power wires. (*Courtesy MidAmerican Energy Co.*)

FIGURE 36.42 Meterman/lineman preparing to insert watthour meter for newly connected customer's residence. (*Courtesy Northwest Lineman College.*)

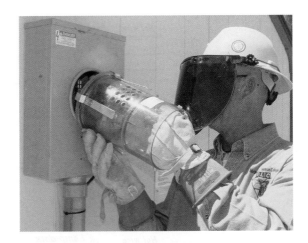

FIGURE 36.43 Two overhead meter socket installations.

Testing Service Wires for Polarity and Ground

Before the service wires can be connected onto the secondary mains, they must be tested with a voltmeter or *lamped out* for polarity. This must be done to avoid connecting the customer ground to either of the *hot* wires of the secondaries. To do so would produce a *short circuit* on the secondary mains. The customer ground wire, usually white or gray or bare, should always be connected to the neutral or grounded wire of the service.

FIGURE 36.44 Lamp test for polarity on two-wire service. The service-drop wire that makes the lamp burn is the grounded wire which should be connected to the neutral or grounded wire of the secondary mains.

FIGURE 36.45 Lamp test for polarity on three-wire service. The service-drop wire that makes the lamp burn is the grounded wire which should be connected to the neutral or grounded wire of the secondary mains.

To perform the polarity, or *lamping-out*, test, an incandescent lamp or voltmeter is used. This should have a voltage rating equal to the voltage across the *hot* wires of the secondary. If the service is a two-wire service, the test is made as shown in Fig. 36.44. The service which is grounded in the customer's premises should be connected to the neutral of the secondary main. To determine which of the service-drop wires is grounded, the lamp is connected from one of the *hot* or *live* secondary main wires to one of the service wires. If the lamp burns, the service-drop wire to which the lamp is then connected is grounded. If the lamp does not burn, that service-drop wire is not grounded. If the service is a three-wire service, the procedure is the same except that three wires of the service must be *lamped out* (see Fig. 36.45). All one needs to remember is that the service-drop wire that makes the lamp burn is the grounded wire and should be connected to the secondary main neutral.

Connecting Service Drop to Secondary Mains

When the service-drop leads are lamped out, they are ready to be permanently connected to the secondary mains.

Connectors are used to obtain a good electrical connection. Taps to secondary mains, service entrance connections, jumper loop connections, etc., are generally made with *compression sleeve type connectors*. The compression-type connector is a form of open-sleeve connector (Figs. 36.46 through 36.48) that is installed with a compression tool.

FIGURE 36.46 Compression sleeve-type connector used for service wire connection. (*Courtesy Homac Mfg. Company.*)

FIGURE 36.47 Insulated compression sleeve type connector used for service wire connection. (*Courtesy Homac Mfg. Company.*)

TYPICAL OVERHEAD RESIDENTIAL SERVICE

FIGURE 36.48 Drawing of service-drop cable connected to customer's service entrance loop with compression sleeve connector. (*Courtesy MidAmerican Energy Company.*)

Connectors are easily and quickly applied. If the conductor has a weatherproof covering, this is first removed and the exposed conductor scraped clean and bright. The two ends are then inserted into the compression-type connector and the proper compression tool is used to crimp the sleeve around the two conductors. Most compression connectors are covered and have inhibitor grease in the tube of the connector when received from the manufacturer. If the compression connector is not covered, it is coated with inhibitor grease and taped so that it will conform to the service-drop wires and secondary mains.

Electrical Drawing Symbols

Diagrams of electric circuits show the manner in which electrical devices are connected. Because it would be impossible to make a pictorial drawing of each electrical device shown in a diagram, the electrical device is represented by a symbol.

A list of some of the most common electrical symbols used follows. It is noted that the symbol is a sort of simplified picture of the electrical device represented.

The reader is encouraged to review IEEE *Standard 315/ANSI Y32.2/CSA Z99, Graphic Symbols for Electrical and Electronics Diagrams.*

Electrical Device	Symbol	Electrical Device	Symbol
Air circuit breaker		Cable termination	
Air circuit breaker drawout type, single pole		Capacitor	
Three pole		Coil, operating	
Air circuit breaker with magnetic-overload device		(Asterisk represents device) (designation	

Electrical Device	Symbol	Electrical Device	Symbol
Air circuit breaker with thermal-overload device		Contact Make-before-break	
Ammeter		Normally closed	
Autotransformer		Normally open	
Battery		Contact (spring-return) Pushbutton operated Circuit closing (make)	
Battery cell		Circuit opening (break)	
		Two circuit	
Contactor Electrically operated with series blowout coil		Lightning arrester with Gap Valve Ground	
Contactor Electrically operated three pole with series blowout coils and auxiliary contacts– one closed, two open		Magnetic blowout coil	
Current transformer		Network protector	
		Motor	
Current transformer with polarity marking		Circuit breaker	C.B.
Fuse		Polarity markings Negative Positive	
Gap horn		Potential transformer	
Gap protective		Potential transformer with polarity markings	
Generator			
Ground connection		Rectifier	

Electrical Device	Symbol	Electrical Device	Symbol
Inductance Fixed		Resister Fixed	
Adjustable		Tapped	
Lamp Red		Variable	
Green		With adjustable contact	
Switch Double throw		Transformer 1-phase with three windings	
Single throw		Transistor PNP-type	(E) (c) (B)
Transformer Single phase two winding		Electron tube Triode	
Transformer connections Three phase Delta Four wire grounded		Rectifier two element	
Open		Voltage regulator General	
Three wire		Induction	
		Single phase	
Scott or T		Three phase	
Wye		Step	
Four wire grounded		Voltmeter	
Three wire ungrounded		Wattmeter	
Zigzag grounded		Wires Connected	
		Crossed	

CHAPTER 38

Single-Line Diagrams

Purpose

Most electric transmission and distribution circuits are three-phase and require three lines to show them schematically in detail. For many functions, it is not necessary to have all the detailed information shown on the schematic diagram. Therefore single-line diagrams are utilized to represent three-phase circuits and equipment. A single-line diagram provides a quick method whereby a lineman or a cableman may analyze the complete facilities in schematic form.

Distribution Substation Single-Line Diagrams

Figure 38.1 is a single-line diagram of a distribution substation supplied by two 69,000-volt three-phase subtransmission circuits designated 66-58-59-1 and 66-58-A-1. The sub-transmission circuits of the 336,400-circular-mil aluminum conductor connect to the high-voltage structure of the substation overhead. Lightning arresters and potential transformers in the substation structure connect to the overhead line terminals. The potential transformers provide inputs for meters, relays, and auxiliary devices.

The subtransmission circuits are switched in and out of service utilizing the 69,000-volt 1,200-amp circuit breakers connected in series with group-operated disconnect switches. The group-operated disconnect switches provide a means of isolating the circuit breakers. The disconnect switches are operated after the circuit breaker contacts have been opened. The substation 69,000-volt bus has a normally closed group-operated disconnect switch, making it possible to manually isolate the bus into two sections. A 69,000-volt group-operated disconnect switch provides a means for connecting a mobile substation to the 69,000-volt bus. The fused potential transformer connected to the bus provides a means of measuring bus voltage.

The two power transformers in the substation (rated 69,000-13,200Y/7,620 volts, 20,000 kVA oil air-cooled, 26,700 kVA forced-oil air-cooled with tap-changing under-load equipment) are each switched in and out of service by a circuit breaker connected to the high-voltage winding and an air circuit breaker in the metal-clad switchgear connected to the low-voltage winding. The 13,200Y/7,620-volt leads from the power transformers connect to the circuit breakers in the metal-clad switchgear through 750-kCMIL copper conductor cables installed underground and terminated with potheads. The current transformers in series with each power transformer air circuit breaker provide a means of measuring the current magnitudes and detecting short circuits with protective relays.

725

FIGURE 38.1 Distribution substation single-line diagram.

Cubicle no. 1 of the metal-clad switchgear has a group-operated disconnect switch and draw-out power fuses that connect to underground cables terminated with potheads providing a source to the house service transformers rated 13,200-480Y/277 volts. Cubicle no. 7 of the metal-clad switchgear contains a bus-tie air circuit breaker, providing a means for sectionalizing the bus for maintenance or to isolate a bus section in case of a bus fault. Switchgear cubicles 3, 4, 8, 9, and 10 each provide a means of switching a distribution circuit in and out of service. Each distribution circuit connects to an air circuit breaker through underground cables. The feeder circuit originating in cubicle no. 4 has a capacitor bank connected to it in the yard of the substation.

Switchgear cubicle no. 5 provides a means for connecting the 13,200-volt bus to a mobile substation with a power transformer rated 69,000-13,200Y/7,620 volts. The mobile substation is utilized to replace one of the power transformers for maintenance work or in the case of a transformer failure.

Distribution Circuit Single-Line Diagram

The single-line diagram of Fig. 38.2 shows the 13,200Y/7,620-volt three-phase four-wire distribution circuits originating at substation 59 in single-line form. A square in the center of the figure with numeral 59 represents the substation. Four circuits originating at the substation extend in different directions to supply electrical power to the customers in the surrounding area. The major physical terrain details—such as Duck Creek, Kimberly Road, Hickory Grove Road, Lombard Street, and Wisconsin Avenue—give the schematic diagram physical significance.

The four circuits originating at substation 59 are designated 13-59-1, 13-59-2, 13-59-3, and 13-59-5. All four circuits exit from the substation through underground cables. Each of the circuits can be connected to an adjacent circuit at the substation perimeter to provide a means of supplying the customer load in case of an exit cable failure, as shown on the schematic tie insert at the left edge of the figure. Circuit 13-59-1 exits from the substation by underground cables that terminate in potheads on the riser pole, as shown on the tie insert schematic. The circle in circuit 13-59-1 with an S in the middle represents single-pole disconnect switches operated with hot-line tools.

The number 3314 is installed near the switches on the cable riser pole to uniquely identify the switches. The single-pole disconnect switches designated 3314 are used to isolate the underground cables from the overhead circuit in case of a cable failure. The circuit extends from the riser pole through single-pole disconnect switches number 3191. Single-pole disconnect switches 3315 are normally open, as designated by NO on the tie insert schematic. Circuit 13-59-1 can be tied to circuit 13-59-5 by closing the normally open single-pole disconnects numbers 3315 and 3319.

Circuit 13-59-1 extends across Duck Creek to Kimberly Road. The circle with an F followed by the number 3 indicates a fused three-phase tap to the main circuit extending north from Kimberly Road. At Pine Street, a capacitor bank number C-279 with the capacity of 1,050 kVAR is connected to the main circuit through switching equipment. The circuit extending north along Pine Street terminates in a group-operated disconnect switch designated by the circle with the letter G in the middle. The group-operated disconnect switch number 3076 is normally open, designated by NO.

Circuit 13-78-1 terminates on the north side of the group-operated disconnect switch. This switch provides a method of connecting circuit 13-78-1 to circuit 13-59-1. The main feeder portion of circuit 13-59-1 extends east on Kimberly Road through group-operated switch 3077 and terminates at group-operated switch 3359, which is

Figure 38.2 Distribution circuit single-line diagram.

Underground loops
1. Five seasons
2 Harrison Homes
3 Georgetown Sq
4 5. Sherwood Forest
6. Teds Mobile Homes
7 Meadowbrook
8. Royal Wood
9. Hetherton Heights
10. Hetherton Heights

Distribution system
Sub 59

13-59-1   13-59-3
13-59-2   13-59-5

728

normally open. Group-operated switch 3077 provides a means for sectionalizing circuit 13-59-1 in case of a sustained fault on the circuit that causes the circuit breaker at substation 59 to remain open.

Group-operated switch 3359 provides a method for connecting circuit 13-59-1 to circuit 13-58-1. The distribution circuit single-line diagram provides the lineman and cableman with a means of understanding switching instructions and locating major pieces of equipment with respect to known landmarks. Circuit 13-59-2 has a voltage regulator installed in it along Hickory Grove Road, southeast of the substation. The voltage regulator is represented by a circle with an arrow through it. The voltage regulator has single-pole disconnect switches in parallel with it for bypassing the regulator and single-pole switches in series with the regulator to isolate it.

Circuit 13-59-5 has a reclosing device in the line along Hickory Grove Road northwest of the substation. The three-phase reclosing device has fused bypass switches in parallel with it. The fused bypass switches are utilized to permit removing the recloser from service for maintenance purposes. The distribution circuit single-line diagram is very valuable for analyzing the circuits to assist the lineman and cableman to complete their work safely and effectively.

CHAPTER 39

Schematic Diagrams

Purpose

The purpose of a schematic diagram is to make it possible to picture the operation of an entire control circuit for a piece of equipment, so that it will be easy to think in terms of an entire circuit instead of a particular component part of it. Once the control schematic diagram is understood, it is easy to apply the information contained on it to some portion of the associated wiring diagram.

Feeder Circuit Control Schematic

The control schematic diagram for a distribution substation feeder circuit breaker installed in a metal-clad switchgear unit (Fig. 39.1) is traced by utilizing the numbers appearing by the component parts of equipment. Each piece of equipment and the purpose it serves are described.

The schematic diagram of Fig. 39.1 is drawn for the condition of all isolating devices open and the circuit breaker in the open position. This is an accepted standard method of drawing schematic control diagrams. The symbol represents a circuit breaker draw-out contact in the switchgear. All equipment shown between the draw-out contacts is located on the feeder circuit breaker.

The control source for closing the feeder circuit breaker is 240 volts alternating current. The lead designations for the control power source are X1 and Y1. Starting with wire X1, a circuit extends to item (1), a knife switch which isolates and deenergizes the control circuit. From the knife switch, the circuit is completed to item (2), which is a low-voltage fuse used for isolating the control circuit for a fault in the control wiring, protecting the control power source and preventing a fault in the wiring of one circuit breaker's control wiring from causing all circuit breakers to be inoperative.

From the fuse, the circuit is completed through wires X2 and 4X2 to item (3), 86, which is a contact on the lockout relay operated by the bus fault relays. The purpose of the 86 contact is to open the closing circuit for the feeder breaker to prevent closing the breaker after a bus fault until the lockout relay has been manually reset. From the lockout relay contact, the circuit is complete through wires 4W and W to item (4), CSO, the control-switch-off contact. This contact on the control switch is closed when the control switch handle is in the off or straight up-and-down position. The purpose of this contact is to keep the 152Y relay coil deenergized until the control switch is returned from the closed position to the off position if the breaker tripped immediately after a close operation.

From the CSO contact, the circuit is completed through wire 2a and the breaker draw-out contact to item (5), 152b, a contact on the circuit breaker auxiliary switch, which is closed when the circuit breaker is open. This contact opens this portion of the

FIGURE 39.1 Unit-type substation, feeder circuit, control schematic diagram.

732

control circuit so that current will not flow through the resistor continuously while the breaker is closed and completes the circuit to energize the 152Y relay when the breaker is in the open position. From the 152b auxiliary switch contact, the circuit is completed through the breaker draw-out contact, wire 1a, and a resistor to item (6), 152Y, closing cutoff relay coil. The purpose of the 152Y closing cutoff relay is to make the control scheme anti-pump, or electrically trip-free. The anti-pumping relay is a device whose function is to prevent multiple breaker closures. From the 152Y relay coil, the circuit is completed through wire Y2, a fuse and a knife switch to wire Y1, the other side of the alternating-current control power source.

If the knife switches are closed, a current will flow through the 152Y closing cutoff relay coil, operating the associated relay contacts, items (7) and (10). Item (7), 152Y, closing cutoff relay contact, completes the circuit through wires W and 1a to establish a second path for the current to flow through the 152Y relay coil. The purpose of this 152Y relay contact is to keep the 152Y closing cutoff relay coil energized when the control switch is turned to the closed position or when the breaker starts to close until the breaker is completely closed.

Starting with wire X1, one side of the alternating-current source, a circuit exists through items (1), (2), (3) and wires 4W and W to item (8), CSC, which is the control-switch-close contact. This contact on the control switch is closed when the control switch handle is turned to the closed position. This contact provides a means of closing the circuit breaker by operating the control switch manually. From the CSC contact, a circuit is completed through wire C to item (9), 152X, the closing relay coil. The purpose of this relay is to energize the circuit breaker closing solenoid. From the closing relay coil, the circuit is complete to item (10), 152Y, which is a contact on the closing cutoff relay. The purpose of this contact is to make the control scheme anti-pump or electrically trip-free. From the closing cutoff relay contact, the circuit is completed through wire Y2, the fuse, and the knife switch to wire Y1, the other side of the alternating-current source. If the CSC contact is closed, the closing relay coil 152X will have a current flow through it, since the closing cutoff relay is energized and the 152Y contact, item (10), is in the closed position. When a current flows through the 152X closing relay coil, the relay will operate its contacts, items (11), (12), and (13). Item (11), 152X, is a closing relay contact which short-circuits the CSC contact to keep the 152X closing relay energized once the breaker closing operation is initiated until the breaker has closed.

Starting with wire X1, one side of the alternating-current source, the circuit is completed through the knife switch, item (1); the fuse, item (2); and wire X2 to item (12), 152X, which is the closing relay contact. The purpose of this contact is to supply a high-current contact capable of energizing the circuit breaker closing coil. From the 152X contact, a circuit extends through the resistor, the rectifier elements, and the circuit breaker draw-out contact to item (14), 152CC, the circuit breaker closing coil. The purpose of this coil is to supply the energy to close the main contacts of the circuit breaker.

From the circuit breaker closing coil, a circuit is completed to the breaker draw-out contact, the copper oxide rectifier, and to item (13), 152X, a closing relay contact which serves the same purpose as item (12). From the closing relay contact, the circuit is established to wire Y2, the other side of the alternating-current source. With the closing relay energized and its associated contacts, items (12) and (13), closed, a current flows through the circuit breaker closing coil, causing the circuit breaker to close its main contacts.

Located on the circuit breaker mechanism is an auxiliary switch which is connected to the linkage which operates the circuit breaker main contacts. The auxiliary switch has contacts which are referred to as "a" contacts and "b" contacts. The auxiliary switch "a" contacts operate the same as the circuit breaker main contacts; that is, when the circuit breaker main contacts are closed, the "a" contacts are closed, and when the circuit breaker main contacts are open, the "a" contacts are open. The auxiliary switch "b" contacts operate opposite to that of the circuit breaker main contacts; that is, when the circuit breaker main contacts are closed, the "b" contacts are open, and when the circuit breaker main contacts are open, the "b" contacts are closed.

Starting with wire Y1, one side of the alternating current, a circuit is established through the knife switch, fuse, wire Y2, and the circuit breaker draw-out contact to item (15), 152aa, a circuit breaker auxiliary "a" switch contact adjusted to close late in the closing cycle. The 152aa switch serves the purpose of cutting off the circuit breaker closing power. The 152aa switch is adjusted to close late in the closing operation to provide adequate power for closing the circuit breaker in case a fault should exist in the distribution line supplied through the breaker. From the 152aa contact, the circuit is completed through the breaker draw-out contact to a point between the resistor and the 152Y closing cutoff relay coil.

With the 152aa auxiliary switch contact closed, the 152Y closing cutoff relay coil is short-circuited and the 152Y relay contacts, items (7) and (10), will be opened. The 152X closing cutoff relay will be deenergized and will open its contacts, items (12) and (13), deenergizing the 152cc circuit breaker closing coil. The 152X closing relay, as described above, was deenergized even though the CSC contact was maintained closed by the operator; therefore, the circuit breaker cannot reclose if it trips out on a closing operation due to a fault until the control switch handle is returned to the off position, closing contact CSO, item (4), and thus reenergizing the 152Y closing cutoff relay coil, item (6). From the Y1 wire of the alternating-current source through the knife switch, the fuse, wire Y2, the breaker draw-out contact, a circuit is established to item (16), 152a, which is an "a" contact on the circuit breaker auxiliary switch. This contact serves the purpose of energizing the red light when the circuit breaker closes. From the 152a contact, a circuit continues through the breaker draw-out contact, wire R, to item (17), R, which is the red indicating light. The red indicating light serves the purpose of informing the operator that the circuit breaker is closed. From the red light, the circuit is completed through wire X2, the fuse, and the knife switch to wire X1, the other side of the alternating-current source.

When the circuit breaker closes and the 152a auxiliary switch contact closes, the red light will be energized, indicating that the circuit breaker is closed. From wire Y1, one side of the alternating-current source, a circuit exists through the knife switch, the fuse, wire Y2, the breaker draw-out contact to item (18), 152b, which is a circuit breaker auxiliary switch "b" contact. The 152b contact serves the purpose of energizing the green light and the reclosing relay. From the 152b contact, the circuit is complete through a circuit breaker draw-out contact, wire G, to item (19), G, which is the green indicating light.

The green indicating light serves the purpose of telling the operator that the circuit breaker is open. From the green light, a circuit extends to item (20), CS CTO, which is the control switch lamp cutout contact. This contact on the control switch is closed except when the control switch handle is turned to the trip position and pulled out, placing the control switch in the lockout position. The purpose of the CS CTO contact is to deenergize the green light so that the green indicating light will not be lighted when the circuit breaker is out of service due to a planned outage.

From the CS CTO contact, the circuit is complete to the X side of the alternating-current source. When the circuit breaker is open and the 152b contact is closed, the green light will be energized, indicating that the circuit breaker is open. From the Y side of the alternating-current source, a circuit is established to item (21), cell switch, which is a contact that is closed when the circuit breaker is in the operating position. The purpose of the cell switch is to energize the amber light.

From the cell switch, the circuit is complete through wire A to item (22), A, the amber light. The purpose of the amber light is to indicate that the circuit breaker is in the operating position. From the amber light, the circuit is complete to the X side of the alternating-current source. When the circuit breaker is in the jacked-in or operating position, the cell switch will be closed, and the amber light will be lighted indicating the breaker is in the operating position.

From the wire P, the positive side of the 24-volt direct-current control power, a circuit extends to item (23), the knife switch. The knife switch can be used to isolate the circuit breaker trip circuit. From item (23), the circuit is completed to item (24), a fuse. The fuse serves the purpose of protecting the direct-current control source from a fault in the trip circuit control wiring of the circuit breaker. From the fuse, the circuit is complete through wire P2 to item (25), CST, the control switch trip contact. The purpose of the CST contact is to initiate a trip operation when the control switch handle is turned to the trip position.

From the CST contact, the circuit is complete through wire T, the breaker draw-out contact, and wire 4, to item (26), 152a, a circuit breaker auxiliary switch "a" contact. The purpose of the circuit breaker auxiliary switch "a" contact is to deenergize the circuit breaker trip coil after the tripping operation has been initiated and the circuit breaker starts to open. From the 152a contact, the circuit is complete to item (28), 152T, which is the circuit breaker trip coil. The circuit breaker trip coil serves the purpose of unlatching the circuit breaker mechanism so that the circuit breaker main contacts will open.

From the 152T trip coil, the circuit is complete to item (27), 152a contact, which serves the same purpose as item (26). From the 152a contact, the circuit extends through wire 5, the breaker draw-out contact, wire N2, the fuse, the knife switch, and wire N to the negative side of the 24-volt direct-current control source. If the control switch handle is turned to the trip position when the circuit breaker is closed, the CST contact will close and the trip coil will be energized, initiating a tripping operation. Item (29), 151, is a trip contact on an overcurrent protective relay.

The overcurrent relay trip contact is connected to wires P2 and T so that when the contact is closed and the circuit breaker is closed, a tripping operation will be initiated. The purpose of the overcurrent relay trip contact is to trip the circuit breaker when a fault exists on the feeder connected to the circuit breaker. Item (30), 86, is the lockout relay trip contact which serves the purpose of tripping the feeder circuit breaker when a bus fault occurs in the substation. The lockout relay trip contact is wired in parallel with the overcurrent relay trip contact.

Starting with the X1 wire of the alternating-current source, a circuit is completed through the knife switch, the fuse, wire X2, wire 4X2, 86 lockout relay contact, and wire 4W to item (31), CSSC, which is the control switch sliding contact. This contact on the control switch closes when the control switch handle is turned to the closed position, and it remains closed until the control switch handle is turned to the trip position. The purpose of the CSSC control switch sliding contact is to open the reclosing circuit to prevent an automatic reclose when the circuit breaker is tripped by the control switch.

From the CSSC contact, a circuit is established to item (32), CSO, an off contact on the control switch. This contact on the control switch is closed when the control switch handle is in the off or the straight up-and-down position. The purpose of the CSO contact is to open the reclosing circuit when the circuit breaker is closed by the operation of the circuit breaker control switch, thus preventing the circuit breaker from reclosing automatically on a fault when the circuit breaker is closed manually.

From the CSO contact, the circuit is complete through wire XC1 to item (33), 143, which is the automatic on/off toggle switch. The purpose of the 143 toggle switch is to deenergize the automatic reclosing relay. From the 143 contact, the circuit is complete through wire XC11 to the reclosing relay. Item (34), LCSW, is the latch check switch contact. The LCSW contact is open except when the circuit breaker is in the full open position. The LCSW is connected between the 152X closing relay and the automatic reclosing relay to prevent an automatic reclosure after a breaker trip operation until the breaker has completely opened. Wire G connects from the 152b, auxiliary switch contact, item (18), to the reclosing relay to energize the relay and initiate a reclosing operation when the circuit breaker opens due to a fault on the line.

The reclosing relay has a timing motor that is connected to wires XC11 and G through cam switches operated by the reclosing relay timing motor. The reclosing relay timing motor is energized after an automatic trip operation if the manual/automatic toggle switch 143 contact is closed. The timing motor operates the reclosing relay cam switches that complete a circuit from wire XC11 to the draw-out contact and the LCSW, item (34). This circuit energizes the closing relay coil, 152X, and initiates an automatic reclosing of the circuit breaker after an automatic trip operation initiated by the overcurrent relay, 151.

Voltage Regulator or Substation Transformer Tap Changer Control Schematic

Distribution substation transformers are supplied by the subtransmission system. The high-voltage windings of most of the transformers have a no-load tap changer and an under-load tap changer.

The purpose of the no-load tap changer is to change the number of turns in the primary winding when the transformer is not energized, so that the number of turns of the winding is in accordance with the general range of voltage that exists on the system at the location of the transformer. For example, if the voltage at a particular site varies from 71,000 to 67,000 volts, the no-load tap changer would probably be set on a tap which is rated 69,000 volts. This means that the actual turns utilized are in proportion to 69,000 volts.

The purpose of the tap-changing under-load equipment is to further change the number of turns in the primary winding to automatically keep the voltage on the secondary of the transformer at the proper value when the primary supply voltage varies to some value other than the value for which the no-load tap changer is adjusted. The under-load tap changer consists of a switching arrangement designed to operate when the transformer is energized and carrying load. This switching arrangement is operated through an insulated mechanism by means of a motor.

The schematic control diagram (Fig. 39.2) illustrates the operation of the control equipment necessary to operate the tap changer motor which supplies the energy to operate the tap-changing under-load switching equipment. Each item of equipment on the schematic control diagram (Fig. 39.2) is numbered. The circuit is traced, each piece of equipment is described, and the purpose that each piece of equipment serves is outlined, utilizing the numbers on the diagram. The circuit elements are shown for the condition

FIGURE 39.2 Unit substation transformer tap changer control, schematic diagram.

737

of the transformer deenergized and all operating devices open. The control schematic for a distribution-line voltage regulator would be similar.

Item (1) is a potential transformer connected to the regulated voltage supply. The purpose of the potential transformer is to reduce the voltage level so that it can be safely connected to relays and other control equipment. The output voltage of the transformer will normally be approximately 120 volts. From terminal V7 on the secondary of the potential transformer, the circuit is completed to item (2), AB1, which is an air circuit breaker. The circuit breaker protects the potential transformer from a short circuit in the control wiring. From item (2), the circuit is completed to item (3), FTS1, which is a test switch. The test switch provides a means of testing the voltage-regulating relay. From item (3), the circuit is completed to item (4), NV, which is the coil of the no-voltage relay. The coil of the no-voltage relay operates its contacts to prevent operation of the tap changer if the regulated voltage supply is removed from the voltage-regulating relay coil. From item (4) the circuit is complete to item (5), FTS1, another test switch, and item (6), AB1, another pole of the air circuit breaker. From item (6), the circuit is complete to terminal VO on the other end of the potential transformer secondary. As long as the potential transformer is energized and the air circuit breaker is closed, the no-voltage relay coil will be energized, and the relay contacts will be closed to permit automatic operation of the tap changer.

From item (3), the circuit is also complete to item (7), P, the voltage-regulating relay coil. The voltage-regulating relay coil actuates the raising and lowering contacts which controls the direction of operation of the tap changer motor. From item (7) the circuit is complete to item (8), the compensator. The compensator elements connect also to the secondary of a current transformer, which has its primary winding connected in series with the load.

The compensator varies the voltage applied to the voltage-regulating relay coil, in accordance with the settings of the resistance and reactance elements, and the current flowing in the secondary of the transformer. Therefore, it is possible to have the regulated voltage at a higher level during times of heavy load conditions to compensate for line voltage drops without readjusting the voltage-regulating relay. From item (8), the circuit is completed to item (5), etc., previously described.

Item (9) is an auxiliary power transformer. This transformer supplies the power to operate the tap changer motor and other equipment. From terminal X on the secondary of the auxiliary power transformer, the circuit is complete to item (10), AB2, an air circuit breaker which protects the auxiliary power transformer from a fault in the tap changer control equipment. From item (10), the circuit is complete through wire TRX to item (11), 152a, a circuit breaker auxiliary switch contact located on the transformer secondary breaker. The 152a switches are closed when the circuit breaker is closed and open when the circuit breaker is open.

The purpose of the auxiliary switch 152a contact is to prevent automatic voltage regulation when the transformer secondary breaker is open, and the transformer is not connected to a load. From item (11), the circuit is completed through wire TC1 to item (12), AM2, an automatic/manual switch contact. The purpose of this contact is to prevent automatic operation of the tap changer controls when the tap changer is being operated by the manual-control switch. From item (12), the circuit is completed to item (13), 123, a contact operated by a cam on the tap changer motor drive. This contact is closed when the tap changer is on a tap position and open when the tap changer is between tap positions.

The purpose of the 123 contact is to deenergize the tap changer control circuit after each tap changer operation is initiated to give nonsequential operation. From item (13) the circuit is complete to item (14), non SEQ, which is the nonsequential contact on the transfer switch. The purpose of this contact is to place the 123 cam switch in the circuit. The circuit is complete from item (14) to item (15) and from item (12) to item (15), SEQ, the sequential contact on the transfer switch. The purpose of this contact is to bypass the 123 cam switch so that the tap changer will keep running once an operation is initiated until the requirements of the voltage-regulating relay are fulfilled.

If the transformer is operated in a network, it is best to have the tap changers set on nonsequential operation to allow first one transformer and then another to operate its tap changer to fulfill the requirements of the voltage-regulating relays. From item (15) the circuit is complete to item (16), NV, the contact on the no-voltage relay previously discussed. The purpose of this contact is to prevent automatic operation of the tap changer if the voltage is removed from the coil of the voltage-regulating relay, thus preventing operation of the tap changer to raise the voltage excessively if the protective devices operate on the regulated potential supply.

From item (16) the circuit is complete to item (17), PL, the contact on the voltage-regulating relay, which closes when the regulated voltage rises above the value for which the voltage-regulating relay is adjusted.

The purpose of this contact is to initiate a tap-changing operation to lower the regulated voltage level. From item (17) the circuit is complete to item (18), HL, the heating element of the timing relay for the tap changer lower operation.

The heating element supplies heat to a bimetal strip which closes a contact after time delay. From item (18) the circuit is complete to item (19), AL2, an auxiliary relay contact, which is a component part of the timing relay. The purpose of this contact is to deenergize the heating element and permit the bimetal strip to return to its original position, opening contact HL1 and closing contact HL2 again. From item (19) the circuit is complete to item (20), LL, the tap changer lower-limit switch. The purpose of this limit switch is to prevent the tap changer motor from operating to change taps in the lower direction after the lowest tap has been reached. From item (20) the circuit is complete to item (21), IS, the interlock switch, which is opened when the crank is inserted in its socket to manually operate the tap changer.

The purpose of the interlock switch is to prevent simultaneous operation of the tap changer electrically and mechanically. From item (21) the circuit is completed to item (22), AB2, another pole of the air circuit breaker. From item (22) the circuit is completed to item (23), Z, the midpoint on the auxiliary power transformer, thus completing a circuit. Therefore, if all the conditions described previously are fulfilled for an automatic lower operation, a current will flow through item (18), HL, causing it to close its contact item (24), HL1, the time-delay relay heating element contact that closes after time delay. The purpose of the HL1 contact is to provide a circuit to energize the lower time-delay relay auxiliary relay coil item (25), AL.

The circuit is then complete from item (17) through items (25) and (24) to item (20), and a current will flow through the auxiliary relay coil, AL, operating its contacts. The purpose of the auxiliary relay is to permit the timing relay to complete its cycle. From item (25) the circuit is complete to item (26), AL1, a contact on the auxiliary relay which closed when the AL coil was energized. The purpose of this contact is to complete a circuit in parallel with the HL1 contact to keep the AL coil energized. When the AL coil was energized, item (19), AL2 contact, opened, deenergizing the HL heating element, which causes contact item (24), HL1, to open, and causes contact HL2, item (27), to close after time delay.

The HL2 contact is a time-delay relay heating element contact, which is closed when the heating element is deenergized. The purpose of this contact is to help energize the relay which energizes the tap changer motor to operate in the lower direction. Item (28), AL3, is another auxiliary relay contact. The purpose of this contact is to energize the tap changer motor relay with the help of item (27), HL2 contact.

From item (10) the circuit is complete to item (29), SL, the tap changer lower motor control relay coil. The purpose of this coil is to operate the relay contacts to energize the tap changer motor. From item (29) the circuit is complete through item (28), AL3 contact, which closed when the timing cycle was halfway completed; item (27), HL2 contact, which opened when the timing cycle was initiated by the voltage-regulating relay and closed when the timing cycle was completed; and item (20), etc., previously discussed.

From item (10) a circuit is established to item (30), TG, the motor thermal guard contact and heater element. The purpose of the thermal guard is to protect the motor if it is overloaded or the limit switches fail, and the motor is stopped by the mechanical stops. From item (30) the circuit is complete to item (31), the tap changer motor, which supplies the power to operate the tap changer. From item (31) the circuit is complete to item (32), SL3 contact on the motor control relay. The purpose of the SL3 contact is to energize the motor so that it will operate the tap changer to a lower tap position. From item (32) the circuit is complete to item (20), etc., as previously described. When the tap changer has moved off position, item (33), 120 cam switch contact, closes.

The purpose of the cam switch is to keep the motor control relay energized until the tap changer is back on position, at which time the 120 cam switch opens again. This prevents a partial tap change due to the voltage-regulating relay opening its contacts or due to the 123 cam switch, item (13), opening if the control is switched to nonsequential operation. This is accomplished through item (34), SR5, a contact on the raise motor control relay which is closed when the raise relay coil is deenergized. The purpose of this contact is twofold: first, to prevent operation of the SL and SR motor control relays simultaneously through the 120 cam switch and, second, to cause the tap changer to run to the next lower tap if it is stopped in a middle position between taps.

Item (35), SL1, a lower motor control relay contact, is opened when the motor is energized to operate the tap changer in the lower direction and closes as the tap change is completed. The purpose of this contact is to assist in placing power on the tap changer motor in the reverse direction to brake the motor when the operation is completed. Item (36), XL, is the lower direction plugging switch. This contact closes when the motor is operated in the direction to lower taps. Therefore, when the lower direction motor control relay coil, SL, is deenergized, power is supplied through contacts XL and SL1, which tends to rotate the motor in the reverse direction, serving the purpose of a brake. When the motor stops, contact XL, item (36), opens, removing power from the motor.

If the motor is energized in the lower direction beyond the lowest tap position, item (20), LL, the lower-limit switch contact, opens, deenergizing the motor, and item (37), 125 cam switch, closes to supply braking power. The purpose of the 125 cam switch, which closes after the tap changer has reached the lowest position, is to apply braking power immediately to prevent operation of the tap changer against the mechanical stop.

Item (38), M17, is a contact on a transfer switch. The purpose of the M17 transfer switch is to energize item (39), 117 cam switch. The 117 cam switch is open only on the odd-numbered tap positions. The purpose of items (39) and (38) is to short-circuit the

120 cam switch on the even-numbered tap positions so that when a tap change is initiated, it will complete two operations instead of one.

Item (40), AM1, is an automatic/manual transfer switch contact. When the transfer switch is operated to manual, contact item (12) is opened and item (40), AM1, is closed. The purpose of this is to permit manual electrical control of the tap changer. From item (29), SL, the lower motor control relay coil, a circuit is completed to item (41), MCL, the manual control switch lower contact. The purpose of this switch is to operate the tap changer electrically by means of the manually operated control switch. From item (41) the circuit is completed through items (40), (21), etc.

From item (10) a circuit is established to item (42), 121, a cam switch which is closed between each tap position. The purpose of this switch is to energize the interposition light and the operation counter. From item (42) a circuit is complete to item (43), RL, a red light. The purpose of the light is to indicate that the tap changer is between positions. From item (43), the circuit is complete to item (22) and the other side of the control power source. From item (42), a circuit is established to item (44), OC, the operation counter. The purpose of the operation counter is to record each tap changer operation to help determine the need for maintenance work and operational inspections.

The tap changer is operated in the raised direction in similar manner as previously described for the lower direction. A circuit exists from the auxiliary transformer through the circuit breaker, item (10), the circuit breaker 152a auxiliary switch contact, item (11), through wire TC1 to item (12), the automatic/manual switch contact, through item (13), 123 cam switch, and item (14), the nonsequential transfer switch contact, or item (15), the sequential switch contact; then through item (16), the no-voltage relay contact, to item (45), PR, the contact on the voltage-regulating relay which closes when the regulated voltage decreases below the value for which the voltage-regulating relay is adjusted.

The purpose of this contact is to initiate a tap-changing operation to raise the regulated voltage level. From item (45) the circuit is complete to item (46), HR, the heating element of the timing relay for the tap changer raise operation. The heating element supplies heat to a bimetal strip which closes a contact after time delay. From item (46) the circuit is complete to item (47), AR2, an auxiliary relay contact which is a component part of the timing relay. The purpose of this contact is to deenergize the heating element and permit the bimetal strip to return to its original position, opening contact HR1 and closing contact HR2 again. From item (47) the circuit is complete to item (48), LR, the tap changer raise limit switch.

The purpose of the limit switch is to prevent a tap changer motor from operating to change taps in the raise direction after the highest tap has been reached. From item (48) the circuit is complete to item (21), the interlock switch, item (22), the air circuit breaker, and item (23), the secondary of the operating transformer. Therefore, if all the conditions described previously are fulfilled for an automatic raise operation, a current will flow through item (46), HR, causing it to close its contact, item (49), HR1, the time-delay relay heating element contact that closes after a predetermined time delay.

The purpose of the HR1 contact is to provide a circuit to energize the raise time-delay relay, auxiliary relay coil item (50), AR. The circuit is complete from item (45) through items (50) and (51) to item (48), and a current will flow through the auxiliary relay coil AR, causing it to operate its contacts. The purpose of the auxiliary relay is to permit the timing relay to complete its cycle. From item (50), the circuit is complete to item (51), AR1, a contact on the auxiliary relay which closed when the AR coil was energized. The purpose of this contact is to complete a circuit in parallel with the HR1 contact to keep the AR coil energized. When the AR coil was energized, item (47), AR2 contact, opened, deenergizing the HR heating element, which opened contact, item (49), HR1, and closed contact HR2, item (52), after time delay. The HR2 contact is the time-delay relay heating element contact which is closed when the heating element is deenergized.

The purpose of this contact is to help energize the relay which energizes the tap changer motor to operate in the raise direction. Item (53), AR3, is another auxiliary relay contact. The purpose of this contact is to energize the tap changer motor relay with the help of item (52), HR2 contact. From item (10), the circuit is complete to item (54), SR, the tap changer raise motor control relay coil. The purpose of this coil is to operate the relay contacts to energize the tap changer motor. From item (54) the circuit is complete through item (53), AR3 contact, which closed when the timing cycle was halfway completed, item (52), HR2 contact, which opened when the timing cycle was initiated by the voltage-regulating relay and closed when the timing cycle was completed, and item (48), etc., as previously discussed.

As previously described, a circuit is established from item (10) through item (30), TG, the motor thermal guard contact and heater element, item (31), M, the tap changer motor, to item (55), SR3 contact on the motor control relay. The purpose of the SR3 contact is to energize the motor so that it will operate the tap changer to raise the regulated voltage. From item (55) the circuit is complete to item (48), LR, etc., as previously described. When the tap changer motor has moved off position, item (33), 120 cam switch contact, closes to keep the raise motor control relay energized until the tap changer is back on position, at which time the 120 cam switch opens. This prevents a partial tap change due to the voltage-regulating relay opening its contact prior to the time the tap changer completes its operation.

Item (56), SR2 raise motor control relay contact, is opened when the motor is energized and closes as the tap change is completed. The purpose of this contact is to assist in placing power on the tap changer motor in the reverse direction to brake the motor when the operation is completed. Item (57), XR, is the raise direction plugging switch. Therefore, when the raise motor control relay coil, SR, is deenergized, power is supplied through contacts XR and SR2, which tends to rotate the motor in the reverse direction, serving the purpose of a brake. When the motor stops, contact XR, item (57), opens, removing the power from the motor.

If the motor is energized in the raise direction beyond the highest tap position, item (48), LR, the raise limit switch contact, opens, deenergizing the motor, and item (58), 124 cam switch contact, closes to supply braking power. The purpose of the 124 cam switch contact, which closes after the tap changer has reached the highest position, is to apply braking power to stop the rotation of the tap changer motor to prevent operation of the tap changer against the mechanical stop.

Item (59), 152b, is a circuit breaker auxiliary switch contact located on the power transformer secondary breaker. This contact is closed when the secondary breaker is open. The purpose of the contact is, if the control power is on, to operate the tap changer in the raise direction and thus to assist the secondary breaker network relays to close the breaker automatically after an outage of the circuit serving as a source to the power transformer, if the transformer is operated in a network system.

CHAPTER 40

Voltage Regulation

*V*oltage regulation on any electrical system is defined as the difference in voltage between periods of no load and periods of full load on the system. Characteristics of the system and the amount of load will determine the extent of the voltage regulation. To provide good electric service to customers served from the distribution system, it is necessary to maintain voltage variations within a very limited range. The voltage variation to residences has been standardized from a high limit of 126 volts to a low limit of 114 volts for a nominal 120-volt secondary service. The term *nominal voltage* designates the line-to-line voltage, as distinguished from the line-to-neutral voltage. It applies to all parts of the system or circuit.

Electrical equipment is designed to operate on a certain voltage, and best operation is obtained when that rated voltage is applied. The effect of low voltage on heating-type appliances such as heaters, toasters, irons, and ranges is an increase in heating time. Low-voltage conditions cause motors to operate at decreased efficiency. They draw more current from the line, have less starting and running torque, and may overheat to the extent that they might experience thermal failure. Incandescent lights operate less efficiently, give less light, and in the case of florescent lights may not operate at all.

High-voltage conditions are also detrimental to the operation of electrical equipment. Heating-type elements, light bulbs, and electron tubes have a shorter life span because of high voltage. Motors and other magnetic equipment will have excess speed or torque. Television sets are particularly susceptible to damage from high voltage.

The best service to customers and the greatest amount of revenue would result from a distribution system which provides rated voltage at every point of the system. Since this is not economically feasible, a compromise is made, and the voltage variation is confined to the standard limits established.

The regulation or control of voltage on a distribution system is usually accomplished by a combination of two different methods. One method utilizes transformer tap-changing equipment, step-voltage regulators, or induction voltage regulators. With this equipment, distribution-system line voltage may be held constant while its supply voltage varies over a certain range. This equipment may also raise or lower the line voltage to compensate for load changes on a circuit. Still another function of this equipment is to provide, in the case of a distribution tie line, some control of kilovar flow over the tie line. To accomplish any of these functions with the above voltage-regulating equipment, an in-phase voltage or approximately in-phase voltage is either added or subtracted from the normal line voltage. Voltage control is accomplished over an approximate ±10% range in easily controlled steps without interruption in service to the customer.

Another voltage control option normally used on distribution systems makes use of capacitors. The control of voltage with capacitors is accomplished by the ability of a capacitor to supply a leading current to a distribution line to counteract the lagging current which the line must supply to magnetize motors, transformers, and other magnetic equipment. The leading current and the lagging current combine to produce a smaller current more nearly in phase with the voltage. The smaller current produces less voltage drop on the line. Capacitors are usually switched on and off in banks, and therefore the control of voltage with capacitors is less flexible than with voltage-regulating equipment. However, capacitors have other advantages, probably the most important of which is their ability to reduce the kVA load on lines, transformers, and generators, making them available for load growth.

The use of voltage-regulating equipment is made necessary because of a characteristic common to all electric circuits. That characteristic is called *impedance*. Impedance is defined as the opposition to the flow of current in an electric circuit. When current flows through an electric circuit, a voltage drop occurs between the input and output terminals of the circuit. This voltage drop is a function of the impedance in the circuit as well as the amount of current flowing in the circuit.

In the circuit having a constant impedance, the voltage drop will be proportional to the current flowing in the circuit. This is defined as:

$$V = IZ$$

where V = voltage drop
      I = current flowing in circuit
      Z = impedance of circuit

The voltage drop through a transformer can be visualized by considering a transformer energized on the primary side with 100% voltage and no connected load on the secondary terminals. For this condition, the voltage on the secondary terminals would be 100% rated voltage also. If a load were now applied to the secondary terminals of the transformer while the primary voltage remained constant, something other than rated voltage would be obtained on the secondary terminals. This change in voltage would be caused by current flowing through the internal impedance of the transformer.

The voltage drop on an overhead transmission or distribution line is probably a little easier to visualize because no transformation ratio is involved as in the case of a transformer. A transmission or distribution line has impedance distributed along its entire length. This impedance is made up of resistance and inductive and capacitive reactance. Each foot of conductor of a line has a certain amount of resistance depending on the conductor size and the material used to make up the conductor (copper, aluminum, etc.). Each 1-ft length of a line has a certain amount of inductive reactance because of the spacing between conductors which is required for electrical clearance. Also, each foot length of line has a certain amount of capacitive reactance due to the conductor size, distance between conductors, and the height of the conductor above ground. The capacitive reactance of a distribution line is small compared to the inductive reactance and is usually neglected. Current flowing through the line impedance causes a voltage drop so that the voltage at the load end of the line is less than the voltage at the source end.

Load current flowing through the impedance of a line produces a voltage drop. To help illustrate the effect of the voltage drop, an electrical representation of a line is shown in Fig. 40.1. Distributed along the line are resistance, inductance, and capacitors.

FIGURE 40.1 Representation of a transmission or distribution line.

To simplify the representation still further, the values of resistance, impedance, and capacitance can be lumped together to form a network, as shown in Fig. 40.2.

If the line is a short transmission line or a distribution feeder or tie line, the capacitive effect of the line can usually be neglected. This results in a still further simplification of the representation of a line, as shown in Fig. 40.3.

The voltage drop, which would occur on this line because of the load current and impedance, is shown in the vector diagram, which assumes a lagging power-factor load (Fig. 40.4).

From the vector diagram (Fig. 40.4) it is apparent that the sending end voltage is of greater magnitude than the receiving end voltage and the difference in the two voltages is the IZ or impedance voltage drop. The IZ component is made up of two parts—the drop due to the current flowing through the resistance and the drop due to the current flowing through the line reactance.

As has previously been noted, capacitance exists between conductors of a line and between the conductors and ground. On a long line that is lightly loaded, the capacitive charging current may exceed the load current and result in the line operating with a leading power factor. In such cases, the receiving end voltage will rise, and it may exceed the voltage at the sending end of the line.

From the vector diagram (Fig. 40.4) several important points can be seen. The first is that the receiving end voltage of a line can be made more nearly equal to the sending end voltage by reducing the current on the line and thereby reducing the impedance drop.

FIGURE 40.2 Representation of a simplified transmission line.

FIGURE 40.3 Further simplification of a short line circuit.

FIGURE 40.4 Vector diagram of voltages on a short line.

V_S = Sending end or source-end voltage

V_R = Receiving end voltage

I_L = Load current

IR = Voltage drop due to the resistance in the line

IX = Voltage drop due to the reactance in the line

This is often done to a distribution feeder when the load on the line becomes so great that excessive voltage drop is encountered. A portion of the load is then transferred to some other line which is not so heavily loaded.

The second point, which is apparent from the diagram, is that the impedance drop may be reduced by decreasing the impedance of the line. To reduce the resistance, wires of a larger cross-sectional area would have to be installed. To reduce the reactance of a line would mean either decreasing the conductor spacing or using a series capacitor, neither of which is too practical.

A third point, which can be seen from the vector diagram, is that the power factor of the load could be improved. This could be done by installing shunt capacitors near the load (Fig. 40.5). This method of voltage improvement to improve the power factor is used quite extensively. For the best effect, the capacitors should be switched off or on as the load changes. This is done with time, temperature, or voltage controls. The effect of adding capacitors to a circuit is shown in Fig. 40.6. A power factor correction example is found in Chap. 41, page 769.

There are times and conditions where the voltage regulation on a line, that is, the difference between the no-load voltage and the full-load voltage, becomes so great that other means must be used to maintain suitable voltages. Under these conditions step-type voltage regulators, tap-changing transformers, or induction voltage regulators are employed. In general, voltage regulation of substation buses is accomplished using under-load tap-changing equipment on transformers, while the voltage regulation on distribution circuits is controlled by shunt capacitors or step-voltage regulators. The step-voltage regulator is primarily an autotransformer with tap-changing under-load switching available to change the ratio between the primary and the secondary windings.

The purpose of under-load tap-changing (LTC) equipment is to change the primary-to-secondary turns ratio of a transformer winding while the transformer is carrying load.

FIGURE 40.5   The physical representation of a 15 kV shunt capacitor bank installed on a distribution pole.

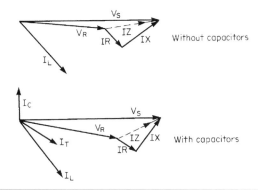

FIGURE 40.6   Effect of installing capacitors.

The *turns ratio* of a transformer is equal to its voltage ratio. That is, a transformer with a turns ratio of 100:1 also has voltage ratio of 100:1. This means that by controlling the turns ratio of a transformer one can regulate the voltage on its "low" side.

The heart of LTC equipment is the automatic control which operates a motor-driven mechanical tap changer, thus maintaining a controlled voltage at the load as the input voltage to the power transformer fluctuates or as the load conditions vary.

As the distribution-line load increases, the voltage at its load end drops due to transformer and line impedances (both reactive and resistive components). This line voltage is monitored at the low-voltage side of the power transformer, and when the voltage changes by a prescribed amount, this automatic control device senses the change and causes the under-load tap-changing equipment to either increase or decrease the turns ratio, thus changing the voltage and bringing it back to its desired value.

There are compensating devices in the automatic control unit that correct for the reactive and resistive loss components of the distribution line. This *line-drop compensating device* raises the voltage at the transformer as the line current increases, thus maintaining a nearly constant voltage at the regulating point on the line.

The typical under-load tap selector (see Fig. 40.7) consists of three (one per phase) dial-type switches (A). Each switch has stationary contacts (B) which are connected to the taps in the transformer winding. These contacts are mounted on an insulating panel in a circle around two concentric slip rings (C). The two movable contact fingers (D) for each switch are mounted as an assembly on the main rotor shaft (E), each finger contacting its own slip ring. The rotor shaft which extends through all three phases of the tap selector is turned by the tap changer motor (F) through reduction gears and switch controls (G). The two fingers move together, stopping either with both on the same contact or with one on each of two adjacent contacts.

Each of the three dial-type tap selector switches is provided with a separate single-pole double-throw switch (H) used to reverse the polarity of the under-load tap-changing windings.

Tap leads from the transformer windings are connected to stationary contacts (see Fig. 40.8) on the tap selector. When the motor operates the selector, it will change taps to vary the transformer turns ratio and thereby regulate the voltage.

Two movable contacts are used on the tap selector so that one can be moved to an adjacent tap while the other contact maintains the power circuit. These movable contacts are connected through a reactor winding to prevent excessive circulating current

FIGURE 40.7   Transformer under-load tap selector diagram.

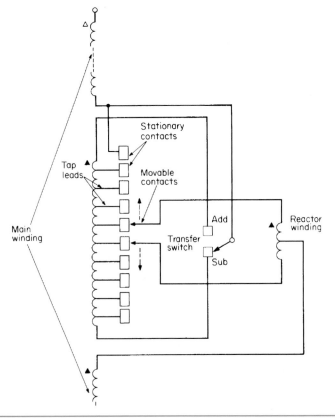

FIGURE 40.8   Under-load tap changer schematic.

through the bridged section of the winding. By designing the winding and reactor to share the circulating current, the bridged position can be used as an intermediate operating position, thus doubling the number of tap-changing steps.

A two-position switch, integral with the tap selector, is used to double the number of steps again. This switch is used to reverse the polarity taken from the tapped section of the transformer, permitting the tapped section to be used twice—one time adding to and the other time subtracting from the number of turns on the main winding.

When the automatic control unit calls for a change of tap position, the drive motor is energized and, through a series of gears, the tap selector is operated. In the first step in a sequence of operation, the movable fingers are slid toward the next stationary contact. The first finger breaks contact, and the current is carried by the second finger (see Fig. 40.9). In the second step, the movable fingers are slid so that the first finger is on the

FIGURE 40.9 Under-load tap changer operation sequence.

next stationary contact while the second finger is still on the original contact. In this step the currents are shared by both fingers and the center-tapped reactor to which the fingers are connected through the slip rings providing for an intermediate voltage between the two stationary contacts. In the third step, the second finger is slid off the original contact and the current is carried by the first finger. In the last step in the sequence of operation, the fingers are slid so that both fingers are on the second stationary contact, thus completing the sequence.

Typical automatic controls which govern operation of the load tap changer consist of a line-drop compensating unit, a voltage-sensing unit, and a time-delay unit.

Line-Drop Compensator

Sometimes it is desirable to maintain a constant voltage at some other point on the circuit than at the source end of the line. This point, for example, may be the center of the load on the circuit. It would be possible to hold a constant voltage at this point by placing a potential transformer on the line and extending a pair of wires back to the voltage-regulating relay for the secondary voltage. This method is not too practical, and a device called a *line-drop compensator* is used instead. The line-drop compensator consists of a resistance and a reactance element wired in series with a current transformer which is in series with one of the regulator terminals, as shown in Fig. 40.10. In this circuit, the values of resistance and reactance are adjusted so that they form a miniature line. Current flow through the current transformer primary winding to the load causes a proportional current to flow through the compensator. This current causes a voltage to appear across the resistance and reactance elements which is proportional to the voltage drop appearing on the line at some previously selected point. The voltages produced across the resistance and reactance elements are then introduced into the voltage-regulating relay circuit. The resistor voltage is applied so as to subtract from the voltage applied to the regulating relay by the potential transformer. The reactance element voltage may be applied so that it either subtracts from or adds to the potential transformer voltage depending on what condition the element is set for. Normally it would be set to subtract.

With this type of equipment, the voltage regulation of a distribution-system feeder for various load conditions is shown graphically in Fig. 40.11. From this figure it can be seen that this method of regulation is a good solution to a regulation problem because it supplies rated voltage or voltage within a very limited range to all customers on the line.

Static Voltage-Regulating Relay

The line-drop compensator unit, the voltage-sensing unit, the time-delay unit, and all auxiliary functions except the final output to the motor can be performed with solid-state

Figure 40.10   Simplified connections of a line-drop compensator.

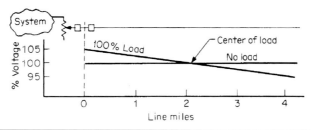

FIGURE 40.11   Regulated line using line-drop compensation.

devices shown on schematic control diagrams (Figs. 40.12 through 40.15). The operation of the control circuits will be described.

The power transformer secondary or distribution-system primary output voltage is stepped down to 120 volts by the control potential supply transformer and is applied to the input of the automatic static-load tap changer control (Fig. 40.12). At this point, a normal test power supply switch permits the input of a test power supply for calibration and maintenance. Past the circuit breaker, a test rheostat is provided to permit variation of the voltage applied to the sensing circuit of the control without changing the transformer output voltage. The test rheostat is left in the "auto" position for normal operation.

A 5:1 ratio transformer (T1) is used to reduce the voltage applied to the line-drop compensator and sensing circuit. This line-drop compensator consists of a tapped line-drop reactance compensating reactor, a compensating polarity switch, and a line-drop resistance compensating rheostat. These devices are fed by the paralleling current transformer, the load current transformer, and transformer T1.

The purpose of the line-drop compensating unit is to develop voltages proportional to the reactive (IX) and resistive (IR) drops in the distribution line fed by the transformer. The final settings for the line-drop compensator may be made by field adjustment; however, initial settings may be determined from line data by using the following expressions:

d = miles from unit to load center or regulating point

r = resistance per conductor from unit to load center, ohms/mile

x = inductive reactance per conductor from unit to load center, ohms/mile

I = load current transformer primary current rating

E = control potential supply transformer ratio

R = resistance dial setting

X = reactance dial setting

For a three-phase balanced unit,

$$R = \frac{rdI}{E} \quad \text{and} \quad X = \frac{rdI}{E}$$

For normal operation, the compensation polarity switch is left in the "normal" position. Reversed reactance compensation is obtained by placing it in the "reverse" position. This reverses the current supplied by the load current transformer in the reactance element.

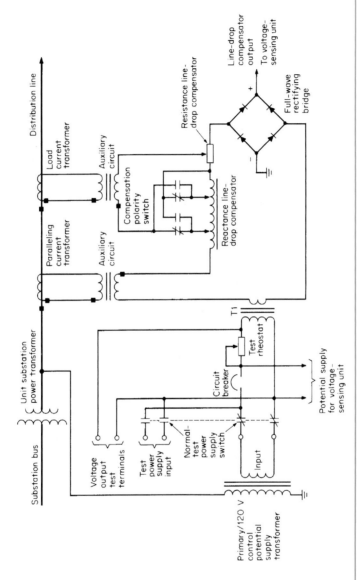

FIGURE 40.12 Line-drop compensating schematic.

Reversed reactance compensation is a method used to reduce the circulating current that might flow when two or more transformers are paralleled. Instead of running toward opposite extreme positions, tap changers having reversed reactance compensation tend to move toward whatever position causes the least amount of transformer circulating current to flow. This is accomplished at some sacrifice to the normal line-drop compensation but is generally satisfactory when units paralleled are not located in close proximity to each other or where the supply is from different sources.

For the paralleling of two transformers located in close proximity and supplied from the same source, the circulating currents are used to control the line-drop compensator through the paralleling current transformer, causing the tap changers to operate in such a direction as to reduce the transformer circulating currents to a minimum.

For normal compensation applications, the voltage developed across the line-drop compensator is subtracted from the output of transformer T1. This forces the output of T1 to be higher for a given voltage into the sensing circuit and thus across the transformer to compensate for the drop created by increased current in the line.

The output of the line-drop compensator is rectified and filtered into a dc voltage by the full-wave bridge rectifier. At this point the voltage-level adjust rheostat provides for transformer voltage-level settings between about 90% and 110% of nominal line voltage.

The adjusted voltage is fed into the voltage-sensing unit which consists of a zener diode bridge network and the control windings of two magnetic pulse amplifiers (Fig. 40.13). The basic elements of the bridge are resistors R1 and R2, zener diodes Z1 and Z2, the minimum-bandwidth rheostat, and the ganged-bandwidth rheostat.

The zener diodes serve as reference devices, and any voltage variations are reflected as differences in voltages across the resistors. The zener diode bridge is in a balanced state when the potential at the junction of R1 and Z1 (point A) is equal to the center of the active portion of the minimum-bandwidth rheostat (point B). As the applied voltage increases, the potential at point B rises with respect to that at point A. As the voltage decreases, the reverse occurs.

The output of the voltage-sensing bridge is fed through the control windings of a pair of magnetic amplifiers (Amp 1 and Amp 2). At balance, the current in the control winding of Amp 1 is from point C to point A, and the current in the control winding of Amp 2 is from point A to point D. Current applied in this direction causes each amplifier to produce negative pulses to the gate of SCR1. Since positive gate voltage is required to make the silicon controlled rectifier (SCR) conduct, there will be no output from the control. At equilibrium the ac windings of the magnetic amplifier are of low impedance; thus, the transformer secondary is short-circuited, and the raise and lower circuits are deenergized.

When the applied input voltage increases, the potential at point B rises with respect to that at point A. When this increase is sufficient to cause the potential at point D to equal that at point A, the current in the control winding of Amp 2 will cut off. When the control current is cut off, the ac winding of Amp 2 changes impedance from low to high. This causes the voltage at the gate of SCR1 to become positive and causes it to conduct on the half-cycle. Since the secondary of transformer T2 is shorted out in only one direction, through diode D1, the voltage builds up across the ac winding of Amp 2 and voltage is applied to the "lower" portion of the relay circuit.

This voltage is applied through diode D4 to the time-delay unit (Fig. 40.14)—the time-delay circuit input voltage is regulated by zener diode Z3—and the time-delay control rheostat varies the amount of time delay. This time-delay unit has a range of about 10 to 100 seconds and is usually set at 30 seconds.

Figure 40.13 LTC voltage-sensing schematic.

FIGURE 40.14   LTC time-delay schematic.

The time-delay circuit consists basically of a unijunction transistor and an RC charging network. When the time-delay capacitor has charged sufficiently to cause the unijunction transistor to conduct, the positive pulse created by this conduction is applied to the gate of SCR2, causing it to conduct. The conduction of SCR2 energizes relay SL which in turn energizes the tap changer operating motor in the direction which will lower the output voltage (Fig. 40.15).

FIGURE 40.15   LTC motor control schematic.

A similar sequence may be followed when the voltage applied to the sensing circuit decreases. In this case, the current through the control winding of Amp 1 will cut off when the potential at point C equals that at point A. (Note that the setting of the band-width rheostat determines the amount of deviation from balance required to initiate a controlling action.)

Since Amp 1 functions when the top of T2 is positive with respect to the bottom, SCR1 is caused to conduct on the half-cycle which energizes the raise portion of the circuit, this time through diode D3. Under these conditions, it will energize relay SR which will in turn operate the drive motor to raise the output voltage.

The tap changer motor is operated from an auxiliary potential supply transformer and has the following operational sequence.

Potential is supplied through the 84I crank-handle interlocking switch. When the crank handle is used to operate the tap changer manually, the 84I contact is open, preventing simultaneous operation of the tap changer manually and electrically. Potential is supplied from this point to the lower/auto/raise switch and to the remote control switch. When the remote switch is activated, the tap changer can be raised, lowered, or set to automatic control from a point outside the tap changer control compartment on the transformer.

The lower/auto/raise switch permits the tap changer to be raised, lowered, or set to automatic control at the tap changer.

In the auto position, the SL and SR contacts of the automatic static control device are employed. If the lower circuit is energized by the sensing unit and the SL relay picks up, the SL contact to the lower terminal of the motor is closed, completing a circuit through the 84/LS lower-end limit switch which cuts off motor operation if the tap changer has advanced to its lowest position. The motor is now energized and will continue to run until it lowers the transformer voltage to satisfy the conditions set by the voltage-sensing unit.

When SL is energized, the contact supplying potential to the SR contacts is opened, preventing simultaneous raise and lower operations. When SL or SR is energized, the power supply to the time-delay unit is cut off, permitting it to reset.

In a similar manner, when the raise circuit is energized by the sensing unit and the SR relay picks up, the SR contact to the raise terminal of the motor is closed, completing a circuit through the 84/LS upper-end limit switch which cuts off motor operation if the tap changer has advanced to its highest position. The motor is again energized and will continue to run until it raises the transformer voltage to satisfy the conditions set by the voltage-sensing unit.

When a tap change occurs, the AG contact is closed, thus causing the operation counter OC to record that an operation has taken place.

Units of Measurement, Electrical Definitions, Electrical Formulas, and Calculations

General

In this chapter, the most common units of measurement, definitions, and electrical formulas with example calculations are provided.

Units of Measurement

Time

1 minute	= 60 seconds		
1 hour	= 60 minutes	= 3,660 seconds	
1 day	= 24 hours	= 1,440 minutes	= 86,400 seconds

60-hertz AC electric system (60 cycles per second)

1 second	= 60 cycles
100 milliseconds	= 6 cycles
250 milliseconds	= 15 cycles
500 milliseconds	= 30 cycles

Length

1 ft	= 12 inches	
1 yard	= 3 ft	= 36 inches
1 mile	= 5,280 ft	= 63,360 inches
1 mile	= 1,760 yards	

Decimal units

milli-	10^{-3}
centi-	10^{-2}
deci-	10^{-1}
kilo-	10^{3}
mega-	10^{6}

Metric equivalents
U.S. to Metric

1 inch	= 0.0254 meter	=	2.54 centimeters	=	25.4 millimeters
2 inches	= 0.0508 meter	=	5.08 centimeters	=	50.8 millimeters
3 inches	= 0.0762 meter	=	7.62 centimeters	=	76.2 millimeters
4 inches	= 0.1016 meter	=	10.16 centimeters	=	101.6 millimeters
5 inches	= 0.1270 meter	=	12.70 centimeters	=	127.0 millimeters
6 inches	= 0.1524 meter	=	15.24 centimeters	=	152.4 millimeters
7 inches	= 0.1778 meter	=	17.78 centimeters	=	177.8 millimeters
8 inches	= 0.2032 meter	=	20.32 centimeters	=	203.2 millimeters
9 inches	= 0.2286 meter	=	22.86 centimeters	=	228.6 millimeters
10 inches	= 0.2540 meter	=	25.40 centimeters	=	254.0 millimeters
11 inches	= 0.2794 meter	=	27.94 centimeters	=	279.4 millimeters
1 foot	= 0.3048 meter	=	30.48 centimeters	=	304.8 millimeters
1 yard	= 0.9144 meter	=	91.44 centimeters	=	914.4 millimeters
1 mile	= 1,609.35 meters	=	1.60935 kilometers		

Metric to U.S.

1 millimeter	= 0.03937 inch					
1 centimeter	= 0.39370 inch					
1 meter	= 1.09361 yards	= 3.28083 feet	= 39.3700 inches			
1 kilometer	= 0.62137 mile					

Temperature

Centigrade to Fahrenheit
$C = (5/9) \times (F - 32)$

0°C	= 32°F
10°C	= 50°F
20°C	= 68°F
30°C	= 86°F
40°C	= 104°F
50°C	= 122°F
60°C	= 140°F
70°C	= 158°F
80°C	= 176°F
90°C	= 194°F
100°C	= 212°F

Fahrenheit to Centigrade
$F = [(9/5) \times C] + 32$

0°F	= −17.8°C
10°F	= −12.2°C
20°F	= −6.7°C
30°F	= −1.1°C
32°F	= 0°C
40°F	= 4.4°C
50°F	= 10.0°C
60°F	= 15.6°C
70°F	= 21.1°C
80°F	= 26.7°C
90°F	= 32.2°C
100°F	= 37.8°C

Definitions

- **Voltage (E)** The electric force of work required to move current through an electric circuit, measured in units of volts.
- **Volt (V)** The unit of measurement of voltage in an electric circuit, denoted by the symbol V.
- **Current (I)** The flow of free electrons in one general direction, measured in units of amps.
- **Amp (A)** The unit of measurement of electric current, denoted by the symbol A.

- **Direct Current (dc)**   Current that flows continually in one direction.

- **Alternating Current (ac)**   Current that flows in a circuit in a positive direction and then reverses itself to flow in a negative direction.

- **Resistance (R)**   The electrical "friction" that must be overcome through a device in order for current to flow when voltage is applied.

- **Ohm (Ω)**   The unit of measurement of resistance in an electric circuit, denoted by the symbol "Ω."

- **Inductance (H)**   The property that causes the current to lag the voltage, measured in units of henrys.

- **Capacitance (F)**   The property that causes the current to lead the voltage, measured in units of farads.

- **Energy**   The amount of electric work (real power) consumed or utilized in an hour. The unit of measurement is the watt-hour.

- **Power (P)**   The combination of electric current and voltage causing electricity to produce work. Power is composed of two components: real power and reactive power.

- **Voltampere (VA)**   The unit of both real and reactive power in an electric circuit.

- **Real Power**   The resistive portion of a load found by taking the cosine (Θ) of the angle that the current and voltage are out of phase.

- **Watt (W)**   The unit of real power in an electric circuit.

- **Reactive Power**   The reactive portion of a load, found by taking the sine (Θ) of the angle that the current and voltage are out of phase.

- **Var (Q)**   The unit of reactive power in an electric circuit.

- **Power Factor (p.f.)**   The ratio of real power to reactive power.

- **Frequency (f)**   The number of complete cycles made per second, measured in units of hertz.

- **Hertz (Hz)**   Units of frequency (equal to 1 cycle per second).

- **Conductors**   Materials that have many free electrons and are good transporters for the flow of electric current.

- **Insulators**   Materials that have hardly any free electrons and inhibit or restrict the flow of electric current.

- **Series Resistive Circuit**   All of the resistive devices are connected to each other so that the same current flows through all of the devices.

- **Parallel Resistive Circuit**   Each resistive device is connected across a voltage source. The current in the parallel path divides and only a portion of the current flows through each of the parallel paths.

Formulas and Calculations

In the material that follows, the most common electrical formulas are given, and their use is illustrated with a problem. In each case, the formula is first stated by using the customary symbols; the same expression is then stated in words; a problem is provided, and, finally, the substitutions are made for the symbols in the formula from which the answer is calculated. Only those formulas that the lineman is apt to use are given. The formulas are divided into three groups: direct-current circuits, alternating-current circuits, and electrical apparatus. A summary of circuit calculations is shown in Table 41.1.

Direct-Current Circuits

Ohm's Law   The formula for Ohm's law is the most fundamental of all electrical formulas. It expresses the relationship that exists in an electric circuit containing resistance only between the current flowing in the resistance, the voltage impressed on the resistance, and the resistance of the circuit.

$$I = \frac{E}{R}$$

where I = current, amps
E = voltage, volts
R = resistance, ohms

Expressed in words, the formula states that current equals voltage divided by resistance, or

$$\text{Amperes} = \frac{\text{volts}}{\text{ohms}}$$

Example   How much direct current will flow through a resistance of 10 ohms if the pressure applied is 120 volts direct current?

Solution $$I = \frac{E}{R} = \frac{120}{10} = 12 \text{ amps}$$

Ohm's law involves three quantities: current, voltage, and resistance. Therefore, when any two are known, the third one can be found. The procedure for solving for current has already been illustrated. To find the voltage required to circulate a given amount of current through a known resistance, Ohm's law is written as

$$E = IR$$

In words, the formula says that the voltage is equal to the current multiplied by the resistance; thus

$$\text{Volts} = \text{amperes} \times \text{ohms}$$

Example   How much direct-current (dc) voltage is required to circulate 5 amps direct current through a resistance of 15 ohms?

Solution $$E = IR = 5 \times 15 = 75 \text{ volts}$$

In like manner, if the voltage and current are known and the value of resistance is to be found, the formula is

$$R = \frac{E}{I}$$

Expressed in words, the formula says that the resistance is equal to the voltage divided by the current; thus

$$\text{Ohms} = \frac{\text{volts}}{\text{amperes}}$$

To Find	Direct Current	Alternating Current	
		Single Phase	**Three Phase**
Amperes when kilovolt-amperes and voltage are known	$\dfrac{\text{Kilovolt-amperes} \times 1{,}000}{\text{Volts}}$	$\dfrac{\text{Kilovolt-amperes} \times 1{,}000}{\text{Volts}}$	$\dfrac{\text{Kilovolt-amperes} \times 1{,}000}{1.73 \times \text{volts}}$
kilovolt-amperes when amperes and voltage are known	$\dfrac{\text{Amperes} \times \text{volts}}{1{,}000}$	$\dfrac{\text{Amperes} \times \text{volts}}{1{,}000}$	$\dfrac{1.73 \times \text{volts} \times \text{amperes}}{1{,}000}$
Amperes when kilowatts, volts, and power factors are known	$\dfrac{\text{Kilowatts} \times 1{,}000}{\text{volts}}$	$\dfrac{\text{Kilowatts} \times 1{,}000}{\text{Volts} \times \text{power factor}}$	$\dfrac{\text{Kilowatts} \times 1{,}000}{1.73 \times \text{volts} \times \text{power factor}}$
Kilowatts when amperes, volts, and power factor are known	$\dfrac{\text{Amperes} \times \text{volts}}{1{,}000}$	$\dfrac{\text{Amperes} \times \text{volts} \times \text{power factor}}{1{,}000}$	$\dfrac{1.73 \times \text{amperes} \times \text{volts} \times \text{power factor}}{1{,}000}$

Table 41.1 Summary of Power Formulas

Example What is the resistance of an electric circuit of 120 volts direct current that causes a current of 5 amps to flow through it?

Solution
$$R = \frac{E}{I} = \frac{120}{5} = 24 \text{ ohms}$$

Series Resistive Circuit

Example The 12-volt battery has three lamps connected in series (Fig. 41.1). The resistance of each lamp is 8 ohms. How much current is flowing in the circuit?

Solution In a series-resistive circuit, all values of resistance are added together, as defined by the formula

$$R_{series} = R_1 + R_2 + \cdots\cdots = R_{total}$$

Therefore: $R = 8 + 8 + 8 = 24$ ohms
And because $I = E/R$
$$= 12/24$$
$$= 0.5 \text{ amps}$$

Parallel Resistive Circuit

Example The 12-volt battery has three lamps connected in parallel (Fig. 41.2). The resistance of each lamp is 8 ohms. How much current is flowing in the circuit?

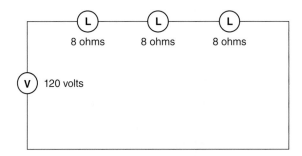

FIGURE 41.1   A series circuit. The same current flows through all the lamps.

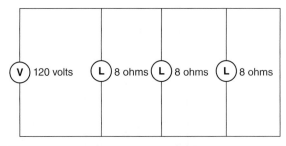

FIGURE 41.2   A parallel circuit. Each lamp is independent of the other lamps and draws its own current.

Solution   In a parallel-resistive circuit, all values of resistance can be added together, as defined by the formula

$$1/R_{\text{parallel}} = 1/R_1 + 1/R_2 + \cdots\cdots + 1/R_n$$

In our three-resistor **Example**

$$
\begin{aligned}
1/R_{\text{parallel}} &= 1/R_1 + 1/R_2 + 1/R_n \\
1/R_{\text{parallel}} &= (R_1 R_2 + R_2 R_3 + R_1 R_3)/(R_1 \times R_2 \times R_3) \\
R_{\text{parallel}} &= (R_1 \times R_2 \times R_3)/(R_1 R_2 + R_2 R_3 + R_1 R_3)
\end{aligned}
$$

Therefore: $R = (8 \times 8 \times 8)/[(8 \times 8) + (8 \times 8) + (8 \times 8)]$
$$
\begin{aligned}
&= 512/192 \\
&= 2.67 \text{ ohms}
\end{aligned}
$$
And because $I = E/R_{\text{parallel}}$
$$
\begin{aligned}
&= 12/2.67 \\
&= 4.5 \text{ amps}
\end{aligned}
$$

Power Formula

The expression for the power drawn by a dc circuit is

$$P = EI$$

where E and I are the symbols for voltage in volts and current in amperes and P is the symbol for power in watts.

Expressed in words, the formula says that the power in watts drawn by a dc circuit is equal to the product of volts and amperes; thus

$$\text{Power} = \text{volts} \times \text{amperes}$$

Example   How much power is taken by a 120-volt dc circuit when the current flowing is 8 amps direct current?

Solution $P = EI = 120 \times 8 = 960 \text{ watts}$

The power formula given above also contains three quantities or terms, watts, volts, and amperes. Therefore, when any two of the three are known, the third one can be found. The procedure for finding the power when the voltage and current are given has already been illustrated. When the power and voltage are given and the current is to be found, the formula is

$$I = \frac{P}{E}$$

In words, the formula states that the current equals the power divided by the voltage; thus

$$\text{Amperes} = \frac{\text{watts}}{\text{volts}}$$

Example   How much direct current would a 1,000-watt load draw when connected to a 120-volt dc circuit?

Solution $I = \dfrac{P}{E} = \dfrac{1,000}{120} = 8.33 \text{ amps}$

In like manner, when the power and current are known, the voltage can be found by writing the formula thus

$$E = \frac{P}{I}$$

Expressed in words, the formula says that the voltage is equal to the power divided by the current; thus

$$\text{Volts} = \frac{\text{watts}}{\text{amperes}}$$

Example   What dc voltage would be required to deliver 660 watts with 6 amps direct current flowing in the circuit?

Solution $E = \dfrac{P}{I} = \dfrac{660}{6} = 110$ volts

Line Loss or Resistance Loss

The formula for computing the power lost in a resistance when current flows through it is

$$P = I^2R$$

where the symbols have the same meaning as in the foregoing formulas.

Expressed in words, the formula says that the power in watts lost in a resistance is equal to the square[1] of the current in amperes multiplied by the resistance in ohms; thus

$$\text{Watts} = \text{amperes squared} \times \text{ohms}$$

Example   Compute the watts lost in a line having a resistance of 4 ohms when 8 amps direct current is flowing in the line.

Solution $P = I^2R = 8 \times 8 \times 4 = 256$ watts

Energy (Electrical Work)

The formula for computing the amount of energy consumed is defined as

$$\text{Energy} = \text{power} \times \text{time}$$

Power is measured in units of watts and time is typically measured in units of hours, so the units for energy will be in watt-hours.

Example   How much energy does a 60-watt load consume in a 10-hour time period? The formula for energy is given by

$$\text{Energy} = \text{Power} \times \text{Time}$$

Solution   Therefore

$$
\begin{aligned}
\text{Energy} &= 60 \text{ watts} \times 10 \text{ hours} \\
&= 600 \text{ watt-hours} \\
&= 0.6 \text{ kilowatt-hours} \\
&= 0.6 \text{ kwh}
\end{aligned}
$$

---

[1]To square means to multiply by itself.

Alternating-Current Circuits

Ohm's Law

Ohm's law is the same for resistance circuits when alternating voltage is applied as when direct voltage is applied, namely

$$I = \frac{E}{R}$$

and

$$E = RI$$

and

$$R = \frac{E}{I}$$

In the preceding equations, E is the effective value of the alternating voltage and I the *effective* value of the alternating current. (See *Direct-Current Circuits* for **Example**s.)

Ohm's Law for Other Than Resistance Circuits

When alternating currents flow in circuits, these circuits may exhibit additional characteristics besides resistance. They may exhibit inductive reactance or capacitive reactance or both. The total opposition offered to the flow of current is then called *impedance* and is represented by the symbol Z. Ohm's law then becomes

$$I = \frac{E}{Z}$$

and

$$E = IZ$$

and

$$Z = \frac{E}{I}$$

where I = current, amps
E = voltage, volts
Z = impedance, ohms

> **Example** Find the impedance of an alternating-current (ac) circuit if a 120-volt alternating voltage source causes 30 amps alternating current to flow.
>
> **Solution**
>
> $$Z = \frac{E}{I} = \frac{120}{30} = 4\text{-ohms impedance}$$

Impedance

The impedance of a series circuit is given by the expression

$$Z = \sqrt{R^2 + (X_L - X_c)^2}$$

where Z = impedance, ohms
R = resistance, ohms
X_L = inductive reactance, ohms
X_c = capacitive reactance, ohms

Example   If an ac series circuit contains a resistance of 5 ohms, an inductive reactance of 10 ohms, and a capacitive reactance of 6 ohms, what is its impedance in ohms?

Solution
$$Z = \sqrt{R^2 + (X_L - X_c)^2}$$
$$= \sqrt{5 \times 5 + (10 - 6)^2}$$
$$= \sqrt{25 + (4 \times 4)}$$
$$= \sqrt{41} = 6.4 \text{ ohms}$$

Note: The values of inductive and capacitive reactance of a circuit depend upon the frequency of the current, the size, spacing, and length of the conductors making up the circuit, etc. For distribution circuits and transmission lines, values may be found in appropriate tables.

In the expression for Z, when neither inductive reactance nor capacitive reactance is present, Z reduces to the value R; thus

$$Z = \sqrt{R^2 + (0 - 0)^2} = \sqrt{R^2} = R$$

Likewise, when only resistance and inductive reactance are present in the circuit, the expression of Z becomes

$$Z = \sqrt{R^2 + (X_L - 0)^2} = \sqrt{R^2 + X_L^2}$$

Example   Compute the impedance of an ac circuit containing 3 ohms of resistance and 4 ohms of inductive reactance connected in series.

Solution
$$Z = \sqrt{R^2 + X_L^2} = \sqrt{3^2 + 4^2}$$
$$= \sqrt{3 \times 3 + 4 \times 4} = \sqrt{9 + 16}$$
$$= \sqrt{25} = 5\text{-ohms impedance}$$

Line Loss or Resistance Loss

The formula for computing the power lost in a resistance or line is

$$P = I^2 R$$

where the symbols have the same meaning as above. Expressed in words, the formula says that the power in watts lost in a resistance is equal to the current in amperes squared multiplied by the resistance in ohms, thus

Watts = amperes squared × ohms

Example   Compute the power lost in a line having a resistance of 3 ohms if 20 amps is flowing in it.

Solution
$$P = I^2 R = 20 \times 20 \times 3 = 1{,}200 \text{ watts}$$

Power Formula

The power formula for single-phase ac circuits is

$$P = E \times I \times \text{pf}$$

where P = power, watts
    E = voltage, volts
     I = current, amps
    pf = power factor of circuit

Expressed in words, the formula says that the power in watts drawn by a single-phase ac circuit is equal to the product of volts, amperes, and power factor. If the power factor is unity, or 1, then the power equals the product of volts and amperes.

Example What is the power drawn by a 20-amp purely resistive load, with a power factor of 1, operating at 120 volts?

Solution
$$\begin{aligned} P &= E \times I \times \text{pf} \\ &= 120 \times 20 \times 1 \\ &= 2{,}400 \text{ watts} \\ &= 2.4 \text{ kilowatts} \\ &= 2.4 \text{ kW} \end{aligned}$$

Example How much power is delivered to a single-phase ac circuit operating at 120 volts if the circuit draws 10 amps at 80 percent power factor?

Solution
$$\begin{aligned} P &= E \times I \times \text{pf} \\ &= 120 \times 10 \times 0.80 \\ &= 960 \text{ watts} \end{aligned}$$

The power formula given previously contains four quantities: P, E, I, and pf. Therefore, when any three are known, the fourth one can be determined. The case where the voltage, current, and power factor are known has already been illustrated. When the voltage, power, and power factor are known, the current can be computed from the expression

$$I = \frac{P}{E \times \text{pf}}$$

Example How many amperes are required to light an ordinary 60-watt, 120-volt incandescent lamp? (Assume that the power factor of a purely resistive load, such as the lamp, is 1.00)

Solution
$$I = \frac{P}{E \times \text{pf}}$$
$$= \frac{60}{120 \times I}$$
$$= 0.5 \text{ amps}$$

Example How much current is drawn by a 10-kW load from a 220-volt ac circuit if the power factor is 0.80?

Solution
$$I = \frac{P}{E \times \text{pf}}$$
$$= \frac{10{,}000}{220 \times 0.80} = 56.8 \text{ amps}$$

In similar manner, if the power, voltage, and current are known, the power factor can be computed from the following formula

$$\text{pf} = \frac{P}{E \times I}$$

Example What is the power factor of a 4-kW load operating at 230 volts if the current drawn is 20 amps?

Solution
$$\text{pf} = \frac{P}{E \times I} = \frac{4,000}{230 \times 20}$$

$$= 0.87 \text{ or } 87\% \text{ power factor}$$

Three-Phase

The power formula for a three-phase ac circuit is

$$P = \sqrt{3} \times E \times I \times \text{pf}$$

where E is the voltage between phase wires and the other symbols have the same meaning as for single-phase. The additional quantity in the three-phase formula is the factor ($\sqrt{3}$), expressed "square root of three," the value of which is

$$\sqrt{3} = 1.73$$

The voltage between phase wires is equal to ($\sqrt{3}$), or 1.73, multiplied by the phase-to-neutral voltage of the three-phase four-wire systems.

Example How much power in watts is drawn by a three-phase load at 230 volts if the current is 10 amps and the power factor is 80%?

Solution
$$P = \sqrt{3} \times E \times I \times \text{pf}$$

$$= \sqrt{3} \times 230 \times 10 \times 0.80$$

$$= 3,183 \text{ watts}$$

The three-phase power formula also contains four quantities, as in the single-phase formula. Therefore, if any three are known, the fourth one can be determined. The case where the voltage, current, and power factor are known was illustrated above. When the power, voltage, and power factor are known, the current in the three-phase line can be computed from the formula

$$I = \frac{P}{\sqrt{3}E \times \text{pf}}$$

Example If a 15-kW three-phase load operates at 2,300 volts and has a power factor of 1, how much current flows in each line of the three-phase circuit?

Solution
$$I = \frac{P}{\sqrt{3}E \times \text{pf}}$$

$$= \frac{15,000}{\sqrt{3} \times 2,300 \times 1.0} = 3.77 \text{ amps}$$

If the only quantity not known is the power factor, this can be found by using the following formula

$$\text{pf} = \frac{P}{\sqrt{3} \times E \times I}$$

Example   Find the power factor of a 100-kW three-phase load operating at 2,300 volts if the line current is 40 amps.

Solution

$$pf = \frac{P}{\sqrt{3} \times E \times I}$$

$$= \frac{100,000}{\sqrt{3} \times 2,300 \times 40}$$

$$= 0.628, \text{ or } 62.8\% \text{ power factor}$$

Voltamperes

Often loads are given in voltamperes instead of watts or in kVA instead of kW. The relationships then become

$$VA = E \times I \qquad\qquad \text{For single-phase}$$

and

$$VA = \sqrt{3} E \times I \qquad\qquad \text{For three-phase}$$

In using these quantities it is not necessary to know the power factor of the load.

Power Factor Correction

To determine the kVARs required to improve the power factor for a known load, the following steps are followed:

Step 1:   Determine the existing KVA load

$$KVA_{existing} = \frac{KW}{pf_{existing}}$$

Step 2:   Determine the existing reactive power load, expressed in kVARs

$$kVAR_{existing} = \sqrt{KVA_{existing}^2 - KW^2}$$

Step 3:   Determine the corrected KVA load

$$KVA_{corrected} = \frac{KW}{pf_{corrected}}$$

Step 4:   Determine the corrected reactive power load, expressed in kVARs

$$kVAR_{corrected} = \sqrt{KVA_{corrected}^2 - KW^2}$$

Step 5:   The required corrective reactive power is equal to the existing reactive power load minus the corrected reactive power load, expressed in kVARs.

$$kVAR_{required} = kVAR_{existing} - kVAR_{corrected}$$

Example For a 40-kW load at 80% power factor, how much kVAR is needed to correct the power factor to 90%?

Solution

Step 1: Determine the kilovolt load in the existing condition:

$$KVA_{existing} = KW/pf_{existing}$$
$$= 40/0.8$$
$$= 50 \text{ kVA}$$

Step 2: Determine the reactive power load in the existing condition:

$$kVAR_{existing} = \sqrt{KVA_{existing}^2 - KW^2}$$
$$= \sqrt{50^2 - 40^2}$$
$$= 30 \text{ kVARs}$$

Having found the existing condition reactive power load, determine what the corrected final values for the total and reactive power will be:

Step 3:
$$KVA_{existing} = KW/pf_{existing}$$
$$= 40/0.9$$
$$= 44.4 \text{ kVA}$$

Step 4:
$$kVAR_{corrected} = \sqrt{KVA^2 - KW^2}$$
$$= \sqrt{44.4^2 - 40^2}$$
$$= 19.37 \text{ kVARs}$$

To determine the amount of reactive power needed on the system to correct the power factor from 80 to 90%, the final step of the problem is to subtract the corrected reactive power value desired from the existing value of reactive power:

Step 5:
$$kVAR_{required} = kVAR_{existing} - kVAR_{corrected}$$
$$= 30 - 19.37$$
$$= 10.63 \text{ kVARs}$$

Line Loss or Resistance Loss (Three-Phase)

The power loss in a three-phase line is given by the expression

$$P = 3 \times I^2R$$

where I = current in each line wire and R = resistance of each line wire. The factor 3 is present so that the loss in a line wire will be taken 3 times to account for the loss in all three wires.

Example If a three-phase line carries a current of 12 amps and each line wire has a resistance of 3 ohms, how much power is lost in the resistance of the line?

Solution $P = 3 \times I^2R = 3 \times 12 \times 12 \times 3 = 1{,}296 \text{ watts}$

Electrical Apparatus

Motors

Direct-Current Motor   The two quantities that are usually desired in an electric motor are its output in horsepower (hp) and its input current rating at full load. The following expression for current is used when the motor is a direct-current motor

$$I = \frac{hp \times 746}{E \times eff}$$

The formula says that the full-load current is obtained by multiplying the horsepower by 746 and dividing the result by the product of voltage and percent efficiency. (The number of watts in 1 hp is 746. A kilowatt equals 1-1/3 horsepower.)

Example   How much current will a 5-hp 230-volt dc motor draw at full load if its efficiency at full load is 90%?

Solution
$$I = \frac{hp \times 746}{E \times eff} = \frac{5 \times 746}{230 \times 0.90} = 18.0 \text{ amps}$$

Alternating-Current Motor, Single-Phase   If the motor is a single-phase ac motor, the expression for current is almost the same, except that it must allow for power factor. The formula for full-load current is

$$I = \frac{hp \times 746}{E \times eff \times pf}$$

Example   How many amperes will a ¼-hp motor take at full load if the motor is rated 110 volts and has a full-load efficiency of 85% and a full-load power factor of 80%?

Solution
$$I = \frac{hp \times 746}{E \times eff \times pf} = \frac{0.25 \times 746}{110 \times 0.85 \times 0.80} = 2.5 \text{ amps}$$

Alternating-Current Motor, Three-Phase   The formula for full-load current of a three-phase motor is the same as for a single-phase motor except that it has the factor in it thus

$$I = \frac{hp \times 746}{\sqrt{3} \times E \times eff \times pf}$$

Example   Calculate the full-load current rating of a 5-hp three-phase motor operating at 220 volts and having an efficiency of 80% and a power factor of 85%.

Solution
$$I = \frac{hp \times 746}{\sqrt{3} \times E \times eff \times pf} = \frac{5 \cdot 746}{\sqrt{3} \times 220 \times 0.80 \times 0.85}$$
$$= 14.4 \text{ amps}$$

Direct-Current Motor   To find the horsepower rating of a dc motor if its voltage and current rating are known, the same quantities are used as for current except the formula is rearranged thus

$$hp = \frac{E \times I \times eff}{746}$$

Example How many horsepower can a 220-volt dc motor deliver if it draws 15 amps and has an efficiency of 90%?

Solution
$$hp = \frac{E \times I \times eff}{746} = \frac{220 \times 15 \times 0.90}{746} = 4.0 \text{ hp}$$

Alternating-Current Motor, Single-Phase The horsepower formula for a single-phase ac motor is

$$hp = \frac{E \times I \times pf \times eff}{746}$$

Example What is the horsepower rating of a single-phase motor operating at 480 volts and drawing 25 amps at a power factor of 88% if it has a full-load efficiency of 90%?

Solution
$$hp = \frac{E \times I \times pf \times eff}{746}$$

$$= \frac{480 \times 25 \times 0.88 \times 0.90}{746}$$

$$= 12.74 \text{ hp}$$

Alternating-Current Motor, Three-Phase The horsepower formula for a three-phase ac motor is

$$hp = \frac{\sqrt{3} \times E \times I \times pf \times eff}{746}$$

Example What horsepower does a three-phase 240-volt motor deliver if it draws 10 amps, has a power factor of 80% and has an efficiency of 85%?

Solution
$$hp = \frac{\sqrt{3} \times E \times I \times pf \times eff}{746}$$

$$= \frac{\sqrt{3} \times 240 \times 10 \times 0.80 \times 0.85}{746} = 3.8 \text{ hp}$$

Alternating-Current Generator

Frequency The frequency of the voltage generated by an ac generator depends on the number of poles in its field and the speed at which it rotates:

$$f = \frac{p \times rpm}{120}$$

where f = frequency, Hz
  p = number of poles in field
  rpm = revolutions of field per minute

Example Compute the frequency of the voltage generated by an alternator having two poles and rotating at 3,600 rpm.

Solution
$$f = \frac{p \times rpm}{120} = \frac{2 \times 3,600}{120} = 60 \text{ hz}$$

Speed   To determine the speed at which an alternator should be driven to generate a given frequency, the following expression is used:

$$\text{rpm} = \frac{f \times 120}{p}$$

Example   At what speed must a four-pole alternator be driven to generate 60 Hz?

Solution

$$\text{rpm} = \frac{f \times 120}{p} = \frac{60 \times 120}{4} = 1,800 \ \text{rpm}$$

Number of Poles   If the frequency and speed of an alternator are known, the number of poles in its field can be calculated by use of the following formula:

$$p = \frac{f \times 20}{\text{rpm}}$$

Example   How many poles does an alternator have if it generates 60 Hz at 1,200 rpm?

Solution

$$p = \frac{f \times 20}{\text{rpm}} = \frac{60 \times 120}{1,200} = 6 \ \text{poles}$$

Transformer (Single-Phase)

Primary Current   The full-load primary current can be readily calculated if the kVA rating of the transformer and the primary voltage are known:

$$I_p = \frac{\text{kVA} \times 1,000}{E_p}$$

where E_p = rated primary voltage and I_p = rated primary current.

Example   Find the rated full-load primary current of a 10-kVA 2,300-volt distribution transformer.

Solution

$$I_p = \frac{\text{kVA} \times 1,000}{E_p} = \frac{10 \times 1,000}{2,300} = 4.3 \ \text{amps}$$

Secondary Current   The expression for secondary current is similar to that for primary current. Thus

$$I_s = \frac{\text{kVA} \times 1,000}{E_s}$$

where I_s = rated secondary current and E_s = rated secondary voltage.

Example   A 3-kVA distribution transformer is rated 2,300 volts primary and 110 volts secondary. What is its full-load secondary current?

Solution

$$I_s = \frac{\text{kVA} \times 1,000}{E_s}$$

$$= \frac{3 \times 1,000}{110} = 27\text{-amps secondary}$$

Summary of Electrical Formulas

For a summary of circuit calculations, see Table 41.1.

Current

1. *Direct current.* To find the current when the voltage and resistance are given:

$$I = \frac{E}{R}$$

2. *Alternating current.* To find the current when the voltage and resistance are given:

$$I = \frac{E}{R}$$

3. *Alternating current.* To find the current when the voltage and impedance are given:

$$I = \frac{E}{Z}$$

4. *Direct current.* To find the current when the voltage and power are given:

$$I = \frac{P}{E}$$

5. *Alternating current, single-phase.* To find the current when the voltage, power, and power factor are given:

$$I = \frac{P}{E \times pf}$$

6. *Alternating current, three-phase.* To find the current when the voltage, power, and power factor are given:

$$I = \frac{P}{\sqrt{3}E \times pf}$$

7. *Alternating current, single-phase.* To find the current when the voltamperes and volts are given:

$$I = \frac{VA}{E}$$

and

$$I = \frac{kVA \times 1,000}{E}$$

8. *Alternating current, three-phase.* To find the current when the voltamperes and volts are given:

$$I = \frac{VA}{\sqrt{3}E}$$

and

$$I = \frac{kVA \times 1,000}{\sqrt{3}E}$$

Voltage

9. *Direct current.* To find the voltage when the current and resistance are given:

$$E = IR$$

10. *Alternating current.* To find the voltage when the current and resistance are given:

$$E = IR$$

11. *Alternating current.* To find the voltage when the current and impedance are given:

$$E = IZ$$

Resistance

12. *Direct current.* To find the resistance when the voltage and current are given:

$$R = \frac{E}{I}$$

13. *Alternating current.* To find the resistance when the voltage and current are given:

$$R = \frac{E}{I}$$

Power Loss

14. *Direct current.* To find the power loss when the current and resistance are given:

$$P = I^2 R$$

15. *Alternating current, single-phase.* To find the power loss when the current and resistance are given:

$$P = I^2 R$$

16. *Alternating current, three-phase.* To find the power loss when the current and resistance are given:

$$P = 3 \times I^2 R$$

Impedance

17. *Alternating current, single-phase.* To find the impedance when the current and voltage are given:

$$Z = \frac{E}{I}$$

18. *Alternating current, single-phase.* To find the impedance when the resistance and inductive reactance are given:

$$Z = \sqrt{R^2 + X_L^2}$$

19. *Alternating current, single-phase.* To find the impedance when the resistance, inductive reactance, and capacitive reactance are given:

$$Z = \sqrt{R^2 + (X_L - X_c)^2}$$

Power

20. *Direct current.* To find the power when the voltage and current are given:

$$P = EI$$

21. *Alternating current, single-phase.* To find the power when the voltage, current, and power factor are given:

$$P = E = I \times pf$$

22. *Alternating current, three-phase.* To find the power when the voltage, current, and power factor are given:

$$P = \sqrt{3} \times I \times pf$$

Voltamperes

23. *Alternating current, single-phase.* To find the voltamperes when the voltage and current are given:

$$VA = E \times I$$

and

$$kVA = \frac{E \times I}{1,000}$$

24. *Alternating current, three-phase.* To find the voltamperes when the voltage and current are given:

$$VA = \sqrt{3}E \times I$$

and

$$kVA = \frac{\sqrt{3}E \times I}{1,000}$$

Power Factor

25. *Alternating current, single-phase.* To find the power factor when the power, volts, and amperes are given:

$$pf = \frac{P}{E \times I}$$

26. *Alternating current, three-phase.* To find the power factor when the power, volts, and amperes are given:

$$pf = \frac{P}{\sqrt{3} \times E \times I}$$

Motor

27. *Direct-current motor.* To find the current when the horsepower of the motor, the voltage, and the efficiency are given:

$$I = \frac{hp \times 746}{E \times eff}$$

28. *Alternating-current motor, single-phase.* To find the current when the horsepower of the motor, the voltage, the efficiency, and the power factor are given:

$$I = \frac{hp \times 746}{E \times pf \times eff}$$

29. *Alternating-current motor, three-phase.* To find the current when the horsepower of the motor, the voltage, the power factor, and the efficiency are given:

$$I = \frac{hp \times 746}{\sqrt{3}E \times pf \times eff}$$

30. *Direct-current motor.* To find the horsepower of the motor when the voltage, current, and efficiency are given:

$$hp = \frac{E \times I \times eff}{746}$$

31. *Alternating-current motor, single-phase.* To find the horsepower of the motor when the voltage, current, power factor, and efficiency are given:

$$hp = \frac{E \times I \times pf \times eff}{746}$$

32. *Alternating-current motor, three-phase.* To find the horsepower of the motor when the voltage, current, power factor, and efficiency are given:

$$hp = \frac{\sqrt{3}E \times I \times pf \times eff}{746}$$

Generator

33. *Alternating-current generator.* To find the frequency of the generator when the number of poles and the rpm are given:

$$f = \frac{p \times rpm}{120}$$

34. *Alternating-current generator.* To find the rpm of the generator when the number of poles and the frequency are given:

$$rpm = \frac{f \times 120}{p}$$

35. *Alternating-current generator.* To find the number of poles of the generator when the frequency and the rpm are given:

$$p = \frac{f \times 120}{rpm}$$

Transformer

36. *Alternating-current transformer, single-phase.*   To find the primary current when the kVA and the primary voltage are given:

$$I_p = \frac{\text{kVA} \times 1,000}{E_p}$$

37. *Alternating-current transformer, single-phase.*   To find the secondary current when the kVA and the secondary voltage are given:

$$I_s = \frac{\text{kVA} \times 1,000}{E_s}$$

Maintenance of Transmission and Distribution Lines

Transmission-Line Inspection

Initial Inspection

All transmission lines should be inspected after construction is completed before energizing the line. Linemen should climb each structure and check the following:

1. Conductor condition.

2. Conductor sag and clearance (horizontal and vertical) to ground, trees, and structures.

3. Insulator conditions.

4. Line hardware for roughness and tightness. Excess inhibitor found should be removed from conductors to prevent corona discharges.

5. Structure vibration and alignment.

6. Guys for anchors that are pulling out, guy wire conditions, missing guy guards.

7. Ground-wire connections and conditions.

8. Ground resistance at each structure.

9. Structure footings for washouts or damage.

10. Obstruction light operations for aircraft warning.

Aerial Inspections

Visual patrols from a helicopter or fixed-wing airplane, especially manufactured for this type of operation with a lineman observer, are normally completed on all high-voltage electric transmission lines several times a year. The lineman observer checks the items listed under initial inspection that can be seen from the air for problems. Aerial patrols are often initiated if a transmission line has an unscheduled outage to locate the problem. Damaged insulators and crossarms, broken or damaged conductors, structures that are leaning or down, and tree clearance problems can easily be seen by an observer in a helicopter or airplane flying along the route of the line.

Regular Inspections

Linemen normally repeat the initial climbing inspection after a line has been in service and at regular intervals thereafter as experience has shown to be necessary. Inspections by linemen walking along the line are normally completed at intervals in between the climbing inspections. These inspections provide close scrutinizing of the high-voltage transmission lines. An instrument capable of measuring radio or TV interference voltages should be used by the linemen while inspecting the line to detect corona discharges and defective insulators.

Distribution-Line Inspection

Linemen work on distribution lines frequently to install facilities for new customers while making modifications required by load increases, and as a result of government agencies' modification of roadways. The lineman should thoroughly check the pole line and equipment each time work is performed. The distribution lines should be inspected at scheduled intervals as experience has shown to be necessary in compliance with the National Electrical Safety Code (NESC) Part 2 Safety Rules for Overhead Lines, Section 21 General requirements, 214. Inspection and Tests of Lines and Equipment unless specified differently by applicable governing authorities. Inspection cycles for lines and equipment vary in length. Typically, 1-year to 10-year inspection cycles are used dependent upon the apparatus. For example, typical inspection cycle for poles varies from a period of 5- to 10-year cycles. Circuits with a record of poor electrical reliability should have special scheduled inspections as soon as possible. Inspectors should follow the route of the line, using a circuit map. Lines located remote from the streets should be inspected by the inspector walking along the circuit. Binoculars, unmanned aerial systems more commonly known as drones, and infrared cameras are tools utilized permitting the inspector to complete a thorough inspection. Operating characteristics and defects of distribution lines and equipment should be recorded in sufficient detail to permit the scheduling of repairs. The lineman can complete the minor repairs as time permits (Fig. 42.1). The installation of sufficient wildlife material guard protection is evaluated (Figs. 42.2 and 42.3).

Common defects or conditions which the inspector should observe, record, and recommend to be repaired are as follows:

1. Poles
   a. Broken or damaged poles needing replacement
   b. Leaning poles in unstable soil
   c. Poles in hazardous locations
   d. Pole ground/ground wire needing repair

2. Guys
   a. Slack, broken, or damaged guys
   b. Guy too close to primary conductors or equipment
   c. Guy insulator not installed as required
   d. Guy anchor needing replacement

3. Pole-top assemblies
   a. Broken, burned, or damaged pins, crossarm and/or crossarm braces requiring replacement
   b. Broken skirt of pin-type insulators

FIGURE **42.1**   The lineman apprentice is learning the technique for the proper repair of a 13.2/7.62 kV pole top pin insulator. (*Courtesy MidAmerican Energy Company.*)

FIGURE **42.2**   The line side of the fused cutouts and arresters are wildlife protected. (*Courtesy MidAmerican Energy Company.*)

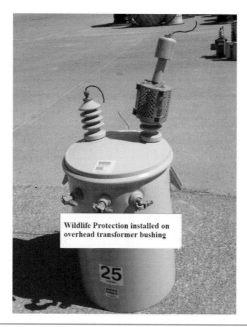

Wildlife Protection installed on overhead transformer bushing

25

FIGURE **42.3**   The 13.2/7.62 kV primary bushing on the 50-kVA pole top overhead transformer is covered with a wildlife protection guard. (*Courtesy MidAmerican Energy Company.*)

4. Line conductors
   a. Too much slack in primary conductors
   b. Floating or loose conductors
   c. Conductors burning in trees
   d. Foreign objects on line
   e. Insufficient vertical clearance over and/or horizontal clearance from:
      (1) Other wires
      (2) Buildings, parks, playgrounds, roads, driveways, loading docks, silos, signs, windmills, and spaces and ways accessible to pedestrians
      (3) Waterways—sailboat-launching area
      (4) Railways
      (5) Airports
      (6) Swimming pools

5. Equipment installations
   a. Equipment leaking oil
   b. Arresters blown or operation of ground lead isolation device
   c. Blown fuses
   d. Switch contacts not properly closed
   e. Equipment ground wire needing repair
   f. Streetlight/mast arm damaged
   g. Wildlife protection adequate

6. Miscellaneous
   a. Unauthorized pole attachments

Overhead Distribution-Line Equipment Maintenance

Capacitor Maintenance

All switched capacitor banks should be inspected and checked for proper operation once each year prior to the time period they are automatically switched on and off. Capacitor-bank oil switches should be maintained on a schedule related with the type of on/off controls installed at each bank. The maximum number of open and close operations between maintenance of the switches should not normally exceed 2,500. Experience has shown that the maintenance schedule in Table 42.1 will normally keep the equipment operating properly.

Type of Control	Maintenance Schedule, Years
Time clock	3
Voltage	3
Dual temperature	3
Temperature only	3
Time clock and temperature	3

TABLE 42.1   Capacitor Switch Control Maintenance Cycle

The capacitor switches are usually removed from the distribution line by the lineman and replaced with a spare unit during the season they are not normally operated. The capacitor switches may be maintained in a distribution shop by the lineman.

Capacitor Precautions

Capacitors and transformers are entirely different in their operation. When a transformer is disconnected from the line on both the primary and secondary sides, it is electrically dead. Unlike the transformer, the capacitor is not dead immediately after it is disconnected from the line. It has the peculiar property of holding its charge as it is essentially an electrical energy storage device. It can hold this charge for a considerable length of time. There is a voltage difference across its terminals after the switch is opened.

Capacitors for use on electrical lines, however, are equipped with an internal discharge resistor. This resistor connected across the capacitor terminals will gradually discharge the capacitor and reduce the voltage across its terminals. After 5 min, the capacitor should be discharged.

To be perfectly safe, however, proceed as follows: Before working on a disconnected capacitor, wait 5 min. Then test the capacitor with a high-voltage tester rated for the circuit voltage. If the voltage is zero, short-circuit the terminals externally using hot-line tools and ground the terminals to the case. Now you can proceed with the work.

Recloser Maintenance

Hydraulically controlled oil circuit reclosers should normally be maintained after 50 operations, or every 2 years. Electronically controlled oil reclosers should be maintained after approximately 150 operations, or every 2 years. The hydraulically or electronically controlled oil reclosers can be maintained in the field or taken to the distribution shop. The dielectric strength of the insulating oil should be preserved by filtering the oil until tests ensure good results. The recloser manufacturer's instructions should be followed to complete the inspection and repairs.

Vacuum reclosers require field testing on a 6-year schedule. An excessive number of recloser operations may make it necessary to reduce the scheduled maintenance time period. High-potential testing of the vacuum interrupter bottles can be completed in the field by the lineman. The electronic controls for the vacuum recloser can be tested in the field by following the manufacturer's instructions.

All line reclosers should be installed with bypass provisions and a means for easy isolation of the equipment for maintenance work.

Pole-Mounted Switch Maintenance

Group-operated or single-pole switches installed in overhead distribution lines should be checked each time they are operated by the lineman for the following:

1. Burned contacts
2. Damaged interrupters or arcing horns
3. Proper alignment
4. Worn parts
5. Defective insulators
6. Adequate lightning arrester protection
7. Proper ground connections
8. Loose hardware

Switches that have not been adequately inspected while being operated in a 5-year period should be scheduled for a maintenance inspection. It may be necessary for the lineman to make temporary line modifications to permit operation, inspection, repairs, or adjustments of the switches.

Underground Distribution System Maintenance

The underground distribution system originates at a riser pole in the overhead distribution lines or in 15-kV URD switchgear. Periodic maintenance activity is required at the riser pole, switchgear, distribution transformer, secondary enclosure, and the primary and secondary cable.

Riser Pole Maintenance

The riser pole for underground distribution circuits should be inspected when overhead lines are inspected and maintained. The inspection should include the disconnect switches and/or fused cutouts, the lightning arresters, and the operation of the arrester ground lead isolation devices, the riser cables and potheads or termination, support of the cables, conduit and/or U-guard, and identification of the circuit and pole conditions.

Switchgear Maintenance

The pad-mount switchgear for underground electric distribution circuits should be inspected annually. The inspection should start with external items such as condition of the pad or foundation. Pads that are tipped or cracked or that have the dirt washed out from under them should be repaired. The paint on metal surfaces of the gear should be in good condition to prevent rusting. The switchgear cabinet doors should be locked and pentahead bolted shut. Cleanliness of the internal areas of the switch-gear is important to maintain the dielectric strength of the insulation. Equipment with insulating oil must be checked for moisture or contamination of the oil. Protective barriers, switches, fuses, and other internal equipment should be inspected in accordance with the manufacturer's instructions. Cable terminations, grounds, and other connections to the switchgear equipment must be kept in good condition. If detected, rodent problems must be resolved. In addition, pad-mounted switchgear safety and clearance labels must be clearly visible and legible. Illegible stickers must be replaced. The labels inform the public to keep out of the transformer and to not obstruct the switchgear.

Thermal measuring hand-held tools provide clues as to potential hotspots. The temperature measurement tools are lightweight and simple to use (Fig. 42.4). Advanced devices provide additional details such as expanded temperature ranges. Even more sophisticated models allow the user to digitally record temperatures at greater distances and download the information to a personal computer for detailed analysis.

Distribution Transformer and Secondary Enclosurer Maintenance

Transformers should be checked for proper loading. Transformer loadings are normally monitored by use of a transformer load management system. The system takes monthly kilowatthour meter readings, used for billing purposes for all customers served from a transformer, and converts them to a value of kVA load that is compared to the kVA rating of the transformer. Overloaded transformers should be replaced with those of adequate capacity to prevent the transformer from failing and to prevent fires.

Pole-type transformers should be inspected visually for defects as a part of the overhead line inspection. Pad-mounted transformers and secondary enclosures are inspected

FIGURE 42.4 A temperature-sensing inspection device being utilized on a metal-enclosed switchgear. Equipment hotspots are detected and repaired. (*Courtesy Raytek Corp.*)

for the items discussed previously for the underground distribution circuit switchgear. In addition, pad-mount transformer and enclosure safety and clearance labels must be clearly visible and legible. Illegible stickers must be replaced. The labels inform the public to keep out of the transformer and to not obstruct the front of the transformer. The transformer cabinet doors and enclosure lids shall be locked and pentahead bolted shut. Annual inspections are preferable.

Underground Cable Maintenance

The terminations of underground cables are inspected as during the maintenance procedures for riser poles, switchgear, and pad-mounted transformers. The route of underground cable circuits should be walked by the lineman on a 2-year cycle. The route of the cable should be inspected for obstacles that could damage the cables, changes in grade of the ground that could cause the cables to be exposed or damaged, and washouts. Cables that fail while in operation must be located and repaired or replaced.

Underground primary distribution cables with solid-dielectric insulation have experienced a high rate of electrical failure after several years of operation because of electrochemical trees. The electrochemical treeing is generally caused by the presence of water in the conductor that penetrates the insulation. It has been successfully demonstrated that the life of the cables can be extended by the cable rejuvenation process. Experienced contractors are available to complete this maintenance for those companies inexperienced in the procedure.

Cable Rejuvenation

An alternative to replacing aged cable is a cable rejuvenation process whereby cable life is extended, water-tree damaged cable insulation mitigated, and cable dielectric strength improved. The cable life extension occurs when an enhancement fluid is pressure injected into the cable and penetrates the cable insulation. The silicone-based fluid fills insulation voids due to electrical stresses that develop in water-tree damaged aged cables. The process has proven effective for high-molecular-weight polyethylene (HMWPE), cross-linked polyethylene (XLPE), and tree-retardant/cross-linked polyethylene (TRXLPE) cables.

An aged relatively splice-free cable is evaluated to determine if the rejuvenation process is logical. Cables are initially evaluated utilizing time-domain reflectivity test equipment. Unacceptable cables contain severely corroded neutrals, poor-quality semiconducting shields, significantly discolored insulation, many air-blocked splices, or badly broken conductor strands. Those cables that are determined to be acceptable for rejuvenation are isolated, re-terminated with specially fitted load-break elbows (Fig. 42.5) or cable-injection adapters (Fig. 42.6), air-tested with nitrogen, and, finally, injected with the silicone fluid (Fig. 42.7).

FIGURE 42.5 A specially fitted load-break elbow being installed on the pad-mount transformer. (*Courtesy Salt River Project/UTILX Corp.*)

FIGURE 42.6 A cable-injection adapter installed on a live-front terminated cable. (*Courtesy Salt River Project/UTILX Corp.*)

FIGURE 42.7 A specially fitted load-break elbow installed on the pad-mount transformer. The silicone fluid canister is attached. The cable is in the process of being injected with the fluid. (*Courtesy Salt River Project/UTILX Corp.*)

The injection process involves both the fluid feed tank (Fig. 42.8) at the start of the cable system and an automatic shutoff vacuum tank at the cable system's end. The process begins by deenergizing the cable and retrofitting the cable terminations with injection ports in a specially designed load-break elbow or in a live-front adapter fitting (Fig. 42.9). The cable is then air-tested with nitrogen (Fig. 42.10) to ensure that the cable is blockage free. The cable sections that pass the air test and are clear of blockage are now ready for the injection of a drying agent to purge any water and dry out the cable. The silicone fluid injection is the next step. The fluid and vacuum tanks are installed. The vacuum tank slowly draws the silicone fluid into the energized cable. When the silicone fluid has successfully saturated the entire length of the cable section's insulation, the vacuum tank automatically shuts off. The tanks are removed, the injection ports are capped, and the cable is now rejuvenated.

FIGURE 42.8 The fluid tank at the start of the cable rejuvenation system. (*Courtesy Salt River Project/UTILX Corp.*)

FIGURE 42.9   The lineman pictured is nearing completion of the installation of the specially fitted cable injection elbow. (*Courtesy Salt River Project/UTILX Corp.*)

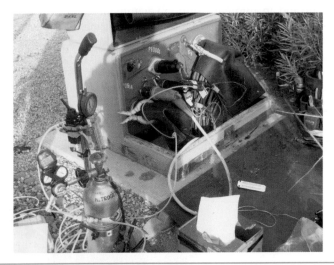

FIGURE 42.10   The nitrogen tester installed and in the process of determining if the cable is blockage free. (*Courtesy Salt River Project/UTILX Corp.*)

Streetlight Maintenance

The lumen output of any electric lamp decreases with use, and accumulated dirt interferes with light reflection and transmission. Streetlight installations are designed such that the average footcandle levels recommended by the *ASA Standards for Roadway Lighting* represent the average illumination delivered on the roadway when the illuminating source is at its lowest output and when the luminaire is in its dirtiest condition. The luminaire refractor should be washed and the lamp replaced when the lumen output of the lamp has fallen to approximately 80% of its initial value, to maintain the designed footcandle level.

Lamp burnout should be a consideration in the maintenance of a streetlight system. The *rated lamp life* is defined as the time required for half of the lamps to fail. The mortality curves for the various types of lamps are relatively flat for the initial 60 to 70% of rated life. Using a 6,000-hour incandescent lamp as an example, only 10% of the lamps will fail in the first 4,000 hours of burning time, while 40% will fail in the next 2,000 hours of burning time.

Proper maintenance of street and highway lighting is important to provide a high degree of service quality. Streetlights should be inspected on a 10-year schedule. Refractors for enclosed-type streetlight luminaires should be cleaned and washed at the time of inspection. The luminaires should be checked for proper level in both the transverse and horizontal directions. If the streetlight is installed on a steep grade, the transverse direction of the luminaire should be leveled parallel to the street grade. Defective or broken refractors or luminaires must be replaced. Many streetlight problems and failures are reported by the public, police, and employees of the concern owning the system. Lamps that fail, defective photocells, and ballasts should be replaced promptly to restore the lighting service (Fig. 42.11). Group replacement of lamps on a 10-year schedule should minimize the number of lamp failures and maintain light output in accordance with established standards. The group replacement can be coordinated with the inspection, cleaning, and washing schedule to minimize the effort required.

FIGURE 42.11   Lineman preforming streetlight maintenance from a single man bucket truck.

Voltage Regulator Maintenance

Distribution voltage regulators should be inspected externally by the inspector while patrolling the overhead lines. Problems such as damaged insulators, leaking oil, or unusual operating noise should be corrected by the lineman as soon as practical. Operation of the voltage regulators should be observed by checking the operation counter and reading the voltage with the aid of recording voltmeters at regular intervals set at various locations along the circuit beyond the regulator. The voltage regulator should be removed from service by the lineman and maintained in accordance with the manufacturer's recommendation as determined by the number of operations of the tap changer or time in service. Normally, the voltage regulators can operate several years between maintenance periods. All manufacturers require that the insulating oil be kept dry and free from carbon to maintain the dielectric strength of the oil.

Maintenance of Equipment Common to Transmission and Distribution Lines

Tree Trimming
See Chap. 35.

Pole Maintenance

All wood poles should be pounded with a hammer by the lineman and visibly inspected before being climbed. A good pole will sound solid when hit with a hammer. A decayed pole will sound hollow or like a drum. The visible inspection should include checking for general condition and position, cracks, shell rot, knots that might weaken the pole, setting depth, soil conditions, hollow spots or woodpecker holes, and burned spots. Poles that reveal possible defects should not be climbed by the linemen until the poles are made safe by repairs or by securing them.

All wood poles should be pressure- or thermal-treated with preservatives when they are new. New treated wood poles should resist decay for 10 to 20 years, depending upon the ambient soil conditions. Scheduled inspection should be based on experience. Inspection cycles for poles vary in length and are typically inspected on a 5- to 10-year cycle. Distribution poles are often replaced before scheduled maintenance is required because of a change in voltage or replacement due to street-widening projects. When a lineman detects defects in wood poles of a particular type and vintage, similar poles installed at the same time should be inspected and maintained, if necessary. A pole hit with a hammer by the lineman will not always reveal decay.

The Electrical Power Research Institute (EPRI) has developed the PoleTest instrument that uses nondestructive test methods to predict the life of a pole. The wood pole strength at ground line and above ground line to pole top is measured. The lineman keys in the wood species and the diameter of the pole on the instrument and then strikes the pole a few times with a marble-size steel ball that has been attached to the pole. A microprocessor in the instrument analyzes the sonic wave patterns passing through the wood pole and indicates the current strength of the pole.

If the instrument indicates the pole strength is deficient or if a pole-testing instrument is not available, the ground around the pole can be excavated to a depth of 12 inches to 18 inches. A probe test can be completed by hand-drilling a 3/8-inch hole below the ground line into the pole butt at a 30° to 45° downward angle to a depth sufficient to reach the center of the pole. The wood shavings are checked for decay. The drilled holes are plugged with a treated 3/8-inch dowel to keep moisture out and to prevent further decay.

Poles that have sufficient decay to weaken them should be reinforced or replaced. External decay of poles should be shaved away. Poles in acceptable condition should have a preservative applied at the ground line and be wrapped with a moisture-proof barrier. The wood poles can be treated internally with chemicals to kill early decay before actual voids form. Poles can be strengthened to delay replacement by placing a section of galvanized steel plate adjacent to the pole. The steel plate is driven into the ground and banded to the pole above the ground line.

Wood poles installed near woodpecker's feeding areas are often damaged by the birds. Woodpecker damage can be prevented by wrapping the pole with extruded sheets of high-density polyethylene with a high-gloss finish. The plastic wrap prevents the woodpeckers from perching on the pole. Lineman can climb the wrapped pole by penetrating the wrap with climber's gaffs. If the linemen must do considerable work on the pole, the pole should be rewrapped when the work is completed to prevent sites of gaff penetration of the plastic wrap from becoming perches for the birds.

Wooden transmission-line poles are usually scheduled for inspection after 20 years of service and treated if pole conditions are adequate. Repeat inspections and treatment are normally scheduled at 8-year to 10-year intervals. Experience indicates that poles properly treated when new and maintained on an adequate schedule should have an average life of 45 years.

Insulator Maintenance

Transmission- and distribution-line inspections will detect damaged insulators. Most insulator damage will result from gun shots, lightning flashovers, contamination flash-overs, and wind damage. Broken, cracked, or damaged insulators should be replaced as soon as they are detected. Defective insulators may cause visible corona and interference voltage propagation. Radio interference conditions can be detected by the lineman by using instruments designed for this purpose while completing the scheduled inspections.

Insulators in severely contaminated atmospheres may require frequent cleaning. These insulators can be washed with high-pressure water, using special hoses and hose nozzles. Procedures developed for safe washing of energized insulators must be followed.

Conductor Maintenance

Conductors can be damaged by aeolian vibration, galloping, sway oscillation, unbalanced loading, lightning discharges, or short-circuit conditions. Inspections of the transmission and distribution lines by the lineman will reveal the need for conductor repairs.

Abrasion to the outside surfaces of the conductor is visible to the inspector. Abrasion damage is a chafing impact wear that accompanies relative movement between a loose tie or other conductor hardware and the conductor or armor rods. Abrasion is a surface damage and can be identified by black deposits on the conductor or tie wire.

Fatigue results in the failure of the conductor strands or tie wire (Fig. 42.12). If the damaged conductor is detected before serious damage or complete failure occurs, it can be repaired by the lineman by installing armor rods or a line splice (Fig. 42.13).

Broken conductor tie wires at an insulator discovered by inspection or as a result of severe radio interference complaints require immediate repairs (Fig. 42.14). The damaged conductor can be repaired even if some of the conductor strands are broken by removing the old tie wires and armor rods, if present, and installing new armor rods or new line guards and retying the assemblies with distribution wraplock ties (Fig. 42.15).

FIGURE 42.12 Excessive conductor damage and broken tie wire at insulator on a distribution circuit. (*Courtesy Preformed Line Products Co.*)

FIGURE 42.13 Damaged conductor being repaired with a conductor splice. Splice will restore original conductivity and strength to all-aluminum, aluminum-alloy, and copper conductors of homogeneous stranding, provided core conductor of ACSR is not damaged. (*Courtesy Preformed Line Products Co.*)

FIGURE 42.14 Picture reveals conductor damage as a result of fatigue breaks. (*Courtesy Preformed Line Products Co.*)

Transmission and distribution lines where experience indicates the conductors have prolonged periods of vibration should have vibration dampers installed to prevent fatigue failure of the conductor. Spiral vibration dampers will prevent conductor fatigue, inner wire fretting, or scoring of the distribution-line insulator glaze.

FIGURE 42.15 Line guard installed on damaged distribution conductor at insulator support. Conductance and strength are completely restored to aluminum or ACSR conductors. Conductor must be retied to insulator to complete repairs. (*Courtesy Preformed Line Products Co.*)

Infrared Scanning of Transmission and Distribution Lines

Transmission and distribution lines can be inspected with infrared scanning equipment to locate hotspots on conductors and equipment caused by loose connections or hardware and other defects. Infrared scanning equipment detects hotspots utilizing the camera's infrared detector to measure thermal temperatures of lines, connections, and equipment. The equipment provides a photographic record as well as a visual display of the hot-spot defects. The infrared scanning equipment can be operated from a helicopter, from a van-type vehicle, or by an operator carrying the equipment. Examples of still photography and infrared images of detected hotspots are shown in Figs. 42.16 through 42.18. The critical hotspots in need of correction are due to an extreme temperature differential (between 25°F and 80°F) of the equipment components.

FIGURE 42.16 Station arresters installed in a substation capacitor bank.

CHAPTER 43

Rope, Knots, Splices, and Gear

Ropes used by linemen and cablemen to build and maintain electric transmission and distribution lines are made from synthetic fibers consisting of nylon, polyester, polypropylene, kevlar, and manila.

Rope Construction

There are several types of rope constructions manufactured. The twisted rope is usually made with three or four strands (Fig. 43.1). A twisted rope is manufactured by first producing the strands, using a specific number of yarns per strand which are placed in a laying machine which controls the strand twist as the finished rope is formed. Braided ropes are mainly 2-in-1 or double-braided, single or hollow braid, 8-strand or 12-strand braided or plaited ropes. The double-braided rope contains a braided core over which a cover is braided, the single or hollow braid consists of a one single braid containing no core, and the 8-strand or 12-strand braided or plaited rope consists of 8 strands or 12 strands, braided in pairs in a maypole fashion on a plaiting machine. Parallel ropes are produced by utilizing parallel yarns as the strength members that are bonded and/or jacketed to form the finished rope. Parallel yarns will provide almost 100% of the available strength from the fibers. Parallel rope is stronger, size for size, than twisted or braided rope, but it has much lower elongation and is stiffer. The 8-strand and 12-strand braided synthetic fiber ropes are commonly used for slings, pilot lines, and hand lines. The 2-in-1 or double-braided synthetic fiber ropes are frequently used for winchlines and slings and for stringing electrical conductors.

Physical Properties

The strengths of the fiber ropes when new are given by the manufacturer or provided in Industry Standards. In particular, the Cordage Institute is an international association of rope, twine, and related manufacturers, their suppliers, and affiliated industries dedicated to the advancement of rope and cordage products. The Cordage Institute International Guideline CI 1401-06 Safer Use of Fiber Rope was published May 2006. Within the Guideline are the concepts of minimum breaking strength, working load, working load limit, and design factor. Minimum Breaking Strength (MBS) is defined as the force that a given rope is required to meet or exceed in a laboratory test when it is new and unused. Working Load (WL) is defined as the weight or force applied to rope or cordage in a given application. Working Load Limit (WLL) is defined as a guideline for the maximum allowable capacity of a rope product and should not be exceeded.

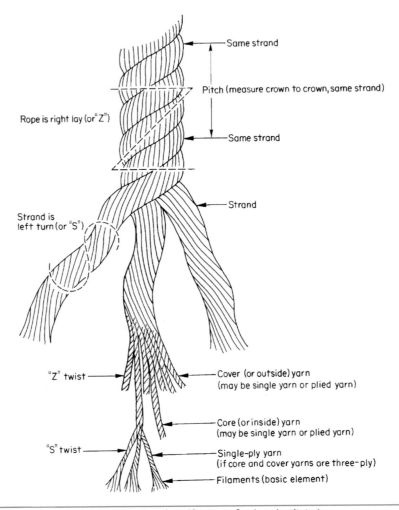

Same strand

Pitch (measure crown to crown, same strand)

Rope is right lay (or "Z")

Same strand

Strand

Strand is
left turn (or "S")

"Z" twist — Cover (or outside) yarn
(may be single yarn or plied yarn)

Core (or inside) yarn
(may be single yarn or plied yarn)

"S" twist — Single-ply yarn
(if core and cover yarns are three-ply)

Filaments (basic element)

FIGURE 43.1   Three-strand rope construction. (*Courtesy Cordage Institute.*)

Design Factor (DF) is defined as the ratio between the MBS and WL (DF = MBS/WL). A design factor in the range of 5:1 to 12:1 is recommended. The concepts are incorporated in Cordage Institute Standards for the various types of rope.

The physical properties for new polyester/polyolefin fiber rope, 3-strand construction with standard terminations from Cordage Institute Standard CI 1302A-96 are provided in Table 43.1. Also, MBS/DF = WLL. For example, a 2-inch rope has a minimum breaking strength of 48,050 lbs. If the design factor 12 is chosen, the 2-inch rope has a working load limit of 48,050/12 = 4,004 lbs. If the design factor 5 is chosen, the 2-inch rope has a working load limit of 48,050/5 = 9,610 lbs. As stated in the guideline "the more severe the application, the higher the design factor needs to be."

POLYESTER/POLYOLEFIN FIBER ROPE STANDARD: CI 1302A-96 V.2

Nominal Size[1]			Physical Properties							Working Load Limits[5]	
Diameter		Size No.	Linear Density[2]		New Rope Min Breaking Strength[3]		Design Factor			Range Lbs (daN)	
Inch	(mm.)	(circ.)	lbs/100[2]	(ktex)	Lbs.	(daN)	Range[4]			1/12	1/5
1/4	(6)	3/4	1.6	(23.8)	1,200	(534)	12–5			100 (44)–240 (107)	
5/16	(8)	1	2.5	(37.2)	1,870	(832)	12–5			156 (69)–374 (166)	
3/8	(10)	1-1/8	3.6	(53.6)	2,700	(1,201)	12–5			225 (100)–540 (240)	
7/16	(11)	1¼	4.8	(71.4)	3,500	(1,557)	12–5			292 (130)–700 (311)	
1/2	(12)	1½	6.2	(92.3)	4,400	(1,957)	12–5			367 (163)–880 (391)	
9/16	(14)	1¾	7.9	(118)	5,200	(2,313)	12–5			433 (193)–1,040 (463)	
5/8	(16)	2	9.5	(141)	6,100	(2,713)	12–5			508 (226)–1,220 (543)	
3/4	(18)	2¼	13.5	(201)	8,400	(3,736)	12–5			700 (311)–1,680 (747)	
7/8	(22)	2¾	18.0	(268)	11,125	(4,948)	12–5			927 (412)–2,225 (990)	
1	(24)	3	21.8	(324)	13,175	(5,860)	12–5			1,098 (488)–2,635 (1,172)	
1-1/16	(26)	3¼	24.5	(365)	14,775	(6,572)	12–5			1,231 (548)–2,955 (1,314)	
1-1/8	(28)	3½	27.1	(403)	16,325	(7,261)	12–5			1,360 (605)–3,265(1,452)	
1¼	(30)	3¾	33.4	(497)	19,900	(8,862)	12–5			1,658 (738)–3,980 (1,770)	
1-5/16	(32)	4	36.5	(543)	21,950	(9,763)	12–5			1,829 (814)–4,390 (1,953)	
1½	(36)	4½	47.0	(699)	28,250	(12,566)	12–5			2,354 (1,047)–5,650 (2,513)	
1-5/8	(40)	5	55.0	(819)	32,950	(14,656)	12–5			2,746(1,221)–6,590 (2,931)	
1¾	(44)	5½	62.0	(923)	36,850	(16,391)	12–5			3,071 (1,366)–7,370 (3,278)	
2	(48)	6	81.0	(1,205)	48,050	(21,373)	12–5			4,004 (1,781)–9,610 (4,275)	
2-1/8	(52)	6½	91.0	(1,354)	53,950	(23,997)	12–5			4,496 (2,000)–10,790 (4,799)	
2¼	(56)	7	101	(1,503)	59,950	(26,666)	12–5			4,996 (2,222)–11,990 (5,333)	

Table 43.1   Polyester/Polyolefin Fiber Rope Standard: CI 1302A-96 V.2 (Continued)

POLYESTER/POLYOLEFIN FIBER ROPE STANDARD: CI 1302A-96 V.2

| Nominal Size[1] | | | Physical Properties | | | | Design Factor | Working Load Limits[5] | |
| Diameter | | Size No. | Linear Density[2] | | New Rope Min Breaking Strength[3] | | | Range Lbs (daN) | |
Inch	(mm.)	(circ.)	lbs/100²	(ktex)	Lbs.	(daN)	Range[4]	1/12	1/5
2½	(60)	7½	124	(1,845)	73,550	(32,715)	12–5	6,129 (2,726)	–14,710 (6,543)
2-5/8	(64)	8	136	(2,024)	80,650	(35,873)	12–5	6,721 (2,989)	–16,130 (7,175)
2¾	(68)	8½	161	(2,396)	95,400	(42,434)	12–5	7,950 (3,536)	–19,080 (8,487)
3	(72)	9	174	(2,589)	102,900	(45,770)	12–5	8,575 (3,814)	–20,580 (9,154)
3¼	(80)	10	212	(3,155)	122,800	(54,621)	12–5	10,233 (4,552)	–24,560 (10,924)
3½	(88)	11	250	(3,720)	144,800	(64,407)	12–5	12,067 (5,367)	–28,960 (12,881)
4	(96)	12	300	(4,465)	171,000	(76,061)	12–5	14,250 (6,338)	–34,200 (15,212)
4¼	(104)	13	345	(5,134)	195,800	(87,092)	12–5	16,317 (7,258)	–39,160 (17,418)
4½	(112)	14	395	(5,878)	224,800	(99,991)	12–5	18,733 (8,333)	–44,960 (19,998)
5	(120)	15	455	(6,771)	254,700	(113,291)	12–5	21,225 (9,441)	–50,940 (22,658)
5-5/16	(128)	16	506	(7,530)	282,600	(125,700)	12–5	23,550 (10,475)	–56,520 (25,140)
5-5/8	(136)	17	562	(8,364)	312,300	(138,911)	12–5	26,025 (11,576)	–62,460 (27,782)
6	(144)	18	635	(9,450)	351,000	(156,125)	12–5	29,250 (13,010)	–70,200 (31,225)

(1) Diameter is approximate and is actually determined by linear density. See Safe Use Guideline CI 1401 (current), page 4.
(2) Linear Density is considered standard. Tolerances are ±10% for diameters 3/16"–5/16" inclusive; ±8% for diameters 3/8"–9/16" inclusive; ±5% for 5/8" diameter and up. The metric unit of linear density is Ktex (kilotex).
(3) New rope Minimum Breaking Strength is based on data from a number of manufacturers and represents a value of 2 standard deviations below the mean, established by regression analysis.
(4) For critical applications, refer to Safe Use Guideline CI 1401 (current) on page 4.
(5) Working Load Limit is determined by dividing the new rope Minimum Breaking Strength by the selected Design Factor. Important information is given in CI 1401 (current) on page 4.
Source: Cordage Institute.

TABLE 43.1 Polyester/Polyolefin Fiber Rope Standard: CI 1302A-96 V.2

Care and Handling of Ropes

A rope's strength will deteriorate slowly or quickly depending on how it is used and/or abused. Rope must be stored properly to prevent deterioration. All ropes should be inspected closely each time before they are used.

Fiber rope must not be kinked. It should be removed carefully from a reel or coil. Kinks can substantially weaken the rope. The rope should be inspected each time before it is used for excessive wear or damage, such as cuts, gouges, badly abraded or glazed spots, and fiber breakage (inside as well as on the surface of the rope); light abrasion is acceptable.

Proper terminations and attachments should be used. Rope splicing can maintain 100% of the rope strength. Knots can reduce rope strength 50% or more and can cause slipping. Avoid running a rope over sharp edges or corners, or around diameters less than eight times the rope diameter. When tying rope to an object that has sharp corners, the rope should be padded to prevent the fibers from being cut. Rope should never be dragged over the ground or over other sharp objects. Dragging one rope over another can result in damage to both ropes. Rope that has become wet should not be permitted to freeze. If a rope has become full of mud or sand, it should be flushed with water and permitted to dry. Rope hand lines and blocks should be gathered neatly and hung in the truck in the proper place when not in use. It is best to place large rope when it is stored, on wood gratings above the floor in a ventilated location. All ropes used in line work should be kept dry. Keep ropes away from harmful chemicals. Avoid excessive exposure of the rope to sunlight and moisture. Keep the ropes away from fire and exposure to temperatures exceeding 150°F. Select the proper rope for the load to be handled. Different ropes have better characteristics for shock and sustained or dynamic loading.

Electrical Conductivity of Ropes

Clean, dry polypropylene rope is an insulator; all other ropes are conductors. As the rope (and this includes polypropylene rope) becomes wet, dirty, and contaminated, its conductivity increases progressively. All insulators conduct electricity if they are not kept clean and dry. All fibers used in ropes, except polypropylene, contain moisture as a part of the composition of the fiber. The moisture regain of fibers varies from 0% for polypropylene to over 7.5% for manila. Polyester fiber has a 0.4% and nylon fiber a 4.5% moisture regain. The fiber moisture content will rise with increased relative humidity. All rope structures will trap water between fibers, yarns, and strands when exposed to rain, mist, puddles, or other sources of moisture which increases the electrical conductivity of the rope. The cordage industry has developed a rope construction that minimizes the conductivity of the finished rope.

Rope Slings

Figure 43.2 provides a pictorial representation of basic sling configurations referred to in Tables 43.2 through 43.10, which give the rated capacity of synthetic and natural fiber rope slings. Spliced-fiber rope slings must be capable of meeting the minimum requirements of Occupational Safety and Health Administration (OSHA) Standard 1910.184.

FIGURE 43.2   Sling configurations.

Various Uses of Rope

Bull Rope
Bull ropes are used for raising or lowering heavy pieces of equipment, for fastening temporary guys, for setting poles, for holding out heavy transformers, and for lowering large limbs and trunks of trees.

Hand Lines
Hand lines are used for raising and lowering light material and tools or for holding small transformers away from a pole while the latter is being raised.

Throw Line
A small-diameter rope is used to throw a line over a crossarm, tree limb, or other support object. The throw line is normally used to pull a larger rope into place for performing a task.

Running Line
The running line is used for pulling in several span lengths of wire at one time.

Eye-and-Eye Sling–Hitch Type / Endless Sling–Hitch Type

Rope Diameter, inch	Eye-and-Eye Vertical	Choker	Two-Leg Bridle or Basket — Horizontal Angle, deg				Endless Vertical	Choker	Basket — Horizontal Angle, deg			
			90°	60°	45°	30°			90°	60°	45°	30°
1/2	1,100	830	2,200	1,900	1,600	1,100	2,000	1,500	4,000	3,500	2,800	2,000
9/16	1,400	1,100	2,800	2,400	2,000	1,400	2,600	2,000	5,200	4,500	3,700	2,600
5/8	1,800	1,400	3,600	3,100	2,500	1,800	3,200	2,400	6,400	5,500	4,500	3,200
3/4	2,600	2,000	5,200	4,500	3,700	2,600	4,600	3,500	9,200	8,000	6,500	4,600
7/8	3,500	2,600	7,000	6,100	4,900	3,500	6,200	4,700	12,400	10,700	8,800	6,200
1	4,400	3,300	8,800	7,600	6,200	4,400	7,900	5,900	15,800	13,700	11,200	7,900
1-1/8	5,700	4,300	11,400	9,900	8,100	5,700	10,100	7,600	20,200	17,500	14,300	10,100
1-1/4	7,000	5,300	14,000	12,100	9,900	7,000	12,400	9,300	24,800	21,500	17,500	12,400
1-5/16	7,700	5,800	15,400	13,300	10,900	7,700	13,700	10,300	27,400	23,700	19,400	13,700
1-1/2	9,700	7,300	19,400	16,800	13,700	9,700	17,400	13,100	34,800	30,100	24,600	17,400
1-5/8	11,500	8,600	23,000	19,900	16,300	11,500	20,500	15,400	41,000	35,500	29,000	20,500
1-3/4	13,200	9,900	26,400	22,900	18,700	13,200	23,600	17,700	47,200	40,900	33,400	23,600
2	16,900	12,700	33,800	29,300	23,900	16,900	30,200	22,700	60,400	52,300	42,700	30,200
2-1/8	19,100	14,300	38,200	33,100	27,000	19,100	34,100	25,600	68,200	59,100	48,200	34,100
2-1/4	21,400	16,100	42,800	37,100	30,300	21,400	38,300	28,700	76,600	66,300	54,200	38,300
2-1/2	26,300	19,700	52,600	45,600	37,200	26,300	46,900	35,200	93,800	81,200	66,300	46,900
2-5/8	28,800	21,600	57,600	49,900	40,700	28,800	51,400	38,600	102,800	89,000	72,700	51,400
3	37,100	27,800	74,200	64,300	52,500	37,100	66,200	49,700	132,400	114,700	93,600	66,200

General Notes:
(a) See Figs. 43.34 and 43.35.
(b) For choker hitch, the angle of choke is 120° or greater.
Source: Occupational Safety and Health Administration "GUIDANCE ON SAFE SLING USE" publication.

TABLE 43.2 Nylon Rope Slings Based on Design Factor = 5 and Rated Loads Expressed in Pounds (lbs)

Rope Diameter, inch	Eye-and-Eye Sling–Hitch Type		Two-Leg Bridle or Basket Horizontal Angle, deg				Endless Sling–Hitch Type		Basket Horizontal Angle, deg			
	Vertical	Choker	90°	60°	45°	30°	Vertical	Choker	90°	60°	45°	30°
1/2	1,000	750	2,000	1,700	1,400	1,000	1,800	1,400	3,600	3,100	2,500	1,800
9/16	1,300	980	2,600	2,300	1,800	1,300	2,300	1,700	4,600	4,000	3,300	2,300
5/8	1,600	1,200	3,200	2,800	2,300	1,600	2,800	2,100	5,600	4,800	4,000	2,800
3/4	2,200	1,700	4,400	3,800	3,100	2,200	4,000	3,000	8,000	6,900	5,700	4,000
7/8	3,000	2,300	6,000	5,200	4,200	3,000	5,400	4,100	10,800	9,400	7,600	5,400
1	4,000	3,000	8,000	6,900	5,700	4,000	7,100	5,300	14,200	12,300	10,000	7,100
1-1/8	5,000	3,800	10,000	8,700	7,100	5,000	8,900	6,700	17,800	15,400	12,600	8,900
1-1/4	6,000	4,500	12,000	10,400	8,500	6,000	10,600	8,000	21,200	18,400	15,000	10,600
1-5/16	6,500	4,900	13,000	11,300	9,200	6,500	11,600	8,700	23,200	20,100	16,400	11,600
1-1/2	8,400	6,300	16,800	14,500	11,900	8,400	15,100	11,300	30,200	26,200	21,400	15,100
1-5/8	9,900	7,400	19,800	17,100	14,000	9,900	17,600	13,200	35,200	30,500	24,900	17,600
1-3/4	11,400	8,600	22,800	19,700	16,100	11,400	20,400	15,300	40,800	35,300	28,800	20,400
2	14,400	10,800	28,800	24,900	20,400	14,400	25,700	19,300	51,400	44,500	36,300	25,700
2-1/8	16,200	12,200	32,400	28,100	22,900	16,200	28,900	21,700	57,800	50,100	40,900	28,900
2-1/4	18,100	13,600	36,200	31,300	25,600	18,100	32,300	24,200	64,600	55,900	45,700	32,300
2-1/2	22,000	16,500	44,000	38,100	31,100	22,000	39,300	29,500	78,600	68,100	55,600	39,300
2-5/8	24,200	18,200	48,400	41,900	34,200	24,200	43,200	32,400	86,400	74,800	61,100	43,200
3	31,200	23,400	62,400	54,000	44,100	31,200	55,700	41,800	111,400	96,500	78,800	55,700

General Notes:
(a) See Figs. 43.34 and 43.35.
(b) For choker hitch, the angle of choke is 120° or greater.
Source: Occupational Safety and Health Administration "GUIDANCE ON SAFE SLING USE" publication.

TABLE 43.3 Polyester Rope Slings Based on Design Factor = 5 and Rated Loads Expressed in Pounds (lbs)

Rope Diameter, inch	Eye-and-Eye Sling–Hitch Type		Two-Leg Bridle or Basket Horizontal Angle, deg				Endless Sling–Hitch Type		Basket Horizontal Angle, deg			
	Vertical	Choker	90°	60°	45°	30°	Vertical	Choker	90°	60°	45°	30°
½	760	570	1,500	1,300	1,100	750	1,400	1,100	2,800	2,400	2,000	1,400
9/16	920	690	1,800	1,600	1,300	900	1,600	1,200	3,200	2,800	2,300	1,699
5/8	1,100	830	2,200	1,900	1,600	1,100	2,000	1,500	4,000	3,500	2,800	2,000
¾	1,500	1,100	3,000	2,600	2,100	1,500	2,700	2,000	5,400	4,700	3,800	2,700
7/8	2,100	1,600	4,200	3,600	3,000	2,100	3,700	2,800	7,400	6,400	5,200	3,700
1	2,600	2,000	5,200	4,500	3,700	2,600	4,600	3,500	9,200	8,000	6,500	4,600
1-1/8	3,200	2,400	6,400	5,500	4,500	3,200	5,700	4,300	11,400	9,900	8,100	5,700
1¼	3,900	2,900	7,800	6,800	5,500	3,900	6,900	5,200	13,800	12,000	9,800	6,900
1-5/16	4,200	3,200	8,400	7,300	5,900	4,200	7,600	5,700	15,200	13,200	10,700	7,600
1½	5,500	4,100	11,000	9,500	7,800	5,500	9,800	7,400	19,600	17,000	13,900	9,800
1-5/8	6,400	4,800	12,800	11,100	9,000	6,400	11,400	8,600	22,800	19,700	16,100	11,400
1¾	7,400	5,600	14,800	12,800	10,500	7,400	13,200	9,900	26,400	22,900	18,700	13,200
2	9,400	7,100	18,800	16,300	13,300	9,400	16,700	12,500	33,400	28,900	23,600	16,700
2-1/8	10,500	7,900	21,000	18,200	14,800	10,500	18,800	14,100	37,600	32,600	26,600	18,800
2¼	11,900	8,900	23,800	20,600	16,800	11,900	21,200	15,900	42,400	36,700	30,000	21,200
2½	14,400	10,800	28,800	24,900	20,400	14,400	25,700	19,300	51,400	44,500	36,300	25,700
2-5/8	16,100	12,100	32,200	27,900	22,800	16,100	28,800	21,600	57,600	49,900	40,700	28,800
3	20,500	15,400	41,000	35,500	29,000	20,500	36,600	27,500	73,200	63,400	51,800	36,600

General Notes:
(a) See Figs. 43.34 and 43.35.
(b) For choker hitch, the angle of choke is 120° or greater.
Source: Occupational Safety and Health Administration "GUIDANCE ON SAFE SLING USE" publication.

TABLE 43.4   Polypropylene Rope Slings Based on Design Factor = 5 and Rated Loads Expressed in Pounds (lbs)

Types I, II, III, and IV

Width, inch	Single-Leg–Hitch Type			Two Leg–Horizontal Angles				Type V
	Vertical	Choker	Vertical Basket	Vertical	60°	45°	30°	Endless Vertical
1	1,100	880	2,200	2,200	1,900	1,600	1,100	2,200
1½	1,600	1,280	3,200	3,200	2,800	2,300	1,600	3,200
1¾	1,900	1,520	3,800	3,800	3,300	2,700	1,900	3,800
2	2,200	1,760	4,400	4,400	3,800	3,100	2,200	4,400
3	3,300	2,640	6,600	6,600	5,700	4,700	3,300	6,600
4	4,400	3,520	8,800	8,800	7,600	6,200	4,400	8,800
5	5,500	4,400	11,000	11,000	9,500	7,800	5,500	11,000
6	6,600	5,280	13,200	13,200	11,400	9,300	6,600	13,200

General Notes:
(a) The rated loads are based on stuffer weave construction webbing with a minimum certified tensile strength of 6,800 lbs per inch of width of the webbing.
(b) Rated loads for Types III and IV slings apply to both tapered and nontapered eye constructions. Rated loads for Type V slings are based on nontapered webbing.
(c) For Type VI slings, consult the manufacturer for rated loads.
(d) For choker hitch, the angle of choke is 120° or greater.
Source: Occupational Safety and Health Administration "GUIDANCE ON SAFE SLING USE" publication.

TABLE 43.5 Rated Load for One-Ply, Class 5 Synthetic Webbing Slings Expressed in Pounds (lbs)

| Width, inch | Types I, II, III, and IV | | | | | | | Type V |
| | Single-Leg-Hitch Type | | | Two Leg-Horizontal Angles | | | | |
	Vertical	Choker	Vertical Basket	Vertical	60°	45°	30°	Endless Vertical
1	2,200	1,760	4,400	4,400	3,800	3,100	2,200	4,400
1½	3,300	2,640	6,600	6,600	5,700	4,700	3,300	6,600
1¾	3,800	3,040	7,600	7,600	6,600	5,400	3,800	7,600
2	4,400	3,520	8,800	8,800	7,600	6,200	4,400	8,800
3	6,600	5,280	13,200	13,200	11,400	9,300	6,600	13,200
4	8,200	6,560	16,400	16,400	14,200	11,600	8,200	16,400
5	10,200	8,160	20,400	20,400	17,700	14,400	10,200	20,400
6	12,300	9,840	24,600	24,600	21,300	17,400	12,300	24,600

General Notes:
(a) The rated loads are based on stuffer weave construction webbing with a minimum certified tensile strength of 6,800 lbs per inch of width of the webbing.
(b) Rated loads for Types III and IV slings apply to both tapered and nontapered eye constructions. Rated loads for Type V slings are based on nontapered webbing.
(c) For Type VI slings, consult the manufacturer for rated loads.
(d) For choker hitch, the angle of choke is 120° or greater.
Source: Occupational Safety and Health Administration "GUIDANCE ON SAFE SLING USE" publication.

TABLE 43.6 Rated Load for Two-Ply, Class 5 Synthetic Webbing Slings Expressed in Pounds (lbs)

| Sling Width, inch | Types I, II, III, and IV | | | | | | | Type V |
| | Single–Leg-Hitch Types | | | | Two Leg–Horizontal Angles | | | Endless Vertical |
	Vertical	Choker	Vertical Basket	Vertical Basket	60°	45°	30°	
1	1,600	1,280	3,200	3,200	2,800	2,300	1,600	3,200
1½	2,300	1,840	4,600	4,600	4,000	3,300	2,300	4,600
1¾	2,700	2,160	5,400	5,400	4,700	3,800	2,700	5,400
2	3,100	2,480	6,200	6,200	5,400	4,400	3,100	6,200
3	4,700	3,760	9,400	9,400	8,100	6,600	4,700	9,400
4	6,200	4,960	12,400	12,400	10,700	8,800	6,200	12,400
5	7,800	6,240	15,600	15,600	13,500	11,000	7,800	15,600
6	9,300	7,440	18,600	18,600	16,100	13,200	9,300	18,600
8	11,750	9,440	21,150	23,600	18,300	15,000	11,750	21,150
10	14,700	11,760	26,450	29,400	22,900	18,700	14,700	26,450
12	17,650	14,120	31,750	35,200	31,750	22,400	17,650	31,750

General Notes:

(a) The rated loads are based on stuffer weave construction webbing with a minimum certified tensile strength of 9,800 lbs per inch of width of the webbing.

(b) Rated loads for Types III and IV slings apply to both tapered and nontapered eye constructions. Rated loads for Type V slings are based on nontapered webbing.

(c) For Type VI slings, consult the manufacturer for rated loads.

(d) For choker hitch, the angle of choke is 120° or greater.

Source: Occupational Safety and Health Administration "GUIDANCE ON SAFE SLING USE" publication.

TABLE 43.7   Rated Load for One-Ply, Class 7 Synthetic Webbing Slings Expressed in Pounds (lbs)

Table 43.8 Rated Loads for Two-Ply, Class 7 Synthetic Webbing Slings Expressed in Pounds (lbs)

| | Types I, II, III, and IV | | | | | | | | Type V |
| | Single-Leg-Hitch Types | | | | Two Leg–Horizontal Angles | | | | |
Sling Width, inch	Vertical	Choker	Vertical Basket	Vertical Basket	Vertical Basket	60°	45°	30°	Endless Vertical
1	3,100	2,480	6,200	6,200	6,200	5,400	4,400	3,100	6,200
1½	4,700	3,760	9,400	9,400	9,400	8,100	6,600	4,700	9,400
1¾	5,400	4,320	10,800	10,800	10,800	9,400	7,600	5,400	10,800
2	6,200	4,960	12,400	12,400	12,400	10,700	8,800	6,200	12,400
3	8,800	7,040	17,600	17,600	17,600	15,200	12,400	8,800	17,600
4	11,000	8,800	22,000	22,000	22,000	19,100	15,600	11,000	22,000
5	13,700	10,960	27,400	27,400	27,400	23,700	19,400	13,700	27,400
6	16,500	13,200	33,000	33,000	33,000	28,600	23,000	16,500	33,000
8	22,750	18,200	42,350	42,350	45,400	36,700	29,900	22,750	42,350
10	28,400	22,720	52,900	52,900	56,800	45,800	37,400	28,400	52,900
12	34,100	27,280	63,500	63,500	68,200	55,000	44,900	34,100	63,500

General Notes:

(a) The rated loads are based on stuffer weave construction webbing with a minimum certified tensile strength of 9,800 lbs per inch of width of the webbing.

(b) Rated loads for Types III and IV slings apply to both tapered and nontapered eye constructions. Rated loads for Type V slings are based on nontapered webbing.

(c) For Type VI slings, consult the manufacturer for rated loads.

(d) For choker hitch, the angle of choke is 120° or greater.

Source: Occupational Safety and Health Administration "GUIDANCE ON SAFE SLING USE" publication.

TABLE 43.9 Rated Load for Four-Ply, Class 7 Synthetic Webbing Slings Expressed in Pounds (lbs)

Sling Width, inch	Single-Leg-Hitch Types			Types I, II, III, and IV			
	Vertical	Choker	Vertical Basket	Vertical Basket	Two Leg—Horizontal Angles		
					60°	45°	30°
1	5,500	4,400	11,000	11,000	9,500	7,800	5,500
2	11,000	8,800	22,000	22,000	19,000	15,600	11,000
3	16,450	13,160	32,900	32,800	28,500	23,300	16,450
4	20,400	16,320	40,800	40,800	35,300	28,800	20,400
5	25,500	20,400	51,000	51,000	44,200	36,100	25,500
6	30,600	24,480	61,200	61,200	53,000	43,300	30,600

General Notes:
(a) The rated loads are based on stuffer weave construction webbing with a minimum certified tensile strength of 9,800 lbs per inch of width of the webbing.
(b) Rated loads for Types III and IV slings apply to both tapered and nontapered eye constructions. Rated loads for Type V slings are based on nontapered webbing.
(c) For Type VI slings, consult the manufacturer for rated loads.
(d) For choker hitch, the angle of choke is 120° or greater.
Source: Occupational Safety and Health Administration "GUIDANCE ON SAFE SLING USE" publication.

Size	Hitch Type			Horizontal Angle, deg		
	Vertical	Choker	Vertical Basket	60°	45°	30°
1	2,600	2,100	5,200	4,500	3,700	2,600
2	5,300	4,200	10,600	9,200	7,500	5,300
3	8,400	6,700	16,800	14,500	11,900	8,400
4	10,600	8,500	21,200	18,400	15,000	10,600
5	13,200	10,600	26,400	22,900	18,700	13,200
6	16,800	13,400	33,600	29,100	23,800	16,800
7	21,200	17,000	42,400	36,700	30,000	21,200
8	25,000	20,000	50,000	43,300	35,400	25,000
9	31,000	24,800	62,000	53,700	43,800	31,000
10	40,000	32,000	80,000	69,300	56,600	40,000
11	53,000	42,400	106,000	91,800	74,900	53,000
12	66,000	52,800	132,000	114,300	93,300	66,000
13	90,000	72,000	180,000	155,900	127,300	90,000

General Notes:
(a) Roundslings are identified by the vertical rated load shown on the sling identification. The size numbers in the first column have been adopted by the Web Sling & Tie Down Association to describe certain polyester roundslings. They are included for references only. Other polyester roundslings may have different vertical rated loads.
(b) Color guidelines for polyester roundsling covers are widely used to indicate the vertical rated load of roundslings; however, this is not followed by some manufacturers. Always select and use roundslings by the rated load as shown on the sling identification; never by color.
(c) For choker hitch, the angle of choke is 120° or greater.
*See Fig. 43.4 for sling configuration descriptions.
Source: Occupational Safety and Health Administration "GUIDANCE ON SAFE SLING USE" publication.

TABLE 43.10   Rated Load for Single-Leg Polyester Roundslings: Endless and Eye-and-Eye Type Expressed in Pounds (lbs)

Safety Line
The safety line is used only for lowering a man to the ground.

Slings
Slings are used for lashing tools or material in place, for attaching blocks and snatch blocks to a pole, for lashing an old pole to a new pole temporarily, and for tying line wires up temporarily.

Knots and Knot Tying
Knots are used for fastening a rope to an object or for joining two ends of a rope. The knot or hitch used must hold the strain to be applied without damaging the rope or the load. The knot used must also be one that can be tied or loosened easily and quickly.

Terms Used in Knot Tying

All knots and hitches are a combination of the three different kinds of bends: the bight, the loop, and the round turn (Fig. 43.3).

For convenience in describing the method of making various knots, the following terms will be used: *standing part, bight,* and *running end.* The standing part is the principal portion, or longest part of the rope; the bight is a loop formed with the rope so that the two parts lie alongside each other; and the running end of a rope is the free end that is used in forming the knot. Figures 43.4 through 43.25 illustrate various kinds of knots.

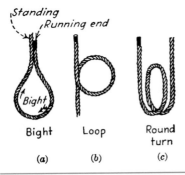

Bight   **Loop**   **Round turn**

(a)   (b)   (c)

FIGURE 43.3   Terms used in knot tying.

FIGURE 43.4   Overhand knot. The overhand knot is the simplest knot made and forms a part of many other knots. This knot is often tied in the end of a rope to prevent the strands from unraveling or as a stop knot to prevent rope from slipping through a block.

FIGURE 43.5   Half hitch. A half hitch is used to throw around the end of an object to guide it or keep it erect while hoisting. A half hitch is ordinarily used with another knot or hitch. A half hitch or two thrown around the standing part of a line after tying a clove hitch makes a very secure knot.

FIGURE 43.6   Two half hitches. This knot is used in attaching a rope for anchoring or snubbing. It is easily and quickly made and easily untied.

FIGURE 43.7   Square knot. The square knot is used to tie ropes together of approximately the same size. It will not slip and can usually be untied even after a heavy strain has been put on it. Linemen use the square knot to bind light leads, lash poles together on changeovers, on slings to raise transformers, and for attaching blocks to poles and crossarms.

FIGURE 43.8 Method of making square knot. (a) Passing left end A over right end B and under. (b) Passing right end B over left end A and under. (c) The completed knot drawn up. (*Courtesy Plymouth Cordage Co.*)

FIGURE 43.9 Granny knot. Care must be taken that the standing and running parts of each rope pass through the loop of the other in the same direction, that is, from above downward, or vice versa; otherwise, a granny knot is made which will not hold.

FIGURE 43.10 Thief knot. In tying the square knot, the standing part of both ropes must cross, as otherwise a useless knot known as the thief knot is formed.

FIGURE 43.11 Single sheet bend. The single sheet bend is used in joining ropes, especially those of unequal size. It is more secure than the square knot but is more difficult to untie. It is made by forming a loop in one end of a rope; the end of the other rope is passed up through the loop and underneath the end and standing part, then down through the loop thus formed.

FIGURE 43.12 The bowline. The bowline is used to place a loop in the end of a line. It will not slip or pull tight. Linemen use the bowline to attach come-alongs (wire grips) to rope, to attach tail lines to hook ladders, and as a loose knot to throw on conductors to hold them in the clear while working on poles.

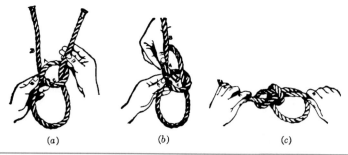

FIGURE 43.13 Method of making bowline knot. (a) Threading the bight from below. (b) Leading around standing part and back through bight C. (c) The completed bowline. (*Courtesy Plymouth Cordage Co.*)

FIGURE 43.14   Running bowline. This knot is used when a hand line or bull rope is to be tied around an object at a point that cannot be safely reached, such as the end of a limb.

FIGURE 43.15   Double bowline. This knot is used to form a loop in the middle of a rope that will not slip when a strain is put upon it.

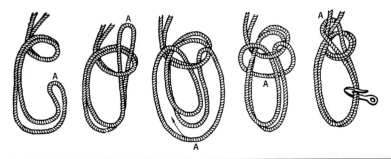

FIGURE 43.16   Bowline on a bight. The bowline on a bight is used to place a loop in a line somewhere away from the end of the rope. It can be used to gain mechanical advantage in a rope guy by doubling back through the bowline on a bight much as a set of blocks. The bowline on a bight also makes a good seat for a man when he is suspended on a rope. To tie this bowline, take the bight of the rope and proceed as with the simple bowline; only instead of tucking the end down through the bight of the knot, carry the bight over the whole and draw up, thus leaving it double in the knot and double in the standing part. The loop under the standing part is single.

FIGURE 43.18   Clove hitch. The clove hitch is used to attach a rope to an object such as a crossarm or pole where a knot that will not slip along the object is desired. Linemen use the clove hitch for side lines, temporary guys, and hoisting steel. To make this hitch, pass the end of the rope around the spar or timber, then over itself, over and around the spar, and pass the end under itself and between the rope and spar as shown.

FIGURE 43.17   Single intermediate bowline. This knot is used in attaching rope to the hook of a block where the end of the rope is not readily available.

FIGURE 43.20   Timber hitch. The timber hitch is used to attach a rope to a pole when the pole is to be towed by hand along the ground in places where it would be impossible to use a truck or its winch line to spot it. The timber hitch is sometimes used to send crossarms aloft. This hitch forms a secure temporary fastening which may be easily undone. It is like the half hitch but is more secure. Instead of the end being passed under the standing part of the rope once, it is wound around the standing part three or four times, as shown in the figure.

FIGURE 43.19   Clove hitch used for lifting.

FIGURE 43.21   Rope timber hitch and half hitch. The timber hitch will not slip under a steady pull but may slip when slack. To make the timber hitch more secure, a single half hitch may be taken a little farther along on the spar.

FIGURE 43.22   Snubbing hitches. Knots used for attaching a rope for anchoring or snubbing purposes.

FIGURE 43.23   This knot is used in attaching one rope to another for snubbing a load.

FIGURE **43.24**   Rolling bend. The rolling bend is often used for attaching the rope to wires that are too large for the wire-pulling grips. It is also used for skidding poles or timber.

FIGURE **43.25**   Blackwall hitch. This hitch consists of a loop with the end of the rope passed under the standing part and across the hook. Under load the hauling part jams the end against the hook. This hitch should be used only where the strain is steady and there is no hazard if the hitch slips. It is exceedingly useful when the hitch must be made quickly or where the hitch must be changed frequently.

Splices

Eye Splice

When it is desirable to have a permanent eye at the end of a rope, such as a hand line, it can be formed by splicing the end of the rope into its side, thereby making an eye or side splice.

The steps in making the eye splice are illustrated and described in Fig. 43.26.

Short Splice

The short splice is sometimes called a *butt splice* because the line is unlaid and the ends are butted together. This splice can be used where it is desirable to splice together two ropes which are not required to pass over a pulley. This splice can be made quickly and is nearly as strong as the rope. As the diameter of the rope is nearly doubled, this type of splice is too bulky to pass through a sheave block.

The steps in making the short splice are illustrated and described in Fig. 43.27.

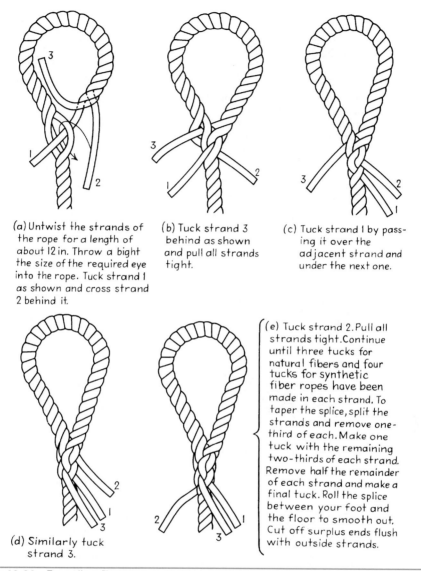

(a) Untwist the strands of the rope for a length of about 12 in. Throw a bight the size of the required eye into the rope. Tuck strand 1 as shown and cross strand 2 behind it.

(b) Tuck strand 3 behind as shown and pull all strands tight.

(c) Tuck strand 1 by passing it over the adjacent strand and under the next one.

(d) Similarly tuck strand 3.

(e) Tuck strand 2. Pull all strands tight. Continue until three tucks for natural fibers and four tucks for synthetic fiber ropes have been made in each strand. To taper the splice, split the strands and remove one-third of each. Make one tuck with the remaining two-thirds of each strand. Remove half the remainder of each strand and make a final tuck. Roll the splice between your foot and the floor to smooth out. Cut off surplus ends flush with outside strands.

FIGURE 43.26   Eye splice, fiber rope.

Long Splice

The long splice is sometimes known as a *running splice*. It is used to splice rope where it is undesirable to increase the diameter of the rope. A good long splice is hard to detect without close examination and will run freely through a sheave or blocks. Using a running splice, a damaged part of a long line can be removed from the line without detracting from its usefulness. The strength will be reduced, however.

The steps in making the long splice are illustrated and described in Fig. 43.28.

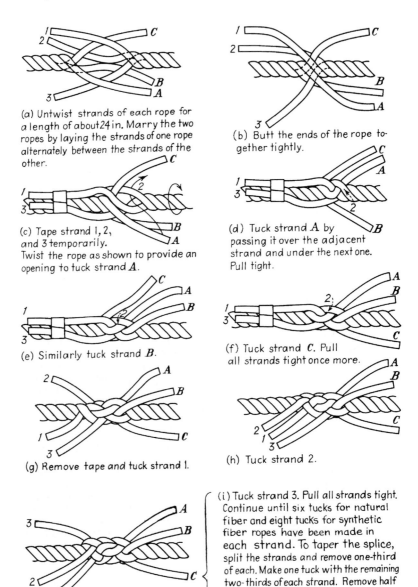

(a) Untwist strands of each rope for a length of about 24 in. Marry the two ropes by laying the strands of one rope alternately between the strands of the other.

(b) Butt the ends of the rope together tightly.

(c) Tape strand 1, 2, and 3 temporarily. Twist the rope as shown to provide an opening to tuck strand A.

(d) Tuck strand A by passing it over the adjacent strand and under the next one. Pull tight.

(e) Similarly tuck strand B.

(f) Tuck strand C. Pull all strands tight once more.

(g) Remove tape and tuck strand 1.

(h) Tuck strand 2.

(i) Tuck strand 3. Pull all strands tight. Continue until six tucks for natural fiber and eight tucks for synthetic fiber ropes have been made in each strand. To taper the splice, split the strands and remove one-third of each. Make one tuck with the remaining two-thirds of each strand. Remove half of the remainder of each strand and make a final tuck. Roll the splice between your foot and the floor to smooth out. Cut off surplus ends flush with outside strands.

FIGURE 43.27 Short splice, fiber rope.

(a) Untwist the strands of each rope 10 to 12 turns, taking care to preserve the natural twist of the rope. Marry the two ropes by laying the strands of one rope alternately between the strands of the other.

(b) Butt the ends of the rope together tightly.

(c) Unlay strand 2 and lay strand *C* in its place until about 6 in. of strand *C* remain.

(d) Tie strands *C* and 2 together with the overhand knot as shown. Similarly tie strands *A* and 1 together.

(e) Work the two overhand knots down into the rope. Unlay strand *B* and lay strand 3 in its place until about 6 in. of strand 3 remain. Tie strands *B* and 3 together in the same manner as the first two pairs of strands.

(f) Untwist each strand a little to make it lay flat and tuck it by passing it over the other strand of the knot and under the adjacent strand. To taper the splice, split the strands and remove one-third of each. Make one tuck with the remaining two-thirds of each strand. Remove half the remainder of each strand and make a final tuck. Roll the splice between your foot and the floor to smooth out. Cut off surplus ends flush with outside strands.

FIGURE **43.28**   Long splice, fiber rope. The long splice is used for permanently joining two ropes which must pass through a close fitting pulley.

In-Line Splice for 12-Strand Braided Rope[1]

The in-line splice is useful to repair synthetic fiber 12-strand braided damaged rope or to join two separate pieces of rope. Start the splicing by securing the ends of both ropes with tape. Lay the ropes side by side and measure one fid length, 20 times rope diameter, from the ends of the rope and place a mark on each rope (Fig. 43.29*a*). Fit rope 1 into the fid and insert through the middle of rope 2 at the mark to start making the splice (Fig. 43.29*b*). Pull the end of rope 1 through rope 2 until both marks meet (Fig. 43.29*c*).

---

[1]*Source:* Yale Cordage, Inc., instructions for splicing braided rope.

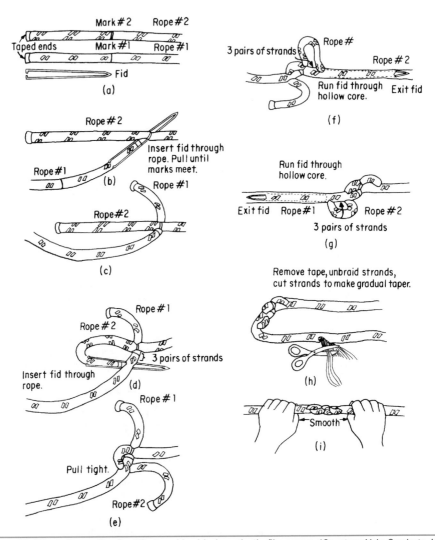

Figure 43.29   In-line splice for 12-strand braided synthetic fiber rope. (*Courtesy Yale Cordage, Inc.*)

Place the end of rope 2 into the fid, and insert the fid and the end of rope 2 through the middle of rope 1 at a point three pairs of strands beyond the marks (Fig. 43.29*d*). Pull rope 2 through rope 1 until the ropes are tight together (Fig. 43.29*e*). Place rope 1 into the fid, and insert the fid and the end of rope 1 into the center of rope 2 at a point three pairs of strands beyond the marks. Run the fid and the end of rope 1 an entire fid length through the center core of rope 2, and at this point push the fid and the end of the rope out through the exterior of rope 2. Pull rope 1 through the exit point on the surface of rope 2 (Fig. 43.29*f*). Place rope 2 into the fid, and insert the fid and the end of rope 2 into the center core of rope 1 at a point three pairs of strands beyond the previous exit point from rope 1. Run the fid and the end of rope 2 an entire fid length through the center core of rope 1, and at this point push the fid out of rope 1 and pull the end of rope 2 through the exit point (Fig. 43.29*g*). Finish the splice by removing the tape from the ends of the ropes and, using a scissors or a knife, trim the rope ends at staggered intervals, near the points the ropes exit from the line (Fig. 43.29*h*). Hold the rope on each side of the splice and smooth the line away from the splice. The tapered ends of the ropes will disappear into the core of the line, leaving a clean splice (Fig. 43.29*i*).

Eye Splice for 12-Strand Braided Rope[2]

Start the splicing by securing the end of the rope with tape. Lay the rope on a flat surface and measure one fid length, 20 times rope diameter, and allowing some extra rope from the end, mark the rope with an *X*. Starting at the *X* mark, form the desired size eye and mark the rope with a dot opposite the *X* (Fig. 43.30*a*). Fit the tapered end of the rope into the fid, and insert the fid and the rope end through the middle of the line at the dot (Fig. 43.30*b*). Pull the end of the rope through the line until the *X* and the dot meet (Fig. 43.30*c*). Insert the fid and the end of the rope back through the middle of the line three pairs of strands beyond the junction of the *X* and the dot. Pull the end of the rope through the exit point (Fig. 43.30*d*). Insert the fid and the end of the rope back into the middle of the line three pairs of strands beyond the previous exit from the line, and run the fid and end of rope down through the center core of the line a fid length. Push the fid and end of the rope through the center core of the line a fid length. Push the fid and end of the rope through the outside of the line. Pull the end of the rope through the exit point (Fig. 43.30*e*). Remove the tape from the end of the rope, and, using a scissors or a knife, trim the rope end at staggered intervals near the point where it exits from the line (Fig. 43.30*f*). Hold the

Figure 43.30   Eye splice for 12-strand braided synthetic fiber rope. (*Courtesy Yale Cordage, Inc.*)

[2]*Source:* Yale Cordage, Inc., instructions for splicing braided rope.

line beyond the eye splice, and smooth the line toward the eye until the tapered end of the rope disappears into the center of the line (Fig. 43.30g). The throat of the eye splice can be whipped if desired (Fig. 43.30h).

Eye Splice for Double-Braided Rope[3]

Start the splice by placing one thin layer of tape on the end of the rope to be spliced. Lay the rope on a flat surface and measure two fid lengths, 20 times the rope diameter, and mark the line with the reference point R. Form a loop of the size desired from reference point R in the line and mark this point X. Tie a slipknot in the line approximately 10 fid lengths, 100 times the rope diameter, beyond point X (Fig. 43.31a). Bend the rope sharply at point X. Spread the braided cover of the line at point X with any sharp tool such as an ice pick, awl, or marlin spike, and pry the core of the rope out through the cover (Fig. 43.31b). Pull the core of the rope out through the cover at point X from the tapered end of the rope. Place one layer of tape on the end of the core. Hold the exposed core and slide the cover as far back as possible toward the tightly tied slipknot. Firmly smooth the cover back from the slipknot toward the taped end until all cover slack is removed. Mark the core of the rope I at the point where it exits from the line (Fig. 43.31c).

FIGURE **43.31**   Eye splice for double-braided synthetic fiber rope. (*Courtesy Yale Cordage, Inc.*)

---

[3]Hybrid double-braided constructed ropes may require special splicing techniques. The user should consult the rope manufacturer for instructions.

Slide the cover of the rope again toward the slipknot to expose more of the rope core. Measure along the core of the rope from mark *I* one-half fid length, and mark the rope core *II*. Measure along the core of the rope from mark *II* 2.5 fid lengths, and mark the core of the rope *III* (Fig. 43.31*d*). Inspection of the braided cover of the rope reveals that half of the strands revolve to the right around the rope and half revolve to the left. Measure one-half fid from reference mark *R* toward the end of the rope, and mark this point *T* and mark a line completely around the braided cover of the rope. Start at point *T* and mark toward the taped end of the rope every fifth right and left strand (single or paired) to the end of the taped braided cover (Fig. 43.31*e*). If the rope has 20 strands in braided cover, start at point *T* and mark fifth right and left strands, then fourth right and left, then fifth, etc., until reaching the end of braided cover. Press prongs of fid into the end of braided cover, and tape over the cover and the fid prongs. Hold the core of the rope lightly at mark *III*, and push the fid with the braided cover attached into the middle of the core at mark *II* and out at mark *III* (Fig. 43.31*f*). Pull the fid and the end of the braided cover from the core at mark *III* until reference point *R* appears. Remove tape and fid from the end of the braided cover of the rope (Fig. 43.32*g*). Start with the last marked pair of the braided cover strands near the end of the rope, and cut and pull the strands out. Proceed to cut and remove each right and left strand at the marks until point *T* is reached (Fig. 43.32*h*). The braided cover from point *T* will have a gradual taper to a point at the end of the rope (Fig. 43.32*i*). Measure one fid length on the braided cover from point *T* toward the slipknot in the rope, and mark this point *Z*. Place the taped end of the core into the fid and insert the fid into the middle of the braided cover at point *T*, and milk the braided cover over the fid and the end of the core of the rope while pulling them through the braided cover exiting at point *Z*, making sure the fid does not catch on any internal core strands (Fig. 43.32*j*). Hold the core tail at the exit point *Z* in one hand and the core with the tapered braid inserted at mark *III* in the other hand, and alternately work the core tail through the exit of the braided cover until the rope core at mark *III* is approximately one diameter of the rope from the braided cover at point *X*. Mark the core tail with a dot at the point it enters the braided cover at point *X* (Fig. 43.32*k*). Smooth out braided cover of eye completely, from mark *T* to the dot on the core, and mark *X* on the braided cover, to get all the slack out of the eye area. The tapered end of the braided cover will disappear into the core of the rope. Pull the core tail out of the braided cover at point *Z* until the dot on the core is exposed. Measure three-fourths of a fid length from the dot on the core toward the end, and cut the core, removing the unneeded end of the core. Cut and remove strands at staggered intervals between the dot on the core and the end of the core to form a gradual taper (Fig. 43.32*l*). Hold the rope at the slipknot and, with the other hand, milk the cover toward the splice, gently at first, then more firmly. The braided cover will slide over the core at mark *III* and *II*. Continue the operation until the reference point *R* on the braided cover is at point *X* on the braided cover (Fig. 43.32*m*). Smooth out the eye during the milking operation to prevent the reduced volume tail from catching in the throat of the splice. Smooth the braided cover from point *T* to point *X* if bunching occurs at mark *I*. It may be necessary to anchor the loop of the slipknot to a fixed object and then use both hands and the weight of the body to extend the cover over the core of the rope. The rope cover can be hammered at point *X* to loosen the strands, and the rope should be flexed and loosened at the crossover point *X* to loosen the strands, and the rope should be flexed and loosened at the crossover point mark *I* during the final burying process. Continue milking the line until all braided cover slack between the knot and the throat of the eye

Smooth out and
cover taper will
disappear.

Pull core tail out until mark
at "X" exposed at Point "Z".

Mark II   Mark III

Mark I   Mark R

Remove tape from cover.

(g)

Mark core tail at "X"

Cover taper

(k)

Cut and remove every 5th strand
(Single or paired).

(h)

Cutting and removing
at mark made on core.

Core taper

3/4 fid

(l)

Cover taper - every 5th strand removed.

Point T at Mark II

Gradual taper

(i)

Beginning at slipknot, "milk" cover by
sliding hand toward loop until core is buried.

Crossover

Bury to Mark R.

(m)

Insert core into cover.

Fid and end of
core come
out at "Z".

(j)

Mechanical tension Crossover Firmly
secured.

Pull

(n)

Pass stitching
through spliced
area near throat
of eye.

Reinsert
at angle

(o)

FIGURE 43.32   Eye splice for double-braided synthetic fiber rope. (*Courtesy Yale Cordage, Inc.*)
(*Continued from Fig. 43.31*)

has been removed. It is helpful to securely anchor the slipknot of larger ropes and attach a small line to the braided core at the crossover, mark *I*, and mechanically apply tension with either a block and tackle or capstan. Tension will reduce the diameter of the core and crossover for easier burying (Fig. 43.32*n*). Finish the eye splice with stitch locking to prevent no-load opening due to mishandling. Nylon or polyester whipping twine approximately the same size as the strands in the rope being spliced should be used for the stitching. Stitch the line, starting at the throat of the eye and in turn two planes perpendicular to each other, making sure the stitching is not pulled too tight. After the stitching is completed, bring the two ends of the twine together through the same opening in the braided cover, and tie the ends together with a square knot. Reinsert the knot and the ends of the stitching twine back into the braided cover between the cover and the core of the rope (Fig. 43.32*o*).

Reduction in Strength due to Knots and Splices
Table 43.11 shows the extent to which the strength of rope is reduced when used in connection with some of the common knots and splices.

Rope Gear

Slings
The simplest sling consists of a short piece of rope whose ends are spliced together to form an endless piece of rope. Slings are used for lashing tools or material in place, for attaching blocks or snatch blocks to a pole, for lashing an old pole to a new pole temporarily, and for tying line wires up temporarily. Slings should be 6 ft to 10 ft long, depending, of course, on their planned use. Slings can also be made with an eye on one end and a dog knot on the other end. Synthetic webbing slings are depicted in Fig. 43.33 and synthetic fiber rope slings are depicted in Fig. 43.34.

Slings are also made of rope, spliced for a hook at one end and a hook or ring at the other.

Safe Loads
Before attempting to lift a load with a rope sling, the weight of the load should be carefully estimated and a sling of the proper size and labeled strength selected. The greatest load can be lifted when all legs of the sling are in a vertical position and the least when

Type or Splice or Knot	Percentage Strength
Straight rope	100
Eye splice	90
Short splice	80
Timber hitch or half hitch	65
Bowline or clove hitch	60
Square knot or sheet bend	50
Overhand knot (half or square knot)	45

TABLE 43.11   Percentage Strength of Spliced or Knotted Rope

Sling made with a triangle fitting on one end and a slotted triangle choker fitting on the other end. It can be used in a vertical, basket, or choker hitch.

Type I

Sling made with a triangle fitting on both ends. It can be used in a vertical or basket hitch only.

Type II

Sling made with flat loop eye on each end with loop eye opening on same plane as sling body. This type of sling is sometimes called a flat eye-and-eye, eye-and-eye, or double-eye sling.

Type III

Sling made with both loop eyes formed as in Type III, except that the loop eyes are turned to form a loop eye which is at a right angle to the plane of the sling body. This type of sling is commonly referred to as a twisted-eye sling.

Type IV

Endless sling, sometimes referred to as a grommet. It is a continuous loop formed by joining the ends of the webbing together.

Type V

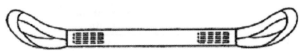

Return-eye (reversed-eye) sling is formed by using multiple widths of webbing held edge-to-edge. A wear pad is attached on one or both sides of the sling body and on one or both sides of the loop eyes to form a loop eye at each end which is at a right angle to the plane of the sling body.

Type VI

43.33   Synthetic Webbing Slings.

Maximum angle
15 deg

Eye splice

Short splice (end-for-end)

Eye and Eye

Endless

Fiber rope thimble (optional)

Angle of choke

Hook (optional)

Choker (with fittings)

Master link

Horizontal angle

Shackle

Multiple leg bridle sling

General note: Ensure that fittings designed for synthetic slings are used.

FIGURE 43.34   Synthetic Fiber Rope Slings.

the legs are nearly horizontal. When a rope sling has been in service for 6 months or more, even though it shows no signs of wear or damage, the loads placed on it should be reduced.

Figure 43.35 illustrates the way the strain on the rope sling is increased when the angle with the horizontal decreases. Thus, at 60° the strain in the legs of the sling is only 578 lbs for a 1,000-lb weight, whereas it is 1,930 lbs for a 1,000-lb load when the angle is reduced to 15°. This is almost twice as great as the weight of the load lifted. At 30° the strain is 1,000 lbs, which is the same as the weight of the load. Sling angles smaller than 30° are therefore not recommended.

If the sling is attached around a sharp corner, it should be carefully padded to prevent cutting.

FIGURE 43.35   Sample loading of parts of lifting slings. For loads other than 1,000 lbs, use direct ratios. Sling angles less than 30° from horizontal are not recommended.

Block and Tackle

Block and tackle are used for applying tension to line conductors when sagging in, for applying tension to guy wires, when hoisting transformers, and for other general-purpose hoisting (Fig. 43.36).

The use of block and tackle has two advantages: (1) the user can stand on the ground and pull downward while hoisting or lifting a load; (2) the manual force applied need only be a fractional part of the load lifted.

Mechanical Advantage

To find the pull required to lift a given weight with a block and tackle, divide the weight by the number of ropes running from the movable block. The lead line or fall line is not to be counted unless the line comes from the moving block. There is always some friction loss around the sheaves. This can be estimated at 10% per sheave and added to the load to be lifted. The sketches in Fig. 43.37 show the various combinations of block and tackle employed to give mechanical advantages from 2 to 7. As mentioned above, the ratio of load to pull on the fall line is given by the number of ropes running from the movable block. The fall line is not included in the calculation of mechanical advantage because the lineman is pulling on the rope from the fixed block. The load that may be

LOAD LIFTED	DIAGRAM OF RIGGING	
2 times safe load on rope (approx.)**	Single 2 parts Single	block block
3 times safe load on rope (approx.)**	Double 3 parts Single	block block
4 times safe load on rope (approx.)**	Double 4 parts Double	block block
5 times safe load on rope (approx.)**	Triple 5 parts Double	block block
6 times safe load on rope (approx.)**	Triple 6 parts Triple	block block
7 times safe load on rope (approx.)**	Quadruple 7 parts Triple	block block

FIGURE 43.36   A three-part block and tackle. (*Courtesy A. B. Chance Co.*)

FIGURE 43.37   Lifting capacity of block and tackle. **Less 20% approximately for friction.

4 part (2 + 2)   5 part (2+3)   6 part (3+3)

FIGURE **43.38**   Four-, five-, and six-part blocks and tackle.

lifted is therefore the mechanical advantages times the safe load on the rope. Normal uses for 4-inch blocks are sagging No. 4 and No. 6 conductors and raising distribution transformers rated up two 37½ kVA single-phase. The 6-inch blocks are used for heavier work up to the limit of the tackle. Blocks having mechanical advantages of four, five, and six parts are illustrated in Fig. 43.38.

CHAPTER 44

Climbing Wood Poles

The ability to climb a pole safely is one of the core competencies required of a lineman. The wood pole climbing principles have remained constant over time. Climbing equipment is designed emphasizing comfort, durability, support, and safety. ASTM F887 *Standard Specifications for Personal Climbing Equipment* is the minimum industry standard.

Utility climbing safety principles incorporate fall arrest protection methods and devices.

Use and Care of Pole-Climbing Equipment[1]

Pole-Climbing Equipment

Pole-climbing equipment consists of a body belt, safety strap, and a pair of climbers.

The equipment allows a workman to climb, stand, or change position on a pole when no other suitable means of support is available. It also allows the free use of both hands while in any position on the pole.

Suggested specifications for pole-climbing equipment are detailed in ASTM F887 *Standard Specifications for Personal Climbing Equipment*. General descriptions of components of the equipment and their use and care follow.

The Body Belt

The body belt consists of a cushion section, a belt section with tongue and buckleends, a tool saddle, and D rings which are attached solidly to the cushion, or on shifting D-ring belts, attached solidly to a D-ring saddle. The body belt usually has provisions made for a holster which will carry one or more tools in addition to the tools which are carried in the tool loops.

Tool loops should be of proper size to prevent the tools from slipping through the loops and falling. There should be no tool loops for 2 inches on either side of the center in the back, in accordance with ASTM F887 *Standard Specifications for Personal Climbing Equipment*.

The belt is marked in "D" sizes. The D size is the distance between the heels of the D rings when the belt is laid flat.

The waist measurement of the belt is found by measuring from the roller on the buckle to the center hole on the tongue end.

Proper measurement for the D-ring size is the distance around behind the body at the point where the belt will be worn between the prominent points of the hip bones, *plus 1 inch.*

---

[1]Reprinted with the permission of the Edison Electric Institute.

829

The proper waist size of body belt is determined by measuring the distance around the body at the point where the belt will be worn. The measurement should be made outside of any clothing normally worn while working.

The body belt should be worn snugly but not too tightly. The end of the strap should always be passed through the keeper and kept clear of the D ring when the belt is being worn.

Manufacturers have standardized on a relationship between D sizes and waist sizes. In the event your measurements do not coincide with these standard sizes, the belt should be ordered by D size, as the waist is adjustable.

The Safety Strap

The safety strap is used for support while working on poles, towers, or platforms. Snap hooks on each end are provided for attachment to the D rings in the body belt.

When climbing poles, under normal conditions, both snaps should be engaged in the *same* D ring for safety. The snap on the double end should have the keeper facing outward, and the other snap should face inward. Right-handed men and some left-handed men usually carry the strap on the left.

When in use, one snap hook should be securely engaged in each D ring, never both snaps in the same D ring. The user should look to be sure that snaps are properly engaged. He should never depend on sound or feel for security.

Safety straps are adjustable for length by means of a buckle in the strap to suit the workman and the size of the pole. When in use, the side of the strap to which the buckle is attached should be next to the pole with the buckle outward.

The Climbers

Climbers are used for ascending, descending, and maintaining the working position on poles when no other means of support is available. The condition, length, and shape of the gaffs of the climbers are of the utmost importance. The gaffs support the workman as he climbs or does his work. Defective gaffs are dangerous and can cause inadvertent electrical contacts or falls from poles.

Climbers are made in adjustable or fixed lengths from 14 inches to 20 inches by ½ inch steps. Gaffs are furnished in either solid or replaceable type. Proper fit requires a leg iron to reach about ½ inch below the inside prominence of the knee joint.

Climbers are secured to the workman's legs by foot and leg straps. These straps should be drawn up to a snug fit, but *not* so tight as to be tiring. High-top shoes with heavy soles and heels should be worn for climbing.

The buckle on the foot strap should lie just outside the shoe lacing. Several types of pads are furnished for the upper end of the leg irons. All leg and foot strap ends should be snugged down in their keepers after buckling, and strap ends should point to the rear and outside.

Before the leg straps are fastened, pull up the pant legs so that they bag at the knees and do not bind. Fold the pant legs snugly against the calf, toward the outside.

This prevents the pant leg from tripping the workman while climbing. With climbers properly adjusted the workman will feel comfortable and confident.

Climbing

Arms and hands should not be bare, but properly protected when climbing. Before ascending a pole, inspect it carefully for unsafe conditions, such as rake, rotted places, nails, tacks, cracks, knots, foreign attachments, pole steps, or ice. Remove rocks and other objects from the ground at the base of the pole. Unauthorized attachments such as signs, radio aerials, or clotheslines should be reported to the workman's supervisor or removed, as company instructions may require.

Inspect the pole as you ascend and descend to avoid placing gaffs in cracks, knots, woodpecker holes, etc., which might cause a fall.

When ascending the pole, keep the arms and body relaxed with the hips, shoulders, and knees a comfortable distance away from the pole. Take it easy; favor short steps (step length should be natural for each workman); use the hands and arms for balance only. Climb with the legs (do not be tempted to pull up with the hands or arms); let the legs take the initiative over the hands (the legs "push" the hands). It is necessary that the gaffs be directed toward the center (or heart) of the pole in a natural manner. The size of the pole and the length of leg between hip and knee will automatically determine the amount of gaff separation on the pole.

The effective leg stroke is that angle or stroke which will cause the gaff to cuteffectively into the pole wood without side thrust of any sort. The effective stroke results when the knee is thrown comfortably away from the pole (without straining the hip), the gaff is aimed at the target (the imaginary line down the center of the pole), and the leg force and travel are made to parallel the climber shank until proper penetration is accomplished. Kicking or slapping the gaffs against the pole should be avoided. The hands and feet should work in coordination; the right hand is raised with the right foot and the left hand with the left foot. Weight should be shifted gradually and easily from one foot to the other.

When descending, the hands are lowered first. Each leg is relaxed and straightened before lowering. When the straightened and relaxed leg is lined up with the center of the pole and the body weight has been shifted above the gaff, drop the gaff into the pole. In descent, the leg is not stroked; it is merely lowered into position with the body weight behind it. The hands and arms take the initiative over the feet. The hands "push" the feet, which is opposite from ascending the pole when the feet push the hands. Keep hips, shoulders, and knees away from the pole. Do not take long steps or try to coast or slide when descending.

When ascending, gaff removal is facilitated by a twisting action of the ankle (outward) and slight prying action of the inside of the footwear against the pole. When descending, the climber gaffs should naturally break out with the outward and lowering movement of the knee. Removal of the climber when the last step to ground is taken is accomplished by a slight twisting and prying action as in ascending.

Wear climbers only when necessary for climbing or working on poles. Always remove them when working on the ground, riding in or driving a vehicle, etc. Gaff guards should be installed when climbers are not being worn.

No one should stand at the base of a pole while a man is ascending, descending, or working on it. There is always a possibility that the workman may fall or drop something. All persons, especially children, should be warned to keep away.

If a second man is to ascend the pole, he should wait until the first man has placed his safety strap in his working position. When descending a pole, one man should remain in his working position with his safety strap in use until the other has reached the ground and is out of the way.

Whenever possible, a slippery pole or one partly coated with snow or ice should be ascended with the gaffs in the slippery side and the hand-holds on the less slippery side.

Under very slippery conditions or when a strong wind is blowing, the safety strap may be placed around the pole and worked upward or downward in ascending or descending.

Do not hold to pins, crossarm braces, insulators, and other hardware in ascending, descending, or changing position on a pole.

Always ascend and descend on the high side of a leaning, raked, or bent pole.

Use of the Safety Strap

In placing the safety strap around a pole for support, the following steps constitute the best procedure.

Place both climber gaffs firmly in the pole at or near the same level. Keep the knees and hips away from the pole and unsnap the single end of the safety strap with the left hand.

Pass the single end of the safety strap to the right hand around the back of the pole and grasp the pole with the left hand. With the right hand, carry the snap hook to the D ring, and engage the snap with the D ring with the keeper facing outward. Both snap keepers should face outward, and the strap should lie flat, without twists, against the pole, with the buckle tongue side out when in use. Always visually check and be sure that the snaps are securely and properly engaged in the D rings before trusting your weight on the safety strap. Never depend on the sound of the snap keeper. It will make the same sound if the snap is engaged in an insecure attachment, such as a plier handle or wire hook.

Keep plier pockets and other objects well clear of D rings to avoid accidentally hooking the safety-belt snap into them. There should be a minimum of 4 inches of clearance between D rings and tool pockets. No wire hooks should be used on body belts.

Always use the belt and safety strap when working aloft on a pole or structure. It is good insurance against falling and provides a safe place to carry tools normally needed. *Never* place the safety strap around the top of a pole above the top crossarm or in any other place where it can accidentally slip off. If it is necessary to place the safety strap high on a new bare pole, place a long through-bolt in the top gain hole to keep the safety strap from slipping off the pole.

To move up or down on a pole with the safety strap in use, hold the pole with one hand, as you release tension on the strap, and move the strap up or down with the other hand.

Before a person is permitted to climb, he is taught the proper method of handling a safety strap.

Practice at the base of a pole until the proper climbing position for the feet, arms, and body is acquired. Place the feet on the ground at about a 90° angle with each other and with the side of the arch of each foot against the sides of the pole. Extend both arms out forward in a horizontal plane from the shoulders and hold on to the back of the pole with the hands. Thrust the hips well back from the pole with the legs and arms straight.

Practice climbing on a medium-size, smooth pole which has not been badly cut by previous climbing. Start using climbers by first confining climbing to a section between the ground and 3 ft above the ground. This permits one to learn to climb without fear of falling. Before starting to climb, inspect the pole from the ground for cracks, knots, holes, tacks, and nails, as the presence of any of these may deflect a gaff and result in an insecure foothold.

Practice placing the safety strap around the pole and fastening it to the D ring while on the pole. Stand on the pole with both legs straight and both gaffs firmly in the pole at or near the same level and at a point approximately 2 ft above the ground level for the first trials. This is the position for placing the strap when in the working position at any location on the pole. Then proceed to place it around the pole as described.

Inspection and Care

The user of climbing equipment should carefully inspect it before each use. Inspect leather or fabric parts for cuts, cracks, tears, enlarged buckle tongue holes, narrowing down due to stretch, or for hard and dry leather. Inspect stitching for broken, ragged, or rotted threads. Inspect metal parts for breaks, cracks, loose attachments, or wear that might affect strength. Care should be taken to determine that the keepers of the snaps are firmly and securely seated in the recess at the end of the hooks and that there is a reasonable resistance to depressing the keeper. Any distortion of the hook may change the tension of the spring so that the keeper is not in firm contact with, or securely seated in, the nose of the hook. If the keeper is not securely seated, it is possible that the D ring may become disengaged by slipping between the nose of the hook and the keeper and be the cause of a serious fall. Check the gaffs and determine whether the cutting edges are properly sharpened and shaped.

Defective equipment should be repaired or replaced, and gaffs properly shaped as soon as possible. If defects affect safety of the equipment, it should be removed from service until repaired or replaced.

Cleaning and Dressing Leather

Leather parts of climbing equipment should be cleaned approximately every 3 months and dressed approximately every 6 months and on return from vacation periods to keep them pliable and in good condition. Cleaning and dressing may be done more often if necessary due to excessive moisture or perspiration. Paint stains should be removed before the paint dries. Never, however, use gasoline or similar solvent on leather, as this will dry out and severely damage the leather.

To Clean Leather   Wipe off surface dirt with a damp sponge.

Next, work up a good lather using a neutral soap, such as Castile, and a clean, moist sponge to remove embedded dirt and perspiration. Wipe off with a clean cloth.

Finally, work up a good lather with saddle soap and a clean sponge, and rub it well into all parts of the leather. Wipe off with a clean cloth.

To Dress Leather   Never dress or oil leather before cleaning as just outlined.

While the leather is still damp after cleaning, apply a good leather dressing or neat's-foot oil, working it into the leather with the hands. Do not use an excess amount of oil, as this will saturate and weaken the leather.

Allow the leather parts to dry in a cool, shady place for 24 hours. Never dry leather near a source of heat, as heat destroys leather.

Remove any excess dressing by rubbing vigorously with a clean, soft cloth. Never use any mineral oil or grease for dressing leather.

Cleaning Nylon Fabric Safety Straps

Since there are differences in the methods of fabricating nylon safety straps, it is recommended that each manufacturer be consulted on the proper cleaning procedure to be used in cleaning the nylon safety straps.

Storage

Store leather equipment away from sources of heat such as stoves, radiators, steam pipes, or open fires. Keep leather goods away from sharp-edged objects. Before storing the climbers, it is good practice to install approved climber gaff guards to protect the gaff points and edges and to protect other objects and persons from accidental injury or damage.

Inspection and Maintenance of Climber Gaffs

Gaffs of climbers should be inspected and checked frequently for length, width, thickness, profile of the point, and sharpness of the cutting edges. If a cutout or any other difficulty is experienced in getting the gaffs to hold in a pole, investigate immediately to determine the cause. The trouble may be that the gaffs have been accidentally damaged since the last inspection.

The gaff gauge can be used for checking climber gaffs and as an aid in shaping the gaff properly.

Be sure to check the instructions issued by the manufacturers of the gauge you are using since there may be differences in the exact procedures to be followed.

The reference line across the gauge indicates the minimum length of the gaff, measured on the underside from the heel of the gaff to the point (Fig. 44.1).

Company requirements as to minimum length of gaffs should be observed.

The TH slots are used to check the thickness of the gaff 1 inch and ½ inch from the point. The face of the gauge should lie flat against the back or ridge of the gaff. If within acceptable limits, the point of the gaff will lie on or between one pair of the reference lines on the gauge, as indicated in Figs. 44.2 and 44.3.

The W slots are used to check the width of the gaff 1 inch and ½ inch from the point, as shown in Figs. 44.4 and 44.5. The face of the gauge should be flat against the back or ridge of the gaff, as in the previous check. If within acceptable limits, the point of the gaff will lie on or between one pair of the reference lines on the gauge.

If necessary to shape the gaff for proper thickness, it should be filed on the flat underside which lies between the gaff and the stirrup. Care should be taken not to cut or notch the leg iron or stirrup with the file. Also, there should be no file marks left on the gaff (Fig. 44.6). A 10-inch mill flat file or an 8-inch smooth knife file will do a nice job,

FIGURE 44.1 Checking for minimum length.

FIGURE 44.2 Checking thickness of gaff 1 inch from the point.

FIGURE 44.3   Checking thickness of gaff ½ inch from the point.

FIGURE 44.4   Checking width of gaff 1 inch from the point.

FIGURE 44.5   Checking width of gaff ½ inch from the point.

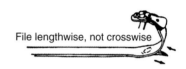

FIGURE 44.6   Filing the gaff.

if kept clean by frequent use of a card file brush. File marks on a gaff, particularly if they are crosswise, may weaken the steel and result in a broken gaff. Final filing, therefore, should be carefully done with light strokes, lengthwise of the gaffs.

If necessary to shape the gaff to proper gauge width, the filing should be done on the two outside rounded surfaces, as shown in Fig. 44.7. Care should be taken not to file the ridge or back edge of the gaff, for it is important to keep it straight, as shown in Fig. 44.8. The back of the gaff gauge can be used as a straightedge to check this.

If the back edge is rounded off, as shown in Fig. 44.9, the gaff will tend to cut out of the pole.

As the filing progresses, the work should be checked with the gaff gauge until the gaff is within the indicated gauge limits.

The point of a gaff should function as a chisel (Fig. 44.8), cutting its way into a pole, instead of as a spike or needle (Fig. 44.10) to be driven into the pole. Spike-pointed gaffs

FIGURE 44.7   Shaping the gaff.

FIGURE 44.8   Correct (chisel point).

FIGURE 44.9   Incorrect (hook-nose point).

FIGURE 44.10   Incorrect (spike point).

will cut out easier than chisel points, if the knee is brought too close to the pole for any reason. Spike points also reduce the cross-sectional area near the tip and may result in gaff breakage.

The chisel point penetrates the pole easier and deeper with less effort and, if properly shaped with the proper gaff angle, will not cut out.

To obtain a chisel point, the flat undersurface should be rounded off toward the point with a slight curve, as shown in Fig. 44.8. This filing should be done with light, even strokes.

If the gauge has a profile recess, use it to check the work, as in Fig. 44.11. If not, use Fig. 44.11 as a guide to the proper profile.

After filing and shaping are completed and the gaff checks with all the proper gauge dimensions, the flat underside of the gaff and the underside of the point should be honed, using a small carborundum pocket stone. Honing removes any burrs on the edges of the gaff and sharpens the edges for easier penetration into the pole. It is this penetration of the wood which supports the workman as he ascends, descends, or does his work on the pole. The edges of the flat underside should be knife-sharp and *not* rounded.

Testing Climber Gaffs for Proper Shape and Sharpness

The Plane Test   The plane test may be used to determine that the climber gaff is properly sharpened to cut into the pole like a chisel.

Place the climber on a soft pine or cedar board, as shown in Fig. 44.12. Holding it upright, but with *no* pressure on the stirrup, push the climber forward along the board. If the gaff is properly shaped and sharpened, and if the gaff angle with the wood is sufficient, the gaff point will dig into the wood and begin to hold within approximately 1 inch.

FIGURE 44.11   Checking the gaff profile.

FIGURE **44.12**   Plane test.

FIGURE **44.13**   Plane test.

If the climber gaff slides along the wood without digging in or merely leaves a mark or groove in the wood as shown in Fig. 44.13, either the gaff is not properly sharpened and shaped, or the gaff angle is too small. The gaff angle is built into the climber by the manufacturer and should be between 9° and 16° with the leg iron placed parallel to the surface of the wood, according to ASTM F887 *Standard Specifications for Personal Climbing Equipment*.

The Cutout Test   The cutout test may be made to determine whether the gaff will cut out of the wood when the angle between the leg iron and the surface of the wood is reduced beyond the critical point.

When the gaff is not properly sharpened or has a hook-nosed point or a convex ridge, it will cut out.

To make the cutout test, jab the climber gaff into the soft pine or cedar board at an angle of about 30° and with sufficient force to penetrate the wood to a depth of ¼ inch or more, as shown in Fig. 44.14. Hold the gaff in place in the wood and bring the leg iron down against the wood surface, applying forward pressure on the stirrup with one hand while holding the leg iron with the other hand. If the gaff is in proper condition, it will tend to cut in deeper rather than cut out.

If the gaff is not properly shaped and sharpened, it will cut out within a horizontal distance of 3 inches or less, leaving a typical gaff cutout mark or groove in the wood.

The Pole Cutout Test   In addition to the above-mentioned tests, the pole cutout test may be used to check the climber gaffs in the field.

To make the test, place the climber on the leg, holding the sleeve with the hand, palm facing the pole, as shown in Fig. 44.15. With the leg at about a 30° angle to the pole and

FIGURE **44.14**   Cutout test.

FIGURE **44.15**   Pole cutout test (step 1).

FIGURE 44.16   Pole cutout test (step 2).

the foot about 12 inches off the ground, lightly jab the gaff into the pole approximately ¼ inch. Keeping enough pressure on the stirrup to keep the gaff in the pole but not so much as to cause the gaff to penetrate any deeper, push the climber and the hand toward the pole by moving the knee until the strap loop of the sleeve is against the pole, as shown in Fig. 44.16. Making certain that the strap loop is held against the pole with pressure from the leg, gradually exert full pressure of the foot straight down on the stirrup without raising the other foot off the ground, to maintain balance if the gaff does not hold.

The point of the gaff should cut into the wood and hold (dig itself in) within a distance of not more than 2 inches, measured from the point of gaff entry into the pole to the bottom of the cut on the pole surface.

Caution: To prevent hand injury from wood splinters, good work gloves should be worn when testing climbers.

Causes of Climber Cutouts

The causes of climber cutouts can be grouped under the following five general headings:

1. Condition of poles

2. Conditions on poles

3. Clothing

4. Climbing practices

5. Climbers

A more detailed listing of causes is given in the following discussion of the five general types. Also typical accidents are cited and preventive measures discussed.

1. *Condition of poles*
   a. *Specific causes*
      (1) Knots
      (2) Knotholes
      (3) Broken poles
      (4) Weather cracks
      (5) Wet creosoted poles
      (6) Excessive gaff cuts

(7) Splinters

(8) Poles coated with ice

(9) Rotted outer surface

(10) Poles hardened by preservative treatments

(11) Crooked or leaning poles

(12) Bird holes

(13) Improperly guyed poles

(14) Pole splintered by lightning

(15) Old gains and drilled holes

(16) Flat surfaces on poles

(17) Small and limber poles

(18) Cemented bird holes

(19) Checks

b. *Examples of accidents*

(1) As a lineman was descending a distribution pole, a climber gaff cut out and he fell to the ground, breaking one leg. The pole surface was badly checked and covered with deep gaff cuts.

(2) After a lineman had unsnapped his safety strap to descend a pole, a piece of the shell of the pole in which one of his gaffs was placed broke off. He fell to the ground, breaking a leg, spraining a wrist, and bruising his chest. The pole was a native cedar and was weathered and checked.

(3) A lineman's gaff struck a knot in a chestnut pole as he was climbing, and he cut out and fell 12 ft. He injured his right leg so severely that he may be permanently crippled.

These are typical instances in which the condition of poles contributes to accidents.

c. *Prevention.*   Prevention of accidents due to condition of poles can be aided in several ways. For example, poles can be purchased to meet better specifications which will help to eliminate some of the defects. A good inspection schedule will help to eliminate other causes. Poles that are to be climbed frequently can perhaps be equipped with steps. Workers themselves should become alert to these hazards and be continually on the lookout for the conditions on the poles which they climb which might contribute to cutouts and injuries.

2. *Conditions on poles*

a. *Specific causes*

(1) Conduit or cable on pole

(2) Strain plates

(3) Nails and tacks

(4) Telephone wires and brackets

(5) Ground wires

(6) Ground wire molding

(7) Signs and posters

(8) Metal pole numbers

(9) Dating nails

(10) Streetlight controls

(11) House knob screws

(12) Insect nests

(13) Metering equipment on pole

(14) Guy hooks, pole plates, and lag screws no longer in use
(15) Tree limbs against the pole
(16) Inadequate climbing space
(17) Telephone company attachments
(18) Excessive number of service drops
(19) Telephone terminal boxes
(20) Vines on poles
(21) Fences attached to poles
(22) Clotheslines and radio aerials attached to poles
(23) Loose and bent pole steps

b. *Examples of accidents*
(1) A lineman climbed a pole to a height of 14 ft, where one gaff cut out and he fell. His right heel struck the curb and was fractured. He also suffered a fracture of the right leg. The pole was excessively gaff-marked, and no steps were provided. Climbing space was restricted by a streetlight riser cable conduit running up the pole and a government mailbox mounted on the lower part of the pole.
(2) A new cedar pole had been set and supply attachments transferred to it. The old chestnut pole was lashed against the new pole pending transfer of the telephone attachments. As the lineman descended the new pole, one of his climber gaffs struck a strain plate and he cut out and fell 12 ft to the ground. He landed on his right side and rolled over on his back. His back was severely injured by the impact of a hammer he carried in his belt.

c. *Prevention.* Preventing cutouts and falls from these causes require the attention of both management and workers. Many of these causes can be removed, or at least minimized, by planning installations and setting construction standards with due regard for the men who must climb the pole, sometimes under very adverse conditions. When a pole is called a *clean pole* by men who climb, it is inferred that hardware, fixtures, and conductors are installed with an eye to adequate and safe climbing space. However, when unsafe conditions are found, they should be properly reported, and extra precautions taken if the pole must be climbed.

3. These cases illustrate how falls can occur because of unsafe conditions of the workman's clothing.
   a. *Specific causes*
   (1) Greasy clothing or gloves
   (2) Worn or loose heels on shoes
   (3) Shoes with nailed soles or heels
   (4) Low-cut shoes
   (5) Loose or ragged clothes
   (6) Wrong-size gloves
   (7) Gloves with holes or torn places
   (8) Soft-soled shoes
   (9) Shoes with poor arch or ankle supports
   (10) Buckles on overalls
   (11) Loose soles on shoes
   (12) High-heel boots

b. *Examples of accidents*
    (1) A lineman climbing a pole cut out and fell about an 18 ft height. The pole was in good condition, but his gloves were greasy. He stated that he lost his hold while he was climbing.
    (2) A lineman was attempting to belt off at his working position. As he reached around the pole to grasp his safety belt, his left gaff cut out and he fell off the pole. Examination revealed that the whole heel had come off the left shoe.

c. *Prevention.* Elimination of cutouts and other accidents due to unsafe conditions of clothing is largely a problem of the worker himself. Nevertheless, management and supervisors can be alerted to recognize such hazards and insist that their workers wear suitable and safe clothing. Many individuals will appreciate having their attention called to any unsafe condition of their work clothing. In some cases, conversely, it may be necessary for supervisors to take stronger action.

4. *Climbing practices*
a. *Specific causes*
    (1) Climbing or descending too fast
    (2) Using low side of pole
    (3) Fatigue
    (4) Gaffs not stuck in pole hard enough
    (5) Climbing too close to pole
    (6) Climbing through or past unprotected conductors
    (7) Not watching placement of gaffs
    (8) Safety belt too loose
    (9) Aiming gaffs at pole at improper angle
    (10) Using too-long steps in climbing or descending
    (11) Not inspecting pole before climbing
    (12) Inattention while ascending or descending
    (13) Improper balancing of body weight on gaffs
    (14) Belting off to pole at wrong position
    (15) Physically unfit for climbing
    (16) Horseplay
    (17) Climbing too close to ropes and hand lines
    (18) Catching material thrown from ground (while working on pole)
    (19) Placing climber gaffs on ground wires
    (20) Failure to get a good handhold
    (21) Holding to braces, etc.
    (22) Climbing in a mechanical manner
    (23) Sliding climber gaffs when climbing (sneaking)
    (24) Showing off (fancy dancing)
b. *Examples of accidents*
    (1) A lineman was climbing a 40-ft cedar pole which was stepped nearly all the way up. He said he was climbing as was customary on a stepped pole; that is, he used the steps for handholds and placed his gaffs between the pole steps. He guessed he cut out at a time when he was reaching for a new handhold with his right hand and fell 15 ft to the ground. In trying to break his fall, he extended both hands. He fractured the radius bone in

both arms near the wrist and suffered minor abrasions on the face. The pole, climbers, gaffs, and straps were in safe condition. This lineman was known as a "fast climber."

(2) A lineman climbed a transformer pole which was raked or leaning. He climbed the low side of the pole. On completing his work, he started to climb down the low side of the pole. About 15 ft from the ground he cut out and fell, receiving a broken right leg, right foot, hip, back, and left foot, and other injuries to his left arm, wrist, and knee.

(3) A lineman had been working for 2 weeks getting transformer name-plate data. He kept a record of the number of poles he climbed each day. The accident occurred late in the afternoon of a very hot day. As he was descending a pole, his right gaff cut out and he fell back 18 ft, fracturing his right leg and permanently disabling his back. He stated that he had climbed more poles that day than ever accomplished by an employee in a single day and that he was just about "all in" before he cut out.

c. *Prevention*

(1) Experienced veteran climbers, as well as new apprentices, need to be reminded from time to time about the pitfalls involved in these unsafe climbing practices.

(2) Foremen should watch the climbing practices of their men and aid them in correcting any unsafe practices that are observed before accidents occur.

(3) Elimination of unsafe climbing practices is based on sound, thorough training of workers in the safe methods of conducting themselves while working on poles.

5. *Climbers.* The importance of the climbers themselves in the prevention of cutout accidents cannot be emphasized too strongly.

a. *Specific causes*

(1) "Duckbill" gaffs
(2) Loose gaffs
(3) Gaffs too short
(4) Dull gaffs
(5) Gaffs improperly sharpened
(6) Straps too tight
(7) Straps too loose
(8) Broken straps
(9) Bent or broken gaffs
(10) Wrong-size climbers
(11) Foreign objects under gaffs
(12) Needlepoint gaffs
(13) Faulty leg irons

b. *Examples of accidents*

(1) A lineman started to climb a pole to attach a guy. About 15 ft above the ground he cut out and fell, breaking his left hip and arm. The pole appeared to be in reasonably good condition. The man's left gaff had

been sharpened in a duckbill shape, and wood splinters were found jammed under the gaff.

(2) A lineman rearranging dead-ends on a pole unfastened his safety belt to swing around to the other side of the pole. When he had almost reached his new working position, his right climber gaff apparently cut out and he fell. He struck the ground on his neck, shoulders, and back, suffering a broken neck and left shoulder blade, fractured ribs, punctured lungs, internal hemorrhages, and many other bruises. He died a few minutes after the accident. The gaff of one climber was found to be loose; otherwise his equipment was in good condition.

(3) As a lineman was preparing to descend to a lower position before removing the line hose from the conductors, he unfastened his safety strap. He then cut out and fell 30 ft to the ground, landing on his hands and arms. Both wrists, both arms, right leg, and lower leg were broken, and he suffered multiple cuts and bruises. He will be disfigured for life and partially disabled. Inspection of his climbers after the accident revealed that the gaffs were very short.

c. *Prevention.* Prevention of accidents due to climber defects can be accomplished with the proper effort.

(1) First. Rigid and frequent inspections of climbers, leg straps, leg iron, etc., with special attention being given to the gaffs, the shape and sharpness of the gaff points, the proper angle of the gaff with the leg iron, etc.

(2) Second. An acceptable gaff gauge should be standardized. It should be simple and easy to use. Minimum lengths of gaffs should be strictly enforced. The smallest deviation below the minimum-length requirement should not be permitted. Gaff shape and size should be similarly dealt with.

(3) Third. There should be a standard procedure for sharpening gaffs. There is universal agreement that gaffs should never be sharpened on ordinary dry grinding wheels, as this destroys the proper heat treatment of the metal. Only the proper type of file should be used. A professional job of precision sharpening should be the goal. Gaffs should always be honed after sharpening to remove file marks which may initiate cracks in the gaff steel and to improve the cutting edges of the point for better penetration of poles and ease in climbing.

(4) Fourth. Workers who use climbing equipment must be thoroughly trained in the fundamentals for proper care and use of the equipment.

(5) Fifth. Workers who use climbing equipment should accept responsibility for inspecting, sharpening, and shaping climbing equipment—and, if necessary, rejecting climbing equipment that is unsafe without waiting for supervisors or safety department personnel to make inspections.

Climbing Poles

Wood Pole Inspection and Testing before Climbing

Precautions must be taken before linemen climb wood poles. All poles should be visually inspected before the lineman starts to climb. The lineman must be particularly aware of the condition of the wood. An unusual angle or buckling at the ground line or along the pole may show that the pole is rotted or broken. Horizontal cracks across the grain of the wood pole may make the pole weak. Vertical cracks seldom weaken the pole. Vertical cracks should be avoided when inserting the climber gaffs into the wood. The lineman should use a hammer to check the soundness of the pole before starting to climb. Pounding the pole with a hammer may show that the pole is shell-rotted or hollow. A solid-wood sound should be heard when the pole is hit with a hammer. A hollow or drum-like sound indicates the pole has internal decay. The surface of the wood pole should not be penetrated by the hammer blows. A shell-rotted or decayed outside layer of the wood pole may cause the lineman's climber gaffs to cut out during climbing. A pole that is shell-rotted or decayed should not be climbed by the lineman using gaffs. Several knots at one location or a single large knot may signify a weak point in the pole. Suspicious poles should be probed with a single bore hole looking for a decayed cross-sectional area. If suspicion remains, a ground-line excavation of 18 inches is advised.

The pole should be set to the proper depth, and the soil must be in condition to support the pole before it is climbed. Soft, wet, or loose soil may permit the pole to shift position, especially if the line tensions change while the lineman is on the pole.

Woodpecker holes or burn spots on the pole may indicate that the pole has been weakened. The lineman should secure the pole before climbing, if inspection reveals that the pole may be weak or if the conductors are to be removed, changing the strain on the pole. The pole can be secured with the line truck boom. A pole may be lashed to a new pole when it will be replaced. Ropes, or guy wires, secured to stationary objects in the proper direction may provide necessary support to permit the lineman to work on the pole. The lineman should use a bucket truck to do the work on a pole that is shell-rotted or decayed, or if the inspection determines that the pole is unstable. If the pole is in an area inaccessible to a bucket truck, then a new pole should be set next to the defective pole and the lineman should climb the new pole and lash the new pole to the defective pole. The lineman should do all his work while he is supported by the new pole. The defective pole should be removed from the line and disposed of.

Wood Pole Fall Restriction Device

The wood pole fall restriction device has been developed for linemen to use while climbing poles (Fig. 44.17). The restriction devices meet or exceed ANSI Z359 *Fall Protection* series of guidelines. The restriction device is manufactured from a resin that provides dielectric protection and makes it easier for the lineman to place himself in the proper position to do the work safely. The restriction device does not damage poles. If the lineman's gaff should cut out while he is climbing or working on a pole, the device will prevent the lineman from falling. The climbing device can be used to lower an injured lineman to the ground safely, simplifying pole-top rescue procedures.

The industry is actively engaged in preventing pole climbing injuries. In addition to the wood pole fall restriction device, additional wood pole fall protection systems for linemen to use while climbing poles are now available. The wood pole fall restriction systems vary in design (Figs. 44.18 through 44.20). One fall restriction system utilizes a three slot dee ring, a threaded removable cleat, a serrated roto-snap carabiner, a cross-over strap, and the traditional safety strap. In the event the lineman's gaffs cut out

FIGURE 44.17   Lineman has climbed a pole, using a wood pole fall restriction device.

FIGURE 44.18   An example of a wood pole fall restriction device. (*Courtesy Buckingham Mfg. Co.*)

FIGURE 44.19   A second example of a wood pole fall restriction device. (*Courtesy Bashlin Industries.*)

FIGURE 44.20   A third example of a wood pole fall restriction device. (*Courtesy J.E. Lortie CO. LTD [Jelco].*)

while climbing the pole, the cross-over strap secures itself to the pole and prevents a fall. A lineman climbing the pole utilizing a wood pole fall restriction device is shown in Figs. 44.21 through 44.24. The lineman is not inhibited by the fall protection system as he climbs the pole.

FIGURE 44.21   The lineman apprentice trainee displays the wood pole fall restriction device attached to his body belt. (*Courtesy MidAmerican Energy Co.*)

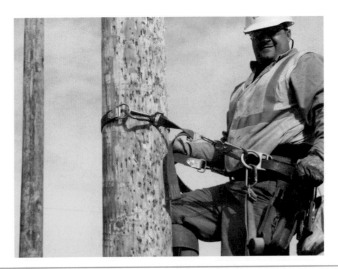

FIGURE 44.22   The lineman apprentice trainee at rest prior to climbing the pole. (*Courtesy MidAmerican Energy Co.*)

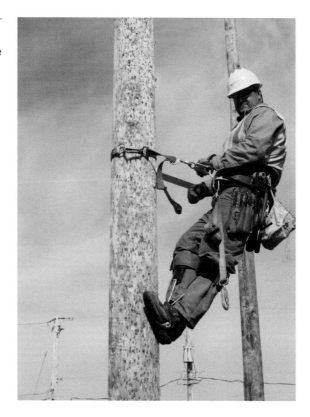

FIGURE 44.23   The lineman apprentice trainee stationary with the wood pole fall restriction device engaged. (*Courtesy MidAmerican Energy Co.*)

FIGURE 44.24   The lineman apprentice trainee suspended from the pole with the wood pole fall restriction device. (*Courtesy MidAmerican Energy Co.*)

Safe Climbing

The need for linemen to climb poles on an everyday basis remains. The fundamentals of climbing incorporating the utilization of the safety strap in conjunction with fall arrest safety devices or other fall protection systems to safely climb the pole is the norm.

Protective Line and Personal Protective Equipment

L inemen and cablemen must use protective line equipment and personal protective equipment (PPE) to safely complete work on energized equipment. The absolute minimum PPE safety rule requirements are defined in Occupational Safety and Health Act (OSHA) rules 1910.132, 1926.950, and 1910.269 (Chap. 46), and the National Electrical Safety Code (NESC). These rules and additional related PPE safety rules enforced by companies are based upon work procedures developed from years of experience and corrective measures adopted from the investigation of serious electrical accidents that all require the use of proper protective equipment to cover energized conductors and equipment as well as grounded surfaces in the work area. Proper procedures, the right tools, and correctly selected and applied protective equipment are all necessary to complete energized electric distribution work safely and efficiently.

Each work project and especially those projects requiring linemen and cablemen to handle energized electrical conductors should be planned in detail. The plan should be developed to ensure the safety of the workers and the reliability of electric service. Hazards to the linemen, the cablemen, and the public should be identified as a part of the planning process. As work procedures are developed, they should include safeguards for all hazards identified. Prior to starting the work, a tailgate conference should be held to communicate the plan and work procedures to all crew members. A tailgate conference is pictured in Fig. 45.1. When the work is initiated, the plan must be followed without shortcuts to ensure safe and reliable completion of the assignment.

Plan the work and work the plan for safety, efficiency, and reliability.

The safest way of completing electric distribution work is to deenergize, isolate, test, and ground the facilities to be worked on. Figure 45.2 illustrates lineman on the pole to be worked. The lineman must correctly identify the circuit to be grounded and obtain from the proper authority assurances that the circuit has been deenergized and isolated. Hold-off tags must be applied to the switches isolating the circuit, and these must be identified with the name of the person in charge of the work. A high-voltage tester mounted on hot-line tools can be used to verify that the circuit is deenergized. The grounding conductors are connected first to a low-resistance ground, usually the neutral conductor, and then to the deenergized phase conductors by the lineman with

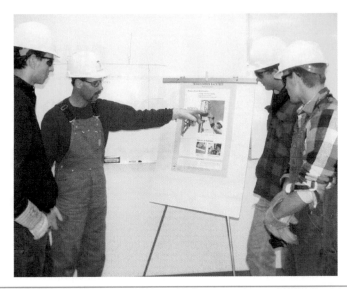

FIGURE 45.1   Supervisor is reviewing work plan with linemen prior to starting the job. The tailgate conference is being held in a conference room prior to driving to the worksite. (*Courtesy Northwest Lineman College.*)

FIGURE 45.2   The linemen are on the pole confirming work order safety requirements prior to proceeding with the job.

FIGURE 45.3

FIGURE 45.3   Linemen are engaged in energized circuit work. The linemen utilize the required PPE and protective line and equipment cover-up for the task required. (*Courtesy A. B. Chance Co.*)

a hot-line tool. Once the work is completed and all personnel are clear of the circuit, the grounds are removed in the reverse sequence.

The public's dependence on continuous electric service to maintain their health, welfare, and safety often makes it impossible to deenergize circuits to complete modifications and perform maintenance. Proper application of the protective equipment is necessary for the lineman to safeguard the work area, that is, to prevent electric shock or flash burn injuries. Correct application of the protective equipment described is essential for injury prevention and the maintenance of reliable electric service. The linemen pictured in Fig. 45.3 are properly protected for the work in progress. The linemen wear hard hats, safety glasses, body belts with safety straps attached to the work platform, rubber gloves with protectors, and rubber sleeves. The work is being performed while the men are standing on an insulated platform. Low primary voltages may be worked directly from the pole by linemen. The higher primary voltages must be worked from an insulated platform attached to the pole or an insulated aerial device. Work areas are guarded by conductor covers, insulated covers, crossarm covers, rubber blankets, and dead-end covers. Energized conductors are elevated and supported clear of the crossarm with hot-line tools.

Safety Rules

The basis for an investor-owned utility company's safety rules are OSHA 29 CFR 1926 Subpart V, Power Transmission; OSHA Federal Register standard numbers 1910.132 and 1910.269; and the National Electrical Safety Code (NESC). Rule 1926.950 is applicable to the construction of electric transmission and distribution lines and equipment. Subpart V was issued in 1970. Rule 1910.132 is applicable whenever it is necessary by reasons of specified hazards. Rule 1910.269 is applicable to the operation and maintenance of electric power generation, control, transformation, transmission, and distribution lines and equipment. The NESC's purpose is for the practical safeguarding of persons during the installation, operation, or maintenance of electric supply and communication lines and associated equipment.

The rules are not intended to be of a conflicting nature. OSHA Rule 1910.269 fills the gap not completely covered by OSHA Rule 1926 or the NESC (latest revision). OSHA Rule 1910.269 is provided in its entirety in Chap. 46.

Protective Line Equipment

Rubber Gloves

The most important article of protection for a lineman or a cableman is a good pair of rubber gloves with the proper dielectric strength for the voltage of the circuit to be worked on. Leather protector gloves must always be worn over the rubber gloves to prevent physical damage to the rubber while work is being performed. When the rubber gloves are not in use, they should be stored in a canvas bag to protect them from mechanical damage or deterioration from ozone generated by sun rays. Rubber gloves should always be given an air test by the lineman or cableman each day before the work is started or if the workman encounters an object that may have damaged the rubber gloves. The lineman pictured in Fig. 45.4 is properly applying an air test to his rubber gloves before putting them on. ASTM F1236 Standard Guide for Visual Inspection of Electrical Protective Rubber Products presents inspection methods, techniques, descriptions, and photographs.

Six classes of rubber gloves are manufactured, as shown in Table 45.1. Standard lengths for Class 00 gloves are 280 mm (11 in.), and 360 mm (14 in.). Standard lengths for Class 0 gloves are 280 mm (11 in), 360 mm (14 in), 410 mm (16 in), and 460 mm (18 in). Standard lengths for Classes 1, 2, and 3 gloves are 360 mm (14 in), 410 mm (16 in), and 460 mm (18 in). Standard lengths for Class 4 gloves are 410 mm (16 in) and 460 mm (18 in). The straight cuff is the default style. The bell cuff and the contour cuff glove options are also available.

FIGURE 45.4 Lineman giving his rubber gloves an air test before starting to work to be sure they do not have a hole in them. (*Courtesy MidAmerican Energy Co.*)

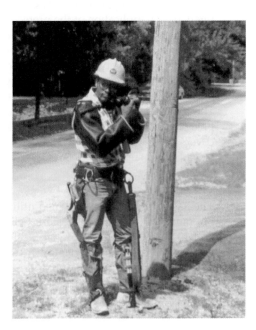

Class of Glove	AC Proof-Test Voltage, rms, V	Maximum AC Use Voltage, ac rms, V	DC Proof-Test Voltage, Average, V	Maximum DC Use Voltage, Average, V
00	2,500	500	10,000	750
0	5,000	1,000	20,000	1,500
1	10,000	7,500	40,000	11,250
2	20,000	17,000	50,000	25,500
3	30,000	26,500	60,000	39,750
4	40,000	36,000	70,000	54,000

TABLE 45.1   Proof-Test/Use Voltage Relationship as Specified by ASTM D120

American National Standards Institute standard ANSI/ASTM D 120, *Rubber Insulating Gloves,* covers manufacturing and testing of rubber insulating gloves for protection of workers from electric shock.

The proof-test voltage of the rubber gloves (Table 45.1) should not be construed to mean the safe voltage on which the glove can be used.

The maximum voltage on which gloves can safely be used depends on many factors including the care exercised in their use; the care followed in handling, storing, and inspecting the gloves in the field; the routine established for periodic laboratory inspection and test; the quality and thickness of the rubber; the design of the gloves; and such other factors as age, usage, or weather conditions.

Inasmuch as gloves are used for personal protection and a serious personal injury may result if they fail while in use, an adequate factor of safety should be provided between the maximum voltage on which they are permitted to be used and the voltage at which they are tested.

It is common practice for the rubber protective equipment manufacturer to prepare complete instructions and regulations to govern in detail the correct and safe use of such equipment. These should include provision for proper-fitting leather protector gloves to protect rubber gloves against mechanical injury and to reduce ozone cutting. Rubber insulating gloves and cover-up blankets should be thoroughly cleaned, inspected, and tested by competent personnel regularly (Figs. 45.4, 45.5, and 45.6).

A procedure for accomplishing this work is as follows:

1. Rubber labels showing the glove size, the glove number, and the test date are cemented to each rubber glove. This label shall be the color specified for each voltage class: Class 00—beige, Class 0—red, Class 1—white, Class 2—yellow, Class 3—green, and Class 4—orange.

2. The rubber gloves are washed in a washing machine, using a detergent to remove all dirt, prior to testing.

3. The rubber gloves are inflated by utilizing glove inflator equipment, checked for air leaks, and inspected. The inspection includes checking the surface of the rubber gloves for defects such as cuts, scratches, blisters, or embedded foreign material. The cuff area of the inside of the gloves is visually inspected after the glove is removed from the inflator. All rubber gloves having surface defects are rejected or repaired.

FIGURE 45.5 Lineman visually checking his inflated rubber gloves for air leaks to detect pin holes.

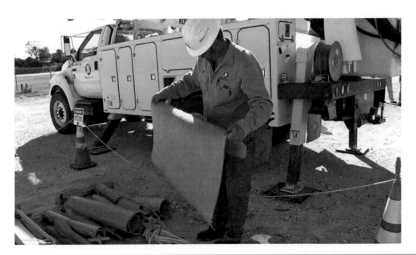

FIGURE 45.6 Rubber insulating blanket being given a thorough inspection by the lineman.

4. After the inflated inspection, the rubber gloves are stored for 24 hours to permit the rubber to relax, thus minimizing the possibility of corona damage while the electrical tests are completed (Fig. 45.7).

5. The rubber gloves are supported in a testing tank by means of plastic clothespins on the supporting rack. A ground electrode is placed inside each glove.

6. The rubber gloves to be tested and the testing tank are filled with water. The height of the water in the glove and in the testing tank depends upon the voltage test to be applied (Fig. 45.8).

Class 00 gloves are tested at 2,500-volt ac or 10,000-volt dc. The flashover clearances at the top edge of the cuff of the gloves during the proof test must be 1½ inch for both the ac and dc test.

FIGURE 45.7 Air-drying and storage rack for rubber gloves. Note rubber label cemented on cuff of glove. (*Courtesy MidAmerican Energy Co.*)

FIGURE 45.8 Rubber gloves in test tank, filled with water to proper level. (*Courtesy MidAmerican Energy Co.*)

Class 0 gloves are tested at 5,000-volt ac or 20,000-volt dc. The flashover clearances at the top edge of the cuff of the gloves during the proof test must be 1½ inch for both the ac and dc test.

Class 1 gloves are tested at 10,000-volt ac or 40,000-volt dc. The flashover clearances at the top edge of the cuff of the gloves during the proof test must be 1½ inch for the ac test and 2 inches for the dc test.

Class 2 gloves are tested at 20,000-volt ac or 50,000-volt dc. The flashover clearances at the top edge of the cuff of the gloves during the proof test must be 2½ inches for the ac test and 3 inches for the dc test.

Class 3 gloves are tested at 30,000-volt ac or 60,000-volt dc. The flashover clearances at the top edge of the cuff of the gloves during the proof test must be 3½ inches for the ac test and 4 inches for the dc test.

Class 4 gloves are tested at 40,000-volt ac or 70,000-volt dc. The flashover clearances at the top edge of the cuff of the gloves during the proof test must be 5 inches for the ac test and 6 inches for the dc test.

7. Sphere gap setting for the test voltage to be applied (Table 45.2). Sphere gap settings will change with variations of temperature and humidity.

Test Voltage ac, Volts	Sphere Gap Setting, mm
2,500	1.25
5,000	2.5
10,000	5.0
20,000	10.0
30,000	15.0
40,000	22.0

TABLE 45.2 Sphere Gap Settings

8. Connect the high-voltage lead of the testing transformer to the metal tank portion of the rubber-glove-testing equipment.

9. Connect the ground electrode in each rubber glove to its associated ground lead.

10. Evacuate all personnel from the enclosed testing area.

11. Close the power-supply-isolating switch to the testing equipment.

12. By means of the remote controls, gradually raise the voltage output of the testing transformer from zero to the proper level at a rate of approximately 1,000 volts/sec. This voltage should be applied to the rubber gloves for a period of 3 minutes.

13. After the high voltage has been applied to the rubber gloves for 3 min, switch the milliammeter to read the leakage current in each ground electrode utilized for each rubber glove. Table 45.3 lists the maximum leakage current for the six classes of rubber gloves.

14. The output voltage of the test transformer is gradually reduced to zero at a rate of approximately 1,000 volts/sec.

15. All rubber gloves which failed during the test or had a current leakage greater than the maximum permissible value are rejected. In some instances, the gloves can be repaired and used if they successfully pass a retest.

16. All rubber gloves which pass the high-voltage test are dried on a drying rack and powdered, and a tag indicating the leakage current recorded for each glove is placed in each glove. The gloves are then returned to the linemen or cablemen for use. In-service care of insulating gloves is specified in ANSI/ASTM F 496-77 standard.

Automatic equipment available can be used to test protective equipment. The test equipment is capable of testing insulated rubber gloves, rubber sleeves, rubber boots, and hard hats (Fig. 45.9). The automatic test equipment is programmed by the operator for the tests to be completed. The tests can be repeated without reprogramming the equipment.

Class of Glove	Proof-Test Voltage, rms, V	Minimum Breakdown Voltage, rms, V	Maximum Leakage Current, Milliamps							
			280-mm (11 inch) Glove		360-mm (14 inch) Glove		410-mm (16 inch) Glove		460-mm (18 inch) Glove	
			50Hz	60Hz	50Hz	60Hz	50Hz	60Hz	50Hz	60Hz
00	2,500	4,000	6.7	8	10	12	B	B	B	B
0	5,000	6,000	6.7	8	10	12	11.7	14	13.3	16
1	10,000	20,000	—	—	11.7	14	13.3	16	15	18
2	20,000	30,000	—	—	13.3	16	15	18	16.7	20
3	30,000	40,000	—	—	15	18	16.7	20	18.3	22
4	40,000	50,000	—	—	—	—	18.3	22	20	24

[A]Proof test current shall be measured to an accuracy of ±1 mA
[B]Not applicable

TABLE 45.3   AC Voltage Requirement Proof Test Currents[A] as Specified by ASTM D120

FIGURE **45.9** Certified technician is removing rubber gloves from automatic testing equipment after tests were completed. (*Courtesy MidAmerican Energy Co.*)

Rubber Sleeves

Rubber sleeves should be worn with rubber gloves to protect the arms and shoulders of the lineman, while he is working on high-voltage distribution circuits, from electrical contacts on the arms or shoulders. Figures 45.3 to 45.5 illustrate the proper use and testing of rubber sleeves by linemen. Rubber insulating sleeves must be treated with care and inspected regularly by the linemen in a manner similar to that described for rubber insulating gloves.

The rubber insulating sleeves should be thoroughly cleaned, inspected, and tested by competent personnel regularly. The inspecting and testing procedures are similar to those described for rubber insulating gloves (Fig. 45.10).

In-service care of insulating sleeves is specified in ANSI/ASTM F 496 standard.

Rubber Insulating Line Hose

Primary distribution conductors can be covered with rubber insulating line hose to protect the lineman from an accidental electrical contact. The line hoses are manufactured in various lengths, with inside-diameter measurements that vary from 1 inch to 1.5 inches and with different test voltage withstand values. The lineman should be sure that the voltage rating of the line hose provides an ample safety factor for the voltage applied to the conductors to be covered.

All line hoses should be cleaned and inspected regularly. A hand crank wringer can be used to spread the line hose to clean and inspect it for cuts or corona damage.

FIGURE **45.10** Rubber insulating sleeve being inspected while inflated after cleaning by washing. (*Courtesy MidAmerican Energy Co.*)

The rubber insulating line hose should be tested for electric withstand in accordance with the specifications at scheduled intervals.

In-service care of insulating line hose and covers is specified in ANSI/ASTM F 478 standard.

Rubber Insulating Insulator Hoods

Pin-type or post-type distribution primary insulators can be covered by hoods. The insulator hood properly installed will overlap the line hose, providing the lineman with complete shielding from the energized conductors.

Insulator hoods, like all other rubber insulating protective equipment, must be treated with care, kept clean, and inspected at regular intervals. Canvas bags of the proper size attached to a hand line should be used to raise and lower the protective equipment when it is to be installed and removed. The lineman pictured in Fig. 45.11 has in the process of installing the handline.

Conductor Covers

A conductor cover, fabricated from high-dielectric polyethylene, clips on and covers conductors up to 2 inches in diameter. A positive air gap is maintained by a swinging latch that can be loosened only by a one-quarter turn with a clamp stick (Fig. 45.12).

FIGURE 45.11   The lineman is in the process of installing the handline.

FIGURE 45.12 Conductor cover fabricated from high-dielectric polyethylene rated 26,000 volts. (*Courtesy A. B. Chance Co.*)

Insulator Covers

Insulator covers are fabricated from high-dielectric polyethylene and are designed to be used in conjunction with two conductor covers. The insulator cover fits over the insulator and locks with a conductor cover on each end. A polypropylene rope swings under the crossarm and hooks with a clamp stick, thus preventing the insulator cover from being removed upward by bumping or wind gusts (Fig. 45.13).

Crossarm Covers

High-dielectric-strength polyethylene crossarm covers are used to prevent tie wires from contacting the crossarm when conductors adjacent to insulators are being tied or untied. It is designed for single- or double-arm construction, with slots provided for the double-arm bolts. Flanges above the slots shield the ends of the double-arm bolts.

Pole Covers

Polyethylene-constructed pole covers are designed to insulate the pole in the area adjacent to high-voltage conductors. The pole covers are available in various lengths. Positive-hold polypropylene rope handles are knotted through holes in the overlap area of the cover (Figs. 45.14 and 45.15).

FIGURE 45.13 Insulator cover fabricated from high-dielectric polyethylene rated 26,000 volts. (*Courtesy A. B. Chance Co.*)

FIGURE 45.14 Pole and conductor covers provide protection from ground to lineman working on insulated platforms. (*Courtesy A. B. Chance Co.*)

FIGURE 45.15 Pole, crossarms, and conductors not being worked on have been covered by the lineman. (*Courtesy A. B. Chance Co.*)

Rubber Insulating Blankets

Odd-shaped conductors and equipment can usually be covered best with rubber insulating blankets.

The rubber insulating blankets are stored in canvas rolls or metal canisters to protect them when they are not in use. The blankets can be held in place by ropes or clamps. In-service care of insulating blankets is specified in ANSI/ASTM F 479 standard.

Ground Mats

Ground mats are utilized to establish an equipotential zone for the lineman to stand on while engaged in construction and maintenance activities on energized equipment. Mats are utilized to eliminate step and touch potentials. The ground mat consists of high-ampacity tinned-copper-braid cable sewn in a grid pattern onto the vinyl/polyester fabric. Mats typically measure 5 ft by 5 ft or 10 ft by 10 ft.

Personal Protective Equipment

Safety Hat

Hard hats, or safety hats, are worn by linemen, cablemen, and groundmen to protect against the impact from falling or moving objects and against accidental electrical contact of the head and energized equipment, as well as to protect the worker from sun rays, cold, rain, sleet, and snow. The first combined impact-resisting and electrical-insulating hat was introduced in 1952. The hat was designed "to roll with the punch" by distributing the force of a blow over the entire head. This is accomplished by a suspension band which holds the hat about an inch away from the head and lets the hat work as a shock absorber (Fig. 45.16).

The hat is made of fiberglass, or plastic material, and has an insulating value of approximately 20,000 volts. New helmets are manufactured to withstand a test of 30,000 volts without failure. The actual voltage that the hat will sustain while being

FIGURE 45.16 Cutaway view of electrical worker's protective helmet showing suspension band which ensures proper crown clearance.

worn depends upon the cleanliness of the hat, weather conditions, the type of electrode contacted, and other variables. The wearing of safety hats by linemen and cablemen has greatly reduced electrical contacts.

Physical injuries to the head have been practically eliminated as a result of workers on the ground wearing protective helmets. The Occupational Safety and Health Act of 1970 and most companies' safety rules require linemen, cablemen, and groundmen to wear safety hats while performing physical work. Specifications for safety hats are found in ANSI Standard Z89.2, *Safety Requirements for Industrial Protective Helmets for Electrical Workers, Class B*.

Safety Glasses and Full-Face Shields

Safety eyewear is a requirement in most, if not all, construction and maintenance situations. The lenses protect against flying particles and UV rays.

Hearing Protection

Certain situations expose the lineman to very loud and sometimes deafening noise levels. In those situations, hearing protection in the way of ear plugs or earmuffs is needed. Hearing protection is also recommended while closing cutouts. Hearing protection is designed to mitigate the effects of noise to desirable levels of sound. Sound levels are measured in units of decibels.

High-Visibility Vests

Many projects involve working in or near roadways (Fig. 45.17). Highly visible shirts or vests alert drivers of the lineman's presence. Properly positioning and barricading the utility vehicle in or just off the traveled roadways is a necessity. Nearly 25,000 serious injuries occur in work zones each year. The importance of barricading worksites properly while working in these situations is critical to the safety of the public as well as the lineman. The lineman should be knowledgeable with traffic control requirements in

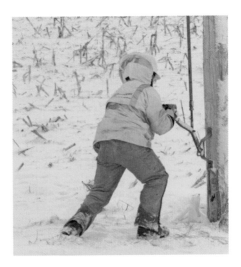

FIGURE 45.17   A lineman is involved in a switching situation in or near the roadway. The lineman is wearing a lime green high-visibility vest. (*Courtesy MidAmerican Energy Company.*)

the jurisdiction worked. Familiarity with rules contained in the National Manual on Uniform Traffic Control Devices (MUTCD) maintained by the Federal Highway Administration (FHWA) aids the lineman in his understanding of the concepts. Permits may be required as well depending upon the agency that controls the highway. These agencies include the Department of Transportation, the County Highway Department, or the City Public Works Department. Information required on the permit includes the nature and duration of the work, proposed positioning and barricading of vehicles, and a diagram of the traffic sign spacing for the work zone.

Clothing

The lineman's clothing requirements are becoming more defined based upon the job's requirements and work situations. Several guidelines pertaining to outerwear, under- wear, winter-wear, and rainwear have been developed. Consideration must be given as to a clothing material's flame resistance, arc resistance, and visibility. Standards written on the topic include ASTM F1506-2000, *Clothing for Electric Arc*, and ASTM F1891-99, *Specification for Arc Resistant Raingear*. Clothing manufacturers utilize fabrics that comply with standards and are comfortable, fashionable, and stylish. The clothing is subjected to rigorous testing including arc exposure and flame resistance. Care must be taken when cleaning in order to maintain the quality and protective properties of the clothing.

Footwear

Comfortable yet durable footwear is required (Fig. 43.18). Slip, trips, and falls are prevented when proper footwear is utilized. Depending upon ground conditions, specific shoes are desired. Spikes or cleats may be worn on snow and ice (Fig. 43.19). Slip-resistant soles ensure safe footing.

FIGURE 45.18 Insulated 10-inch dark brown pole climber lineman boot. (*Courtesy Hall Safety Equipment Corp.*)

FIGURE 45.19 Lineman wearing ice cleats over the lineman boot. (*Courtesy MidAmerican Energy Co.*)

FIGURE 45.20   A standard design shock-absorbing lanyard. The lanyard is 6 ft in length. The tubular nylon exterior may expand 42 inches maximum while the interior material stretches to reduce the arresting force to less than 900 lbs. (*Courtesy Buckingham Manufacturing Company, Inc.*)

Lanyards and Full-Body Harnesses

Falls are a leading cause of injury for the lineman. The industry has implemented solutions to minimize or eliminate the severity of fall-related injuries experienced. ANSI standards A10.14 and Z359.1, ASTM standard F887, and OSHA standards, in particular 1910.269, have been written addressing the topic in more detail. Prior to 1910.269, linemen were only required to utilize a lanyard when working from the bucket. After this standard was implemented, a majority of climbing situations dictated shock-absorbing lanyards (Fig. 45.20) designed for specific fall arrest forces, deceleration distances, and minimum breaking strengths. The full-body harness and associated connecting straps are now required in more situations (Figs. 45.21*a* and *b*). Safety systems are in place for fall restraint, fall arrest, work positioning, suspension, and retrieval. Shock-absorbing lanyards reduce the impact of a free fall from elevated work positions. The equipment is designed with worker's comfort and safety in mind. The equipment is lightweight

(a)   (b)

FIGURE 45.21   (a) The front view of an "H"-style full-body harness. (b) Rear view. The harness is made from nylon webbing and features interlocking leg strap buckles with leather chafes. The harness shown has a Velcro chest strap and a web dorsal loop. (*Courtesy Buckingham Manufacturing Company, Inc.*)

FIGURE 45.22 The linemen illustrated are working on a 13,200Y/7,620-volt energized distribution circuit, utilizing proper protective equipment and tools. The linemen are wearing rubber gloves, rubber sleeves, insulated hard hats, safety glasses, and body belts with safety strap secured. They are working on an insulated platform. Hot-line tools are in use to maintain proper clearances. The protective cover-up equipment installed includes conductor covers, crossarm covers, pole covers, dead-end insulator covers, and rubber blankets. (*Courtesy A. B. Chance Co.*)

and durable. Back and total body support are included. Unnecessary chafing is minimized. As with any piece of safety equipment, the harnesses, lanyards, body belts, and straps are to be thoroughly examined before use.

Summary

Protective line equipment and PPE are required while working on energized distribution conductors. Application of the protective equipment should be planned as with all other aspects of work. When the plan has been established, it should be followed without shortcuts. The safe way of completing the work is the most efficient and reliable method. The skilled lineman and cableman is an expert in the application and use of protective equipment. Figure 45.22 pictures two skilled linemen utilizing protective equipment to good advantage while working on energized distribution circuits.

CHAPTER 46

OSHA 29 CFR 1910.269

OSHA Regulations (Standards—29 CFR): Electric Power Generation, Transmission, and Distribution—1910.269

- Standard Number: 1910.269
- Standard Title: Electric Power Generation, Transmission, and Distribution
- Part Number: 1910
- Part Title: Occupational Safety and Health Standards
- Subpart Number: R
- Subpart Title: Special Industries
- Appendix: A, B, C, D, E, F, G

1910.269(a) *General.*

1910.269(a)(1) Application.

1910.269(a)(1)(i) This section covers the operation and maintenance of electric power generation, control, transformation, transmission, and distribution lines and equipment. These provisions apply to:

1910.269(a)(1)(i)(A) Power generation, transmission, and distribution installations, including related equipment for the purpose of communication or metering that are accessible only to qualified employees;

Note to paragraph (a)(1)(i)(A): The types of installations covered by this paragraph include the generation, transmission, and distribution installations of electric utilities, as well as equivalent installations of industrial establishments. Subpart S of this part covers supplementary electric generating equipment that is used to supply a workplace for emergency, standby, or similar purposes only. (See paragraph (a)(1)(i)(B) of this section.)

1910.269(a)(1)(i)(B) Other installations at an electric power generating station, as follows:

1910.269(a)(1)(i)(B)(1) Fuel and ash handling and processing installations, such as coal conveyors,

1910.269(a)(1)(i)(B)(2) Water and steam installations, such as penstocks, pipelines, and tanks, providing a source of energy for electric generators, and

1910.269(a)(1)(i)(B)(3) Chlorine and hydrogen systems;

1910.269(a)(1)(i)(C) Test sites where employees perform electrical testing involving temporary measurements associated with electric power generation, transmission, and distribution in laboratories, in the field, in substations, and on lines, as opposed to metering, relaying, and routine line work;

1910.269(a)(1)(i)(D) Work on, or directly associated with, the installations covered in paragraphs (a)(1)(i)(A) through (a)(1)(i)(C) of this section; and

1910.269(a)(1)(i)(E) Line-clearance tree trimming performed for the purpose of clearing space around electric power generation, transmission, or distribution lines or equipment and on behalf of an organization that operates, or that controls the operating procedures for, those lines or equipment, as follows:

1910.269(a)(1)(i)(E)(1) Entire § 1910.269 of this part, except paragraph (r)(1) of this section, applies to line-clearance tree-trimming covered by the introductory text to paragraph (a)(1)(i)(E) of the section when performed by qualified employees (those who are knowledgeable in the construction and operation of the electric power generation, transmission, or distribution equipment involved, along with the associated hazards).

1910.269(a)(1)(i)(E)(2) Paragraphs (a)(2), (a)(3), (b), (c), (g), (k), (p), and (r) of this section apply to line-clearance tree trimming covered by the introductory text to paragraph (a)(1)(i)(E) of this section when performed by line-clearance tree trimmers who are not qualified employees.

1910.269(a)(1)(ii) Notwithstanding paragraph (a)(1)(i) of this section, § 1910.269 of this part does not apply:

1910.269(a)(1)(ii)(A) To construction work, as defined in § 1910.12 of this part, except for line-clearance tree trimming and work involving electric power generation installations as specified in § 1926.950(a)(3) of this chapter; or

1910.269(a)(1)(ii)(B) To electrical installations, electrical safety-related work practices, or electrical maintenance considerations covered by Subpart S of this part.

Note 1 to paragraph (a)(1)(ii)(B): The Occupational Safety and Health Administration considers work practices conforming to §§ 1910.332 through 1910.335 as complying with the electrical safety-related work-practice requirements of § 1910.269 identified in Table 1 of Appendix A-2 to this section, provided that employers are performing the work on a generation or distribution installation meeting §§ 1910.303 through 1910.308. This table also identifies provisions in § 1910.269 that apply to work by qualified persons directly on, or associated with, installations of electric power generation, transmission, and distribution lines or equipment, regardless of compliance with §§ 1910.332 through 1910.335.

Note 2 to paragraph (a)(1)(ii)(B): The Occupational Safety and Health Administration considers work practices performed by qualified persons and conforming to § 1910.269 as complying with §§ 1910.333(c) and 1910.335.

1910.269(a)(1)(iii) This section applies in addition to all other applicable standards contained in this Part 1910. Employers covered under this section are not exempt from complying with other applicable provisions in Part 1910 by the operation of § 1910.5(c). Specific references in this section to other sections of Part 1910 are for emphasis only.

1910.269(a)(2) Training.

1910.269(a)(2)(i) All employees performing work covered by this section shall be trained as follows:

1910.269(a)(2)(i)(A) Each employee shall be trained in, and familiar with, the safety-related work practices, safety procedures, and other safety requirements in this section that pertain to his or her job assignments.

1910.269(a)(2)(i)(B) Each employee shall also be trained in and familiar with any other safety practices, including applicable emergency procedures (such as pole-top and manhole rescue), that are not specifically addressed by this section but that are related to his or her work and are necessary for his or her safety.

1910.269(a)(2)(i)(C) The degree of training shall be determined by the risk to the employee for the hazard involved.

1910.269(a)(2)(ii) Each qualified employee shall also be trained and competent in:

1910.269(a)(2)(ii)(A) The skills and techniques necessary to distinguish exposed live parts from other parts of electric equipment,

1910.269(a)(2)(ii)(B) The skills and techniques necessary to determine the nominal voltage of exposed live parts,

1910.269(a)(2)(ii)(C) The minimum approach distances specified in this section corresponding to the voltages to which the qualified employee will be exposed and the skills and techniques necessary to maintain those distances,

1910.269(a)(2)(ii)(D) The proper use of the special precautionary techniques, personal protective equipment, insulating and shielding materials, and insulated tools for working on or near exposed energized parts of electric equipment, and

1910.269(a)(2)(ii)(E) The recognition of electrical hazards to which the employee may be exposed and the skills and techniques necessary to control or avoid these hazards.

Note to paragraph (a)(2)(ii): For the purposes of this section, a person must have the training required by paragraph (a)(2)(ii) of this section to be considered a qualified person. **1910.269(a)(2)(iii)** Each line-clearance tree trimmer who is not a qualified employee shall also be trained and competent in:

1910.269(a)(2)(iii)(A) The skills and techniques necessary to distinguish exposed live parts from other parts of electric equipment,

1910.269(a)(2)(iii)(B) The skills and techniques necessary to determine the nominal voltage of exposed live parts, and

1910.269(a)(2)(iii)(C) The minimum approach distances specified in this section corresponding to the voltages to which the employee will be exposed and the skills and techniques necessary to maintain those distances.

1910.269(a)(2)(iv) The employer shall determine, through regular supervision and through inspections conducted on at least an annual basis, that each employee is complying with the safety-related work practices required by this section.

1910.269(a)(2)(v) An employee shall receive additional training (or retraining) under any of the following conditions:

1910.269(a)(2)(v)(A) If the supervision or annual inspections required by paragraph (a)(2)(iv) of this section indicate that the employee is not complying with the safety-related work practices required by this section, or

1910.269(a)(2)(v)(B) If new technology, new types of equipment, or changes in procedures necessitate the use of safety-related work practices that are different from those which the employee would normally use, or

1910.269(a)(2)(v)(C) If he or she must employ safety-related work practices that are not normally used during his or her regular job duties.

Note to paragraph (a)(2)(v)(C): The Occupational Safety and Health Administration considers tasks that are performed less often than once per year to necessitate retraining before the performance of the work practices involved.

1910.269(a)(2)(vi) The training required by paragraph (a)(2) of this section shall be of the classroom or on-the-job type.

1910.269(a)(2)(vii) The training shall establish employee proficiency in the work practices required by this section and shall introduce the procedures necessary for compliance with this section.

1910.269(a)(2)(viii) The employer shall ensure that each employee has demonstrated proficiency in the work practices involved before that employee is considered as having completed the training required by paragraph (a)(2) of this section.

Note 1 to paragraph (a)(2)(viii): Though they are not required by this paragraph, employment records that indicate that an employee has successfully completed the required training are one way of keeping track of when an employee has demonstrated proficiency.

Note 2 to paragraph (a)(2)(viii): For an employee with previous training, an employer may determine that that employee has demonstrated the proficiency required by this paragraph using the following process:

1910.269(a)(2)(viii)(1) Confirm that the employee has the training required by paragraph (a)(2) of this section,

1910.269(a)(2)(viii)(2) Use an examination or interview to make an initial determination that the employee understands the relevant safety-related work practices before he or she performs any work covered by this section, and

1910.269(a)(2)(viii)(3) Supervise the employee closely until that employee has demonstrated proficiency as required by this paragraph.

1910.269(a)(3) Information transfer.

1910.269(a)(3)(i) Before work begins, the host employer shall inform contract employers of:

1910.269(a)(3)(i)(A) The characteristics of the host employer's installation that are related to the safety of the work to be performed and are listed in paragraphs (a)(4)(i) through (a)(4)(v) of this section;

Note to paragraph (a)(3)(i)(A): This paragraph requires the host employer to obtain information listed in paragraphs (a)(4)(i) through (a)(4)(v) of this section if it does not have this information in existing records.

1910.269(a)(3)(i)(B) Conditions that are related to the safety of the work to be performed, that are listed in paragraphs (a)(4)(vi) through (a)(4)(viii) of this section, and that are known to the host employer;

Note to paragraph (a)(3)(i)(B): For the purposes of this paragraph, the host employer need only provide information to contract employers that the host employer can obtain from its existing records through the exercise of reasonable diligence. This paragraph does not require the host employer to make inspections of worksite conditions to obtain this information.

1910.269(a)(3)(i)(C) Information about the design and operation of the host employer's installation that the contract employer needs to make the assessments required by this section; and

Note to paragraph (a)(3)(i)(C): This paragraph requires the host employer to obtain information about the design and operation of its installation that contract employers need to make required assessments if it does not have this information in existing records.

1910.269(a)(3)(i)(D) Any other information about the design and operation of the host employer's installation that is known by the host employer, that the contract employer requests, and that is related to the protection of the contract employer's employees.

Note to paragraph (a)(3)(i)(D): For the purposes of this paragraph, the host employer need only provide information to contract employers that the host employer can obtain from its existing records through the exercise of reasonable diligence. This paragraph does not require the host employer to make inspections of worksite conditions to obtain this information.

1910.269(a)(3)(ii) Contract employers shall comply with the following requirements:

1910.269(a)(3)(ii)(A) The contract employer shall ensure that each of its employees is instructed in the hazardous conditions relevant to the employee's work that the contract employer is aware of as a result of information communicated to the contract employer by the host employer under paragraph (a)(3)(i) of this section.

1910.269(a)(3)(ii)(B) Before work begins, the contract employer shall advise the host employer of any unique hazardous conditions presented by the contract employer's work.

1910.269(a)(3)(ii)(C) The contract employer shall advise the host employer of any unanticipated hazardous conditions found during the contract employer's work that the host employer did not mention under paragraph (a)(3)(i) of this section. The contract employer shall provide this information to the host employer within 2 working days after discovering the hazardous condition.

1910.269(a)(3)(iii) The contract employer and the host employer shall coordinate their work rules and procedures so that each employee of the contract employer and the host employer is protected as required by this section.

1910.269(a)(4) Existing characteristics and conditions. Existing characteristics and conditions of electric lines and equipment that are related to the safety of the work to be performed shall be determined before work on or near the lines or equipment is started. Such characteristics and conditions include, but are not limited to:

1910.269(a)(4)(i) The nominal voltages of lines and equipment,

1910.269(a)(4)(ii) The maximum switching-transient voltages,

1910.269(a)(4)(iii) The presence of hazardous induced voltages,

1910.269(a)(4)(iv) The presence of protective grounds and equipment grounding conductors,

1910.269(a)(4)(v) The locations of circuits and equipment, including electric supply lines, communication lines, and fire-protective signaling circuits,

1910.269(a)(4)(vi) The condition of protective grounds and equipment grounding conductors,

1910.269(a)(4)(vii) The condition of poles, and

1910.269(a)(4)(viii) Environmental conditions relating to safety.

1910.269(b) *Medical services and first aid.* The employer shall provide medical services and first aid as required in § 1910.151. In addition to the requirements of § 1910.151, the following requirements also apply:

1910.269(b)(1) First-aid training. When employees are performing work on, or associated with, exposed lines or equipment energized at 50 volts or more, persons with first-aid training shall be available as follows:

1910.269(b)(1)(i) For field work involving two or more employees at a work location, at least two trained persons shall be available. However, for line-clearance tree trimming operations performed by line-clearance tree trimmers who are not qualified employees, only one trained person need be available if all new employees are trained in first aid within 3 months of their hiring dates.

1910.269(b)(1)(ii) For fixed work locations such as substations, the number of trained persons available shall be sufficient to ensure that each employee exposed to electric shock can be reached within 4 minutes by a trained person. However, where the existing number of employees is insufficient to meet this requirement (at a remote substation, for example), each employee at the work location shall be a trained employee.

1910.269(b)(2) First-aid supplies. First-aid supplies required by § 1910.151(b) shall be placed in weatherproof containers if the supplies could be exposed to the weather.

1910.269(b)(3) First-aid kits. The employer shall maintain each first-aid kit, shall ensure that it is readily available for use, and shall inspect it frequently enough to ensure that expended items are replaced. The employer also shall inspect each first aid kit at least once per year.

1910.269(c) *Job briefing.*

1910.269(c)(1) Before each job.

1910.269(c)(1)(i) In assigning an employee or a group of employees to perform a job, the employer shall provide the employee in charge of the job with all available information that relates to the determination of existing characteristics and conditions required by paragraph (a)(4) of this section.

1910.269(c)(1)(ii) The employer shall ensure that the employee in charge conducts a job briefing that meets paragraphs (c)(2), (c)(3), and (c)(4) of this section with the employees involved before they start each job.

1910.269(c)(2) Subjects to be covered. The briefing shall cover at least the following subjects: hazards associated with the job, work procedures involved, special precautions, energy-source controls, and personal protective equipment requirements.

1910.269(c)(3) Number of briefings.

1910.269(c)(3)(i) If the work or operations to be performed during the work day or shift are repetitive and similar, at least one job briefing shall be conducted before the start of the first job of each day or shift.

1910.269(c)(3)(ii) Additional job briefings shall be held if significant changes, which might affect the safety of the employees, occur during the course of the work.

1910.269(c)(4) Extent of briefing.

1910.269(c)(4)(i) A brief discussion is satisfactory if the work involved is routine and if the employees, by virtue of training and experience, can reasonably be expected to recognize and avoid the hazards involved in the job.

1910.269(c)(4)(ii) A more extensive discussion shall be conducted:

1910.269(c)(4)(ii)(A) If the work is complicated or particularly hazardous, or

1910.269(c)(4)(ii)(B) If the employee cannot be expected to recognize and avoid the hazards involved in the job.

Note to paragraph (c)(4): The briefing must address all the subjects listed in paragraph (c)(2) of this section.

1910.269(c)(5) Working alone. An employee working alone need not conduct a job briefing. However, the employer shall ensure that the tasks to be performed are planned as if a briefing were required.

1910.269(d) *Hazardous energy control (lockout/tagout) procedures.*

1910.269(d)(1) Application. The provisions of paragraph (d) of this section apply to the use of lockout/tagout procedures for the control of energy sources in installations for the purpose of electric power generation, including related equipment for communication or metering. Locking and tagging procedures for the deenergizing of electric energy sources which are used exclusively for purposes of transmission and distribution are addressed by paragraph (m) of this section.

Note to paragraph (d)(1): Installations in electric power generation facilities that are not an integral part of, or inextricably commingled with, power generation processes or equipment are covered under § 1910.147 and Subpart S of this part.

1910.269(d)(2) General.

1910.269(d)(2)(i) The employer shall establish a program consisting of energy control procedures, employee training, and periodic inspections to ensure that, before any employee performs any servicing or maintenance on a machine or equipment where the unexpected energizing, start up, or release of stored energy could occur and cause injury, the machine or equipment is isolated from the energy source and rendered inoperative.

1910.269(d)(2)(ii) The employer's energy control program under paragraph (d)(2) of this section shall meet the following requirements:

1910.269(d)(2)(ii)(A) If an energy isolating device is not capable of being locked out, the employer's program shall use a tagout system.

1910.269(d)(2)(ii)(B) If an energy isolating device is capable of being locked out, the employer's program shall use lockout, unless the employer can demonstrate that the use of a tagout system will provide full employee protection as follows:

1910.269(d)(2)(ii)(B)(1) When a tagout device is used on an energy isolating device which is capable of being locked out, the tagout device shall be attached at the same location that the lockout device would have been attached, and the employer shall demonstrate that the tagout program will provide a level of safety equivalent to that obtained by the use of a lockout program.

1910.269(d)(2)(ii)(B)(2) In demonstrating that a level of safety is achieved in the tagout program equivalent to the level of safety obtained by the use of a lockout program, the employer shall demonstrate full compliance with all tagout-related provisions of this standard together with such additional elements as are necessary to provide the equivalent safety available from the use of a lockout device. Additional means to be considered as part of the demonstration of full employee protection shall include the implementation of additional safety measures such as the removal of an isolating circuit element, blocking of a controlling switch, opening of an extra disconnecting device, or the removal of a valve handle to reduce the likelihood of inadvertent energizing.

1910.269(d)(2)(ii)(C) After November 1, 1994, whenever replacement or major repair, renovation, or modification of a machine or equipment is performed, and whenever new machines or equipment are installed, energy isolating devices for such machines or equipment shall be designed to accept a lockout device.

1910.269(d)(2)(iii) Procedures shall be developed, documented, and used for the control of potentially hazardous energy covered by paragraph (d) of this section.

1910.269(d)(2)(iv) The procedure shall clearly and specifically outline the scope, purpose, responsibility, authorization, rules, and techniques to be applied to the control of hazardous energy, and the measures to enforce compliance including, but not limited to, the following:

1910.269(d)(2)(iv)(A) A specific statement of the intended use of this procedure;

1910.269(d)(2)(iv)(B) Specific procedural steps for shutting down, isolating, blocking and securing machines or equipment to control hazardous energy;

1910.269(d)(2)(iv)(C) Specific procedural steps for the placement, removal, and transfer of lockout devices or tagout devices and the responsibility for them; and

1910.269(d)(2)(iv)(D) Specific requirements for testing a machine or equipment to determine and verify the effectiveness of lockout devices, tagout devices, and other energy control measures.

1910.269(d)(2)(v) The employer shall conduct a periodic inspection of the energy control procedure at least annually to ensure that the procedure and the provisions of paragraph (d) of this section are being followed.

1910.269(d)(2)(v)(A) The periodic inspection shall be performed by an authorized employee who is not using the energy control procedure being inspected.

1910.269(d)(2)(v)(B) The periodic inspection shall be designed to identify and correct any deviations or inadequacies.

1910.269(d)(2)(v)(C) If lockout is used for energy control, the periodic inspection shall include a review, between the inspector and each authorized employee, of that employee's responsibilities under the energy control procedure being inspected.

1910.269(d)(2)(v)(D) Where tagout is used for energy control, the periodic inspection shall include a review, between the inspector and each authorized and affected employee, of that employee's responsibilities under the energy control procedure being inspected, and the elements set forth in paragraph (d)(2)(vii) of this section.

1910.269(d)(2)(v)(E) The employer shall certify that the inspections required by paragraph (d)(2)(v) of this section have been accomplished. The certification shall identify the machine or equipment on which the energy control procedure was being used, the date of the inspection, the employees included in the inspection, and the person performing the inspection.

Note to paragraph (d)(2)(v)(E): If normal work schedule and operation records demonstrate adequate inspection activity and contain the required information, no additional certification is required.

1910.269(d)(2)(vi) The employer shall provide training to ensure that the purpose and function of the energy control program are understood by employees and that the knowledge and skills required for the safe application, usage, and removal of energy controls are acquired by employees. The training shall include the following:

1910.269(d)(2)(vi)(A) Each authorized employee shall receive training in the recognition of applicable hazardous energy sources, the type and magnitude of energy available in the workplace, and in the methods and means necessary for energy isolation and control.

1910.269(d)(2)(vi)(B) Each affected employee shall be instructed in the purpose and use of the energy control procedure.

1910.269(d)(2)(vi)(C) All other employees whose work operations are or may be in an area where energy control procedures may be used shall be instructed about the procedures and about the prohibition relating to attempts to restart or reenergize machines or equipment that are locked out or tagged out.

1910.269(d)(2)(vii) When tagout systems are used, employees shall also be trained in the following limitations of tags:

1910.269(d)(2)(vii)(A) Tags are essentially warning devices affixed to energy isolating devices and do not provide the physical restraint on those devices that is provided by a lock.

1910.269(d)(2)(vii)(B) When a tag is attached to an energy isolating means, it is not to be removed without authorization of the authorized person responsible for it, and it is never to be bypassed, ignored, or otherwise defeated.

1910.269(d)(2)(vii)(C) Tags must be legible and understandable by all authorized employees, affected employees, and all other employees whose work operations are or may be in the area, in order to be effective.

1910.269(d)(2)(vii)(D) Tags and their means of attachment must be made of materials which will withstand the environmental conditions encountered in the workplace.

1910.269(d)(2)(vii)(E) Tags may evoke a false sense of security, and their meaning needs to be understood as part of the overall energy control program.

1910.269(d)(2)(vii)(F) Tags must be securely attached to energy isolating devices so that they cannot be inadvertently or accidentally detached during use.

1910.269(d)(2)(viii) Retraining shall be provided by the employer as follows:

1910.269(d)(2)(viii)(A) Retraining shall be provided for all authorized and affected employees whenever there is a change in their job assignments, a change in machines, equipment, or processes that present a new hazard or whenever there is a change in the energy control procedures.

1910.269(d)(2)(viii)(B) Retraining shall also be conducted whenever a periodic inspection under paragraph (d)(2)(v) of this section reveals, or whenever the employer has reason to believe, that there are deviations from or inadequacies in an employee's knowledge or use of the energy control procedures.

1910.269(d)(2)(viii)(C) The retraining shall reestablish employee proficiency and shall introduce new or revised control methods and procedures, as necessary.

1910.269(d)(2)(ix) The employer shall certify that employee training has been accomplished and is being kept up to date. The certification shall contain each employee's name and dates of training.

1910.269(d)(3) Protective materials and hardware.

1910.269(d)(3)(i) Locks, tags, chains, wedges, key blocks, adapter pins, self-locking fasteners, or other hardware shall be provided by the employer for isolating, securing, or blocking of machines or equipment from energy sources.

1910.269(d)(3)(ii) Lockout devices and tagout devices shall be singularly identified; shall be the only devices used for controlling energy; may not be used for other purposes; and shall meet the following requirements:

1910.269(d)(3)(ii)(A) Lockout devices and tagout devices shall be capable of withstanding the environment to which they are exposed for the maximum period of time that exposure is expected.

1910.269(d)(3)(ii)(A)(1) Tagout devices shall be constructed and printed so that exposure to weather conditions or wet and damp locations will not cause the tag to deteriorate or the message on the tag to become illegible.

1910.269(d)(3)(ii)(A)(2) Tagout devices shall be so constructed as not to deteriorate when used in corrosive environments.

1910.269(d)(3)(ii)(B) Lockout devices and tagout devices shall be standardized within the facility in at least one of the following criteria: color, shape, size. Additionally, in the case of tagout devices, print and format shall be standardized.

1910.269(d)(3)(ii)(C) Lockout devices shall be substantial enough to prevent removal without the use of excessive force or unusual techniques, such as with the use of bolt cutters or metal cutting tools.

1910.269(d)(3)(ii)(D) Tagout devices, including their means of attachment, shall be substantial enough to prevent inadvertent or accidental removal. Tagout device attachment means shall be of a non-reusable type, attachable by hand, self-locking, and non-releasable with a minimum unlocking strength of no less than 50 lbs and shall have the general design and basic characteristics of being at least equivalent to a one-piece, all-environment-tolerant nylon cable tie.

1910.269(d)(3)(ii)(E) Each lockout device or tagout device shall include provisions for the identification of the employee applying the device.

1910.269(d)(3)(ii)(F) Tagout devices shall warn against hazardous conditions if the machine or equipment is energized and shall include a legend such as the following: Do Not Start, Do Not Open, Do Not Close, Do Not Energize, Do Not Operate.

Note to paragraph (d)(3)(ii)(F): For specific provisions covering accident prevention tags, see § 1910.145.

1910.269(d)(4) Energy isolation. Lockout and tagout device application and removal may only be performed by the authorized employees who are performing the servicing or maintenance.

1910.269(d)(5) Notification. Affected employees shall be notified by the employer or authorized employee of the application and removal of lockout or tagout devices. Notification shall be given before the controls are applied and after they are removed from the machine or equipment.

Note to paragraph (d)(5): See also paragraph (d)(7) of this section, which requires that the second notification take place before the machine or equipment is reenergized.

1910.269(d)(6) Lockout/tagout application. The established procedures for the application of energy control (the lockout or tagout procedures) shall include the following elements and actions, and these procedures shall be performed in the following sequence:

1910.269(d)(6)(i) Before an authorized or affected employee turns off a machine or equipment, the authorized employee shall have knowledge of the type and magnitude of the energy, the hazards of the energy to be controlled, and the method or means to control the energy.

1910.269(d)(6)(ii) The machine or equipment shall be turned off or shut down using the procedures established for the machine or equipment. An orderly shutdown shall be used to avoid any additional or increased hazards to employees as a result of the equipment stoppage.

1910.269(d)(6)(iii) All energy isolating devices that are needed to control the energy to the machine or equipment shall be physically located and operated in such a manner as to isolate the machine or equipment from energy sources.

1910.269(d)(6)(iv) Lockout or tagout devices shall be affixed to each energy isolating device by authorized employees.

1910.269(d)(6)(iv)(A) Lockout devices shall be attached in a manner that will hold the energy isolating devices in a "safe" or "off" position.

1910.269(d)(6)(iv)(B) Tagout devices shall be affixed in such a manner as will clearly indicate that the operation or movement of energy isolating devices from the "safe" or "off" position is prohibited.

1910.269(d)(6)(iv)(B)(1) Where tagout devices are used with energy isolating devices designed with the capability of being locked out, the tag attachment shall be fastened at the same point at which the lock would have been attached.

1910.269(d)(6)(iv)(B)(2) Where a tag cannot be affixed directly to the energy isolating device, the tag shall be located as close as safely possible to the device, in a position that will be immediately obvious to anyone attempting to operate the device.

1910.269(d)(6)(v) Following the application of lockout or tagout devices to energy isolating devices, all potentially hazardous stored or residual energy shall be relieved, disconnected, restrained, or otherwise rendered safe.

1910.269(d)(6)(vi) If there is a possibility of reaccumulation of stored energy to a hazardous level, verification of isolation shall be continued until the servicing or maintenance is completed or until the possibility of such accumulation no longer exists.

1910.269(d)(6)(vii) Before starting work on machines or equipment that have been locked out or tagged out, the authorized employee shall verify that isolation and deenergizing of the machine or equipment have been accomplished. If normally energized parts will be exposed to contact by an employee while the machine or equipment is deenergized, a test shall be performed to ensure that these parts are deenergized.

1910.269(d)(7) Release from lockout/tagout. Before lockout or tagout devices are removed and energy is restored to the machine or equipment, procedures shall be followed and actions taken by the authorized employees to ensure the following:

1910.269(d)(7)(i) The work area shall be inspected to ensure that nonessential items have been removed and that machine or equipment components are operationally intact.

1910.269(d)(7)(ii) The work area shall be checked to ensure that all employees have been safely positioned or removed.

1910.269(d)(7)(iii) After lockout or tagout devices have been removed and before a machine or equipment is started, affected employees shall be notified that the lockout or tagout devices have been removed.

1910.269(d)(7)(iv) Each lockout or tagout device shall be removed from each energy isolating device by the authorized employee who applied the lockout or tagout device. However, if that employee is not available to remove it, the device may be removed under the direction of the employer, provided that specific procedures and training for such removal have been developed, documented, and incorporated into the employer's energy control program. The employer shall demonstrate that the specific procedure provides a degree of safety equivalent to that provided by the removal of the device by the authorized employee who applied it. The specific procedure shall include at least the following elements:

1910.269(d)(7)(iv)(A) Verification by the employer that the authorized employee who applied the device is not at the facility;

1910.269(d)(7)(iv)(B) Making all reasonable efforts to contact the authorized employee to inform him or her that his or her lockout or tagout device has been removed; and

1910.269(d)(7)(iv)(C) Ensuring that the authorized employee has this knowledge before he or she resumes work at that facility.

1910.269(d)(8) Additional requirements.

1910.269(d)(8)(i) If the lockout or tagout devices must be temporarily removed from energy isolating devices and the machine or equipment must be energized to test or position the machine, equipment, or component thereof, the following sequence of actions shall be followed:

1910.269(d)(8)(i)(A) Clear the machine or equipment of tools and materials in accordance with paragraph (d)(7)(i) of this section;

1910.269(d)(8)(i)(B) Remove employees from the machine or equipment area in accordance with paragraphs (d)(7)(ii) and (d)(7)(iii) of this section;

1910.269(d)(8)(i)(C) Remove the lockout or tagout devices as specified in paragraph (d)(7)(iv) of this section;

1910.269(d)(8)(i)(D) Energize and proceed with the testing or positioning; and

1910.269(d)(8)(i)(E) Deenergize all systems and reapply energy control measures in accordance with paragraph (d)(6) of this section to continue the servicing or maintenance.

1910.269(d)(8)(ii) When servicing or maintenance is performed by a crew, craft, department, or other group, they shall use a procedure which affords the employees a level of protection equivalent to that provided by the implementation of a personal lockout or tagout device. Group lockout or tagout devices shall be used in accordance with the procedures required by paragraphs (d)(2)(iii) and (d)(2)(iv) of this section including, but not limited to, the following specific requirements:

1910.269(d)(8)(ii)(A) Primary responsibility shall be vested in an authorized employee for a set number of employees working under the protection of a group lockout or tagout device (such as an operations lock);

1910.269(d)(8)(ii)(B) Provision shall be made for the authorized employee to ascertain the exposure status of all individual group members with regard to the lockout or tagout of the machine or equipment;

1910.269(d)(8)(ii)(C) When more than one crew, craft, department, or other group is involved, assignment of overall job-associated lockout or tagout control responsibility shall be given to an authorized employee designated to coordinate affected work forces and ensure continuity of protection; and

1910.269(d)(8)(ii)(D) Each authorized employee shall affix a personal lockout or tagout device to the group lockout device, group lockbox, or comparable mechanism when he or she begins work and shall remove those devices when he or she stops working on the machine or equipment being serviced or maintained.

1910.269(d)(8)(iii) Procedures shall be used during shift or personnel changes to ensure the continuity of lockout or tagout protection, including provision for the orderly transfer of lockout or tagout device protection between off-going and on-coming employees, to minimize their exposure to hazards from the unexpected energizing or start-up of the machine or equipment or from the release of stored energy.

1910.269(d)(8)(iv) Whenever outside servicing personnel are to be engaged in activities covered by paragraph (d) of this section, the on-site employer and the outside employer shall inform each other of their respective lockout or tagout procedures, and each employer shall ensure that his or her personnel understand and comply with restrictions and prohibitions of the energy control procedures being used.

1910.269(d)(8)(v) If energy isolating devices are installed in a central location and are under the exclusive control of a system operator, the following requirements apply:

1910.269(d)(8)(v)(A) The employer shall use a procedure that affords employees a level of protection equivalent to that provided by the implementation of a personal lockout or tagout device.

1910.269(d)(8)(v)(B) The system operator shall place and remove lockout and tagout devices in place of the authorized employee under paragraphs (d)(4), (d)(6)(iv), and (d)(7)(iv) of this section.

1910.269(d)(8)(v)(C) Provisions shall be made to identify the authorized employee who is responsible for (that is, being protected by) the lockout or tagout device, to transfer responsibility for lockout and tagout devices, and to ensure that an authorized employee requesting removal or transfer of a lockout or tagout device is the one responsible for it before the device is removed or transferred.

Note to paragraph (d): Lockout and tagging procedures that comply with paragraphs (c) through (f) of § 1910.147 will also be deemed to comply with paragraph (d) of this section if the procedures address the hazards covered by paragraph (d) of this section.

1910.269(e) *Enclosed spaces.* This paragraph covers enclosed spaces that may be entered by employees. It does not apply to vented vaults if the employer makes a determination that the ventilation system is operating to protect employees before they enter the space. This paragraph applies to routine entry into enclosed spaces in lieu of the permit space entry requirements contained in paragraphs (d) through (k) of § 1910.146. If, after the employer takes the precautions given in paragraphs (e) and (t) of this section, the hazards remaining in the enclosed space endanger the life of an entrant or could interfere with an entrant's escape from the space, then entry into the enclosed space shall meet the permit-space entry requirements of paragraphs (d) through (k) of § 1910.146.

1910.269(e)(1) *Safe work practices.* The employer shall ensure the use of safe work practices for entry into, and work in, enclosed spaces and for rescue of employees from such spaces.

1910.269(e)(2) Training. Each employee who enters an enclosed space or who serves as an attendant shall be trained in the hazards of enclosed-space entry, in enclosed-space entry procedures, and in enclosed-space rescue procedures.

1910.269(e)(3) Rescue equipment. Employers shall provide equipment to ensure the prompt and safe rescue of employees from the enclosed space.

1910.269(e)(4) Evaluating potential hazards. Before any entrance cover to an enclosed space is removed, the employer shall determine whether it is safe to do so by checking for the presence of any atmospheric pressure or temperature differences and by evaluating whether there might be a hazardous atmosphere in the space. Any conditions making it unsafe to remove the cover shall be eliminated before the cover is removed.

Note to paragraph (e)(4): The determination called for in this paragraph may consist of a check of the conditions that might foreseeably be in the enclosed space. For example, the cover could be checked to see if it is hot and, if it is fastened in place, could be loosened gradually to release any residual pressure. An evaluation also needs to be made of whether conditions at the site could cause a hazardous atmosphere, such as an oxygen-deficient or flammable atmosphere, to develop within the space.

1910.269(e)(5) Removing covers. When covers are removed from enclosed spaces, the opening shall be promptly guarded by a railing, temporary cover, or other barrier designed to prevent an accidental fall through the opening and to protect employees working in the space from objects entering the space.

1910.269(e)(6) Hazardous atmosphere. Employees may not enter any enclosed space while it contains a hazardous atmosphere, unless the entry conforms to the permit-required confined spaces standard in § 1910.146.

1910.269(e)(7) Attendants. While work is being performed in the enclosed space, an attendant with first-aid training shall be immediately available outside the enclosed space to provide assistance if a hazard exists because of traffic patterns in the area of the opening used for entry. The attendant is not precluded from performing other duties outside the enclosed space if these duties do not distract the attendant from: monitoring employees within the space or ensuring that it is safe for employees to enter and exit the space.

Note to paragraph (e)(7): See paragraph (t) of this section for additional requirements on attendants for work in manholes and vaults.

1910.269(e)(8) Calibration of test instruments. Test instruments used to monitor atmospheres in enclosed spaces shall be kept in calibration and shall have a minimum accuracy of ±10%.

1910.269(e)(9) Testing for oxygen deficiency. Before an employee enters an enclosed space, the atmosphere in the enclosed space shall be tested for oxygen deficiency with a direct-reading meter or similar instrument, capable of collection and immediate analysis of data samples without the need for offsite evaluation. If continuous forced-air ventilation is provided, testing is not required provided that the procedures used ensure that employees are not exposed to the hazards posed by oxygen deficiency.

1910.269(e)(10) Testing for flammable gases and vapors. Before an employee enters an enclosed space, the internal atmosphere shall be tested for flammable gases and vapors with a direct-reading meter or similar instrument capable of collection and immediate analysis of data samples without the need for off-site evaluation. This test shall be performed after the oxygen testing and ventilation required by paragraph (e)(9) of this section demonstrate that there is sufficient oxygen to ensure the accuracy of the test for flammability.

1910.269(e)(11) Ventilation, and monitoring for flammable gases or vapors. If flammable gases or vapors are detected or if an oxygen deficiency is found, forced-air ventilation shall be used to maintain oxygen at a safe level and to prevent a hazardous concentration of flammable gases and vapors from accumulating. A continuous monitoring program to ensure that no increase in flammable gas or vapor concentration above safe levels occurs may be followed in lieu of ventilation if flammable gases or vapors are initially detected at safe levels.

Note to paragraph (e)(11): See the definition of "hazardous atmosphere" for guidance in determining whether a specific concentration of a substance is hazardous.

1910.269(e)(12) Specific ventilation requirements. If continuous forced-air ventilation is used, it shall begin before entry is made and shall be maintained long enough for the employer to be able to demonstrate that a safe atmosphere exists before employees are allowed to enter the work area. The forced-air ventilation shall be so directed as to ventilate the immediate area where employees are present within the enclosed space and shall continue until all employees leave the enclosed space.

1910.269(e)(13) Air supply. The air supply for the continuous forced-air ventilation shall be from a clean source and may not increase the hazards in the enclosed space.

1910.269(e)(14) Open flames. If open flames are used in enclosed spaces, a test for flammable gases and vapors shall be made immediately before the open flame device is used and at least once per hour while the device is used in the space. Testing shall be conducted more frequently if conditions present in the enclosed space indicate that once per hour is insufficient to detect hazardous accumulations of flammable gases or vapors.

Note to paragraph (e)(14): See the definition of "hazardous atmosphere" for guidance in determining whether a specific concentration of a substance is hazardous.

Note to paragraph (e): Entries into enclosed spaces conducted in accordance with the permit-space entry requirements of paragraphs (d) through (k) of § 1910.146 are considered as complying with paragraph (e) of this section.

1910.269(f) *Excavations.* Excavation operations shall comply with Subpart P of Part 1926 of this chapter.

1910.269(g) *Personal protective equipment.*

1910.269(g)(1) General. Personal protective equipment shall meet the requirements of Subpart I of this part.

Note to paragraph (g)(1) of this section: Paragraph (h) of § 1910.132 sets employer payment obligations for the personal protective equipment required by this section, including, but not limited to, the fall protection equipment required by paragraph (g)(2) of this section, the electrical protective equipment required by paragraph (l)(3) of this section, and the flame-resistant and arc rated clothing and other protective equipment required by paragraph (l)(8) of this section.

1910.269(g)(2) Fall protection.

1910.269(g)(2)(i) Personal fall arrest systems shall meet the requirements of Subpart I of this part.

1910.269(g)(2)(ii) Personal fall arrest equipment used by employees who are exposed to hazards from flames or electric arcs, as determined by the employer under paragraph (l)(8)(i) of this section, shall be capable of passing a drop test equivalent to that required by paragraph (g)(2)(iii)(L) of this section after exposure to an electric arc with a heat energy of 40 ± 5 cal/cm^2.

1910.269(g)(2)(iii) Body belts and positioning straps for work-positioning equipment shall meet the following requirements:

1910.269(g)(2)(iii)(A) Hardware for body belts and positioning straps shall meet the following requirements:

1910.269(g)(2)(iii)(A)(1) Hardware shall be made of drop-forged steel, pressed steel, formed steel, or equivalent material.

1910.269(g)(2)(iii)(A)(2) Hardware shall have a corrosion-resistant finish.

1910.269(g)(2)(iii)(A)(3) Hardware surfaces shall be smooth and free of sharp edges.

1910.269(g)(2)(iii)(B) Buckles shall be capable of withstanding an 8.9-kilonewton (2,000-pound-force) tension test with a maximum permanent deformation no greater than 0.4 millimeters (0.0156 inches).

1910.269(g)(2)(iii)(C) D rings shall be capable of withstanding a 22-kilonewton (5,000-pound-force) tensile test without cracking or breaking.

1910.269(g)(2)(iii)(D) Snaphooks shall be capable of withstanding a 22-kilonewton (5,000-pound-force) tension test without failure.

Note to paragraph (g)(2)(iii)(D): Distortion of the snaphook sufficient to release the keeper is considered to be tensile failure of a snaphook.

1910.269(g)(2)(iii)(E) Top grain leather or leather substitute may be used in the manufacture of body belts and positioning straps; however, leather and leather substitutes may not be used alone as a load-bearing component of the assembly.

1910.269(g)(2)(iii)(F) Plied fabric used in positioning straps and in load-bearing parts of body belts shall be constructed in such a way that no raw edges are exposed and the plies do not separate.

1910.269(g)(2)(iii)(G) Positioning straps shall be capable of withstanding the following tests:

1910.269(g)(2)(iii)(G)(1) A dielectric test of 819.7 volts, AC, per centimeter (25,000 volts per foot) for 3 minutes without visible deterioration;

1910.269(g)(2)(iii)(G)(2) A leakage test of 98.4 volts, AC, per centimeter (3,000 volts per foot) with a leakage current of no more than 1 mA;

Note to paragraphs (g)(2)(iii)(G)(1) and (g)(2)(iii)(G)(2): Positioning straps that pass direct-current tests at equivalent voltages are considered as meeting this requirement.

1910.269(g)(2)(iii)(G)(3) Tension tests of 20 kilonewtons (4,500 pounds-force) for sections free of buckle holes and of 15 kilonewtons (3,500 pounds-force) for sections with buckle holes;

1910.269(g)(2)(iii)(G)(4) A buckle-tear test with a load of 4.4 kilonewtons (1,000 pounds-force); and

1910.269(g)(2)(iii)(G)(5)

A flammability test in accordance with Table R-2.

Test Method	Criteria for Passing the Test
• Vertically suspend a 500-mm (19.7-inch) length of strapping supporting a 100-kg (220.5-lb) weight. • Use a butane or propane burner with a 76-mm (3-inch) flame. • Direct the flame to an edge of the strapping at a distance of 25 mm (1 inch). • Remove the flame after 5 seconds. • Wait for any flames on the positioning strap to stop burning.	• Any flames on the positioning strap shall self extinguish. • The positioning strap shall continue to support the 100-kg (220.5-lb) mass.

TABLE **R-2** Flammability Test

1910.269(g)(2)(iii)(H) The cushion part of the body belt shall contain no exposed rivets on the inside and shall be at least 76 millimeters (3 inches) in width.

1910.269(g)(2)(iii)(I) Tool loops shall be situated on the body of a body belt so that the 100 millimeters (4 inches) of the body belt that is in the center of the back, measuring from D ring to D ring, is free of tool loops and any other attachments.

1910.269(g)(2)(iii)(J) Copper, steel, or equivalent liners shall be used around the bars of D rings to prevent wear between these members and the leather or fabric enclosing them.

1910.269(g)(2)(iii)(K) Snaphooks shall be of the locking type meeting the following requirements:

1910.269(g)(2)(iii)(K)(1) The locking mechanism shall first be released, or a destructive force shall be placed on the keeper, before the keeper will open.

1910.269(g)(2)(iii)(K)(2) A force in the range of 6.7 N (1.5 lbf) to 17.8 N (4 lbf) shall be required to release the locking mechanism.

1910.269(g)(2)(iii)(K)(3) With the locking mechanism released and with a force applied on the keeper against the face of the nose, the keeper may not begin to open with a force of 11.2 N (2.5 lbf) or less and shall begin to open with a maximum force of 17.8 N (4 lbf).

1910.269(g)(2)(iii)(L)

Body belts and positioning straps shall be capable of withstanding a drop test as follows:

1910.269(g)(2)(iii)(L)(1) The test mass shall be rigidly constructed of steel or equivalent material with a mass of 100 kg (220.5 lbm). For work-positioning equipment used by employees weighing more than 140 kg (310 lbm) fully equipped, the test mass shall be increased proportionately (that is, the test mass must equal the mass of the equipped worker divided by 1.4).

1910.269(g)(2)(iii)(L)(2) For body belts, the body belt shall be fitted snugly around the test mass and shall be attached to the test structure anchorage point by means of a wire rope.

1910.269(g)(2)(iii)(L)(3) For positioning straps, the strap shall be adjusted to its shortest length possible to accommodate the test and connected to the test-structure anchorage point at one end and to the test mass on the other end.

1910.269(g)(2)(iii)(L)(4) The test mass shall be dropped an unobstructed distance of 1 meter (39.4 inches) from a supporting structure that will sustain minimal deflection during the test.

1910.269(g)(2)(iii)(L)(5) Body belts shall successfully arrest the fall of the test mass and shall be capable of supporting the mass after the test.

1910.269(g)(2)(iii)(L)(6) Positioning straps shall successfully arrest the fall of the test mass without breaking, and the arrest force may not exceed 17.8 kilonewtons (4,000 pounds-force). Additionally, snaphooks on positioning straps may not distort to such an extent that the keeper would release.

Note to paragraph (g)(2)(iii) of this section: When used by employees weighing no more than 140 kg (310 lbm) fully equipped, body belts and positioning straps that conform to American Society of Testing and Materials Standard Specifications for Personal Climbing Equipment, ASTM F887-12e1, are deemed to be in compliance with paragraph (g)(2)(iii) of this section.

1910.269(g)(2)(iv) The following requirements apply to the care and use of personal fall protection equipment.

1910.269(g)(2)(iv)(A) Work-positioning equipment shall be inspected before use each day to determine that the equipment is in safe working condition. Work-positioning equipment that is not in safe working condition may not be used.

Note to paragraph (g)(2)(iv)(A): Appendix F to this section contains guidelines for inspecting work-positioning equipment.

1910.269(g)(2)(iv)(B) Personal fall arrest systems shall be used in accordance with Subpart I of this part.

Note to paragraph (g)(2)(iv)(B): Fall protection equipment rigged to arrest falls is considered a fall arrest system and must meet the applicable requirements for the design and use of those systems. Fall protection equipment rigged for work positioning is considered work-positioning equipment and must meet the applicable requirements for the design and use of that equipment.

1910.269(g)(2)(iv)(C) The employer shall ensure that employees use fall protection systems as follows:

1910.269(g)(2)(iv)(C)(1) Each employee working from an aerial lift shall use a travel restraint system or a personal fall arrest system.

1910.269(g)(2)(iv)(C)(2) Except as provided in paragraph (g)(2)(iv)(C)(3) of this section, each employee in elevated locations more than 1.2 meters (4 ft) above the ground on poles, towers, or similar structures shall use a personal fall arrest system, work-positioning equipment, or fall restraint system, as appropriate, if the employer has not provided other fall protection meeting Subpart D of this part.

1910.269(g)(2)(iv)(C)(3) Until March 31, 2015, a qualified employee climbing or changing location on poles, towers, or similar structures need not use fall protection equipment, unless conditions, such as, but not limited to, ice, high winds, the design of the structure (for example, no provision for holding on with hands), or the presence of contaminants on the structure, could cause the employee to lose his or her grip or footing. On and after April 1, 2015, each qualified employee climbing or changing location on poles, towers, or similar structures must use fall protection equipment unless the employer can demonstrate that climbing or changing location with fall protection is infeasible or creates a greater hazard than climbing or changing location without it.

Note 1 to paragraphs (g)(2)(iv)(C)(2) and (g)(2)(iv)(C)(3): These paragraphs apply to structures that support overhead electric power transmission and distribution lines and equipment. They do not apply to portions of buildings, such as loading docks, or to electric equipment, such as transformers and capacitors. Subpart D of this part contains the duty to provide fall protection associated with walking and working surfaces.

Note 2 to paragraphs (g)(2)(iv)(C)(2) and (g)(2)(iv)(C)(3): Until the employer ensures that employees are proficient in climbing and the use of fall protection under paragraph (a)(2)(viii) of this section, the employees are not considered "qualified employees" for the purposes of paragraphs (g)(2)(iv)(C)(2) and (g)(2)(iv)(C)(3) of this section. These paragraphs require unqualified employees (including trainees) to use fall protection any time they are more than 1.2 meters (4 ft) above the ground.

1910.269(g)(2)(iv)(D) On and after April 1, 2015, work positioning systems shall be rigged so that an employee can free fall no more than 0.6 meters (2 ft).

1910.269(g)(2)(iv)(E) Anchorages for work-positioning equipment shall be capable of supporting at least twice the potential impact load of an employee's fall, or 13.3 kilonewtons (3,000 pounds-force), whichever is greater.

Note to paragraph (g)(2)(iv)(E): Wood-pole fall-restriction devices meeting American Society of Testing and Materials Standard Specifications for Personal Climbing Equipment, ASTM F887-12e1, are deemed to meet the anchorage-strength requirement when they are used in accordance with manufacturers' instructions.

1910.269(g)(2)(iv)(F) Unless the snaphook is a locking type and designed specifically for the following connections, snaphooks on work-positioning equipment may not be engaged:

1910.269(g)(2)(iv)(F)(1) Directly to webbing, rope, or wire rope;

1910.269(g)(2)(iv)(F)(2) To each other;

1910.269(g)(2)(iv)(F)(3) To a D ring to which another snaphook or other connector is attached;

1910.269(g)(2)(iv)(F)(4) To a horizontal lifeline; or

1910.269(g)(2)(iv)(F)(5) To any object that is incompatibly shaped or dimensioned in relation to the snaphook such that accidental disengagement could occur should the connected object sufficiently depress the snaphook keeper to allow release of the object.

1910.269(h) *Portable ladders and platforms.*

1910.269(h)(1) General. Requirements for portable ladders contained in Subpart D of this part apply in addition to the requirements of paragraph (h) of this section, except as specifically noted in paragraph (h)(2) of this section.

1910.269(h)(2) Special ladders and platforms. Portable ladders used on structures or conductors in conjunction with overhead line work need not meet § 1910.23(c)(4) and (9). Portable ladders and platforms used on structures or conductors in conjunction with overhead line work shall meet the following requirements:

1910.269(h)(2)(i) In the configurations in which they are used, portable ladders and platforms shall be capable of supporting without failure at least 2.5 times the maximum intended load.

1910.269(h)(2)(ii) Portable ladders and platforms may not be loaded in excess of the working loads for which they are designed.

1910.269(h)(2)(iii) Portable ladders and platforms shall be secured to prevent them from becoming dislodged.

1910.269(h)(2)(iv) Portable ladders and platforms may be used only in applications for which they are designed.

1910.269(h)(3) Conductive ladders. Portable metal ladders and other portable conductive ladders may not be used near exposed energized lines or equipment. However, in specialized high-voltage work, conductive ladders shall be used when the employer demonstrates that nonconductive ladders would present a greater hazard to employees than conductive ladders.

1910.269(i) *Hand and portable power equipment.*

1910.269(i)(1)

General. Paragraph (i)(2) of this section applies to electric equipment connected by cord and plug. Paragraph (i)(3) of this section applies to portable and vehicle-mounted generators used to supply cord- and plug-connected equipment. Paragraph (i)(4) of this section applies to hydraulic and pneumatic tools.

1910.269(i)(2) Cord- and plug-connected equipment. Cord- and plug-connected equipment not covered by Subpart S of this part shall comply with one of the following instead of § 1910.243(a)(5):

1910.269(i)(2)(i) The equipment shall be equipped with a cord containing an equipment grounding conductor connected to the equipment frame and to a means for grounding the other end of the conductor (however, this option may not be used where the introduction of the ground into the work environment increases the hazard to an employee); or

1910.269(i)(2)(ii)

The equipment shall be of the double-insulated type conforming to Subpart S of this part; or

1910.269(i)(2)(iii) The equipment shall be connected to the power supply through an isolating transformer with an ungrounded secondary of not more than 50 volts.

1910.269(i)(3) Portable and vehicle-mounted generators. Portable and vehicle mounted generators used to supply cord- and plug-connected equipment covered by paragraph (i)(2) of this section shall meet the following requirements:

1910.269(i)(3)(i) The generator may only supply equipment located on the generator or the vehicle and cord- and plug connected equipment through receptacles mounted on the generator or the vehicle.

1910.269(i)(3)(ii) The non-current-carrying metal parts of equipment and the equipment grounding conductor terminals of the receptacles shall be bonded to the generator frame.

1910.269(i)(3)(iii) For vehicle-mounted generators, the frame of the generator shall be bonded to the vehicle frame.

1910.269(i)(3)(iv) Any neutral conductor shall be bonded to the generator frame.

1910.269(i)(4) Hydraulic and pneumatic tools.

1910.269(i)(4)(i) Safe operating pressures for hydraulic and pneumatic tools, hoses, valves, pipes, filters, and fittings may not be exceeded.

Note to paragraph (i)(4)(i): If any hazardous defects are present, no operating pressure is safe, and the hydraulic or pneumatic equipment involved may not be used. In the absence of defects, the maximum rated operating pressure is the maximum safe pressure.

1910.269(i)(4)(ii) A hydraulic or pneumatic tool used where it may contact exposed energized parts shall be designed and maintained for such use.

1910.269(i)(4)(iii) The hydraulic system supplying a hydraulic tool used where it may contact exposed live parts shall provide protection against loss of insulating value, for the voltage involved, due to the formation of a partial vacuum in the hydraulic line. *Note to paragraph (i)(4)(iii)*: Use of hydraulic lines that do not have check valves and that have a separation of more than 10.7 meters (35 ft) between the oil reservoir and the upper end of the hydraulic system promotes the formation of a partial vacuum.

1910.269(i)(4)(iv) A pneumatic tool used on energized electric lines or equipment, or used where it may contact exposed live parts, shall provide protection against the accumulation of moisture in the air supply.

1910.269(i)(4)(v) Pressure shall be released before connections are broken, unless quick acting, self-closing connectors are used.

1910.269(i)(4)(vi) Employers must ensure that employees do not use any part of their bodies to locate, or attempt to stop, a hydraulic leak.

1910.269(i)(4)(vii) Hoses may not be kinked.

1910.269(j) *Live-line tools.*

1910.269(j)(1) Design of tools. Live-line tool rods, tubes, and poles shall be designed and constructed to withstand the following minimum tests:

1910.269(j)(1)(i) If the tool is made of fiberglass-reinforced plastic (FRP), it shall withstand 328,100 volts per meter (100,000 volts per foot) of length for 5 minutes, or

Note to paragraph (j)(1)(i): Live-line tools using rod and tube that meet ASTM F711-02 (2007), Standard Specification for Fiberglass-Reinforced Plastic (FRP) Rod and Tube Used in Live Line Tools, are deemed to comply with paragraph (j)(1) of this section.

1910.269(j)(1)(ii) If the tool is made of wood, it shall withstand 246,100 volts per meter (75,000 volts per foot) of length for 3 minutes, or

1910.269(j)(1)(iii) The tool shall withstand other tests that the employer can demonstrate are equivalent.

1910.269(j)(2) Condition of tools.

1910.269(j)(2)(i) Each liveline tool shall be wiped clean and visually inspected for defects before use each day.

1910.269(j)(2)(ii) If any defect or contamination that could adversely affect the insulating qualities or mechanical integrity of the live-line tool is present after wiping, the tool shall be removed from service and examined and tested according to paragraph (j)(2)(iii) of this section before being returned to service.

1910.269(j)(2)(iii) Live-line tools used for primary employee protection shall be removed from service every 2 years, and whenever required under paragraph (j)(2)(ii) of this section, for examination, cleaning, repair, and testing as follows:

1910.269(j)(2)(iii)(A) Each tool shall be thoroughly examined for defects.

1910.269(j)(2)(iii)(B) If a defect or contamination that could adversely affect the insulating qualities or mechanical integrity of the live-line tool is found, the tool shall be repaired and refinished or shall be permanently removed from service. If no such defect or contamination is found, the tool shall be cleaned and waxed.

1910.269(j)(2)(iii)(C) The tool shall be tested in accordance with paragraphs (j)(2) (iii)(D) and (j)(2)(iii)(E) of this section under the following conditions:

1910.269(j)(2)(iii)(C)(1) After the tool has been repaired or refinished; and

1910.269(j)(2)(iii)(C)(2) After the examination if repair or refinishing is not performed, unless the tool is made of FRP rod or foam-filled FRP tube and the employer can demonstrate that the tool has no defects that could cause it to fail during use.

1910.269(j)(2)(iii)(D) The test method used shall be designed to verify the tool's integrity along its entire working length and, if the tool is made of fiberglass-reinforced plastic, its integrity under wet conditions.

1910.269(j)(2)(iii)(E) The voltage applied during the tests shall be as follows:

1910.269(j)(2)(iii)(E)(1) 246,100 volts per meter (75,000 volts per foot) of length for 1 minute if the tool is made of fiberglass, or

1910.269(j)(2)(iii)(E)(2) 164,000 volts per meter (50,000 volts per foot) of length for 1 minute if the tool is made of wood, or

1910.269(j)(2)(iii)(E)(3) Other tests that the employer can demonstrate are equivalent.

Note to paragraph (j)(2): Guidelines for the examination, cleaning, repairing, and in service testing of live-line tools are specified in the Institute of Electrical and Electronics Engineers' IEEE Guide for Maintenance Methods on Energized Power Lines, IEEE Std 516-2009.

1910.269(k) *Materials handling and storage.*

1910.269(k)(1) General. Materials handling and storage shall comply with applicable material handling and material-storage requirements in this part, including those in Subpart N of this part.

1910.269(k)(2) Materials storage near energized lines or equipment.

1910.269(k)(2)(i) In areas to which access is not restricted to qualified persons only, materials or equipment may not be stored closer to energized lines or exposed energized parts of equipment than the following distances, plus a distance that provides for the maximum sag and side swing of all conductors and for the height and movement of material-handling equipment:

1910.269(k)(2)(i)(A) For lines and equipment energized at 50 kilovolts or less, the distance is 3.05 meters (10 ft).

1910.269(k)(2)(i)(B) For lines and equipment energized at more than 50 kilovolts, the distance is 3.05 meters (10 ft) plus 0.10 meter (4 inches) for every 10 kilovolts over 50 kilovolts.

1910.269(k)(2)(ii) In areas restricted to qualified employees, materials may not be stored within the working space about energized lines or equipment.

Note to paragraph (k)(2)(ii): Paragraphs (u)(1) and (v)(3) of this section specify the size of the working space.

1910.269(l) *Working on or near exposed energized parts.* This paragraph applies to work on exposed live parts, or near enough to them to expose the employee to any hazard they present.

1910.269(l)(1) General.

1910.269(l)(1)(i) Only qualified employees may work on or with exposed energized lines or parts of equipment.

1910.269(l)(1)(ii) Only qualified employees may work in areas containing unguarded, uninsulated energized lines or parts of equipment operating at 50 volts or more.

1910.269(l)(1)(iii) Electric lines and equipment shall be considered and treated as energized unless they have been deenergized in accordance with paragraph (d) or (m) of this section.

1910.269(l)(2) At least two employees.

1910.269(l)(2)(i) Except as provided in paragraph (l)(2)(ii) of this section, at least two employees shall be present while any employees perform the following types of work:

1910.269(l)(2)(i)(A) Installation, removal, or repair of lines energized at more than 600 volts,

1910.269(l)(2)(i)(B) Installation, removal, or repair of deenergized lines if an employee is exposed to contact with other parts energized at more than 600 volts,

1910.269(l)(2)(i)(C) Installation, removal, or repair of equipment, such as transformers, capacitors, and regulators, if an employee is exposed to contact with parts energized at more than 600 volts,

1910.269(l)(2)(i)(D) Work involving the use of mechanical equipment, other than insulated aerial lifts, near parts energized at more than 600 volts, and

1910.269(l)(2)(i)(E) Other work that exposes an employee to electrical hazards greater than, or equal to, the electrical hazards posed by operations listed specifically in paragraphs (l)(2)(i)(A) through (l)(2)(i)(D) of this section.

1910.269(l)(2)(ii) Paragraph (l)(2)(i) of this section does not apply to the following operations:

1910.269(l)(2)(ii)(A) Routine circuit switching, when the employer can demonstrate that conditions at the site allow safe performance of this work,

1910.269(l)(2)(ii)(B) Work performed with live-line tools when the position of the employee is such that he or she is neither within reach of, nor otherwise exposed to contact with, energized parts, and

1910.269(l)(2)(ii)(C) Emergency repairs to the extent necessary to safeguard the general public.

1910.269(l)(3) Minimum approach distances.

1910.269(l)(3)(i) The employer shall establish minimum approach distances no less than the distances computed by Table R-3 for ac systems or Table R-8 for dc systems.

For phase-to-phase system voltages of 50 V to 300 V:[1] MAD = avoid contact	
For phase-to-phase system voltages of 301 V to 5 kV:[1] $$MAD = M + D, \text{ where}$$ $D = 0.02$ m .... $M = 0.31$ m for voltages up to 750 V and 0.61 m otherwise	the electrical component of the minimum approach distance. the inadvertent movement factor.
For phase-to-phase system voltages of 5.1 kV to 72.5 kV:[1,4] $$MAD = M + AD, \text{ where}$$ $M = 0.61$ m .... $A =$ the applicable value from Table R-5 .... $D =$ the value from Table R-4 corresponding to the voltage and exposure or the value of the electrical component of the minimum approach distance calculated using the method provided in Appendix B to this section	the inadvertent movement factor. the altitude correction factor. the electrical component of the minimum approach distance.
For phase-to-phase system voltages of more than 72.5 kV, nominal:[2,4] $$MAD = 0.3048(C + a)V_{LG}TA + M, \text{ where}$$ $C = 0.01$ for phase-to-ground exposures that the employer can demonstrate consist only of air across the approach distance (gap), 0.01 for phase-to-phase exposures if the employer can demonstrate that no insulated tool spans the gap and that no large conductive object is in the gap, or 0.011 otherwise $V_{LG} =$ phase-to-ground rms voltage, in kV $T =$ maximum anticipated per-unit transient overvoltage; for phase-to-ground exposures, T equals T_{LG}, the maximum per-unit transient overvoltage, phase-to-ground, determined by the employer under paragraph (l)(3)(ii) of this section; for phase-to-phase exposures, T equals $1.35T_{LG} + 0.45$ $A =$ altitude correction factor from Table R-5 $M = 0.31$ m, the inadvertent movement factor $a =$ saturation factor, as follows:	

Phase-to-Ground Exposures

$V_{Peak} = T_{L-G}V_{L-G}\sqrt{2}$	635 kV or less	635.1 to 915 kV	915.1 to 1,050 kV	More than 1,050 kV
a	0	$(V_{Peak}-635)/140,000$	$(V_{Peak}-645)/135,000$	$(V_{Peak}-675)/125,000$

Phase-to-Phase Exposures[3]

$V_{Peak} = (1.35T_{L-G} + 0.45)V_{L-G}\sqrt{2}$	630 kV or less	630.1 to 848 kV	848.1 to 1,131 kV	1,131.1 to 1,485 kV	More than 1,485 kV
a	0	$(V_{Peak}-630)/155,000$	$(V_{Peak}-633.6)/152,207$	$(V_{Peak}-628)/153,846$	$(V_{Peak}-350.5)/203,666$

---

[1]Employers may use the minimum approach distances in Table R-6. If the worksite is at an elevation of more than 900 meters (3,000 ft), see footnote 1 to Table R-6.

[2]Employers may use the minimum approach distances in Table R-7, except that the employer may not use the minimum approach distances in Table R-7 for phase-to-phase exposures if an insulated tool spans the gap or if any large conductive object is in the gap. If the worksite is at an elevation of more than 900 meters (3,000 ft), see footnote 1 to Table R-7. Employers may use the minimum approach distances in Table 14 through Table 21 in Appendix B to this section, which calculated MAD for various values of T, provided the employer follows the notes to those tables.

[3]Use the equations for phase-to-ground exposures (with V_{Peak} for phase-to-phase exposures) unless the employer can demonstrate that no insulated tool spans the gap and that no large conductive object is in the gap.

[4]Until March 31, 2015, employers may use the minimum approach distances in Table 6 through Table 13 in Appendix B to this section.

Table R-3   AC Live-Line Work Minimum Approach Distance [The minimum approach distance (MAD; in meters) shall conform to the following equations.]

1910.269(l)(3)(ii) No later than April 1, 2015, for voltages over 72.5 kilovolts, the employer shall determine the maximum anticipated per-unit transient overvoltage, phase-to-ground, through an engineering analysis or assume a maximum anticipated per-unit transient overvoltage, phase-to-ground, in accordance with Table R-9. When the employer uses portable protective gaps to control the maximum transient overvoltage, the value of the maximum anticipated per-unit transient overvoltage, phase-to-ground, must provide for five standard deviations between the statistical sparkover voltage of the gap and the statistical withstand voltage corresponding to the electrical component of the minimum approach distance. The employer shall make any engineering analysis conducted to determine maximum anticipated per unit transient overvoltage available upon request to employees and to the Assistant Secretary or designee for examination and copying.

Note to paragraph (l)(3)(ii): See Appendix B to this section for information on how to calculate the maximum anticipated per-unit transient overvoltage, phase-to-ground, when the employer uses portable protective gaps to reduce maximum transient overvoltages.

1910.269(l)(3)(iii) The employer shall ensure that no employee approaches or takes any conductive object closer to exposed energized parts than the employer's established minimum approach distance, unless:

1910.269(l)(3)(iii)(A) The employee is insulated from the energized part (rubber insulating gloves or rubber insulating gloves and sleeves worn in accordance with paragraph (l)(4) of this section constitutes insulation of the employee from the energized part upon which the employee is working provided that the employee has control of the part in a manner sufficient to prevent exposure to uninsulated portions of the employee's body), or

1910.269(l)(3)(iii)(B) The energized part is insulated from the employee and from any other conductive object at a different potential, or

1910.269(l)(3)(iii)(C) The employee is insulated from any other exposed conductive object in accordance with the requirements for live-line barehand work in paragraph (q)(3) of this section.

1910.269(l)(4) Type of insulation

1910.269(l)(4)(i) When an employee uses rubber insulating gloves as insulation from energized parts (under paragraph (l)(3)(iii)(A) of this section), the employer shall ensure that the employee also uses rubber insulating sleeves. However, an employee need not use rubber insulating sleeves if:

1910.269(l)(4)(i)(A) Exposed energized parts on which the employee is not working are insulated from the employee; and

1910.269(l)(4)(i)(B) When installing insulation for purposes of paragraph (l)(4)(i)(A) of this section, the employee installs the insulation from a position that does not expose his or her upper arm to contact with other energized parts.

1910.269(l)(4)(ii) When an employee uses rubber insulating gloves or rubber insulating gloves and sleeves as insulation from energized parts (under paragraph (l)(3)(iii)(A) of this section), the employer shall ensure that the employee:

1910.269(l)(4)(ii)(A) Puts on the rubber insulating gloves and sleeves in a position where he or she cannot reach into the minimum approach distance, established by the employer under paragraph (l)(3)(i) of this section; and

1910.269(l)(4)(ii)(B) Does not remove the rubber insulating gloves and sleeves until he or she is in a position where he or she cannot reach into the minimum approach distance, established by the employer under paragraph (l)(3)(i) of this section.

1910.269(l)(5) Working position.

1910.269(l)(5)(i) The employer shall ensure that each employee, to the extent that other safety-related conditions at the worksite permit, works in a position from which a slip or shock will not bring the employee's body into contact with exposed, uninsulated parts energized at a potential different from the employee's.

1910.269(l)(5)(ii) When an employee performs work near exposed parts energized at more than 600 volts, but not more than 72.5 kilovolts, and is not wearing rubber insulating gloves, being protected by insulating equipment covering the energized parts, performing work using live-line tools, or performing live-line barehand work under paragraph (q)(3) of this section, the employee shall work from a position where he or she cannot reach into the minimum approach distance, established by the employer under paragraph (l)(3)(i) of this section.

1910.269(l)(6) Making connections. The employer shall ensure that employees make connections as follows:

1910.269(l)(6)(i) In connecting deenergized equipment or lines to an energized circuit by means of a conducting wire or device, an employee shall first attach the wire to the deenergized part;

1910.269(l)(6)(ii) When disconnecting equipment or lines from an energized circuit by means of a conducting wire or device, an employee shall remove the source end first; and

1910.269(l)(6)(iii) When lines or equipment are connected to or disconnected from energized circuits, an employee shall keep loose conductors away from exposed energized parts.

1910.269(l)(7) Conductive articles. When an employee performs work within reaching distance of exposed energized parts of equipment, the employer shall ensure that the employee removes or renders nonconductive all exposed conductive articles, such as keychains or watch chains, rings, or wrist watches or bands, unless such articles do not increase the hazards associated with contact with the energized parts.

1910.269(l)(8) Protection from flames and electric arcs.

1910.269(l)(8)(i) The employer shall assess the workplace to identify employees exposed to hazards from flames or from electric arcs.

1910.269(l)(8)(ii) For each employee exposed to hazards from electric arcs, the employer shall make a reasonable estimate of the incident heat energy to which the employee would be exposed.

Note 1 to paragraph (l)(8)(ii): Appendix E to this section provides guidance on estimating available heat energy. The Occupational Safety and Health Administration will deem employers following the guidance in Appendix E to this section to be in compliance with paragraph (l)(8)(ii) of this section. An employer may choose a method of calculating incident heat energy not included in Appendix E to this section if the chosen method reasonably predicts the incident energy to which the employee would be exposed.

Note 2 to paragraph (l)(8)(ii): This paragraph does not require the employer to estimate the incident heat energy exposure for every job task performed by each employee. The employer may make broad estimates that cover multiple system areas provided the employer uses reasonable assumptions about the energy-exposure distribution throughout the system and provided the estimates represent the maximum employee exposure for those areas. For example, the employer could estimate the heat energy just outside a substation feeding a radial distribution system and use that estimate for all jobs performed on that radial system.

1910.269(l)(8)(iii) The employer shall ensure that each employee who is exposed to hazards from flames or electric arcs does not wear clothing that could melt onto his or her skin or that could ignite and continue to burn when exposed to flames or the heat energy estimated under paragraph (l)(8)(ii) of this section.

Note to paragraph (l)(8)(iii) of this section: This paragraph prohibits clothing made from acetate, nylon, polyester, rayon and polypropylene, either alone or in blends, unless the employer demonstrates that the fabric has been treated to withstand the conditions that may be encountered by the employee or that the employee wears the clothing in such a manner as to eliminate the hazard involved.

1910.269(l)(8)(iv) The employer shall ensure that the outer layer of clothing worn by an employee, except for clothing not required to be arc rated under paragraphs (l)(8)(v)(A) through (l)(8)(v)(E) of this section, is flame resistant under any of the following conditions:

1910.269(l)(8)(iv)(A) The employee is exposed to contact with energized circuit parts operating at more than 600 volts,

1910.269(l)(8)(iv)(B) An electric arc could ignite flammable material in the work area that, in turn, could ignite the employee's clothing,

1910.269(l)(8)(iv)(C) Molten metal or electric arcs from faulted conductors in the work area could ignite the employee's clothing, or

Note to paragraph (l)(8)(iv)(C): This paragraph does not apply to conductors that are capable of carrying, without failure, the maximum available fault current for the time the circuit protective devices take to interrupt the fault.

1910.269(l)(8)(iv)(D) The incident heat energy estimated under paragraph (l)(8)(ii) of this section exceeds 2.0 cal/cm^2.

1910.269(l)(8)(v) The employer shall ensure that each employee exposed to hazards from electric arcs wears protective clothing and other protective equipment with an arc rating greater than or equal to the heat energy estimated under paragraph (l)(8)(ii) of this section whenever that estimate exceeds 2.0 cal/cm^2. This protective equipment shall cover the employee's entire body, except as follows:

1910.269(l)(8)(v)(A) Arc-rated protection is not necessary for the employee's hands when the employee is wearing rubber insulating gloves with protectors or, if the estimated incident energy is no more than 14 cal/cm^2, heavy-duty leather work gloves with a weight of at least 407 gm/m^2 (12 oz/yd^2),

1910.269(l)(8)(v)(B) Arc-rated protection is not necessary for the employee's feet when the employee is wearing heavy-duty work shoes or boots,

1910.269(l)(8)(v)(C) Arc-rated protection is not necessary for the employee's head when the employee is wearing head protection meeting § 1910.135 if the estimated incident energy is less than 9 cal/cm^2 for exposures involving single-phase arcs in open air or 5 cal/cm^2 for other exposures,

1910.269(l)(8)(v)(D) The protection for the employee's head may consist of head protection meeting § 1910.135 and a face shield with a minimum arc rating of 8 cal/cm^2 if the estimated incident-energy exposure is less than 13 cal/cm^2 for exposures involving single-phase arcs in open air or 9 cal/cm^2 for other exposures, and

1910.269(l)(8)(v)(E) For exposures involving single phase arcs in open air, the arc rating for the employee's head and face protection may be 4 cal/cm^2 less than the estimated incident energy.

Note to paragraph (l)(8): See Appendix E to this section for further information on the selection of appropriate protection.

1910.269(l)(8)(vi) Dates.

1910.269(l)(8)(vi)(A) The obligation in paragraph (l)(8)(ii) of this section for the employer to make reasonable estimates of incident energy commences January 1, 2015.

1910.269(l)(8)(vi)(B) The obligation in paragraph (l)(8)(iv)(D) of this section for the employer to ensure that the outer layer of clothing worn by an employee is flame-resistant when the estimated incident heat energy exceeds 2.0 cal/cm^2 commences April 1, 2015.

1910.269(l)(8)(vi)(C) The obligation in paragraph (l)(8)(v) of this section for the employer to ensure that each employee exposed to hazards from electric arcs wears the required arc-rated protective equipment commences April 1, 2015.

1910.269(l)(9) Fuse handling. When an employee must install or remove fuses with one or both terminals energized at more than 300 volts, or with exposed parts energized at more than 50 volts, the employer shall ensure that the employee uses tools or gloves rated for the voltage. When an employee installs or removes expulsion-type fuses with one or both terminals energized at more than 300 volts, the employer shall ensure that the employee wears eye protection meeting the requirements of Subpart I of this part, uses a tool rated for the voltage, and is clear of the exhaust path of the fuse barrel.

1910.269(l)(10) Covered (noninsulated) conductors. The requirements of this section that pertain to the hazards of exposed live parts also apply when an employee performs work in proximity to covered (noninsulated) wires.

1910.269(l)(11) Non-current-carrying metal parts. Non-current-carrying metal parts of equipment or devices, such as transformer cases and circuit-breaker housings, shall be treated as energized at the highest voltage to which these parts are exposed, unless the employer inspects the installation and determines that these parts are grounded before employees begin performing the work.

1910.269(l)(12) Opening and closing circuits under load.

1910.269(l)(12)(i) The employer shall ensure that devices used by employees to open circuits under load conditions are designed to interrupt the current involved.

1910.269(l)(12)(ii) The employer shall ensure that devices used by employees to close circuits under load conditions are designed to safely carry the current involved.

1910.269 (m) *Deenergizing lines and equipment for employee protection.*

1910.269(m)(1) *Application.* Paragraph (m) of this section applies to the deenergizing of transmission and distribution lines and equipment for the purpose of protecting employees. See paragraph (d) of this section for requirements on the control of hazardous energy sources used in the generation of electric energy. Conductors and parts of electric equipment that have been deenergized under procedures other than those required by paragraph (d) or (m) of this section, as applicable, shall be treated as energized.

1910.269(m)(2) General.

1910.269(m)(2)(i) If a system operator is in charge of the lines or equipment and their means of disconnection, the employer shall designate one employee in the crew to be in charge of the clearance and shall comply with all of the requirements of paragraph (m)(3) of this section in the order specified.

1910.269(m)(2)(ii) If no system operator is in charge of the lines or equipment and their means of disconnection, the employer shall designate one employee in the crew to be in charge of the clearance and to perform the functions that the system operator would otherwise perform under paragraph (m) of this section. All of the requirements

Nominal Voltage (kV) Phase-to-Phase	Phase-to-Ground Exposure D (m)	Phase-to-Phase Exposure D (m)
5.1 to 15.0	0.04	0.07
15.1 to 36.0	0.16	0.28
36.1 to 46.0	0.23	0.37
46.1 to 72.5	0.39	0.59

TABLE R-4   Electrical Component of the Minimum Approach Distance at 5.1 to 72.5 kV [D; In meters]

Altitude Above Sea Level (m)	A
0 to 900	1.00
901 to 1,200	1.02
1,201 to 1,500	1.05
1,501 to 1,800	1.08
1,801 to 2,100	1.11
2,101 to 2,400	1.14
2,401 to 2,700	1.17
2,701 to 3,000	1.20
3,001 to 3,600	1.25
3,601 to 4,200	1.30
4,201 to 4,800	1.35
4,801 to 5,400	1.39
5,401 to 6,000	1.44

TABLE R-5   Altitude Correction Factor

of paragraph (m)(3) of this section apply, in the order specified, except as provided in paragraph (m)(2)(iii) of this section.

1910.269(m)(2)(iii) If only one crew will be working on the lines or equipment and if the means of disconnection is accessible and visible to, and under the sole control of, the employee in charge of the clearance, paragraphs (m)(3)(i), (m)(3)(iii), and (m)(3)(v) of this section do not apply. Additionally, the employer does not need to use the tags required by the remaining provisions of paragraph (m)(3) of this section.

1910.269(m)(2)(iv) If two or more crews will be working on the same lines or equipment, then:

1910.269(m)(2)(iv)(A) The crews shall coordinate their activities under paragraph (m) of this section with a single employee in charge of the clearance for all of the crews and follow the requirements of paragraph (m) of this section as if all of the employees formed a single crew, or

	Distance			
	Phase-to-Ground Exposure		Phase-to-Phase Exposure	
Nominal Voltage (kV) Phase-to-Phase	m	ft	m	ft
0.050 to 0.300[2]	Avoid Contact		Avoid Contact	
0.301 to 0.750[2]	0.33	1.09	0.33	1.09
0.751 to 5.0	0.63	2.07	0.63	2.07
5.1 to 15.0	0.65	2.14	0.68	2.24
15.1 to 36.0	0.77	2.53	0.89	2.92
36.1 to 46.0	0.84	2.76	0.98	3.22
46.1 to 72.5	1.00	3.29	1.20	3.94

[1]Employers may use the minimum approach distances in this table provided the worksite is at an elevation of 900 meters (3,000 ft) or less. If employees will be working at elevations greater than 900 meters (3,000 ft) above mean sea level, the employer shall determine minimum approach distances by multiplying the distances in this table by the correction factor in Table R-5 corresponding to the altitude of the work.
[2]For single-phase systems, use voltage-to-ground.

TABLE R-6   Alternative Minimum Approach Distances for Voltages of 72.5 kV and Less[1]

	Phase-to-Ground Exposure		Phase-to-Phase Exposure	
Voltage Range Phase to Phase (kV)	m	ft	m	ft
72.6 to 121.0	1.13	3.71	1.42	4.66
121.1 to 145.0	1.30	4.27	1.64	5.38
145.1 to 169.0	1.46	4.79	1.94	6.36
169.1 to 242.0	2.01	6.59	3.08	10.10
242.1 to 362.0	3.41	11.19	5.52	18.11
362.1 to 420.0	4.25	13.94	6.81	22.34
420.1 to 550.0	5.07	16.63	8.24	27.03
550.1 to 800.0	6.88	22.57	11.38	37.34

[1]Employers may use the minimum approach distances in this table provided the worksite is at an elevation of 900 meters (3,000 ft) or less. If employees will be working at elevations greater than 900 meters (3,000 ft) above mean sea level, the employer shall determine minimum approach distances by multiplying the distances in this table by the correction factor in Table R-5 corresponding to the altitude of the work.

TABLE R-7   Alternative Minimum Approach Distances for Voltages of More Than 72.5 kV[1 2 3]

Maximum Anticipated Per-Unit Transient Overvoltage	Distance (m) Maximum Line-to-Ground Voltage (kV)				
	250	400	500	600	750
1.5 or less	1.12	1.60	2.06	2.62	3.61
1.6	1.17	1.69	2.24	2.86	3.98
1.7	1.23	1.82	2.42	3.12	4.37
1.8	1.28	1.95	2.62	3.39	4.79

[1]The distances specified in this table are for air, bare-hand, and live-line tool conditions. If employees will be working at elevations greater than 900 meters (3,000 ft) above mean sea level, the employer shall determine minimum approach distances by multiplying the distances in this table by the correction factor in Table R-5 corresponding to the altitude of the work.

TABLE R-8   DC Live-Line Minimum Approach Distance with Overvoltage Factor[1] (In meters)

Voltage Range (kV)	Type of Current (ac or dc)	Assumed Maximum Per-Unit Transient Overvoltage
72.6 to 420.0	ac	3.5
420.1 to 550.0	ac	3.0
550.1 to 800.0	ac	2.5
250 to 750	dc	1.8

TABLE R-9   Assumed Maximum Per-Unit Overvoltage

1910.269(m)(2)(iv)(B) Each crew shall independently comply with paragraph (m) of this section and, if there is no system operator in charge of the lines or equipment, shall have separate tags and coordinate deenergizing and reenergizing the lines and equipment with the other crews.

1910.269(m)(2)(v) The employer shall render any disconnecting means that are accessible to individuals outside the employer's control (for example, the general public) inoperable while the disconnecting means are open for the purpose of protecting employees.

1910.269(m)(3) Deenergizing lines and equipment.

1910.269(m)(3)(i) The employee that the employer designates pursuant to paragraph (m)(2) of this section as being in charge of the clearance shall make a request of the system operator to deenergize the particular section of line or equipment. The designated employee becomes the employee in charge (as this term is used in paragraph (m)(3) of this section) and is responsible for the clearance.

1910.269(m)(3)(ii) The employer shall ensure that all switches, disconnectors, jumpers, taps, and other means through which known sources of electric energy may be supplied to the particular lines and equipment to be deenergized are open. The employer shall render such means inoperable, unless its design does not so permit, and then ensure that such means are tagged to indicate that employees are at work.

1910.269(m)(3)(iii) The employer shall ensure that automatically and remotely controlled switches that could cause the opened disconnecting means to close are also tagged at the points of control. The employer shall render the automatic or remote control feature inoperable, unless its design does not so permit.

1910.269(m)(3)(iv) The employer need not use the tags mentioned in paragraphs (m)(3)(ii) and (m)(3)(iii) of this section on a network protector for work on the primary feeder for the network protector's associated network transformer when the employer can demonstrate all of the following conditions:

1910.269(m)(3)(iv)(A) Every network protector is maintained so that it will immediately trip open if closed when a primary conductor is deenergized;

1910.269(m)(3)(iv)(B) Employees cannot manually place any network protector in a closed position without the use of tools, and any manual override position is blocked, locked, or otherwise disabled; and

1910.269(m)(3)(iv)(C) The employer has procedures for manually overriding any network protector that incorporate provisions for determining, before anyone places a network protector in a closed position, that: The line connected to the network protector is not deenergized for the protection of any employee working on the line; and (if the line connected to the network protector is not deenergized for the protection of any employee working on the line) the primary conductors for the network protector are energized.

1910.269(m)(3)(v) Tags shall prohibit operation of the disconnecting means and shall indicate that employees are at work.

1910.269(m)(3)(vi) After the applicable requirements in paragraphs (m)(3)(i) through (m)(3)(v) of this section have been followed and the system operator gives a clearance to the employee in charge, the employer shall ensure that the lines and equipment are deenergized by testing the lines and equipment to be worked with a device designed to detect voltage.

1910.269(m)(3)(vii) The employer shall ensure the installation of protective grounds as required by paragraph (n) of this section.

1910.269(m)(3)(viii) After the applicable requirements of paragraphs (m)(3)(i) through (m)(3)(vii) of this section have been followed, the lines and equipment involved may be considered deenergized.

1910.269(m)(3)(ix) To transfer the clearance, the employee in charge (or the employee's supervisor if the employee in charge must leave the worksite due to illness or other emergency) shall inform the system operator and employees in the crew; and the new employee in charge shall be responsible for the clearance.

1910.269(m)(3)(x) To release a clearance, the employee in charge shall:

1910.269(m)(3)(x)(A) Notify each employee under that clearance of the pending release of the clearance;

1910.269(m)(3)(x)(B) Ensure that all employees under that clearance are clear of the lines and equipment;

1910.269(m)(3)(x)(C) Ensure that all protective grounds protecting employees under that clearance have been removed; and

1910.269(m)(3)(x)(D) Report this information to the system operator and then release the clearance.

1910.269(m)(3)(xi) Only the employee in charge who requested the clearance may release the clearance, unless the employer transfers responsibility under paragraph (m)(3)(ix) of this section.

1910.269(m)(3)(xii) No one may remove tags without the release of the associated clearance as specified under paragraphs (m)(3)(x) and (m)(3)(xi) of this section.

1910.269(m)(3)(xiii) The employer shall ensure that no one initiates action to reenergize the lines or equipment at a point of disconnection until all protective grounds have been removed, all crews working on the lines or equipment release their clearances, all employees are clear of the lines and equipment, and all protective tags are removed from that point of disconnection.

1910.269(n) *Grounding for the protection of employees.*

1910.269(n)(1) Application. Paragraph (n) of this section applies to grounding of generation, transmission, and distribution lines and equipment for the purpose of protecting employees. Paragraph (n)(4) of this section also applies to protective grounding of other equipment as required elsewhere in this section.

Note to paragraph (n)(1): This paragraph covers grounding of generation, transmission, and distribution lines and equipment when this section requires protective grounding and whenever the employer chooses to ground such lines and equipment for the protection of employees.

1910.269(n)(2) General. For any employee to work transmission and distribution lines or equipment as deenergized, the employer shall ensure that the lines or equipment are deenergized under the provisions of paragraph (m) of this section and shall ensure proper grounding of the lines or equipment as specified in paragraphs (n)(3) through (n)(8) of this section. However, if the employer can demonstrate that installation of a ground is impracticable or that the conditions resulting from the installation of a ground would present greater hazards to employees than working without grounds, the lines and equipment may be treated as deenergized provided that the employer establishes that all of the following conditions apply:

1910.269(n)(2)(i) The employer ensures that the lines and equipment are deenergized under the provisions of paragraph (m) of this section.

1910.269(n)(2)(ii) There is no possibility of contact with another energized source.

1910.269(n)(2)(iii) The hazard of induced voltage is not present.

1910.269(n)(3) Equipotential zone. Temporary protective grounds shall be placed at such locations and arranged in such a manner that the employer can demonstrate will prevent each employee from being exposed to hazardous differences in electric potential. *Note to paragraph (n)(3)*: Appendix C to this section contains guidelines for establishing the equipotential zone required by this paragraph. The Occupational Safety and Health Administration will deem grounding practices meeting these guidelines as complying with paragraph (n)(3) of this section.

1910.269(n)(4) Protective grounding equipment.

1910.269(n)(4)(i) Protective grounding equipment shall be capable of conducting the maximum fault current that could flow at the point of grounding for the time necessary to clear the fault.

1910.269(n)(4)(ii) Protective grounding equipment shall have an ampacity greater than or equal to that of No. 2 AWG copper.

1910.269(n)(4)(iii) Protective grounds shall have an impedance low enough so that they do not delay the operation of protective devices in case of accidental energizing of the lines or equipment.

Note to paragraph (n)(4): American Society for Testing and Materials Standard Specifications for Temporary Protective Grounds to Be Used on De-Energized Electric Power Lines and Equipment, ASTM F855-09, contains guidelines for protective grounding equipment. The Institute of Electrical Engineers Guide for Protective Grounding of Power Lines, IEEE Std 1048-2003, contains guidelines for selecting and installing protective grounding equipment.

1910.269(n)(5) Testing. The employer shall ensure that, unless a previously installed ground is present, employees test lines and equipment and verify the absence of nominal voltage before employees install any ground on those lines or that equipment.

1910.269(n)(6) Connecting and removing grounds.

1910.269(n)(6)(i) The employer shall ensure that, when an employee attaches a ground to a line or to equipment, the employee attaches the ground-end connection first and then attaches the other end by means of a live-line tool. For lines or equipment operating at 600 volts or less, the employer may permit the employee to use insulating equipment other than a live-line tool if the employer ensures that the line or equipment is not energized at the time the ground is connected or if the employer can demonstrate that each employee is protected from hazards that may develop if the line or equipment is energized.

1910.269(n)(6)(ii) The employer shall ensure that, when an employee removes a ground, the employee removes the grounding device from the line or equipment using a live-line tool before he or she removes the ground-end connection. For lines or equipment operating at 600 volts or less, the employer may permit the employee to use insulating equipment other than a live-line tool if the employer ensures that the line or equipment is not energized at the time the ground is disconnected or if the employer can demonstrate that each employee is protected from hazards that may develop if the line or equipment is energized.

1910.269(n)(7) Additional precautions. The employer shall ensure that, when an employee performs work on a cable at a location remote from the cable terminal, the cable is not grounded at the cable terminal if there is a possibility of hazardous transfer of potential should a fault occur.

1910.269(n)(8) Removal of grounds for test. The employer may permit employees to remove grounds temporarily during tests. During the test procedure, the employer shall ensure that each employee uses insulating equipment, shall isolate each employee from any hazards involved, and shall implement any additional measures necessary to protect each exposed employee in case the previously grounded lines and equipment become energized.

1910.269(o) *Testing and test facilities.*

1910.269(o)(1) Application. Paragraph (o) of this section provides for safe work practices for high-voltage and high-power testing performed in laboratories, shops, and substations, and in the field and on electric transmission and distribution lines and equipment. It applies only to testing involving interim measurements using high voltage, high power, or combinations of high voltage and high power, and not to testing involving continuous measurements as in routine metering, relaying, and normal line work.

Note to paragraph (o)(1): OSHA considers routine inspection and maintenance measurements made by qualified employees to be routine line work not included in the scope of paragraph (o) of this section, provided that the hazards related to the use of intrinsic high-voltage or high-power sources require only the normal precautions associated with routine work specified in the other paragraphs of this section. Two typical examples of such excluded test work procedures are "phasing-out" testing and testing for a "no-voltage" condition.

1910.269(o)(2) General requirements.

1910.269(o)(2)(i) The employer shall establish and enforce work practices for the protection of each worker from the hazards of high-voltage or high-power testing at all test areas, temporary and permanent. Such work practices shall include, as a minimum, test area safeguarding, grounding, the safe use of measuring and control circuits, and a means providing for periodic safety checks of field test areas.

1910.269(o)(2)(ii) The employer shall ensure that each employee, upon initial assignment to the test area, receives training in safe work practices, with retraining provided as required by paragraph (a)(2) of this section.

1910.269(o)(3) Safeguarding of test areas.

1910.269(o)(3)(i) The employer shall provide safeguarding within test areas to control access to test equipment or to apparatus under test that could become energized as part of the testing by either direct or inductive coupling and to prevent accidental employee contact with energized parts.

1910.269(o)(3)(ii) The employer shall guard permanent test areas with walls, fences, or other barriers designed to keep employees out of the test areas.

1910.269(o)(3)(iii) In field testing, or at a temporary test site not guarded by permanent fences and gates, the employer shall ensure the use of one of the following means to prevent employees without authorization from entering:

1910.269(o)(3)(iii)(A) Distinctively colored safety tape supported approximately waist high with safety signs attached to it,

1910.269(o)(3)(iii)(B) A barrier or barricade that limits access to the test area to a degree equivalent, physically and visually, to the barricade specified in paragraph (o)(3)(iii)(A) of this section, or

1910.269(o)(3)(iii)(C) One or more test observers stationed so that they can monitor the entire area.

1910.269(o)(3)(iv) The employer shall ensure the removal of the safeguards required by paragraph (o)(3)(iii) of this section when employees no longer need the protection afforded by the safeguards.

1910.269(o)(4) Grounding practices.

1910.269(o)(4)(i) The employer shall establish and implement safe grounding practices for the test facility.

1910.269(o)(4)(i)(A) The employer shall maintain at ground potential all conductive parts accessible to the test operator while the equipment is operating at high voltage.

1910.269(o)(4)(i)(B) Wherever ungrounded terminals of test equipment or apparatus under test may be present, they shall be treated as energized until tests demonstrate that they are deenergized.

1910.269(o)(4)(ii) The employer shall ensure either that visible grounds are applied automatically, or that employees using properly insulated tools manually apply visible grounds, to the high-voltage circuits after they are deenergized and before any employee performs work on the circuit or on the item or apparatus under test. Common ground connections shall be solidly connected to the test equipment and the apparatus under test.

1910.269(o)(4)(iii) In high-power testing, the employer shall provide an isolated ground-return conductor system designed to prevent the intentional passage of current, with its attendant voltage rise, from occurring in the ground grid or in the earth. However, the employer need not provide an isolated ground-return conductor if the employer can demonstrate that both of the following conditions exist:

1910.269(o)(4)(iii)(A) The employer cannot provide an isolated ground-return conductor due to the distance of the test site from the electric energy source, and

1910.269(o)(4)(iii)(B) The employer protects employees from any hazardous step and touch potentials that may develop during the test.

Note to paragraph (o)(4)(iii)(B): See Appendix C to this section for information on measures that employers can take to protect employees from hazardous step and touch potentials.

1910.269(o)(4)(iv) For tests in which using the equipment grounding conductor in the equipment power cord to ground the test equipment would result in greater hazards to test personnel or prevent the taking of satisfactory measurements, the employer may use a ground clearly indicated in the test set-up if the employer can demonstrate that this ground affords protection for employees equivalent to the protection afforded by an equipment grounding conductor in the power supply cord.

1910.269(o)(4)(v) The employer shall ensure that, when any employee enters the test area after equipment is deenergized, a ground is placed on the high-voltage terminal and any other exposed terminals.

1910.269(o)(4)(v)(A) Before any employee applies a direct ground, the employer shall discharge high capacitance equipment through a resistor rated for the available energy.

1910.269(o)(4)(v)(B) A direct ground shall be applied to the exposed terminals after the stored energy drops to a level at which it is safe to do so.

1910.269(o)(4)(vi) If the employer uses a test trailer or test vehicle in field testing, its chassis shall be grounded. The employer shall protect each employee against hazardous touch potentials with respect to the vehicle, instrument panels, and other conductive parts accessible to employees with bonding, insulation, or isolation.

1910.269(o)(5) Control and measuring circuits.

1910.269(o)(5)(i) The employer may not run control wiring, meter connections, test leads, or cables from a test area unless contained in a grounded metallic sheath and terminated in a grounded metallic enclosure or unless the employer takes other precautions that it can demonstrate will provide employees with equivalent safety.

1910.269(o)(5)(ii) The employer shall isolate meters and other instruments with accessible terminals or parts from test personnel to protect against hazards that could arise should such terminals and parts become energized during testing. If the employer provides this isolation by locating test equipment in metal compartments with viewing windows, the employer shall provide interlocks to interrupt the power supply when someone opens the compartment cover.

1910.269(o)(5)(iii) The employer shall protect temporary wiring and its connections against damage, accidental interruptions, and other hazards. To the maximum extent possible, the employer shall keep signal, control, ground, and power cables separate from each other.

1910.269(o)(5)(iv) If any employee will be present in the test area during testing, a test observer shall be present. The test observer shall be capable of implementing the immediate deenergizing of test circuits for safety purposes.

1910.269(o)(6) Safety check.

1910.269(o)(6)(i) Safety practices governing employee work at temporary or field test areas shall provide, at the beginning of each series of tests, for a routine safety check of such test areas.

1910.269(o)(6)(ii) The test operator in charge shall conduct these routine safety checks before each series of tests and shall verify at least the following conditions:

1910.269(o)(6)(ii)(A) Barriers and safeguards are in workable condition and placed properly to isolate hazardous areas;

1910.269(o)(6)(ii)(B) System test status signals, if used, are in operable condition;

1910.269(o)(6)(ii)(C) Clearly marked test-power disconnects are readily available in an emergency;

1910.269(o)(6)(ii)(D) Ground connections are clearly identifiable;

1910.269(o)(6)(ii)(E) Personal protective equipment is provided and used as required by Subpart I of this part and by this section; and

1910.269(o)(6)(ii)(F) Proper separation between signal, ground, and power cables.

1910.269(p) *Mechanical equipment.*

1910.269(p)(1) General requirements.

1910.269(p)(1)(i) The critical safety components of mechanical elevating and rotating equipment shall receive a thorough visual inspection before use on each shift.

Note to paragraph (p)(1)(i): Critical safety components of mechanical elevating and rotating equipment are components for which failure would result in free fall or free rotation of the boom.

1910.269(p)(1)(ii) No motor vehicle or earthmoving or compacting equipment having an obstructed view to the rear may be operated on off-highway jobsites where any employee is exposed to the hazards created by the moving vehicle, unless:

1910.269(p)(1)(ii)(A) The vehicle has a reverse signal alarm audible above the surrounding noise level, or

1910.269(p)(1)(ii)(B) The vehicle is backed up only when a designated employee signals that it is safe to do so.

1910.269(p)(1)(iii) Rubber-tired self-propelled scrapers, rubber-tired front-end loaders, rubber-tired dozers, wheel-type agricultural and industrial tractors, crawler-type tractors, crawler-type loaders, and motor graders, with or without attachments, shall have rollover protective structures that meet the requirements of Subpart W of Part 1926 of this chapter.

1910.269(p)(1)(iv) The operator of an electric line truck may not leave his or her position at the controls while a load is suspended, unless the employer can demonstrate that no employee (including the operator) is endangered.

1910.269(p)(2) Outriggers.

1910.269(p)(2)(i) Mobile equipment, if provided with outriggers, shall be operated with the outriggers extended and firmly set, except as provided in paragraph (p)(2)(iii) of this section.

1910.269(p)(2)(ii) Outriggers may not be extended or retracted outside of the clear view of the operator unless all employees are outside the range of possible equipment motion.

1910.269(p)(2)(iii) If the work area or the terrain precludes the use of outriggers, the equipment may be operated only within its maximum load ratings specified by the equipment manufacturer for the particular configuration of the equipment without outriggers.

1910.269(p)(3) Applied loads. Mechanical equipment used to lift or move lines or other material shall be used within its maximum load rating and other design limitations for the conditions under which the mechanical equipment is being used.

1910.269(p)(4) Operations near energized lines or equipment.

1910.269(p)(4)(i) Mechanical equipment shall be operated so that the minimum approach distances, established by the employer under paragraph (l)(3)(i) of this section, are maintained from exposed energized lines and equipment. However, the insulated portion of an aerial lift operated by a qualified employee in the lift is exempt from this requirement if the applicable minimum approach distance is maintained between the uninsulated portions of the aerial lift and exposed objects having a different electrical potential.

1910.269(p)(4)(ii) A designated employee other than the equipment operator shall observe the approach distance to exposed lines and equipment and provide timely warnings before the minimum approach distance required by paragraph (p)(4)(i) of this section is reached, unless the employer can demonstrate that the operator can accurately determine that the minimum approach distance is being maintained.

1910.269(p)(4)(iii) If, during operation of the mechanical equipment, that equipment could become energized, the operation also shall comply with at least one of paragraphs (p)(4)(iii)(A) through (p)(4)(iii)(C) of this section.

1910.269(p)(4)(iii)(A) The energized lines or equipment exposed to contact shall be covered with insulating protective material that will withstand the type of contact that could be made during the operation.

1910.269(p)(4)(iii)(B) The mechanical equipment shall be insulated for the voltage involved. The mechanical equipment shall be positioned so that its uninsulated portions cannot approach the energized lines or equipment any closer than the minimum approach distances, established by the employer under paragraph (l)(3)(i) of this section.

1910.269(p)(4)(iii)(C) Each employee shall be protected from hazards that could arise from mechanical equipment contact with energized lines or equipment. The measures used shall ensure that employees will not be exposed to hazardous differences in electric potential. Unless the employer can demonstrate that the methods in use protect each employee from the hazards that could arise if the mechanical equipment contacts the energized line or equipment, the measures used shall include all of the following techniques:

1910.269(p)(4)(iii)(C)(1) Using the best available ground to minimize the time the lines or electric equipment remain energized,

1910.269(p)(4)(iii)(C)(2) Bonding mechanical equipment together to minimize potential differences,

1910.269(p)(4)(iii)(C)(3) Providing ground mats to extend areas of equipotential, and

1910.269(p)(4)(iii)(C)(4) Employing insulating protective equipment or barricades to guard against any remaining hazardous electrical potential differences.

Note to paragraph (p)(4)(iii)(C): Appendix C to this section contains information on hazardous step and touch potentials and on methods of protecting employees from hazards resulting from such potentials.

1910.269(q) *Overhead lines and live-line barehand work.* This paragraph provides additional requirements for work performed on or near overhead lines and equipment and for live-line barehand work.

1910.269(q)(1) General.

1910.269(q)(1)(i) Before allowing employees to subject elevated structures, such as poles or towers, to such stresses as climbing or the installation or removal of equipment may impose, the employer shall ascertain that the structures are capable of sustaining the additional or unbalanced stresses. If the pole or other structure cannot withstand the expected loads, the employer shall brace or otherwise support the pole or structure so as to prevent failure.

Note to paragraph (q)(1)(i): Appendix D to this section contains test methods that employers can use in ascertaining whether a wood pole is capable of sustaining the forces imposed by an employee climbing the pole. This paragraph also requires the employer to ascertain that the pole can sustain all other forces imposed by the work employees will perform.

1910.269(q)(1)(ii) When a pole is set, moved, or removed near an exposed energized overhead conductor, the pole may not contact the conductor.

1910.269(q)(1)(iii) When a pole is set, moved, or removed near an exposed energized overhead conductor, the employer shall ensure that each employee wears electrical protective equipment or uses insulated devices when handling the pole and that no employee contacts the pole with uninsulated parts of his or her body.

1910.269(q)(1)(iv) To protect employees from falling into holes used for placing poles, the employer shall physically guard the holes, or ensure that employees attend the holes, whenever anyone is working nearby.

1910.269(q)(2) Installing and removing overhead lines. The following provisions apply to the installation and removal of overhead conductors or cable (over-head lines).

1910.269(q)(2)(i) When lines that employees are installing or removing can contact energized parts, the employer shall use the tension-stringing method, barriers, or other equivalent measures to minimize the possibility that conductors and cables the employees are installing or removing will contact energized power lines or equipment.

1910.269(q)(2)(ii) For conductors, cables, and pulling and tensioning equipment, the employer shall provide the protective measures required by paragraph (p)(4)(iii) of this section when employees are installing or removing a conductor or cable close enough to energized conductors that any of the following failures could energize the pulling or tensioning equipment or the conductor or cable being installed or removed:

1910.269(q)(2)(ii)(A) Failure of the pulling or tensioning equipment,

1910.269(q)(2)(ii)(B) Failure of the conductor or cable being pulled, or

1910.269(q)(2)(ii)(C) Failure of the previously installed lines or equipment.

1910.269(q)(2)(iii) If the conductors that employees are installing or removing cross over energized conductors in excess of 600 volts and if the design of the circuit interrupting devices protecting the lines so permits, the employer shall render inoperable the automatic-reclosing feature of these devices.

1910.269(q)(2)(iv) Before employees install lines parallel to existing energized lines, the employer shall make a determination of the approximate voltage to be induced in the new lines, or work shall proceed on the assumption that the induced voltage is hazardous. Unless the employer can demonstrate that the lines that employees are installing are not subject to the induction of a hazardous voltage or unless the lines are treated as energized, temporary protective grounds shall be placed at such locations and arranged in such a manner that the employer can demonstrate will prevent exposure of each employee to hazardous differences in electric potential.

Note 1 to paragraph (q)(2)(iv): If the employer takes no precautions to protect employees from hazards associated with involuntary reactions from electric shock, a hazard exists if the induced voltage is sufficient to pass a current of 1 milliampere through a 500-ohm resistor. If the employer protects employees from injury due to involuntary reactions from electric shock, a hazard exists if the resultant current would be more than 6 milliamperes.

Note 2 to paragraph (q)(2)(iv): Appendix C to this section contains guidelines for protecting employees from hazardous differences in electric potential as required by this paragraph.

1910.269(q)(2)(v) Reel-handling equipment, including pulling and tensioning devices, shall be in safe operating condition and shall be leveled and aligned.

1910.269(q)(2)(vi) The employer shall ensure that employees do not exceed load ratings of stringing lines, pulling lines, conductor grips, load-bearing hardware and accessories, rigging, and hoists.

1910.269(q)(2)(vii) The employer shall repair or replace defective pulling lines and accessories.

1910.269(q)(2)(viii) The employer shall ensure that employees do not use conductor grips on wire rope unless the manufacturer specifically designed the grip for this application.

1910.269(q)(2)(ix) The employer shall ensure that employees maintain reliable communications, through two-way radios or other equivalent means, between the reel tender and the pulling rig operator.

1910.269(q)(2)(x) Employees may operate the pulling rig only when it is safe to do so.

Note to paragraph (q)(2)(x): Examples of unsafe conditions include: employees in locations prohibited by paragraph (q)(2)(xi) of this section, conductor and pulling line hang-ups, and slipping of the conductor grip.

1910.269(q)(2)(xi) While a power-driven device is pulling the conductor or pulling line and the conductor or pulling line is in motion, the employer shall ensure that employees are not directly under overhead operations or on the crossarm, except as necessary for the employees to guide the stringing sock or board over or through the stringing sheave.

1910.269(q)(3) Live-line barehand work. In addition to other applicable provisions contained in this section, the following requirements apply to live-line bare-hand work:

1910.269(q)(3)(i) Before an employee uses or supervises the use of the live-line barehand technique on energized circuits, the employer shall ensure that the employee completes training conforming to paragraph (a)(2) of this section in the technique and in the safety requirements of paragraph (q)(3) of this section.

1910.269(q)(3)(ii) Before any employee uses the liveline barehand technique on energized high-voltage conductors or parts, the employer shall ascertain the following information in addition to information about other existing conditions required by paragraph (a)(4) of this section:

1910.269(q)(3)(ii)(A) The nominal voltage rating of the circuit on which employees will perform the work,

1910.269(q)(3)(ii)(B) The clearances to ground of lines and other energized parts on which employees will perform the work, and

1910.269(q)(3)(ii)(C) The voltage limitations of equipment employees will use.

1910.269(q)(3)(iii) The employer shall ensure that the insulated equipment, insulated tools, and aerial devices and platforms used by employees are designed, tested, and made for live-line barehand work.

1910.269(q)(3)(iv) The employer shall ensure that employees keep tools and equipment clean and dry while they are in use.

1910.269(q)(3)(v) The employer shall render inoperable the automatic-reclosing feature of circuit-interrupting devices protecting the lines if the design of the devices permits.

1910.269(q)(3)(vi) The employer shall ensure that employees do not perform work when adverse weather conditions would make the work hazardous even after the employer implements the work practices required by this section. Additionally, employees may not perform work when winds reduce the phase-to-phase or phase-to-ground clearances at the work location below the minimum approach distances specified in paragraph (q)(3)(xiv) of this section, unless insulating guards cover the grounded objects and other lines and equipment.

Note to paragraph (q)(3)(vi): Thunderstorms in the vicinity, high winds, snow storms, and ice storms are examples of adverse weather conditions that make liveline barehand work too hazardous to perform safely even after the employer implements the work practices required by this section.

1910.269(q)(3)(vii) The employer shall provide and ensure that employees use a conductive bucket liner or other conductive device for bonding the insulated aerial device to the energized line or equipment.

1910.269(q)(3)(vii)(A) The employee shall be connected to the bucket liner or other conductive device by the use of conductive shoes, leg clips, or other means.

1910.269(q)(3)(vii)(B) Where differences in potentials at the worksite pose a hazard to employees, the employer shall provide electrostatic shielding designed for the voltage being worked.

1910.269(q)(3)(viii) The employer shall ensure that, before the employee contacts the energized part, the employee bonds the conductive bucket liner or other conductive device to the energized conductor by means of a positive connection. This connection shall remain attached to the energized conductor until the employee completes the work on the energized circuit.

1910.269(q)(3)(ix) Aerial lifts used for live-line barehand work shall have dual controls (lower and upper) as follows:

1910.269(q)(3)(ix)(A) The upper controls shall be within easy reach of the employee in the bucket. On a two-bucket-type lift, access to the controls shall be within easy reach of both buckets.

1910.269(q)(3)(ix)(B) The lower set of controls shall be near the base of the boom and shall be designed so that they can override operation of the equipment at any time.

1910.269(q)(3)(x) Lower (ground-level) lift controls may not be operated with an employee in the lift except in case of emergency.

1910.269(q)(3)(xi) The employer shall ensure that, before employees elevate an aerial lift into the work position, the employees check all controls (ground level and bucket) to determine that they are in proper working condition.

1910.269(q)(3)(xii) The employer shall ensure that, before employees elevate the boom of an aerial lift, the employees ground the body of the truck or barricade the body of the truck and treat it as energized.

1910.269(q)(3)(xiii) The employer shall ensure that employees perform a boom-current test before starting work each day, each time during the day when they encounter a higher voltage, and when changed conditions indicate a need for an additional test.

1910.269(q)(3)(xiii)(A) This test shall consist of placing the bucket in contact with an energized source equal to the voltage to be encountered for a minimum of 3 minutes.

1910.269(q)(3)(xiii)(B) The leakage current may not exceed 1 microampere per kilovolt of nominal phase-to-ground voltage.

1910.269(q)(3)(xiii)(C) The employer shall immediately suspend work from the aerial lift when there is any indication of a malfunction in the equipment.

1910.269(q)(3)(xiv) The employer shall ensure that employees maintain the minimum approach distances, established by the employer under paragraph (l)(3)(i) of this section, from all grounded objects and from lines and equipment at a potential different from that to which the live-line barehand equipment is bonded, unless insulating guards cover such grounded objects and other lines and equipment.

1910.269(q)(3)(xv) The employer shall ensure that, while an employee is approaching, leaving, or bonding to an energized circuit, the employee maintains the minimum approach distances, established by the employer under paragraph (l)(3)(i) of this section, between the employee and any grounded parts, including the lower boom and portions of the truck and between the employee and conductive objects energized at different potentials.

1910.269(q)(3)(xvi) While the bucket is alongside an energized bushing or insulator string, the employer shall ensure that employees maintain the phase-to ground minimum approach distances, established by the employer under paragraph (l)(3)(i) of this section, between all parts of the bucket and the grounded end of the bushing or insulator string or any other grounded surface.

1910.269(q)(3)(xvii) The employer shall ensure that employees do not use handlines between the bucket and the boom or between the bucket and the ground. However, employees may use nonconductive-type handlines from conductor to ground if not supported from the bucket. The employer shall ensure that no one uses ropes used for live-line barehand work for other purposes.

1910.269(q)(3)(xviii) The employer shall ensure that employees do not pass uninsulated equipment or material between a pole or structure and an aerial lift while an employee working from the bucket is bonded to an energized part.

1910.269(q)(3)(xix) A nonconductive measuring device shall be readily accessible to employees performing live-line barehand work to assist them in maintaining the required minimum approach distance.

1910.269(q)(4) Towers and structures. The following requirements apply to work performed on towers or other structures that support overhead lines.

1910.269(q)(4)(i) The employer shall ensure that no employee is under a tower or structure while work is in progress, except when the employer can demonstrate that such a working position is necessary to assist employees working above.

1910.269(q)(4)(ii) The employer shall ensure that employees use tag lines or other similar devices to maintain control of tower sections being raised or positioned, unless the employer can demonstrate that the use of such devices would create a greater hazard to employees.

1910.269(q)(4)(iii) The employer shall ensure that employees do not detach the loadline from a member or section until they safely secure the load.

1910.269(q)(4)(iv) The employer shall ensure that, except during emergency restoration procedures, employees discontinue work when adverse weather conditions would make the work hazardous in spite of the work practices required by this section.

Note to paragraph (q)(4)(iv): Thunderstorms in the vicinity, high winds, snow storms, and ice storms are examples of adverse weather conditions that make this work too hazardous to perform even after the employer implements the work practices required by this section.

1910.269(r) *Line-clearance tree trimming operations.* This paragraph provides additional requirements for line-clearance tree-trimming operations and for equipment used in these operations.

1910.269(r)(1) Electrical hazards. This paragraph does not apply to qualified employees.

1910.269(r)(1)(i) Before an employee climbs, enters, or works around any tree, a determination shall be made of the nominal voltage of electric power lines posing a hazard to employees. However, a determination of the maximum nominal voltage to which an employee will be exposed may be made instead, if all lines are considered as energized at this maximum voltage.

1910.269(r)(1)(ii) There shall be a second line-clearance tree trimmer within normal (that is, unassisted) voice communication under any of the following conditions:

1910.269(r)(1)(ii)(A) If a line-clearance tree trimmer is to approach more closely than 3.05 meters (10 ft) to any conductor or electric apparatus energized at more than 750 volts or

1910.269(r)(1)(ii)(B) If branches or limbs being removed are closer to lines energized at more than 750 volts than the distances listed in Table R-5, Table R-6, Table R-7, and Table R-8 or

1910.269(r)(1)(ii)(C) If roping is necessary to remove branches or limbs from such conductors or apparatus.

1910.269(r)(1)(iii) Line-clearance tree trimmers shall maintain the minimum approach distances from energized conductors given in Table R-5, Table R-6, Table R-7, and Table R-8.

1910.269(r)(1)(iv) Branches that are contacting exposed energized conductors or equipment or that are within the distances specified in Table R-5, Table R-6, Table R-7, and Table R-8 may be removed only through the use of insulating equipment.

Note to paragraph (r)(1)(iv): A tool constructed of a material that the employer can demonstrate has insulating qualities meeting paragraph (j)(1) of this section is considered as insulated under paragraph (r)(1)(iv) of this section if the tool is clean and dry.

1910.269(r)(1)(v) Ladders, platforms, and aerial devices may not be brought closer to an energized part than the distances listed in Table R-5, Table R-6, Table R-7, and Table R-8. **1910.269(r)(1)(vi)** Line-clearance tree-trimming work may not be performed when adverse weather conditions make the work hazardous in spite of the work practices required by this section. Each employee performing line-clearance tree trimming work in the aftermath of a storm or under similar emergency conditions shall be trained in the special hazards related to this type of work.

Note to paragraph (r)(1)(vi): Thunderstorms in the immediate vicinity, high winds, snow storms, and ice storms are examples of adverse weather conditions that are presumed to make line-clearance tree trimming work too hazardous to perform safely.

1910.269(r)(2) Brush chippers.

1910.269(r)(2)(i) Brush chippers shall be equipped with a locking device in the ignition system.

1910.269(r)(2)(ii) Access panels for maintenance and adjustment of the chipper blades and associated drive train shall be in place and secure during operation of the equipment. **1910.269(r)(2)(iii)** Brush chippers not equipped with a mechanical infeed system shall be equipped with an infeed hopper of length sufficient to prevent employees from contacting the blades or knives of the machine during operation.

1910.269(r)(2)(iv) Trailer chippers detached from trucks shall be chocked or otherwise secured.

1910.269(r)(2)(v) Each employee in the immediate area of an operating chipper feed table shall wear personal protective equipment as required by Subpart I of this part.

1910.269(r)(3) Sprayers and related equipment.

1910.269(r)(3)(i) Walking and working surfaces of sprayers and related equipment shall be covered with slip-resistant material. If slipping hazards cannot be eliminated, slip-resistant footwear or handrails and stair rails meeting the requirements of Subpart D of this part may be used instead of slip-resistant material.

1910.269(r)(3)(ii) Equipment on which employees stand to spray while the vehicle is in motion shall be equipped with guardrails around the working area. The guardrail shall be constructed in accordance with Subpart D of this part.

1910.269(r)(4) Stump cutters.

1910.269(r)(4)(i) Stump cutters shall be equipped with enclosures or guards to protect employees.

1910.269(r)(4)(ii) Each employee in the immediate area of stump grinding operations including the stump cutter operator) shall wear personal protective equipment as required by Subpart I of this part.

1910.269(r)(5) Gasoline-engine power saws. Gasoline-engine power saw operations shall meet the requirements of § 1910.266(e) and the following:

1910.269(r)(5)(i) Each power saw weighing more than 6.8 kilograms (15 lbs, service weight) that is used in trees shall be supported by a separate line, except when work is performed from an aerial lift and except during topping or removing operations where no supporting limb will be available.

1910.269(r)(5)(ii) Each power saw shall be equipped with a control that will return the saw to idling speed when released.

1910.269(r)(5)(iii) Each power saw shall be equipped with a clutch and shall be so adjusted that the clutch will not engage the chain drive at idling speed.

1910.269(r)(5)(iv) A power saw shall be started on the ground or where it is otherwise firmly supported. Drop starting of saws over 6.8 kilograms (15 lbs), other than chain saws, is permitted outside of the bucket of an aerial lift only if the area below the lift is clear of personnel.

Note to paragraph (r)(5)(iv): Paragraph (e)(2)(vi) of § 1910.266 prohibits drop starting of chain saws.

1910.269(r)(5)(v) A power saw engine may be started and operated only when all employees other than the operator are clear of the saw.

1910.269(r)(5)(vi) A power saw may not be running when the saw is being carried up into a tree by an employee.

1910.269(r)(5)(vii) Power saw engines shall be stopped for all cleaning, refueling, adjustments, and repairs to the saw or motor, except as the manufacturer's servicing procedures require otherwise.

1910.269(r)(6) Backpack power units for use in pruning and clearing.

1910.269(r)(6)(i) While a backpack power unit is running, no one other than the operator may be within 3.05 meters (10 ft) of the cutting head of a brush saw.

1910.269(r)(6)(ii) A backpack power unit shall be equipped with a quick shutoff switch readily accessible to the operator.

1910.269(r)(6)(iii) Backpack power unit engines shall be stopped for all cleaning, refueling, adjustments, and repairs to the saw or motor, except as the manufacturer's servicing procedures require otherwise.

1910.269(r)(7) Rope.

1910.269(r)(7)(i) Climbing ropes shall be used by employees working aloft in trees. These ropes shall have a minimum diameter of 12 millimeters (0.5 inch) with a minimum breaking strength of 10.2 kilonewtons (2,300 lbs). Synthetic rope shall have elasticity of not more than 7%.

1910.269(r)(7)(ii) Rope shall be inspected before each use and, if unsafe (for example, because of damage or defect), may not be used.

1910.269(r)(7)(iii) Rope shall be stored away from cutting edges and sharp tools. Rope contact with corrosive chemicals, gas, and oil shall be avoided.

1910.269(r)(7)(iv) When stored, rope shall be coiled and piled, or shall be suspended, so that air can circulate through the coils.

1910.269(r)(7)(v) Rope ends shall be secured to prevent their unraveling.

1910.269(r)(7)(vi) Climbing rope may not be spliced to effect repair.

1910.269(r)(7)(vii) A rope that is wet, that is contaminated to the extent that its insulating capacity is impaired, or that is otherwise not considered to be insulated for the voltage involved may not be used near exposed energized lines.

1910.269(r)(8) Fall protection. Each employee shall be tied in with a climbing rope and safety saddle when the employee is working above the ground in a tree, unless he or she is ascending into the tree.

1910.269(s) *Communication facilities.*

1910.269(s)(1) Microwave transmission.

1910.269(s)(1)(i) The employer shall ensure that no employee looks into an open waveguide or antenna connected to an energized microwave source.

1910.269(s)(1)(ii) If the electromagnetic-radiation level within an accessible area associated with microwave communications systems exceeds the radiation-protection guide specified by § 1910.97(a)(2), the employer shall post the area with warning signs containing the warning symbol described in § 1910.97(a)(3). The lower half of the warning symbol shall include the following statements, or ones that the employer can demonstrate are equivalent: "Radiation in this area may exceed hazard limitations and special precautions are required. Obtain specific instruction before entering."

1910.269(s)(1)(iii) When an employee works in an area where the electromagnetic radiation could exceed the radiation protection guide, the employer shall institute measures that ensure that the employee's exposure is not greater than that permitted by that guide. Such measures may include administrative and engineering controls and personal protective equipment.

1910.269(s)(2) Power-line carrier. The employer shall ensure that employees perform power-line carrier work, including work on equipment used for coupling carrier current to power line conductors, in accordance with the requirements of this section pertaining to work on energized lines.

1910.269(t) *Underground electrical installations.* This paragraph provides additional requirements for work on underground electrical installations.

1910.269(t)(1) Access. The employer shall ensure that employees use a ladder or other climbing device to enter and exit a manhole or subsurface vault exceeding 1.22 meters (4 ft) in depth. No employee may climb into or out of a manhole or vault by stepping on cables or hangers.

1910.269(t)(2) Lowering equipment into manholes.

1910.269(t)(2)(i) Equipment used to lower materials and tools into manholes or vaults shall be capable of supporting the weight to be lowered and shall be checked for defects before use.

1910.269(t)(2)(ii) Before anyone lowers tools or material into the opening for a manhole or vault, each employee working in the manhole or vault shall be clear of the area directly under the opening.

1910.269(t)(3) Attendants for manholes and vaults.

1910.269(t)(3)(i) While work is being performed in a manhole or vault containing energized electric equipment, an employee with first-aid training shall be available on the surface in the immediate vicinity of the manhole or vault entrance to render emergency assistance.

1910.269(t)(3)(ii) Occasionally, the employee on the surface may briefly enter a manhole or vault to provide nonemergency assistance.

Note 1 to paragraph (t)(3)(ii): Paragraph (e)(7) of this section may also require an attendant and does not permit this attendant to enter the manhole or vault.

Note 2 to paragraph (t)(3)(ii): Paragraph (l)(1)(ii) of this section requires employees entering manholes or vaults containing unguarded, uninsulated energized lines or parts of electric equipment operating at 50 volts or more to be qualified.

1910.269(t)(3)(iii) For the purpose of inspection, housekeeping, taking readings, or similar work, an employee working alone may enter, for brief periods of time, a manhole or vault where energized cables or equipment are in service if the employer can demonstrate that the employee will be protected from all electrical hazards.

1910.269(t)(3)(iv) The employer shall ensure that employees maintain reliable communications, through two-way radios or other equivalent means, among all employees involved in the job.

1910.269(t)(4) Duct rods. The employer shall ensure that, if employees use duct rods, the employees install the duct rods in the direction presenting the least hazard to employees. The employer shall station an employee at the far end of the duct line being rodded to ensure that the employees maintain the required minimum approach distances.

1910.269(t)(5) Multiple cables. When multiple cables are present in a work area, the employer shall identify the cable to be worked by electrical means, unless its identity is obvious by reason of distinctive appearance or location or by other readily apparent means of identification. The employer shall protect cables other than the one being worked from damage.

1910.269(t)(6) Moving cables. Except when paragraph (t)(7)(ii) of this section permits employees to perform work that could cause a fault in an energized cable in a manhole or vault, the employer shall ensure that employees inspect energized cables to be moved for abnormalities.

1910.269(t)(7) Protection against faults.

1910.269(t)(7)(i) Where a cable in a manhole or vault has one or more abnormalities that could lead to a fault or be an indication of an impending fault, the employer shall deenergize the cable with the abnormality before any employee may work in the manhole or vault, except when service-load conditions and a lack of feasible alternatives require that the cable remain energized. In that case, employees may enter the manhole or vault provided the employer protects them from the possible effects of a failure using shields or other devices that are capable of containing the adverse effects of a fault. The employer shall treat the following abnormalities as indications of impending faults unless the employer can demonstrate that the conditions could not lead to a fault: Oil or compound leaking from cable or joints, broken cable sheaths or joint sleeves, hot localized surface temperatures of cables or joints, or joints swollen beyond normal tolerance.

1910.269(t)(7)(ii) If the work employees will perform in a manhole or vault could cause a fault in a cable, the employer shall deenergize that cable before any employee works in the manhole or vault, except when service-load conditions and a lack of feasible alternatives require that the cable remain energized. In that case, employees may enter the manhole or vault provided the employer protects them from the possible effects of a failure using shields or other devices that are capable of containing the adverse effects of a fault.

1910.269(t)(8) Sheath continuity. When employees perform work on buried cable or on cable in a manhole or vault, the employer shall maintain metallic-sheath continuity, or the cable sheath shall be treated as energized.

1910.269(u) *Substations.* This paragraph provides additional requirements for substations and for work performed in them.

1910.269(u)(1) Access and working space. The employer shall provide and maintain sufficient access and working space about electric equipment to permit ready and safe operation and maintenance of such equipment by employees.

Note to paragraph (u)(1): American National Standard National Electrical Safety Code, ANSI/IEEE C2-2012 contains guidelines for the dimensions of access and working space about electric equipment in substations. Installations meeting the ANSI provisions comply with paragraph (u)(1) of this section. The Occupational Safety and Health Administration will determine whether an installation that does not conform to this ANSI standard complies with paragraph (u)(1) of this section based on the following criteria:

(1) Whether the installation conforms to the edition of ANSI C2 that was in effect when the installation was made,

(2) Whether the configuration of the installation enables employees to maintain the minimum approach distances, established by the employer under paragraph (l)(3)(i) of this section, while the employees are working on exposed, energized parts, and

(3) Whether the precautions taken when employees perform work on the installation provide protection equivalent to the protection provided by access and working space meeting ANSI/IEEE C2-2012.

1910.269(u)(2) Draw-out-type circuit breakers. The employer shall ensure that, when employees remove or insert draw-out-type circuit breakers, the breaker is in the open position. The employer shall also render the control circuit inoperable if the design of the equipment permits.

1910.269(u)(3) Substation fences. Conductive fences around substations shall be grounded. When a substation fence is expanded or a section is removed, fence sections shall be isolated, grounded, or bonded as necessary to protect employees from hazardous differences in electric potential.

Note to paragraph (u)(3): IEEE Std 80-2000, IEEE Guide for Safety in AC Substation Grounding, contains guidelines for protection against hazardous differences in electric potential.

1910.269(u)(4) Guarding of rooms and other spaces containing electric supply equipment.

1910.269(u)(4)(i) Rooms and other spaces in which electric supply lines or equipment are installed shall meet the requirements of paragraphs (u)(4)(ii) through (u)(4)(v) of this section under the following conditions:

1910.269(u)(4)(i)(A) If exposed live parts operating at 50 to 150 volts to ground are within 2.4 meters (8 ft) of the ground or other working surface inside the room or other space,

1910.269(u)(4)(i)(B) If live parts operating at 151 to 600 volts to ground and located within 2.4 meters (8 ft) of the ground or other working surface inside the room or other space are guarded only by location, as permitted under paragraph (u)(5)(i) of this section, or

1910.269(u)(4)(i)(C) If live parts operating at more than 600 volts to ground are within the room or other space, unless:

1910.269(u)(4)(i)(C)(1) The live parts are enclosed within grounded, metal-enclosed equipment whose only openings are designed so that foreign objects inserted in these openings will be deflected from energized parts, or

1910.269(u)(4)(i)(C)(2) The live parts are installed at a height, above ground and any other working surface, that provides protection at the voltage on the live parts corresponding to the protection provided by a 2.4-meter (8-ft) height at 50 volts.

1910.269(u)(4)(ii) Fences, screens, partitions, or walls shall enclose the rooms and other spaces so as to minimize the possibility that unqualified persons will enter.

1910.269(u)(4)(iii) Unqualified persons may not enter the rooms or other spaces while the electric supply lines or equipment are energized.

1910.269(u)(4)(iv) The employer shall display signs at entrances to the rooms and other spaces warning unqualified persons to keep out.

1910.269(u)(4)(v) The employer shall keep each entrance to a room or other space locked, unless the entrance is under the observation of a person who is attending the room or other space for the purpose of preventing unqualified employees from entering.

1910.269(u)(5) Guarding of energized parts.

1910.269(u)(5)(i) The employer shall provide guards around all live parts operating at more than 150 volts to ground without an insulating covering unless the location of the live parts gives sufficient clearance (horizontal, vertical, or both) to minimize the possibility of accidental employee contact.

Note to paragraph (u)(5)(i): American National Standard National Electrical Safety Code, ANSI/IEEE C2-2002 contains guidelines for the dimensions of clearance distances about electric equipment in substations. Installations meeting the ANSI provisions comply with paragraph (u)(5)(i) of this section. The Occupational Safety and Health Administration will determine whether an installation that does not conform to this ANSI standard complies with paragraph (u)(5)(i) of this section based on the following criteria:

(1) Whether the installation conforms to the edition of ANSI C2 that was in effect when the installation was made,

(2) Whether each employee is isolated from energized parts at the point of closest approach; and

(3) Whether the precautions taken when employees perform work on the installation provide protection equivalent to the protection provided by horizontal and vertical clearances meeting ANSI/IEEE C2-2002.

1910.269(u)(5)(ii) Except for fuse replacement and other necessary access by qualified persons, the employer shall maintain guarding of energized parts within a compartment during operation and maintenance functions to prevent accidental contact with energized parts and to prevent dropped tools or other equipment from contacting energized parts.

1910.269(u)(5)(iii) Before guards are removed from energized equipment, the employer shall install barriers around the work area to prevent employees who are not working on the equipment, but who are in the area, from contacting the exposed live parts.

1910.269(u)(6) Substation entry.

1910.269(u)(6)(i) Upon entering an attended substation, each employee, other than employees regularly working in the station, shall report his or her presence to the employee in charge of substation activities to receive information on special system conditions affecting employee safety.

1910.269(u)(6)(ii) The job briefing required by paragraph (c) of this section shall cover information on special system conditions affecting employee safety, including the location of energized equipment in or adjacent to the work area and the limits of any deenergized work area.

1910.269(v) *Power generation.* This paragraph provides additional requirements and related work practices for power generating plants.

1910.269(v)(1) Interlocks and other safety devices.

1910.269(v)(1)(i) Interlocks and other safety devices shall be maintained in a safe, operable condition.

1910.269(v)(1)(ii) No interlock or other safety device may be modified to defeat its function, except for test, repair, or adjustment of the device.

1910.269(v)(2) Changing brushes. Before exciter or generator brushes are changed while the generator is in service, the exciter or generator field shall be checked to determine whether a ground condition exists. The brushes may not be changed while the generator is energized if a ground condition exists.

1910.269(v)(3) Access and working space. The employer shall provide and maintain sufficient access and working space about electric equipment to permit ready and safe operation and maintenance of such equipment by employees.

Note to paragraph (v)(3) of this section: American National Standard National Electrical Safety Code, ANSI/IEEE C2-2012 contains guidelines for the dimensions of access and working space about electric equipment in substations. Installations meeting the ANSI provisions comply with paragraph (v)(3) of this section. The Occupational Safety and Health Administration will determine whether an installation that does not conform to this ANSI standard complies with paragraph (v)(3) of this section based on the following criteria:

(1) Whether the installation conforms to the edition of ANSI C2 that was in effect when the installation was made;

(2) Whether the configuration of the installation enables employees to maintain the minimum approach distances, established by the employer under paragraph (l)(3)(i) of this section, while the employees are working on exposed, energized parts, and;

(3) Whether the precautions taken when employees perform work on the installation provide protection equivalent to the protection provided by access and working space meeting ANSI/IEEE C2-2012.

1910.269(v)(4) Guarding of rooms and other spaces containing electric supply equipment.

1910.269(v)(4)(i) Rooms and other spaces in which electric supply lines or equipment are installed shall meet the requirements of paragraphs (v)(4)(ii) through (v)(4)(v) of this section under the following conditions:

1910.269(v)(4)(i)(A) If exposed live parts operating at 50 to 150 volts to ground are within 2.4 meters (8 ft) of the ground or other working surface inside the room or other space,

1910.269(v)(4)(i)(B) If live parts operating at 151 to 600 volts to ground and located within 2.4 meters (8 ft) of the ground or other working surface inside the room or other space are guarded only by location, as permitted under paragraph (v)(5)(i) of this section, or

1910.269(v)(4)(i)(C) If live parts operating at more than 600 volts to ground are within the room or other space, unless:

1910.269(v)(4)(i)(C)(1) The live parts are enclosed within grounded, metal-enclosed equipment whose only openings are designed so that foreign objects inserted in these openings will be deflected from energized parts, or

1910.269(v)(4)(i)(C)(2) The live parts are installed at a height, above ground and any other working surface, that provides protection at the voltage on the live parts corresponding to the protection provided by a 2.4-meter (8-ft) height at 50 volts.

1910.269(v)(4)(ii) Fences, screens, partitions, or walls shall enclose the rooms and other spaces so as to minimize the possibility that unqualified persons will enter.

1910.269(v)(4)(iii) Unqualified persons may not enter the rooms or other spaces while the electric supply lines or equipment are energized.

1910.269(v)(4)(iv) The employer shall display signs at entrances to the rooms and other spaces warning unqualified persons to keep out.

1910.269(v)(4)(v) The employer shall keep each entrance to a room or other space locked, unless the entrance is under the observation of a person who is attending the room or other space for the purpose of preventing unqualified employees from entering.

1910.269(v)(5) Guarding of energized parts.

1910.269(v)(5)(i) The employer shall provide guards around all live parts operating at more than 150 volts to ground without an insulating covering unless the location of the live parts gives sufficient clearance (horizontal, vertical, or both) to minimize the possibility of accidental employee contact.

Note to paragraph (v)(5)(i): American National Standard National Electrical Safety Code, ANSI/IEEE C2-2002 contains guidelines for the dimensions of clearance distances about electric equipment in substations. Installations meeting the ANSI provisions comply with paragraph (v)(5)(i) of this section. The Occupational Safety and Health Administration will determine whether an installation that does not conform to this ANSI standard complies with paragraph (v)(5)(i) of this section based on the following criteria:

(1) Whether the installation conforms to the edition of ANSI C2 that was in effect when the installation was made;

(2) Whether each employee is isolated from energized parts at the point of closest approach; and

(3) Whether the precautions taken when employees perform work on the installation provide protection equivalent to the protection provided by horizontal and vertical clearances meeting ANSI/IEEE C2-2002.

1910.269(v)(5)(ii) Except for fuse replacement and other necessary access by qualified persons, the employer shall maintain guarding of energized parts within a compartment during operation and maintenance functions to prevent accidental contact with energized parts and to prevent dropped tools or other equipment from contacting energized parts.

1910.269(v)(5)(iii) Before guards are removed from energized equipment, the employer shall install barriers around the work area to prevent employees who are not working on the equipment, but who are in the area, from contacting the exposed live parts.

1910.269(v)(6) Water or steam spaces. The following requirements apply to work in water and steam spaces associated with boilers:

1910.269(v)(6)(i) A designated employee shall inspect conditions before work is permitted and after its completion. Eye protection, or full face protection if necessary, shall be worn at all times when condenser, heater, or boiler tubes are being cleaned.

1910.269(v)(6)(ii) Where it is necessary for employees to work near tube ends during cleaning, shielding shall be installed at the tube ends.

1910.269(v)(7) Chemical cleaning of boilers and pressure vessels. The following requirements apply to chemical cleaning of boilers and pressure vessels:

1910.269(v)(7)(i) Areas where chemical cleaning is in progress shall be cordoned off to restrict access during cleaning. If flammable liquids, gases, or vapors or combustible materials will be used or might be produced during the cleaning process, the following requirements also apply:

1910.269(v)(7)(i)(A) The area shall be posted with signs restricting entry and warning of the hazards of fire and explosion; and

1910.269(v)(7)(i)(B) Smoking, welding, and other possible ignition sources are prohibited in these restricted areas.

1910.269(v)(7)(ii) The number of personnel in the restricted area shall be limited to those necessary to accomplish the task safely.

1910.269(v)(7)(iii) There shall be ready access to water or showers for emergency use.

Note to paragraph (v)(7)(iii): See § 1910.141 for requirements that apply to the water supply and to washing facilities.

1910.269(v)(7)(iv) Employees in restricted areas shall wear protective equipment meeting the requirements of Subpart I of this part and including, but not limited to, protective clothing, boots, goggles, and gloves.

1910.269(v)(8) Chlorine systems.

1910.269(v)(8)(i) Chlorine system enclosures shall be posted with signs restricting entry and warning of the hazard to health and the hazards of fire and explosion.

Note to paragraph (v)(8)(i): See Subpart Z of this part for requirements necessary to protect the health of employees from the effects of chlorine.

1910.269(v)(8)(ii) Only designated employees may enter the restricted area. Additionally, the number of personnel shall be limited to those necessary to accomplish the task safely.

1910.269(v)(8)(iii) Emergency repair kits shall be available near the shelter or enclosure to allow for the prompt repair of leaks in chlorine lines, equipment, or containers.

1910.269(v)(8)(iv) Before repair procedures are started, chlorine tanks, pipes, and equipment shall be purged with dry air and isolated from other sources of chlorine.

1910.269(v)(8)(v) The employer shall ensure that chlorine is not mixed with materials that would react with the chlorine in a dangerously exothermic or other hazardous manner.

1910.269(v)(9) Boilers.

1910.269(v)(9)(i) Before internal furnace or ash hopper repair work is started, overhead areas shall be inspected for possible falling objects. If the hazard of falling objects exists, overhead protection such as planking or nets shall be provided.

1910.269(v)(9)(ii) When opening an operating boiler door, employees shall stand clear of the opening of the door to avoid the heat blast and gases which may escape from the boiler.

1910.269(v)(10) Turbine generators.

1910.269(v)(10)(i) Smoking and other ignition sources are prohibited near hydrogen or hydrogen sealing systems, and signs warning of the danger of explosion and fire shall be posted.

1910.269(v)(10)(ii) Excessive hydrogen makeup or abnormal loss of pressure shall be considered as an emergency and shall be corrected immediately.

1910.269(v)(10)(iii) A sufficient quantity of inert gas shall be available to purge the hydrogen from the largest generator.

1910.269(v)(11) Coal and ash handling.

1910.269(v)(11)(i) Only designated persons may operate railroad equipment.

1910.269(v)(11)(ii) Before a locomotive or locomotive crane is moved, a warning shall be given to employees in the area.

1910.269(v)(11)(iii) Employees engaged in switching or dumping cars may not use their feet to line up drawheads.

1910.269(v)(11)(iv) Drawheads and knuckles may not be shifted while locomotives or cars are in motion.

1910.269(v)(11)(v) When a railroad car is stopped for unloading, the car shall be secured from displacement that could endanger employees.

1910.269(v)(11)(vi) An emergency means of stopping dump operations shall be provided at railcar dumps.

1910.269(v)(11)(vii) The employer shall ensure that employees who work in coal- or ash-handling conveyor areas are trained and knowledgeable in conveyor operation and in the requirements of paragraphs (v)(11)(viii) through (v)(11)(xii) of this section.

1910.269(v)(11)(viii) Employees may not ride a coal- or ash-handling conveyor belt at any time. Employees may not cross over the conveyor belt, except at walkways, unless the conveyor's energy source has been deenergized and has been locked out or tagged in accordance with paragraph (d) of this section.

1910.269(v)(11)(ix) A conveyor that could cause injury when started may not be started until personnel in the area are alerted by a signal or by a designated person that the conveyor is about to start.

1910.269(v)(11)(x) If a conveyor that could cause injury when started is automatically controlled or is controlled from a remote location, an audible device shall be provided that sounds an alarm that will be recognized by each employee as a warning that the conveyor will start and that can be clearly heard at all points along the conveyor where personnel may be present. The warning device shall be actuated by the device starting the conveyor and shall continue for a period of time before the conveyor starts that is long enough to allow employees to move clear of the conveyor system. A visual warning may be used in place of the audible device if the employer can demonstrate that it will provide an equally effective warning in the particular circumstances involved. However, if the employer can demonstrate that the system's function would be seriously hindered by the required time delay, warning signs may be provided in place of the audible warning device. If the system was installed before January 31, 1995, warning signs may be provided in place of the audible warning device until such time as the conveyor or its control system is rebuilt or rewired. These warning signs shall be clear, concise, and legible and shall indicate that conveyors and allied equipment may be started at any time, that danger exists, and that personnel must keep clear. These warning signs shall be provided along the conveyor at areas not guarded by position or location.

1910.269(v)(11)(xi) Remotely and automatically controlled conveyors, and conveyors that have operating stations which are not manned or which are beyond voice and visual contact from drive areas, loading areas, transfer points, and other locations on the conveyor path not guarded by location, position, or guards shall be furnished with emergency stop buttons, pull cords, limit switches, or similar emergency stop devices. However, if the employer can demonstrate that the design, function, and operation of the conveyor do not expose an employee to hazards, an emergency stop device is not required.

1910.269(v)(11)(xi)(A) Emergency stop devices shall be easily identifiable in the immediate vicinity of such locations.

1910.269(v)(11)(xi)(B) An emergency stop device shall act directly on the control of the conveyor involved and may not depend on the stopping of any other equipment.

1910.269(v)(11)(xi)(C) Emergency stop devices shall be installed so that they cannot be overridden from other locations.

1910.269(v)(11)(xii) Where coal-handling operations may produce a combustible atmosphere from fuel sources or from flammable gases or dust, sources of ignition shall be eliminated or safely controlled to prevent ignition of the combustible atmosphere.

Note to paragraph (v)(11)(xii): Locations that are hazardous because of the presence of combustible dust are classified as Class II hazardous locations. See § 1910.307.

1910.269(v)(11)(xiii) An employee may not work on or beneath overhanging coal in coal bunkers, coal silos, or coal storage areas, unless the employee is protected from all hazards posed by shifting coal.

1910.269(v)(11)(xiv) An employee entering a bunker or silo to dislodge the contents shall wear a body harness with lifeline attached. The lifeline shall be secured to a fixed support outside the bunker and shall be attended at all times by an employee located outside the bunker or facility.

1910.269(v)(12) Hydro plants and equipment. Employees working on or close to water gates, valves, intakes, forebays, flumes, or other locations where increased or decreased water flow or levels may pose a significant hazard shall be warned and shall vacate such dangerous areas before water flow changes are made.

1910.269(w) *Special conditions.*

1910.269(w)(1) Capacitors. The following additional requirements apply to work on capacitors and on lines connected to capacitors.

Note to paragraph (w)(1): See paragraphs (m) and (n) of this section for requirements pertaining to the deenergizing and grounding of capacitor installations.

1910.269(w)(1)(i) Before employees work on capacitors, the employer shall disconnect the capacitors from energized sources and short circuit the capacitors. The employer shall ensure that the employee short circuiting the capacitors waits at least 5 minutes from the time of disconnection before applying the short circuit,

1910.269(w)(1)(ii) Before employees handle the units, the employer shall short circuit each unit in series-parallel capacitor banks between all terminals and the capacitor case or its rack. If the cases of capacitors are on ungrounded substation racks, the employer shall bond the racks to ground.

1910.269(w)(1)(iii) The employer shall short circuit any line connected to capacitors before the line is treated as deenergized.

1910.269(w)(2) Current transformer secondaries. The employer shall ensure that employees do not open the secondary of a current transformer while the transformer is energized. If the employer cannot deenergize the primary of the current transformer before employees perform work on an instrument, a relay, or other section of a current transformer secondary circuit, the employer shall bridge the circuit so that the current transformer secondary does not experience an open-circuit condition.

1910.269(w)(3) Series street lighting.

1910.269(w)(3)(i) If the open-circuit voltage exceeds 600 volts, the employer shall ensure that employees work on series street lighting circuits in accordance with paragraph (q) or (t) of this section, as appropriate.

1910.269(w)(3)(ii) Before any employee opens a series loop, the employer shall deenergize the street lighting transformer and isolate it from the source of supply or shall bridge the loop to avoid an open-circuit condition.

1910.269(w)(4) Illumination. The employer shall provide sufficient illumination to enable the employee to perform the work safely.

1910.269(w)(5) Protection against drowning.

1910.269(w)(5)(i) Whenever an employee may be pulled or pushed, or might fall, into water where the danger of drowning exists, the employer shall provide the employee with, and shall ensure that the employee uses, a U.S. Coast Guard approved personal flotation device.

1910.269(w)(5)(ii) The employer shall maintain each personal flotation device in safe condition and shall inspect each personal flotation device frequently enough to ensure that it does not have rot, mildew, water saturation, or any other condition that could render the device unsuitable for use.

1910.269(w)(5)(iii) An employee may cross streams or other bodies of water only if a safe means of passage, such as a bridge, is available.

1910.269(w)(6) Employee protection in public work areas.

1910.269(w)(6)(i) Traffic-control signs and traffic-control devices used for the protection of employees shall meet § 1926.200(g)(2) of this chapter.

1910.269(w)(6)(ii) Before employees begin work in the vicinity of vehicular or pedestrian traffic that may endanger them, the employer shall place warning signs or flags and other traffic-control devices in conspicuous locations to alert and channel approaching traffic.

1910.269(w)(6)(iii) The employer shall use barricades where additional employee protection is necessary.

1910.269(w)(6)(iv) The employer shall protect excavated areas with barricades.
1910.269(w)(6)(v) The employer shall display warning lights prominently at night.
1910.269(w)(7) Backfeed. When there is a possibility of voltage backfeed from sources of cogeneration or from the secondary system (for example, backfeed from more than one energized phase feeding a common load), the requirements of paragraph (l) of this section apply if employees will work the lines or equipment as energized, and the requirements of paragraphs (m) and (n) of this section apply if employees will work the lines or equipment as deenergized.

1910.269(w)(8) Lasers. The employer shall install, adjust, and operate laser equipment in accordance with § 1926.54 of this chapter.

1910.269(w)(9) Hydraulic fluids. Hydraulic fluids used for the insulated sections of equipment shall provide insulation for the voltage involved.

1910.269(x) *Definitions.*

Affected employee. An employee whose job requires him or her to operate or use a machine or equipment on which servicing or maintenance is being performed under lockout or tagout, or whose job requires him or her to work in an area in which such servicing or maintenance is being performed.

Attendant. An employee assigned to remain immediately outside the entrance to an enclosed or other space to render assistance as needed to employees inside the space.

Authorized employee. An employee who locks out or tags out machines or equipment in order to perform servicing or maintenance on that machine or equipment. An affected employee becomes an authorized employee when that employee's duties include performing servicing or maintenance covered under this section.

Automatic circuit recloser. A self-controlled device for automatically interrupting and reclosing an alternating-current circuit, with a predetermined sequence of opening and reclosing followed by resetting, hold closed, or lockout.

Barricade. A physical obstruction such as tapes, cones, or A-frame type wood or metal structures that provides a warning about, and limits access to, a hazardous area.

Barrier. A physical obstruction that prevents contact with energized lines or equipment or prevents unauthorized access to a work area.

Bond. The electrical interconnection of conductive parts designed to maintain a common electric potential.

Bus. A conductor or a group of conductors that serve as a common connection for two or more circuits.

Bushing. An insulating structure that includes a through conductor or that provides a passageway for such a conductor, and that, when mounted on a barrier, insulates the conductor from the barrier for the purpose of conducting current from one side of the barrier to the other.

Cable. A conductor with insulation, or a stranded conductor with or without insulation and other coverings (single conductor cable), or a combination of conductors insulated from one another (multiple-conductor cable).

Cable sheath. A conductive protective covering applied to cables.

Note to the definition of "cable sheath": A cable sheath may consist of multiple layers one or more of which is conductive.

Circuit. A conductor or system of conductors through which an electric current is intended to flow.

Clearance (between objects). The clear distance between two objects measured surface to surface.

Clearance (for work). Authorization to perform specified work or permission to enter a restricted area.

Communication lines. (See Lines; (1) Communication lines.)

Conductor. A material, usually in the form of a wire, cable, or bus bar, used for carrying an electric current.

Contract employer. An employer, other than a host employer, that performs work covered by this section under contract.

Covered conductor. A conductor covered with a dielectric having no rated insulating strength or having a rated insulating strength less than the voltage of the circuit in which the conductor is used.

Current-carrying part. A conducting part intended to be connected in an electric circuit to a source of voltage. Non-current-carrying parts are those not intended to be so connected.

Deenergized. Free from any electrical connection to a source of potential difference and from electric charge; not having a potential that is different from the potential of the earth.

Note to the definition of "deenergized": The term applies only to current-carrying parts, which are sometimes energized (alive).

Designated employee (designated person). An employee (or person) who is assigned by the employer to perform specific duties under the terms of this section and who has sufficient knowledge of the construction and operation of the equipment, and the hazards involved, to perform his or her duties safely.

Electric line truck. A truck used to transport personnel, tools, and material for electric supply line work.

Electric supply equipment. Equipment that produces, modifies, regulates, controls, or safeguards a supply of electric energy.

Electric supply lines. (See Lines; (2) Electric supply lines.)

Electric utility. An organization responsible for the installation, operation, or maintenance of an electric supply system.

Enclosed space. A working space, such as a manhole, vault, tunnel, or shaft, that has a limited means of egress or entry, that is designed for periodic employee entry under normal operating conditions, and that, under normal conditions, does not contain a hazardous atmosphere, but may contain a hazardous atmosphere under abnormal conditions.

Note to the definition of "enclosed space": The Occupational Safety and Health Administration does not consider spaces that are enclosed but not designed for employee entry under normal operating conditions to be enclosed spaces for the purposes of this section. Similarly, the Occupational Safety and Health Administration does not consider spaces that are enclosed and that are expected to contain a hazardous atmosphere to be enclosed spaces for the purposes of this section. Such spaces meet the definition of permit spaces in § 1910.146, and entry into them must conform to that standard.

Energized (alive, live). Electrically connected to a source of potential difference, or electrically charged so as to have a potential significantly different from that of earth in the vicinity.

Energy isolating device. A physical device that prevents the transmission or release of energy, including, but not limited to, the following: a manually operated electric circuit breaker, a disconnect switch, a manually operated switch, a slide gate, a slip blind, a line valve, blocks, and any similar device with a visible indication of the position of the device. (Push buttons, selector switches, and other control-circuit-type devices are not energy isolating devices.)

Energy source. Any electrical, mechanical, hydraulic, pneumatic, chemical, nuclear, thermal, or other energy source that could cause injury to employees.

Entry (as used in paragraph (e) of this section). The action by which a person passes through an opening into an enclosed space. Entry includes ensuing work activities in that space and is considered to have occurred as soon as any part of the entrant's body breaks the plane of an opening into the space.

Equipment (electric). A general term including material, fittings, devices, appliances, fixtures, apparatus, and the like used as part of or in connection with an electrical installation.

Exposed, Exposed to contact (as applied to energized parts). Not isolated or guarded.

Fall restraint system. A fall protection system that prevents the user from falling any distance.

First-aid training. Training in the initial care, including cardiopulmonary resuscitation (which includes chest compressions, rescue breathing, and, as appropriate, other heart and lung resuscitation techniques), performed by a person who is not a medical practitioner, of a sick or injured person until definitive medical treatment can be administered.

Ground. A conducting connection, whether planned or unplanned, between an electric circuit or equipment and the earth, or to some conducting body that serves in place of the earth.

Grounded. Connected to earth or to some conducting body that serves in place of the earth.

Guarded. Covered, fenced, enclosed, or otherwise protected, by means of suitable covers or casings, barrier rails or screens, mats, or platforms, designed to minimize the possibility, under normal conditions, of dangerous approach or inadvertent contact by persons or objects.

Note to the definition of "guarded": Wires that are insulated, but not otherwise protected, are not guarded.

Hazardous atmosphere. An atmosphere that may expose employees to the risk of death, incapacitation, impairment of ability to self-rescue (that is, escape unaided from an enclosed space), injury, or acute illness from one or more of the following causes:

(1) Flammable gas, vapor, or mist in excess of 10% of its lower flammable limit (LFL);

(2) Airborne combustible dust at a concentration that meets or exceeds its LFL;

Note to the definition of "hazardous atmosphere" (2): This concentration may be approximated as a condition in which the dust obscures vision at a distance of 1.52 meters (5 ft) or less.

(3) Atmospheric oxygen concentration below 19.5% or above 23.5%;

(4) Atmospheric concentration of any substance for which a dose or a permissible exposure limit is published in Subpart G, Occupational Health and Environmental Control, or in Subpart Z, Toxic and Hazardous Substances, of this part and which could result in employee exposure in excess of its dose or permissible exposure limit;

Note to the definition of "hazardous atmosphere" (4): An atmospheric concentration of any substance that is not capable of causing death, incapacitation, impairment of ability to self-rescue, injury, or acute illness due to its health effects is not covered by this provision.

(5) Any other atmospheric condition that is immediately dangerous to life or health.

Note to the definition of "hazardous atmosphere" (5): For air contaminants for which the Occupational Safety and Health Administration has not determined a dose or permissible exposure limit, other sources of information, such as - Safety Data Sheets that comply with the Hazard Communication Standard, § 1910.1200, published information, and internal documents can provide guidance in establishing acceptable atmospheric conditions.

High-power tests. Tests in which the employer uses fault currents, load currents, magnetizing currents, and line dropping currents to test equipment, either at the equipment's rated voltage or at lower voltages.

High-voltage tests. Tests in which the employer uses voltages of approximately 1,000 volts as a practical minimum and in which the voltage source has sufficient energy to cause injury.

High wind. A wind of such velocity that one or more of the following hazards would be present:

(1) The wind could blow an employee from an elevated location,

(2) The wind could cause an employee or equipment handling material to lose control of the material, or

(3) The wind would expose an employee to other hazards not controlled by the standard involved.

Note to the definition of "high wind": The Occupational Safety and Health Administration normally considers winds exceeding 64.4 kilometers per hour (40 miles per hour), or 48.3 kilometers per hour (30 miles per hour) if the work involves material handling, as meeting this criteria, unless the employer takes precautions to protect employees from the hazardous effects of the wind.

Host employer. An employer that operates, or that controls the operating procedures for, an electric power generation, transmission, or distribution installation on which a contract employer is performing work covered by this section.

Note to the definition of "host employer": The Occupational Safety and Health Administration will treat the electric utility or the owner of the installation as the host employer if it operates or controls operating procedures for the installation. If the electric utility or installation owner neither operates nor controls operating procedures for the installation, the Occupational Safety and Health Administration will treat the employer that the utility or owner has contracted with to operate or control the operating procedures for the installation as the host employer. In no case will there be more than one host employer.

Immediately dangerous to life or health (IDLH). Any condition that poses an immediate or delayed threat to life or that would cause irreversible adverse health effects or that would interfere with an individual's ability to escape unaided from a permit space.

Note to the definition of "immediately dangerous to life or health": Some materials-hydrogen fluoride gas and cadmium vapor, for example-may produce immediate transient effects that, even if severe, may pass without medical attention, but are followed by sudden, possibly fatal collapse 12–72 hours after exposure. The victim "feels normal" from recovery from transient effects until collapse. Such materials in hazardous quantities are considered to be "immediately" dangerous to life or health.

Insulated. Separated from other conducting surfaces by a dielectric (including air space) offering a high resistance to the passage of current.

Note to the definition of "insulated": When any object is said to be insulated, it is understood to be insulated for the conditions to which it normally is subjected. Otherwise, it is, for the purpose of this section, uninsulated.

Insulation (cable). Material relied upon to insulate the conductor from other conductors or conducting parts or from ground.

Isolated. Not readily accessible to persons unless special means for access are used.
Line-clearance tree trimmer. An employee who, through related training or on-the-job experience or both, is familiar with the special techniques and hazards involved in line-clearance tree trimming.

Note 1 to the definition of "line-clearance tree trimmer": An employee who is regularly assigned to a line-clearance tree-trimming crew and who is undergoing on-the-job training and who, in the course of such training, has demonstrated an ability to perform duties safely at his or her level of training and who is under the direct supervision of a line-clearance tree trimmer is considered to be a line-clearance tree trimmer for the performance of those duties.

Note 2 to the definition of "line-clearance tree trimmer": A line-clearance tree trimmer is not considered to be a "qualified employee" under this section unless he or she has the training required for a qualified employee under paragraph (a)(2)(ii) of this section. However, under the electrical safety-related work practices standard in Subpart S of this part, a line-clearance tree trimmer is considered to be a "qualified employee." Tree trimming performed by such "qualified employees" is not subject to the electrical safety-related work practice requirements contained in §§ 1910.331 through 1910.335 when it is directly associated with electric power generation, transmission, or distribution lines or equipment. (See § 1910.331 for requirements on the applicability of the electrical safety-related work practice requirements contained in §§ 1910.331 through 1910.335 to line-clearance tree trimming performed by such "qualified employees," and see the note following § 1910.332(b)(3) for information regarding the training an employee must have to be considered a qualified employee under §§ 1910.331 through 1910.335.)

Line-clearance tree trimming. The pruning, trimming, repairing, maintaining, removing, or clearing of trees, or the cutting of brush, that is within the following distance of electric supply lines and equipment:

(1) For voltages to ground of 50 kilovolts or less-3.05 meters (10 ft);

(2) For voltages to ground of more than 50 kilovolts-3.05 meters (10 ft) plus 0.10 meters (4 inches) for every 10 kilovolts over 50 kilovolts.

NOTE TO DEFINITION OF "LINE-CLEARANCE TREE TRIMMING": This section applies only to line-clearance tree trimming performed for the purpose of clearing space around electric power generation, transmission, or distribution lines or equipment and on behalf of an organization that operates, or that controls the operating procedures for, those lines or equipment. See paragraph (a)(1) of this section. Tree trimming performed on behalf of a homeowner or commercial entity other than an organization that operates, or that controls the operating procedures for, electric power generation, transmission, or distribution lines or equipment is not directly associated with an electric power generation, transmission, or distribution installation and is outside the scope of this section. In addition, tree trimming that is not for the purpose of clearing space around electric power generation, transmission, or distribution lines or equipment is not directly associated with an electric power generation, transmission, or distribution installation and is outside the scope of this section. Such tree trimming may be covered by other applicable standards. See, for example, §§1910.268 and 1910.331 through 1910.335.

Lines. (1) Communication lines. The conductors and their supporting or containing structures which are used for public or private signal or communication service, and which operate at potentials not exceeding 400 volts to ground or 750 volts between any two points of the circuit, and the transmitted power of which does not exceed 150 watts. If the lines are operating at less than 150 volts, no limit is placed on the transmitted power of the system. Under certain conditions, communication cables may include communication circuits exceeding these limitations where such circuits are also used to supply power solely to communication equipment.

Note to the definition of "communication lines": Telephone, telegraph, railroad signal, data, clock, fire, police alarm, cable television, and other systems conforming to this definition are included. Lines used for signaling purposes, but not included under this definition, are considered as electric supply lines of the same voltage.

Lines. (2) Electric supply lines. Conductors used to transmit electric energy and their necessary supporting or containing structures. Signal lines of more than 400 volts are always supply lines within this section, and those of less than 400 volts are considered as supply lines, if so run and operated throughout.

Manhole. A subsurface enclosure that personnel may enter and that is used for installing, operating, and maintaining submersible equipment or cable.

Minimum approach distance. The closest distance an employee may approach an energized or a grounded object.

Note to the definition of "minimum approach distance": Paragraph (l)(3)(i) of this section requires employers to establish minimum approach distances.

Personal fall arrest system. A system used to arrest an employee in a fall from a working level.

Qualified employee (qualified person). An employee (person) knowledgeable in the construction and operation of the electric power generation, transmission, and distribution equipment involved, along with the associated hazards.

Note 1 to the definition of "qualified employee (qualified person)": An employee must have the training required by (a)(2)(ii) of this section to be a qualified employee.

Note 2 to the definition of "qualified employee (qualified person)": Except under (g)(2)(iv)(C)(2) and (g)(2)(iv)(C)(3) of this section, an employee who is undergoing on-the-job training and who has demonstrated, in the course of such training, an ability to perform duties safely at his or her level of training and who is under the direct supervision of a qualified person is a qualified person for the performance of those duties.

Statistical sparkover voltage. A transient overvoltage level that produces a 97.72% probability of sparkover (that is, two standard deviations above the voltage at which there is a 50% probability of sparkover).

Statistical withstand voltage. A transient overvoltage level that produces a 0.14% probability of sparkover (that is, three standard deviations below the voltage at which there is a 50% probability of sparkover).

Switch. A device for opening and closing or for changing the connection of a circuit. In this section, a switch is manually operable, unless otherwise stated.

System operator. A qualified person designated to operate the system or its parts.

Vault. An enclosure, above or below ground, that personnel may enter and that is used for installing, operating, or maintaining equipment or cable.

Vented vault. A vault that has provision for air changes using exhaust flue stacks and low-level air intakes operating on pressure and temperature differentials that provide for airflow that precludes a hazardous atmosphere from developing.

Voltage. The effective (root mean square, or rms) potential difference between any two conductors or between a conductor and ground. This section expresses voltages in nominal values, unless otherwise indicated. The nominal voltage of a system or circuit is the value assigned to a system or circuit of a given voltage class for the purpose of convenient designation. The operating voltage of the system may vary above or below this value.

Work-positioning equipment. A body belt or body harness system rigged to allow an employee to be supported on an elevated vertical surface, such as a utility pole or tower leg, and work with both hands free while leaning.

[59 FR 40672, Aug. 9, 1994; 59 FR 51672, Oct. 12, 1994; 79 FR 20633-20659, July 10, 2014]

Appendix A to §1910.269—Flow Charts

This appendix presents information, in the form of flow charts, that illustrates the scope and application of § 1910.269. This appendix addresses the interface between § 1910.269 and Subpart S of this Part (Electrical), between § 1910.269 and § 1910.146 (Permit required confined spaces), and between § 1910.269 and § 1910.147 (The control of hazardous energy (lockout/tagout)). These flow charts provide guidance for employers trying to implement the requirements of § 1910.269 in combination with other General Industry Standards contained in Part 1910. Employers should always consult the relevant standards, in conjunction with this appendix, to ensure compliance with all applicable requirements.

Appendix A-1 to §1910.269—Application of §1910.269 and Subpart S of this Part to Electrical Installations.

[1]This chart applies to electrical installation design requirements only. See Appendix A-2 for electrical safety-related work practices. Supplementary electric generating equipment that is used to supply a workplace for emergency, standby, or similar purposed only is not considered an electric power generation installation.

Appendix A-2 to §1910.269—Application of §1910.269 and Subpart S of this Part to Electrical Safety-Related Work Practices.[1]

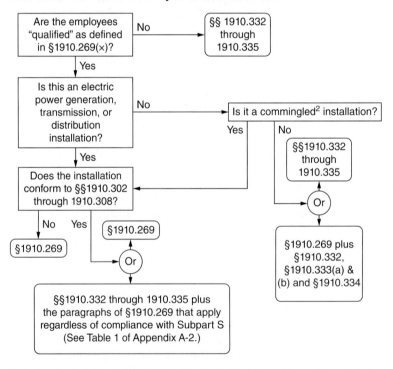

[1]This flowchart applies only to the electrical safety-related work practice and training requirements in §1910.269 and §§ 1910.332 through 1910.335.

[2]See §1910.269(a)(1)(ii)(B) and 1910.331 (b) and (c)(1).

[3]This means commingled to the extent that the electric power generation, transmission, or distribution installation poses the greater hazard.

Compliance with Subpart S Will Comply with These Paragraphs of §1910.269[1]	Paragraphs That Apply Regardless of Compliance with Subpart S[2]
(d), electric-shock hazards only	(a)(2), (a)(3) and (a)(4).
(h)(3)	(b)
(i)(2) and (i)(3)	(c)
(k)	(d), for other than electric-shock hazards.
(l)(1) through (l)(5), (l)(7), and (l)(10) through (l)(12)	(e)
(m)	(f)
(p)(4)	(g)
(s)(2)	(h)(1) and (h)(2).
(u)(1) and (u)(3) through (u)(5)	(i)(4)
(v)(3) through (v)(5)	(j)
(w)(1) and (w)(7)	(l)(6), (l)(8) and (l)(9).
	(n)
	(o)
	(p)(1) through (p)(3). (q)
	(r) (s)(1)
	(t)
	(u)(2) and (u)(6)
	(v)(1), (v)(2), and (v)(6) through (v)(12).
	(w)(2) through (w)(6), (w)(8), and (w)(9).

[1]If the electrical installation meets the requirements of §§ 1910.302 through 1910.308 of this part, then the electrical installation and any associated electrical safety-related work practices conforming to §§ 1910.332 through 1910.335 of this part are considered to comply with these provisions of § 1910.269 of this part.

[2]These provisions include electrical safety and other requirements that must be met regardless of compliance with subpart S of this part.

APPENDIX A.2, TABLE 1   Electrical Safety-Related Work Practices in §1910.269

Appendix A-3 to §1910.269—Application of §1910.269 and Subpart S of this Part to Tree Trimming

[1]3.05 meters (10 ft) plus 0.10 meters (4 inches) for every 10 kilovolts over 50 kilovolts.
Note: Paragraph (t) of § 1910.269 contains additional requirements for work in manholes and underground vaults.

Appendix A-4 to §1910.269—Application of §§1910.147, 1910.269, and 1910.333 to Hazardous Energy Control Procedures (Lockout/Tagout).

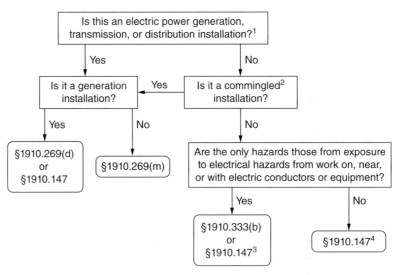

[1]If a generation, transmission, or distribution installation conforms to §§ 1910.302 through 1910.308, the lockout and tagging procedures of § 1910.333(b) may be followed for electric-shock hazards.

[2]This means commingled to the extent that the electric power generation, transmission, or distribution installation poses the greater hazard.

[3]Paragraphs (b)(2)(iii)(D) and (b)(2)(iv)(B) of § 1910.333 still apply.

[4]Paragraph (b) of § 1910.333 applies to any electrical hazards from work on, near, or with electric conductors and equipment.

Appendix A-5 to §1910.269—Application of §§1910.146 and 1910.269 to Permit-Required Confined Spaces.

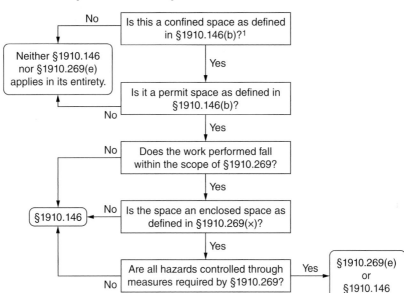

[1]See § 1910.146(c) for general nonentry requirements that apply to all confined spaces.
Note: Paragraph(t) of § 1910.269 contains additional requirements for work in manholes and underground vaults.

Appendix B to §1910.269—Working on Exposed Energized Parts

I. Introduction

Electric utilities design electric power generation, transmission, and distribution installations to meet National Electrical Safety Code (NESC), ANSI C2, requirements. Electric utilities also design transmission and distribution lines to limit line outages as required by system reliability criteria[1] and to withstand the maximum overvoltages impressed on the system. Conditions such as switching surges, faults, and lightning can cause overvoltages. Electric utilities generally select insulator design and lengths and the clearances to structural parts so as to prevent outages from contaminated line insulation and during storms. Line insulator lengths and structural clearances have, over the years, come closer to the minimum approach distances used by workers. As minimum approach distances and structural clearances converge, it is increasingly important that system designers and system operating and maintenance personnel understand the concepts underlying minimum approach distances.

The information in this appendix will assist employers in complying with the minimum approach-distance requirements contained in § 1910.269(l)(3) and (q)(3). Employers must use the technical criteria and methodology presented in this appendix in establishing minimum approach distances in accordance with § 1910.269(l)(3)(i)

---

[1] Federal, State, and local regulatory bodies and electric utilities set reliability requirements that limit the number and duration of system outages.

and Table R-3 and Table R-8. This appendix provides essential background information and technical criteria for the calculation of the required minimum approach distances for live-line work on electric power generation, transmission, and distribution installations.

Unless an employer is using the maximum transient overvoltages specified in Table R-9 for voltages over 72.5 kilovolts, the employer must use persons knowledgeable in the techniques discussed in this appendix, and competent in the field of electric transmission and distribution system design, to determine the maximum transient overvoltage.

II. General

A. Definitions   The following definitions from § 1910.269(x) relate to work on or near electric power generation, transmission, and distribution lines and equipment and the electrical hazards they present.

Exposed. Not isolated or guarded.

Guarded. Covered, fenced, enclosed, or otherwise protected, by means of suitable covers or casings, barrier rails or screens, mats, or platforms, designed to minimize the possibility, under normal conditions, of dangerous approach or inadvertent contact by persons or objects.

Note to the definition of "guarded": Wires that are insulated, but not otherwise protected, are not guarded.

Insulated. Separated from other conducting surfaces by a dielectric (including air space) offering a high resistance to the passage of current.

Note to the definition of "insulated": When any object is said to be insulated, it is understood to be insulated for the conditions to which it normally is subjected. Otherwise, it is, for the purpose of this section, uninsulated.

Isolated. Not readily accessible to persons unless special means for access are used. Statistical sparkover voltage. A transient overvoltage level that produces a 97.72% probability of sparkover (that is, two standard deviations above the voltage at which there is a 50% probability of sparkover).

Statistical withstand voltage. A transient overvoltage level that produces a 0.14% probability of sparkover (that is, three standard deviations below the voltage at which there is a 50% probability of sparkover).

B. Installations Energized at 50 to 300 Volts   The hazards posed by installations energized at 50 to 300 volts are the same as those found in many other workplaces. That is not to say that there is no hazard, but the complexity of electrical protection required does not compare to that required for high voltage systems. The employee must avoid contact with the exposed parts, and the protective equipment used (such as rubber insulating gloves) must provide insulation for the voltages involved.

C. Exposed Energized Parts Over 300 Volts AC   Paragraph (l)(3)(i) of § 1910.269 requires the employer to establish minimum approach distances no less than the distances computed by Table R-3 for ac systems so that employees can work safely without risk of sparkover.[2]

Unless the employee is using electrical protective equipment, air is the insulating medium between the employee and energized parts. The distance between the employee and an energized part must be sufficient for the air to withstand the maximum transient

---

[2]Sparkover is a disruptive electric discharge in which an electric arc forms and electric current passes through air.

overvoltage that can reach the worksite under the working conditions and practices the employee is using. This distance is the minimum air insulation distance, and it is equal to the electrical component of the minimum approach distance.

Normal system design may provide or include a means (such as lightning arrestors) to control maximum anticipated transient overvoltages, or the employer may use temporary devices (portable protective gaps) or measures (such as preventing automatic circuit breaker reclosing) to achieve the same result. Paragraph (l)(3)(ii) of § 1910.269 requires the employer to determine the maximum anticipated per-unit transient overvoltage, phase-to-ground, through an engineering analysis or assume a maximum anticipated per-unit transient overvoltage, phase-to-ground, in accordance with Table R-9, which specifies the following maximums for ac systems:

72.6 to 420.0 kilovolts-3.5 per unit

420.1 to 550.0 kilovolts-3.0 per unit

550.1 to 800.0 kilovolts-2.5 per unit

See paragraph IV.A.2, later in this appendix, for additional discussion of maximum transient overvoltages.

D. Types of Exposures   Employees working on or near energized electric power generation, transmission, and distribution systems face two kinds of exposures: Phase-to-ground and phase-to-phase. The exposure is phase-to-ground: (1) With respect to an energized part, when the employee is at ground potential or (2) with respect to ground, when an employee is at the potential of the energized part during live-line bare-hand work. The exposure is phase-to-phase, with respect to an energized part, when an employee is at the potential of another energized part (at a different potential) during live-line bare-hand work.

III. Determination of Minimum Approach Distances for AC Voltages Greater Than 300 Volts

A. Voltages of 301 to 5,000 Volts   Test data generally forms the basis of minimum air insulation distances. The lowest voltage for which sufficient test data exists is 5,000 volts, and these data indicate that the minimum air insulation distance at that voltage is 20 millimeters (1 inch). Because the minimum air insulation distance increases with increasing voltage, and, conversely, decreases with decreasing voltage, an assumed minimum air insulation distance of 20 millimeters will protect against sparkover at voltages of 301 to 5,000 volts. Thus, 20 millimeters is the electrical component of the minimum approach distance for these voltages.

B. Voltages of 5.1 to 72.5 Kilovolts   For voltages from 5.1 to 72.5 kilovolts, the Occupational Safety and Health Administration bases the methodology for calculating the electrical component of the minimum approach distance on Institute of Electrical and Electronic Engineers (IEEE) Standard 4-1995, Standard Techniques for High-Voltage Testing. Table 1 lists the critical sparkover distances from that standard as listed in IEEE Std 516-2009, IEEE Guide for Maintenance Methods on Energized Power Lines.

To use this table to determine the electrical component of the minimum approach distance, the employer must determine the peak phase-to-ground transient overvoltage and select a gap from the table that corresponds to that voltage as a withstand voltage rather than a critical sparkover voltage. To calculate the electrical component

60 Hz Rod-to-Rod Sparkover (kV Peak)	Gap Spacing from IEEE Std 4-1995 (cm)
25	2
36	3
46	4
53	5
60	6
70	8
79	10
86	12
95	14
104	16
112	18
120	20
143	25
167	30
192	35
218	40
243	45
270	50
322	60

Source: IEEE Std 516-2009.

TABLE 1   Sparkover Distance for Rod-to-Rod Gap

of the minimum approach distance for voltages between 5 and 72.5 kilovolts, use the following procedure:

1. Divide the phase-to-phase voltage by the square root of 3 to convert it to a phase-to-ground voltage.

2. Multiply the phase-to-ground voltage by the square root of 2 to convert the rms value of the voltage to the peak phase-to-ground voltage.

3. Multiply the peak phase-to-ground voltage by the maximum per-unit transient overvoltage, which, for this voltage range, is 3.0, as discussed later in this appendix. This is the maximum phase-to-ground transient overvoltage, which corresponds to the withstand voltage for the relevant exposure.[3]

[3]The withstand voltage is the voltage at which sparkover is not likely to occur across a specified distance. It is the voltage taken at the 3s point below the sparkover voltage, assuming that the sparkover curve follows a normal distribution.

Step	Maximum System Phase-to-Phase Voltage (kV)			
	15	36	46	72.5
1. Divide by √3	8.7	20.8	26.6	41.9
2. Multiply by √2	12.2	29.4	37.6	59.2
3. Multiply by 3.0	36.7	88.2	112.7	177.6
4. Divide by 0.85	43.2	103.7	132.6	208.9
5. Interpolate from Table 1	3+(7.2/10)*1	14+(8.7/9)*2	20+(12.6/23)*5	35+(16.9/26)*5
Electrical component of MAD (cm)	3.72	15.93	22.74	38.25

TABLE 2   Calculating the Electrical Component of MAD 751 V to 72.5 kV

4. Divide the maximum phase-to-ground transient overvoltage by 0.85 to determine the corresponding critical sparkover voltage. (The critical sparkover voltage is 3 standard deviations (or 15%) greater than the withstand voltage.)

5. Determine the electrical component of the minimum approach distance from Table 1 through interpolation.

Table 2 illustrates how to derive the electrical component of the minimum approach distance for voltages from 5.1 to 72.5 kilovolts, before the application of any altitude correction factor, as explained later.

C. Voltages of 72.6 to 800 Kilovolts   For voltages of 72.6 kilovolts to 800 kilovolts, this section bases the electrical component of minimum approach distances, before the application of any altitude correction factor, on the following formula:

Equation 1-For Voltages of 72.6 kV to 800 kV

$$D = 0.3048(C + a) \text{ VL-GT}$$

where D = Electrical component of the minimum approach distance in air in meters;

C = A correction factor associated with the variation of gap sparkover with voltage;

a = A factor relating to the saturation of air at system voltages of 345 kilovolts or higher;[4]

VL-G = Maximum system line-to-ground rms voltage in kilovolts-it should be the "actual" maximum, or the normal highest voltage for the range (for example, 10% above the nominal voltage); and

T = Maximum transient overvoltage factor in per unit.

---

[4]Test data demonstrates that the saturation factor is greater than 0 at peak voltages of about 630 kilovolts. Systems operating at 345 kilovolts (or maximum system voltages of 362 kilovolts) can have peak maximum transient overvoltages exceeding 630 kilovolts. Table R-3 sets equations for calculating a based on peak voltage.

In Equation 1, C is 0.01: (1) For phase-to-ground exposures that the employer can demonstrate consist only of air across the approach distance (gap) and (2) for phase-to-phase exposures if the employer can demonstrate that no insulated tool spans the gap and that no large conductive object is in the gap. Otherwise, C is 0.011.

In Equation 1, the term a varies depending on whether the employee's exposure is phase-to-ground or phase-to-phase and on whether objects are in the gap. The employer must use the equations in Table 3 to calculate a. Sparkover test data with insulation spanning the gap form the basis for the equations for phase-to-ground exposures, and sparkover test data with only air in the gap form the basis for the equations for phase-to-phase exposures. The phase-to-ground equations result in slightly higher values of a, and, consequently, produce larger minimum approach distances, than the phase-to-phase equations for the same value of VPeak.

In Equation 1, T is the maximum transient overvoltage factor in per unit. As noted earlier, § 1910.269(l)(3)(ii) requires the employer to determine the maximum anticipated per-unit transient overvoltage, phase-to-ground, through an engineering analysis or assume a maximum anticipated per-unit transient overvoltage, phase-to-ground, in accordance with Table R-9. For phase-to-ground exposures, the employer uses this value, called TL-G, as T in Equation 1. IEEE Std 516-2009 provides the following formula to calculate the phase-to-phase maximum transient overvoltage, TL-L, from TL-G:

$$TL\text{-}L = 1.35TL\text{-}G + 0.45$$

For phase-to-phase exposures, the employer uses this value as T in Equation 1.

D. Provisions for Inadvertent Movement The minimum approach distance must include an "adder" to compensate for the inadvertent movement of the worker relative to an energized part or the movement of the part relative to the worker. This "adder" must

Phase-to-Ground Exposures			
VPeak = TL-GVL-G $\sqrt{2}$	635 kV or less 0	635.1 to 915 kV (VPeak-635)/140,000	915.1 to 1,050 kV (VPeak-645)/135,000
a			
VPeak = TL-GVL-G$\sqrt{2}$	More than 1,050 kV		
a	(VPeak-675)/125,000		
Phase-to-Phase Exposures 1			
VPeak = (1.35TL-G + 0.45) VL-G$\sqrt{2}$	630 kV or less 0	630.1 to 848 kV (VPeak-630)/155,000	848.1 to 1,131 kV (VPeak-633.6)/152,207
a			
VPeak = (1.35TL-G + 0.45) VL-G$\sqrt{2}$	1,131.1 to 1,485 kV (VPeak-628)/153,846	More than 1,485 kV (VPeak-350.5)/203,666	
a			

[1]Use the equations for phase-to-ground exposures (with VPeak for phase-to-phase exposures) unless the employer can demonstrate that no insulated tool spans the gap and that no large conductive object is in the gap.

TABLE 3 Equations for Calculating the Surge Factor, a

account for this possible inadvertent movement and provide the worker with a comfortable and safe zone in which to work. Employers must add the distance for inadvertent movement (called the "ergonomic component of the minimum approach distance") to the electrical component to determine the total safe minimum approach distances used in live-line work.

The Occupational Safety and Health Administration based the ergonomic component of the minimum approach distance on response time-distance analysis. This technique uses an estimate of the total response time to a hazardous incident and converts that time to the distance traveled. For example, the driver of a car takes a given amount of time to respond to a "stimulus" and stop the vehicle. The elapsed time involved results in the car's traveling some distance before coming to a complete stop. This distance depends on the speed of the car at the time the stimulus appears and the reaction time of the driver.

In the case of live-line work, the employee must first perceive that he or she is approaching the danger zone. Then, the worker responds to the danger and must decelerate and stop all motion toward the energized part. During the time it takes to stop, the employee will travel some distance. This is the distance the employer must add to the electrical component of the minimum approach distance to obtain the total safe minimum approach distance.

At voltages from 751 volts to 72.5 kilovolts,[5] the electrical component of the minimum approach distance is smaller than the ergonomic component. At 72.5 kilovolts, the electrical component is only a little more than 0.3 meters (1 ft). An ergonomic component of the minimum approach distance must provide for all the worker's unanticipated movements. At these voltages, workers generally use rubber insulating gloves; however, these gloves protect only a worker's hands and arms. Therefore, the energized object must be at a safe approach distance to protect the worker's face. In this case, 0.61 meters (2 ft) is a sufficient and practical ergonomic component of the minimum approach distance.

For voltages between 72.6 and 800 kilovolts, employees must use different work practices during energized line work. Generally, employees use live-line tools (hot sticks) to perform work on energized equipment. These tools, by design, keep the energized part at a constant distance from the employee and, thus, maintain the appropriate minimum approach distance automatically.

The location of the worker and the type of work methods the worker is using also influence the length of the ergonomic component of the minimum approach distance. In this higher voltage range, the employees use work methods that more tightly control their movements than when the workers perform work using rubber insulating gloves. The worker, therefore, is farther from the energized line or equipment and must be more precise in his or her movements just to perform the work. For these reasons, this section adopts an ergonomic component of the minimum approach distance of 0.31 m (1 ft) for voltages between 72.6 and 800 kilovolts.

Table 4 summarizes the ergonomic component of the minimum approach distance for various voltage ranges.

---

[5]For voltages of 50 to 300 volts, Table R-3 specifies a minimum approach distance of "avoid contact." The minimum approach distance for this voltage range contains neither an electrical component nor an ergonomic component.

Voltage Range (kV)	Distance	
	m	ft
0.301 to 0.750	0.31	1.0
0.751 to 72.5	0.61	2.0
72.6 to 800	0.31	1.0

Note: The employer must add this distance to the electrical component of the minimum approach distance to obtain the full minimum approach distance.

TABLE 4   Ergonomic Component of Minimum Approach Distance

The ergonomic component of the minimum approach distance accounts for errors in maintaining the minimum approach distance (which might occur, for example, if an employee misjudges the length of a conductive object he or she is holding), and for errors in judging the minimum approach distance. The ergonomic component also accounts for inadvertent movements by the employee, such as slipping. In contrast, the working position selected to properly maintain the minimum approach distance must account for all of an employee's reasonably likely movements and still permit the employee to adhere to the applicable minimum approach distance. (See Figure 1.) Reasonably likely movements include an employee's adjustments to tools, equipment, and working positions and all movements needed to perform the work. For example, the employee should be able to perform all of the following actions without straying into the minimum approach distance:

Adjust his or her hardhat,

maneuver a tool onto an energized part with a reasonable amount of overreaching or underreaching,

FIGURE 1   Maintaining the Minimum Approach Distance.

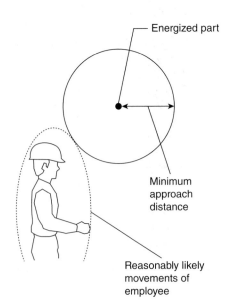

Energized part

Minimum approach distance

Reasonably likely movements of employee

reach for and handle tools, material, and equipment passed to him or her, and adjust tools, and replace components on them, when necessary during the work procedure.

The training of qualified employees required under § 1910.269(a)(2), and the job planning and briefing required under § 1910.269(c), must address selection of a proper working position.

E. Miscellaneous Correction Factors   Changes in the air medium that forms the insulation influences the strength of an air gap. A brief discussion of each factor follows.

1. Dielectric strength of air. The dielectric strength of air in a uniform electric field at standard atmospheric conditions is approximately 3 kilovolts per millimeter.[6] The pressure, temperature, and humidity of the air, the shape, dimensions, and separation of the electrodes, and the characteristics of the applied voltage (wave shape) affect the disruptive gradient.

2. Atmospheric effect. The empirically determined electrical strength of a given gap is normally applicable at standard atmospheric conditions (20°C, 101.3 kilopascals, 11 grams/cubic centimeter humidity). An increase in the density (humidity) of the air inhibits sparkover for a given air gap. The combination of temperature and air pressure that results in the lowest gap sparkover voltage is high temperature and low pressure. This combination of conditions is not likely to occur. Low air pressure, generally associated with high humidity, causes increased electrical strength. An average air pressure generally correlates with low humidity. Hot and dry working conditions normally result in reduced electrical strength. The equations for minimum approach distances in Table R-3 assume standard atmospheric conditions.

3. Altitude. The reduced air pressure at high altitudes causes a reduction in the electrical strength of an air gap. An employer must increase the minimum approach distance by about 3% per 300 meters (1,000 ft) of increased altitude for altitudes above 900 meters (3,000 ft). Table R-5 specifies the altitude correction factor that the employer must use in calculating minimum approach distances.

IV. Determining Minimum Approach Distances

A. Factors Affecting Voltage Stress at the Worksite

1. System voltage (nominal). The nominal system voltage range determines the voltage for purposes of calculating minimum approach distances. The employer selects the range in which the nominal system voltage falls, as given in the relevant table, and uses the highest value within that range in per-unit calculations.

---

[6] For the purposes of estimating arc length, § 1910.269 generally assumes a more conservative dielectric strength of 10 kilovolts per 25.4 millimeters, consistent with assumptions made in consensus standards such as the National Electrical Safety Code (IEEE C2-2012). The more conservative value accounts for variables such as electrode shape, wave shape, and a certain amount of overvoltage.

2. Transient overvoltages. Operation of switches or circuit breakers, a fault on a line or circuit or on an adjacent circuit, and similar activities may generate transient overvoltages on an electrical system. Each overvoltage has an associated transient voltage wave shape. The wave shape arriving at the site and its magnitude vary considerably.

In developing requirements for minimum approach distances, the Occupational Safety and Health Administration considered the most common wave shapes and the magnitude of transient overvoltages found on electric power generation, transmission, and distribution systems. The equations in Table R-3 for minimum approach distances use per-unit maximum transient overvoltages, which are relative to the nominal maximum voltage of the system. For example, a maximum transient overvoltage value of 3.0 per unit indicates that the highest transient overvoltage is 3.0 times the nominal maximum system voltage.

3. Typical magnitude of overvoltages. Table 5 lists the magnitude of typical transient overvoltages.

4. Standard deviation-air-gap withstand. For each air gap length under the same atmospheric conditions, there is a statistical variation in the breakdown voltage. The probability of breakdown against voltage has a normal (Gaussian) distribution. The standard deviation of this distribution varies with the wave shape, gap geometry, and atmospheric conditions. The withstand voltage of the air gap is three standard deviations (3s) below the critical sparkover voltage. (The critical sparkover voltage is the crest value of the impulse wave that, under specified conditions, causes sparkover 50% of the time. An impulse wave of three standard deviations below this value, that is, the withstand voltage, has a probability of sparkover of approximately 1 in 1,000.)

5. Broken Insulators. Tests show reductions in the insulation strength of insulator strings with broken skirts. Broken units may lose up to 70% of their withstand capacity. Because an employer cannot determine the insulating capability of a broken unit without testing it, the employer must consider damaged units in an insulator to have no insulating value. Additionally, the

Cause	Magnitude (per unit)
Energized 200-mile line without closing resistors	3.5
Energized 200-mile line with one-step closing resistor	2.1
Energized 200-mile line with multistep resistor	2.5
Reclosing with trapped charge one-step resistor	2.2
Opening surge with single restrike	3.0
Fault initiation unfaulted phase	2.1
Fault initiation adjacent circuit	2.5
Fault clearing	1.7 to 1.9

TABLE 5   Magnitude of Typical Transient Overvoltages

presence of a live-line tool alongside an insulator string with broken units may further reduce the overall insulating strength. The number of good units that must be present in a string for it to be "insulated" as defined by § 1910.269(x) depends on the maximum overvoltage possible at the worksite.

B. Minimum Approach Distances Based on Known, Maximum-Anticipated Per-Unit Transient Overvoltages

1. Determining the minimum approach distance for AC systems. Under § 1910.269(l)(3)(ii), the employer must determine the maximum anticipated per-unit transient overvoltage, phase-to-ground, through an engineering analysis or must assume a maximum anticipated per-unit transient overvoltage, phase-to-ground, in accordance with Table R-9. When the employer conducts an engineering analysis of the system and determines that the maximum transient overvoltage is lower than specified by Table R-9, the employer must ensure that any conditions assumed in the analysis, for example, that employees block reclosing on a circuit or install portable protective gaps, are present during energized work. To ensure that these conditions are present, the employer may need to institute new livework procedures reflecting the conditions and limitations set by the engineering analysis.

2. Calculation of reduced approach distance values. An employer may take the following steps to reduce minimum approach distances when the maximum transient overvoltage on the system (that is, the maximum transient overvoltage without additional steps to control overvoltages) produces unacceptably large minimum approach distances:

   Step 1. Determine the maximum voltage (with respect to a given nominal voltage range) for the energized part.

   Step 2. Determine the technique to use to control the maximum transient overvoltage. (See paragraphs IV.C and IV.D of this appendix.) Determine the maximum transient overvoltage that can exist at the worksite with that form of control in place and with a confidence level of 3s. This voltage is the withstand voltage for the purpose of calculating the appropriate minimum approach distance.

   Step 3. Direct employees to implement procedures to ensure that the control technique is in effect during the course of the work.

   Step 4. Using the new value of transient overvoltage in per unit, calculate the required minimum approach distance from Table R-3.

C. Methods of Controlling Possible Transient Overvoltage Stress Found on a System

1. Introduction. There are several means of controlling overvoltages that occur on transmission systems. For example, the employer can modify the operation of circuit breakers or other switching devices to reduce switching transient overvoltages. Alternatively, the employer can hold the overvoltage to an acceptable level by installing surge arresters or portable protective gaps on the system. In addition, the employer can change the transmission system to minimize the effect of switching operations. Section 4.8 of IEEE Std 516-2009 describes various ways of controlling, and thereby reducing, maximum transient overvoltages.

2. Operation of circuit breakers.[7] The maximum transient overvoltage that can reach the worksite is often the result of switching on the line on which employees are working. Disabling automatic reclosing during energized line work, so that the line will not be reenergized after being opened for any reason, limits the maximum switching surge overvoltage to the larger of the opening surge or the greatest possible fault-generated surge, provided that the devices (for example, insertion resistors) are operable and will function to limit the transient overvoltage and that circuit breaker restrikes do not occur. The employer must ensure the proper functioning of insertion resistors and other overvoltage-limiting devices when the employer's engineering analysis assumes their proper operation to limit the overvoltage level. If the employer cannot disable the reclosing feature (because of system operating conditions), other methods of controlling the switching surge level may be necessary.

Transient surges on an adjacent line, particularly for double circuit construction, may cause a significant overvoltage on the line on which employees are working. The employer's engineering analysis must account for coupling to adjacent lines.

3. Surge arresters. The use of modern surge arresters allows a reduction in the basic impulse-insulation levels of much transmission system equipment. The primary function of early arresters was to protect the system insulation from the effects of lightning. Modern arresters not only dissipate lightning-caused transients, but may also control many other system transients caused by switching or faults.

The employer may use properly designed arresters to control transient overvoltages along a transmission line and thereby reduce the requisite length of the insulator string and possibly the maximum transient overvoltage on the line.[8]

4. Switching Restrictions. Another form of overvoltage control involves establishing switching restrictions, whereby the employer prohibits the operation of circuit breakers until certain system conditions are present. The employer restricts switching by using a tagging system, similar to that used for a permit, except that the common term used for this activity is a "hold-off" or "restriction." These terms indicate that the restriction does not prevent operation, but only modifies the operation during the livework activity.

D. Minimum Approach Distance Based on Control of Maximum Transient Overvoltage at the Worksite   When the employer institutes control of maximum transient overvoltage at the worksite by installing portable protective gaps, the employer may calculate the minimum approach distance as follows:

---

[7]The detailed design of a circuit interrupter, such as the design of the contacts, resistor insertion, and breaker timing control, are beyond the scope of this appendix. The design of the system generally accounts for these features. This appendix only discusses features that can limit the maximum switching transient overvoltage on a system.

[8]Surge arrester application is beyond the scope of this appendix. However, if the employer installs the arrester near the work site, the application would be similar to the protective gaps discussed in paragraph IV.D of this appendix.

Step 1. Select the appropriate withstand voltage for the protective gap based on system requirements and an acceptable probability of gap sparkover.[9]

Step 2. Determine a gap distance that provides a withstand voltage[10] greater than or equal to the one selected in the first step.[11]

Step 3. Use 110% of the gap's critical sparkover voltage to determine the phase-to-ground peak voltage at gap sparkover (VPPG Peak).

Step 4. Determine the maximum transient overvoltage, phase-to-ground, at the worksite from the following formula:

$$T = \frac{V_{PPG\ Peak}}{V_{L\text{-}G}\sqrt{2}}$$

Step 5. Use this value of T[12] in the equation in Table R-3 to obtain the minimum approach distance. If the worksite is no more than 900 meters (3,000 ft) above sea level, the employer may use this value of T to determine the minimum approach distance from Table 7 through Table 14.

Note: All rounding must be to the next higher value (that is, always round up).

Sample protective gap calculations.

Problem: Employees are to perform work on a 500-kilovolt transmission line at sea level that is subject to transient overvoltages of 2.4 p.u. The maximum operating voltage of the line is 550 kilovolts. Determine the length of the protective gap that will provide the minimum practical safe approach distance. Also, determine what that minimum approach distance is.

Step 1. Calculate the smallest practical maximum transient overvoltage (1.25 times the crest phase-to-ground voltage):[13]

$$550\ kV \times \frac{\sqrt{2}}{\sqrt{3}} \times 1.25 = 561\ kV$$

This value equals the withstand voltage of the protective gap.

---

[9]The employer should check the withstand voltage to ensure that it results in a probability of gap flashover that is acceptable from a system outage perspective. (In other words, a gap sparkover will produce a system outage. The employer should determine whether such an outage will impact overall system performance to an acceptable degree.) In general, the withstand voltage should be at least 1.25 times the maximum crest operating voltage.

[10]The manufacturer of the gap provides, based on test data, the critical sparkover voltage for each gap spacing (for example, a critical sparkover voltage of 665 kilovolts for a gap spacing of 1.2 meters). The withstand voltage for the gap is equal to 85% of its critical sparkover voltage.

[11]Switch steps 1 and 2 if the length of the protective gap is known.

[12]IEEE Std 516-2009 states that most employers add 0.2 to the calculated value of T as an additional safety factor.

[13]To eliminate sparkovers due to minor system disturbances, the employer should use a withstand voltage no lower than 1.25 p.u. Note that this is a practical, or operational, consideration only. It may be feasible for the employer to use lower values of withstand voltage.

Step 2. Using test data for a particular protective gap, select a gap that has a critical sparkover voltage greater than or equal to:

$$561 \text{ kV} \div 0.85 = 660 \text{ kV}$$

For example, if a protective gap with a 1.22-m (4.0-ft) spacing tested to a critical sparkover voltage of 665 kilovolts (crest), select this gap spacing.

Step 3. The phase-to-ground peak voltage at gap sparkover (VPPG Peak) is 110% of the value from the previous step:

$$665 \text{ kV} \times 1.10 = 732 \text{ kV}$$

This value corresponds to the withstand voltage of the electrical component of the minimum approach distance.

Step 4. Use this voltage to determine the worksite value of T:

$$T = \frac{732}{564} = 1.7 \text{ p.u.}$$

Step 5. Use this value of T in the equation in Table R-3 to obtain the minimum approach distance, or look up the minimum approach distance in Table 7 through Table 14:

$$\text{MAD} = 2.29 \text{ m (7.6 ft)}.$$

E. Location of Protective Gaps

1. Adjacent structures. The employer may install the protective gap on a structure adjacent to the worksite, as this practice does not significantly reduce the protection afforded by the gap.

2. Terminal stations. Gaps installed at terminal stations of lines or circuits provide a level of protection; however, that level of protection may not extend throughout the length of the line to the worksite. The use of substation terminal gaps raises the possibility that separate surges could enter the line at opposite ends, each with low enough magnitude to pass the terminal gaps without sparkover. When voltage surges occur simultaneously at each end of a line and travel toward each other, the total voltage on the line at the point where they meet is the arithmetic sum of the two surges. A gap installed within 0.8 km (0.5 mile) of the worksite will protect against such intersecting waves. Engineering studies of a particular line or system may indicate that employers can adequately protect employees by installing gaps at even more distant locations. In any event, unless using the default values for T from Table R-9, the employer must determine T at the worksite.

3. Worksite. If the employer installs protective gaps at the worksite, the gap setting establishes the worksite impulse insulation strength. Lightning strikes as far as 6 miles from the worksite can cause a voltage surge greater than the gap withstand voltage, and a gap sparkover can occur. In addition, the gap can sparkover from overvoltages on the line that exceed the withstand voltage of the gap. Consequently, the employer must protect employees from hazards resulting from any sparkover that could occur.

F. Disabling Automatic Reclosing   There are two reasons to disable the automatic-reclosing feature of circuit-interrupting devices while employees are performing live-line work:

To prevent reenergization of a circuit faulted during the work, which could create a hazard or result in more serious injuries or damage than the injuries or damage produced by the original fault;

To prevent any transient overvoltage caused by the switching surge that would result if the circuit were reenergized.

However, due to system stability considerations, it may not always be feasible to disable the automatic-reclosing feature.

V. Minimum Approach-Distance Tables

A. Legacy Tables   Employers may use the minimum approach distances in Table 6 through Table 13 until March 31, 2015.*

B. Alternative Minimum Approach Distances   Employers may use the minimum approach distances in Table 14 through Table 21 provided that the employer follows the notes to those tables.

[79 FR 20665-20677, July 10, 2014]

---

*Table 6 through Table 13 not included.

T (p.u.)	Phase-to-Ground Exposure		Phase-to-Ground Exposure	
	m	ft	m	ft
1.5	0.67	2.2	0.84	2.8
1.6	0.69	2.3	0.87	2.9
1.7	0.71	2.3	0.90	3.0
1.8	0.74	2.4	0.93	3.1
1.9	0.76	2.5	0.96	3.1
2.0	0.78	2.6	0.99	3.2
2.1	0.81	2.7	1.01	3.3
2.2	0.83	2.7	1.04	3.4
2.3	0.85	2.8	1.07	3.5
2.4	0.88	2.9	1.10	3.6
2.5	0.90	3.0	1.13	3.7
2.6	0.92	3.0	1.16	3.8
2.7	0.95	3.1	1.19	3.9
2.8	0.97	3.2	1.22	4.0
2.9	0.99	3.2	1.24	4.1
3.0	1.02	3.3	1.27	4.2
3.1	1.04	3.4	1.30	4.3
3.2	1.06	3.5	1.33	4.4
3.3	1.09	3.6	1.36	4.5
3.4	1.11	3.6	1.39	4.6
3.5	1.13	3.7	1.42	4.7

TABLE 14   AC Minimum Approach Distances—72.6 to 121.0 kV

T (p.u.)	Phase-to-Ground Exposure		Phase-to-Ground Exposure	
	m	ft	m	ft
1.5	0.74	2.4	0.95	3.1
1.6	0.76	2.5	0.98	3.2
1.7	0.79	2.6	1.02	3.3
1.8	0.82	2.7	1.05	3.4
1.9	0.85	2.8	1.08	3.5
2.0	0.88	2.9	1.12	3.7
2.1	0.90	3.0	1.15	3.8
2.2	0.93	3.1	1.19	3.9
2.3	0.96	3.1	1.22	4.0
2.4	0.99	3.2	1.26	4.1
2.5	1.02	3.3	1.29	4.2
2.6	1.04	3.4	1.33	4.4
2.7	1.07	3.5	1.36	4.5
2.8	1.10	3.6	1.39	4.6
2.9	1.13	3.7	1.43	4.7
3.0	1.16	3.8	1.46	4.8
3.1	1.19	3.9	1.50	4.9
3.2	1.21	4.0	1.53	5.0
3.3	1.24	4.1	1.57	5.2
3.4	1.27	4.2	1.60	5.2
3.5	1.30	4.3	1.64	5.4

TABLE 15   AC Minimum Approach Distances—121.1 to 145.0 kV

T (p.u.)	Phase-to-Ground Exposure		Phase-to-Ground Exposure	
	m	ft	m	ft
1.5	0.81	2.7	1.05	3.4
1.6	0.84	2.8	1.09	3.6
1.7	0.87	2.9	1.13	3.7
1.8	0.90	3.0	1.17	3.8
1.9	0.94	3.1	1.21	4.0
2.0	0.97	3.2	1.25	4.1
2.1	1.00	3.3	1.29	4.2
2.2	1.03	3.4	1.33	4.4
2.3	1.07	3.5	1.37	4.5
2.4	1.10	3.6	1.41	4.6
2.5	1.13	3.7	1.45	4.8
2.6	1.17	3.8	1.49	4.9
2.7	1.20	3.9	1.53	5.0
2.8	1.23	4.0	1.57	5.2
2.9	1.26	4.1	1.61	5.3
3.0	1.30	4.3	1.65	5.4
3.1	1.33	4.4	1.70	5.6
3.2	1.36	4.5	1.76	5.8
3.3	1.39	4.6	1.82	6.0
3.4	1.43	4.7	1.88	6.2
3.5	1.46	4.8	1.94	6.4

TABLE 16   AC Minimum Approach Distances—145.1 to 169.0 kV

T (p.u.)	Phase-to-Ground Exposure		Phase-to-Ground Exposure	
	m	ft	m	ft
1.5	1.02	3.3	1.37	4.5
1.6	1.06	3.5	1.43	4.7
1.7	1.11	3.6	1.48	4.9
1.8	1.16	3.8	1.54	5.1
1.9	1.21	4.0	1.60	5.2
2.0	1.25	4.1	1.66	5.4
2.1	1.30	4.3	1.73	5.7
2.2	1.35	4.4	1.81	5.9
2.3	1.39	4.6	1.90	6.2
2.4	1.44	4.7	1.99	6.5
2.5	1.49	4.9	2.08	6.8
2.6	1.53	5.0	2.17	7.1
2.7	1.58	5.2	2.26	7.4
2.8	1.63	5.3	2.36	7.7
2.9	1.67	5.5	2.45	8.0
3.0	1.72	5.6	2.55	8.4
3.1	1.77	5.8	2.65	8.7
3.2	1.81	5.9	2.76	9.1
3.3	1.88	6.2	2.86	9.4
3.4	1.95	6.4	2.97	9.7
3.5	2.01	6.6	3.08	10.1

TABLE 17   AC Minimum Approach Distances—169.1 to 242.0 kV

T (p.u.)	Phase-to-Ground Exposure		Phase-to-Ground Exposure	
	m	ft	m	ft
1.5	1.37	4.5	1.99	6.5
1.6	1.44	4.7	2.13	7.0
1.7	1.51	5.0	2.27	7.4
1.8	1.58	5.2	2.41	7.9
1.9	1.65	5.4	2.56	8.4
2.0	1.72	5.6	2.71	8.9
2.1	1.79	5.9	2.87	9.4
2.2	1.87	6.1	3.03	9.9
2.3	1.97	6.5	3.20	10.5
2.4	2.08	6.8	3.37	11.1
2.5	2.19	7.2	3.55	11.6
2.6	2.29	7.5	3.73	12.2
2.7	2.41	7.9	3.91	12.8
2.8	2.52	8.3	4.10	13.5
2.9	2.64	8.7	4.29	14.1
3.0	2.76	9.1	4.49	14.7
3.1	2.88	9.4	4.69	15.4
3.2	3.01	9.9	4.90	16.1
3.3	3.14	10.3	5.11	16.8
3.4	3.27	10.7	5.32	17.5
3.5	3.41	11.2	5.52	18.1

TABLE 18   AC Minimum Approach Distances—242.1 to 362.0 kV

T (p.u.)	Phase-to-Ground Exposure		Phase-to-Ground Exposure	
	m	ft	m	ft
1.5	1.53	5.0	2.40	7.9
1.6	1.62	5.3	2.58	8.5
1.7	1.70	5.6	2.75	9.0
1.8	1.78	5.8	2.94	9.6
1.9	1.88	6.2	3.13	10.3
2.0	1.99	6.5	3.33	10.9
2.1	2.12	7.0	3.53	11.6
2.2	2.24	7.3	3.74	12.3
2.3	2.37	7.8	3.95	13.0
2.4	2.50	8.2	4.17	13.7
2.5	2.64	8.7	4.40	14.4
2.6	2.78	9.1	4.63	15.2
2.7	2.93	9.6	4.87	16.0
2.8	3.07	10.1	5.11	16.8
2.9	3.23	10.6	5.36	17.6
3.0	3.38	11.1	5.59	18.3
3.1	3.55	11.6	5.82	19.1
3.2	3.72	12.2	6.07	19.9
3.3	3.89	12.8	6.31	20.7
3.4	4.07	13.4	6.56	21.5
3.5	4.25	13.9	6.81	22.3

TABLE 19   AC Minimum Approach Distances—362.1 to 420.0 kV

T (p.u.)	Phase-to-Ground Exposure		Phase-to-Ground Exposure	
	m	ft	m	ft
1.5	1.95	6.4	3.46	11.4
1.6	2.11	6.9	3.73	12.2
1.7	2.28	7.5	4.02	13.2
1.8	2.45	8.0	4.31	14.1
1.9	2.62	8.6	4.61	15.1
2.0	2.81	9.2	4.92	16.1
2.1	3.00	9.8	5.25	17.2
2.2	3.20	10.5	5.55	18.2
2.3	3.40	11.2	5.86	19.2
2.4	3.62	11.9	6.18	20.3
2.5	3.84	12.6	6.50	21.3
2.6	4.07	13.4	6.83	22.4
2.7	4.31	14.1	7.18	23.6
2.8	4.56	15.0	7.52	24.7
2.9	4.81	15.8	7.88	25.9
3.0	5.07	16.6	8.24	27.0

TABLE 20 AC Minimum Approach Distances—420.1 to 550.0 kV

T (p.u.)	Phase-to-Ground Exposure		Phase-to-Ground Exposure	
	m	ft	m	ft
1.5	3.16	10.4	5.97	19.6
1.6	3.46	11.4	6.43	21.1
1.7	3.78	12.4	6.92	22.7
1.8	4.12	13.5	7.42	24.3
1.9	4.47	14.7	7.93	26.0
2.0	4.83	15.8	8.47	27.8
2.1	5.21	17.1	9.02	29.6
2.2	5.61	18.4	9.58	31.4
2.3	6.02	19.8	10.16	33.3
2.4	6.44	21.1	10.76	35.3
2.5	6.88	22.6	11.38	37.3

Notes to Table 14 through Table 21:

The employer must determine the maximum anticipated per-unit transient overvoltage, phase-to-ground, through an engineering analysis, as required by § 1910.269(l)(3)(ii), or assume a maximum anticipated per-unit transient overvoltage, phase-to-ground, in accordance with Table R-9.

For phase-to-phase exposures, the employer must demonstrate that no insulated tool spans the gap and that no large conductive object is in the gap.

The worksite must be at an elevation of 900 meters (3,000 ft) or less above sea level.

TABLE 21 AC Minimum Approach Distances—550.1 to 800.0 kV

Appendix C to §1910.269—Protection from Step and Touch Potentials

I. Introduction

Current passing through an impedance impresses voltage across that impedance. Even conductors have some, albeit low, value of impedance. Therefore, if a "grounded"[14] object, such as a crane or deenergized and grounded power line, results in a ground fault on a power line, voltage is impressed on that grounded object. The voltage impressed on the grounded object depends largely on the voltage on the line, on the impedance of the faulted conductor, and on the impedance to "true," or "absolute," ground represented by the object. If the impedance of the object causing the fault is relatively large, the voltage impressed on the object is essentially the phase-to-ground system voltage. However, even faults to grounded power lines or to well grounded transmission towers or substation structures (which have relatively low values of impedance to ground) can result in hazardous voltages.[15] In all cases, the degree of the hazard depends on the magnitude of the current through the employee and the time of exposure. This appendix discusses methods of protecting workers against the possibility that grounded objects, such as cranes and other mechanical equipment, will contact energized power lines and that deenergized and grounded power lines will become accidentally energized.

II. Voltage-Gradient Distribution

A. Voltage-Gradient Distribution Curve   Absolute, or true, ground serves as a reference and always has a voltage of 0 volts above ground potential. Because there is an impedance between a grounding electrode and absolute ground, there will be a voltage difference between the grounding electrode and absolute ground underground-fault conditions. Voltage dissipates from the grounding electrode (or from the grounding point) and creates a ground potential gradient. The voltage decreases rapidly with increasing distance from the grounding electrode. A voltage drop associated with this dissipation of voltage is a ground potential. Figure 1 is a typical voltage-gradient distribution curve (assuming a uniform soil texture).

B. Step and Touch Potentials   Figure 1 also shows that workers are at risk from step and touch potentials. Step potential is the voltage between the feet of a person standing near an energized grounded object (the electrode). In Figure 1, the step potential is equal to the difference in voltage between two points at different distances from the electrode (where the points represent the location of each foot in relation to the electrode). A person could be at risk of injury during a fault simply by standing near the object.

Touch potential is the voltage between the energized grounded object (again, the electrode) and the feet of a person in contact with the object. In Figure 1, the touch potential is equal to the difference in voltage between the electrode (which is at a distance of 0 meters) and a point some distance away from the electrode (where the point represents the location of the feet of the person in contact with the object). The touch potential could be nearly the full voltage across the grounded object if that object is

---

[14]This appendix generally uses the term "grounded" only with respect to grounding that the employer intentionally installs, for example, the grounding an employer installs on a deenergized conductor. However, in this case, the term "grounded" means connected to earth, regardless of whether or not that connection is intentional.

[15]Thus, grounding systems for transmission towers and substation structures should be designed to minimize the step and touch potentials involved.

FIGURE 1 Typical Voltage–Gradient Distribution Curve.

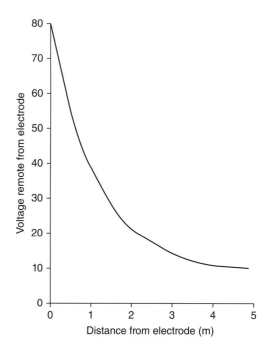

grounded at a point remote from the place where the person is in contact with it. For example, a crane grounded to the system neutral and that contacts an energized line would expose any person in contact with the crane or its uninsulated load line to a touch potential nearly equal to the full fault voltage.

Figure 2 illustrates step and touch potentials.

FIGURE 2 Step and Touch Potentials.

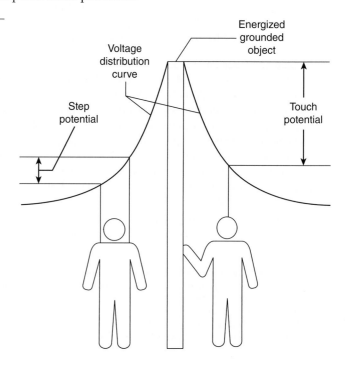

III. Protecting Workers from Hazardous Differences in Electrical Potential

A. Definitions   The following definitions apply to section III of this appendix:

Bond. The electrical interconnection of conductive parts designed to maintain a common electric potential.

Bonding cable (bonding jumper). A cable connected to two conductive parts to bond the parts together.

Cluster bar. A terminal temporarily attached to a structure that provides a means for the attachment and bonding of grounding and bonding cables to the structure.

Ground. A conducting connection between an electric circuit or equipment and the earth, or to some conducting body that serves in place of the earth.

Grounding cable (grounding jumper). A cable connected between a deenergized part and ground. Note that grounding cables carry fault current and bonding cables generally do not. A cable that bonds two conductive parts but carries substantial fault current (for example, a jumper connected between one phase and a grounded phase) is a grounding cable.

Ground mat (grounding grid). A temporarily or permanently installed metallic mat or grating that establishes an equipotential surface and provides connection points for attaching grounds.

B. Analyzing the Hazard   The employer can use an engineering analysis of the power system under fault conditions to determine whether hazardous step and touch voltages will develop. The analysis should determine the voltage on all conductive objects in the work area and the amount of time the voltage will be present. Based on the analysis, the employer can select appropriate measures and protective equipment, including the measures and protective equipment outlined in Section III of this appendix, to protect each employee from hazardous differences in electric potential. For example, from the analysis, the employer will know the voltage remaining on conductive objects after employees install bonding and grounding equipment and will be able to select insulating equipment with an appropriate rating, as described in paragraph III.C.2 of this appendix.

C. Protecting Workers on the Ground   The employer may use several methods, including equipotential zones, insulating equipment, and restricted work areas, to protect employees on the ground from hazardous differences in electrical potential.

1. An equipotential zone will protect workers within it from hazardous step and touch potentials. (See Figure 3.) Equipotential zones will not, however, protect employees located either wholly or partially outside the protected area. The employer can establish an equipotential zone for workers on the ground, with respect to a grounded object, through the use of a metal mat connected to the grounded object. The employer can use a grounding grid to equalize the voltage within the grid or bond conductive objects in the immediate work area to minimize the potential between the objects and between each object and ground. (Bonding an object outside the work area can increase the touch potential to that object, however.) Section III.D of this appendix discusses equipotential zones for employees working on deenergized and grounded power lines.

2. Insulating equipment, such as rubber gloves, can protect employees handling grounded equipment and conductors from hazardous touch potentials. The insulating equipment must be rated for the highest voltage that can be

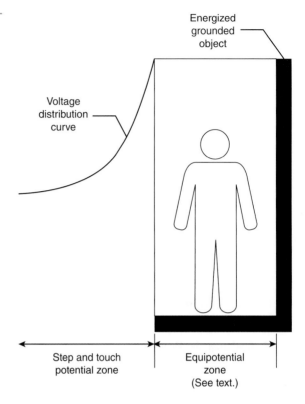

Figure 3   Protection from Ground-Potential Gradients.

impressed on the grounded objects under fault conditions (rather than for the full system voltage).

3. Restricting employees from areas where hazardous step or touch potentials could arise can protect employees not directly involved in performing the operation. The employer must ensure that employees on the ground in the vicinity of transmission structures are at a distance where step voltages would be insufficient to cause injury. Employees must not handle grounded conductors or equipment likely to become energized to hazardous voltages unless the employees are within an equipotential zone or protected by insulating equipment.

D. Protecting Employees Working on Deenergized and Grounded Power Lines   This Section III.D of Appendix C establishes guidelines to help employers comply with requirements in § 1910.269(n) for using protective grounding to protect employees working on deenergized power lines. Paragraph (n) of § 1910.269 applies to grounding of transmission and distribution lines and equipment for the purpose of protecting workers. Paragraph (n)(3) of § 1910.269 requires temporary protective grounds to be placed at such locations and arranged in such a manner that the employer can demonstrate will prevent exposure of each employee to hazardous differences in electric potential.[16] Sections III.D.1 and III.D.2

---

[16]The protective grounding required by § 1910.269(n) limits to safe values the potential differences between accessible objects in each employee's work environment. Ideally, a protective grounding system would create a true equipotential zone in which every point is at the same electric potential. In practice, current passing through the grounding and bonding elements creates potential differences. If these potential differences are hazardous, the employer may not treat the zone as an equipotential zone.

of this appendix provide guidelines that employers can use in making the demonstration required by § 1910.269(n)(3). Section III.D.1 of this appendix provides guidelines on how the employer can determine whether particular grounding practices expose employees to hazardous differences in electric potential. Section III.D.2 of this appendix describes grounding methods that the employer can use in lieu of an engineering analysis to make the demonstration required by § 1910.269(n)(3). The Occupational Safety and Health Administration will consider employers that comply with the criteria in this appendix as meeting § 1910.269(n)(3).

Finally, Section III.D.3 of this appendix discusses other safety considerations that will help the employer comply with other requirements in § 1910.269(n). Following these guidelines will protect workers from hazards that can occur when a deenergized and grounded line becomes energized.

1. *Determining safe body current limits. This Section III.D.1 of Appendix C provides* guidelines on how an employer can determine whether any differences in electric potential to which workers could be exposed are hazardous as part of the demonstration required by § 1910.269(n)(3).

Institute of Electrical and Electronic Engineers (IEEE) Standard 1048-2003, IEEE Guide for Protective Grounding of Power Lines, provides the following equation for determining the threshold of ventricular fibrillation when the duration of the electric shock is limited:

$$I = \frac{116}{\sqrt{t}},$$

where I is the current through the worker's body, and t is the duration of the current in seconds. This equation represents the ventricular fibrillation threshold for 95.5% of the adult population with a mass of 50 kilograms (110 lbs) or more. The equation is valid for current durations between 0.0083 to 3.0 seconds. To use this equation to set safe voltage limits in an equipotential zone around the worker, the employer will need to assume a value for the resistance of the worker's body. IEEE Std 1048-2003 states that "total body resistance is usually taken as 1000 Ω for determining . . . body current limits." However, employers should be aware that the impedance of a worker's body can be substantially less than that value. For instance, IEEE Std 1048-2003 reports a minimum hand-to-hand resistance of 610 ohms and an internal body resistance of 500 ohms. The internal resistance of the body better represents the minimum resistance of a worker's body when the skin resistance drops near zero, which occurs, for example, when there are breaks in the worker's skin, for instance, from cuts or from blisters formed as a result of the current from an electric shock, or when the worker is wet at the points of contact.

Employers may use the IEEE Std 1048-2003 equation to determine safe body current limits only if the employer protects workers from hazards associated with involuntary muscle reactions from electric shock (for example, the hazard to a worker from falling as a result of an electric shock). Moreover, the equation applies only when the duration of the electric shock is limited. If the precautions the employer takes, including those required by applicable standards, do not adequately protect employees from hazards associated with involuntary reactions from electric shock, a hazard exists if the induced voltage is sufficient to pass a current of 1 milliampere through a 500-ohm resistor. (The 500-ohm resistor the represents resistance of an employee. The 1-milliampere current is the threshold of perception.) Finally, if the employer protects employees from injury due to involuntary reactions from electric shock, but the duration of the electric shock is unlimited (that is, when the fault current at the work location will be insufficient to trip the devices protecting the circuit), a hazard exists if the resultant current would be more than 6 milliamperes (the recognized let-go threshold for workers[17]).

2. *Acceptable methods of grounding for employers that do not perform an engineering determination.* The grounding methods presented in this section of this appendix ensure that differences in electric potential are as low as possible and, therefore, meet § 1910.269(n)(3) without an engineering determination of the potential differences. These methods follow two principles: (i) The grounding method must ensure that the circuit opens in the fastest available clearing time, and (ii) the grounding method must ensure that the potential differences between conductive objects in the employee's work area are as low as possible.

Paragraph (n)(3) of § 1910.269 does not require grounding methods to meet the criteria embodied in these principles. Instead, the paragraph requires that protective grounds be "placed at such locations and arranged in such a manner that the employer can demonstrate will prevent exposure of each employee to hazardous differences in electric potential." However, when the employer's grounding practices do not follow these two principles, the employer will need to perform an engineering analysis to make the demonstration required by § 1910.269(n)(3).

i. *Ensuring that the circuit opens in the fastest available clearing time.* Generally, the higher the fault current, the shorter the clearing times for the same type of fault. Therefore, to ensure the fastest available clearing time, the grounding method must maximize the fault current with a low impedance connection to ground. The employer accomplishes this objective by grounding the circuit conductors to the best ground available at the worksite. Thus, the employer must ground to a grounded system neutral conductor, if one is present. A grounded system neutral has a direct connection to the system ground at the source, resulting in an extremely low impedance to ground. In a substation, the employer may instead ground to the substation grid, which also has an extremely low impedance to the system ground and, typically, is connected to a grounded system neutral when one is present. Remote system grounds, such as pole and tower grounds, have a higher impedance to the system ground than grounded system neutrals and substation grounding grids; however, the employer may use a remote ground when lower impedance grounds are not available. In the absence of a grounded system neutral, substation grid, and remote ground, the employer may use a temporary driven ground at the worksite.

In addition, if employees are working on a three-phase system, the grounding method must short circuit all three phases. Short circuiting all phases will ensure faster clearing and lower the current through the grounding

---

[17] Electric current passing through the body has varying effects depending on the amount of the current. At the let-go threshold, the current overrides a person's control over his or her muscles. At that level, an employee grasping an object will not be able to let go of the object. The let-go threshold varies from person to person; however, the recognized value for workers is 6 milliamperes.

cable connecting the deenergized line to ground, thereby lowering the voltage across that cable. The short circuit need not be at the worksite; however, the employer must treat any conductor that is not grounded at the worksite as energized because the ungrounded conductors will be energized at fault voltage during a fault.

ii. *Ensuring that the potential differences between conductive objects in the employee's work area are as low as possible.* To achieve as low a voltage as possible across any two conductive objects in the work area, the employer must bond all conductive objects in the work area. This section of this appendix discusses how to create a zone that minimizes differences in electric potential between conductive objects in the work area.

The employer must use bonding cables to bond conductive objects, except for metallic objects bonded through metal-to-metal contact. The employer must ensure that metal-to-metal contacts are tight and free of contamination, such as oxidation, that can increase the impedance across the connection. For example, a bolted connection between metal lattice tower members is acceptable if the connection is tight and free of corrosion and other contamination. Figure 4 shows how to create an equipotential zone for metal lattice towers.

Wood poles are conductive objects. The poles can absorb moisture and conduct electricity, particularly at distribution and transmission voltages. Consequently, the employer must either: (1) Provide a conductive platform, bonded to a grounding cable, on which the worker stands or (2) use cluster bars to bond wood poles to the grounding cable. The employer must ensure that employees install the cluster bar below, and close to, the worker's feet. The inner portion of the wood pole is more conductive than the outer shell, so it is important that the cluster bar be in conductive contact with a metal spike or nail

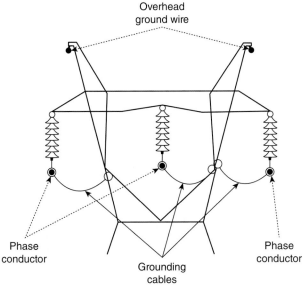

Overhead ground wire

Phase conductor

Phase conductor

Grounding cables

Notes:
1. Employers must ground overhead ground wires that are within reach of the employee.
2. The grounding cable must be as short as practicable; therefore, the attachment points between the grounding cable and the tower may be different from that shown in the figure.

FIGURE 4   Equipotential Zone for Metal Lattice Tower.

that penetrates the wood to a depth greater than or equal to the depth the worker's climbing gaffs will penetrate the wood. For example, the employer could mount the cluster bar on a bare pole ground wire fastened to the pole with nails or staples that penetrate to the required depth. Alternatively, the employer may temporarily nail a conductive strap to the pole and connect the strap to the cluster bar. Figure 5 shows how to create an equipotential zone for wood poles.

For underground systems, employers commonly install grounds at the points of disconnection of the underground cables. These grounding points are typically remote from the manhole or underground vault where employees will be working on the cable. Workers in contact with a cable grounded at a remote location can experience hazardous potential differences if the cable becomes energized or if a fault occurs on a different, but nearby, energized cable. The fault current causes potential gradients in the earth, and a potential difference will exist between the earth where the worker is standing and the earth where the cable is grounded. Consequently, to create an equipotential zone for the worker, the employer must provide a means of connecting the deenergized cable to ground at the worksite by

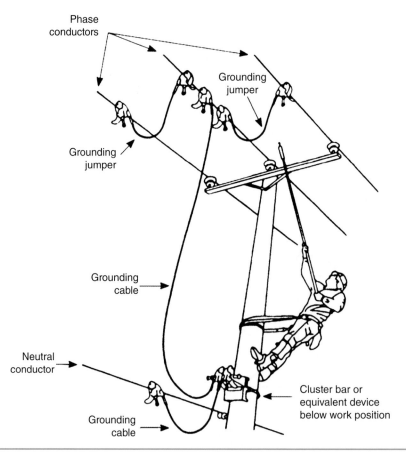

Figure 5   Equipotential Grounding for Wood Poles. *(Reprinted with permission from Hubbell Power Systems, Inc. (Hubbell) OSHA revised the figure from Hubbell's original.)*

having the worker stand on a conductive mat bonded to the deenergized cable. If the cable is cut, the employer must install a bond across the opening in the cable or install one bond on each side of the opening to ensure that the separate cable ends are at the same potential. The employer must protect the worker from any hazardous differences in potential any time there is no bond between the mat and the cable (for example, before the worker installs the bonds).

3. *Other safety-related considerations.* To ensure that the grounding system is safe and effective, the employer should also consider the following factors:[18]

   i. *Maintenance of grounding equipment.* It is essential that the employer properly maintain grounding equipment. Corrosion in the connections between grounding cables and clamps and on the clamp surface can increase the resistance of the cable, thereby increasing potential differences. In addition, the surface to which a clamp attaches, such as a conductor or tower member, must be clean and free of corrosion and oxidation to ensure a low-resistance connection. Cables must be free of damage that could reduce their current-carrying capacity so that they can carry the full fault current without failure. Each clamp must have a tight connection to the cable to ensure a low resistance and to ensure that the clamp does not separate from the cable during a fault.

   ii. *Grounding cable length and movement.* The electromagnetic forces on grounding cables during a fault increase with increasing cable length. These forces can cause the cable to move violently during a fault and can be high enough to damage the cable or clamps and cause the cable to fail. In addition, flying cables can injure workers. Consequently, cable lengths should be as short as possible, and grounding cables that might carry high fault current should be in positions where the cables will not injure workers during a fault.

Appendix D to § 1910.269—Methods of Inspecting and Testing Wood Poles

I. Introduction

When employees are to perform work on a wood pole, it is important to determine the condition of the pole before employees climb it. The weight of the employee, the weight of equipment to be installed, and other working stresses (such as the removal or retensioning of conductors) can lead to the failure of a defective pole or a pole that is not designed to handle the additional stresses.[19] For these reasons, it is essential that, before an employee climbs a wood pole, the employer ascertain that the pole is capable of sustaining the stresses of the work. The determination that the pole is capable of sustaining these stresses includes an inspection of the condition of the pole.

If the employer finds the pole to be unsafe to climb or to work from, the employer must secure the pole so that it does not fail while an employee is on it. The employer

---

[18]This appendix only discusses factors that relate to ensuring an equipotential zone for employees. The employer must consider other factors in selecting a grounding system that is capable of conducting the maximum fault current that could flow at the point of grounding for the time necessary to clear the fault, as required by § 1910.269(n)(4)(i). IEEE Std 1048-2003 contains guidelines for selecting and installing grounding equipment that will meet § 1910.269(n)(4)(i).

[19]A properly guyed pole in good condition should, at a minimum, be able to handle the weight of an employee climbing it.

can secure the pole by a line truck boom, by ropes or guys, or by lashing a new pole alongside it. If a new one is lashed alongside the defective pole, employees should work from the new one.

II. Inspecting Wood Poles

A qualified employee should inspect wood poles for the following conditions:[20]

A. *General condition.* Buckling at the ground line or an unusual angle with respect to the ground may indicate that the pole has rotted or is broken.

B. *Cracks.* Horizontal cracks perpendicular to the grain of the wood may weaken the pole. Vertical cracks, although not normally considered to be a sign of a defective pole, can pose a hazard to the climber, and the employee should keep his or her gaffs away from them while climbing.

C. *Holes.* Hollow spots and woodpecker holes can reduce the strength of a wood pole.

D. *Shell rot and decay.* Rotting and decay are cutout hazards and possible indications of the age and internal condition of the pole.

E. *Knots.* One large knot or several smaller ones at the same height on the pole may be evidence of a weak point on the pole.

F. *Depth of setting.* Evidence of the existence of a former ground line substantially above the existing ground level may be an indication that the pole is no longer buried to a sufficient depth.

G. *Soil conditions.* Soft, wet, or loose soil around the base of the pole may indicate that the pole will not support any change in stress.

H. *Burn marks.* Burning from transformer failures or conductor faults could damage the pole so that it cannot withstand changes in mechanical stress.

III. Testing Wood Poles

The following tests, which are from § 1910.268(n)(3), are acceptable methods of testing wood poles:

A. *Hammer test.* Rap the pole sharply with a hammer weighing about 1.4 kg (3 lbs), starting near the groundline and continuing upwards circumferentially around the pole to a height of approximately 1.8 meters (6 ft). The hammer will produce a clear sound and rebound sharply when striking sound wood. Decay pockets will be indicated by a dull sound or a less pronounced hammer rebound. Also, prod the pole as near the ground line as possible using a pole prod or a screwdriver with a blade at least 127 millimeters (5 inches) long. If substantial decay is present, the pole is unsafe.

B. *Rocking test.* Apply a horizontal force to the pole and attempt to rock it back and forth in a direction perpendicular to the line. Exercise caution to avoid causing power lines to swing together. Apply the force to the pole either by pushing it with a pike pole or pulling the pole with a rope. If the pole cracks during the test, it is unsafe.

---

[20]The presence of any of these conditions is an indication that the pole may not be safe to climb or to work from. The employee performing the inspection must be qualified to make a determination as to whether it is safe to perform the work without taking additional precautions.

Appendix E to § 1910.269—Protection from Flames and Electric Arcs

I. Introduction

Paragraph (l)(8) of § 1910.269 addresses protecting employees from flames and electric arcs. This paragraph requires employers to: (1) Assess the workplace for flame and electric-arc hazards (paragraph (l)(8)(i)); (2) estimate the available heat energy from electric arcs to which employees would be exposed (paragraph (l)(8)(ii)); (3) ensure that employees wear clothing that will not melt, or ignite and continue to burn, when exposed to flames or the estimated heat energy (paragraph (l)(8)(iii)); and (4) ensure that employees wear flame-resistant clothing[21] and protective clothing and other protective equipment that has an arc rating greater than or equal to the available heat energy under certain conditions (paragraphs (l)(8)(iv) and (l)(8)(v)). This appendix contains information to help employers estimate available heat energy as required by § 1910.269(l)(8)(ii), select protective clothing and other protective equipment with an arc rating suitable for the available heat energy as required by § 1910.269(l)(8)(v), and ensure that employees do not wear flammable clothing that could lead to burn injury as addressed by §§ 1910.269(l)(8)(iii) and (l)(8)(iv).

II. Assessing the Workplace for Flame and Electric-Arc Hazards

Paragraph (l)(8)(i) of § 1910.269 requires the employer to assess the workplace to identify employees exposed to hazards from flames or from electric arcs. This provision ensures that the employer evaluates employee exposure to flames and electric arcs so that employees who face such exposures receive the required protection. The employer must conduct an assessment for each employee who performs work on or near exposed, energized parts of electric circuits.

A. Assessment Guidelines Sources of electric arcs. Consider possible sources of electric arcs, including:

- Energized circuit parts not guarded or insulated,
- Switching devices that produce electric arcs in normal operation,
- Sliding parts that could fault during operation (for example, rack-mounted circuit breakers), and
- Energized electric equipment that could fail (for example, electric equipment with damaged insulation or with evidence of arcing or overheating).

Exposure to flames. Identify employees exposed to hazards from flames. Factors to consider include:

- The proximity of employees to open flames, and
- For flammable material in the work area, whether there is a reasonable likelihood that an electric arc or an open flame can ignite the material.

---

[21] Flame-resistant clothing includes clothing that is inherently flame resistant and clothing chemically treated with a flame retardant. (See ASTM F1506-10a, Standard Performance Specification for Flame Resistant Textile Materials for Wearing Apparel for Use by Electrical Workers Exposed to Momentary Electric Arc and Related Thermal Hazards, and ASTM F1891-12 Standard Specification for Arc and Flame Resistant Rainwear.)

Probability that an electric arc will occur. Identify employees exposed to electric-arc hazards. The Occupational Safety and Health Administration will consider an employee exposed to electric-arc hazards if there is a reasonable likelihood that an electric arc will occur in the employee's work area, in other words, if the probability of such an event is higher than it is for the normal operation of enclosed equipment. Factors to consider include:

- For energized circuit parts not guarded or insulated, whether conductive objects can come too close to or fall onto the energized parts,

- For exposed, energized circuit parts, whether the employee is closer to the part than the minimum approach distance established by the employer (as permitted by § 1910.269(l)(3)(iii)).

- Whether the operation of electric equipment with sliding parts that could fault during operation is part of the normal operation of the equipment or occurs during servicing or maintenance, and

- For energized electric equipment, whether there is evidence of impending failure, such as evidence of arcing or overheating.

B. Examples   Table 1 provides task-based examples of exposure assessments.

Task		Is Employee Exposed to Flame or Electric Arc Hazard?
Normal operation of enclosed equipment, such as closing or opening a switch.	The employer properly installs and maintains enclosed equipment, and there is no evidence of impending failure.	No.
	There is evidence of arcing or overheating	Yes.
	Parts of the equipment are loose or sticking, or the equipment otherwise exhibits signs of lack of maintenance.	Yes.
Servicing electric equipment, such as racking in a circuit breaker or replacing a switch		Yes.
Inspection of electric equipment with exposed energized parts.	The employee is not holding conductive objects and remains outside the minimum approach distance established by the employer.	No.
	The employee is holding a conductive object, such as a flashlight, that could fall or otherwise contact energized parts (irrespective of whether the employee maintains the minimum approach distance).	Yes.
	The employee is closer than the minimum approach distance established by the employer (for example, when wearing rubber insulating gloves or rubber insulating gloves and sleeves).	Yes.
Using open flames, for example, in wiping cable splice sleeves		Yes.

TABLE 1   Example Assessments for Various Tasks

III. Protection against Burn Injury

A. Estimating Available Heat Energy Calculation methods. Paragraph (l)(8)(ii) of § 1910.269 provides that, for each employee exposed to an electric-arc hazard, the employer must make a reasonable estimate of the heat energy to which the employee would be exposed if an arc occurs. Table 2 lists various methods of calculating values of available heat energy from an electric circuit. The Occupational Safety and Health Administration does not endorse any of these specific methods. Each method requires the input of various parameters, such as fault current, the expected length of the electric arc, the distance from the arc to the employee, and the clearing time for the fault (that is, the time the circuit protective devices take to open the circuit and clear the fault). The employer can precisely determine some of these parameters, such as the fault current and the clearing time, for a given system. The employer will need to estimate other parameters, such as the length of the arc and the distance between the arc and the employee, because such parameters vary widely.

The amount of heat energy calculated by any of the methods is approximately inversely proportional to the square of the distance between the employee and the arc. In other words, if the employee is very close to the arc, the heat energy is very high; but if the employee is just a few more centimeters away, the heat energy drops substantially. Thus, estimating the distance from the arc to the employee is key to protecting employees.

The employer must select a method of estimating incident heat energy that provides a reasonable estimate of incident heat energy for the exposure involved. Table 3 shows which methods provide reasonable estimates for various exposures.

Selecting a reasonable distance from the employee to the arc. In estimating available heat energy, the employer must make some reasonable assumptions about how far the employee will be from the electric arc. Table 4 lists reasonable distances from the employee to the electric arc. The distances in Table 4 are consistent with national consensus standards, such as the Institute of Electrical and Electronic Engineers' National Electrical Safety Code, ANSI/IEEE C2-2012, and IEEE Guide for Performing Arc-Flash Hazard Calculations, IEEE Std 1584b-2011. The employer is free to use other reasonable

1. Standard for Electrical Safety Requirements for Employee Workplaces, NFPA 70E-2012, Annex D, "Sample Calculation of Flash Protection Boundary."
2. Doughty, T.E., Neal, T.E., and Floyd II, H.L., "Predicting Incident Energy to Better Manage the Electric Arc Hazard on 600 V Power Distribution Systems," Record of Conference Papers IEEE IAS 45th Annual Petroleum and Chemical Industry Conference, September 28–30, 1998.
3. Guide for Performing Arc-Flash Hazard Calculations, IEEE Std 1584-2002, 1584a-2004 (Amendment 1 to IEEE Std 1584-2002), and 1584b-2011 (Amendment 2: Changes to Clause 4 of IEEE Std 1584-2002).*
4. ARCPRO, a commercially available software program developed by Kinectrics, Toronto, ON, CA.

*This appendix refers to IEEE Std 1584-2002 with both amendments as IEEE Std 1584b-2011.

TABLE 2 Methods of Calculating Incident Heat Energy from an Electric Arc

Incident-Energy Calculation Method	600 V and Less[2]			601 V to 15 kV[2]			More than 15 kV		
	1Φ	3Φa	3Φb	1Φ	3Φa	3Φb	1Φ	3Φa	3Φb
NFPA 70E-2012 Annex D (Lee Equation)	Y-C	Y	N	Y-C	Y-C	N	N[3]	N[3]	N[3]
Doughty, Neal, and Floyd	Y-C	Y	Y	N	N	N	N	N	N
IEEE Std 1584b-2011	Y	Y	Y	Y	Y	Y	N	N	N
ARCPRO	y	N	N	Y	N	N	Y	Y[4]	Y[4]

Key:

1Φ: Single-phase arc in open air.

3Φa: Three-phase arc in open air.

3Φb: Three-phase arc in an enclosure (box).

Y: Acceptable; produces a reasonable estimate of incident heat energy from this type of electric arc.

N: Not acceptable; does not produce a reasonable estimate of incident heat energy from this type of electric arc.

 Y-C: Acceptable; produces a reasonable, but conservative, estimate of incident heat energy from this type of electric arc.

Notes:

1. Although the Occupational Safety and Health Administration will consider these methods reasonable for enforcement purposes when employers use the methods in accordance with this table, employers should be aware that the listed methods do not necessarily result in estimates that will provide full protection from internal faults in transformers and similar equipment or from arcs in underground manholes or vaults.

2. At these voltages, the presumption is that the arc is three-phase unless the employer can demonstrate that only one phase is present or that the spacing of the phases is sufficient to prevent a multiphase arc from occurring.

3. Although the Occupational Safety and Health Administration will consider this method acceptable for purposes of assessing whether incident energy exceeds 2.0 cal/cm[2], the results at voltages of more than 15 kilovolts are extremely conservative and unrealistic.

4. The Occupational Safety and Health Administration will deem the results of this method reasonable when the employer adjusts them using the conversion factors for three-phase arcs in open air or in an enclosure, as indicated in the program's instructions.

TABLE 3   Selecting a Reasonable Incident-Energy Calculation Method[1]

distances, but must consider equipment enclosure size and the working distance to the employee in selecting a distance from the employee to the arc. The Occupational Safety and Health Administration will consider a distance reasonable when the employer bases it on equipment size and working distance.

Selecting a reasonable arc gap. For a single-phase arc in air, the electric arc will almost always occur when an energized conductor approaches too close to ground. Thus, an employer can determine the arc gap, or arc length, for these exposures by the dielectric strength of air and the voltage on the line. The dielectric strength of air is approximately 10 kilovolts for every 25.4 millimeters (1 inch). For example, at 50 kilovolts, the arc gap would be $50 \div 10 \times 25.4$ (or 50×2.54), which equals 127 millimeters (5 inches).

Class of Equipment	Single-Phase Arc mm (inches)	Three-Phase Arc mm (inches)
Cable	*NA	455 (18)
Low voltage MCCs and panel boards	NA	455 (18)
Low-voltage switchgear	NA	610 (24)
5-kV switchgear..	NA	910 (36)
15-kV switchgear..	NA	910 (36)
Single conductors in air (up to 46 kilovolts), work with rubber insulating gloves	380 (15)	NA
Single conductors in air, work with live-line tools and live-line bare-hand work	MAD − (2 × kV × 2.54)	NA
	(MAD − (2 × kV /10))[†]	

*NA = not applicable.
[†]The terms in this equation are:
MAD = The applicable minimum approach distance, and kV = The system voltage in kilovolts.

TABLE 4   Selecting a Reasonable Distance from the Employee to the Electric Arc

For three-phase arcs in open air and in enclosures, the arc gap will generally be dependent on the spacing between parts energized at different electrical potentials. Documents such as IEEE Std 1584b-2011 provide information on these distances. Employers may select a reasonable arc gap from Table 5, or they may select any other reasonable arc gap based on sparkover distance or on the spacing between (1) live parts at different potentials or (2) live parts and grounded parts (for example, bus or conductor spacings in equipment). In any event, the employer must use an estimate that reasonably resembles the actual exposures faced by the employee.

Making estimates over multiple system areas. The employer need not estimate the heat-energy exposure for every job task performed by each employee. Paragraph (l)(8)(ii) of § 1910.269 permits the employer to make broad estimates that cover multiple system areas provided that: (1) The employer uses reasonable assumptions about the energy-exposure distribution throughout the system, and (2) the estimates represent the maximum exposure for those areas. For example, the employer can use the maximum fault current and clearing time to cover several system areas at once.

Incident heat energy for single-phase-to ground exposures. Table 6 and Table 7 provide incident heat energy levels for open air, phase-to-ground electric-arc exposures typical for overhead systems.[22] Table 6 presents estimates of available energy for employees using rubber insulating gloves to perform work on overhead systems

---

[22] The Occupational Safety and Health Administration used metric values to calculate the clearing times in Table 6 and Table 7. An employer may use English units to calculate clearing times instead even though the results will differ slightly.

Class of Equipment	Single-Phase Arc mm (inches)	Three-Phase Arc mm[1] (inches)
Cable	NA[2]	13 (0.5)
Low voltage MCCs and panel boards	NA	25 (1.0)
Low-voltage switchgear	NA	32 (1.25)
5-kV switchgear	NA	104 (4.0)
15-kV switchgear	NA	152 (6.0)
Single conductors in air, 15 kV and less	51 (2.0)	Phase conductor spacing
Single conductor in air, more than 15 kV	Voltage in kV × 2.54 .... (Voltage in kV × 0.1), but no less than 51 mm (2 inches)	Phase conductor spacing

[1]*Source:* IEEE Std 1584b-2011.
[2]NA = not applicable.

TABLE 5   Selecting a Reasonable Arc Gap

operating at 4 to 46 kilovolts. The table assumes that the employee will be 380 millimeters (15 inches) from the electric arc, which is a reasonable estimate for rubber insulating glove work. Table 6 also assumes that the arc length equals the sparkover distance for the maximum transient overvoltage of each voltage range.[23] To use the table, an employer would use the voltage, maximum fault current, and maximum clearing time for a system area and, using the appropriate voltage range and fault-current and clearing time values corresponding to the next higher values listed in the table, select the appropriate heat energy (4, 5, 8, or 12 cal/cm^2) from the table. For example, an employer might have a 12,470-volt power line supplying a system area. The power line can supply a maximum fault current of 8 kiloamperes with a maximum clearing time of 10 cycles. For rubber glove work, this system falls in the 4.0-to-15.0-kilovolt range; the next-higher fault current is 10 kA (the second row in that voltage range); and the clearing time is under 18 cycles (the first column to the right of the fault current column). Thus, the available heat energy for this part of the system will be 4 cal/cm^2 or less (from the column heading), and the employer could select protection with a 5-cal/cm^2 rating to meet § 1910.269(l)(8)(v). Alternatively, an employer could select a base incident-energy value and ensure that the clearing times for each voltage range and fault current listed in the table do not exceed the corresponding clearing time specified in the table. For example, an employer that provides employees with arc-flash protective equipment rated at 8 cal/cm^2 can use the table to determine if any system area exceeds 8 cal/cm^2

[23]The Occupational Safety and Health Administration based this assumption, which is more conservative than the arc length specified in Table 5, on Table 410-2 of the 2012 NESC.

Voltage Range (kV)**	Fault Current (kA)	Maximum Clearing Time (Cycles)			
		4 cal/ cm²	5 cal/ cm²	8 cal/ cm²	12 cal/ cm²
4.0 to 15.0	5	46	58	92	138
	10	18	22	36	54
	15	10	12	20	30
	20	6	8	13	19
15.1 to 25.0	5	28	34	55	83
	10	11	14	23	34
	15	7	8	13	20
	20	4	5	9	13
25.1 to 36.0	5	21	26	42	62
	10	9	11	18	26
	15	5	6	10	16
	20	4	4	7	11
36.1 to 46.0	5	16	20	32	48
	10	7	9	14	21
	15	4	5	8	13
	20	3	4	6	9

Notes:

*This table is for open-air, phase-to-ground electric-arc exposures. It is not for phase-to-phase arcs or enclosed arcs (arc in a box).

†The table assumes that the employee will be 380 mm (15 in.) from the electric arc. The table also assumes the arc length to be the sparkover distance for the maximum transient overvoltage of each voltage range (see Appendix B to § 1910.269), as follows:

    4.0 to 15.0 kV 51 mm (2 in.)
    15.1 to 25.0 kV 102 mm (4 in.)
    25.1 to 36.0 kV 152 mm (6 in.)
    36.1 to 46.0 kV 229 mm (9 in.)

‡The Occupational Safety and Health Administration calculated the values in this table using the ARCPRO method listed in Table 2.

**The voltage range is the phase-to-phase system voltage.

TABLE 6 Incident Heat Energy for Various Fault Currents, Clearing Times, and Voltages of 4.0 to 46.0 KV: Rubber Insulating Glove Exposures Involving Phase-to-Ground Arcs in Open Air Only *†‡

Voltage Range (kV)**	Fault Current (kA)	Maximum Clearing Time (Cycles)			
		4 cal/ cm²	5 cal/ cm²	8 cal/ cm²	12 cal/ cm²
4.0 to 15.0	5	197	246	394	591
	10	73	92	147	220
	15	39	49	78	117
	20	24	31	49	73
15.1 to 25.0	5	197	246	394	591
	10	75	94	150	225
	15	41	51	82	122
	20	26	33	52	78
25.1 to 36.0	5	138	172	275	413
	10	53	66	106	159
	15	30	37	59	89
	20	19	24	38	58
36.1 to 46.0	5	129	161	257	386
	10	51	64	102	154
	15	29	36	58	87
	20	19	24	38	57
46.1 to 72.5	20	18	23	36	55
	30	10	13	20	30
	40	6	8	13	19
	50	4	6	9	13
72.6 to 121.0	20	10	12	20	30
	30	6	7	11	17
	40	4	5	7	11
	50	3	3	5	8
121.1 to 145.0	20	12	15	24	35
	30	7	9	15	22
	40	5	6	10	15
	50	4	5	8	11
145.1 to 169.0	20	12	15	24	36
	30	7	9	15	22
	40	5	7	10	16
	50	4	5	8	12
169.1 to 242.0	20	13	17	27	40
	30	8	10	17	25
	40	6	7	12	17
	50	4	5	9	13

TABLE 7   Incident Heat Energy for Various Fault Currents, Clearing Times, and Voltages: Live-Line Tool Exposures Involving Phase-to-Ground Arcs in Open Air Only *†‡# (Continued)

Voltage Range (kV)**	Fault Current (kA)	Maximum Clearing Time (Cycles)			
		4 cal/ cm²	5 cal/ cm²	8 cal/ cm²	12 cal/ cm²
242.1 to 362.0	20	25	32	51	76
	30	16	19	31	47
	40	11	14	22	33
	50	8	10	16	25
362.1 to 420.0	20	12	15	25	37
	30	8	10	15	23
	40	5	7	11	16
	50	4	5	8	12
420.1 to 550.0	20	23	29	47	70
	30	14	18	29	43
	40	10	13	20	30
	50	8	9	15	23
550.1 to 800.0	20	25	31	50	75
	30	15	19	31	46
	40	11	13	21	32
	50	8	10	16	24

Notes:

*This table is for open-air, phase-to-ground electric-arc exposures. It is not for phase-to-phase arcs or enclosed arcs (arc in a box).

†The table assumes the arc length to be the sparkover distance for the maximum phase-to-ground voltage of each voltage range (see Appendix B to this section). The table also assumes that the employee will be the minimum approach distance minus twice the arc length from the electric arc.

‡The Occupational Safety and Health Administration calculated the values in this table using the ARCPRO method listed in Table 2.

#For voltages of more than 72.6 kV, employers may use this table only when the minimum approach distance established under § 1910.269(l)(3)(i) is greater than or equal to the following values:

72.6 to 121.0 kV 1.02 m.
121.1 to 145.0 kV 1.16 m.
145.1 to 169.0 kV 1.30 m.
169.1 to 242.0 kV 1.72 m.
242.1 to 362.0 kV 2.76 m.
362.1 to 420.0 kV 2.50 m.
420.1 to 550.0 kV 3.62 m.
550.1 to 800.0 kV 4.83 m.

**The voltage range is the phase-to-phase system voltage.

TABLE 7   Incident Heat Energy for Various Fault Currents, Clearing Times, and Voltages: Live-Line Tool Exposures Involving Phase-to-Ground Arcs in Open Air Only * † ‡ #

by checking the clearing time for the highest fault current for each voltage range and ensuring that the clearing times do not exceed the values specified in the 8-cal/cm² column in the table.

Table 7 presents similar estimates for employees using live-line tools to perform work on overhead systems operating at voltages of 4 to 800 kilovolts. The table assumes that the arc length will be equal to the sparkover distance[24] and that the employee will be a distance from the arc equal to the minimum approach distance minus twice the sparkover distance.

The employer will need to use other methods for estimating available heat energy in situations not addressed by Table 6 or Table 7. The calculation methods listed in Table 2 and the guidance provided in Table 3 will help employers do this. For example, employers can use IEEE Std 1584b-2011 to estimate the available heat energy (and to select appropriate protective equipment) for many specific conditions, including lower voltage, phase-to-phase arc, and enclosed arc exposures.

B. Selecting Protective Clothing and Other Protective Equipment Paragraph (l)(8)(v) of § 1910.269 requires employers, in certain situations, to select protective clothing and other protective equipment with an arc rating that is greater than or equal to the incident heat energy estimated under § 1910.269(l)(8)(ii). Based on laboratory testing required by ASTM F1506-10a, the expectation is that protective clothing with an arc rating equal to the estimated incident heat energy will be capable of preventing second-degree burn injury to an employee exposed to that incident heat energy from an electric arc. Note that actual electric-arc exposures may be more or less severe than the estimated value because of factors such as arc movement, arc length, arcing from reclosing of the system, secondary fires or explosions, and weather conditions. Additionally, for arc rating based on the fabric's arc thermal performance value[25] (ATPV), a worker exposed to incident energy at the arc rating has a 50% chance of just barely receiving a second-degree burn. Therefore, it is possible (although not likely) that an employee will sustain a second-degree (or worse) burn wearing clothing conforming to § 1910.269(l)(8)(v) under certain circumstances. However, reasonable employer estimates and maintaining appropriate minimum approach distances for employees should limit burns to relatively small burns that just barely extend beyond the epidermis (that is, just barely a second degree burn). Consequently, protective clothing and other protective equipment meeting § 1910.269(l)(8)(v) will provide an appropriate degree of protection for an employee exposed to electric-arc hazards.

---

[24] The dielectric strength of air is about 10 kilovolts for every 25.4 millimeters (1 inch). Thus, the employer can estimate the arc length in millimeters to be the phase-to-ground voltage in kilovolts multiplied by 2.54 (or voltage (in kilovolts) × 2.54).

[25] ASTM F1506-10a defines "arc thermal performance value" as "the incident energy on a material or a multilayer system of materials that results in a 50% probability that sufficient heat transfer through the tested specimen is predicted to cause the onset of a second-degree skin burn injury based on the Stoll [footnote] curve, cal/cm²." The footnote to this definition reads: "Derived from: Stoll, A. M., and Chianta, M. A., 'Method and Rating System for Evaluations of Thermal Protection,' Aerospace Medicine, Vol 40, 1969, pp. 1232–1238 and Stoll, A. M., and Chianta, M. A., 'Heat Transfer through Fabrics as Related to Thermal Injury,' Transactions-New York Academy of Sciences, Vol 33(7), Nov. 1971, pp. 649–670."

Paragraph (l)(8)(v) of § 1910.269 does not require arc-rated protection for exposures of 2 cal/cm² or less. Untreated cotton clothing will reduce a 2-cal/cm² exposure below the 1.2- to 1.5-cal/cm² level necessary to cause burn injury, and this material should not ignite at such low heat energy levels. Although § 1910.269(l)(8)(v) does not require clothing to have an arc rating when exposures are 2 cal/cm² or less, § 1910.269(l)(8)(iv) requires the outer layer of clothing to be flame resistant under certain conditions, even when the estimated incident heat energy is less than 2 cal/cm², as discussed later in this appendix.

Additionally, it is especially important to ensure that employees do not wear undergarments made from fabrics listed in the note to § 1910.269(l)(8)(iii) even when the outer layer is flame resistant or arc rated. These fabrics can melt or ignite easily when an electric arc occurs. Logos and name tags made from non-flame-resistant material can adversely affect the arc rating or the flame resistant characteristics of arc-rated or flame resistant clothing. Such logos and name tags may violate § 1910.269(l)(8)(iii), (l)(8)(iv), or (l)(8)(v).

Paragraph (l)(8)(v) of § 1910.269 requires that arc-rated protection cover the employee's entire body, with limited exceptions for the employee's hands, feet, face, and head. Paragraph (l)(8)(v)(A) of § 1910.269 provides that arc-rated protection is not necessary for the employee's hands under the following conditions:

For any estimated incident heat energy	When the employee is wearing rubber insulating gloves with protectors.
If the estimated incident heat energy does not exceed 14 cal/cm²	When the employee is wearing heavy-duty leather work gloves with a weight of at least 407 gm/m² (12 oz/yd²).

Paragraph (l)(8)(v)(B) of § 1910.269 provides that arc-rated protection is not necessary for the employee's feet when the employee is wearing heavy-duty work shoes or boots. Finally, § 1910.269(l)(8)(v)(C), (l)(8)(v)(D), and (l)(8)(v)(E) require arc-rated head and face protection as follows:

		Minimum Head and Face Protection	
Exposure	None*	Arc-Rated Faceshield with a Minimum Rating of 8 cal/cm²*	Arc-Rated Hood or Faceshield with Balaclava
Single-phase, open air	2-8 cal/cm²	9-12 cal/cm²	13 cal/cm² or higher†.
Three-phase	2-4 cal/cm²	5-8 cal/cm²	9 cal/cm² or higher†.

*These ranges assume that employees are wearing hardhats meeting the specifications in § 1910.135 or § 1926.100(b)(2), as applicable.

†The arc rating must be a minimum of 4 cal/cm² less than the estimated incident energy. Note that § 1910.269(l)(8)(v)(E) permits this type of head and face protection, with a minimum arc rating of 4 cal/cm² less than the estimated incident energy, at any incident energy level.

‡Note that § 1910.269(l)(8)(v) permits this type of head and face protection at any incident energy level.

IV. Protection against Ignition

Paragraph (l)(8)(iii) of § 1910.269 prohibits clothing that could melt onto an employee's skin or that could ignite and continue to burn when exposed to flames or to the available heat energy estimated by the employer under § 1910.269(l)(8)(ii). Meltable fabrics, such as acetate, nylon, polyester, and polypropylene, even in blends, must be avoided. When these fibers melt, they can adhere to the skin, thereby transferring heat rapidly, exacerbating burns, and complicating treatment. These outcomes can result even if the meltable fabric is not directly next to the skin. The remainder of this section focuses on the prevention of ignition.

Paragraph (l)(8)(v) of § 1910.269 generally requires protective clothing and other protective equipment with an arc rating greater than or equal to the employer's estimate of available heat energy. As explained earlier in this appendix, untreated cotton is usually acceptable for exposures of 2 cal/cm² or less.[26] If the exposure is greater than that, the employee generally must wear flame-resistant clothing with a suitable arc rating in accordance with § 1910.269(l)(8)(iv) and (l)(8)(v). However, even if an employee is wearing a layer of flame-resistant clothing, there are circumstances under which flammable layers of clothing would be uncovered, and an electric arc could ignite them. For example, clothing ignition is possible if the employee is wearing flammable clothing under the flame-resistant clothing and the underlayer is uncovered because of an opening in the flame-resistant clothing. Thus, for purposes of § 1910.269(l)(8)(iii), it is important for the employer to consider the possibility of clothing ignition even when an employee is wearing flame-resistant clothing with a suitable arc rating.

Under § 1910.269(l)(8)(iii), employees may not wear flammable clothing in conjunction with flame-resistant clothing if the flammable clothing poses an ignition hazard.[27] Although outer flame-resistant layers may not have openings that expose flammable inner layers, when an outer flame-resistant layer would be unable to resist breakopen,[28] the next (inner) layer must be flame-resistant if it could ignite.

Non-flame-resistant clothing can ignite even when the heat energy from an electric arc is insufficient to ignite the clothing. For example, nearby flames can ignite an employee's clothing; and, even in the absence of flames, electric arcs pose ignition hazards beyond the hazard of ignition from incident energy under certain conditions. In addition to requiring flame-resistant clothing when the estimated incident energy exceeds 2.0 cal/cm², § 1910.269(l)(8)(iv) requires flame-resistant clothing when: The employee is exposed to contact with energized circuit parts operating at more than 600 volts (§ 1910.269(l)(8)(iv)(A)), an electric arc could ignite flammable material in the work area that, in turn, could ignite the employee's clothing (§ 1910.269(l)(8)(iv)(B)), and molten metal or electric arcs from faulted conductors in the work area could ignite the employee's clothing (§ 1910.269(l)(8)(iv)(C)). For example, grounding conductors can become a source of heat energy if they cannot carry fault current without failure. The employer must consider these possible

---

[26]See § 1910.269(l)(8)(iv)(A), (l)(8)(iv)(B), and (l)(8)(iv)(C) for conditions under which employees must wear flame-resistant clothing as the outer layer of clothing even when the incident heat energy does not exceed 2 cal/cm².

[27]Paragraph (l)(8)(iii) of § 1910.269 prohibits clothing that could ignite and continue to burn when exposed to the heat energy estimated under paragraph (l)(8)(ii) of that section.

[28]Breakopen occurs when a hole, tear, or crack develops in the exposed fabric such that the fabric no longer effectively blocks incident heat energy.

sources of electric arcs[29] in determining whether the employee's clothing could ignite under § 1910.269(l)(8)(iv)(C).

Appendix F to § 1910.269—Work-Positioning Equipment Inspection Guidelines

I. Body Belts
Inspect body belts to ensure that:

A. The hardware has no cracks, nicks, distortion, or corrosion;

B. No loose or worn rivets are present;

C. The waist strap has no loose grommets;

D. The fastening straps are not 100% leather; and

E. No worn materials that could affect the safety of the user are present.

II. Positioning Straps
Inspect positioning straps to ensure that:

A. The warning center of the strap material is not exposed;

B. No cuts, burns, extra holes, or fraying of strap material is present;

C. Rivets are properly secured;

D. Straps are not 100% leather; and

E. Snaphooks do not have cracks, burns, or corrosion.

III. Climbers
Inspect pole and tree climbers to ensure that:

A. Gaffs are at least as long as the manufacturer's recommended minimums (generally 32 and 51 millimeters (1.25 and 2.0 inches) for pole and tree climbers, respectively, measured on the underside of the gaff);

Note: Gauges are available to assist in determining whether gaffs are long enough and shaped to easily penetrate poles or trees.

B. Gaffs and leg irons are not fractured or cracked;

C. Stirrups and leg irons are free of excessive wear;

D. Gaffs are not loose;

E. Gaffs are free of deformation that could adversely affect use;

F. Gaffs are properly sharpened; and

G. There are no broken straps or buckles.

[79 FR 20691, July 10, 2014]

---

[29]Static wires and pole grounds are examples of grounding conductors that might not be capable of carrying fault current without failure. Grounds that can carry the maximum available fault current are not a concern, and employers need not consider such grounds a possible electric arc source.

Appendix G to § 1910.269—Reference Documents

The references contained in this appendix provide information that can be helpful in understanding and complying with the requirements contained in § 1910.269. The national consensus standards referenced in this appendix contain detailed specifications that employers may follow in complying with the more performance-based requirements of § 1910.269. Except as specifically noted in § 1910.269, however, the Occupational Safety and Health Administration will not necessarily deem compliance with the national consensus standards to be compliance with the provisions of § 1910.269.

ANSI/SIA A92.2-2009, *American National Standard for Vehicle-Mounted Elevating and Rotating Aerial Devices.*

ANSI Z133-2012, *American National Standard Safety Requirements for Arboricultural Operations-Pruning, Trimming, Repairing, Maintaining, and Removing Trees, and Cutting Brush.*

ANSI/IEEE Std 935—1989, *IEEE Guide on Terminology for Tools and Equipment to Be Used in Live Line Working.*

ASME B20.1-2012, *Safety Standard for Conveyors and Related Equipment.*

ASTM D120-09, *Standard Specification for Rubber Insulating Gloves.*

ASTM D149-09 (2013), *Standard Test Method for Dielectric Breakdown Voltage and Dielectric Strength of Solid Electrical Insulating Materials at Commercial Power Frequencies.*

ASTM D178-01 (2010), *Standard Specification for Rubber Insulating Matting.*

ASTM D1048-12, *Standard Specification for Rubber Insulating Blankets.*

ASTM D1049-98 (2010), *Standard Specification for Rubber Insulating Covers.*

ASTM D1050-05 (2011), *Standard Specification for Rubber Insulating Line Hose.*

ASTM D1051-08, *Standard Specification for Rubber Insulating Sleeves.*

ASTM F478-09, *Standard Specification for In-Service Care of Insulating Line Hose and Covers.*

ASTM F479-06 (2011), *Standard Specification for In-Service Care of Insulating Blankets.*

ASTM F496-08, *Standard Specification for In-Service Care of Insulating Gloves and Sleeves.*

ASTM F711-02 (2007), *Standard Specification for Fiberglass-Reinforced Plastic (FRP) Rod and Tube Used in Live Line Tools.*

ASTM F712-06 (2011), *Standard Test Methods and Specifications for Electrically Insulating Plastic Guard Equipment for Protection of Workers.*

ASTM F819-10, *Standard Terminology Relating to Electrical Protective Equipment for Workers.*

ASTM F855-09, *Standard Specifications for Temporary Protective Grounds to Be Used on De-energized Electric Power Lines and Equipment.*

ASTM F887-12[e1], *Standard Specifications for Personal Climbing Equipment.*

ASTM F914/F914M-10, *Standard Test Method for Acoustic Emission for Aerial Personnel Devices without Supplemental Load Handling Attachments.*

ASTM F1116-03 (2008), *Standard Test Method for Determining Dielectric Strength of Dielectric Footwear.*

ASTM F1117-03 (2008), *Standard Specification for Dielectric Footwear.*

ASTM F1236-96 (2012), *Standard Guide for Visual Inspection of Electrical Protective Rubber Products.*

ASTM F1430/F1430M-10, *Standard Test Method for Acoustic Emission Testing of Insulated and Non-Insulated Aerial Personnel Devices with Supplemental Load Handling Attachments.*

ASTM F1505-10, *Standard Specification for Insulated and Insulating Hand Tools.*

ASTM F1506-10a, *Standard Performance Specification for Flame Resistant and Arc Rated Textile Materials for Wearing Apparel for Use by Electrical Workers Exposed to Momentary Electric Arc and Related Thermal Hazards.*

ASTM F1564-13, *Standard Specification for Structure-Mounted Insulating Work Platforms for Electrical Workers.*

ASTM F1701-12, *Standard Specification for Unused Polypropylene Rope with Special Electrical Properties.*

ASTM F1742-03 (2011), *Standard Specification for PVC Insulating Sheeting.*

ASTM F1796-09, *Standard Specification for High Voltage Detectors-Part 1 Capacitive Type to be Used for Voltages Exceeding 600 Volts AC.*

ASTM F1797-09[e1], *Standard Test Method for Acoustic Emission Testing of Insulated and Non-Insulated Digger Derricks.*

ASTM F1825-03 (2007), *Standard Specification for Clampstick Type Live Line Tools.*

ASTM F1826-00 (2011), *Standard Specification for Live Line and Measuring Telescoping Tools.*

ASTM F1891-12, *Standard Specification for Arc and Flame Resistant Rainwear.*

ASTM F1958/F1958M-12, *Standard Test Method for Determining the Ignitability of Non-flame-Resistant Materials for Clothing by Electric Arc Exposure Method Using Mannequins.*

ASTM F1959/F1959M-12, *Standard Test Method for Determining the Arc Rating of Materials for Clothing.*

IEEE Stds 4-1995,4a-2001(Amendment to IEEE Standard Techniques for High-Voltage Testing), *IEEE Standard Techniques for High-Voltage Testing.*

IEEE Std 62-1995, *IEEE Guide for Diagnostic Field Testing of Electric Power Apparatus-Part 1: Oil Filled Power Transformers, Regulators, and Reactors.*

IEEE Std 80-2000, *Guide for Safety in AC Substation Grounding.*

IEEE Std 100-2000, *The Authoritative Dictionary of IEEE Standards Terms Seventh Edition.*

IEEE Std 516-2009, *IEEE Guide for Maintenance Methods on Energized Power Lines.*

IEEE Std 524-2003, *IEEE Guide to the Installation of Overhead Transmission Line Conductors.*

IEEE Std 957-2005, *IEEE Guide for Cleaning Insulators.*

IEEE Std 1048-2003, *IEEE Guide for Protective Grounding of Power Lines.*

IEEE Std 1067-2005, *IEEE Guide for In-Service Use, Care, Maintenance, and Testing of Conductive Clothing for Use on Voltages up to 765 kV AC and ±750 kV DC.*

IEEE Std 1307-2004, *IEEE Standard for Fall Protection for Utility Work.*

IEEE Stds 1584-2002,1584a-2004 Amendment 1 to IEEE Std 1584-2002), and 1584b-2011 (Amendment 2: Changes to Clause 4 of IEEE Std 1584-2002), *IEEE Guide for Performing Arc-Flash Hazard Calculations.*

IEEE C2-2012, *National Electrical Safety Code.*

NFPA 70E-2012, *Standard for Electrical Safety in the Workplace.*

[79 FR 20691-20692, July 10, 2014]

Resuscitation

Introduction

Emergencies may require the lineman to rescue and resuscitate a coworker. If cardio-pulmonary resuscitation (CPR) is required, the recommended procedure is to sequentially alternate 30 chest compressions and 2 rescue breaths as explained in Chap. 48.

These chapters will present the steps you should take in an emergency by offering a review and explanation of what you have learned in professionally given first aid, CPR, and pole-top rescue classes. It is not intended as a replacement for taking the *Occupational Safety and Health Act* (OSHA)—required American Red Cross, American Heart Association, United States Bureau of Mines, or equivalent training that can be verified by documentary evidence.

To be prepared for an emergency, you should take a review course yearly and practice the skills on a mannequin in the presence of a trained instructor. To emphasize, the cardiopulmonary resuscitation (CPR) recommended procedure is to sequentially alternate 30 chest compressions and 2 rescue breaths as explained in Chap. 48.

Be Prepared for an Emergency—The Best Way to Care for an Emergency Is to Prevent It—Think Safety

Have and **use** PPE.

Inspect rubber gloves daily and have them electrically tested as per industry recommendations.

Be aware of what is going on around you.

Keep your mind on what you are doing.

Do not use any substance that will impair your ability to act and to react.

Wear clothes made of 100% cotton and rated at 11-lb weave.

Avoid wearing the orange traffic vest when not directing traffic.

Wear clothing that is nonconductive and flame-resistant.

Medical Services and First Aid

The employer shall provide medical services and first aid as required in § 1910.151. For fieldwork involving two or more employees at a work location, at least two trained persons shall be available. For fixed work locations such as substations, the number of trained persons available shall be sufficient to ensure that each employee exposed to electric shock can be reached within 4 minutes by a trained person. However, where the existing number of employees is insufficient to meet this requirement (at a remote substation, for example), each employee at the work location shall be a trained employee.

First Aid Kits and Supplies

First aid kits and supplies required by § 1910.151(b) shall be in a weatherproof container with individual sealed packages for each type of item. The contents shall be checked by the employer before being sent out on each job and at least weekly thereafter to see that expended items are replaced.

First aid kits are for your use and protection on the job. When they are taken off of the truck for unauthorized use and an emergency occurs, the missing supplies could mean the difference between an incident of small proportions and a life-threatening incident.

The employer is responsible for the first aid kits and so are you. You need to make sure they are present, and the supplies are not out of date or soiled. Make sure the appropriate company personnel knows if supplies need to be replaced.

Trucks may be stocked with an artificial electronic defibrillator (AED). Step by step instructions on use of the AED are illustrated on the lid of the device. All crew members are expected to be familiar with when and how to use the AED.

This is your responsibility. Your life could depend upon it.

You Have an Emergency—What Do You Do?

Be prepared.

Remain calm.

If the victim is on a pole or in a bucket, shout at your victim to see if they are conscious and can respond.

If they cannot respond, immediately send someone to call 911 or your local emergency system.

Use common sense.

We do not want two victims.

Do not approach the victim until it is safe for you to do so.

Assess your victim.

Use protective barriers.

Give rescue breathing, CPR, or Heimlich maneuver as needed (these skills will be reviewed for you in the following pages).

Effects of Electric Shock on the Human Body

The effect of electric shock on a human being is rather unpredictable and may manifest itself in a number of ways:

1. *Asphyxia.* Electric shock may cause a cessation of respiration (asphyxia). Electric current passing through the body may temporarily paralyze (or destroy) either the nerves or the area of the brain which controls respiration.

2. *Burns—contact and flash.* Contact burns are a common result of electric current passing through the body. The burns are generally found at the points where the current entered and left the body and vary in severity, the same as thermal burns. The seriousness of these burns may not be immediately evident because their appearance may not indicate the depth to which they have penetrated.

In some accidents there is a flash or electric arc, the rays and heat from which may damage the eyes or result in thermal burns to exposed parts of the body.

3. *Fibrillation*. Electric shock may disturb the natural rhythm of the heartbeat. When this happens, the muscles of the heart are thrown into a twitching or trembling state, and the actions of the individual muscle fibers are no longer coordinated. The pulse disappears, and circulation ceases. This condition is known as ventricular fibrillation and is serious.

4. *Muscle spasm*. A series of erratic movements of a limb or limbs may occur owing to alternating contractions and relaxations of the muscles. This muscle spasm action on the muscles of respiration may be a factor in the stoppage of breathing.

Rescue

Because a person may receive electric shock in many different locations—on the ground, in buildings, on poles, or on steel structures—it is neither possible nor desirable to lay down definite methods of rescue. However, certain facts should be remembered.

Freeing the Victim

Because of the muscle spasm at the time of shock, most victims are thrown clear of contact. However, in some instances (usually low voltage) the victim is still touching live equipment. In either situation, the rescuer must be extremely careful not to get in contact with the live equipment or to touch the victim while he or she is still in contact. The rescuer should "free" the victim as soon as possible so that artificial respiration can be applied without hazard. This may involve opening switches or cutting wires so that equipment within reach is deenergized or using rubber gloves or other approved insulation to move the victim out of danger.

If the victim is to be lowered from a pole, tower, or other structure, a hand line of adequate strength, tied or looped around him and placed over a crossarm or tower member, is a simple and satisfactory method. Procedures for lowering a lineman from a pole or electric line truck bucket are illustrated in Chap. 49.

Artificial Respiration Methods

Early methods of artificial respiration involved applying heat by building a fire on the victim's stomach, beating the victim with a whip, inverting the victim by suspending him by his feet, or rolling the victim over a barrel. Modern methods of artificial respiration include the Schafer prone method, the chest-pressure-arm-lift method, the back-pressure-arm-lift method, and the mouth-to-mouth method. Artificial respiration should be used when breathing stops as a result of electric shock, exposure to gases, drowning, or physical injury. If the heart has stopped functioning as a result of cardiac arrest or ventricular fibrillation, heart-lung resuscitation should be used with mouth-to-mouth resuscitation. Heart-lung resuscitation methods are illustrated in Chap. 48.

Applying Artificial Respiration

To be successful, artificial respiration must be applied within the shortest time possible to a victim who is not breathing. Without oxygen, the brain begins to die. The first 4 minutes are critical. Irreversible brain damage begins. The best chance of survival is for the victim to receive rescue breathing, CPR immediately, and advanced medical care from an emergency medical system (EMS).

Normally, the type of resuscitation in which the victim is placed on the ground or floor is the best because the victim can be given additional treatment (stoppage of bleeding, wrapping in blanket to keep warm, etc.) and the rescuer can be easily relieved without an interruption of the breathing rhythm or cycle. However, pole-top resuscitation or any adaptation of it should be started if there will be any appreciable delay in getting the victim into the prone position.

In choosing the type of resuscitation to be used, the rescuer must consider the obvious injuries suffered by the victim. Broken ribs, for example, might make inadvisable the use of certain types of resuscitation; burns on the arms or on the face might exclude other types. However, no time should be lost in searching for injuries; artificial respiration should be started at once.

The Need for Speed

Artificial respiration must be started at once. Time is the most important single factor in successful artificial respiration. The human brain can exist for only a few minutes, probably not more than 4 or 5, without oxygenated blood; therefore, every second counts. The greater the delay, the smaller the chance of successful recovery. There should be no delay to loosen clothing, warm the victim, get him down from the pole, or move him to a more comfortable position. However, an immediate check of the victim's mouth should be made by a quick pass of the fingers (with a protective barrier such as rubber gloves) to remove tobacco, chewing gum, etc. Do not remove dentures unless they are loose. They help to support the facial structure.

Use Protective Barriers

Today, we run the risk of ourselves being exposed and exposing the victim to diseases. To prevent the transmission of disease, use protective barrier masks.

Resuscitation masks vary in size, shape, and features. The most effective are ones made of transparent, pliable material that allows you to make a tight seal on the person's face and have a one-way valve for releasing a person's exhaled air.

Use medical latex gloves when available.

Assess Your Victim

You must act immediately. Send someone to call 911 or your local emergency system immediately. Determine and use the best method of rescuing the victim. If the injured person is on a pole or in a bucket, make sure the area is safe for you to approach and then begin pole-top or bucket rescue procedures.

Shout at the victim on the pole to see if he can respond and is conscious. If he cannot respond, and the area is safe to reach the victim, determine if your victim is breathing. If not breathing, give two rescue breaths before you lower him to the ground (Fig. 47.1). Use protective barriers if possible.

Lower Victim to Ground—Open the Airway—Determine If Victim Is Breathing

To open the airway, you should tilt the head back by placing one hand on the victim's forehead. Using your other hand, place two fingers under the bony part of the lower jaw near the chin. Tilt head back and lift jaw. Avoid closing the victim's mouth. Avoid pushing on soft parts under the chin (Fig. 47.2).

The head-tilt/chin-lift will remove the tongue away from the back of the throat. Many times a victim will begin breathing on his or her own once you have done this (Fig. 47.3).

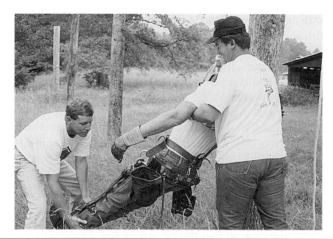

FIGURE 47.1   Lower the victim to the ground using the pole-top rescue techniques. Lay victim on back. Open the airway. Use the head-tilt/chin-lift method.

FIGURE 47.2   Proper method to be used to open the victim's airway.

FIGURE 47.3   Look for the rise and fall of the chest, listen and feel for escaping air for 3 to 5 sec.

If Not Breathing, Give Two Rescue Breaths

If the victim is not breathing, give two rescue breaths. Remember to use a protective barrier, if possible. A rescue breath is one breath given slowly for approximately 1½ seconds to 2 seconds. If you find that air will not enter or that the chest is not rising and falling, you should reposition the head and attempt giving a breath. Improper chin and head positioning is the most common cause of difficulty with ventilations. If after you have repositioned the airway and air is not going in, you may have a foreign object trapped in the airway. You should begin airway obstruction maneuvers (Heimlich maneuver). While maintaining the head-tilt/chin-lift, the rescuer should take a deep breath and seal their lips around the victim's mouth, creating an airtight seal. The rescuer gently pinches the nose closed with the thumb and index finger of the hand on the forehead, thereby preventing air from escaping through the victim's nose (Fig. 47.4).

Gently blow into the victim's mouth for 1½ seconds to 2 seconds. Look for the chest to rise. Release the nose so that the air can escape. Pause only long enough for you to take a breath, and then repeat the procedure for a second rescue breath.

Check for a Pulse

After you have given two rescue breaths, check for a pulse at the carotid artery (Fig. 47.5). To find the carotid artery, find the Adam's apple, slide your fingers down into the groove closest to you. Feel the pulse for approximately 5 seconds to 10 seconds. If a pulse is present, but the victim is not breathing, begin rescue breathing. If a pulse is absent, begin CPR.

FIGURE 47.4 The rescuer gently pinches the nose closed with the thumb and index finger of the hand on the forehead.

FIGURE 47.5 To find the carotid artery, find the Adam's apple and slide your fingers down into the groove closest to you. Feel for 5 seconds to 10 seconds for a pulse.

Mouth-to-Nose Rescue Breathing

The Journal of the American Medical Association recommends that mouth-to-nose breathing be used when:

- It is impossible to ventilate through the victim's mouth.
- The mouth cannot be opened.
- The mouth is seriously injured.
- A tight mouth-to-mouth seal is difficult to achieve.

Keep the victim's head tilted back with one hand on the forehead and use the other hand to lift the victim's lower jaw (as in the head-tilt/chin-lift) and close the mouth. The rescuer takes a deep breath and then breathes into the victim's nose. The rescuer removes his mouth and allows the air to escape.

If a Pulse Is Present But the Victim Is Not Breathing, Begin Rescue Breathing

Give the victim one rescue breath about every 5 seconds. Do this for about 1 minute (12 breaths).

Recheck the pulse and breathing about every minute.

You should continue rescue breathing as long as a pulse is present and the victim is not breathing, or until one of the following things happen:

- The victim begins to breathe without your help.
- The victim has no pulse (begin CPR).
- Another trained rescuer takes over for you.
- You are too tired to continue.

Review

Recognize that you have an emergency. Call for help immediately.

Assess your victim for unresponsiveness.

Make sure the scene is safe.

Begin pole-top or bucket rescue.

Check for breathing. If the victim is not breathing, give two breaths.

Lower the victim to the ground.

Open the airway and use head-tilt/chin-lift.

If the victim is not breathing, give two rescue breaths. (Remember to use a protective barrier if available.)

Check for a pulse.

If a pulse is present and breathing is absent, begin rescue breathing (one breath every 5 seconds).

If there is no pulse, start CPR.

Airway Obstruction—Conscious Adult

A choking victim may be making the universal signal for an obstructed airway. The chocking sign is one or both hands clutching the throat. Ask the victim if they are choking and if you can help him or her (Fig. 47.6). If they are coughing forcefully, the rescuer

FIGURE 47.6   Universal signal for an obstructed airway is one or both hands clutching the throat.

should not interfere with the victim's own attempt to expel the foreign object, but stay with them and encourage them to continue coughing. If the victim has a weak, ineffective cough or is making high-pitched noises while inhaling or is unable to speak, breathe, or cough, this should be treated as a complete airway obstruction. Stand behind the victim. Place your leg between the victim's legs; this will help you control the victim if he passes out.

Wrap your arms around the victim's waist, make a fist with one hand, the thumb side of the fist is placed against the victim's abdomen just slightly above the navel and well below the tip of the breastbone (Fig. 47.7).

The fist is grasped with the other hand and pressed into the victim's abdomen with a quick upward thrust (Fig. 47.8). Each new thrust should be a separate and distinct

FIGURE 47.7   The thumb should be folded into the fist so that there is a flat surface placed on the abdomen.

FIGURE 47.8   The fist is pressed into the victim's abdomen with a quick upward thrust.

movement administered with the intent of relieving the obstruction. The thrusts should be repeated and continued until the object is expelled from the airway or the victim becomes unconscious.

Conscious Adult Choking Victim Who Becomes Unconscious

If the victim becomes unconscious, lower victim to the floor or ground. The rescuer straddles the victim and places the heel of one hand slightly above the navel and well below the tip of the breastbone. The second hand is placed directly on top of the fist. The rescuer presses into the abdomen with a quick upward motion (Fig. 47.9).

Give up to five abdominal thrusts. The victim's mouth should be opened, and a finger sweep done after you have given five abdominal thrusts (Heimlich maneuvers). Give the victim a rescue breath, if air does not go in, repeat the sequence of five abdominal thrusts, finger sweep, and attempt to ventilate. Repeat as long as necessary.

How to Do a Finger Sweep

A finger sweep is done by opening the mouth—grasp both the tongue and the lower jaw between the thumb and fingers of one hand, lift jaw. Insert index finger of other hand into the mouth, running it along the inside of the cheek and deep into throat to the base of the tongue. Use a "hooking" action to dislodge any object that might be there. Remove the object. Be careful not to force the object deeper into the airway.

What to Do When the Object Is Cleared

If the object is cleared and the victim is conscious, the victim may want to leave the area and go to the restroom, etc. Do not let them leave by themselves. The throat has been irritated and may swell and the airway may close. Stay with the victim until EMS personnel arrive.

If You Are Alone and Choking

If you are alone and choking, you can perform the Heimlich maneuver on yourself (Figs. 47.10 and 47.11). If this is unsuccessful, the upper abdomen should be pressed

FIGURE 47.9   The rescuer straddles the unconscious victim and presses into the abdomen with a quick upward motion.

FIGURE **47.10** Applying Heimlich maneuver to yourself.

FIGURE **47.11** The lineman is utilizing the closest source available to apply the Heimlich maneuver to himself. The bed of the line truck is being used by the lineman to apply the Heimlich maneuver.

quickly over any firm surface such as the back of a chair, side of a table, or porch railing. Several thrusts may be needed to clear the airway.

What to Do If Your Victim Is Pregnant or Large

If you are not able to get your arms around the waist of a choking victim who is conscious, you must give chest thrusts.

Chest thrusts are given by getting behind the victim, the victim can be sitting or standing. Place your arms under the victim's armpits and around the chest. Place the thumb side of your fist on the middle of the breastbone. Be sure that your fist is centered right on the breastbone and not on the ribs. Make sure that your fist is not near the lower tip of the breastbone.

Grasp your fist with your other hand and give backward thrusts.

Give thrusts until the obstruction is cleared or until the person loses consciousness.

You should think of each thrust as a separate attempt to dislodge the object.

Review

Ask if victim is choking. Ask for permission to help.

If victim is coughing forcefully, encourage victim to continue coughing.

If victim has obstructed airway and is conscious, do abdominal thrusts until object is expelled or victim becomes unconscious.

If victim becomes unconscious, lower to floor. Give five abdominal thrusts.

Do a finger sweep.

Give a rescue breath, if air still does not go in, repeat sequence of five thrusts, finger sweep, and breath.

If alone, give Heimlich maneuver (abdominal thrust) to yourself, or lean over firm surface.

Heart-Lung Resuscitation

Cardiopulmonary Resuscitation (CPR)

Step A: Kneel facing the victim's chest. Find the correct hand position by sliding your fingers up the rib cage to the breastbone. *Step B:* Place your middle finger in the notch and the index finger next to it on the lower end of the breastbone. *Step C:* Place your other hand beside the two fingers. *Step D:* Place hand used to locate notch on top of the other hand. Lace the fingers together. Keep fingers off the chest. Position the shoulders over the hands, with the elbows locked and the arms straight (Fig. 48.1).

There are varying sizes and shapes of hands. Arthritic or injured hands and wrists will not allow some rescuers to lace their fingers together. An acceptable alternative hand position is to grasp the wrist of the hand on the chest with the hand that has been locating the lower end of their sternum. Another acceptable alternative is to hold your fingers up off the chest. Use the method that works best for you. Just remember to keep your fingers off the chest. Use the heel of your hand to do the compressions (Fig. 48.2). Compress the breastbone 1½ inches to 2 inches. Give 30 compressions. Count 1 and 2 and 3 and 4. Give two rescue breaths after the 30th compression. This cycle of 30 compressions and 2 breaths is consistent with ensuring the blood continues circulation within the body. Do 4 cycles of 30 compressions and 2 breaths. Check the pulse at the end of the fourth cycle.

When you press down, the weight of your upper body creates the force you need to compress the chest. It is important to keep your shoulders directly over your hands and your elbows locked so you use the weight of your upper body to compress the chest. Push with the weight of your upper body to compress the chest. Push with the weight of your upper body, not with your arm muscles. Push straight down; don't rock (Figs. 48.3 through 48.6).

If the thrust is not in a straight downward direction, the torso has a tendency to roll; part of the force is lost, and the chest compression may be less effective.

Once you have started CPR, you must continue until one of the following happens:

- Another trained person takes over CPR for you.
- EMS personnel arrive and take over the care of the victim.
- You are too exhausted and unable to continue.
- The scene becomes unsafe.

A

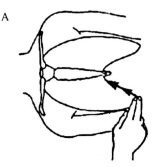

Kneel facing the victim's chest. Find the correct hand position by sliding your fingers up the rib cage to the breastbone.

B

Place your middle finger in the notch and the index finger next to it on the lower end of the breastbone.

C

Place your other hand beside the two fingers.

D

Place hand used to locate notch on top of the other hand. Lace the fingers together. Keep fingers off the chest. Position shoulders over hands, with elbows locked and arms straight.

FIGURE 48.1   Steps A, B, C, and D illustrate procedures to properly position hands prior to applying CPR.

FIGURE 48.2 Alternate positions for hands while applying CPR.

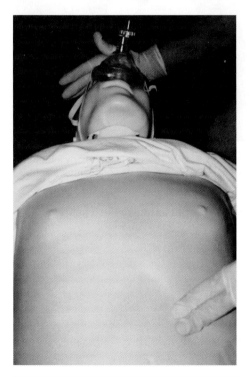

FIGURE 48.3 Correctly tilting the head to lift the chin while positioning finger along the injured lineman's rib cage.

FIGURE 48.4 The hands are being properly positioned on the victim prior to beginning the chest compressions.

Figure 48.5   The rescuer is using the weight of his upper body to compress the chest.

Figure 48.6   The rescuer has properly interlaced his fingers while providing chest compressions.

In the case of several rescuers, the rescue tasks are split amongst the team (Figs. 48.7 through 48.13).

Review

Call for advance help—911.

Determine that victim has no pulse.

Find correct hand compression.

Do 4 cycles of 30 compressions and 2 breaths.

Recheck pulse for 5 seconds.

If pulse is not present, continue with CPR.

Why Do I Do That and Why Did They Change?

1. *We used to give CPR for a minute and then go call for help.*
   Why has it been changed to—"Call for help immediately and then begin CPR?"
   It was found that trained single rescuers often performed CPR much longer than 1 min. Thereby, the 911 call to an EMS system or an advanced cardiac life support (ACLS) care was seriously delayed. It is important for the victim to receive the advanced life-saving medical support as soon as possible.

FIGURE 48.7   The three linemen have come to the aid of their injured coworker.

FIGURE 48.8   Two rescue linemen are accessing the victim while the third lineman is away to request 911 assistance.

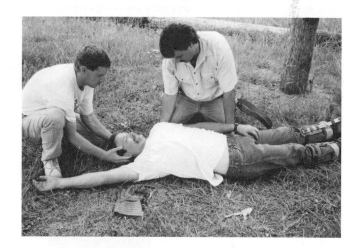

FIGURE 48.9   The third rescue lineman has returned to the injured lineman. The injured lineman has been placed on his side while the rescuers determine the proper course of action.

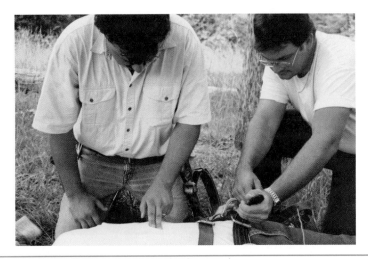

FIGURE 48.10   The injured lineman's belts are being loosened.

FIGURE 48.11   It has been determined that CPR is necessary. Each of the lineman has assumed a rescue task for the injured lineman.

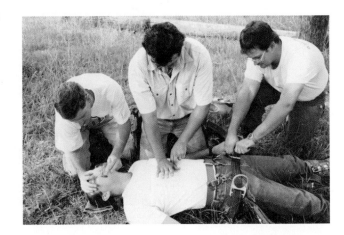

FIGURE 48.12   The chest compressions are being provided while the lineman providing rescue breaths is in position.

FIGURE 48.13 Rescue breaths have been given and chest compressions are resumed.

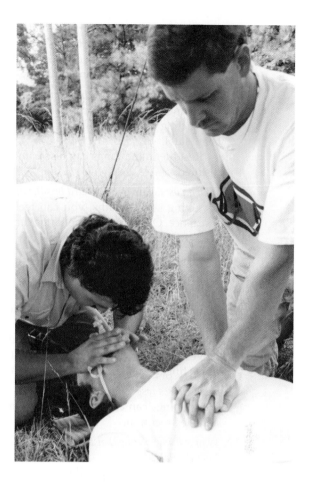

2. *Why do we use the head-tilt/chin-lift now instead of the old way of pushing up on the neck?*
   Using the head-tilt/chin-lift method helps to minimize further damage to a victim. It is more efficient in opening and sealing the airway.

3. *Why do we ask if a victim is choking when it appears obvious that they are?*
   An allergic reaction can also cause the airway to close. In such a case, the victim needs advanced medical support immediately and all of the abdominal thrusts in the world is not going to open the airway.

4. *What if the jaw or mouth is injured, the mouth is shut too tight, or your mouth is too small?*
   You should do mouth-to-nose breathing.

5. *What if a victim is breathing on his own but is unconscious?*
   It is vital that professional emergency personnel be called immediately. Victims who may be breathing on their own may still go into shock if counteractive measures are not begun immediately. There is a point when shock becomes irreversible, and the victim will die.

6. *Why has the time taken during ventilation for filling the lungs of an adult increased to 1½ seconds to 2 seconds per breath?*
   The increased time further decreases the likelihood of the stomach filling with the air and the victim then vomiting.

7. *What's the big deal if a victim vomits?*
   When an unconscious victim vomits, stomach contents may get into the lungs, obstructing breathing, which may hamper rescue breathing attempts, thus being fatal.
   Residue in the lungs can also cause complications after breathing has been restored.

8. *Why do we no longer give hard breaths but slow breaths?*
   By giving slow breaths, you can feel when the lungs are full and you meet resistance. Overinflating the lungs can cause air to go into the stomach.

9. *Do we remove dentures?*
   No. The dentures should be removed only if they cannot be kept in place.

10. *What if the victim has something in his or her mouth like chewing gum, tobacco, or vomit?*
    If foreign material or vomit is visible in the mouth, it should be removed quickly. Liquids or semiliquids should be wiped out with the index and middle fingers. The fingers should be covered by a protective barrier such as rubber gloves or by a piece of cloth. Solid material should be extracted with a hooked index finger. Remember to use *protective barriers whenever possible.*

11. *What causes respiratory arrest?*
    Many things can cause respiratory arrest: shock, allergic reactions, drowning, and electrocution.

12. *Why is it imperative that two rescue breaths be given before you bring the victim down from the pole?*
    Extensive soft-tissue swelling may develop rapidly and close the airway. The heart may stop beating soon after breathing has stopped. The heart will continue to beat in some instances after breathing has stopped. Giving rescue breathing before bringing the victim down helps delay the heart stopping and brain damage from lack of oxygen.

13. *Why do you stop breathing after an electrocution?*
    In electrocution, depending upon the path of the current, several things may occur. Your victim may continue to have a heart rhythm, but not be breathing due to:

    a. The electrical current having passed through the brain and disrupting the area of the brain that controls breathing.
    b. The diaphragm and chest muscles which control breathing may have been damaged.
    c. The respiratory muscles may have been paralyzed.

    If artificial respiration is not started immediately, the heart will stop beating.

14. *I know CPR and I am good at my skills, why was I not able to save my victim? Did I do something wrong?*
    There are some instances when all the skills you possess will not be able to save the victim. With your skills the victim had a chance, whereas if you did

nothing, the victim had no chance. A hand-to-hand pathway is more likely to be fatal than a hand-to-foot or a foot-to-foot.

Electricity in the body causes organs (the heart, etc.) and bones to explode, rupture, and break as well as cause tissue destruction. When the organ has ruptured, there is very little if anything you would have been able to do. This is why it is so important that you use safe work procedures and protective equipment to prevent accidents.

15. *How do I know that I am giving adequate ventilations?*
    a. You should be able to observe the rise and fall of the chest with each breath you give.
    b. You should be able to hear and feel the escape of air during the passive exhalation of your victim.

16. *What causes a victim to vomit when they are not even breathing?*
    a. During rescue breathing, you may not have the airway opened properly. You are forcing air into the stomach and not into the lungs.
    b. The breath you give is given too long, at too fast or at too much pressure. The air is forced into the stomach.

17. *Is it a problem if the stomach is full of air?*
    a. A stomach filled with air reduces lung volume by elevating the diaphragm.
    b. The victim may also have the stomach contents forced up by the air in the stomach.

18. *What do I do if the victim starts to vomit?*
    Turn the victim onto one side, wipe out his or her mouth with a protective barrier. Reassess your victim to see if they are breathing on their own. If not breathing, continue with the rescue procedure you are using (i.e., rescue breathing, CPR). If you are alone and must leave an unconscious victim, position the victim on his or her side in case the victim vomits while you are gone.

19. *How can someone get enough oxygen out of the air we exhale? Didn't I use all of the oxygen?*
    The air we breathe has approximately 21% oxygen. We use about 5% of the oxygen. That means that we exhale air containing approximately 16% of oxygen. This shows you that there is enough oxygen in the air you breathe out to support life until advanced medical care arrives.

20. *Do chest compressions circulate the same amount of blood as the heart does normally?*
    Under the best of conditions, chest compressions only circulate approximately 1/3 of the normal blood flow. Body tissues are receiving only the barest minimum of oxygen required for short-term survival. That is why it is so important to get professional help immediately.

21. *What does CPR stand for and what does it do?*
    CPR stands for cardiopulmonary resuscitation. Cardio refers to the heart and pulmonary refers to the lungs. CPR is a combination of chest compressions and rescue breathing. The purpose of CPR is to keep the lungs supplied with oxygen when breathing has stopped, and to keep blood circulating with oxygen to the brain, heart, and other parts of the body.

22. *Do we ever slap an adult choking victim on the back?*
   No. You could lodge the item more securely.

23. *Why does a victim who is going to be given CPR need to be on a level surface?*
   When the head is elevated above the heart, blood flow to the brain is further reduced or eliminated.

24. *What do I do if my hands come off the victim's chest or move from the correct compression point when I am giving CPR?*
   You must go up the rib cage and find the correct hand position again.

25. *Does the victim's chest need to be bared to perform compressions?*
   No. Clothing should remain on unless it interferes with finding the proper location for chest compressions. Do not waste time or delay the starting of the compressions by removing clothing.

26. *Can or should you do CPR on someone who has a pacemaker?*
   Yes. The pacemaker is placed to the side of the heart and directly below the breastbone, it will not get in the way of chest compressions.

27. *Do I give CPR to a victim who has a very slow or weak pulse?*
   No. Never perform CPR on a victim who has a pulse; it can cause serious medical complications.

28. *What if a victim regains a pulse and resumes breathing during or following resuscitation?*
   The rescuer should continue to help maintain an open airway. The rescuer should then place the patient in the recovery position. To place the victim in the recovery position, the victim is rolled onto his or her side so that the head, shoulders, and torso move simultaneously without twisting. If the victim has sustained trauma or trauma is suspected, the victim should not be moved.

29. *How hard do I press when doing the head-tilt/chin-lift maneuver?*
   The fingers must not press deeply into the soft tissue under the chin, put them on the bone. Pressing into the soft tissue might obstruct the airway. The thumb should not be used for lifting the chin.

30. *If there are several people who can help the victim, should one of them take off the tool belt and hooks?*
   Yes. Smoldering shoes, belts, and hooks should be removed to prevent further thermal damage and possible injury to the rescuer. NEVER remove adhered particles of charred clothing from the body. It will pull the skin off with it. If you are the only rescuer, do not take time to remove items unless it interferes with the ability to administer effective compressions. It is critical to begin CPR immediately.

Pole-Top and Bucket Truck Rescue

Time Is Critical

You may have to help a lineman on a pole reach the ground safely when the lineman:

- Becomes ill
- Is injured
- Loses consciousness

You must know:

- When the lineman needs help
- When and why time is critical
- The approved method of lowering

Basic Steps in Pole-Top Rescue

- Evaluate the situation.
- Provide for your protection.
- Climb to rescue position.
- Determine the injured lineman's condition.

Then, if necessary:

- Call or arrange for help from emergency medical services.
- Give first aid.
- Lower the injured lineman.
- Give follow-up care until help arrives.

Evaluate the Situation

Call to the lineman on the pole. If the lineman does not answer or appears stunned or dazed, do the following:

- Prepare to aid the lineman.

Time is extremely important.

Provide for Your Protection
Your safety is vital to the rescue. Use personal tools and rubber gloves (also rubber sleeves, if required).

- Checklist of equipment and site conditions
  - Extra rubber goods
  - Live-line tools
  - Physical condition of the pole
    - Damaged conductors or equipment?
    - Fire on the pole?
    - Broken pole?
    - Hand-line on pole and in good condition?

Climb to the Rescue Position
Climb carefully and position yourself to:

- Ensure your safety.
- Clear the injured lineman from the hazard.
- Determine the injured lineman's condition.
- Render aid as required.
- Start mouth-to-mouth, if required.
- Lower the injured lineman, if necessary.

The best position to render assistance will usually be slightly above the injured lineman.

Determine the Injured Lineman's Condition
He may be:

- Conscious
- Unconscious, but breathing
- Unconscious and not breathing
- Unconscious, not breathing, and heart stopped

If the injured lineman is conscious:

- Call for help from emergency medical services, if necessary.
- Time may no longer be critical.
- Give necessary first aid on the pole.
- Reassure the injured lineman.
- Help the lineman descend the pole.
- Give additional first aid on the ground.

If the injured lineman is unconscious, but breathing:

- Call for help from emergency medical services.
- Watch the lineman closely in case the lineman's breathing stops.
- Lower the lineman to the ground.
- Give first aid on the ground.

If the injured lineman is unconscious and not breathing: While the lineman is still on the pole:

- Call for help from emergency medical services.
- Provide an open airway.
- Give the lineman two rescue breaths.
- If the lineman responds, continue mouth-to-mouth until the lineman is breathing without help.
- If the lineman does not respond to the first two rescue breaths, check skin color, check for pupil dilation.
- If the lineman's pupils' contract and skin color is good, continue mouth-to-mouth until he is breathing without help.
  - Then, help the lineman descend the pole.
  - Watch the lineman closely, the lineman's breathing may stop again.
  - Give any additional first aid needed until help arrives.
  - If the lineman's pupils do not contract and skin color is bad, the lineman's **Heart Has Stopped**.
    - Prepare to lower the lineman immediately.
    - Give the lineman two rescue breaths just before lowering.
    - Lower the lineman to the ground.
    - Start cardiopulmonary resuscitation (CPR) until help arrives.

The method of lowering an injured lineman is:

- Safe
- Simple
- Available

Equipment needed:

- ½-inch hand-line

Procedure

- Position the hand-line.
- Tie the injured lineman.

- Remove the slack in the hand-line.
- Take firm grip on fall line.
- Cut the injured lineman's safety strap.
- Lower the injured lineman.

Rescuer

- Call or arrange for help from emergency medical services to be called.
- Position the hand-line over the crossarm or other part of the structure.
- Position the hand-line for a clear path to the ground (usually the best position is 2 or 3 feet from the pole).
- Wrap the short end of the hand-line around the fall line twice—two wraps around the fall line.
- Tie the hand-line around the lineman's chest using three half-hitch knots.
- Tie the injured lineman as follows:
  - Pass the hand-line around the injured lineman high on the chest.
  - Tie three half-hitch knots in front, near one armpit.
  - Snug the knots.
- Remove the slack in the hand-line.
- If only one rescuer—rescuer removes the slack while on pole.
- If two rescuers—the lineman on the ground removes the slack.
- **Important**—Give the injured lineman **TWO** quick breaths…if necessary, then
- Take firm grip on fall line.
  - If only one rescuer—the rescuer holds the fall line with one hand.
  - If two rescuers—the lineman on the ground holds the fall line.
- Cut the injured lineman's safety strap on the side opposite the desired swing.
  - Caution: Do not cut your own safety strap.

Lower the Injured Lineman (Fig. 49.1)

- One rescuer—the rescuer guides the load-line with one hand and controls the rate of descent with the other hand.
- Two rescuers—the lineman on the pole guides the load-line, the lineman on the ground controls the rate of descent.

If a conscious lineman is being assisted in climbing down, the only difference is that enough slack is fed into the line to permit the lineman climbing freedom.

One-Man Rescue versus Two-Man Rescue

These rescues differ mainly in control of the fall line.

Remember, the approved method of lowering an injured lineman is:

- Call or arrange for help from emergency medical services to be called.
- Position the hand-line.

Figure 49.1   The lineman on the ground safely assisted the injured lineman to the ground.

- Tie the injured lineman.
- Remove the slack in the hand-line.
- Take a firm grip on the fall line.
- Cut the injured lineman's safety strap.
- Lower the injured lineman.

If the pole does not have a crossarm, the rescuer places the hand-line over the fiber-glass bracket insulator support, or some other substantial piece of pole line hardware, such as a secondary rack, neutral bracket, or guy wire attachment, strong enough to support the weight of the injured lineman. The short end of the hand-line is wrapped around the fall line twice and tied around the injured lineman's chest using three half-hitch knots. The injured lineman's safety strap is cut and the rescuer lowers the injured lineman to the ground.

After the lineman is eased to ground, the hand-line is removed from the lineman (Fig. 49.2). Lay the lineman on their back and observe if the lineman is conscious (Fig 49.3). If the lineman is conscious, time may no longer be critical. Give necessary first aid until help arrives.

FIGURE 49.2   The hand-line is being removed from the injured lineman.

FIGURE 49.3   The injured lineman is being evaluated by his rescuers.

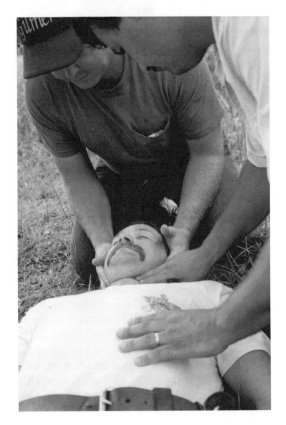

If the injured lineman is unconscious and not breathing, provide an open airway. Give the lineman two rescue breaths. If the lineman responds, continue mouth-to-mouth resuscitation until the lineman is breathing without help. If no pulse is present, the lineman's pupils do not contract, and the lineman's skin color is not normal, **the lineman's heart has stopped**. Restore circulation. Use CPR.

Bucket Truck Rescue

Time Is Critical

- Equip a portion of the insulated boom of the truck with rope blocks designed for hot-line work.
- The strap is placed around the insulated boom approximately 10 ft from the bucket to support the blocks.
- The blocks are held taut on the boom from the strap to the bottom of the boom.
- The rescuer on the ground evaluates the conditions when an emergency arises.
- Call for help from emergency medical services.
- The bucket is lowered using the lower controls.
- Obstacles in the path of the bucket must be avoided.
- The hook on the rope blocks is engaged in a ring on the lineman's safety strap.
- The safety strap is released from the boom of the truck.
- The rope blocks are drawn taut by the rescuer on the ground.
- The unconscious lineman is raised out of the bucket with rope blocks.
- The rescuer eases the injured lineman to the ground.
- Care should be taken to protect the lineman from further injury.
- Release the rope blocks from the lineman.
- The basket or bucket on the aerial device may be constructed to tilt after being released, eliminating the need for special rigging, to remove an injured lineman from the basket or bucket.
- Lay the injured lineman on their back and observe if the lineman is conscious.
- If the lineman is conscious, time may no longer be critical.
- Give necessary first aid.
- If the lineman is unconscious and not breathing, provide an open airway.
- Give the lineman two rescue breaths.
- If the lineman responds, continue mouth-to-mouth resuscitation until the lineman is breathing without help.
- If no pulse is present, the lineman's pupils do not contract, and the lineman's skin color is not normal, the lineman's **heart has stopped**.
- Restore circulation. Use CPR.
- After every 30 chest compressions, give 2 breaths.
- Continue CPR until emergency medical services arrive.

Self-Testing Questions and Exercises

The chapter number and page number for the answer to the question or a description of the exercise requested are indicated for each item.

1. Describe the electron theory. Chap. 1, p. 1

2. Define an electric current. Chap. 1, p. 1

3. What makes a good electrical conductor? Chap. 1, p. 2

4. What makes a good insulator. Chap. 1, p. 2

5. Name three examples of good insulators? Chap. 1, p. 2

6. What is the unit of electric current? Chap. 1, p. 5

7. What instrument is used to measure electric current? Chap. 1, p. 5

8. What is required in order for current to flow in a circuit? Chap. 1, p. 6

9. What does the voltmeter measure in volts? Chap. 1, p. 7

10. What is the unit of electric power? Chap. 1, p. 8

11. A wattmeter is essentially the combination of what two instruments? Chap. 1, p. 9

12. What is the formula for Ohm's law? Chap. 1, p. 11

13. Define an alternating current. Chap. 1, p. 12

14. What is the unit of an alternating-current frequency? Chap. 1, p. 13

15. What unit is inductance measured in? Chap. 1, page 20

16. What unit is capacitance measured in? Chap. 1, p. 20

17. In a circuit, what is a diode semiconductor used for? Chap. 1, p. 27

18. What are the distinguishing characteristics of transmission lines? Chap. 2, p. 39

19. How do the voltage levels of alternating-current (ac) transmission lines compare? Chap. 2, p. 39

20. When is a direct-current transmission line used? Chap. 2, pp. 39–40

21. List the functions that can be accomplished in an electric substation. Chap. 2, p. 41 and Chap. 3, p. 62

22. List the component parts of the electric distribution system. Chap. 2, p. 42

23. Describe a typical distribution feeder circuit. Chap. 2, p. 42

24. What is a primary circuit? Chap. 2, p. 47

25. What is the purpose of the distribution transformer installed on an electric distribution circuit? Chap. 2, p. 47

26. Describe an electric service. Chap. 2, p. 52

27. What is the purpose of an electric substation? Chap. 2, p. 41 and Chap. 3, p. 62

28. What types of equipment can be found in a typical substation? Chap. 3, p. 63

29. What is a power transformer used for in an electric substation? Chap. 3, p. 63

30. What function is accomplished with circuit breakers installed in electric substations? Chap. 3, p. 66

31. Why are disconnect switches installed in series with circuit breakers? Chap. 3, p. 67

32. What function is accomplished with protective relays installed in a substation? Chap. 3, p. 74

33. What normally limits the capacity of a transmission line? Chap. 4, p. 86

34. What are the advantages of using dc transmission for underground circuits rather than ac transmission? Chap. 4, p. 91

35. What is the relationship between the phase-to-neutral voltage and the phase-to-phase voltage of a three-phase four-wire system? Chap. 5, p. 99

36. Provide an image of how the secondary windings of the substation transformer would be connected to provide a source for a three-phase four-wire distribution circuit Chap. 5, p. 100

37. Which conductor of a three-phase four-wire distribution system is grounded? Chap. 5, p. 102

38. List the three-phase secondary voltages normally used. Chap. 5, p. 102

39. What are harmonic voltages? Chap. 5, p. 104

40. How are wood poles classified? Chap. 7, p. 120

41. What types of lines are commonly constructed with wood pole structures? Chap. 7, p. 120

42. What are the advantages of using crossarms on a subtransmission structure? Chap. 7, p. 126

43. What types of steel structures are used for transmission lines? Chap. 8, p. 149

44. List the advantages of monitoring devices? Chap. 9, p. 164

45. List some of the disaster response assessment data utilized to access a storm situation. Chap. 10, p. 180

46. What is the minimum recommended depth for setting a pole in soil? Chap. 11, p. 197

47. What is the rule for determining the setting depth for a pole to be installed in soil? Chap. 11, p. 197

48. Where should guys be installed? Chap. 12, p. 225

49. When is it necessary to insulate a guy? Chap. 12, p. 233

50. Describe an insulator. Chap. 13, p. 257

51. Where are double crossarms used? Chap. 13, p. 261

52. What types of circuits permit the installation of post-type insulators? Chap. 13, p. 270

53. The polymer suspension insulator is used to address what issue? Chap. 13, p. 274

54. Where are strain insulators used? Chap. 13, p. 273

55. What is the difference between a strain insulator and a suspension insulator? Chap. 13, p. 274

56. What factors must be taken into account when selecting line conductors? Chap. 14, pp. 291 and 300

57. What is the purpose of a distribution transformer? Chap. 15, p. 323

58. What equipment is used with a self-protected type of distribution transformer? Chap. 15, p. 323

59. Draw a schematic diagram of a single-phase distribution transformer connected to a primary circuit and a load through a switch. Chap. 15, p. 327

60. Describe the operation of a single-phase distribution transformer, first without connected load, and second with the switch closed and the load connected. Chap. 15, pp. 324–329

61. How is the capacity of a distribution transformer determined? Chap. 15, p. 330

62. Draw a schematic diagram of a distribution transformer secondary winding connected for 120-volt two-wire operation and 120/240-volt three-wire operation. Chap. 15, pp. 331–332

63. How can the polarity of a transformer be determined? Chap. 15, pp. 334–336

64. Why are some distribution transformers manufactured with taps on the primary windings? Chap. 15, p. 337

65. Describe the theory of operation of an elementary lightning arrester. Chap. 16, pp. 354, 356

66. Describe the operation of a metal-oxide valve-type lightning arrester. Chap. 16, pp. 357, 360

67. What is the basic insulation level of a 15 kV system? Chap. 16, p. 360

68. What is the purpose of installing a fuse in a circuit? Chap. 17, p. 371

69. Describe the operation of an expulsion fuse cutout. Chap. 17, pp. 372–373

70. How do you know when the fuse in an indicating enclosed cutout is blown? Chap. 17, p. 373

71. Refer to the time-current curve (Fig. 17.16) and determine the minimum clearing time for a 15K fuse link if 100 amps of current flowed through the fuse link. Chap. 17, pp. 381–382

72. Describe the operation of a solid-material fuse. Chap. 17, p. 382

73. Describe the operation of a current-limiting fuse. Chap. 17, pp. 388–389

74. What is likely to happen if a disconnect switch is opened while current is flowing through the switch? Chap. 18, p. 403

75. Describe the operation of an oil circuit recloser. Chap. 18, pp. 404–408

76. What is the purpose of using a voltage regulator in a distribution circuit? Chap. 19, p. 411

77. Describe the operation of a distribution step voltage regulator. Chap. 19, pp. 415–416

78. Why must the voltage regulator be in the neutral position to operate a voltage regulator bypass switch with the circuit energized? Chap. 19, pp. 419–420

79. What is the order of operations for transmission-line tower erection? Chap. 20, p. 421

80. Why must the surface of high-voltage transmission-line conductors be protected from scratches while they are being installed? Chap. 21, p. 433

81. Why is it necessary to effectively ground conductors while they are being installed? Chap. 21, p. 434

82. Describe the procedures to be followed to reconductor distribution lines. Chap. 21, p. 440

83. When should conductors be sagged? Chap. 22, p. 443

84. Describe how a lineman can use a transit to complete sagging operations. Chap. 22, pp. 447–449

85. Why are armor rods installed on some line conductors at the point of stress? Chap. 22, p. 462

86. Why are spacers installed on bundled-conductor transmission lines? Chap. 22, p. 464

87. What types of splices can be used on line conductors under tension? Chap. 23, p. 465

88. When can bolted connectors be used to join line conductors? Chap. 23, p. 465

89. What types of work can be completed to maintain lines while energized with hot-line tools? Chap. 24, p. 481

90. Describe the hot-line tool called a wire tong. Chap. 24, pp. 484–485

91. Describe the method used to phase out two circuits while they are energized before connecting them. Chap. 24, p. 505

92. Describe the procedure for testing insulators in a string with a high-voltage tester to determine if one of the insulators is defective. Chap. 24, pp. 506–507

93. What methods can be used to perform live-line maintenance while using an insulated aerial platform? Chap. 25, p. 511

94. How deep should a ground rod be driven? Chap. 26, p. 524

95. If the normal ground rod installation does not establish an acceptable value of ground resistance, what methods can be used to reduce the ground resistance? Chap. 26, p. 524

96. Protective grounds can guard the lineman from what types of hazards? Chap. 27, p. 537

97. How can a high-voltage circuit be tested to be sure it is deenergized? Chap. 27, pp. 539–540

98. What is the proper sequence for installing protective grounds? Chap. 27, p. 540

99. What are the requirements necessary for a lineman to establish an adequate protective ground? Chap. 27, pp. 540–542

100. Where should protective grounds be installed to establish the "man-shorted-out" concept that provides the best protection for the lineman? Chap. 27, pp. 543, 545

101. Describe the best procedures for establishing a temporary protective ground on a cable in a dead-front-type pad-mounted distribution transformer. Chap. 27, pp. 547–550

102. Define the following street-lighting terms: lamp, luminaire, lumen, footcandle, ballast, mast arm. Chap. 28, p. 551

103. What types of lamps are in common use for streetlighting? Chap. 28, p. 561

104. What types of streetlight lamps require a ballast? Chap. 28, p. 551

105. How is the light emitted from a streetlight controlled? Chap. 28, pp. 563–565

106. Name two common streetlight distribution patterns. Chap. 28, p. 564

107. Name the parts of an underground system. Chap. 29, p. 569

108. In general, the top of the duct run should not be less than how many feet below the street surface? Chap. 29, p. 570

109. What is the advantage of compact sector cable construction? Chap. 29, p. 582

110. What is the purpose of having shielding over the insulation of a high-voltage cable? Chap. 29, p. 583

111. Why are potheads installed on cables? Chap. 29, pp. 590–591

112. What purpose do vaults serve in an underground system? Chap. 31, p. 616

113. What precautions must be taken before cables are installed in a duct? Chap. 32, pp. 619, 623

114. Describe precautions that must be taken to remove the insulation shielding from a cable to prepare the cable for splicing. Chap. 33, pp. 629, 633–634

115. Describe the types of underground residential distribution systems normally installed. Chap. 34, p. 640

116. What are the voltages normally used for underground residential distribution circuits? Chap. 34, p. 644

117. Describe the triplexed-type cable assemblies used for underground residential distribution secondaries and services. Chap. 34, p. 645

118. Identify the component parts of transformers used for underground electric distribution systems. Chap. 34, pp. 646, 648–649

119. What is the minimum depth allowable for direct-buried cables operating in the range of 601 to 22,000 volts? Chap. 34, p. 655

120. What requirements are necessary to install electric supply cables and communication cables at the same depth with no deliberate separation? Chap. 34, pp. 655–657

121. What is the maximum distance permitted between the points where effectively grounded electric cables are bonded to communication cable shields if they are installed with random separation? Chap. 34, p. 663

122. What are the objectives of line clearance tree trimming? Chap. 35, p. 669

123. What are the fundamentals essential for safe and competent tree-trimming operations? Chap. 35, p. 670

124. Where should distribution transformers be installed? Chap. 36, p. 691

125. Describe the proper location of grounds for various types of distribution transformer secondary wiring. Chap. 36, p. 692

126. Describe the procedure to be followed if two single-phase distribution transformers are to be connected in parallel. Chap. 36, p. 697

127. When is it desirable to determine the phase sequence of an ac circuit? Chap. 36, pp. 701–703, 705

128. What is the minimum height allowable for attaching a low-voltage service to a building? Chap. 36, p. 713

129. Describe a method for testing the service wires to avoid connecting the customer's ground wire to either of the hot wires of a secondary. Chap. 36, pp. 717–718

130. Draw the symbol of a single-pole draw-out type of air circuit breaker. Chap. 37, p. 721

131. Draw the symbol for a cable termination. Chap. 37, p. 721

132. Draw the symbol for a variable resister. Chap. 37, p. 723

133. Draw the symbol for a ground connection. Chap. 37, p. 722

134. Draw the symbol for a single phase, two winding transformer. Chap. 37, p. 723

135. Draw the symbol for the three-phase wye-connected four-wire grounded transformer connection. Chap. 37, p. 723

136. Describe the components and significant features shown on the distribution circuit single-line diagram in Fig. 38.2. Chap. 38, pp. 727, 729

137. Define voltage regulation on an electrical system. Chap. 40, p. 743

138. What are the high limit and the low limit of the voltages of an electric service to a residence? Chap. 40, p. 743

139. What is the effect of low voltage on residential electrical equipment? Chap. 40, p. 743

140. What is the effect of high voltage on residential electrical equipment? Chap. 40, p. 743

141. How can the voltage on an electric distribution system be controlled? Chap. 40, pp. 743–744

142. What is the purpose of a line-drop compensator on a voltage regulator? Chap. 40, p. 750

143. What is the formula for Ohm's law? Chap. 41, p. 760

144. What is the formula for calculating the impedance of an alternating-current series circuit? Chap. 41, pp. 765–766

145. What is the formula for calculating power for a single-phase alternating-current circuit? Chap. 41, pp. 767–768

146. What is the formula for calculating power for a three-phase alternating-current circuit? Chap. 41, pp. 768–769

147. What is the horsepower rating of a single-phase motor operating at 480 volts and drawing 25 amps at a power factor of 88% if it has a full-load efficiency of 90%? Chap. 41, p. 771

148. What items are important to check as part of the initial inspection of transmission lines before they are energized? Chap. 42, p. 779

149. How can most transmission lines be inspected quickly after the line experiences a sustained unscheduled outage? Chap. 42, p. 779

150. What conditions normally occur when insulators are defective? Chap. 42, p. 791

151. What common problems or defective conditions should a lineman check for when inspecting distribution circuits? Chap. 42, pp. 780 and 782

152. What procedures are normally followed to maintain capacitor bank switches? Chap. 42, pp. 782–783

153. What items are important to check when inspecting switches? Chap. 42, pp. 783–784

154. What items should be checked when inspecting and maintaining underground distribution switchgear? Chap. 42, p. 784

155. How can distribution transformer loading be checked? Chap. 42, p. 784

156. What work should be performed by the lineman to keep streetlights properly maintained? Chap. 42, pp. 788–789

157. Using a piece of rope, tie an example of the following knot: two half hitches Chap. 43, p. 810

158. Describe the proper inspection procedures that should be executed prior to climbing a pole. Chap. 44, pp. 843–844

159. Describe the proper methods of inspecting and caring for climbing equipment. Chap. 44, p. 833

160. Describe procedures for maintaining climber gaffs. Chap. 44, pp. 834–838

161. What is the safest way of completing electric distribution work? Chap. 45, p. 849

162. Itemize the protective equipment that should be used by the lineman or cableman when working on energized electric circuits. Chap. 45, p. 851

163. How often should rubber gloves be given an air test by the lineman or cableman? Chap. 45, p. 852

164. How soon must resuscitation be started after a person stops breathing to prevent brain damage? Chap. 47, p. 979

165. Describe the proper assessment procedure prior to performing mouth-to-mouth resuscitation. Chap. 47, p. 978

166. Describe the procedures necessary to perform CPR. Chap. 48, pp. 987, 990

167. What is the main difference in one-man versus two-man pole top rescue? Chap. 49, p. 1000

Index

Note: Page numbers referencing figures are followed by an "*f*"; page numbers referencing tables are followed by a "*t*".